HOME FOOD SYSTEMS

Design and layout: Merole Berger, K. A. Schell, Linda Jacopetti.

Photography: T. L. Gettings, Douglas C. Green, Christie Tito, Margaret Smyser, Sara Bell (coordinator).

Cover: K. A. Schell (design), Fred Matlack (construction), Mitchell T. Mandel and T. L. Gettings (photography).

Illustration: Jack Crane (decorative), Frank Rohrbach (technical).

Product evaluation: Diana Branch, with Anna Winkler, Don Buss, Joyce Hubler, Janice Kay, Keith Marks, Judy Rittenhouse.

Test kitchens: JoAnn Benedick, Linda Gilbert, Anita Hirsch, Tom Ney.

Chapter editors: Corliss Bachman (Fish Gardening); Tom Dybdahl (Sprouting, Canning, Juicing, Drying, Freezing and Cold Storage, Backyard Animals); Thomas Stoneback (Mushrooms); Roger Yepsen (Grains, Beans, Tongue Foods, The Home Dairy, Planning the Home).

HOME FOOD SYSTEMS

Rodale's catalog of methods and tools for producing, processing, and preserving naturally good foods.

Corliss A. Bachman, writer

Tom Dybdahl, writer and editor

Catherine A. High, secretary and assistant researcher

Jean Polak, editorial coordinator and researcher

Thomas Stoneback, project coordinator

Roger B. Yepsen, Jr., writer and editor

Edited by Roger B. Yepsen, Jr.

 Rodale Press, Emmaus, Pennsylvania

Printed in the United States of America on recycled paper, containing a high percentage of de-inked fiber.

Library of Congress Cataloging in Publication Data
Main entry under title:

Home food systems.

    Includes bibliographical references and index.
    1. Food, Natural. 2. Canning and preserving.
3. Kitchen utensils. 4. Livestock. 5. Vegetable gardening. 6. Organic gardening. I. Bachman, Corliss A. II. Yepsen, Roger B.
TX369.H65   641.3′02   80-23806

ISBN 0-87857-325-9 hardcover
ISBN 0-87857-320-8 paperback

Lowest figure indicates number of this printing:
2 4 6 8 10 9 7 5 3 1 hardcover
2 4 6 8 10 9 7 5 3 1 paperback

Props for some photographs supplied by La Belle Cuisine Fine Cookware, Allentown, Pennsylvania.

Please note that the prices of products and foods stated herein can only serve as a suggestion of those you will find by the time this book reaches you.

Contents

What is a home food system? A series of steps that takes a food from the garden (or nesting box or barnyard or tree) to the plate. Some systems are very simple: dipping water from spring to mouth; sampling tomatoes and peas and berries on a walk through the garden; nibbling at sheep sorrel that grows wild at one's feet.

Most food systems are more involved. A kitchen comes betwixt living plant or animal and mouth: harvested food may be stored, or it may be milled and cooked for the sake of palatability and variety. Or so it was in the kitchen of a hundred years ago, when a kitchen was still a workplace on a par with the home workshop, being just as filled with tools, storage space, and the serious air of important work to be done—work that is now largely done for us in factories. A food system, when performed in a factory, becomes streamlined and perverted.

Grains (page 18), the subject of our first chapter, are the origin of a few food systems, most of them usurped decades ago by the factory. There, the grain is diddled with to make it more obedient—stripped of its vitamin E for maximum shelf life, stripped of its bran to make it white and pliant in baked goods and pasta. The story is told in

Grain profiles, beginning with

tell of the personality and nutritional contribution of 12 grains. One of the most rewarding food systems starts with the flint-hard grains and ends with warm loaves of whole grain bread.

You don't save much by milling your own flour, but many bakers wouldn't depend on those five-pound supermarket bags for anything.

Grains become sweet when sprouted, a fact not overlooked by brewers and bakers who have traditionally depended upon

"It's likely that store-bought pasta isn't as deficient as store-bought bread," opens

Pasta, 59,

but "pasta-makers can feel good in knowing the product of the hands (or their high-chrome machines) is better than money can buy."

Grains don't have to be ground into a fine powder of course, and other possibilities are given in

Storage is important in keeping grain and flour in quantity.

Those of us who take an interest in grains are in a small minority, judging by their absence from the supermarket, so we suggest

Beans (page 86) are among the humblest of foods. They defy those who would cook them up in a minute; their flavor does not call to mind such culinary adjectives as succulent and piquant; and with few exceptions they come in sparrowlike colors that vary from subtle to drab. It's hard for a book to talk up foods that seem to enjoy pretending they are little stones. These little stones happen to fuel a great number of the earth's people, particularly those of modest means, gaining beans the unfortunate nickname

Many cultures have evolved a great variety of ways of combining beans with grains for optimum nutrition, as we demonstrate with recipes in

Fourteen best-known beans are profiled alphabetically, beginning with

Many varieties aren't easy to find, and we offer help in

The soybean has the talent of changing personality, from a rather bland bean to tofu, soymilk, tempeh, and other, less-known foods from the East. These soyfoods may be the biggest thing to invade the American diet since pizza.

Sprouting (page 116) is the simple process of a seed beginning to grow. Nevertheless, sprouts are struggling to escape their exotic Eastern heritage and gain acceptance as a basic, nutritious food. Even their looks are against them: a stringlike mass of sprouts looks more appropriate for a biology lab than the supper table. To promote their benefits, both to your health and wealth, we begin with:

Sprouts have definitely arrived, and are increasingly easy to find in health food stores, produce stands, and even some supermarkets. But in less time than it takes to drive to the store, you can learn

Once you've got the basic process down, you may start looking for new sprout challenges.

With a little water and warmth, sprouts will grow in almost any device, including a sprouter. We confirmed that by testing nine of them in

Juicing (page 136) is an easily overlooked way of processing food. It's not a second-rate way of eating; indeed, the natural goodness of a food is distilled in its juice. Even if you don't have a surplus of produce, you can still find

There's juice in almost every fruit and vegetable. We take a closer look at ten of them, as well as at what they're reputed to do for your body.

Apple juice is a popular beverage, though most of us know it only as a clear, pasteurized liquid that's far removed from its origins. You can learn how to make the real thing, and enjoy its varied forms, in

No matter how hard you squeeze, you won't get much liquid from a carrot. You need a juicer. There are many on the market. First we tell you how the different kinds are supposed to work, and then how 19 of them actually performed.

Lest you think a juicer is a one-talent appliance, don't miss

When your garden or orchard won't quit, you'll need to go beyond drinking juices to

Drying (page 160) was the principal method of food preservation until this century. But it's such a basic, low-cost way of keeping food that it's being rediscovered.

Almost any food—except very fatty fish or meat—can be dried if you know

Once you've dried the food, you need to keep it that way until you plan to use it.

Electric dehydrators make food drying quick and easy. Most of them look somewhat similar, until you get inside. After a look at the mechanics of dryers, we review 13 of them.

The cheapest way to dry food is with the sun. But it's a fickle star—bright one day, hidden the next. We reveal what you can do to minimize these variations, and make a success of

There's no trick that will enable you to dry food during a rainstorm. So as a final resort, there's

Canning (page 190) evokes contradictory feelings in many of us. The nostalgic satisfaction of seeing (or remembering) home-grown, handpicked food stored up for the cold times is combined with darker fears that lethal botulism may be hiding in Aunt Helen's green beans. But canning may not be worthy of either such praise or such terror; it's simply a time-tested, low-cost way of preserving your food. Contrary to some reports,

In fact, we trace canning from its origin to its resurgence.

Most people's first concern is for safety.

To get started, you need some basic

The canning process may seem complicated, but it is just a series of simple steps done in a logical order. Collect your food, and we'll tell you

Canning is an old process, but new approaches are making it even better. It's

Finally, when you've mastered the basics, you can move beyond fruits and vegetables to

You can even bring home what is usually a commercial process.

Grandmother's only canning tools may have been her two hands and a sharp knife, but today there are various devices that claim they're indispensable. Learn the real truth about:

Freezing and Cold Storage (page 226) preserve much of America's food. Virtually every home has a refrigerator, and nearly half have freezers. Most of the time, we just drop food into these appliances, knowing it will keep but not really knowing why.

It is hard to imagine modern life without

The surest sign we take them for granted is the way we use them: badly. You can save money by

Freezers grow in popularity every year, largely because they're unexcelled for maintaining the quality of stored food. You pay for that quality in energy bills, however, so it's worthwhile to learn all you can about freezing and

Before the birth of refrigerators and freezers, people kept food cold with ice, streams, springs, and air, or by stashing it in Mother Earth. These elementary methods still work just fine, and they don't cost much.

Tongue Foods (page 246) tells how to make your own toppings, condiments, herbs and seasonings, and snacks, rather than rely on commercial versions with their salt, sugar, and additives.

Condiments, 253,

gives recipes for chutney, horseradish, mayonnaise, ketchup, mustard, tomato sauce and paste, chili sauce, and spice powders, plus the

Curious truth about the peppercorn, 256.

As the spurious reputation of chocolate spreads, more people turn to

Carob, 257.

Not many people take the time and trouble, but you *can* make your own

Essential oils and extracts, 258.

Homemade cider, free of preservatives, will make a good and inexpensive vinegar:

Cider and the mother, 259.

Crackers, 260,

if made at home, won't look just like store-bought, but are quick and easy to bake.

Fine jams and jellies can be made without sugar.

Is honey really a health food? Is maple sugar any better for you than table sugar? Read

We apply different standards to food and drink: even the sternest of label readers are often careless about what they pour down, says

in its review of sane alternatives to soda and booze:

Each year a crop goes largely unnoticed, perhaps because the plants are often the size of a house:

The Home Dairy (page 288) may or may not involve a live-in goat, cow, or sheep. You can use store-bought milk, but the savings drop and may nearly disappear altogether in some cases. Why bother making your own milk, cream, butter, yogurt, and cheese? One good reason stares at you from the labels on dairy foods: additives. Start from scratch and avoid them. (Start from the lactating animal, and avoid pesticides and medications that can wind up in milk.) You might start off with

Big savings can be had by making yogurt: commercial yogurt is double your costs. Here is a food particularly suited to home production, ranking with sprouts and bread, and we devote a number of pages to yogurt and other fermented dairy foods.

Once you've gotten yogurt down pat, the next challenges are

and a variety of

A bit trickier, and more time-consuming is

They may be jowls on Nutrition's face, but we can't ignore

The first choice of dairying homesteaders is

The next section quotes a farmer as saying, "You know, that barn full of cows is my jail," but big families with a deep thirst and many idle hands may want to invest time and love in

Backyard Animals (page 332) says that many animals—even some rather large ones—can be comfortably raised in a small space. The benefits include: companionship, food, and perhaps some wool, skins, or down. Even slaughtering has its rewards, as you'll discover in

The biggest hurdle to raising animals may not be your small backyard, but a city ordinance.

The most popular backyard animal is a versatile fowl that provides eggs, meat, and a reliable alarm clock.

Waterfowl make friendly, loyal neighbors, and they'll liven your life and your menu.

Pigeons are usually considered pests, but they're prolific breeders that can earn you a profit with their gourmet meat.

The Japanese emperor eats quail at his ceremonial dinners, and you can raise them on your back porch and eat them anytime.

You may be hesitant about turning your backyard into a pigpen. But given half a chance, hogs will stay quite clean, and they're amazingly efficient at transforming scraps into bacon, ham, and pork chops.

Nobody likes to kill a cute, furry animal. But if you've got the heart for it, four does will provide you with five pounds of tender, white meat every week.

Sheep are not the smartest animals, but they may be a smart investment. They'll give you wool, skins, and meat in exchange for a minimal shelter and a little grazing land.

Your garden will profit if you get into worms—the natural composters.

Even if you don't grow your own meat, there are ways of avoiding the chemical-laden cuts stocked by most supermarkets. We end with tips for

Fish Gardening (page 362) is the most avant-garde of this book's food production systems, an intriguing way of raising a fresh supply of protein for the family. Fish is a good food, as we explain in

The condition of the water in which fish live influences the success of fish gardening more than anything else. We show how interrelated environmental factors affect fish in

A step-by-step guide to understanding a backyard fish system starts with

We outline the events in a fish gardening season from setting up to hauling in the catch.

For advice on keeping a pool system operating smoothly, see

Profiles of several of the fish best suited to home fish gardening begin with

Basements are good for something else besides Ping-Pong tables and power tools. Fish can grow indoors, too, as we explain in

What's all that clucking and splashing? We report about experiments to combine fish and fowl in

Natural bodies of water usually support fish of some sort. If tended and coaxed, though, the water will produce a much greater harvest, as discussed under

Several organizations concerned with fish gardening have been working with resource-conserving solar structures and wind power. Accounts of their efforts are in

A whole chapter on

Mushrooms (page 388)? The chapter subhead bills the mushroom as "home-grown protein from an unlikely source," and it's a fact that mushrooms produce more of that nutrient in a given space than any other crop.

After taking the trouble to grow your own mushrooms, you'll want to treat them considerately in the pan.

Nutrition aside, mushrooms are a delicacy, and growing them shares some of the mystique of wine-making.

Perhaps the finest of mushrooms are free for the picking—providing you know both where to look and what to leave untouched on the forest floor. The former may be a closely guarded secret among local mushroom hunters; the latter is best learned at the side of an experienced forager, with the aid of a good, illustrated field guide.

We weren't overly impressed with the kits we reviewed, but simple growing shelves can be built of lumber and nails.

Even the best-laid outlines of editors fail to embrace a few wandering ideas. An idea is no less valuable for its vagrancy, and so it is that the thirteenth chapter,

Planning the Home (page 398), accounts for the bits and pieces left in the files after the first dozen had left for the typesetter. Much of the lees in the editorial barrel can be drawn off in the name of

"One cook's pet appliance is another's nuisance." So opens our next generously broad section, named

From the kitchen we next go to the green spaces about the house and yard.

This chapter doesn't quit at the back door or at the far end of the garden. One home system goes beyond the yard—the plumbing. Whether you are hooked up to the city main or drink sweet water from your own well, you'll want to read

Appropriately, the last section talks of "making the least home waste, then making the most of it."

Grains

The whole grain philosophy

Tastes in clothes and furnishings follow the dictates of fashion. This is common knowledge. But consider for a moment that the same is true of food. Our food has strayed as far from its origins as a flimsy $300 dress has from the wool shirt.

The humble grain just wasn't sexy and streamlined enough for this century. While we moderns didn't invent white flour, we brought the art of refining grains to unimaginable heights. Styrofoam bread stands at the zenith of a centuries-long quest for blandness and whiteness in grain products. Until the 20th century, millers hadn't been able to achieve the blinding white flour we take for granted today; rice has never been as fluffy and disturbingly cheerful as our commercial products; and modern-day pasta is unmatched in history as an unobtrusive blotter for rich sauces.

No grain, it seems, has come through time unscathed. In addition to wheat and rice, we have lime-treated white corn flour (or masa), pearled (hulled) barley, refined rye flour, and commercial breakfast cereals.

How is it humans acquired a taste for bland, nutritionally inferior foods, and an eye for paleness?

RY

The snob appeal of refined grain

Like so much else in our world, the taste for refined foods had its roots in ancient Greece and Rome. It was there that a new method of milling made refinement possible. The manual labor of thorough grinding and sifting required to produce white flour made it quite costly. And the break made from it was lighter and more delicate in flavor and texture than whole grain breads. Both the cost of white bread and its delicacy endeared it to the privileged classes because this set them apart from the plebian multitudes who could never afford such things. Before long, Hippocrates added medical blessing to social discrimination: mistakenly praising white bread's better digestibility, he caused centuries of confusion on the subject.

Once the myth of white flour's superiority was established, it was never dispelled. In medieval England, grades of bread were noted as punctiliously as social classes. Only the wealthy and noble ate bread made from well-sifted wheat flour.

A whiter shade of pale.

The bulk of the populace received the unintended benefit of coarser breads, in grades that descended down from whole wheat loaves and bran breads, through breads made with heavier grains like rye, barley, and oats, to breads in which beans and peas were main ingredients.

Throughout most of history, white wheat flour was so costly to refine that only the wealthy could afford it. But suddenly in the 19th century, with the development of a new milling process, the bread of the rich was available to everyone. Using steel rollers instead of massive stone wheels, the new process actually produced flour more "refined" than the royalty of earlier times could buy. Instead of crushing the wheat germ, the rollers squeezed it out from the kernel, so it, along with the bran, could be removed more completely; the resulting flour was whiter and smoother than ever.

This water-powered grain mill in Kashmir, India, shares basic principles with a countertop mill. The mill is operated 24 hours a day by a single miller. He naps on a mat a few feet away, and makes his meals in a tiny alcove.

Practical considerations added to the appeal of white flour: without bran and wheat germ, it was less likely to turn rancid than whole grain flour; also, white dough was easier to knead than whole wheat dough, housewives discovered, and it rose more predictably.

Lest you think the preference for refined grains is an aberration of Western history, consider rice. As soon as technologically possible, the Oriental countries that rely on rice began to refine it by polishing off the outer layer of that grain, for the same aesthetic and practical reasons that motivated the refiners of wheat flour.

The smooth texture and whiter-than-white appearance of roller-milled flour came cheaply, it seemed, but in actuality those who made it a dietary staple had to pay a heavy price. Early in this century, the poorer parts of this country saw an increase in pellegra, a disease that progresses from a skin condition to a general breakdown of the digestive and nervous system. Its cause, scientists discovered, was the lack of niacin, a B vitamin. Home bakers of white bread were heavily represented among pellegra's victims.

The solution seemed obvious— put back the nutrients that had been removed by refinement. It sounded like a perfect idea, but like many easy ways out it had its drawbacks. "Enrichment" simply means putting back four of the most obviously important nutrients: the mineral iron and the B vitamins thiamine, riboflavin, and niacin. This provided a defense against pellegra, anemia, and other deficiency diseases, but it also

In 1972, about 38 percent of all bread baked was white bread. By 1977 that total had slipped to 29 percent. For 1982, the Commerce Department projects that only 23 percent of all bread will be white, as consumers turn to bran breads and other baked goods that have come to be considered more "healthful." . . . To combat the slump, Campbell Taggart [a major U.S. baker] plans a big ad campaign based on the slogan, "It's not called the staff of life for nothing." Much of the effort will be aimed at boosting the image of white bread.

Wall Street Journal.

created an illusion—that enriched white flour, and later, enriched white rice, were the nutritional equivalents of the original whole grains. Which they are not, not by a long shot.

What enrichment does, to be more accurate, is put a patch on the devastation inflicted by refinement. Whole wheat, for instance, has a whole range of B vitamins, all of which are decimated by refinement. Only *three* are returned by enrichment. Whole grains in general supply small but significant amounts of minerals; aside from iron, enrichment forgets about them. We require trace minerals like chromium, manganese, copper, zinc, selenium, and magnesium in very, very small amounts, but those small amounts play crucial roles in bodily processes that range from the action of heart muscle to the enzyme systems on which life depends. Whole grains are an important source of these minerals; enriched, refined flour is a poor source. And enrichment does absolutely nothing to restore what has been ruthlessly extracted from whole grain—fiber. Removing fiber from grain is a crime against nature.

The role of fiber—the part of food that passes virtually undigested through the body—in sickness and health has been a major focus of nutritional research during the past ten years. Strong evidence has accumulated that a lack of fiber in the modern American diet is in part responsible for our high rates of "the diseases of civilization"—illnesses like bowel cancer, heart disease, diabetes, obesity, ulcer, gallstones, and varicose veins.

Some speculate that the human body is adapted by nature to deal with foods in which digestible matter and fiber are mixed together, as they are in fresh vegetables, fruits, and whole grains. When we take fiber from food—as we do when we refine grains—we are committing a crime against nature, creating an artificial food that is wrong for the body.

Another benefit of fiber is that it serves as a valuable "natural barrier" to eating and digestion. Whole wheat bread, which has twice the fiber of white bread, actually takes longer to eat, so that you can feel satisfied with fewer calories (something to remember if you're watching your weight). Good bread passes more slowly from the stomach to the small intestine, sparing the duodenum the heavy load of ulcer-promoting acid that may accompany the digestion of white bread. Foods with natural fiber, like whole grains, are absorbed into

(Continued on page 20)

(Continued from page 19)
the bloodstream more slowly and evenly than refined foods. Blood sugar and insulin levels remain steadier and, according to a widely held theory, diabetes is less likely.

In the large intestine, fiber adds bulk to the stool. This, scientists believe, protects against bowel ailments (including cancer) and indirectly prevents the development of varicose veins. The possible protection that fiber offers against heart disease and gallstones is not fully understood, but it has been suggested that its effect on cholesterol absorption is involved.

Because it leans heavily on processed foods like white sugar and white flour, the modern American diet is woefully short of fiber. It is hard not to notice that the diseases of civilization increased here during the same period that our fiber intake declined.

Whole grains as the first step in reforming your diet

To go a step further, when you make more room on your table for whole grains, displacing some meat and refined foods, you'll accomplish, at a single stroke, a major reformation of your diet. You'll be consuming less fats (in particular, less saturated fats), fewer calories, and more trace minerals, as well as more fiber. If you are concerned about pesticides and other chemicals in your food, you can appreciate the fact that grains have far less of these toxic substances than meat or animal products.

Practical illustrations of the difference whole grain can make are not hard to find. During World War I, when shortages of meat forced the Danes to switch to a diet in which

A funny thing happened on the way to the supermarket shelf

Wheat is the most widespread and versatile of cereal grains. The whole grain is a rich source of protein, calcium, carbohydrates, iron, and B vitamins.

Into the mill. The milling process strips the wheat berry of its bran and germ to produce a soft, light bread that you can chew with your tongue.

Federal laws require that B vitamins and iron be returned to the white flour to enrich it, but the result is a paint-by-numbers portrait of whole wheat flour.

Double, double, toil and trouble;
Fire burn, and caldron bubble.

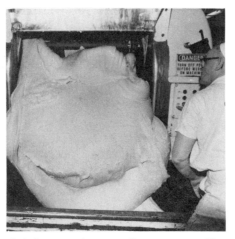

A bakery employee warily eyes a quaking mass of living dough.

Loaves trundle off into the oven. At no point are they touched by the human hand.

coarse bran bread and barley porridge figured prominently, mortality rates dropped significantly. England's experience during World War II was similar: the mass consumption of "national flour" high in bran was accompanied by a sharp decrease in the number of deaths from diabetes. In those modern countries where whole grains still make a substantial contribution to the diet, the diseases of civilization are generally far less troublesome than here.

Our disdain for grain is quite a recent development. In 1910, Ameri-

(Continued on page 22)

Today's baker pushes buttons instead of pummeling dough.

Through advertising, we've been taught to judge our bread as we judge our toilet paper—with a squeeze.

Amaranth

Different varieties grown for nutritious leaves, high-protein seeds, or luxuriant flowers. For centuries, a significant crop in scattered areas of both the old and new worlds. Now gaining recognition as an important food-producing plant suitable for gardens.

Appearance: Can grow to four or five feet high, has red or golden seedheads. Leaves may be magenta-tinged. Plants valued as ornamentals.

Uses: Toasted seeds can be milled into flour or boiled for cereal. Mexicans pop seeds and stick them together with honey or molasses for a version of popcorn balls. Flour lacks the gluten to make raised bread. Makes fine flatbread, such as Indian chapatis or Mexican tortillas. Mixes well with other flours; use in breads, muffins, and pancakes. Boil like rice for gruel or porridge. Leaves of young plants or potherb varieties are cooked like other green leafy vegetables; won't wilt or turn bitter in hot weather.

High in protein—especially lysine, which is lacking in most grains. Mixed with whole wheat it makes a complete protein.

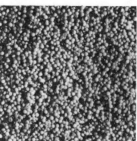
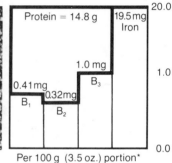

Per 100 g (3.5 oz.) portion*

Barley

A hardy grain; may have been the first crop ever consciously cultivated. Can grow in climates from the tropics to the Arctic Circle.

Appearance: Short, stubby kernels with a hard outer shell. White, translucent pearl barley has been polished to remove most of the endosperm. Whole grain brown barley has only a single outer layer removed.

Uses: Gives flavor and thickness to soups and stews. Flour and flakes have a light, nutty flavor brought out by toasting. The low-gluten flour won't make raised bread, but loaves have a cakelike consistency and delightful sweetness. Flakes or whole grains make a chewy, hearty breakfast cereal.

Whole grain barley has twice as much calcium as pearl barley, and more iron, protein, phosphorus, and potassium.

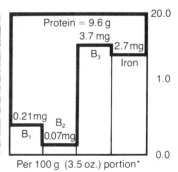

Per 100 g (3.5 oz.) portion*

*B₁ is thiamine, B₂ is riboflavin, B₃ is niacin.
Source: *Composition of Foods,* USDA Handbook No. 8.

Buckwheat

Not a true cereal; its brothers are such common weeds as dock and sorrel.

Appearance: Produces dark, three-cornered seeds that resemble tiny beechnuts. The hard, fibrous hull can be removed only with a special buckwheat huller.

Uses: In the U.S. used most often in pancakes, biscuits, and muffins. As toasted kasha, it's a diet staple in Eastern Europe and the Ukraine. Far Easterners make it into soba (noodles). Blossoms are the source of a popular dark honey. The flour is light or dark, depending on whether the hull has been removed prior to grinding. Dark flour is far superior nutritionally. For tasty cereal, cook the whole groat like rice.

Has a protein content superior to true cereal grains because it is high in lysine, an essential amino acid.

Protein = 11.7 g
B$_3$ = 4.4 mg
Iron = 3.1 mg
B$_1$ = 0.6 mg
B$_2$ = 0.0 mg

Per 100 g (3.5 oz.) portion*

Corn

A staple food of this continent when Columbus landed.

Appearance: Plump, round kernels. Most common varieties in this country are yellow or white.

Uses: Can be ground into meal, exploded to create popcorn, and crushed to extract corn oil. Use meal to make unraised breads and pones. Good as breakfast food in corn flakes and cornmeal mush, and as a cooked, whole grain at other meals. Often added to pumpernickel recipes. Corn flour (masa) makes Mexican tortilla.

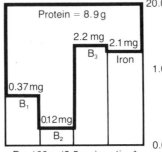

Protein = 8.9 g
B$_3$ = 2.2 mg
Iron = 2.1 mg
B$_1$ = 0.37 mg
B$_2$ = 0.12 mg

Per 100 g (3.5 oz.) portion*

*B$_1$ is thiamine, B$_2$ is riboflavin, B$_3$ is niacin.
Source: *Composition of Foods, USDA Handbook No. 8.*

(Continued from page 21)

cans consumed a hefty average of 300 pounds of grain per person per year, twice what we eat now. Grains used to be our chief source of protein, supplying an average of 38 grams per day—more than half the Recommended Dietary Allowance. In the early 1900s, we appreciated a wide variety of grains, including respectable quantities of rye, barley, cornmeal, and corn flour.

To bring a variety of whole grains into your own diet, you might take some inspiration from cultures that have developed recipes over thousands of years. Whole wheat bread is a good place to start your acquaintance with whole grains, but why stop there? *Real* rye bread (as opposed to the pale packaged variety) is a dark, heavy loaf made from whole rye flour. It was familiar to everyone except the wealthy in medieval England. Today, a dark pumpernickel made from rye is the standard in Russia.

In Ethiopia, bread means *injera:*

(Continued on page 24)

Saffron Rice

1½   teaspoons saffron threads
3   tablespoons and 3½ cups boiling water
6   tablespoons butter
2   cups brown rice
1   2-inch piece cinnamon stick
4   whole cloves
¼   teaspoon cardamom seeds, crushed finely
1   cup raisins
½   cup toasted almond slivers

Soak the saffron threads in the 3 tablespoons of boiling water for about 10 minutes.

Meanwhile, melt the butter in a heavy skillet. Add the cinnamon and cloves and coat with the butter. Over low heat, saute lightly for about 5 minutes. Be careful not to burn the spices.

Add the rice to the skillet and stir until coated evenly with the butter mixture. Cook over low heat, for about 5 minutes. Add the remaining 3½ cups of water and the cardamom seeds. Bring to a boil over high heat. Add the saffron and its soaking water; stir gently, then cover and reduce the heat.

Simmer for 20 to 25 minutes, or until the rice is tender. Drain off any remaining liquid. Remove clove and cinnamon stick.

Stir in the raisins and almonds and serve at once.

Yield: 8 servings

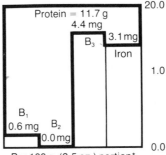

THE BOOK OF WHOLE GRAINS

Marlene Anne Bumgarner. 1976. St. Martin's Press, New York, NY 10010. 334 p. paperback $5.95.

Offers a fine variety of recipes, arranged by grain. The seldom-treated millet and sorghum both get chapters. The book is refreshing on two counts: recipes are simple and practical, and the text is free of health-food hype.

RY

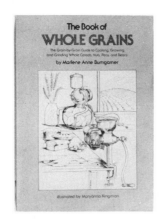

Kenke

This is yet another version of cornmeal mush, which in parts of Africa is called mealie meal porridge. Kenke is made from fermented corn dough and steamed in corn husks. In this version, yogurt is used to introduce a culture which is allowed to grow for 8 to 24 hours before cooking the dough.

4 cups whole cornmeal
4 cups boiling water
2 tablespoons yogurt
2 teaspoons salt
2 cups boiling water

Stir boiling water into cornmeal, stirring rapidly to prevent lumps. Cool to lukewarm and stir in 2 tablespoons plain yogurt. Cover with a damp cloth and leave overnight.

When ready to prepare kenke, add 2 teaspoons salt to the dough and stir dough into 2 cups rapidly boiling water, stirring constantly until mixture has thickened. Spoon onto squares of aluminum foil, folding sides and ends to make rectangles. Place on rack inside large kettle of boiling water; steam for 2 hours. Cool slightly, until dough easily pulls free from foil. Serve hot or cold.

Yield: 12 kenke

Tuo Zaafi

This stiff millet porridge is the staple meal of many North Africans. We like to shred apples over a big bowlful, and sometimes we also add chopped nuts and raisins.

1 cup millet, coarsely ground
3 cups water
½ to 1 teaspoon salt

Mix meal with 1 cup cold water. Bring remaining water to a boil; stir meal/water mixture and pour into boiling water, stirring constantly to prevent lumps. Lower heat and cook, stirring frequently, until mixture thickens. Use a bamboo paddle or wet wooden spoon to prevent sticking. Serve with honey or brown sugar and milk.

Polenta

This is the staple dish of northern Italy, and although it begins much the same as our own Southern cornmeal mush, somehow it comes out differently. The cooking of polenta is still a daily ritual in many Italian kitchens. Traditionally made in a copper pot, it now is often stirred in a large stainless steel pan sitting on a modern stove, but the old criterion for "doneness" hasn't changed: simmer and stir until the spoon will stand by itself in the mixture.

Grind 1 cup of corn to medium coarseness. Bring to a boil 3 cups chicken stock, then slowly stir in corn. Reduce heat and cook slowly, stirring often, until thick, about 20 minutes. Spoon into serving dish or bowls and top with butter and grated cheese. Polenta is usually served with meat.

Polenta may also be spooned into a mold and allowed to cool slightly before unmolding onto a plate, or it may be refrigerated in a loaf pan, sliced while still cold, and fried until crisp before serving.

■

Tabouli

4 cups boiling water
1¼ cups bulgur
¾ cup green onion or onions, chopped
3 medium-size tomatoes, chopped
1½ cups parsley, chopped
1 cucumber, chopped
½ cup lemon juice
¼ cup oil
  fresh mint, finely chopped, to taste

Pour the boiling water over the bulgur and let stand 1 hour until the grain is light and fluffy. Drain and press out excess water. Add remaining ingredients and chill for about 1 hour. Serve on salad greens.

Yield: 8 servings

(Continued from page 22)

"made from millet." Millet may be overlooked in this country, but elsewhere is highly regarded, as befits this most ancient of grain foods. Millions in Africa and Asia eat it daily. North Africans make *tuo zaafi,* a porridge, from millet. (See accompanying recipes for this and other italicized dishes.) Try millet as the principle ingredient of a hearty, nourishing soup.

Barley, whose history reaches about as far back as millet also makes a fine soup; it is the basis of Scotch broth, a Polish soup called krupnik, and a vegetarian version of the classic Polish stuffed cabbage. Another Eastern European staple is kasha, made from buckwheat and produced in Pennsylvania and central New York State. Kasha quickly cooks up into an unusually flavorful breakfast cereal.

If you're fond of Mexican food, make your own tortillas (page 30) and tamales from undegerminated cornmeal. With cornmeal on hand, you'll also want to try *polenta,* a cornmeal mush that is as important in northern Italy as pasta is in southern Italy. A yogurt culture gives *kenke,* an African version of cornmeal mush, the added tang of fermentation.

Over half the people of the world lean heavily on rice for nourishment. From simple brown rice, cooked to accompany Chinese or Japanese food, branch out to pilau, the favored rice dish of the Near East, and to *saffron rice,* which is highly popular in the Middle and Far East as well. Nutritionally, rice goes well with beans, as in the black bean soup (feijoada) served with rice in Brazil.

If "wheat" just means bread to you, bulgur will open your eyes to possibilities that are taken for granted in the Near and Middle East. Since it is precooked and dried, bulgur can be prepared quickly and easily. Hot, it makes a hearty side dish, breakfast cereal, or the foundation of a casserole. Cold, it can become *tabouli,* a minty, refreshing salad.

Carl Sherman

Making bread

Baking demystified

Authors of books on baking share a nasty trait: they seem intent upon intimidating the novice with iron-clad rules and a bewildering array of recipes.

Perhaps bakers, in becoming authors, automatically turn dictatorial and wordy. (We at Rodale Press are no exception when it comes to authors' quirks. According to one recipe published some time ago, "When the dough leaves the palm clean when held for a slow count of 25, it is ready to set aside for rising." Dally to 30, and you court disaster?) Rules and recipes tend to obscure the fact that all breads spring from a very few basic recipes. All breads, by definition, are cooked flour or meal of one combination or another. Three principles are used to loft the ingredients into something other than a warm brick: fermentation (with yeast), interaction between acidic and basic ingredients (baking soda), or heat (producing the puffed breads of India and the Middle East). Botched bread can usually be traced to a broken principle. Yeast may be annihilated by cold kitchen drafts. Baking soda loses its power with time. And a novice's puris and pitas may refuse to puff up unless the heat is just right.

Most recipes are bossy, pedantic, and humorless. They don't suggest that foodstuffs are fun to play with. So, instead of following recipes, take inspiration from them as you can. Experiment with times, amounts, and techniques. Yeast, oil, sweeteners, herbs, salt, kneading, pans, and an oven are all refinements that we in our affluence take for necessary, but good-tasting, nutritious bread can be made with just flour and water and a frying pan over the fire. I can make a whole wheat tortilla for a sandwich in five minutes—from ingredients and cold stove to my plate.

My bread-making took a giant step forward the day I abandoned measuring devices in favor of my own judgment. My hands now remember what the dough of a successful loaf feels like, so that I can duplicate the proper plasticity and texture by adding a handful of this and a pour of that.

Violating recipes is good exercise. I have violated so many that my imagination is liberated to the point where everything on the kitchen shelves is a potential bread ingredient.

Read the gentle recipe that follows, and put your hands to work. Discover that home baking combines the best of sport and craft: through playful, relaxing exercise, you make a handsome and valuable product.

RY

Starting out simply

Three months ago, I had never baked bread. Since then, I have baked all the bread our family eats, and had a few extra loaves to give away besides.

There's no miracle in the story. I simply found out how easy it can be to make bread, along with how satisfying it is and how good homemade bread tastes. I'm still just a beginner and wouldn't presume to give advice to an experienced baker, but if you've never made bread, or tried a time or two and failed, my experience may be useful.

The breakthrough was finding a recipe for no-knead, whole wheat bread. I'd always assumed that baking bread was a complicated, time-consuming task, worth the effort only if you had hours to kill. But you can make whole wheat bread that tastes great and only takes about 20 minutes of actual preparation time.

This bread originated in County Cork, Ireland, and is baked and served daily at Fitness House, Rodale Press' employee eatery. It's drawn more praise from readers of *The Rodale Cookbook* than any other single recipe.

Fitness House Bread

7½ cups whole wheat flour
6 teaspoons dry yeast
4 cups warm water
1 tablespoon honey
4 tablespoons molasses (unsulphured)
1 teaspoon butter (for greasing pans)

It's best to warm the whole wheat flour by placing it in a large bowl in a warm oven for about 20 minutes.

Sprinkle the yeast over 1 cup of warm water and add the honey. Let this mixture work while you get the rest of the ingredients and utensils ready.

Butter 2 large loaf pans (or 3 small ones), taking care to grease the corners of the pans well.

Mix the molasses with 1 cup of warm water.

Remove the warmed flour from the oven, and combine it with the yeast and molasses mixtures. Then add the 2 cups of warm water, and mix the flour and liquid thoroughly. The dough should be quite sticky.

Turn the dough into the buttered pans and let it rise to at least ⅓ more of its size in bulk. This should take 45 minutes to an hour. When the bread is nearly finished rising, preheat the oven to 375° F. Then put the loaves in and bake for 45 minutes or so, until the crust is brown.

Remove the pans from the oven and cool them for about 10 minutes. Then loosen the loaves, turn them out onto a rack and cool further before slicing.

Yield: 2 large loaves, or 3 small ones

That's all there is to it. So if you think bread-making is too difficult for you, try this recipe. It will change your mind.

By baking three or four loaves at a time, I only need to bake once a week to meet our family needs. We've never had such good bread, and for so little cost. The sense of satisfaction and the lovely aroma throughout the house are free.

TD

The ingredients for Fitness House bread are few.

Stirring is a lot easier than kneading.

The dough rises handsomely.

This moist bread tastes good with or without butter.

The inside story

Mix the ingredients, then knead the dough. After it rises, punch it down and let it rise again. Bake it for an hour or so, let it cool, and you've produced a loaf of bread.

But what is really happening? Why is bread kneaded? What makes bread rise? Or fall? Why is some bread heavier than other bread?

Making bread may seem like magic, but it's not. There's a fairly simple explanation for most of what happens, and with a little understanding you can usually figure out why things went right—or wrong.

The basic ingredients of yeast bread are water, flour, and yeast. (Continued on page 26)

Millet

The primary food grain in China before the introduction of rice. Still a staple in North China, India, and Ethiopia.

Appearance: Tiny, round kernels that look more like seeds than grains. Swells hugely when cooked: pound for pound supplies more servings than any other grain.

Uses: Mostly fed to birds in the U.S. Boiled like rice, makes excellent breakfast cereal and main dishes. Thickens and flavors soups and stews. Has too little gluten to make raised bread, but can be used in Indian chapatis. Add half a cup of whole grain to a loaf of wheat bread for crunchier texture.

High in essential amino acids, and richer in iron than any other grain except amaranth. Unlike most grains, has a high alkaline content, making it easy to digest.

Protein = 9.9 g

0.73mg B_1	0.38mg B_2	2.3 mg B_3	6.8 mg Iron

20.0 / 1.0 / 0.0

Per 100 g (3.5 oz.) portion*

Oats

Native to Central Asia, this good animal food followed the horse to Europe. The Scots are responsible for making oatmeal a common food in northern Europe, Ireland, and the U.S.

Appearance: A long, light brown grain seldom seen in whole grain form. Rolled or cut oats retain both bran and germ.

Uses: An ingredient of granola and cookies. Used in place of bread crumbs to bind loaves or patties of meat, fish, or nuts. Combined with wheat or rye flour, oat flour makes an excellent loaf. Longer-cooking, steel-cut or rolled oats make a more nutritious oatmeal than highly processed quick-cooking varieties.

One of the highest among cereal grains in protein, calcium, and thiamine (vitamin B_1).

Protein = 14.2 g

0.6 mg B_1	0.14mg B_2	1.0 mg B_3	4.5 mg Iron

20.0 / 1.0 / 0.0

Per 100 g (3.5 oz.) portion*

*B_1 is thiamine, B_2 is riboflavin, B_3 is niacin.
Source: *Composition of Foods,* USDA Handbook No. 8.

(Continued from page 25)

Some kind of sweetener and fat are usually added, but they're really not essential.

When the three main ingredients are mixed together, several things happen. The water enables the proteins in the flour to start bonding together and form gluten. The water also wets the starch granules in the flour, causing them to expand, and it helps disperse the yeast through the dough.

Gluten is the glue which holds the dough together, trapping the carbon dioxide given off by yeast and thereby enabling the bread to rise. It is made from flour proteins found in the endosperm (the starchy part of the wheat).

Yeast consists of millions of fast-growing plant cells that feed on the sugars in the dough. As the yeast grows, it produces alcohol as well as carbon dioxide gas, but this bakes off. Yeast helps give bread its flavor and adds to its nutritive value.

The rising time and the height to which the bread will rise also depend on the gluten. Hard wheat flours—made from wheat which has hard, flinty kernels—are generally high in gluten and make good bread. Because white flour is made entirely from the wheat endosperm, it is higher in gluten than whole wheat flour, and makes a lighter (though less nutritious) bread.

Flour with too little gluten, or with weak gluten, does not rise properly. If the gluten does not hold together, the bread will not rise at all. If it is too weak, the dough may rise too rapidly, since the gluten stretches too easily.

Most recipes suggest mixing the yeast with the water and sweetener and letting this mixture stand for 20 mintues or so before adding it to the flour. This not only allows the yeast to begin to work, it also "proofs" it—lets you know whether it is active. If the yeast is not fully active, or if you use too little, the dough will not rise well, and will yield heavy bread with a firm crumb. Too much yeast will make the bread rise too far, giving it a large coarse grain and a yeasty flavor.

The purpose of kneading is to thoroughly mix the dough, and to stretch or develop the gluten. In a way, the gluten in unkneaded flour is like short, stiff bands of taffy. These bands cannot expand very far without breaking. Kneading the dough gradually pulls these bands, so they can stretch further and more uniformly without breaking. In addition, it mixes these bands more evenly throughout the bread.

It is possible to make excellent bread without kneading, as with our Fitness House bread recipe. Given sufficient gluten and yeast, the bread will rise to some extent. But without the stretching and mixing that kneading provides, the bread will be heavier and tend to have an uneven texture.

Bread should be kneaded until it feels smooth and elastic—anywhere from 5 to 15 minutes. Improperly kneaded bread rises poorly. If you knead too little, the gluten remains stiff and holds the bread down. Too much kneading breaks the gluten, and the bread falls. A simple way to tell when the bread has been kneaded enough is to poke it with your finger. If your finger leaves an impression in the dough, it means the gluten is well stretched. At an earlier stage, the elasticity in the gluten would cause the impression to disappear rapidly.

After kneading, the bread is set aside to rise. This rising process works best at a temperature between 75 and 86° F. At higher temperatures, the yeast will grow more rapidly, but other undesirable microorganisms are more likely to grow as well. A standard practice is to let dough rise until it doubles in size. That can take anywhere from 30 minutes to several hours, depending upon the temperature, the kind of bread you are making, and the amount of yeast you used.

When the dough has risen, it should be knocked back or punched down. This lets the carbon dioxide gas escape, mixes the yeast cells more evenly throughout the dough, and gives them access to fresh food and oxygen. The second rise results in a more evenly grained bread and a lighter loaf.

The dough continues to rise during the first 15 minutes or so of baking. The yeast is still producing carbon dioxide, and other gases in the bread are expanding because of the heat. When the temperature in the loaf reaches about 140° F., the yeast cells die. If the oven is too hot, the

(Continued on page 28)

Rice

The most widely consumed cereal grain in the world. Americans eat so little rice that despite producing only 1 percent of the world's crop, we are the leading rice exporters.

Appearance: Long, medium, or short kernels covered with a green-brown husk. Brown rice has only the husk removed; white rice has lost its germ and several outer layers.

Uses: Can be boiled as a basic food and served alone or with meat or vegetables. Ingredient of soups and casseroles. In Asia, made into sweet dishes, noodles, and wine. Flour is used primarily for Oriental foods: small pastries, sweet cakes, and noodles. Can be used with rye and oat flour to make wheatless bread. Cooked whole grain is good for breakfast or as an accompaniment to main dishes.

Brown rice is a good source of carbohydrates, with substantial amounts of protein, minerals, and B-vitamins. Milling away the hull to make white rice takes away much of the nutritive value.

  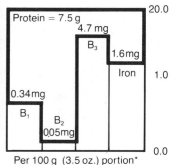

Protein = 7.5 g

B_1 0.34 mg
B_2 0.05 mg
B_3 4.7 mg
Iron 1.6 mg

Per 100 g (3.5 oz.) portion*

Rye

Perhaps the hardiest grain of all. Grows at temperatures as low as 40° F., and can overwinter at −40°F.

Appearance: Has dark brown kernels that are longer and thinner than those of wheat. Especially attractive to ergot, a virulent fungus that causes a nervous disorder known as St. Anthony's Fire.

Uses: Flour has less gluten than wheat, so it makes a heavier bread. Bread color ranges from pale to dark brown; German pumpernickel, made with unrefined rye flour and molasses, is the blackest and heaviest form. Can be boiled like rice for a very substantial cereal.

Lower in protein than wheat, but a good source of potassium, phosphorus, and iron.

  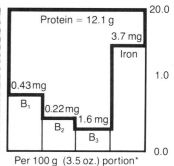

Protein = 12.1 g

B_1 0.43 mg
B_2 0.22 mg
B_3 1.6 mg
Iron 3.7 mg

Per 100 g (3.5 oz.) portion*

Sorghum

Unknown to many people, despite being the major cereal grain of Africa.

Appearance: Roundish seeds, slightly smaller than peppercorns, that are brown with yellow and red coloration mixed in. Lighter varieties—sometimes called yellow endosperm sorghum—are the best tasting.

Uses: A principal feed grain in southwestern U.S. Boiled seeds can be eaten whole, ground into meal for cereal, or milled into flour. Syrup is made from sweet sorghum stalks. Flour is delicious in pancakes and cookies. Chinese and Japanese eat sorghum boiled like rice; Africans cook freshly ground meal into porridge or brew the grain into liquor.

Protein = 11.0 g
Iron 4.4 mg
B_3 3.9 mg
B_1 0.38 mg
B_2 0.15 mg
20.0 — 1.0 — 0.0

Per 100 g (3.5 oz.) portion*

Triticale

The youngest grain—a man-made hybrid that combines the productivity of wheat with the ruggedness of rye. Has high nutritional value.

Appearance: Gray brown, oval-shaped kernels, larger than wheat grains and plumper than rye grains.

Uses: Flour makes raised bread with a ryelike flavor, but is generally mixed with wheat flour to strengthen the gluten and produce a higher loaf. Triticale's gluten can be worn out by too much kneading. Whole or cracked, it makes excellent cooked breakfast cereal.

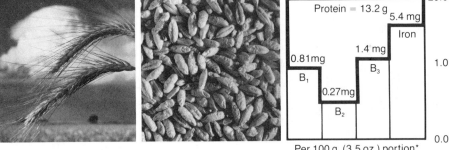
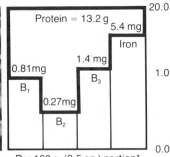

Protein = 13.2 g
Iron 5.4 mg
B_3 1.4 mg
B_1 0.81 mg
B_2 0.27 mg
20.0 — 1.0 — 0.0

Per 100 g (3.5 oz.) portion*

*B_1 is thiamine, B_2 is riboflavin, B_3 is niacin.
Source: *Composition of Foods,* USDA Handbook No. 8.

(Continued from page 27)

yeast will die sooner, resulting in heavier bread. If it is too cool, the bread may rise too far and either develop a bubbly appearance or even collapse.

After the first quarter hour of baking, the gluten proteins begin to thicken and the starch particles harden. The internal structure of the bread becomes firm, and a crust develops as the surface dries. During baking, the moisture content of the bread decreases from about 45 percent to 35 percent.

In 30 minutes to an hour, you have a finished loaf of bread.

—TD

Melt-in-Your-Mouth Biscuits

    1   cup whole wheat flour
    ¼   teaspoon salt (optional)
    2   teaspoons baking powder
    ⅔   cup light cream
  2½   teaspoons sour cream

Sift together flour, salt, baking powder, and gradually stir in the light cream with a fork. Then add the sour cream. Drop 6 portions about 3 inches apart on an ungreased baking sheet.
Bake at 450°F. for 12 to 15 minutes.
Yield: 6 biscuits

Breads of different shapes

Breads from other lands are not only novel, but usually easier and quicker to make than the yeasted loaf of the Western world. Recipes can be found in modern and good bread cookbooks.

Chapatis and tortillas are fastest, requiring little finesse and no kneading. Just mix flour and water, combine with the hands, roll out, and cook a couple of minutes on each side in a dry skillet. These flat breads usually are made as part of an ethnic meal, but they can be kept on hand as sandwich bread. Heat them until pliable just before you're ready to add the peanut butter and jelly. Tomato and melted cheese go well, too. Or, use them as the structural component of a handfilling, multilayered dagwood.

Quick-baking pita bread (pocket bread) is more demanding, as it involves yeast and high heat (500° F.). Unfortunately, our experience suggests that batches may not come out with a well-developed air space—and it's no fun trying to stuff sprouts and mayo into a pocketless pita. Properly hollow pita loaves are a boom to lunch toters tired of sandwiches that disassemble in transit, because they provide a sandwich with not only a top and bottom but walls as well.

A traditional Middle East accompaniment to pita is hummus, a combination of sesame seed butter (tahini) and chickpeas. For a recipe, see the Beans chapter, page 88.

(Continued on page 31)

Flat breads cook quickly in a pan, and unlike oven breads, making them won't render the kitchen unlivable in summer.

86 steps

You're 86 steps and $305.75 away from this home-built baking center, designed at Rodale Press. A built-in proofing box, heated by a rheostat-controlled light bulb, encourages yeast to do its best work. The butcher block top extends over the sides so that food processing equipment can be clamped on, and is 34 inches high, 2 inches below normal to facilitate kneading and rolling. The center rolls about on heavy-duty, 3-inch casters. Drawers and cabinets provide storage space for baking utensils that always seem to lose themselves at baking time. There's even an electrical receptacle on one end for countertop appliances. You can build a baking center with less work and fewer dollars, but this prototype may inspire you to come up with your own design.

Wheat

The most widespread and versatile grain, a staple food for half of the world's population. As early as 4000 B.C., the Egyptians were making light, yeasted wheat bread.

Appearance: Produces short, rounded kernels of varying shades of brown. Darker (red) wheats produce flour for bread and pastry; lighter, white wheat is used primarily in pastries and processed breakfast cereals.

Uses: Milled into flour to make bread, pastries, cookies, crackers, pasta, and cereal. Primarily outside the U.S., it's cracked, cooked, then combined with meat or vegetables for a main dish. Produces the best bread-making flour of any grain. Bran and germ are removed during the commercial milling of white flour. Wheat flakes can be boiled for breakfast cereal. Add whole grains to rice and burgers for a chewy texture. Wheat can be sprouted for sweeter flour, or to make malt for baking.

On average, white flour has 21 percent less protein, 55 percent less calcium, 77 percent less phosphorus, and 74 percent less potassium than whole wheat.

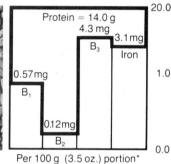

Protein = 14.0 g
B_1 = 0.57 mg
B_2 = 0.12 mg
B_3 = 4.3 mg
Iron = 3.1 mg

Per 100 g (3.5 oz.) portion*

2 loaves, from grain to bread, in 2 hours 20 minutes, for 40¢ each.

Flour Tortillas

This recipe introduces soy flour for its ability to complement the protein in wheat. If the tortillas are to be filled with beans or topped with cheese, there is little or no disadvantage to using straight whole wheat flour.

½ cup soy flour
½ cup water
1 cup whole wheat pastry flour

Mix soy flour and water together well. Blend in the whole wheat flour. Knead the dough until smooth and shiny. Divide into 6 small balls.

Sprinkle a sheet of waxed paper heavily with whole wheat flour. Lightly flatten one of the balls of dough and lay on the waxed paper. Sprinkle with flour and lay another sheet of waxed paper on top. Roll paper thin.

Heat a cast iron skillet over low-medium heat. When hot, lay the tortilla in the pan and bake until lightly brown, then turn to brown the other side. Tortillas should remain soft and flexible for use in such foods as burritos (page 88), or they can be cooked till crisp for use as a chip.

Yield: 6 tortillas

For variety, try substituting cornmeal for up to half the flour, or masa harina (a pale corn flour) for any or all the flour.

A tortilla press of wood.

Pita

Pockets can be filled with avocado dip, salad, cheese, alfalfa sprouts, a combination of cooked vegetables, or mozzarella and spaghetti sauce for an inside-out pizza (known in some parts of North America as a stromboli).

Pennsylvania's Trexlertown Velodrome offers spectators its alternative to the hot dog. Into a whole wheat pita are stuffed mayo, cucumber, carrot, onion, green pepper, and muenster cheese, topped off with alfalfa sprouts.

¼ teaspoon honey
2 tablespoons lukewarm water
2 teaspoons dry yeast
3½ cups whole wheat flour
1¼ cup lukewarm water
2 tablespoons oil (olive or other if preferred)
¼ cup cornmeal

Add ¼ teaspoon honey to 2 tablespoons lukewarm water and sprinkle yeast over surface. Set aside for 2 or 3 minutes, then stir to dissolve yeast completely. Set aside again for 5 minutes until mixture is working.

Place 3 cups flour in a bowl. Combine remaining lukewarm water with oil, add to yeast mixture, and pour into center of flour, stirring until dough forms one mass and leaves the sides of the bowl. Turn dough out onto lightly floured board and knead for 8 to 10 minutes, using approximately ⅓ cup flour in the kneading process. Lightly oil surface of dough, cover and leave in a warm place free of drafts for 45 minutes to an hour, or until dough has doubled.

Punch down dough, form into 4 balls about 2½ inches in diameter. Cover and let them rest for 30 minutes. Roll out each ball to a diameter of 8 inches, with a thickness of ⅛ inch. Sprinkle cornmeal on baking sheet wherever breads are to be placed, and lay them 2 inches apart on the baking sheet. Cover and let rise 30 minutes.

Preheat oven to 500° F. Bake breads 4 minutes on lowest rack of an electric oven (if you have a gas oven, bake the breads 4 to 5 minutes on the floor of the oven), then bake them 3 to 4 minutes on the rack about 3 inches above the lowest rack (or floor of oven). Breads should puff up about 2 inches while on the lower shelf, and after being moved up they will turn brown and cook through, remaining puffed. If you're not serving them immediately, wrap cooked breads in foil and set aside. When unwrapped, these breads will have fallen somewhat, but the pocket will remain. They may be broken apart and eaten with butter, or cut in half and filled.

Yield: 4 8-inch breads

Chapatis

Chapatis are one of several breads made in the Indian home. They look much like Mexican tortillas, and the preceding recipe will serve. Chapatis are best eaten warm and straight from the pan, but can be stored and then reheated briefly to restore their flavor and suppleness.

Whole wheat flour works well, but you might try whole wheat pastry flour for a more delicate bread. In Kashmir in the north of India, chapatis are also made from bean, corn, or rice flours. These work in flat breads because without leavening there's no need to call on the elasticity of wheat's gluten.

(Continued from page 29)

A bagel is a donut without the burden of oil: bagels are boiled in water prior to being baked, while donuts are boiled in oil at a temperature that makes further cooking unnecessary. Because of their small size and the hole in the middle, bagels cook in a third the time of loaves. They're also more fun to make than bread—something like the relaxing exercise of shaping clay into non-objective shapes—and more fun to eat than either clay or bread.

Commercial bagels come light and dark, but darker is not necessarily more wholesome. Pumpernickel and rye are attractive sell words, and are more descriptive of color than flavor or ingredients. A dark shade is likely the work of caramel—burnt sugar. Redress this misfortune by making your own honest bagels.

RY

Bagels

6   cups whole wheat flour, or 4 cups whole wheat and 2 cups rye
2   cups water
2   tablespoons dry yeast
¼   cup oil
¼   cup molasses or honey
2   tablespoons caraway seeds (optional)
3   eggs

Place the flour in a mixing bowl. Dissolve the yeast in water, then stir in the oil and molasses or honey. (Add caraway seeds if you choose.)

Stir this liquid into the flour. Beat the eggs, and stir in.

Place the dough on a floured board and knead for 10 minutes. Return the dough to the bowl and allow to rise until doubled in volume.

Preheat oven to 350° F.

Knead the risen dough a minute or two, then remove apple-sized chunks, one at a time, and roll them into cylinders of ¾ to 1 inch in diameter. Precision isn't necessary, but at this stage, any deviations from perfect cylinder will be greatly magnified when the bagel rises. Cut cylinders in lengths of 6 to 8 inches, and connect their ends by pinching to make what looks like a pale, emaciated bagel.

Place a pot with 4 inches of water on to boil. Place dough rings in 1 or several at a time. They'll sink, then swim for air. When they appear at the surface, flip them over for a minute or two. Then place them on oiled cookie sheets.

(At this stage you can first dip one side in a bath of egg yolk and then on a plate of chopped, fried garlic or onion, or sesame seeds.)

Bake for 10 minutes, then check to see how the bagels are doing. They're done when tanned golden on top, but can be left in longer if you prefer a tougher bagel.

Bagels freeze well. Let them cool and throw off water vapor before bagging and freezing.

Yield: about 2 dozen

Cakes from the pan

Pancakes can be thought of as our version of the Mexican tortilla and the Indian chapati. The similarity is slight. While the two foreign foods are important to the diets of their nations, we relegate our flat bread to the role of an occasional breakfast indulgence. Pancakes *are* the star of one of America's few culinary traditions: going out for a Sunday brunch of pale, syrup-smothered discs. This ultimate carbohydrate feast challenges church as a Sabbath draw.

The pancake can be more than an over-sized confection, and without giving it the tiresome brewer's yeast-and-soyflour treatment. Just consider for a moment the pancake's potential. Use anything other than white flour, and you'll find this mild-mannered food takes on a robust character, a new personality that invites the complement of other than the standard butter and syrup. Cakes of whole wheat, rye, or buckwheat move up to main-dish status with toppings and accompaniments such as applesauce, yogurt, or a layer of cheese (melted on the flip side by putting the lid on for the last minute of cooking).

Potato pancakes can be made pretty and sweet, but a good oniony or garlicky batter cries out for yogurt or sour cream, or a tart fruit compote. More-conventional people might enjoy ground pepper, mustard, ketchup, a dill pickle, or other condiments customarily used to give hamburger a personality.

Most cookbooks add little to the literature on the most-neglected meal of the day. No wonder it's pancakes as usual, year in and year out. An exception is Diana Scezny Greene's *Sunrise* (a $5.95 paperback from Crossing Press, 1980). This cookbook challenges the reader's preconceptions with dozens of original recipes. Here is one for a homemade pancake mix that, like the store-bought versions, awaits only egg, liquid, and frying pan.

(Continued on page 34)

The dough hook

Get set to forget all you know about mixing and kneading dough when you use a dough hook. No matter what your recipe calls for, kneading takes only three minutes. This is an especially helpful tool for large batches of bread (it has a capacity of seven or eight loaves) but it's a time-saver no matter what volume you're mixing.

All ingredients are mixed in the basin, liquids first. After a total of three minutes for mixing and kneading, the first raising takes place right in the basin. After it has risen, the hook is turned until the dough forms a ball. We anticipated having a problem getting the hook out of the dough but it came right out, leaving a perfect dough ball.

Dough mixers are sold by Garden Way ($27.95), Charlotte, VT

05445, and Cumberland General Store ($29.95), Route 3, Crossville, TN 38555. Attractive, heavier-duty mixers occasionally turn up at yard sales and flea markets.

Baking: Potomac, Maryland

For Betsy Sussman, baking has become an important part of supporting her family of four. Two of those four are often on hand to help. The fourth, writer Vic Sussman, hides in vain from the smell of baking, and blames his six-mile-per-day running habit on Betsy's success with bread.

Baking is one of several important activities that allow the family to live on a modest income in the high-priced outskirts of Washington, D.C.

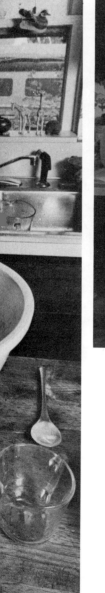

FLOURLESS BREAD?

Bread made without flour sounds a little suspect. It may look odd, too, and has a rather unique taste and texture. There's a knack to making flourless bread at home, we've found—a knack that escapes us—but at least two commercial bakeries are enjoying success with a recipe they believe has an almost divine origin. It comes from the *Gospel of Peace*, a holy book of the first-century Essene Community, and some attribute it to Jesus.

The recipe is translated thusly: "Moisten the grain, and the spirit of water will enter and cause the germ of life to sprout. Then crush your grain and form thin loaves as did your fathers before you."

That's all. Just sprout the wheat, grind it in a meat grinder, pat it into a loaf, and bake it. You've got bread—or something quite like it—with no flour.

Through a call to the Lifestream Natural Foods people of Richmond, British Columbia, we learned the secret is to bake slowly at low temperatures and carefully controlled humidity. But they wouldn't say just how to grind the sprouts, or what baking times or temperatures they use. The method remains as arcane as the ingredients are simple: sprouted wheat and unchlorinated water.

A taste of their bread is enough to make one experiment in the kitchen. Our efforts at Rodale turned out a low, chewy loaf that was something less than a taste treat, and didn't belie the fact that its ingredients were humble.

TD and RY

Crab Grass Cereal

Crab grass seeds are plentiful nearly everywhere there are lawns and may be ground or run through a food processor to create a very coarse meal.

3 cups water
1 cup crab grass seeds, coarsely ground

Boil water in a pot over high heat. Stir in crab grass meal and boil for 30 seconds. Cover pot tightly and reduce heat. Cook 15 minutes.

Serve with berry topping or honey, and coconut milk.

(Continued from page 31)
Pancake Mix

2 cups hard wheat flour
⅓ cup soy flour
½ cup wheat germ
⅓ cup powdered milk
2 tablespoons baking powder
1 teaspoon powdered kelp (or ¾ teaspoon sea salt)
1 tablespoon lecithin granules

Combine all above ingredients in a medium bowl. Mix thoroughly and store, refrigerated, in a tightly lidded glass jar. This mix will make 3 batches of pancakes for 2 persons.

2 eggs
1 cup whole milk (or 1 cup water and ⅓ cup powdered instant non-fat milk)
2½ tablespoons cold-pressed safflower oil

To make the pancakes, measure the above ingredients into a small bowl. Combine with 1 cup mix.

Novel flours for variety and enhanced nutritional value

We're all familiar with wheat bread, and most of us know rye bread as an aromatic, somewhat darker variation. But bread can be made from a variety of grains and even peas and beans, either as flour, meal, or flakes. The list that follows should stimulate your wheat-stifled imagination. Wheat will likely be an ingredient in most recipes, as only this grain contains enough gluten to loft a loaf into a familiar shape and texture.

Pea and bean flours can make wheat-based bread a better food. While whole wheat is high in several minerals, protein, and carbohydrates, its weak spot is lysine, one of the amino acids necessary to make a complete protein. Lysine is a "limiting" amino acid, which means that when it is used up, the body cannot make use of the additional amounts of the other amino acids. Add bean or pea flour to increase the lysine, and the loaf will offer more usable protein. You'll like the variety that high-protein flours give to your baking. Experiment a little and find out what combinations work well and taste best to you. Remember that some of these flours, especially in larger quantities, can make a considerable difference in baking and handling qualities.

Amaranth. A mild-tasting flour that works well in quick breads using baking powder or soda.

Barley. Used with wheat to give bread a sweet, cakelike character.

Bean and chickpea. Bread with 15 percent pea flour has an increased protein value of about 15 percent and increased fiber, with little change in taste, baking quality, or cost. Any green or yellow tinge is largely bleached out in baking. The humble navy bean—especially when roasted—makes a flour that mixes well with wheat at levels up to 15 percent.

The flours can be made at home. Work with about ¼ cup of dried peas or beans at a time, sifting out the coarse particles and regrinding them with the next batch. (Some stone mills aren't suited for grinding seeds and beans.) Minimum orders of 25 pounds of these flours are available from Regina Nut Products, 5213 Second Avenue, Brooklyn, NY 11232.

Buckwheat. Has a strong, distinctive taste which is mellowed by other flours.

Corn. The meal, more coarsely ground than flour, gives a distinctive corn flavor to breads, muffins, and crackers.

Millet. Millet's sweet, nutty flavor is brought out by light toasting before milling.

Oat. A bland, sweet taste. Contains an anti-oxidant that keeps bread fresh longer. Use flakes, or flour available ready-ground as Scotch oatmeal.

Peanut. In tests with cookies, peanut flour behaved like wheat flour and could be used at levels up to 30 percent. With this much peanut flour, the cookies had twice as much protein as those made with wheat flour only. Tasters reported that peanut flour did not affect the flavor of the cookies, though it did cause increased browning and an overall darker color.

Potato. Dried cooked potato can be milled and combined with wheat flour.

Rice. Rice flour can be used to supplement wheat flour in most recipes. It is used in Chinese and Indian noodles, Scottish short bread, and Sri Lankan crepelike *hoppers* (page 90).

Rye. Used with wheat flour for rye bread and pumpernickel. Excellent for sourdough breads. Pure rye loaves are heavy.

Sorghum. In complement with grain flours, boosts protein. The strong flavor may come through unpleasantly unless used sparingly. (See recipe, page 76.)

Soy. According to some experimenters, flour from sprouted soybeans makes a loaf that tastes and looks better. Soybeans are oily and few store mills can handle them; the beans even clog some burr mills. If your mill isn't touted as being up to soybeans, you can buy the flour from the following: Walnut Acres, Penns Creek, PA 17862; Shiloh Farms Inc., P.O. Box 97, Sulphur Springs, AR 72768; and Sam Wylde Flour Company Inc., 6901 Fox Avenue, Box 24724, Seattle, WA 98124.

Sunflower. According to tests, bread with 15 percent sunflower seed flour has 32 percent more protein than white wheat bread. You can make this novel flour at home with either a blender or a mill. The blender may work best, as sunflower seeds contain considerable oil and tend to clog mills. Sunflower flour is available from Walnut Acres, Penns Creek, PA 17862; Hopkinsville Milling Co., P.O. Box 669, Hopkinsville, KY 42240; and Green Earth Enterprises, 1505 Newton Street, Los Angeles, CA 90021.

Triticale. Hybrid cross between wheat and rye, with pleasant nutty taste. Makes a slightly heavier loaf than whole wheat.

Wheat. Pastry flour is very finely milled. Gluten flour is high in stretchy protein and rises well, but much of the grain's goodness is sacrificed for aesthetics.

Grow your own yeast

Yeast is one of the easiest of house plants to grow. You'll save money growing your own and probably get better results as well because

(Continued on page 37)

MAKING BREADS WITH HOME-GROWN YEASTS AND HOME-GROUND GRAINS

Phyllis Hobson. 1975.
Garden Way Publishing, Charlotte,
VT 05445. 44 p. paperback $2.95.

An attractive little book that takes baking all the way back to growing grain in the garden and propagating yeast. Ms. Hobson even offers methods for putting wild yeasts to work, but cautions that she hasn't had any luck. The recipes are simple and hearty (calling for white flour in only one instance), and range from basic whole-grain breads to such exotics as potato donuts and skillet-puffed wheat cereal.

RY

Growing your own wheat—One-fourth of an acre of rich, fertile land should produce at least five bushels of wheat, more than a year's supply for the average family. Wheat requires well-fertilized soil, so prepare the plot in advance by plowing under barn manure, then sowing a cover crop of rye. In the spring, work under this cover crop and till the soil to a fine bed.

Planting—On this prepared bed, broadcast (by hand or with a broadcast seeder) one peck of seed to the one-fourth acre. Walk with a medium pace as you seed. You can leave the seed uncovered, but it will germinate better if you can roll it down.

The time to plant wheat depends on the variety, which depends on the climate and the soil. Here in the midwest, we plant hard winter wheat in September and harvest it the following July. But some wheats are planted in the spring and harvested in the fall. Your local County Agent will know the variety and time for planting in your area. He can also tell you where to obtain seed.

Once the seed is planted, it's up to the sun and the rain. Wheat will sprout within a week and soon will look like bright green grass. In every stage, wheat is one of the most attractive crops, so it would not be unpleasant near the house.

As it matures, the green stalks grow two or two-and-one-half feet tall, then turn golden. When it has passed the peak of its golden color and turns tan, the wheat is ready for harvest. You can test it by tasting a grain. It should be hard, yet soft enough to be dented with pressure from a fingernail. And it should have a nut-like flavor.

Harvesting—If you let your wheat reach this past-its-golden-prime color, you can skip the step of shocking it in the field to dry, although this is the most picturesque stage. If you want to shock your wheat, cut the stalks, tie them into bundles and stand them in the sun to dry for a week to 10 days.

Either way, you're now ready for threshing. You can vent your frustrations on it, as the European peasants used to do, or you can do it in half the time with half the effort by beating the kernels directly into a barrel.

As you cut each bundle, hold it, head down, in the barrel and hit it several times sharply against the sides to knock off the kernels of wheat. When most of the kernels have fallen into the barrel, break or cut off the end of the stalks and toss them to the chickens, which will finish threshing them for you. Save the stalks. They're called straw, and it makes fine barn bedding for the winter ahead.

If the wheat in the barrel is "dirty" (mixed with chaff), you'll need to winnow it. You can make a big production of this with a sheet of canvas and a fanning mill; you can toss the grains of wheat into the air on a windy day, or you can do as we do—simply pick the weed seeds out of each cupful just before we grind it into flour, which is just before we make bread.

Brick-oven baking

In a brick oven, food is cooked by heat that is stored in the bricks. Cooking starts with building a substantial fire in the oven with hardwood. When the bricks reach a temperature of about 700° F., the fire, coals, and ashes are removed. Food is then cooked by the heat in the bricks. I've found my oven will hold temperatures for four hours at 350° to 375° F., three additional hours at 300°, and two more hours at 250°.

After putting out the fire, remove the coals and ashes, and sweep the oven floor clean. Allow the door to remain open to cool the oven down to about 300°. Close the door and in ten minutes the temperature should be up to 400°—and you're ready for cooking.

I've cooked entire meals in my

oven at once, including breads, roasts, puddings, and other "delicate" foods. One big advantage of an outdoor brick oven is that you can be preparing your food while the oven is heating up, yet your kitchen won't be overheated. Because the oven holds such a steady heat, you can plan your dinner so everything is ready at once, quite a switch from charcoal-grill cooking.

The first step is to pour a concrete slab in the shape of a half circle, with a diameter of 6½ feet. The slab should be 18 inches above ground level, and extend below the frost line. I made mine 30 inches below ground level.

Once the slab has cured, begin to stack bricks on it in the shape of the oven, using no mortar. Try to get the flat type of fire brick, as opposed to the conventional rectangular shape: flat bricks will stack much easier. When stacking the bricks on the very outside of the oven, you might have to chip off a corner every now and then to keep the shape. If you are handy with adobe, you could even make your own bricks.

The illustrated oven measures 5 feet across, 44 inches deep on the outside and 30 by 22 inches on the inside. Bricks are stacked in a beehive shape. Make the walls as thick and solid as you can for thermal mass to store the heat to cook the food. The more bricks, the longer the oven will hold an even temperature. An opening in the top, about the size of a coffee can, serves as the draft vent and chimney. Leave a door opening of about 2 feet, or big enough to put in a roast.

A Kashmiri baker makes the morning batch of bagellike breads in a wood-fired tandoor.

GALLAGHER'S EXPENSES

114 clay fire bricks at 25 cents each	$28.50
Iron door	$30.00
Regular cement and mortar	$15.00
Total	$73.50

A door and metal frame can be made at a local welding shop for modest cost. Or, make your own of wood: you will have to remove it when burning the fire to heat the oven, and it must have an angle-iron frame to fit into. The metal frame also provides support for the bricks that lay directly on top of the door.

With all bricks in place, apply concrete to the outside, pressing it well into the cracks and chinks between the bricks. Cement a coffee can in the chimney opening in the top.

Allow the oven to cure for four or five days. Occasionally wet it down with a hose to help it dry evenly and prevent cracking. After it has cured, seal the kiln with a coat of white cement or mortar, mixed to a soupy consistency, with a utility brush or large paintbrush.

Allow it to cure a few more days, and you're ready to go.

William Gallagher

(Continued from page 35)
the yeast will be fresher and more active.

You're likely familiar with packaged dry or cake yeast; the recipe below makes a third form: a yeast sponge, used by our grandmothers for breadmaking. Sometimes these starters were kept for years at a time, and even passed down from one generation to another.

Yeast Starter

5	potatoes
4	tablespoons honey
1	tablespoon ground ginger
2	cups whole wheat flour
1	package dry yeast (or 1 cup homemade yeast mixture)

Peel the potatoes, cover them with water, and boil until soft. Don't discard the water; reheat it to boiling as you mash the potatoes. Mix honey, ginger, and flour with the mashed potatoes, then pour a pint of boiling potato water over the mixture and beat until smooth.

When the mixture is lukewarm, start it by adding 1 package of dry yeast that has been dissolved in a small amount of warm water according to package instruction. Allow the mixture to rise for an hour or so, then store it in the refrigerator. (Be sure the storage container is large enough to accommodate the yeast as it rises.)

This recipe makes about 5 cups of starter. Each cup is roughly equal to 1 package of dry yeast, or 1 yeast cake.

Yeast starter must be used or refreshed every two weeks, so it is wise to make only as much as you plan to use in that time, plus another cup for the new batch of starter. Should you need to refresh the starter, put it into a clean dish and wash out the refrigerator container thoroughly. Return the cup of starter back to the container along with fresh potato-honey-flour food. Let it stand for one hour, then refrigerate.

Even if you don't bake regularly, you can still make your own yeast. The following hops yeast cakes keep several months. (Hops help prevent the growth of bacteria, and can usually be purchased from stores carrying beer-making supplies.)

Yeast Cakes*

1	medium potato
¼	cup hops
3	cups water
1	package dry yeast (or 1 yeast cake)
¼	cup whole wheat flour
½	cup cornmeal

Cook the potato and hops in 3 cups water until the potato is well done. Drain the potato, keeping the liquid and discarding the hops. Add enough cold water to the liquid to make 2 ½ cups. Then add the yeast, and let the mixture set for 1 or 2 hours, until bubbly. Mash the potato, and mix it into the liquid along with the flour and cornmeal, stirring to make a stiff paste.

After cooling the mixture, roll into cakes, using 2 tablespoons of paste for each cake. Dip cakes into cornmeal and roll until easily handled. Dry cakes for several days on a wire rack, turning daily. When cakes are completely dry, store in a closed container in a cool place.

For longer-lasting cakes, stir more cornmeal into the yeast mixture until it becomes a hard dough. Make the dough into small, thin cakes and dry without baking or cooking. These cakes should keep for several months.

TD

*This recipe is reprinted from *Making Breads with Home-grown Yeasts and Home-ground Grains* by Phyllis Hobson (Garden Way Publishing Co., 1975). See review on page 35.

BAKERY ON WHEELS

Nan is an Indian bread baked on the inside of a clay oven known as a tandoor (see photo, page 36). In the Kashmiri capital of Srinigar, street vendors bake a potato-filled nan in a small, portable tandoor encased with a steel barrel. A slim pancake of dough is sprinkled with chopped green and red chilis, then kneaded into a ball and shaped again into a pancake, on which is placed a dollop of mashed potato. The dough is brought up around the dollop to form a ball within a ball. Then the whole thing is flattened between the palms, sprinkled with water, and slapped against the inside of the tandoor. When done, the nans are removed with the aid of two iron pokers. At home, you might experiment with a large clay flower pot, bermed with earth and topped with a stone lid.

Mills

Five reasons to buy a grain mill

A number of the mills we tested are expensive (many electric models cost more than $300 and good hand mills are now in the $200 range), but you may consider the price a bargain if you are serious about optimum taste and nutrition.

1) Superior flour. With a mill, you know the flour is fresh and that nothing synthetic or artificial has been added to it. Connoisseurs of baked goods may find that fresh flour tastes better (although our tongues failed to detect a difference between breads baked with fresh-ground and six-week-old flour). The oil in whole wheat flour is sensitive to heat, and while little degradation occurs over six weeks if the flour is refrigerated, the fresher the better.

Unless you buy flour from a health food store that has a mill, you generally can't tell how old it is. It may have been in the store for several weeks, plus the time taken by processing and shipping. Preservatives are often added to increase shelf life.

Vitamin E, primarily in the oil-rich wheat germ, begins to deteriorate shortly after the flour is ground and most of it is gone in several months. After several weeks of storage, there will be some vitamin A loss in yellow cornmeal. The folic acid, a B-complex vitamin, in flour decreases during the first four to six weeks of storage, then stabilizes at about 80 percent of its original content. Under adverse storage conditions, such as high temperature, nutrient loss and spoilage would be more rapid. Again, the fresher the flour, the better it is for you.

2) Convenience. Whenever you want to bake or cook, you can have freshly ground whole-grain flour at hand, rather than having to make a trip to the store. Properly stored, grains will keep for years.

3) Variety. A mill grants access to a wider variety of flours, meals, and cereals. It is usually easy to find whole wheat flour, but rye, barley, and buckwheat flour can involve a

hunt. How many stores stock soy or amaranth flour? What about finely ground flours for pastry? You can also use a mill to split beans and peas for soups, to make cereals, and—with many mills—to hull sunflower seeds.

4) Satisfaction. As with any area of home food production, there are other, less-tangible benefits. Milling your own flour gives you control of the whole bread-making process and the security and satisfaction that come from taking charge of an important part of your life.

5) A better diet. A mill in the house could encourage better eating habits. Once you've purchased a grain mill and discovered what it can do, you may start eating more whole grains. That's not a bad idea; Americans ate an average of 107 pounds of wheat per capita in 1977, and twice that amount wouldn't have been too much considering that they'd have eaten less of other, nutritionally inferior foods.

So if you're thinking about buy-

ing a mill, don't think in terms of dollars alone. The benefits and rewards from owning and using a grain mill cannot be measured simply in economic terms, although some mill manufacturers would have you believe the savings are enormous.

How much can you save by owning a mill?

The advertised claims vary a good deal. One company says its mill will save an "average family" $201.60 each year. Another promises annual savings of $601.32 for a family with three to five children. Still a third gives elaborate figures to show that "one family" saved as much as $392.40 in a year. Based on these amounts, you absolutely can't afford to *not* buy a mill.

A closer look at these claims suggests they are misleading. The manufacturer's figures are based on a comparison between the cost of buying your bread and cereal in the store and the cost of grinding and preparing your own. In this light, the figures are fairly accurate. But what the manufacturers ignore is a more relevant perspective: comparing the cost of baking bread from store-bought flour with baking bread from home-ground flour. The real savings come not from *grinding* your own flour or cereal but from *processing* it into food. If you're already baking your bread from store-bought flour, buying a grain mill to grind your flour will save little. A check with health food stores and mail order houses showed that whole grain flour costs only about five cents more per pound than buying the grains themselves, and in some cases only two cents more. So if you add anything for the expense of operating the mill, the difference is negligible. (In some specific situations, a grain mill could result in considerable savings: if you aren't making your own bread and cereal and a mill would provide the incentive to do so, it might pay for itself; and if you live near a grain farmer who'll sell to you in bulk, you'll probably save some money.)

There are still plenty of reasons why it makes sense to grind your own flour and cereal. But if you're in the market for a mill, look carefully at the money-saving claims made by the manufacturer. Chances are they won't apply to you.

TD

For people in many Third World nations, the mechanics of grinding grain haven't changed over the centuries.

Hand mills

A hand mill can grind grains, nuts, and seeds just like an electric mill, and some people are drawn to the idea of using muscle rather than buying electricity.

But few people realize the amount of time and energy involved in grinding by hand. It is heavy work no matter which mill you buy. Production is slow: it takes anywhere from 15 to 50 minutes of actual grinding time (not including rest
(Continued on page 40)

Atlas

$225 East, $240 West. Steel grinding surface. Grinds a pound of medium flour in 5 min. 75 lbs. W 20" × D 20" × H 20".

The Atlas hand mill is designed more for rugged farm use than for grinding fine kitchen flour, as its 75-pound cast-iron construction attests. However, it will produce flour suitable for heavy breads, and a lighter flour can be had by sifting away the coarse bran. Its three sets of burr plates will grind dry and oily grains, nuts, coffee, and spices.

For all but the finest settings, the mill is easy to crank. A 20-inch-diameter flywheel weighing 24 pounds gives real momentum to the stroke. It has a splendid 7½-inch handle, long enough to get a firm grip with both hands side-by-side. Two people can even crank the mill, one hand each. There's no accommodation for V-belt drive from the flywheel, which is unusual for a mill this size.

The Atlas' production rate is quick (if you aren't after fine flour) because the grains are cracked before being ground and because the grind is more coarse than average. Quite a bit of flour remains in the grinding chamber and it is hard to clean

out. Dismantling requires a special tool.

You'll definitely want a permanent spot for the Atlas. Seventy-five pounds of cast iron isn't easy to tote around. Its base requires bolting; no clamps are provided.

In-Tec Equipment Co., Box 123, D.V. Station, Dayton, OH 45406.

C. S. Bell #2 Grist Mill

$49.95. Steel grinding surface. Grinds a pound of coarse meal in 10 min. 30 lbs. W 8" × D 15" × H 16".

The C. S. Bell #2 mill is a sturdy hand grist mill intended for rugged homestead chores. It has one permanent set of cone-shaped steel burrs which will grind a variety of materials—whole dry grains, corn, dried chicken bones, soybeans, and animal feeds—into a coarse, mealy texture. It is *not* designed to produce fine flour.

We found the Bell #2 mill relatively easy to operate. Once the cast-iron base is bolted to its permanent working surface, work can begin. No clamps are provided. The demand for arm power is less than that for many of the other mills we used since you are not grinding to a powder-fine texture. A counterweight is built into the handle to offer a flywheel effect, but we noticed little or no advantage and it got in the way of a two-handed stroke.

Our new mill never produced flour fine enough for bread without sifting. The manufacturer explained that with time the steel burr cones wear and eventually grind finer, but we didn't use the mill long enough to notice this change. We were able to produce a ready-to-use texture of

corn and soy meals.

The Bell is the only mill recommended for grinding your own bone meal. Only fish and chicken bones should be used and they should be dried first. The company does not recommend this mill for peanut butter or for other moist grains and nuts.

The C. S. Bell Co., P.O. Box 291, Tiffin, OH 44883.

(Continued from page 39)

stops) to produce the 2½ pounds of fine flour called for by an average two-loaf bread recipe.

You can't stand meekly by the mill, weight on your heels and one hand on the crank. You'll need a work shirt to soak up the sweat and sneakers for traction. The home miller must grasp the handle with both hands and put the body's weight into each stroke. With practice you learn to pace yourself, to integrate grinding with other chores so that the necessary time-outs are productive. Like jogging, the first 100 yards are the roughest.

A coarser flour than usual is easier and faster to grind, but fine flour is best for sandwich breads that are sliced thinly and have to hold together. Bread from coarser flour rises well and has more whole-grain flavor, but tends to be more crumbly. With pancakes, on the other hand, coarse flour produces a flavorful cake that rises well without baking powder, while fine flour makes a flatter, heavier cake.

Wheat is one of the hardest grains to grind, and the hand miller may favor buckwheat, millet, and oat flour. They can be milled very finely with much less effort. (At a wide setting we could produce oat flakes that resembled rolled oats, although they weren't as thin as Quaker's and we had to sift off the flour.)

Should you buy a hand mill instead of an electric one? If your only goal is to save on your energy bill, the answer is no. Electric mills draw quite a bit of power, but because they're so quick, grinding a pound of grain into flour will cost you less than ½¢ of electricity.

Of course, choosing a hand mill does mean one less electric appliance to plug in. A hand mill means less reliance on complex and fallible technology.

You may try to get the best of both worlds by buying an electric mill equipped with a hand crank. If you plan to do much hand grinding, beware. The cranks work, but turning them is very difficult; they're really an afterthought, designed primarily for back-up during a power failure. A machine designed to run at 1,750 RPM's won't be very efficient at 60 RPM's.

Our views on hand mills changed as we worked with them. During the first week of tests, we were ready to ship them back: hand grinding was a hot, tiring chore. By the start of the second week, however, we felt different. The work was still arduous, but we were becoming equal to it and found the grinding a source of exercise and enjoyment. There was a real satisfaction in taking hard grain and turning it into pure, fine flour with just the energy from our own muscles.

If those rewards appeal to you, a hand mill will be a great buy. If you just want a quick source of fresh, unadulterated flour, go with an electric model.

We tried eight different hand mills, ranging in price from $45 to $240. They all grind grain, but there are considerable differences in ease and efficiency. We tested for: 1) grinding speed and flour texture; 2) ease and comfort of grinding; 3) time required for assembly and cleaning; 4) general design and safety features; and 5) ability to perform advertised functions, such as grinding oily beans or making peanut butter.

As with electric mills, stones generally wear longer and produce finer flour than burr plates. But our tests revealed an exception: the Diamant. Its steel milling plates produce as fine a flour as stones at a high rate of production.

Hand mills differ from most electric mills in that their rotating stones or steel plates are held in place by a single center support. They have a wide range of adjust-

Corona King Convertible

$45.50. Steel and stone grinding surfaces. Grinds a pound of fine flour in 16 min., 30 sec. 10 lbs. W 19" × D 8" × H 14".

Versatility is the most outstanding feature of the cast-iron Corona convertible mill, handling dry and oily grains, beans, spices, and nuts. The mill comes equipped with interchangable stone and steel grinding surfaces and two augers which match the feed rate with the capacity of the milling surface.

Steel burrs generally don't grind as fine as stones, but they handle a variety of jobs: producing an evenly textured whole grain cereal; grinding pea, bean, and soy flours; and turning sesame seeds into tahini and peanuts into flour and peanut butter.

Changing from steel plates to stones takes a screwdriver and about 3 minutes. Corona stones produce as fine a flour as you could want but the stroke is rugged. We wished the mill had a handle large enough for a two-hand grip. At any other than a fine setting the stones tend to wobble and produce uneven-textured flour.

For most jobs, the 9½-inch hand crank offers a comfortable stroke without rough spots—but only if the mill is securely fastened. The Corona King comes with its own clamp, but ours had an uneven base which made it impossible to clamp the mill tightly. To keep it from swiveling, we mounted it to a wooden table top with two wood screws (holes are provided in the base). There are warnings against tighten-

ing the clamp with anything other than your fingers. Heed them: cast-iron snaps easily.

Adjusting texture is simple and the setting doesn't tend to wander. No flour pan is provided but there is plenty of room for your own bowl. The mill comes apart—and goes back together—easily after you've practiced it once or twice. Production slows down when cleaning the last ¼ cup out of the hopper but if you have the patience, the mill will grind all but the last ounce or two.

R & R Mill Co., 45 W. First North, Smithfield, UT 84335.

ment, but when opened for a coarser grind they tend to wobble, producing flour with an uneven texture. With sunflower seeds, we could get the meats out, but they were in bits and pieces. If you want cracked grains, seeds, and cereals, you'll do best with an interchangeable mill designed to grind different textures while butted up close to the stationary plate.

With most of the mills, we discovered that 60 RPM's was a natural cranking speed. Curiously, this rate was independent of the strength of the tester, and kept fairly constant even when the grinding adjustment was changed. Sixty RPM's provided sufficient momentum to keep grinding without tiring too quickly.

DB

Electric mills

There are two basic types of electric grain mills: stone and steel burr plate mills. Stones grind finer, produce less heat at finer settings, and wear better. Steel burr plates are more versatile, and some can grind not only flour but also shells, roots, bark, dried bones, spices, and animal feed. If the mill gets gummed up, the burr plates can be washed and then dried in an oven. Stones must *never* be washed.

Most stone mills use man-made grinding stones (an exception is the Meadows Mills Company, which uses natural granite stone) fabricated from bauxite particles held together with a natural clay. These usually come with a good warranty.

With stones, you are limited to grinding dry materials of 12 percent moisture or less, and you should follow the manufacturer's directions explicitly to avoid trouble. Wet or oily materials, such as soybeans and peanuts, tend to coat the stones, making them useless. Stones can be cleaned by running through a handful of popcorn at a wide setting, but this isn't a green light to grinding moist or oily things.

If you want cracked grains or cereals, stone mills produce a more uneven and powdery cereal than burr mills. The flour that's produced with cracked cereal has to be sifted off to

Diamant Domestic Mill (Type D 525)

$275 East, $285 West. Steel and stone grinding surfaces. Grinds a pound of fine flour in 5 min. 55 lbs. W 16" × D 20" × H 20".

The cast-iron Danish Diamant is an impressive mill in performance and appearance. It turned relatively easily, and was one of the most efficient flour producers we tested. This is partly due to an auger with teeth which cracks the grain before pushing it into the grinding chamber. The steel burr plates produced flour as fine as any stones could grind. The mill will grind animal feeds and seeds, as well as grains and beans for people.

A sturdy, permanent mounting is necessary; the mill comes with four bolts to hold it secure. A 16-pound flywheel with a 16-inch diameter gives a stroke with good momentum and a comfortable reach. Unfortunately, the handle wouldn't quite accommodate a two-hand grip. Cranking is most comfortable at just over the usual 60 RPMs. We found that while a faster pace was more tiresome, the ample flow of flour kept us cranking.

It's an attractive machine, covered with bright green metallic paint, and careful design and craftsmanship are evident in its performance. But the paint on the hopper of our test mill began to flake off, and chips could easily have ended up in the flour.

With three sets of metal plates and one set of man-made stones available at extra cost, the Diamant ranks high in versatility. Good alignment means the grind is even and allows no whole or cracked kernels to escape into the flour. A coarse setting with the fine plates rubs bran away in large shavings that can be sifted off and used as bran in other recipes. Coarser plates produce a uniformly textured cereal or cracked wheat. Plates can be changed in about 3 minutes with only the help of a large screwdriver. Cleaning is not difficult.

The manufacturer's literature recommends the mill for grinding oily foods like soybeans and peanuts. We had success producing soy flour but the mill clogged when we tried to make peanut butter.

In-Tec Equipment Co., Box 123, D.V. Station, Dayton, OH 45406.

keep the cereal from getting too gummy.

Grinding can have an effect on nutritional value of grains. A finer grind exposes more surface area, so vitamins and minerals can be more readily assimilated during digestion.

Grain mill manufacturers sometimes boast about the low flour temperature during grinding, reasoning that chemical reactions happen more easily at high temperatures. High temperatures are more of a worry during storage, when oxidation (the reaction which turns germ oil rancid) has a chance to get under way. Home millers don't have to worry about storage if they grind just enough flour for each batch of bread.

Mills should be kept clean to discourage bugs and molds. Some mills are easier to tidy up than others. While wood cabinets make for attractive mills, untreated wood absorbs moisture and traps flour in the cross grain. Formica and metal housings present fewer problems. In high-speed cabinet mills, steam can cause flour to cake or turn to paste, making cleaning a harder task. Steam can be avoided by backing off to a coarser grind that will cool the mill down.

The grinding times we recorded show that high-speed cabinet mills grind more quickly than others, but they don't reflect the time required for stopping to dump the flour or adjust the stone as the drive shaft lengthens with heat and brings the stones too close together.

DB

I have always found that the coarser my ship bread, the healthier my crew is.

A sea captain of more than thirty years' experience, quoted by Sylvester Graham in *A Treatise on Bread and Bread-Making*, 1837. Reprinted by Lee Foundation for Nutritional Research, Milwaukee, Wisconsin.

EXCALIBUR KIT

The Excalibur kit is the same design as the preassembled model. Refer to the Excalibur evaluation (page 45) for technical details.

One Excalibur owner of our acquaintance was shocked the day the kit arrived. "I just about died when I saw all the pieces." I felt the same way upon opening the box. But assembly isn't difficult, if your awe at piles of little parts makes you a careful assembler. The only two prerequisites for taking on a project like this are the ability to recognize parts as they are called for (the kit's illustrations help with that) and the confidence to go ahead and make a move.

Assembly took me a little over six hours, including finishing the wooden cabinet. Those hours were spread over three days to allow the coats of finish to dry. I encountered a few hitches. My first was that a piece of the cabinet didn't fit. A quick call to the manufacturer revealed it was mismarked, not the wrong size.

Once I had glued up all the parts, I experienced hitch number two: something didn't fit quite right, and the glue was setting fast. I had positioned a piece improperly but fortunately caught the mistake in time. The moral: No matter how confident you feel, humble yourself enough to be thorough.

The third problem visited me when installing the capaciter. A quick-connecting electrical gismo wouldn't hold, and I finally had to solder. In spite of, or perhaps because of, these complications, I became fond of this mechanical contraption over the quiet hours of assembly. It occurred to me that a novice like me might be more likely to succeed at this project than a pro: when you know what you're doing, it's easy to skip details and that means potential trouble.

DB

Hi Life

$46 to $54. Stone grinding surface. Grinds a pound of fine flour in 14 min. 15 lbs. W 19" × D 10" × H 5".

The Hi Life, also sold as the Great Northern and the GF3 Stone Hand Mill, has an attractively rugged look with its large grinding stones, ceramic hopper, and steel stand.

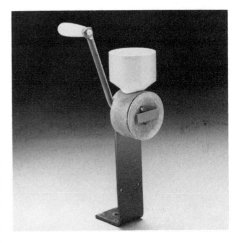

The mill will produce an extremely fine flour, although the stroke is difficult. A standard fine grind required reasonable effort. We could get a good overlapping two-hand grip, but the nut attaching the handle to the machine rubbed against the outside of the hand. The Hi Life's stand is longer than the crank, so the mill can be mounted permanently on a wall. Because it stands so high on a table, a large rectangular pan is needed to catch the wide spray of falling flour. You can buy conversion kits for running the mill with a 1/3- or 1/2-horsepower motor or with a bicycle, but we did not test these.

The mill does a good job with dry grains and can handle soybeans at a coarse setting. The grain hopper empties almost completely, and the mill comes apart easily for cleaning.

Tempco Products, Inc., 564 W. 800 South, Bountiful, UT 84010.

Retsel Little Arc

$47.50. Stone grinding surface. Grinds a pound of fine flour in 20 min., 30 sec. 8 lbs. W 10" × D 12" × H 16".

The Little Arc was one of the easiest mills to crank, and one of the slowest. We felt comfortable cranking with just one arm for the extended periods of time necessary to grind powder-fine flour. The production rate is reasonable for a standard fine flour, however, and the stroke is comfortable enough, especially at a faster-than-average pace. A slower pace revealed a hitch in the stroke.

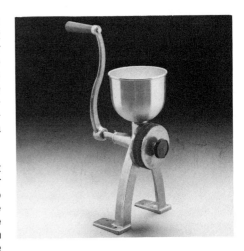

The cast-aluminum base is lightweight and strong. Two legs with wide feet offer good stability. No clamps are provided, so the mill needs to be screwed down. The compact design makes for easy storage but poses a problem when trying to catch the flour: the legs are so close to the stones that we couldn't position a receptacle to catch all the flour. The guard provided to funnel flour into a bowl is a makeshift cardboard device, held in place with rubber bands.

The short cast-aluminum auger feeds grain efficiently at a rate that accommodates an easy stroke and completely empties the grain bin. What little cleaning is necessary can be accomplished by simply twisting off the adjustment nut on the front, removing the stone and handle, and blowing out the grain and flour.

As with all mills having a center pin mounting, the rotating stone has a tendency to wobble just slightly at a wide setting. This admits some coarse particles into the flour, yielding an uneven texture. If it's just cracked cereal you want, sift away the fine flour; if left in, the flour will make a creamier cereal. At a fine setting, there is no wobble, just a steady stream of fine flour.

Retsel Corp., P.O. Box 291, McCammon, ID 83250.

Samap

$99.95. Stone grinding surface. Grinds a pound of fine flour in 14 min. 18 lbs. W 10" × D 17" × H 11".

The Samap hand mill is unique in appearance, function, and performance. Though based on the design of ancient grist mills, it has a contemporary look indicative of its French origin. Body construction is of magnesite cement, in which stone granules are imbedded. It is light enough to be put away after each use.

The Samap's grinding arm is on the top, calling for a horizontal cranking motion shared by no other muscle-powered mill. This took getting used to (some testers never did), but with practice we found a comfortable stance—feet spaced far apart with one side to the mill while bending over it slightly. Cranking is a full-bodied motion.

A stationary table is necessary to prevent even light horizontal cranking from pushing it across the floor. The table should be heavy and low enough for you to bend over the mill. No bolts, screws, or extra clamps are needed to hold the mill firmly on a table top or counter ledge; its own two clamps do the trick, and felt beneath the mill protects table tops.

The rate of feed from hopper to grinding chamber can be adjusted, another unique feature. If the stroke is too difficult, a clockwise turn of a knob will make the going easier (but slow down production as well). To grind fine flour from hard, red winter wheat we set the feed adjustment as slow as possible, and the stroke was still difficult. Going to either a slightly coarser grind or a softer grain made the stroke more comfortable.

The Samap's peculiar grinding action produces a flaky rather than a granular flour.

Often what looks like a coarse chunk of bran is really a flat bran flake which feels fine to the touch. In our test kitchen, however, we found no differences in its baking qualities for pancakes and bread.

The company's literature says the mill will handle oilseeds, yet soybeans clogged the mill and glazed the stones. Fortunately, the mill comes apart easily for cleaning.

The handle fits across the hopper, leaving a very small opening to fill with grain. Dumping from a bag works all right, but when using a cup, bowl, or large scoop, grain spills easily over the side and into the flour in the catch basin. This moatlike basin is attractive, but a nuisance to empty. A brush is necessary for collecting the flour.

Miracle Exclusives, 16 W. 40th Street, New York, NY 10018.

STOVE-TOP BREAD

It's possible to make bread without an oven by steaming it. Using your favorite bread recipe, make enough dough for two loaves. Divide the dough into three equal parts and place them into three one-pound coffee cans that have been wiped with liquid lecithin to prevent sticking. If you're making yeast bread, set the cans aside in a warm spot to allow the dough to rise.

Next, cover the cans tightly with aluminum foil. Get out your big roasting pan, one with a tight-fitting lid, and place a rack in its bottom. Pour one inch of water into the bottom of the pan, and place the three coffee cans on the rack. Cover the roaster and set it on top of the stove. Over medium heat, cook the bread for about three hours. Make sure there is always an inch of water in the bottom of the roaster or the bread won't steam properly.

Remove bread from cans to cool; slice, and return to the cans where they can be stored topped with their airtight plastic lids.

Lana Littell

. . . **The sense of taste is so easily vitiated, that we can very easily become reconciled to the most offensive gustatory qualities, and even learn to love them.**

Sylvester Graham in *A Treatise on Bread and Bread-Making*, 1837. Reprinted by Lee Foundation for Nutritional Research, Milwaukee, Wisconsin.

HOMINY

We used hominy very freely during the winter. We could not buy it as we can now, but had to "pound" it for ourselves. Every neighborhood, but not every farm, had a hominy mortar. We would take an oak log, about four or five feet long, a foot or 15 inches in diameter, and by cutting, boring, and burning would excavate a bowl in one end that would hold a peck or more of corn. We would fasten an iron wedge, that we used to split rails with, into the end of a handle to use as a pestle or pounder, put about two quarts of smooth, hard grained "flint" corn in the mortar, pour in hot water enough to soften the hulls of the corn, and pound it lightly, so as not to break the grains, until the hulls were loose, and we had a good hominy, which we ate boiled and fried after it got cold. Hominy mills had not been invented.

My uncle, on an adjoining farm, fastened a "spring-pole" overhead in the shop, [from] the end of which the pestle or "beater" was suspended, and it made the work of pounding much easier.

Pounding hominy was an evening job. If the men were notified that some was needed, they would on one evening select and shell half a bushel or a bushel of corn, and the next evening carry it to the mortar and pound it.

John Jay Janney's Virginia, copyright © 1978 by Edmund Derby Haigler, is published by EPM Publications, Inc., 1003 Turkey Run Road, McLean, VA 22101.

THE LEE MILL

We asked the Lee Engineering Company for a mill to use in our tests, but they declined. So, we didn't include a Lee mill in our regular evaluations. By coincidence, Jim Osborne, the home miller profiled on page 54, owns a Lee, and this report is based on his experience and our observation.

The Lee mill is a strange beast. It grinds flour, not between two stones, but by slamming the grain against a single, stationary stone.

The mill has a circular, Carborundum stone with a rotating stainless steel feed disc inside it. The grain falls into the rapidly spinning disc, where centrifugal action throws it against the stone ring, which grinds it into flour.

The Lee mill is not particularly fast or quiet. It took 20 minutes to grind three pounds of flour, and the noise level was that of a vacuum cleaner. But neither of these drawbacks is serious because you can fill the hopper, turn on the mill, and retreat to silence for the 20 minutes it takes to grind a load.

The unique design makes for minimal maintenance, and cleaning is quick and easy. Since the stone isn't grinding against another stone, there's virtually no wear, and the flour comes out cool. Prices are very moderate for electric mills; they start at $110, and the most expensive model sells for $190.

A disadvantage of the Lee is that it can't handle oily seeds, nuts, beans or small grains (millet and amaranth).

Lee Engineering Co., 2023 W. Wisconsin Avenue, P.O. Box 652, Milwaukee, WI 53201.

All Grain A-22

$179. Stone grinding surface. Grinds 10 lbs. fine flour in 1 hour, 19 min. 19 lbs. W 12" × D 18½" × H 14".

The All Grain receives a lot of flack from competitors for using a flow of air to both cool flour and deliver it to the catch bin. They claim that this increased exposure to oxygen causes rapid oxidation (chemical breakdown) of the flour, so that the flour is nutritionally inferior. We took a long hard look at the accusation, and have come to a different conclusion. But first, let's study the mill and how it works.

The All Grain has a ⅓-horsepower motor housed in stainless steel, powering horizontal stones (vertical is conventional). Gravity feeds grain from a small 4-cup hopper into the grinding chamber. From there it is blown into a huge polyethylene bucket with a 17½-pound flour capacity (there's no stopping in the middle of things to empty this one).

At first glance, we judged the flour to be the finest we had been able to mill at home without sifting—but the truth was elusive. The flour is blown over a relatively wide area and separates into piles of different textures. The finest, light white endosperm falls first, collecting around the exit spout. It feels like talcum powder. On the far side of the bin you'll find the browner, heavier whole wheat flour. Mix them together, and the product is quite like any other fine whole wheat flour.

The All Grain makes a lot of noise—for a long time. It was the slowest in the 10-pound grinding test. But you needn't bother with close tending and you can hear when it's ready for another hopperful of grain.

Cleanup wasn't the burden endured by owners of high-speed cabinet models, un-less someone wasn't careful to make sure the lid on the flour bin was secure.

Now for our opinion on nutrition and the All Grain. It is true that oxidation is destructive to the nutritional value of flour. But chemical reactions take time to occur, and home millers likely grind up batches of flour as needed for bread. So between milling and baking, there's little time for the flour to degrade. Also, it's a fact that flour entering the All Grain bin is cooler than with any of the cabinet mills. Because of its design, we couldn't get a temperature probe inside the All Grain's grinding chamber. But even if the temperature were extremely high, flour would be cooled within seconds. Although we did not conduct nutritional analyses, our staff biochemist advises us that nothing critical is taking place in that short amount of time to degrade the flour.

All Grain Distributing Co., 3333 S. 900 East, Salt Lake City, UT 84106.

VERNA MAE'S HEART SPEAKS

Of all the things my father taught me, I am thankful that I learned from him the enjoyment one could obtain from work. I did not know until I was grown that there were people who did not like to work, and not until my children were grown did I realize that some folks thought it was shameful to do manual labor. (I must admit, though, I do not like to do housework. I like for a house to be a home, a comfortable place to live, not a "show room.")

I wish I could pass on to my grandchildren how I feel about growing our own food. All good things come from God. But you seem so close to Him, one with nature, when you plant the tiny seeds, in faith that they will grow. Later there is the joy of gathering and storing away these results of your partnership with nature. The food you grow yourself tastes much better and seems a lot cleaner to me. It may not be "untouched by human

hands," but at least you know whose hands they were.

Mike was once helping me to dig a mess of potatoes for supper and asked, "Granny, why do you put your potatoes in the ground? Is it so your chickens can't get at them? Mother gets hers at the market."

Verna Mae Slone, *What My Heart Wants to Tell* (New Republic Books, 1979).

Excalibur

$259 completed, $189 kit. Stone grinding surface. Grinds 10 lbs. of fine flour in 38 min. 39½ lbs. W 7⅛" × D 17" × H 11".

The Excalibur can be bought either in kit form or fully assembled. Since the kit is the same design as the finished model, we'll discuss the features of the ready-made mill as representative of both and discuss separately the kit's assembly.

The Excalibur is a high-speed mill with an enclosed cabinet. What stands out most at first sight is its size—long and narrow, about half the width of other cabinet models, and likely to park easier on your counter.

With the convenience of its small size comes the inconvenience of tending. The grain hopper holds slightly over 4 cups of grain and the flour bin holds about 6 cups of flour. Because grinding goes quite quickly, you have to remain on hand to empty the flour bin and fill the hopper about every 2 minutes. You'll be able to hear when the mill is hungry because the grain feed slows down as the hopper empties. Grain is delivered to the stones by gravity feed; as grain passes through, there is less weight on the remaining grain and the flow slows considerably. In fact, emptying the hopper completely requires some dabbling with a brush to scoot the remaining kernels into the feed hole. We think that the hopper design could stand some improvement.

The Excalibur produces flour as fine as you wish, as well as coarse flours or cracked wheat (though the latter is pro-

duced with some flour). The on-off switch and lever for adjusting texture are both located at the back of the mill.

The Excalibur wins praise at cleanup time. Its stationary stone can be removed with a small tug, providing access around and between the stones. With the cabinet as small as it is, getting at all the corners still requires some patience.

The Excalibur does have a hand crank, but the 1:1 ratio coupled with a stone intended for 1,725 RPMs means that production is slow. The crank is intended only for real emergencies.

Excalibur Flour Mills, 5711 Florin-Perkins Road, Sacramento, CA 95828.

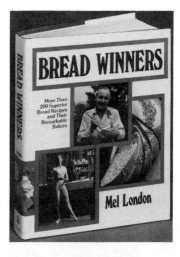

BREAD WINNERS
Mel London. 1979.
Rodale Press, 33 E. Minor Street, Emmaus, PA 18049. 365 p. hardcover $14.95.

Mel London, an extremely affable man, has made this cookbook from the recipes of 45 friends and acquaintances across the country. Because the reader is introduced to each contributor, this is one cookbook that's fun to sit down and read.

Corinne lives in Yerington, Nevada, and writes a weekly column for the Mason Valley News— *"The Only Newspaper in the World That Gives a Damn about Yerington." She's baked in just about every kind of oven. Her first was a kerosene stove with an oven that sat over the burner. She's also baked in the old iron variety wood stove (and thinks the results are the best of all). Her daughter also bakes bread and Corinne tells me that she uses bread pans that are now in their fourth generation and have never been washed.*

ARROWHEAD MILLS: HEREFORD, TEXAS

Deaf Smith County, in the high plains of the Western Texas panhandle, is famous for its hard, red winter wheat. The long nine-month growing season, the dry climate, and the mineral-rich soil give this wheat a high nutritional quality.

In 1960, a local farmer named Frank Ford started a milling business in Hereford. With an old pick-up truck and trailer, and a 30-inch stone mill, he began selling whole wheat flour and cornmeal. It was seven years before Arrowhead Mills began to break even.

Today the old mill is still working, and Frank Ford's operation has become a multi-million dollar business that operates out of a 20-acre complex. The company distributes over 250

whole grain products under such brand names as Olde Mill, Deaf Smith, The Simpler Life, and Arrowhead Mills.

Ford believes the reason for his success is the quality of his products. He's tried to act as a partner with his land, rather than its master, maintaining the quality of his food by leaving its natural goodness alone. His motto: "We treat food with respect, not with chemicals."

In addition to whole grains and flours, Arrowhead Mills now markets oils, cereals, seeds, beans, nuts, peanut butter, whole grain pasta, vitamins, and frozen dinners. A recent addition to their product line is Powerhouse 32, a complete, all-vegetable protein food containing such potent

substances as chia seeds, rose hip powder, fenugreek powder, and dulse.

In the early 1970s, Ford became a born-again Christian and Scripture references crept into his writings. But he wears the change gracefully, and most of the references come as gentle reminders that we need to care for our bodies, our souls, and our environment.

You can write for a catalog and order natural products directly from Arrowhead Mills. But because of shipping expenses, the minimum order is 300 pounds. Write Arrowhead Mills, Box 866, Hereford, TX 79045.

TD

Grove's Mill

Grove's Mill, located west of Lewisburg, Pennsylvania, is a rarity—a water-powered commercial mill that grinds locally grown grain and corn into flour and feed. The mill stream exits the creek about a quarter mile upstream, and meanders to the mill in a controlled current. This is not a museum for buses of school kids; the Groves, father and son, do business out of an office in the mill. Farmers bring in grain and it is ground rough, into meal, or into refined white flour. A farmer loads bags of meal as Mr. Grove the younger looks on, then checks the grind at one venerable machine and stitches the bags shut with another. The Grove's Roller Mills flour is retailed locally in five-pound bags, and it's available at the mill as well.

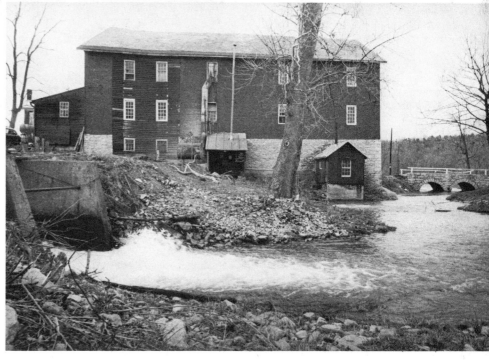

Clinicians recently discovered what the ancients well knew, that roughage was an important element of diet. ITT Continental Baking Company met this need with a loaf that promised added fiber. The government has insisted that the company identify the ingredient more plainly. It is sawdust. We have come to that.

John Hess. *Saturday Review.*

The Mill Wheel

The mill wheels are clapping; the
*   brook turns them round, clip,*
*   clap!*
By day and by night is the grain
*   being ground, clip, clap!*
The miller is jolly and ever alert,
That we may have bread and be
*   glad like a bird,*
clip, clap, clip, clap, clip, clap!

How busy the wheels are in turning
*   the stone, clip, clap!*
And grinding so finely the grain we
*   have grown, clip, clap!*
The baker the flour for the baking
*   will use,*
And make us a roll, or a cake if we
*   choose,*
clip, clap, clip, clap, clip, clap!

Traditional German Song

Garden Way Kit

$169. Stone grinding surface. Grinds 10 lbs. of flour in 34 min., 15 sec. 32½ lbs. W 9¾″ × D 17″ × 11¾″.

We assembled the Garden Way kit when it first came on the market and can't speak for the new and improved instructions now included. Even with our handicap, the kit went together in less than 3 hours, excluding the finish.

The cabinet pieces, of hardwood-core birch veneer plywood, were precut and fit together just right, but lacked the craftsmanship of the Excalibur cabinet. Some of the edges and drill holes were chipped slightly (most often in places where they wouldn't show) and the wood components didn't match quite as well as on the Excalibur.

We found other indications that the Garden Way mill was newly on the market. An oversight in design kept the hopper from being tightly closed, while placing stress on lid hinges which, secured with tiny brads and not screws, soon came loose.

The cabinet is bottomless, and its two sides carry the weight of the mill. This allows access to the flour pan, but within a few months the sides had bowed slightly inward, making the flour pan fit awfully snugly. Perhaps our not putting any kind of finish on the mill had something to do with it.

A benefit of assembling your own grain mill is that you become knowledgeable enough to make most of your own repairs

and adjustments. Our first batch of flour had an uneven texture, revealing that the stones were out of alignment. After some fiddling, we finally got it right.

The hopper fed well down to the last ¼ cup, which had to be pushed into the feed hole by hand or with a brush. The texture adjustment is conveniently located in the front, so that you can easily note your favorite positions for cornmeal, wheat flour, and cracked cereal. The on-off switch is located on the mill's electric cord, rather than hiding away on the back of the cabinet.

Garden Way Catalog, Ferry Road, Charlotte, VT 05445.

THE GIRL WHO STEPPED ON BREAD: A FABLE

[Through much of history, bread was considered so essential to human life that it was almost sacred. One of the myths that grew from this belief was that it was wicked to walk on bread.]

I suppose you have heard about the girl who stepped on bread in order not to get her shoes dirty, and how badly she fared . . . .

She was a poor child, but proud and arrogant; she had what is commonly called a bad character. When she was very little it had given her pleasure to tear the wings off flies, so they forever after would have to crawl. If she caught a dung beetle, she would stick a pin through its body; then place a tiny piece of paper where the poor creature's legs could grab hold of it; and watch the insect twist and turn the paper, round and round, in the vain hope that, with its help, it could pull itself free of the pin.

"Look, the dung beetle is reading," little Inger—that was her name—would scream and laugh. "Look, it is turning over the page."

She did not improve as she grew up; in fact, she became worse. She was pretty, and that was probably her misfortune; otherwise, the world would have treated her rougher.

"A strong brine is needed to scrub that head," her own mother said about her. "You stepped on my apron when you were small, I am afraid you will step on my heart when you grow older."

And she did!

A job was found for her as a maid in a house out in the country. The family she worked for was very distinguished and wealthy. Both her master and mistress treated her kindly, more as if she were their daughter than their servant. Pretty she was and prettily was she dressed, and prouder and prouder she became.

After she had been in service for a year her mistress said to her, "You should go and visit your parents, little Inger."

She went, but it was because she wanted to show off her fine dresses. When she came to the entrance of her village, near the little pond where the young men and girls were gossiping, she saw her mother sitting on a stone. The woman was resting, for she had been in the forest gathering wood, and a whole bundle of faggots lay beside her.

Inger was ashamed that she—who was so finely dressed—should have a mother who wore rags and had to collect sticks for her fire. The girl turned around and walked away, with irritation but no regret.

Half a year passed and her mistress said again, "You should go home for the day and visit your old parents. Here is a big loaf of white bread you can take along. I am sure they will be very happy to see you."

Inger dressed in her very best clothes and put on her new shoes. She lifted her skirt a little as she walked and was very careful where she trod, so that she would not dirty or spoil her finery. That one must not hold against her, but when the path became muddy, and finally a big puddle blocked her way, she threw the bread into it rather than get her shoes wet. As she stepped on the bread, it sank deeper and deeper into the mud, carrying her with it, until she disappeared. At last, all that could be seen were a few dark bubbles on the surface of the puddle.

Excerpt from "The Girl Who Stepped on Bread" in *The Complete Fairy Tales & Stories* by Hans Christian Andersen, translated by Erik Christian Haugaard. Copyright © 1974 by Erik Christian Haugaard. Reprinted by permission of Doubleday & Company, Inc.

Golden Grain Grinder

$299 Deluxe, $269 Standard, $189 Nugget. Stone grinding surface. Grinds 10 lbs. of fine flour in 8 min., 20 sec. 43½ lbs. W 12½" × D 19¾" × H 15".

If you're looking for a nice piece of furniture to add to your kitchen, consider this mill in its deluxe Early American version, available in either a maple or walnut finish. Decorative spindles at the back of the cabinet hide the motor but allow full ventilation for cooling.

The nice-looking outsides surround a capable, serious machine—the fastest we worked with in the 10-pound grinding test. Responsible for this speed are the ¾-horsepower motor, preslicing metal inserts in the stones, and a large flour bin. The texture adjustment, located on the rear of the motor, allows for grinding a wide range of textures of hard dry grains. The manufacturer suggests the mill will grind soybeans, but only to a medium texture or a little finer if mixed with some dry grain.

The mill is heavy—over the 50-pound UPS limit, so we had our sample unit delivered by bus. Because we're only six miles from the bus station, this wasn't any trouble for us. But the weight means finding a permanent spot for this one. A spot on the countertop or under a cupboard may make the on-off and texture controls hard to get at.

The Golden's flour bin is worth an extra mention, not only because of its generous 25-cup capacity but also because it is made of heavy-gauge stainless steel. The container is easy to clean and comes in handy for other household chores; on many mills, the cabinet front and drawer are one, ruling out a second use.

Cheaper than the Deluxe are the Golden Grain Standard model and the Golden Nugget. The Deluxe and Standard differ mostly in cabinet design, while the Nugget is much smaller. We tested neither.

Kuest Enterprises, Box 110, Filer, ID 83328.

THE SEDUCED TONGUE

Everyone knows that people eat the food they like, and avoid the foods they do not like. This fact lies behind our making tempting dishes for the convalescent, as well as our tendency to overeat at meals prepared by a superb cook. We have, however, largely ignored the nutritional consequences of this eating for palatability. It used to be true that when we ate what we liked, we ate what our bodies needed, both quantitatively and qualitatively. But today the ability of the food technologist to separate palatability from nutritional value means that taste is no longer a guide to good nutrition.

We may soon be presented with something that has the texture, taste, smell, and cooking properties of beefsteak but containing neither protein, vitamins, nor any other nutrients. And we shall eat it because we like it. More important, we are already presented with very palatable sweets, chocolate, ice cream, biscuits, cakes, and drinks, which supply us with little but carbohydrates and calories. As a result, we eat these instead of, or as well as, some of our other foods, so that they may displace part of our diet and also add to it. There is a reason to believe that some of the diseases of civilization—obesity, dental decay, myocardial infarction, peptic ulceration, and diabetics—are at least in part caused by the fact that our diet, as well as sometimes showing a desirable increase in its protein content, shows a persistently high intake of carbohydrates, much of which comes from sugar.

John Yudkin. The Lancet.

Granzow

$279. Stone grinding surface. Grinds 10 lbs. flour in 41 min., 30 sec. 37 lbs. W 12" × D 16¹/2" × H 16¹/4".

The Granzow is a mill of a different sort. Relatively new to the market, this high-speed mill lacks the familiar boxy cabinet common to most such mills. Grain is gravity-fed from a large 8-cup hopper, through the stones and into the catcher of your choice—either an enclosed heavy plastic flour bin or, for greater capacity, a plastic bag held in place with a huge rubber band.

The design is special in that the grinding chamber opens for cleaning by a simple flip of a clasp. There before you lies the mill's innards—the grinding face of both stones awaiting cleanup. With the Granzow, you can actually see the parts that do the work. The Granzow has precutting teeth that serve to crack the grain before it touches the stone.

The Granzow has a stainless steel cover around the ½-horsepower motor. Although it is very attractive and makes the body of the mill easy to clean, we're afraid the motor isn't vented properly. In our 10-pound wheat test, enough heat was generated to cause the plastic cover around the base of the motor to melt a bit out of shape.

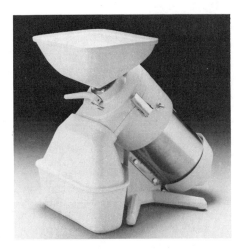

A second problem is that flour can hide in a spot that can't be reached without removing the stationary stone. This cache provides a potential home for weevils, and fooling with the stone is apt to throw the stones out of alignment. The manufacturer helps by providing a small brush for getting behind the stone, but we couldn't reach quite all of it.

Granzow Grain Mill Co., 2516 E. Jackson, Phoenix, AZ 85034.

CORN PONES

Bob Rodale, chairman of the board of Rodale Press, calls them corn pones, but you may know them as journeycakes or johnnycakes: a quickly made bread substitute that packs all the fiber, nutrition, and good taste of corn.

The crisp consistency of corn pones harkens back to the pioneer days. Strong teeth and healthy jaws are a must to get to the natural, nutlike flavor. You can't just bite off a piece, chew twice and swallow as you would bread. Eating them is good exercise for the jaws and gums.

According to Bob they also have a mouth-cleaning effect. No matter what you eat during a meal, have a corn pone or two for dessert and you'll leave the table with a

MAIL-ORDER NATURAL FOODS: PENNS CREEK, PENNSYLVANIA

When Paul and Betty Keene started Walnut Acres in 1946, they began modestly. They bought a 100-acre farm near Penns Creek, Pennsylvania, farmed the land with horses and natural fertilizer, and sold apple butter that was cooked outdoors in an iron pot.

Today Walnut Acres sells hundreds of items, ranging from peanut flour to watercress dressing to chicken stew. Last year, the company grossed over $2 million, and had as many as 100 employees in peak season.

Despite the growth, Paul Keene's philosophy hasn't changed. Every issue of his mail-order price list claims: "We are not trying to supply the whole world. We just want to be small, careful, and trustworthy."

Many items on the list have special notes. The peanut butter-honey-sesame spread is said to be "crunchy and likewise toothsome." The Holland Honey Raisin Date Loaf is "tasty and

sweet," but it "gets gummy gradually." Under the black walnut pieces it warns: "strongly flavored—it's better not to order if you don't know the taste."

Much of what Walnut Acres sells is raised on the Keene farm, which has increased to 500 acres. They grow vegetables, fruits, and grains, as well as beef cattle and chickens. They also market their own soups, dressings, sauces, breads, and cookies. The price list indicates whatever items are organically grown; grown without chemical pesticides; free of added salt or sweetener; and those containing "materials some persons may find undesirable," such as baking soda, cornstarch, or dark sugars.

The Walnut Acres commitment to quality extends to storage as well. To keep their grain safe from bugs and mold, they refrigerate it—an expensive but highly effective practice.

TD

The wheat was cut a few days before complete ripeness to prevent its shattering out when bound. After it was all cut, one went through the field picking up sheaves, two at a time, standing them apart at the bottoms, both for stability and to allow air to circulate throughout to dry things, and leaning the tops together. A sheaf at each end, and a cap sheaf spread across the top with stems all pointing downward to shed the rain, completed that part. It was a sweaty, scratchy job to stand up hundreds of shocks on a steaming day.

Later came the pitching onto a wagon by fork (careful—you can lose a load of slippery sheaves before you know it!), the hauling into the barn, the pitching into a mow, to be pitched back out later into the big threshing machine when it and the crew got to your farm. And all that wasn't nearly so long ago that memory must stretch itself to recall.

Walnut Acres' newsletter.

clean, satisfying taste in your mouth. Bob likes to serve them up with a dip of unsalted chunky peanut butter from Walnut Acres and quartered Granny Smith or MacIntosh apples. On less formal nights he'll hand you the apple and a sharp paring knife, and set the peanut butter can on the table.

How did Bob happen upon this unusual food? "Corn is a food I like to eat regularly, and in significant quantities, because it's so rich in fiber and so low in cost," he explains. "But most important, corn has an enjoyable sweet taste that seldom gets boring. After trying dozens of variations of corn products and much kitchen experimenting I settled on this simple corn pone recipe. I went through recipes for cornmeal mush, cornbread, hush puppies, corn pancakes, parched corn, popcorn and others."

Here is the Rodale recipe. Put three cups of good, white cornmeal in a mixing bowl. (Either white or yellow meal can be used to make corn pones depending on your taste preference, but try to get a meal that is soft and floury, not one with gritty particles.) Add ¼ cup of sesame seeds to improve both protein quality and flavor, if you wish. You can also add some caraway seeds. Put slightly more than ¼ cup of corn oil in a mixing cup. Heat two cups of water to boiling. When the water is hot, pour the oil and about 1½ cups of water into the dry mixture, and mix with a large spoon. Exactly how much water you add depends on the moisture content of the meal, plus your personal taste. You want to end up with a stiff batter that can be formed easily with your hands without clinging aggressively to your fingers. Too much water will leave you with slightly soggy cakes, even after baking; add too little water, and they'll fall apart in your hands as you form them.

The batter will have to cool before you go onto the next step. Meanwhile, preheat the oven to 350°F. Oil a cookie sheet and also your hands. Take a small handful of comfortably cooled batter and form it quickly into a flat cake. Bake for about 40 minutes, more or less to compensate for your oven's quirks and the moisture content of the batter. Use the oven exhaust fan or keep your oven door ajar during the last 10 minutes of baking, and the corn pones will be crisper.

TS

Kitchen Aid

$209.95 mixer, $89.95 attachment. Steel grinding surface. Grinds 10 lbs. flour in 17 min., 45 sec. 4 lbs. (mill only). W 4" × D 7" × H 8¼".

The Kitchen Aid grain mill is an attachment for the firm's multi-purpose kitchen mixers. It's a steel burr mill to be used only with dry grains. A peek inside revealed the best-made burrs we've seen so far: the metal burrs were cut evenly, assuring an even grind. Because grain follows the path of least resistance, uneven burrs permit coarser grain than intended to filter through the larger spaces; thus, a very fine flour is impossible to produce.

However, even with its precise milling, the Kitchen Aid mill is not intended for grinding powder-fine whole wheat flour. The burrs are milled to produce a medium-textured flour that is acceptable for many purposes but not ideal for pastries. A very fine flour is possible with millet, barley, or roasted buckwheat groats. A medium setting produces an evenly textured cereal, but some flour is produced when cracking grains.

The Kitchen Aid folks have set a strict 10-cup grinding limit on this mill (good for three 1-pound loaves) with a 45-minute waiting period between grinds to avoid overheating the motor. Because the power unit was originally designed for mixing, blending, and kneading, this otherwise rugged machine can't handle the continuous heavy-duty load that milling requires. Here's a likely reason they didn't design it to grind any finer: the finer you go, the more power is needed.

Mounting the attachment is simple. You just slip it into the front of the mixer and throw a lever to lock it in place. The mill itself may be fully disassembled for cleaning and goes back together in just a minute. Although the unit is steel and fully washable (if then dried thoroughly by popping in a warm oven for a few minutes), it is not recommended for moist or oily grains. We went ahead and tried grinding sunflower seeds, and sure enough, the mill clogged. Cleaning up wasn't difficult but we'll stick with dry grains. No nuts, soybeans, or peanuts for this one.

Kitchen Aid Division, Hobart Corp., Troy, OH 45374.

Bowler on Bread

White "dead" flour makes tasteless stuff, more like blotting paper than real bread, only fit for tasteless and ignorant people.
White flour makes white faces.

Commander Geoffrey Bowler, Royal Navy, in *Natural Bread* (Rodale Press, 1944). Out of print.

A train food system? On India's Tamil Nadu Express, we found this miller at work in the kitchen car, rotating the huge stone pestle at 60 per. Clearly, Indians put a premium on fresh food, even when traveling.

Marathon Uni Mill

$310. Stone grinding surface. Grinds 10 lbs. of fine flour in 15 min. 48 lbs. W 10³/₄" × D 16⁷/₈" × H 13⁵/₈".

The Marathon Uni Mill is a high-speed stone mill enclosed in a hardwood-core cabinet with a Formica finish. We found it to be the most convenient cabinet mill because the on-off switch and texture controls are located up front. Another good point is its versatility—the Marathon is one of the few stone mills recommended for grinding soybeans. The mill's appetite for soybeans and its considerable speed can be attributed to metal inserts in the stones which shear grain before it goes between the stones.

As with all high-speed mills, the Marathon gets a pasty build-up with extended use. The Formica cabinet proved to be a real asset at cleanup time.

Because of its high speed and small 4³/₄-cup grain hopper, you'll not want to wander too far from the mill. Fortunately the flour bin was larger, holding 14 cups, and didn't need emptying as often as the hopper needed filling. The hopper was not efficient at emptying the last ½ to ¼ cup. The laggard grains needed a little prodding by hand or with a brush.

During one of our many conversations with the Grover Company, we learned there is a reason for that small hopper. It

was explained that the heat of operation causes the motor's drive shaft to swell. That closes the gap between the stones so that, at a fine setting, they could come together and stall out the motor. Therefore, with each additional batch of grain to the hopper, the stones should be readjusted to compensate for the swelling shaft—a tip that isn't shared in the directions or by other manufacturers.

The Grover Company, 2111 S. Industrial Park Avenue, Tempe, AZ 85282.

RYE MADNESS

Clusters of people gathered by the fountains and in the streets. They were not talking excitedly now. The voices were hushed. "Le pain tue!" ("The bread kills!") It was a whisper louder than any shout.

They were nearly all screaming and shouting that long night, as they were brought into the hospital on stretchers, from trucks, cars, wagons, any form of transportation that could be pressed into use. There was the man and wife, both bleeding and disheveled, who had been chasing each other around the kitchen table with knives. There was the woman with delusions of grandeur, who insisted that she was a baroness and screamed that she was being persecuted when they dragged her through the hospital doors. There was the woman who was absolutely certain that her three children had been drawn and quartered and were hanging from the rafters of the attic to be made into sausages. There was the man who clutched his head because he was sure that red snakes were eating his brain. There was the man cringing and twisting his body in contortions because there were bandits with huge donkey ears chasing him. There was the seven-year-old child

THE CARROT STORY

J. I. Rodale had a simple formula for eating: "Do not eat anything that has gone through a factory." In 1953, he wrote this story to show what is taken out of our food before we get it on our table.

The first thing done is: the green top is cut off and thrown away. It goes into the garbage pail and eventually is burned up in the city incinerator, fed to pigs, or dumped out at sea. What a terrible blunder this is! Nutritionally the green top has far greater value than the pulpy bottom part which grows underground. The greens contain vitamin K, for example, a vitamin which is completely lacking in the carrot itself.

Many people are not aware that by far the largest portion of the minerals resides in the skin and in the area immediately under it. They therefore scrape the skin off, thinking it is better for their health if they do so. They are afraid of a little dirt. A good washing is

all that is necessary.

The next step is the unpardonable crime of cooking the carrot. Whole boiled carrots will retain 90 percent of vitamin C and much of the minerals, but slicing before cooking results in destruction of the vitamin C and the niacin portion of the vitamin B complex. Salt in the cooking water lowers the vitamin C retention in sliced or quartered carrots. It also causes mineral loss in the water. Copper utensils will destroy vitamin C. Soda in the water destroys both thiamine (of the vitamin B) and vitamin C. Boiling carrots can cause loss of 25 percent of their thiamine. If carrots are frozen and you thaw them slowly, you lose vitamin C. If you sieve carrots hot, you lose 15 percent of their vitamin C. If you sieve them cold, the C loss is 5 percent.

In many kitchens the cooking methods are so careless that the water that has drained out of the carrots, a

precious essence full of minerals, is recklessly thrown down the sink drain.

It is not enough that man destroys so much of the nutritional percentage of the carrot; he is not yet satisfied. At the last moment he decides not to eat it all and leaves a small portion, sometimes not so small, to find its way into the garbage pail and to the city incinerator where it is burned up.

It is not the so-called fast pace that is killing people, but a degenerating and weakening of the body due in part to an emasculation of its food. When one sees what is being done to our food supply, a case can be built for not only the carrot but our bread, fruits, and other foods. It is a sad note that the NUTRITIONISTS have eyes but cannot see. *They* are the ones who could set things right by recognizing conditions for what they are.

J.I. Rodale

whose every toy changed suddenly into a fantastic, indescribable beast. There was the woman screaming that her son was dead, although he stood beside her and was helping her up the hospital stairs. There was the woman shouting, "You've come to kill me, to poison me!" There was the man who saw the hospital attendants as giant fish with gaping mouths, ready to eat him alive. There was the woman surrounded by the dead everywhere.

The delusions were not all horrifying. Some heard giant celestial choruses singing in the heavens above. Others saw flaming, gorgeous bouquets of flowers growing suddenly out of their hands and feet. Still others saw colors of the most breathtaking beauty everywhere, surrounding everything. Some saw in the most ordinary things—a thimble, a fingernail, a shoe—the whole essence of the world and the universe, a revelation they had never had before, a great religious mysticism.

But most were in terror. As one doctor approached a victim on her bed, she screamed: "No, Doctor. Do not approach me. Do not come near me! Look at the fire roaring from my fingers!"

John G. Fuller. *The Day of St. Anthony's Fire* (Macmillan Co., 1968), describing the outbreak of ergotism that occurred in Pont-Saint-Esprit, France, in August 1951, after the inhabitants ate infected rye bread. Reprinted by permission of International Creative Management, copyright © 1968 by John G. Fuller.

Miller Boy Model 1

$197.95. Stone grinding surface. Grinds 10 lbs. of fine flour in 32 min., 10 sec. 36 lbs. W 8″ × D 14″ × H 14″.

The Miller Boy high-speed cabinet mill is available in three forms. Our test unit is the economy model; there's nothing fancy about it, with a particle board cabinet painted red, a tiny 2-cup grain hopper, and a small but sturdy 8-cup stainless steel flour bin. The more expensive Model IV offers a fancier cabinet and a slightly larger bin capacity, but the milling components are the same. A third Miller Boy, bearing the name Independent, is simply a sawed-off Miller Boy without a motor. You can power it with your own electric motor or rig it up with a sprocket and chain to pedal power. It comes equipped with its own 4:1 hand crank.

All Miller Boy designs use an auger instead of gravity to feed grain into the stone. The auger also precracks the grain to take some of the burden off the stones. This mill is suggested for only dry grains, not soybeans, and will grind the usual range of fine flour to cracked cereal. However the stones do not open wide enough for hulling sunflower seeds— unless you have some pretty small seeds. The texture adjustment and on-off switch made their way halfway to the front of the mill, an advantage over rear-mounted controls.

Magic Valley Industries, Inc., Box 10, Filer, ID 83328.

Retsel Mill Master

$277.40. Stone grinding surface. Grinds 10 lbs. flour in 31 min. 38½ lbs. W 11½″ × D 13½″ × H 17⅝″.

The Retsel Mill Master has an appealing calm about it. It is geared down to grind slowly, delivering flour at a steady, cool 115° F. The hopper is large enough to keep the mill busy without constant attention, though you may just want to watch the flour as it gently falls from the stones into your bowl. There's no dust flying, so no need for a cabinet. It may not win a beauty contest, but it requires no more cleaning than a light dusting. The stones can easily be removed for cleaning.

The Mill Master grinds more slowly than high-speed cabinet mills, but because there were no stops for dumping the receptacle or adjusting the stone setting during the 10-pound grind, the *real* grinding time, 31 minutes, is comparable with higher-speed mills. Retsel's mid-sized model, the Mil-Rite ($227.50), has half the production rate and is recommended for the average family; the small Mil-Maid, which grinds at one-sixth the Mill Master's speed, is $179.50.

These Retsel Mills take the hand grinding adaptation seriously. A gear can be removed from the Mil-Rite so you're not turning the motor too—an exclusive feature. And the crank is extra long for added leverage. Most high-speed mills have a direct drive handle with a 1:1 ratio, which is extremely slow; since the Retsel mill is made to be productive at slow speeds, cranking by hand is much more effective.

Retsel Corp., Box 47, McCammon, ID 83250.

Home grain mill: Belmont Village, Vermont

Vermonter Jim Osborne used to indulge his taste for fresh, whole grain bread by driving 20 miles up the road to a health food store in Middlebury. "I'd stop at the store, and by the time I'd lost a chess game, my flour would be ground," he says.

After a year, Jim got tired of making the trip regularly. "I was doing a lot of baking, and the travel got to be more and more inconvenient. If I bought more flour, to save on trips, it wouldn't be fresh by the time I used it up." He decided to buy a home mill.

He didn't know which brand to purchase, so he talked to his chess-playing miller, who recommended a Lee. "When he promised to buy the mill back from me if I didn't like it, I made up my mind pretty fast," Jim said.

That was in 1970. Since then Jim's Lee mill (an S-600 model) has ground several thousand pounds of flour. He and his friend Mary Jane Rand had been grinding about 30 pounds of wheat berries a month, but that amount doubled in recent months when his sister's family moved in. Several friends also come over to grind flour for themselves.

Although Jim uses his mill for a variety of grains—rye, corn, triticale, buckwheat—90 percent of his grinding is wheat. "We make all our own

Jim Osborne and his Lee S-600.

bread, along with muffins, cookies, pancakes, and pie crusts. Whole wheat flour is great in thickeners, gravies, and sauces as well. And once you get used to the taste of freshly ground flour, you won't settle for anything else." Jim purchases his wheat berries from a co-op at $16 for 100 pounds, and stores them in gar-

bage cans. The cool Vermont weather keeps bugs in check.

Can owning a mill make a change in one's diet? Jim thinks so. "I've always liked fresh, whole grain products," he says. "But because we've got the mill, and it's so easy to have fresh flour, we think of more ways to use it."

Malt

Baking with diastatic malt

Professional bakers in Europe have an ingredient in their yeast breads that is virtually unknown to home bakers in the United States. It is a unique yeast food and bread improver called diastatic malt.

You may have noticed that many yeast bread recipes call for small (sometimes not so small) amounts of refined sugar, even for those breads which are not meant to be sweet. The purpose of sugar is to feed the yeast and increase its gassing power. Sugar, because of its ability to caramelize, also makes the crust

brown. But these are the sum of sugar's virtues in bread, and we all know about its manifold vices.

People who are concerned about nutrition often substitute honey or molasses, which perform the same functions of feeding the yeast and browning the crust. More effective still is diastatic malt. Not only does it do everything sugar does, but it has other significant qualities as well. Diastatic malt is full of enzymes and vitamins which increase bread's nutritional value. In addition, the catalytic action of these enzymes on the yeast

and flour improves both flavor and appearance of the loaf, fosters a finer texture, and helps the bread stay fresh. Is it any wonder that travelers remark about the superiority of bread in Europe?

The search

Just what is diastatic malt anyway? That was my question when I first learned about it in a letter from my daughter who was apprenticing in a Swiss bakery. I remembered seeing cans of Blue Ribbon Malt on grocery shelves when I was a child, but I had no idea what malt really was or where it came from. I thumbed through every cookbook I had, but found no reference to diastatic malt. Most cookbooks don't mention malt of any kind, except possibly a passing reference to malted milk. Even the dictionary referred to malt only as "germinated grain used in brewing and distilling."

I took the dictionary's lead and began reading books on beer making. I discovered that malt is sprouted grain (usually barley) that has been roasted and ground. It is then dissolved in water and filtered to remove the husks and bran, after which the liquid is reduced to a syrup or dehydrated to make a powder. I also learned that *diastatic* malt, unlike conventional malt, has been dried and processed at a low temperature (under 170° F.) so that its special enzymes are not destroyed by heat. Much later I learned from my daughter that these enzymes have the power to transform the starch in flour into maltose and dextrin—yeast foods which assist in the fermentation process. These enzymes also help in the production of soluble proteins for the yeast's use.

After checking several sources I was finally able to find diastatic malt syrup at a beer supply store, and thus began an adventure in baking bread with malt rather than sugar. The bread was great. The problem was that the malt syrup was so thick and unmanageable it was a nuisance to use.

At about that time my husband happened to be reading a book on Mesopotamia. In it was a detailed description of how the people of those ancient times made malt for their alcoholic beverages by sprouting barley kernels and then drying them in the sun. That is when I made the critical connection. Malt was not so exotic after all. I had been sprouting mung beans in my kitchen for years. Why not sprout barley as well and carry it two steps further—dry the sprouts and grind them? There would be no need to filter out the husks and bran since my malt was to be made into bread rather than beer.

The problem was where to find the grain. Hulled barley would not sprout and there was no unhulled barley anywhere. I searched through

health food stores in New York City and wrote many letters to organic grain suppliers, but to no avail. Finally, in an encyclopedia, I learned that malt can be made from wheat as well as barley, and that the enzyme action is the same.

For several years now I have been making my own diastatic malt with wheat berries. They are available at any health food store. I am convinced that diastatic malt makes the subtle difference between a good bread and a great bread.

Used in place of sugar, honey, or molasses in any yeast bread recipe, the action of the diastatic malt is so powerful that only a small amount is needed. One teaspoonful will be enough for a batch of dough yielding three to four loaves. A little more won't hurt, but use restraint. Once you have made your own diastatic malt and see how easy it is, you may be tempted to simply dump it into the dough, believing that "more is better." It is true that diastatic malt is richer in nutrients than the grain it was made from. However, if used in excess it will overwhelm the yeast (give it indigestion, so to speak). This will cause a breakdown in the texture of the loaf during baking and will yield a sweet, sticky fiasco.

For those who do not wish to make their own, Schiff Bio Food Products, Inc. makes a dehydrated diastatic malt called Dimalt. It is available in many health food stores, but ask for it by its trade name, for many clerks are unfamiliar with diastatic malt per se.

The method

Place one cup of wheat berries in a wide-mouth glass jar and cover the top with a piece of nylon net or cheesecloth, secured with a rubber band. Pour four cups of tepid water into the jar through the net and let the grain soak for about 12 hours or overnight. Drain off the water from the swollen grain. (Save the water for bread liquid, soup stock, or for watering your house plants—it's full of water-soluble vitamins and minerals.) Pour more tepid water into the jar, shake gently, and drain thoroughly. This rinsing and draining keeps mold from forming on the sprouts. Keep the jar near your kitchen sink and repeat the rinsing, shaking, and

draining three times a day for two days, or until the little shoots are about the same length as the grains. (There will also be tiny white rootlets.) The temperature of your kitchen will determine the length of time required.

When the sprouts have reached their proper length, rinse and drain once again and arrange the sprouts evenly in thin layers on two large baking sheets. Place them in an oven at a temperature no higher than 150° F. The sprouts should be dry in eight hours or less. (They'll give off a delightfully sweet, earthy fragrance as they dry.) Or you can air-dry them by placing the baking sheets in a warm place, preferably in the sun, for several days until they are thoroughly dry. Then grind the dried sprouts to a fine meal or flour in an electric grinder or blender. This amount will yield one cup of diastatic malt—enough for up to 150 loaves of bread. Store the malt in a tightly closed glass jar in the refrigerator or freezer. It will keep indefinitely.

Jane Nordstrom

Hot Malt Beverage

Use 1 heaping teaspoonful of powdered malt for each cup of water, and simmer the water and malt together for about 10 minutes. Strain before serving.

Hamburger Buns

1 tablespoon active dry yeast
1 teaspoon wheat malt (or honey)
¼ cup warm water
4 tablespoons oil
1 egg, beaten
3 cups whole wheat flour
¾ cup milk
1 tablespoon sesame seeds

Soften the yeast and malt (or honey) in the water. When the mixture bubbles (gives a spongy appearance), add the oil and egg. Blend thoroughly and allow to rise for 10 minutes.

Mix whole wheat flour, milk, and yeast mixture together to make a soft dough. Turn dough out onto a floured board, and knead until it is smooth and elastic. Shape into round, flat buns (about 3 to 3½ inches in diameter), and place on greased cookie sheet, allowing space between. Brush the tops of the buns with cold water and sprinkle with sesame seeds. Cover the buns with a damp cloth and set in a warm place to rise until double in bulk. Bake at 375° F. for 15 to 20 minutes.

Yield: 10 buns

Malt Bread

2 cups warm water
1 teaspoon diastatic malt
2 tablespoons or 2 packages yeast
2 tablespoons vegetable oil
5 cups whole wheat flour

Combine the water, malt, and yeast, and let them sit until bubbly. Add the oil and enough flour to make a soft dough, and knead it until it is smooth and elastic. Put the dough in a greased bowl in a warm place to rise until double in volume.

Punch down the dough, knead it briefly, and then let it rise again. After the second rising, punch it down and divide the dough into 2 pieces, and let it rest, covered, for about 10 minutes. Form it into loaves, place them in 2 medium-size greased pans, and let them rise to about 1½ times the original volume. Bake the loaves at 350°F. for 35 to 40 minutes or until they sound hollow when rapped. The bread should have a smooth texture and the unadorned sweet, earthy flavor of the whole grain.

Yield: 2 medium size loaves

A grain glossary

Amaranth. See illustrated grain profiles.

Barley. See illustrated grain profiles, *hull-less barley, pearl barley.*

Bran. The outer part of the cereal grain, rich in B vitamins and composed mainly of indigestible cellulose lulose material.

Brewing. A method for preparing liquids from grains. This is done by steeping or boiling the grain in water, and then, as in the case of beer, allowing the liquid to ferment.

Broomcorn. A variety of sorghum whose seed head tassels are used to make brooms.

Brown rice. Rice from which only the fibrous husk has been removed, leaving the bran intact.

Buckwheat. See illustrated grain profiles, *kasha.*

Bulgur. Wheat that has been steamed, then dried and cracked. Usually it is lightly milled to remove the outer bran.

Corn. See illustrated grain profiles, *field corn, grits, hominy, Indian corn, maize, popcorn, sweet corn.*

Cous-cous. A Middle Eastern and North African dish made with bulgur, served over a meat broth.

Cracked wheat. Wheat that has been broken into small pieces by very coarse milling, but not precooked. (See also *bulgur, cous-cous.*)

Durum wheat. A hard spring wheat whose flour is used primarily to make pasta. The starch granules in durum wheat are so hard that the flour cannot be used for making bread.

Emmer. An ancient, hard red wheat, now found only in grain glossaries.

Endosperm. The inner, starch portion of the grain kernel, which contains most of the protein. White flour is made from the wheat endosperm.

Enriched flour. White flour to which three B vitamins (thiamine, niacin, and riboflavin) and iron have been added. This practice was begun in 1942 to prevent beri-beri and pellagra, two nervous system diseases encouraged by the popularity of white flour.

Farina. Starch obtained from wheat

Short on flavor but long on fiber, broomcorn has yet to be replaced by synthetic bristles.

other than durum wheat. Or, a name for a breakfast cereal made from fine wheat meal.

Field corn. Corn that is dried and then used primarily as animal food. It is coarser than sweet corn.

Flax. A member of the herb family used primarily for making linen. Flax seed can be sprouted, and is fed to animals. The seeds are the source of linseed oil.

Germ. The embryo or sprouting section of the seed, often separated because it contains fat which limits the storage of flours. It is especially rich in thiamine and vitamin E.

Gluten. The protein part of wheat, which can be isolated from flour by washing out the starch. It is what gives cohesiveness to dough. Powdered gluten, labeled as gluten flour, is available in natural food stores.

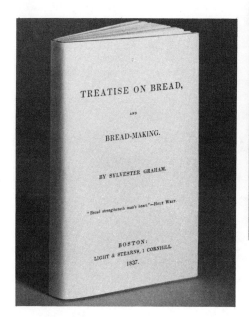

The traditional method of separating the hulls from rice.

Graham flour. A flour in which the bran is finely ground. It is named for Sylvester Graham, a miller of whole meal flour who advocated its use in his 1837 book, *Treatise on Bread and Bread Making.*

Grits. Coarsely ground, hulled grain, especially corn.

Groats. Oats from which the husks have been entirely removed. Or, any hulled grain broken into large pieces.

Hard wheat. Wheats used to make bread flour. They can be either winter or spring wheats, and are higher in protein and gluten than soft wheats.

Hominy. Field corn that has been soaked, precooked, and dehulled (and in most cases, degermed). Corn grits are ground hominy.

Hull or husk. The outer, inedible cellulose covering of the cereal grain. In some grains, such as wheat, it is loosely attached; in others, such as barley, it is firmly attached.

Hull-less barley. A variety of barley from which the hull can be removed easily, leaving most of the bran intact. Little is grown in the United States because it is unsuitable for malting.

Indian corn. Synonymous with maize, as used outside the United States. In the United States, decorative corn with multicolored kernels.

Kasha. Cooked buckwheat. A diet staple in Eastern Europe and the Ukraine.

Maize. The more common name for corn outside the United States.

Malt. A sweet extract of sprouted barley or wheat. Sprouted grain is soaked in hot water to dissolve the malt. The solution can then be concentrated into a syrup or evaporated to dryness.

Middlings. Small fragments of groats resulting from processing. Middlings are a by-product of hulling, milling, and rolling.

(Continued on page 58)

(Continued from page 57)

Millet. See illustrated grain profiles.

Milling. The process of converting grain into flour.

Miso. A Japanese food made by fermenting rice with soybeans and salt.

Oatmeal. Common term used for a breakfast cereal made from ground or rolled oats.

Oats. See illustrated grain profiles, *rolled oats, Scotch oats, steel-cut oats.*

Pearl barley. Barley from which most of the bran layer has been removed, so-called because of its white, smooth appearance. Because the bran layer is fused to the starchy core, the barley must be partially ground to remove it.

Polished rice. Rice from which the outer bran layers and the germ have been removed. Sometimes called white rice. When white rice is polished with a mixture of glucose and talc, it is labeled coated rice and should be rinsed before eating.

Popcorn. A special strain of corn with a very hard hull. When the popcorn is heated, the moisture inside turns to steam, causing the kernel to explode.

Pumpernickel. A dark bread or flour made from whole rye berries. The bread is traditionally baked at low temperatures for up to 12 hours. Commercial pumpernickel is further darkened with caramel coloring.

Rice. See illustrated grain profiles, *brown rice, polished rice, sweet rice, white rice, wild rice.*

Rolled oats. A lightly steamed, flattened oat. It is used as a breakfast cereal, and in cookies, breads, and muffins.

Rye. See illustrated grain profiles, *pumpernickel.*

Saracen corn. Another name for buckwheat.

Scotch oats. Groats that have been cut into various size pieces, often roasted to give them a nutty flavor. They cook faster than whole groats because the exposed starchy core absorbs water easily.

Semolina. The starch obtained from the endosperm of durum wheat. Usually used to make pasta.

Soft wheats. Wheats used mainly for pastry or cookies. Compared to hard wheats, they have less gluten and protein.

Sorghum. See illustrated grain profiles, *broomcorn, sweet sorghum.*

Soy flour. A high-protein flour from soybeans, which has a somewhat bitter taste. Add in small amounts to standard recipes.

Spelt. A rare variety of wheat which does not thresh easily and is used as as an animal food.

Spring wheat. Wheat that is planted in the spring and harvested in the fall. It can be either hard or soft wheat.

Sprout. A germinated seed or grain.

Steel-cut oats. Oats cut with rotary-type cutters to speed cooking. They may simply be cut (scotch oats) or cut and then rolled.

Sweet corn. The type of corn usually eaten by humans. Its genetic structure prevents the natural sugars from changing into starch before the corn matures. (See field corn.)

Sweet rice. Short grain, waxy rice which is very glutinous. It cooks into a sweet sticky mass and is excellent for rice pudding.

Sweet sorghum. A variety containing juice in its stalk that can be made into syrup, molasses, or sugar. The stalks are used for fodder and silage. Also called sorgo.

Threshing. Any procedure to separate the grain or seed from the straw or plant.

Triticale. See illustrated grain profiles.

Traditionally, winnowing is accomplished by letting a breeze blow hulls away from a shallow wicker basket.

This Moslem grandfather's contribution to the family homestead is threshing the rice harvest.

Unbleached flour. White flour which has had the bran and germ removed, but which, as the name implies, has not been chemically bleached. Sometimes, the germ is returned to the flour.

Wheat. See illustrated grain profiles, bulgur, cous-cous, cracked wheat, durum wheat, farina, graham flour, hard wheat, kasha, semolina, soft wheats, spring wheat, whole wheat, winter wheat.

Wheat pilaf. A type of processed cracked wheat used as a substitute for rice. (See *bulgur, cous-cous.*)

White flour. Flour made entirely from the wheat endosperm, with both the bran and germ removed. It is usually bleached, most commonly with chlorine dioxide.

White rice. Rice from which the bran and germ have been removed. (See *polished rice.*)

Whole wheat. Wheat from which only the outer hull, or husk, has been removed.

Wild rice. A native North American grass that grows in marshy areas and produces a long, thin, greenish grain. Little is grown and it is difficult to harvest, so it has become a gourmet food.

Winnowing. Separating the grain from the chaff by wind power.

Winter wheat. Wheat that is planted in late autumn and begins to grow before the winter snows. It then lies dormant until spring, and is ready for harvest by early or mid-summer. Winter wheat is favored by pretzel makers.

Pasta

You make pasta at home for the same good reasons you make bread: to enjoy fresher, more-flavorful food of known ingredients and superior nutrition. It's likely that store-bought pasta isn't as deficient as store-bought bread, so pasta-making isn't accorded the same importance in most households.

In fact, homemade pasta was until recently the sole domain of Italian grandmothers and Pennsylvania Dutch women.

Now, home pasta-making lags only a little behind bread-making in popularity. Just as it is now fashionable to scorn snow white bread, pasta-makers can feel good in knowing the product of their hands (or their high-chrome machine) is better than money can buy.

Pasta has as many supposed histories as it does shapes. The creation of the food is often attributed to the pasta-loving Italians, but it is said that Marco Polo, the Venetian merchant-traveler, borrowed the idea on a voyage to China.

One account has Polo sending a sailor ashore for fresh water. By a well in the courtyard of a large home, he noticed a young girl cooking strange long strings. She invited him to sample the strings, and he found them delicious. The sailor went back to the ship with both water and a recipe. Back home, the edible strings caught on in a big way, and came to be named after the sailor who discovered them, Spaghetti.

Another version of the same story credits the sailor with actually inventing spaghetti. The Chinese maiden was outside making bread when the sailor happened by and lured her away from her baking. The pair returned later to find that leaves had fallen into the bread dough. The gallant sailor rushed to the rescue and attempted to extract the leaves by forcing the dough through a wicker basket. The thin strands which came out dried in the sun, and thus pasta was born of their affair.

Unfortunately for romantics, it couldn't have happened that way. As a matter of fact, pasta was commonplace in Europe long before 1270, when Marco Polo made his now-famous journey. The Spaghetti Museum in Italy (yes, there *is* one) has proof that there were regulations for the size and shape of various

(Continued on page 60)

(Continued from page 59)

pastas as early as 1200. And on the walls of a 4th century B.C. Etruscan tomb, not far from Rome, were iound bas-reliefs of pasta-making utensils. Undoubtedly, pasta-making evolved independently in China, in Europe, and the Middle East.

The origins of macaroni are somewhat clearer, being relatively modern history. Although handed a poor deal by life, a beneficent Neapolitan by the name of Chicho devoted his modest funds to developing a new affordable food for the masses. He experimented with simple wheat dough and primitive machines in his tiny garret, and happened to extrude it into tubes. He also developed a special sauce to suit this novel pasta. Meanwhile, across the alley from Chicho's garret, the cunning Javanella was watching. She saw him working far into the night, watched his lonely experiments, and took note of his sauce ingredients. She cooked up some tubes herself. When her husband praised the new dish, Javanella declared the recipe had been revealed to her by an angel in a dream.

It so happened that the husband worked in the kitchen of King Frederick, and asked the chef if his wife could prepare the dish for the king. The chef was only too happy to comply with someone who had communicated with an angel. So Javanella made the macaroni for King Frederick, who promptly exclaimed that it was marvelous. Not only that, he gave Javanella 100 gold pieces in thanks. Soon princes and lesser nobles were enjoying this new food. Javanella's fame spread.

But Chicho, slaving away in his garret, heard none of it. He could return to the world only when he had perfected the economical yet tasty dish he so wanted to share.

That day came. He left his quarters to impress a prince with the culinary invention, only to be stopped cold by a smell, a familiar smell—his secret sauce. How could this be? He traced the smell until he found a woman in the very act of pouring the sauce over tubes of pasta. She explained that the dish was a favorite of the king, and he appreciated the tubes and sauce so much that he had bestowed a fortune upon its creator, a neighbor by the name of

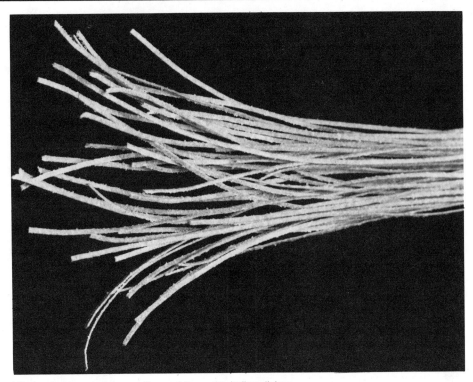

Whole wheat pasta brings a flavor of its own to Italian dishes.

Javanella.

Devastated, Chicho slunk back to his garret and smashed all his equipment. It is said that his ghost still haunts the street, and that on Halloween he can be heard relentlessly rolling his pasta, while Javanella stirs the sauce and the devil grates the cheese.

The name of this food is explained by a happier tale. When a 13th century king (perhaps Frederick?) was given his first taste, he cried out, "Ma caroni!"—how very dear.

By the 1700s, Naples was the world capital of pasta. Noodles were everywhere, drying in the open air, and sold hot in the streets as a fast food to be eaten with the fingers.

When Thomas Jefferson came home from his years as Ambassador to France, he brought America's first "machinetta" with him. But pasta didn't catch on with the rest of America until the late 1800s, when floods of Italians began immigrating to the United States. Soon bohemian artists of New York's Greenwich Village were attracted by the aroma of spaghetti cooking right next door in Little Italy.

Manufacture of pasta in America was helped considerably by the efforts of Dr. Mark Carleton, an agronomist for the USDA. In 1898 Carle-

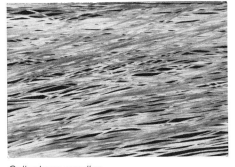

Cellophane noodles.

ton went to Russia to find a rust-resistant, drought-resistant strain of wheat to rescue America's diseased crop. He returned with durum wheat, and zealously encouraged farmers in the plains states to grow this variety—which happens to have the hardness necessary for good pasta. When World War I cut off imports, American pasta manufacturing got a further boost.

In Japan, the noodle is second only to rice in the diet. On breaks between meals Japanese commonly have steaming bowls of hot noodles in the winter or chilled noodles in the summer. Their thick whole wheat noodles, *udon,* and skinny *somen* are served in broth. Japanese also make *soba* from half buckwheat flour and half wheat flour. In Thailand shoppers and theater-goers buy fried rice noodles from street vendors. The Chinese celebrate the New Year with noodles, because their length symbolizes longevity. *Mein* have been a staple in northern China for at least 6,000 years. Americans are most familiar with the Chinese *chao mein* (chow mein to us), meaning fried noodles. The Chinese also make tangled bundles of noodles from flours other than wheat. Rice noodles are made from rice flour and water. Mung bean flour noodles are known as cellophane noodles, because they become transparent when cooked. The Chinese prepare the noodles by first soaking them in hot water and then steaming or stir-frying them. Or they deep-fry the noodles, which turns them into a puffy, cloudlike mass.

Not all pastas are created equal

Traditionally, commercial pasta has been made from semolina, a highly refined flour milled from the starchy endosperm of hard durum wheat. This golden grain gives commercial pasta its characteristic yellow color. An exceptionally high amount of gluten develops elasticity when the dough is kneaded, enabling the pasta to maintain its shape when cooked.

All-purpose flours or regular whole wheat flours have a lower gluten content, so homemade pasta is likely not to hold together quite as well. Enthusiasts may want to search out whole durum wheat to grind into fine flour at home. For that extra effort, the pasta will have both the gluten of hard wheat and the nutritional advantages of the whole grain (missing in commercial pasta). Otherwise, standard whole wheat flour, kneaded well, will produce pasta more than satisfactorily.

The nutritional superiority of whole grain pasta over commercial semolina pasta parallels that of whole grain bread over white bread: Enrichment of commercial pasta replaces some B vitamins, but the small yet significant amounts of fiber and minerals (including calcium, phosphorus, iron, potassium, and magnesium) remain on the mill floor.

Pasta becomes more fun to eat when you learn that, contrary to common belief, spaghetti and noodles are not fattening foods dieters need shun. Nutritionists are encouraging people to include more whole foods, such as grains, fruits, and vegetables, in their diets. Whole grain pasta is one of those basic, healthful foods. It offers a good proportion of protein, carbohydrates, and fat—about 70 percent carbohydrates, 13 percent protein, and only about 2 percent fat.

These carbohydrates (sugars and starches in their natural state) are digested more slowly and are more satisfying than highly processed carbohydrates.

The protein in wheat has a varied distribution of essential amino acids, but is low in lysine, which is necessary for complete protein. Combining pasta with cheese or beans (as in pasta fagioli) rectifies this. And recipes for pasta made at home usually call for eggs, both because egg-based dough is easier to handle and because egg offers complete protein.

A main dish of pasta also helps make a meal low in fat. Whole wheat flour's fat content is a mere 2 percent. Adding eggs to a pasta recipe puts it up to 5 percent—but compare that to lean ground beef at 10 percent fat. Semolina pasta has the slimmest fat content—only 1 percent. But that's because the valuable germ is squeezed out during the milling process.

Pasta's calorie count is often assumed to be higher than it is. A quarter-pound portion (fresh, before cooking) of flour-and-water pasta has about 200 calories. Eggs in the dough add only 15 or 20 more calories per serving. Either way, it's a modest amount for a main dish. That serving of unadorned pasta supplies about 50 more calories—but 50 percent more protein—than two slices of bread. You needn't worry about gaining weight from eating pasta if you remember that the thing to beware is not the pasta itself, but rich sauces with hidden calories.

According to federal food-labeling regulations, commercial *macaroni* products may be prepared from just flour and water, but *noodles* must contain fresh, dried, or frozen whole eggs or egg yolks. Recently, however, egg-less Oriental noodles have come on the market. These noodles have been exempt from the egg requirements, much to the dismay of pasta manufacturers who must still use costly eggs for their noodles, American-style.

By law, commercial macaroni and noodle products may also contain disodium phosphate, salt (often in *large* quantities), glyceryl monostearate (a fat used as protective coating), gum gluten, and seasonings such as onion, celery, garlic, and bay leaf.

Sometimes, even less desirable substances find their way into the commercial pasta box. Analyses by Consumer Reports revealed 69 insect fragments or larvae and animal (probably rodent) hairs per ounce of one brand of imported spaghetti.

Adornment

Italians don't hide good pasta by smothering it with sauce. They use just a little, to keep the pasta from sticking together and to add aroma, flavor, and color. Conversely, the standard American spaghetti sauce is tomato sauce with a hefty dose of meat—the spaghetti serving only as an excuse for lapping up the rich topping.

The simplest way to enjoy pasta, especially spaghetti or noodles, is with garlic fried in olive oil. By all means don't abandon tomato sauces. Just don't overwhelm the pasta. A good way to let the personality of the noodles come through is to use a fresh, uncooked sauce made with tomatoes and other garden vegetables.

Pesto is one of the most basic Italian pasta sauces. Although pesto can fall prey to modern technology and be whipped up in a blender or food processor, the traditional practice of grinding the herbs in a mortar and pestle preserves more of the delicate aromatic oils. Fresh basil leaves and garlic cloves are ground in the mortar, then olive oil and grated Parmesan are beaten in slowly. Pesto frequently includes pine nuts.

Pasta by hand and by machine

If you can knead bread dough or stir up a good granola, you can make pasta. Without a machine, strands vary in width in a decidedly uncommercial way, but this roughness will come to symbolize good flavor—fresh noodles taste better than their perfect-looking cousins from the box. Use whole grain flour, and the pasta will be a lot better for you, too. An extra advantage of homemade pasta, for people who want to limit their sodium consumption, is that you can avoid the salt added to commercial brands.

Making pasta doesn't take all day, either. With the help of a pasta machine, a meal's worth of noodles takes only about half an hour to make, once you're familiar with the process. Even without a machine the whole process is no more than an hour's work.

You can create a batch of homemade pasta with just five utensils: measuring cup, fork, rolling pin, sharp knife, and large wooden pastry

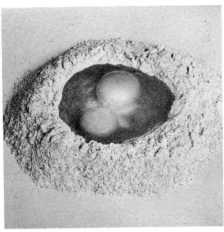

Place the flour in a mound on the board, and form a well in the center. Drop the eggs into the well.

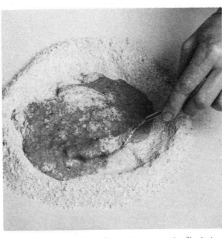

Beat the eggs with a fork, then gradually bring in the flour as you mix.

Work the dough into a ball with your hands. The dough should pick up only as much flour as it is able to. If the dough seems dry, wet your hands and then work with the ball. It should be stiff, but not unmanageable. Cover the dough with an inverted bowl and let it rest a few minutes while you clean all the bits of dough off the board and your hands.

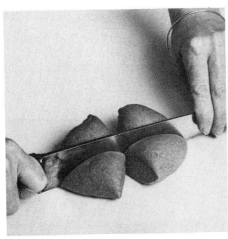

Divide the dough into smaller pieces for rolling. Set aside and cover all the pieces but one.

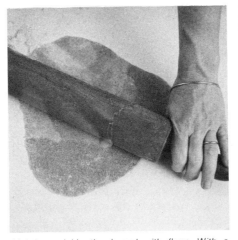

Lightly sprinkle the board with flour. With a rolling pin, flatten the dough into a smooth oval. Stretch the dough from the center outward, working with all portions uniformly. Instead of making the dough thinner by pressing it, stretch it. Wrap an end around the rolling pin, and move your hands out from the center so that you pull the dough wider as you stretch it away from you. Work quickly, but don't rush or give up before the dough is as thin as you can make it. Roll and stretch until the sheet is translucent. Let that piece rest in a warm, dry place while you repeat the process with the remainder of the dough.

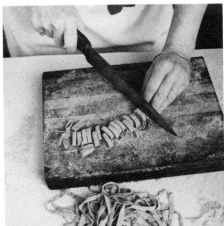

When the dough is dry and supple as leather (but before it becomes brittle), loosely roll each piece like a jelly roll. Cut noodles the desired width with a large knife.

board (or a sheet of Formica or plastic board or even the dining room table covered with an old sheet).

Many pasta-makers insist the best rolling pin for pasta is long and very thin—the "broomstick" type. Choose whatever you can manipulate well.

Cooking is an inexact science, and pasta-making is no exception. The proportions of flour and egg may change each time, depending on the size of the eggs, the type of flour, the humidity and temperature of the kitchen, and other mysteries. Don't hesitate to use a little more or less of something to make the dough look and feel right. You'll want a general, reliable formula to start from, though.

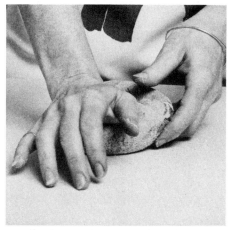

To knead the dough, push it away from you with the heels and palms of your hands, then fold the outer edge over toward you. Continue that motion, giving the dough a quarter turn each time. Knead at least ten minutes, until the dough is hard and smooth. It should not stick to the board.

To straighten the noodles, carefully unroll them one at a time. Or gently lift a pile of noodles into the air with your fingertips, then drop them. Think of it as tossing a salad. The noodles should uncurl themselves. If you want perfectly straight noodles, you can cut them to almost spaghetti-thinness from a flat sheet of dough with a pizza or pastry wheel. It takes longer that way, but then you needn't roll and unroll the noodles, and they will be as uniform as those made by machine.

Use ¾ cup flour and 1 large egg for 2 servings. To make more, simply add an egg for each additional ¾ cup flour. One and a half cups flour and 2 eggs make almost 1 pound of fresh, uncooked pasta. One-fourth that amount is an average main course serving size. Of course, each cook's idea of "one serving" varies, as do peoples' appetites. (Some pasta dough recipes call for 1 cup flour per egg, but they usually also include water or oil. It's wise to start with just ¾ cup flour, so that more can be added during rolling to prevent sticking, without risk of the pasta becoming tough and coarse.)

If pasta is to be cooked the same day, let it dry in the open air for about an hour. Lay it flat or, if you don't have enough space, hang the pasta over the backs of chairs or on a broomstick, like laundry on a line, as the Italians do.

To keep fresh pasta, store it in airtight plastic bags or containers. It will last about a week in the refrigerator, or almost indefinitely in the freezer. Frozen pasta shouldn't be thawed before cooking, or it will get mushy when boiled.

Pasta may also be dried completely for another several hours, until it is quite brittle, and stored in tightly capped jars in a cool place. Very thin pasta curls up when it dries. To keep it nice and straight for storing in tall jars, place a heavy object (the rolling pin will do for one) on both ends of the noodles until they dry.

Although wheat flour is a necessary pasta ingredient because of the rigidity it gives to the dried forms, you can supplement it with other flours for variety and enhanced nutrition. Half and half buckwheat flour and whole wheat makes a flavorful version of Japanese *soba*. Mixing in half soy flour helps complement the protein in wheat.

Although flour and eggs are all that is needed to make a good homemade pasta, many recipes include a tablespoon or so of water, oil, or cream. These ingredients make the dough easier to knead.

You can add spinach, beets, carrot, or tomato to pasta dough to give it a bright, appetizing color. Spinach noodles are green and have a smooth, creamy texture. The taste is mild—guaranteed to fool people who shy away from spinach. Besides livening the appearance of your pasta, adding low-calorie spinach supplies it with one of the richest sources of Vitamin A.

To make 4 servings of spinach noodles, use ½ pound fresh spinach with 2 cups flour and 2 eggs. Remove stems, and cook or steam the spinach until it wilts. Drain and squeeze well to remove as much moisture as possible, and chop finely. Beat the spinach well with the eggs before drawing in flour. The dough may be stickier than usual; dust with additional flour as necessary. Knead, roll, and cut the dough as you would basic pasta dough.

Pureed beets, strained tomato juice, or pureed cooked carrots make red, pink, or orange pasta. Use the same general method, always beating the coloring agent with the eggs first.

(Continued on page 66)

Raw Mixed Vegetable Sauce

4   tablespoons olive oil
1   clove garlic, finely minced
¼   cup red or Bermuda onion, finely chopped
¼   cup red pepper, finely chopped
¼   cup green pepper, finely chopped
¼   cup baby zucchini, unpeeled and finely chopped
5   medium tomatoes, peeled, seeded, and coarsely chopped
½   cup young sprouts (lentil, mung, or alfalfa).
2 to 4 tablespoons fresh basil or parsley

Marinate all ingredients except herbs and sprouts 6 hours or more at room temperature. Toss hot pasta with the sauce, then sprinkle with sprouts and herbs.

Yield: sauce for 4 servings

Raw sauces are very perishable and should be used within a few days. After that period they should be brought to a boil, cooled, and frozen.

Broccoli flowerettes are a crunchy mate for steaming pasta. Saute steamed broccoli in a little chicken broth before tossing them with noodles. Or team pasta and peas, then pour on a cheese sauce.

Pasta topography

The long history and great importance of pasta are suggested by its many named varieties. Most are made from the same simple ingredients. It's just the shapes and sizes that change.

Basically, pasta comes in four styles. Cords and strings are long, thin, and solid—such as spaghetti. Spaghettini is a little finer, and vermicelli is finer still.

Tubular forms include macaroni, mostaccioli, rigatoni, and ziti, differing in length and diameter.

Flat pasta ribbons are just plain noodles to us, but the Italians have a name for several widths: linguine, fettuccine, tagliatelle, and the popular wide lasagne.

Filled forms are stuffed—usually with a meat, cheese, or spinach mixture—then rolled, folded, or pressed together. Ravioli are stuffed little squares. Cappelletti look like hats, and cannelloni like big pipes.

Fresh pasta will be done in only three to five minutes of boiling, while completely dried or commercial pasta may take up to eight or ten minutes.

If you're willing to tamper with tradition, you can save energy (and water-soluble vitamins and minerals lost by whole grain dough during boiling) by cooking the pasta directly in a soup or broth. Expect about five cups of liquid to be absorbed per pound of dry pasta, less for fresh. You'll likely have to add water to rich soups.

Or, cook fresh pasta as you would rice or oatmeal. Use just as much water as it will absorb. This will vary depending on the quantity cooked, size of the pot and the degree of dryness of the pasta. A good estimate is four cups liquid per pound of freshly made pasta. Bring the water to a boil, preferably in a wide, shallow pot. Add the pasta very slowly without stopping the boiling. Keep it at a slow, uncovered boil, stirring often with a wooden spoon or chopstick until the water is almost absorbed. Then tilt the pot with the lid on to drain off the remaining water and serve immediately on warmed plates.

Yet another option is to cook pasta directly in a thin tomato sauce or the juice drained from canned tomatoes. Add herbs, a touch of garlic, and olive oil, then cook and serve from the same pot.

Make lazy lasagne by eliminating the precooking, draining, and drying of the noodles. Prepare the dough and cut out the strips to the desired length. Simply layer the ingredients as usual, without precooking.

A pasta rake (also called a spaghetti spoon) helps untangle long noodles when removing them from the cooking pot.

To make ravioli, roll out two large sheets of dough. Spoon filling mixture onto one at regular intervals. Then cover that with the second sheet of dough, gently press down the spaces between the lumps of filling, and cut out squares with a pasta crimper. Neatness counts.

Homemade pasta can be just as varied and interesting in shape as storebought.

The chitarra

Anyone who knows how to wield a hammer could easily make a *chitarra*, an amusing Italian kitchen instrument. It's named a chitarra, Italian for guitar, because that's what it looks like. Thin, strong wires are strung close together over a rectangular wooden frame.

A sheet of dough is pressed through the wires with a rolling pin, and falls through as thin noodles. They are gently scooped into a pile and served as nests of noodles.

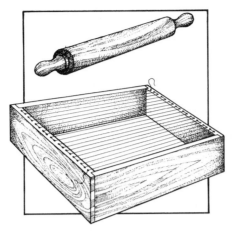

(Continued from page 63)

A novice reports

After watching a friend make homemade pasta I could hardly wait to try it. It looked so easy. Just two ingredients—flour and eggs—and *voila!* you have lovely fresh noodles.

Well, my first pasta-making session certainly couldn't have been filmed for a television gourmet show. But in the end the noodles tasted great.

Careful as I was to keep the eggs dammed in while I beat them, some forced a path through the flour and made a golden rivulet across the pastry board before I could stem the flow. At that point, with my one hand full of sticky egg and flour, and my other hand buttressing the other side of the mound, which threatened to burst at any moment, I wished I had an extra hand or two. But the cameras weren't watching, so I scooped the whole mess together and worked it into a respectable lump of dough.

The pasta machine I borrowed was a treat to use. Things proceeded smoothly as I kneaded and thinned some dough between the rollers—until I got to the last (thinnest) setting on the machine. Then the edges of my beautiful strip of dough started to disintegrate! When I doubled the dough over to try to remedy the situation, it got shirred. Not knowing how else to salvage the dough, I squeezed it back into a ball and started over at the widest roller setting. (Yes, if you overwork the dough it does get tough, but I wanted to come out with some respectable-looking noodles.)

This time, I ended up with a long, thin strip that looked just the way it should have—because I quit while I was ahead at the next-to-last setting. The other pieces of dough ran through the pasta-maker in even, uniform sheets.

Then came the part that was most fun. I cranked the dough through the noodle cutter, and out came lovely, perfectly straight noodles, cool on my hand as I caught them up. The tiny spaghettis were even niftier. Watching them tumble out of the machine was an unexpected delight.

Just to see how well I could do it, I tried kneading and rolling part of the

PASTA INTERNATIONAL
Gertrude Harris. 1978.
101 Productions, 834 Mission Street,
San Francisco, CA 94103. 185 p.
paperback $4.95.

If you'd welcome guidance from a practiced pasta-maker before attempting your own batch, do spend an hour browsing through *Pasta International.*

In her careful step-by-step explanation of how to make hand-rolled pasta, Gertrude Harris treats the practice as an art—but one whose secrets she is glad to share. She includes instructions for making noodles in manual and electric pasta machines and food processors. All are explicit yet conversational.

As befits the title Ms. Harris weaves a long and interesting history of pasta through Europe, China, Africa, and the United States. She briefly discusses and suggests uses for several dozen types of pasta eaten all around the world. In addition to the illustrations scattered liberally throughout the text, the inside front and back covers are an interesting find for those who never knew exactly what linguine or mostaccioli or gnocchi looks like. Pictured there are sketches of the shapes and relative sizes of 66 forms of pasta, each with its Italian or Oriental name.

The bulk of *Pasta International* is a collection of recipes, grouped by the main ingredient of their sauces and accompaniments, and includes pasta with vegetables and herbs, pasta with fruit and nuts, pasta with poultry, and dumplings and filled pasta.

CB

Cheese Pancakes with Fine Noodles

I am told that when these pancakes are eaten, the diners agree that "Baghdad set-el-Beled"—Baghdad is the ancestor of all countries.

½   cup very fine egg noodles
2   large eggs
    salt and freshly ground white pepper to taste
½   cup crumbled or grated white cheese, such as feta or Syrian halumin
4   tablespoons unsalted butter

Cook the noodles in boiling salted water until tender. Drain and set aside.

Beat the eggs and add the salt, pepper, cheese and noodles; mix thoroughly.

In a large wide frying pan or on a griddle, melt the butter. When it begins to foam, pour 2 tablespoons of the egg mixture into the butter for each pancake, making only as many pancakes at a time as the pan will comfortably hold. Turn the pancakes when the undersides are golden and cook second sides until nicely browned. Serve hot with fruit preserves or syrup, if desired.

Yield: 4 servings

dough by hand. Kneading pasta takes a lot of energy. I had even slipped some white flour in with the whole wheat, thinking it would make the mixture easier to work with. I don't believe it made much difference. Pushing around that pasta dough, that solid elastic mass, really gave my arm muscles a workout. When it was time for the dough to rest, I was ready for a rest, too.

Rolling and stretching pasta dough is no light chore either. It's a far cry from rolling out a soft, cooperative pie crust. But it wasn't *that* bad. It was simple to cut noodles into quarter-inch or even eighth-inch widths. The ones I made by hand were not quite as thin as those from the machine, but they were just as good tasting.

So after a few setbacks, my first batch of homemade noodles turned out to be a rousing success. Pasta-making is so enjoyable, it might become habit-forming. A week later, in my second batch, I added spinach for pretty Pasta Verde.

And that time the eggs didn't even spill out of the flour.

Pasta machines

If there's a task to be done by hand, chances are someone has already marketed a gadget to relieve the hands of anything but pushing a button or turning a crank. Those who would make pasta are courted by a good share of tools and machine. They're fun to use, and they speed things along. Some will make fancy shapes. But when you get right down to it, a trusty rolling pin and knife will suffice.

After you have mixed the dough to the proper consistency, the machine takes over, kneading and rolling the dough between two sleek, steel rollers. A knob regulates the distance between rollers. To knead, you pass a small piece of dough through the rollers at the widest setting about 15 times, doubling the strip over each time and turning it so all parts of the dough are worked. The rollers should be dusted with flour frequently to avoid sticking.

Then, to gradually stretch the dough into a paper-thin sheet, you click the knob down through each of the settings, running the dough through about five times at each thickness without doubling.

(Continued on page 68)

THE GREAT COOKS' GUIDE TO PASTA AND NOODLE DISHES

1977.
Random House, 201 E. 50th Street,
New York, NY 10022. 54 p.
paperback $1.95.

Run-of-the-mill cooks should not mistake the ambiguous title. *The Great Cooks' Guide to Pasta and Noodle Dishes* was written *by* culinary experts, not *for* them. As the book's cover explains, "America's leading food authorities share their home-tested recipes and expertise on cooking equipment and techniques."

Nine pages of the slim volume are devoted to discussion of homemade pasta, the machines and gadgets to make it, directions for cooking, and a few quick paragraphs about sauces and grated cheese.

The rest of the book consists of recipes developed by cookbook authors, food writers, and restaurateurs. In addition to traditional Italian fare, *Guide* includes some good old American casseroles, a couple Jewish favorites, and a couple Oriental-style dishes.

—CB

Fresh Herb Pasta

This is a delicious pasta. The addition of fragrant fresh herbs—mint, basil, chives, thin scallions, Italian parsley—to the dough gives a lively flavor to the noodles. The dish can be served as a main course with a little olive oil and some grated cheese. Sauces should not be used, as they would smother the fragrances of the herbs.

1 to 2 cups fresh herbs, minced (choose one or two: mint, basil, chives, very small scallions, Italian parsley)
4 cups unbleached flour
6 eggs
salt
4 tablespoons olive oil
1 teaspoon peanut oil
Parmesan or Swiss cheese, freshly grated (optional)

Dry the minced herbs thoroughly in a paper towel. If you prefer, pound the herbs in a mortar.

Place the flour in a large bowl. Make a well in the center. Add the eggs, 1 tablespoon of salt, and 2 tablespoons of the olive oil and gradually work the flour into the eggs, mixing well.

Add the herbs and place on a counter or table and knead for about 20 minutes. Flour your hands and the work surface as often as necessary to prevent the dough from sticking. Keep pushing the dough away from you with the heel of your hand and gathering it back into a mass, until it is smooth and elastic. Make a ball of the dough, cover with a towel and let it rest for 1 hour.

Divide the ball into 3 parts. Roll each part through a pasta machine, following the manufacturer's instructions, to make thin sheets. Let the sheets of dough rest on a floured surface for 30 minutes, then pass them through the machine's cutting rollers to cut them into strips. Let them fall loosely on a floured tray. Sprinkle with flour.

Add salt and the peanut oil to a large pot filled about ¾ full with water and bring to a boil. Add the pasta and cook, uncovered, for 5 to 10 minutes, depending on how tender you want it to be. Drain the pasta well.

Pour the drained pasta into a large bowl, add the remaining 2 tablespoons of olive oil, and toss with two forks.

Serve with a bowl of cheese, if you please.

Yield: 6 servings

A WORLD OF PASTA

*Maria Luisa Scott and Jack Denton
Scott. 1978.
McGraw-Hill Book Company, 1221
Avenue of the Americas, New York,
NY 10036. 226 p. hardcover $12.95.*

While most any big fat cookbook will
include the better-known pasta recipes, *A
World of Pasta* devotes fewer than 30 of
its 200 recipes to Italian specialties. The
rest are from the cuisines of 50 nations:
France, Germany, Japan, China, Poland,
Switzerland, Yugoslavia, Belgium, Sweden, the Ukraine, Lebanon, Algeria,
Tunisia, Puerto Rico, the United States,
Mexico, India, Thailand, Australia, and so
on. The recipes are not intimidatingly exotic even though the Scotts introduce all
sorts of shapes and varieties of pasta to
inspire experimenting. The Scotts discovered the following pasta soup in Syria on
one of their dozen trips around the world.

CB

Fettuccelle and Yogurt Soup

    5   cups chicken broth
    1   cup fettuccelle (¼-inch-wide
         noodles), broken up
    2   tablespoons butter
    1   medium size onion, finely chopped
         salt and pepper to taste
    2   tablespoons fresh mint leaves,
         minced, or 1 teaspoon of dried,
         crushed mint
    2   tablespoons fresh parsley, minced
    1½   cups yogurt
    2   egg yolks, beaten

*In a saucepan, bring the chicken
broth to a boil. Lower heat and add the
noodles. Cook noodles about 8 minutes
or until they are al dente. Meanwhile, in a
frypan over medium heat, melt the butter,
add the onion, sprinkle with salt and
pepper, and saute until soft. Stir in the
mint and parsley. Cook 1 minute and then
stir into the saucepan with the noodles. In
a bowl stir the yogurt until smooth. Add
the egg yolks and mix well. Stir about ½
cup of the hot soup, a spoonful at a time
into the yogurt bowl. Pour this into the
soup pot, stirring constantly. Taste for
seasoning. Heat just to a simmer. Do not
boil.*

Yield: 4 to 6 servings

(Continued from page 67)

Cutting the noodles is merely a
matter of cranking dough through the
cutting rollers. The dough *must not*
be too damp, or what is supposed to
be a couple dozen individual noodles
will come out looking like soggy corrugated cardboard.

The Atlas pasta machine is a
dependable product of sound construction. It operates smoothly and
produces neat, supple pasta with
minimal effort. One person can easily
handle feeding the dough into the
machine, cranking, and gathering up
the finished noodles. Because whole
wheat pasta is not as elastic as
semolina pasta, the dough may
begin to get holes in it if thinned to
the finest setting of six. Whole wheat
dough becomes nicely translucent on
#5 or even #4. The cutters make
¼-inch and ¹/₁₆-inch noodles. Other
accessories are available, including
ones for curly lasagne and superfine
"angel's hair." The Italian-made Atlas
costs about $40. (The $37 Ampia,
made by the same manufacturer, has
the same construction, but the cutters are not removable.)

The CAD pasta machine looks
very much like the Atlas in materials
and design, but just isn't the same
top-notch quality. When the clamp is
positioned properly, the crank bangs
into the clamp on the way around.
Shift the clamp to avoid this, and the
crank has the nasty habit of falling
out onto the floor every few minutes.

The knob on the CAD that
changes the roller settings has no
number markings. Instead, grooves

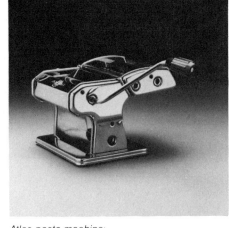

Atlas pasta machine:

in the plastic knob indicate dough
thickness. The CAD kneads and cuts
acceptably. It makes ¼-inch and
¹/₁₆-inch noodles, and costs about
$20.

The most sophisticated gadget
available for making pasta in the
home is the Bialetti electric pasta
machine. It not only kneads, rolls,
and cuts, but also mixes the flour and
eggs into dough.

The machine is about the size of
a toaster oven, and costs approximately $250. It is not complicated to
operate. With the spoked kneading
spindle in place, you put the flour
and eggs inside the plastic kneading
box, close the lid, and set the timer
for its six minute maximum. The
motor turns the spindle, which churns
up the ingredients. The instructions
recommend setting the timer for
another six minutes and adding a
little water or flour until the dough is a
good consistency. The Bialetti won't
magically create one huge blob of
dough, but it does make well-mixed,

CAD pasta machine.

Bialetti electric pasta machine.

moist particles that can be easily shaped into a ball.

The Bialetti's kneading and cutting attachments work like those of the manual machines, except that they are electrically powered. It is possible to stretch dough—even whole wheat—much thinner than you might dare to try with a hand-crank pasta machine. The manufacturer claims that, unlike steel rollers, the Bialetti's nylon rollers produce pasta with a texture to which sauce will cling more readily.

Cutting the sheet of dough is accurate and speedy. The speed may be encouraged by the fact that the Bialetti is extremely noisy, and you want to hurry up and shut the thing off.

Despite its considerable noise, the Bialetti is enjoyable to use. If you feel like it, you can stretch dough literally tissue-paper thin, or make spaghetti five feet long—diversions that may or may not justify for you the purchase price.

Lillo spaghetti machine.

These and similar pasta machines turn out yards of wispy noodles that come close to perfect-looking commercial varieties. But those skinny strips of dough just aren't spaghetti. By definition, spaghetti is round in cross-section and between 0.06 and 0.11 of an inch in diameter, by F.D.A. regulation. The only way to get these super-thin cylinders is to force the dough through dies. Wouldn't it be nice, for authenticity's sake, to have a home machine that could make round spaghetti?

We found such a device, but it is bound to disappoint pasta-makers, novice and expert alike. The Lillo spaghetti machine may seem like a bargain at about $12, compared with others, but it won't knead the dough, and takes considerably more exertion. The dough goes into a hopper at the top of the machine and must be pushed down by hand. When the crank is turned, an auger takes up part of the dough and forces it out through the die.

The Lillo *does* work. If your dough is nice and pliable, and you keep on cranking, spaghetti will ooze out slowly. The plump round strands look more special than noodles, just as slim but flat. But the time and effort needed to make the spaghetti is exorbitant. Extruding only half a pound of pasta takes at least 20 minutes of steady cranking, and that doesn't include mixing and kneading.

Except for the crank and clamp, the Lillo is all plastic, and the parts can be washed. That's fortunate because a significant amount of dough gets stuck inside the machine. The heat generated by the friction from all the heavy cranking makes the wasted dough stick inside with a vengeance. The tiny globs of dough remaining in the dies have to be picked out with a toothpick.

The Lillo comes with additional attachments for ¼-inch noodles, lasagne noodles, and others.

If you knead and roll your dough by hand, the Matfer French rolling cutter will cut it, just like 24 pastry wheels working all at once. This simple tool is nothing more than a rolling pin with thin stainless steel discs that can be spaced variably for noodles of 6mm., 12mm., and 15mm., widths. To change gauge, unscrew the handles, slip off the discs and wooden spacers, and reassemble with a dif-

ferent set of spacers. This task isn't difficult, but takes up enough time that you wouldn't want to switch sizes in the middle of one day's pasta-making.

Fortunately the $37 rolling cutter is well made and hefty, as you must press hard on the dough to cut it. Even then the noodles aren't completely separated and have to be peeled apart.

A cavatelli maker cuts, flutes and curls strips of dough, and all you do is crank.

Half the appeal of pasta is in its many whimsical shapes. The Vitantonio cavatelli maker lets you make novel sea shells for just over $20. When cranked, the simple device flattens a finger-wide strip of dough between wooden rollers, cuts off a chunk, and flutes and curls it against a curved piece of aluminum—all in about two seconds. Shells fly out in such succession that it's wise to have a bowl underneath to catch them.

The little square pillows of ravioli are easy to make if you have a simple $1 or $2 pasta crimper. Perhaps one is already in your kitchen disguised as a pastry wheel. The crimper cuts and seals ravioli in one easy motion (and can also make curly-edged noodles).

A round stamp with a zig-zag edge is marketed to cut agnolotti, the round version of ravioli, one at a time. Cutting circles from a large sheet of dough will leave some waste. This stamp also sells for only a dollar or two.

Faster ravioli production can be had with a $7 wooden rolling pin of scooped-out squares, which seals dozens of the pockets at once. You roll out two large sheets of dough, spoon filling onto one (being careful about spacing), and cover with the second sheet of dough. The ravioli

(Continued on page 70)

(Continued from page 69)

pin rolls over and seals the small squares. They must be cut apart separately.

The Italian metal ravioli press we tested looks like a tray for small, shallow ice cubes. You place a sheet of dough on the tray and gently press into the indentations with your thumb. Give each dent a little filling and top with a second sheet of dough. By rolling over the layers with a rolling pin, 36 ravioli squares are cut by the tray's crimping edges. The six-inch rolling pin that comes with the $8

Raviol-Wit.

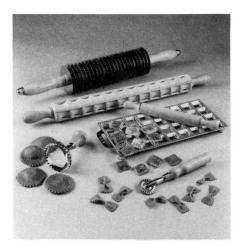

Specialized tools for particular pasta products. Clockwise: French roller cutter, ravioli rolling pin, ravioli press with pin, pasta crimper, and round stamp.

press is silly. You can get very little force behind it to cut the dough. Your own pin will do much better.

The sturdy Raviol-Wit crank ravioli maker automates the process one step further by filling the squares at the same time it cuts and seals the dough. To use the Raviol-Wit you first need to roll two long four-inch-wide strips of dough with the hand-held cutter provided. Then you lay the pieces of dough into a V-shaped trough at the top of the machine, one piece flopped over each side. A dollop of filling goes into the trough between the layers of dough. When you turn the crank the dough and filling are drawn down between two rollers,

one with indents and zig-zagged cutting edges. Each ravioli pouch takes only as much filling as it can hold, then is sealed and passes out through the bottom of the rollers. The rest of the filling remains in the trough for the next ravioli. Two squares are formed side by side, in as long a strip as there is dough. The pieces are not separated completely, but can be gently pulled apart. The Italian Raviol-Wit costs about $25.

You can probably purchase the ravioli makers from your local hotel and restaurant supply store or gourmet shop.

CB

Atlas Pasta Machine and the Ampia Pasta Machine; Kalkus-Hirco, 5714 W. Cermak Road, Cicero, IL 60650.
CAD Pasta Machine; Aviatex Corp., 6515/A Corporate Drive, Houston, TX 77036.
Bialetti Electric Pasta Machine; Coffee Imports International, 275 Barneveld Avenue, San Fracisco, CA 94124.
Lillo Pasta Machine; James W. Ernst Associates, Box 37, Rosenhayn, NJ 08352.
Matfer French Rolling Cutter; Euromarket Designs Inc., 190 Northfield Road, Northfield, IL 60093.
Vitantonio Cavatelli Maker; Vitantonio Manufacturing Co., 34355 Vokes Drive, Eastlake, OH 44094.

Inspired grains

Grains needn't be atomized into flour for palatability, of course. They can be sprouted, popped, cut and rolled to make oatmeal, cracked to make bulgur, or soaked and eaten raw. We start this section at the day's beginning, with thoughts on breakfast cereals.

Meals and mushes

Meals and mushes are eaten only for breakfast because they'd be laughed right off the table at any other meal. In appearance, these dishes rank just ahead of tripe and overripe brie. Worse, they lack the seductive flavor and aroma that's liberated when grains are baked into bread or muffins. But these grain foods have great potential.

That tired workhorse, oatmeal, can be spelled with flakes of other grains. Grains are rolled flat (and

sometimes embossed with a pattern) to make them cook faster. Wheat, rye, and triticale flakes are available at natural food stores and co-ops; so are soy flakes, the most nutritious and least appetizing of the bunch. Excepting soy, these grains are quite similar to the tongue and nose, and vary more in color and texture.

The art in cooking meals and mushes lies in the optional ingredients—administer them as you would medicine to a fading patient. Here's a

real challenge for the kitchen experimenter, with a real reward. Divine a pleasing combination of oatmeal adjuncts, and you've got yourself an excellent food.

Anything on the shelves is a potential ally because grains can be lead in several directions: sweet, pungent, rich, lean. If melted cheese or peanut butter tastes good on oatmeal bread, then it's only logical that it'll be just as good perched atop oatmeal. Heed: that's prejudice

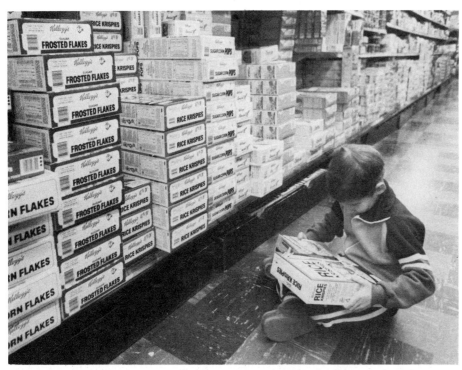

Many lifetimes of technological and creative genius have been lavished on developing and marketing breakfast cereals. Like summer-reading paperbacks, these cereals rely on a flashy package to compensate for the pap within.

Sunrise

A Breakfast Cookbook
Using Whole Grains
And Natural Foods

By Diana Scesny Green

making you gag at this suggestion, not any innate knowledge of the proper order of things. Other ingredients to explore are tahini, tamari, hummus (recipe, page 88), cider,

yogurt, and any fruit you happen to have on hand. Oatmeal is traditionally seasoned with cinnamon, but a pinch or two of cumin per serving will subtly alter the dish's personality.

Devote a few mornings to such experimentation, and you'll bust apart the myth that meals and mush are dull grandma foods.

Inspiration for another breakfast mush comes from India in the soft, domed form of uppuma. Rather than something to eat at breakfast, it looks like a flying saucer as fabricated in mush by Claes Oldenburg. Those maintaining an oatmeal habit will appreciate the change this spicy recipe can bring to their mornings.

The recipe is a variation of the uppuma served at the Woodlands Hotel in Madras, courtesy of the manager, S. Gopal Bhat.

Uppuma

2   tablespoons butter
1   small onion, finely chopped
1   green chili, finely chopped
¼   teaspoon black mustard seeds
1   cup cream of wheat (see below) or bulgur
2   cups boiling water
    chopped cashews

An unrefined cream of wheat can be ground from toasted grain. Commercial cream of wheat has been refined, then enriched, and is paler and not so wholesome. A recipe follows.

Fry the mustard seeds, onion, and chili in the butter, then add the cream of wheat. Stir over low heat to incorporate the other ingredients with the grain.

Add the water and continue stirring. When the cereal is cooked, remove it from the stove and add chopped cashews. Put on the lid and allow to sit for a minute or two before serving.

The cereal can be plopped unceremoniously in the bowl, or formed into a dome with the aid of an ice cream scoop. Season with a squeeze of lemon juice and tamari.

Yield: 2 or 3 servings

Lineage of the rolled oat

Unless this item is the very first in this book to attract your attention, you know that our purpose is to show that most foods can be processed at home.

Oatmeal is an exception. Although some mills produce the famil-

(Continued on page 72)

Diana Greene, author of Sunrise (reviewed below) says "Any cereal grain can be turned into a cream cereal. The grain need only be toasted ahead of time, then ground into fine particles in a steel or stone mill. Starting with a cold, dry, cast-iron skillet, cover the bottom with a layer of whole berries or kernels. Then stir constantly over medium heat until browned. Berries and rice grains may 'pop,' but do not be alarmed. Such popped or parched wheat is a snack all by itself. Transfer the toasted grain to a bowl and let cool thoroughly. Then grind in a grain mill on a setting for meal. If the cereal is ground too fine, then the cooked cereal will be pasty."

Here is her recipe.

Cream of Wheat

2   cups water
⅔   cup powdered milk (instant)
¼   teaspoon sea salt
½   cup toasted, ground hard red spring wheat

Bring first three to a simmer. Whisk in the wheat, and stirring constantly, bring to a boil. Then simmer, covered, for 15 minutes.

Sprinkle ¼ cup raisins and 2 tablespoons sunflower seeds on top. Steam for an additional 2 or 3 minutes.

Yield: 2 servings

Reprinted with permission of The Crossing Press, Trumansburg, New York 14886.

SUNRISE
Diana Greene. 1980.
The Crossing Press, R.D. 3,
Trumansburg, NY 14886. paperback
$5.95.

The author lives in Alaska, no doubt a place that would stimulate one to eat a good breakfast. Her recipes will appeal to all but the most frantic of breakfast inhalers. We would like to share here more than a couple of her recipes to round out our cereal section, but her publisher understandably did not want us to print the heart of the book. Recommended.

Growth of the stand-up quickie breakfast

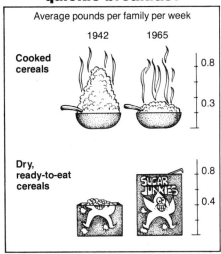

Average pounds per family per week

(From Food Consumption Group, Agricultural Research Service, USDA)

(Continued from page 71)

iar flakes, a good deal of flour comes along with it and must be sieved off. Most of you, like most of us on the book staff, will continue to buy oatmeal at the store.

How is this deceptively complex food processed? We've gleaned our answer from *Chemistry and Technology of Food and Food Products,* a heavy-duty treatise published by Wiley Interscience, a division of John Wiley and Sons, Inc., in New York.

Oats are first separated from the company of such curious impurities as mustard and onion seed (*that would be a rude awakening to groggy oatmeal eaters*), smut, and rosebuds. Separating and sizing grains requires a battery of some 30 machines. Oats are heated to bring out a superior flavor, then hulled between two stones or, more recently, with impact hullers. This refining takes away little more than the fibrous hull, leaving the bran and germ. Oatmeal, therefore, is as whole a food as brown rice. The finished product of hulling is known as a groat. Groats are processed to make them cook faster by one of a few methods. For traditional slow-cooking oatmeal, the whole groats are partially cooked with live steam; the moisture content thereby increased will mean less flour is produced in the subsequent rolling.

Quick-cooking oats are steel-cut with rotary cutters into graded sizes.

The smaller the piece, the quicker the oatmeal will cook. Each flake you see is approximately one-third of a whole groat. The pieces are then steamed as are the slower-cooking variety. (This was once the last step for oatmeal. So called steel-cut Scotch oats were not rolled to decrease cooking time. You still can buy oats that have been spared the roller.)

Next, the whole grain or grain fragment is rolled into the familiar flake. Grain drops from a hopper to the maws of a large steel roller. Quick oats are squeezed thinner, and thus are rendered quicker still.

RY

Granola

Granola is a prebaked cereal of cut and rolled grains, with any number of embellishments for flavor and nutrition, plus oil and a liquid of some sort to keep it all together. The cereal is baked at low heat until golden brown and crisp.

Granola can either be one of the least expensive of breakfasts or a fancy and somewhat costly food. The difference lies in what's added to the grain-oil-liquid foundation: almonds, pumpkin seeds, and dried fruit cost plenty.

Our recipe calls for apple juice to add moisture and a touch of sweetness, thereby avoiding the potentially cloying effect of honey, found in most granola recipes.

Good natural food stores and co-ops may stock flaked grains other than oat, such as wheat, rye, triticale,

and soy. Soy flakes don't win any friends when boiled like oatmeal, but are fine if toasted in a granola recipe. Soy also complements the protein in the grain flakes.

Fancy Granola

    3 cups oat flakes
 1½ cups dry coconut shreds,
         unsweetened
    ½ cup raw wheat germ
    1 cup sunflower seeds
    ¼ cup sesame seeds
    ¼ cup oil
    ½ cup apple juice or cider
    1 cup blanched almonds, slivered
    1 cup raisins (optional)

Preheat oven to 225°F.

In a large mixing bowl, combine rolled oats, coconut, wheat germ, sunflower seeds, and sesame seeds. Toss ingredients together thoroughly.

Add oil to dry ingredients, stirring until well-mixed. Add the apple juice a little at a time, mixing until crumbly.

Pour mixture into a large, heavy, shallow baking pan which has been lightly brushed with oil. Spread mixture evenly to edges of pan.

Place pan on middle rack of a preheated oven and bake for 1½ hours, stirring every 15 minutes. Add 1 cup slivered almonds and continue to bake for ½ hour longer, or until mixture is thoroughly dry and light brown in color. Cereal should feel crisp to the touch.

Turn oven off and allow cereal to cool in oven. If raisins are to be added to cereal, do so at this point.

Remove cereal from oven. When cool, put the cereal in a tightly covered container and store in a cool, dry place.

Serve plain or with fresh fruit.

Yield: 8 cups (about 24 1-ounce servings)

Rodale Press is committed to reviving the reputation of amaranth

What is amaranth, and why would a publishing company spend considerable amounts of time, effort, and money developing and promoting it?

Amaranth is a cereallike annual herb with large seed heads ranging in color from white to brilliant magenta. Growing wild, it is known to us as pigweed. Long ago it was the most important crop of the flourishing Aztec civilization in Mexico. The grain both sustained life and served an integral part of the Indian religious ceremonies. Then in the 1500s the Spaniard Cortez invaded Mexico, massacred the Axtecs, plundered the golden riches of their capital city, and burned and trampled the fields of amaranth. The grain was all but forgotten, even though through different historical circumstances it might today rank with wheat and corn.

Rodale's interest in amaranth began in the 1970s with the receipt of a letter from Dr. John Robson, then director of the Human Nutrition Program at the University of Michigan. He asked for help in exploring the plant's tremendous potential as a food crop, both for America and the Third World. The grain produced in the seed heads is rich in vegetable protein, and comes closer to complete protein than any other grain; the leaves can also be eaten raw or cooked, much like spinach. This convinced Rodale Press that the grain was worth a long look, and seed was procured from Mexico.

The first amaranth was planted at the Organic Gardening and Farming Research Center in the summer of 1974. While the growth of amaranth was carefully monitored, researchers evaluated its nutritive value.

During the second summer of amaranth trials the OGF farm compared seed yields of different species. 1975 also was the first year that Rodale Press readers were invited to help research amaranth's possibilities. People from all around the country conducted growth trials, keeping careful records of weather and dates of sowing, germination and flowering, height measurements, and yields of vegetative and seed harvest.

More reader-researchers joined up the next summer, while cultivation research went on at the Research Center. In 1977, a small notice in *Organic Gardening* magazine brought in 13,500 requests from readers wanting to participate as amaranth reader-researchers.

Begun in 1977, Rodale's germplasm collection now includes about 650 varieties, including samples from every major area of the world where the genus is found. Breeding with grain amaranth was begun in 1978, with the aim of finding a high-yielding amaranth variety that is resistant to disease and insects, shatter-resist-

ant, threshes cleanly and easily, produces white or golden grain, and is bountiful in high-quality protein and starch.

Those properties are essential if amaranth is to become a food crop for America. Rodale planted a two-

acre pilot field in 1979, and for the 1980s will concentrate on using standard cultivation machinery to handle the crop.

At the Rodale test kitchens, amaranth is ground into flour, popped like popcorn, and boiled into

(Continued on page 74)

The adobe oven

Mud-and-brick adobe ovens are endemic to the Southwest, but the process is not: building a fire in a massive oven that will hold heat for hours (and take hours to heat), then removing the fire and putting in the loaves to bake. Bread is still baked this way around the world—wherever technology or the worship of convenience have yet to take hold.

The adobe pictured here must

be heated two or three hours with a wood fire. The coals are then raked out, and the oven floor is cleaned with a damp mop. The vent at the top must be plugged with a damp cloth to keep in the heat. A thermometer indicates when the oven has cooled to a standard baking temperature of 300 to 325° F. Loaves are placed upon the oven floor with a long-handled wooden paddle.

(Continued from page 73)

a cereal in developing recipes for the novel food, including amaranth salad dressings, pancakes, muffins, crepes, crackers, and breads. Amaranth flour alone has too little gluten for yeasted breads, but can be blended with wheat flour for a bread that rises well and takes advantage of amaranth's complement of wheat protein.

To encourage other companies to get into the act, the Test Kitchens sent sample kits of amaranth seed, flour, and confections to food and snack makers. The good word has been spread further through two amaranth seminars in 1977 and 1979.

Sauteed Amaranth

1 tablespoon salad oil
1 to 2 cloves garlic
1 onion, sliced
1 green pepper, chopped
1 cup mung bean sprouts
1 tablespoon tamari
1 tablespoon water
4 cups amaranth greens

Heat oil in heavy skillet. Add garlic, onion, and green pepper, and stir fry for a few minutes. Stir in amaranth and sprouts.

Mix tamari and water and add to vegetables. Cover and cook over medium heat just until amaranth is tender, for 3 to 5 minutes.

Yield: 4 servings

Alegria

4½ teaspoons honey
1½ teaspoons molasses
1 tablespoon butter
1 cup popped amaranth
    oil to grease pan

In a saucepan, combine honey, molasses, and butter. Cook over medium heat for 4 to 7 minutes, stirring constantly until the mixture turns golden brown.

Add 1 cup popped amaranth and stir with a wooden spoon until all grains are coated with the syrup.

Press coated grains into a greased 8-inch square pan. Cut into squares or bars and allow to cool.

Yield: 10 to 12 bars

Sunflower seeds, raisins, or chopped peanuts, pumpkin seeds, walnuts and pecans can be added for flavor or texture.

The net result of Rodale's energetic commitment to amaranth? It'll probably be a few years before amaranth bread appears on your grocer's shelf, but health food companies are now considering amaranth's talents. The real test is interesting the consumer in this forgotten food.

How can you incorporate the forgotten grain in your diet? Because amaranth has very little (if any) gluten, it must be combined with wheat flour for a yeast dough. The Rodale Test Kitchens have used successfully a 50:50 ratio of raw amaranth flour to whole wheat flour in a sweet dough. (Remember, the wheat complements amaranth's protein pattern).

Breads made entirely from raw amaranth flour are very heavy and compact due to the lack of gluten. However, all-amaranth muffins can be leavened with baking powder; the rising time is so short that no gluten is necessary for a successful product. The flavor is improved by adding an acidic ingredient (such as lemon or vinegar) to the dough. This softens the crumb and thereby lightens the texture.

When used in products with shorter baking times (such as pan-

Amaranth Corn Muffins

⅓ cup soy flour
⅔ cup corn flour
1 cup amaranth flour
¼ cup honey
3 teaspoons baking powder
1 cup cranberries, blueberries, or
    raspberries, coarsely chopped
1 cup milk
⅓ cup melted butter
2 eggs

Preheat oven to 400° F. Grease 12 muffin cups.

Combine flours with baking powder in a large bowl. Add honey to berries and add to flour mixture. In a separate container, measure milk in 2-cup measure. Add melted butter and eggs; beat with fork to mix thoroughly. Make a well in the center of the flour mixture. Pour in the milk mixture all at once; stir quickly with fork just until dry ingredients are moistened. Do not beat — the batter should be lumpy. Quickly fill muffin cups with batter, about ⅔ full.

Bake 25 minutes or until golden brown.

Yield: 12 muffins

cakes, waffles, and cookies) some amaranth flours add a raw, earthy flavor. Our test kitchen tasters find this objectionable. By lightly toasting the flour in a 325° F. oven for about 15 minutes, it takes on a nutty, toasted flavor. This treatment is usually unnecessary in products with a longer baking time.

Amaranth seed can be popped into small white kernels similar to popcorn. The Press has found that a wok or heavy skillet works best. The pan is heated until very hot and then approximately one tablespoon of seeds is poured in. To prevent burning the seeds must be agitated constantly. A small pastry brush can be used to keep them moving. When popping has subsided, remove the pan from the heat and empty the seeds into a bowl. Do as many batches as necessary. One cup of amaranth seeds will yield three to four cups of popped amaranth.

An electric corn popper will also work but the percentage of popped seeds is lower. An electric skillet works fairly well.

The seeds can be eaten freshly popped. Or, use them in confections bound with sorghum, molasses or honey; granolas; cheese spreads; peanut butter balls: fudge; fillings for sweet rolls; breadings for chicken and fish; garnishes for vegetables; and toppings for casseroles and desserts. Popped amaranth can be ground into a meal for baking.

Wild rice, the gourmet grain

To the Ojibway people of Canada, wild rice is a sacred food, as central to the tribe's religion and culture as bread and wine to the Christians. These natives of the Lake of the Woods region call their most important staple *Manitou gi ti gahn*, which means "the plant the Great Spirit gave us," and *manomin*, "delicacy of the Great Spirit." Seventeenth-century French explorers dubbed this strange new delicacy *folle avoine*, "crazy oats."

Besides corn, wild rice is the only well-known cereal grain native to the western hemisphere. But while corn has been hybridized into a cash crop used for cattle feed, starches, sweeteners, cooking oil, and even auto fuel, wild rice remains a misunderstood gourmet luxury. In fact, wild rice is neither a type of rice nor is it entirely wild.

Wild rice is an annual aquatic grass that belongs to its own genus, *Zizania,* and is only distantly related to rice, oats, wheat, and other cereal grains of the grass family. In August and September, its ripe seeds fall from the grain head into the three-foot river and lake shallows where the plant grows. The seeds remain dormant in the bottom sediment until the following April, when they sprout submerged leaves. In May, one or two floating leaves appear followed by erect leaves in June. In July, the flower-bearing stem rises from the water, growing two to four feet above its surface before its seeds ripen again in August. Several species of wild rice grow in the eastern half of North America, from Louisiana and Texas to Manitoba and Quebec. But the variety that produces grains large enough to harvest commercially occurs only in northern Minnesota and nearby portions of Ontario, Manitoba, and Wisconsin.

Although wild rice is much tougher than domesticated grains, each year reseeding itself and growing without human attention until harvest, it has probably benefited from human contact. That's because the plants spread very slowly on their own. Their seeds sink quickly and are completely digested by foraging waterfowl, leading scientists to believe that much of the wild rice growing in lakes and streams today was originally planted by Indians. Legend supports that theory. The birchbark scrolls of the Ojibway say that wherever their ancestors travelled, they sowed the seeds of *manomin.*

Wild rice isn't rice. And it may not be completely wild. But nearly everyone who tastes it seems to agree with the Ojibway that it's a divine delicacy. It looks and tastes like no other grain. The long dark kernels have a nutty flavor and herbal aroma that add exotic appeal to foods. Just as a few slices of celery or pepper in soup will enhance its flavor, only a small amount of wild rice grains or flour will change the character of breads and casseroles. Food industry researchers have conducted taste tests with wild rice and found that it

has nearly the same insatiable appeal as candy, explaining why wild rice fanciers will pay from $7.50 to $20 per pound.

There are several reasons why wild rice costs so much. The most obvious is its scarcity. In spite of attempts to grow it elsewhere, the prized long-grain varieties thrive only in about 34,000 acres of upper Midwestern and Canadian wetlands. The yearly harvest from such a limited area is enough to provide each American with a stingy allotment of about 2/5 ounce. Nor is the wild rice plant as bountiful as domesticated grains. A good stand yields only about 800 to 1,000 pounds per acre, a paltry amount compared to the 4,000- to 6,000-pound yield from an acre of conventional rice. Unlike domesticated cereals, wild rice also gives up its grain slowly, ripening gradually over two weeks. During that period many of the seeds "shatter," or fall from the seed head, before they can be harvested. When traditional Indian harvesting methods were more prevalent than they are today, that tendency to shatter made gathering the grains a task of skill and devotion. Indian families would camp on shore near their favorite wild rice stands and make several passes through the plants with a canoe. While one person propelled the canoe with a pole, the other would gently flail the grain into it with two cedar sticks. After about four hours of hot work, the two would return to shore with perhaps 50 pounds of grain. Repeated trips in a canoe might catch 10 to 60 percent of the ripe seeds. And after another long process of drying, roasting, and separating the grains from the chaff, an acre of rice would provide a family with about 40 pounds of grain to be stored for the winter.

It's no wonder, then, that wild rice has always had high monetary value. But the nicest surprise about wild rice is that such a good-tasting food also has high nutritional value. Its protein content ranges between 12.4 and 15 percent, higher than that of any other common cereal grain except oats, which averages about 14 percent. And the quality of wild rice protein surpasses that of oats. It has a better balance of the essential amino acids, especially lysine and methionine. Wild rice is also lower in

fat than most other grains and, like other whole grains, it is a good source of B vitamins and minerals.

Wild rice may never replace potatoes and wheat as a staple food but it should become more abundant in the future. Enterprising farmers and University of Minnesota researchers have developed shatter-resistant varieties that can be grown in artificial paddies and harvested with a combine just like conventional rice. Today the farmers and the Indians each harvest about the same amount of wild rice acreage. But because mechanical methods in shatter-resistant plantings are so efficient—yielding about 280 pounds of processed rice from an acre—only about 20 to 30 percent of the crop is still harvested by Indians. The Indians, too, are abandoning tradition in an attempt to hang on to a share of the wild rice market. Some Minnesota tribes now plant and harvest wild rice using the same methods and machinery as the farmers. In Canada other tribes use motorized pontoon boats with screened-in scoops to gather lake rice. It's already unlikely that the wild rice you buy in those convenient supermarket boxes grew naturally or was harvested in the traditional way.

Progress has its price. The commercial wild rice is just as nutritious as its lakeside cousins but it must be grown with pesticides. Farmers use malathion to protect their crop from the wild rice worm. And, since intensive production makes the plants suceptible to leaf blight, the fungicide Diathane M-45 is sometimes used.

So far, pesticide residues haven't shown up in lab tests of commercial wild rice, perhaps because the chemicals are destroyed by the heat of processing. But if you don't like to eat food grown with pesticides and you do like wild rice, you may want to try growing your own. It's tricky but not impossible if you are lucky enough to have a pond or stream with the right environment.

Wild rice is a finicky plant that needs clean, gently flowing water. It's more sensitive to pollution than most aquatic plants and won't grow in waters with dissolved salts or sulphates. Nor does wild rice compete very well with cattails, reeds, and other shoreline vegetation.

(Continued on page 76)

(Continued from page 75)

It grows best in limestone regions or farming areas where the waters are fertile and rich in minerals. At the same time, though, the water must be clear enough to allow sunlight to reach the plant's emerging leaves in spring. And it must be at least an inch deep to support those young, floating leaves in April and May. If the water rises more than six inches at that stage, the plants will die. In later stages of growth, wild rice can live in three feet of water. It rarely reaches maturity from greater depths.

Wild rice seeds won't germinate unless they are kept wet. They also need a period of dormancy at freezing or near-freezing temperatures and must be deprived of oxygen. If all those things happen, the seeds will sprout in spring when temperatures rise and the seeds are again exposed to oxygen dissolved in the water.

All that may sound complicated, but those are exactly the conditions Mother Nature has created in the special niche where wild rice grows.

During the winter the seeds buried in mud stay moist. A heavy layer of ice keeps their temperature at about 32° F. and prevents oxygen from reaching the mud. When the ice breaks up in spring it tears away any weeds that grew in the shallows the year before, naturally cultivating and aerating the wild rice seedbed. The ligher weed seeds also tend to float away while the heavy rice seeds remain. Through the rest of the growing season, the gentle currents and stable water levels of the interconnected lakes of northern Minnesota and Canada maintain good growing conditions. If you own riverfront or lakefront land in an area that has those conditions, you might try seeding it with wild rice.

A bushel of seed will be sufficient for an acre. In smaller areas a large handful can be scattered over an area about six feet on a side. Plant either in fall after the harvest or in spring soon after the ice breakup. If many ducks frequent the area in fall, spring planting is better. Otherwise, plant in fall, as storing seed through the winter can be tricky. For a plant-ing that will reseed itself, you should use seeds of shattering wild rice varieties rather than the commercial non-shattering types. (One source is Wildlife Nurseries, P.O. Box 2724, Oshkosh, WI 54901.) The best spot to plant in small areas is near an inlet or outlet. Plantings in small landlocked ponds often fail. Bays and protected portions of lakes are better than areas fringing deep water.

For more information, consult Wild-Rice by William G. Dore (publication 1393, Canada Department of Agriculture, Research Branch. Order from Supply and Services Canada, Publications Centre, 45 Sacred Heart Street, Hull, Quebec K1A 0S9.) The University of Minnesota offers Agronomy Fact Sheet No. 33, Seeding Time, Method, and Rate for Wild Rice Grown as a Field Crop; and Home Economics—Family Living Fact Sheet No. 21, Wild Rice—How It Grows, How to Cook It (order from Bulletin Room, 3 Coffey Hall, 1420 Eckles Avenue, University of Minnesota, St. Paul, MN 55108.)

Dan Looker

Meet sorghum and bulgur

Sorghum Crepes with Vegetables

1½ cups sorghum flour
3 eggs
2 cups milk, scalded and cooled
2 tablespoons butter, melted
   butter for the pan

Beat the eggs and mix them into the flour. Add the milk and butter, blend well with a wire whisk.

Cover the bowl and let the batter stand at room temperature for an hour or two.

Heat a crepe pan and add a bit of butter to oil the pan. The first crepe or two will probably stick, so butter the pan well. Take the pan from the heat when it is very hot, pour in just enough batter to cover it, swirling lightly to distribute the batter evenly. (For a 6-inch pan, use about ⅓ cup of batter.)

Cook over medium heat about 1 minute, then loosen the edges gently with a spatula. Continue to cook until the crepe is a golden brown and then turn and cook on the other side until brown. After the first crepe or two you will need very little butter to oil the pan.

Vegetable Filling

3 tablespoons butter
4 cups mushrooms, sliced
1 teaspoon dried chervil
½ cup onion, finely chopped
2 cloves garlic, minced
4 cups zucchini, bite-size slices
½ cup sweet red chili pepper
¼ teaspoon dried dill
   pinch dried basil
2 cups Swiss cheese, grated

Melt the butter in a skillet and add sliced mushrooms. Over medium-high heat, cook the mushrooms until soft, stirring frequently. Sprinkle with chervil and stir.

Add the onion and garlic to the mushrooms and cook briefly over medium-high heat until wilted. Stir in the zucchini and pepper and cook until the zucchini are done but still crisp. Sprinkle with the dill and basil and cook for 1 minute more.

Spoon some of the vegetable mixture onto each crepe, top with cheese and roll up. Arrange the stuffed crepes in a lightly oiled baking dish and heat in a preheated 350°F. oven for about 15 minutes. Serve immediately.

Yield: 4 servings

Bulgur Mexicana

We asked Fran Wilson, Chef Manager of Rodale's Fitness House Kitchen, for her favorite grain recipe and she offered this bulgur dish.

3 tablespoons butter
1 cup bulgur
1 medium size onion, chopped
1 cup celery, thinly sliced
½ green or red bell pepper, diced
1 teaspoon chili powder
¾ teaspoon ground cumin
2¼ cups regular-strength broth

Cook bulgur and onion until onion is limp and bulgur is golden. Stir in the celery, pepper, and spices and cook 2 minutes. Pour in broth, bring to a boil, cover, reduce heat, and simmer 15 minutes, or until all liquid is absorbed. To serve, mound bulgur mixture into portions. Offer condiments, each in a separate bowl, to spoon over: 1 cup each shredded cheddar cheese, yogurt, alfalfa sprouts, and roasted, chopped peanuts; ¼ to ½ cup sliced green onion; 3 small tomatoes (diced); and taco sauce.

Yield: 4 to 6 servings

Raw grains

There's good reason to believe that most foods are at their nutritional best with little or no cooking. How about grains? When dry, they approach pebbles in hardness and are always cooked before eating.

Almost always, that is. While cooking is indeed necessary if mealtime is minutes away, grains can be simply soaked in an equal amount of water, at room temperature, for 18 to 24 hours. The shorter time yields chewy grain, the longer, a texture like that of cooked grain. The yield will be twice the dry measure.

What happens if you oversoak? The liquid may ferment, with unpredictable and likely unappetizing results. The grains may sprout. Neither means catastrophe, but it's best to pour off any remaining liquid at the end of the recommended soaking period and refrigerate. Soaked raw grains will keep for several days.

Now that you have a bowl full of cold, wet grain, what do you do with it?

Illustration from Feasting on Raw Foods.

FEASTING ON RAW FOODS
Charles Gerras, editor. 1980.
Rodale Press, 33 E. Minor Street,
Emmaus, PA 18049. 320 p. hardcover
$12.95

Cooking does more to food than warm it up: the composition changes, and not often for the better. *Feasting on Raw Foods* questions our habitual roasting, baking, boiling, and frying, and suggests that only raw foods make it to the table with their goodness intact.

In the safe and predictable world of cookbooks, this comes across as unalloyed radicalism and, in fact, the dishes that showed up here at the Press for taste tests were at times a challenge to the tasters. But this book should prove fascinating to anyone interested in exploring the frontiers of nutrition. We print here a raw grain recipe that will please all but the most habit-bound.

Soak thoroughly rinsed grains for 18 to 24 hours in an equal amount of water. Cover the soaking container to keep out curious fruit flies, and keep at room temperature. Do not be upset if grains produce little sprouts. To store soaked grain, strain and place in the refrigerator.

RY

Curried Rye and Tofu

    1 pound tofu, coarsely mashed
2½ cups soaked rye
    3 green onions, thinly sliced
    2 tablespoons parsley, minced
    2 seeded green peppers (1 chopped
         and 1 cut into 6 rings)
       Oil and Vinegar Dressing (see
         recipe below)
    6 lettuce leaves

Mix the first 4 ingredients with chopped peppers. Add dressing and toss until blended. Cover and refrigerate several hours.

Place a mound on each lettuce leaf and garnish with a pepper ring.

Yield: 6 servings

Oil and Vinegar Dressing

⅓ cup oil
⅓ cup rice vinegar
  1 teaspoon curry powder
  1 garlic clove, peeled and minced

Place ingredients in a jar with a lid; cover and shake well to mix. May be prepared a day in advance.

Yield: about 1 cup

Keeping it fresh

If you bake bread regularly and have your own mill, it makes sense to buy grains in bulk. You'll pay less for grain per pound, you'll make fewer trips to the market, and you'll have freshly ground flour whenever you get the urge to bake. But, bugs and mold like grains too.

An average family of four consumes about 560 pounds of grain in a year. The major price break usually comes at about 25 pounds. A local health food store sells organic hard wheat for 50 cents a pound in 2 pound bags, 45 cents a pound in 5 pound bags, and 35 cents a pound for quantities from 25 to 300 pounds. A mail order house sells wheat in 1 pound lots for 40 cents a pound, and 25 pound bags at 23 cents a pound, but buying larger quantities reduces the price only to 22 cents a pound. So if you buy more than 25 pounds, you may save in convenience, but you probably won't save much on the cost.

Making large quantity purchases may simplify your life, but it can create a new problem: storage. Grains are not difficult to store, but they are vulnerable to insects, ro-

(Continued on page 78)

(Continued from page 77)
dents, and mold. You need to protect your grain against these pests.

The two most important factors in storing grain are the temperature and moisture content of the grain. Ideal storage conditions are a temperature of 40° F., and a moisture content of 12 percent or less.

These ideals may not be easy to attain. Grain is normally dried before being sold, but it can pick up moisture during shipping or storage. As for temperature, unless you live in an unusual house, it is likely that the refrigerator is the only place with a constant temperature of 40° F.

Where to store?

This doesn't mean you must store your grain in the refrigerator. What it means is that you should store the grain in a relatively cool, dry place, and that the cooler and drier it is, the longer the grain will keep.

The reason cooler temperatures are best is that they help preserve the grain and inhibit pest activity. Insects don't flourish at temperatures much below 60° F.; mites can't grow if it's below 40° F.

The best storage spot depends on the house. The basement may be coolest, but it may also be too moist. If you store grain in the kitchen, keep it away from appliances or hot water pipes. Make sure the grain is in a well-ventilated place and not exposed to direct sunlight.

The grain itself should be stored in dry, airtight containers. Jars, some

metal cans, or plastic bags will work fine. If you're using a metal can, don't place it directly on a concrete floor: it may sweat and cause the grain to mold.

Plastic, paper, or cloth bags are not good storage containers if rodents are around. Rats can chew through plastic, wood, and even aluminum to get to the stored grain.

If you keep grains within the safe storage period, you probably won't have any trouble. But bugs are persistent pests, and if you regularly store grain, they may crack your defenses one day. Or you may have the misfortune to purchase grain that is already infested.

No-bug grain

When your grain gets bugged, or if you just want to make sure it doesn't, you can sterilize it by heating or freezing. Keeping the grain at 0° F. for three or four days will do the job, as will heating it in a 130 to 150° F. oven for 20 to 30 minutes.

If you heat the grain, spread it no more than ¾-inch deep in shallow pans. Leave the oven door slightly open to allow air circulation and prevent overheating. Stir the grain periodically to be sure it is all exposed to the heat. If you plan to sprout the grain, don't heat it, or you could affect its germination ability. (This procedure also works well for drying grain.)

After sterilizing the grain, sift it with a screen to remove dead insects. Once it's clean, an airtight container should prevent re-infestation.

Buggy grain, grotesquely magnified.

Moisture and mold

The other danger is moisture. As the table shows, grains that are low in moisture can be stored for considerable periods, even at temperatures as high as 80 or 90° F. But if the grain gets too moist—especially with warm temperatures—mold can begin to grow.

Not all molds are harmful, but most of them are distasteful. If you get moldy grain, you may be able to reclaim some of it. Wash the grain in lukewarm water, then dry it immediately, preferably in the oven, and use it as soon as possible. (Be-

SAFE STORAGE LIFE OF GRAINS
(CONSERVATIVE ESTIMATES, IN DAYS)

Grain temp. °F	Moisture content (percent)				
	9.5	11	12.5	14	15.5
50	2048	1024	512	256	128
60	1024	512	256	128	64
70	512	256	128	64	32
80	256	128	64	32	16
90	128	64	32	16	8
100	64	32	16	8	4

cause of the problems moisture can cause, it is generally not advisable to wash your grain. It should be clean when purchased.)

Another danger under warm, wet conditions is that the grain may begin to sprout. The sprouted grain will begin to spoil after a week or so and infect its neighbors. If this happens, remove the sprouted grain and dry the rest before storing it again.

Nutrition losses

Storage does affect the nutritive value of the grain. Under good storage conditions, the nutrient losses are minimal, but it makes sense to buy no more than one year's supply at a time.

Most studies have shown only small changes in the protein or mineral content of grains during normal storage. There is some vitamin loss. Wheat with a ·12 percent moisture level lost 12 percent of its thiamine (vitamin B_1) over a five-month period. Wheat with a 17 percent moisture level lost 30 percent of its thiamine in the same period.

Some nutrition loss is inevitable, of course, since it is impossible to get "fresh" grain regularly. Because grain is harvested only during a limited time, it must be stored somewhere—and you can make as good a storage environment in your home as is likely to be found on a farm, or in a warehouse or store.

Ironically, it is also possible—especially if you grow your own grain—to not store it long enough. After harvest, the grain needs to dry for about a month. If you use it while it is still "green," it may clog your mill.

Storing flour

If you don't have a mill, and buy your flour already ground, it is best not to buy in bulk. Flour rapidly loses a significant portion of its nutritive value when stored, so it should be used as quickly as possible. If you're lucky enough to live near a health food store with a mill, you may be able to have grain ground to order when you plan to bake. Try to avoid commercially ground flour; there's no way to determine just how old it is.

Flour, like grain, is best stored in dry, airtight containers. Pests and molds like flour as much as they like grains, and will invade it if given the chance.

The primary factor affecting the storagability of flour is its fat content. The fat can turn rancid and spoil the flour. Whole wheat flour will keep up to two months at room temperature, and from three to six months in the refrigerator. Whole grain rye and barley flour will store for similar periods. White flour will keep longer because much of the fat is found in the wheat germ, which is removed in milling.

Soybeans are high in fat, so soy flour will keep only one to three weeks at room temperature. Corn and oats have a higher fat content than wheat, and these whole grain flours should be used quickly or stored in the refrigerator.

The best container for storing large amounts of grain is probably a garbage can or a barrel lined with a polyethylene plastic bag. The bag will make it easy to seal in the grain, and the polyethylene will not affect it. You can also purchase plastic containers; just check the label to make sure they're FDA approved for food storage.

With large quantities, it is usually not practical to sterilize the grain by heating or freezing (unless you live in a cold climate and it's winter). In that case, one alternative is dry ice.

Even pests need oxygen to survive, so they can be destroyed by removing the oxygen from the container. Dry ice, which is solid carbon dioxide, makes this a fairly simple job. You should be able to get some from a dairy that makes ice cream, a firm that rents freezer lockers, or a local hospital. Drugs are sometimes shipped packed in dry ice.

To sterilize your supply, spread crushed dry ice over the bottom of the container. You'll need about four ounces for every ten gallons of grain. Pour the grain over the dry ice, and let the ice sublime before you seal your container. This should take about 30 minutes. If the grain is stored in an airtight container you should have no trouble with insects.

There are other natural methods for controlling pests or moisture, but they all have some drawbacks. Nitrogen can be used instead of carbon dioxide, but it is more expensive than dry ice, and handling gas under pressure can be risky.

DIATOMACEOUS EARTH

For as long as people have been storing grain, they have been struggling to keep bugs out of it. The Chinese and Egyptians had some success controlling insects by mixing dust with their grain. They may have learned this trick from birds or other animals who take dust baths to combat ticks and fleas. Diatomaceous earth is one of the latest products in this line of natural insecticides. It is a dust milled from the shells of microscopic, one-celled algae called diatoms. As the diatoms died, their shells piled up, leaving deposits at the bottom of lakes and lagoons which have since dried up.

Making diatomaceous earth is a slow process; each inch takes about 1,000 years to produce. But in several areas of the western United States, there are deposits more than 1,000 feet deep extending over several square miles.

Diatomaceous earth kills on contact. The tiny, sharp particles pierce the outer waxy coating of the insect, causing death by dehydration some 12 hours later. When mixed with your grain, it will kill any bugs that are already feasting or that dare to venture in, although eggs are not affected. Since it kills only on contact, every kernel must be coated for best results.

You don't have to worry about washing it off your grain, either. The dust will not affect the taste or texture of bread, and it may even improve nutrition. Rats fed a diet of 5 percent diatomaceous earth for 90 days had no ill effects or toxic build-up. In fact, they were heavier and healthier than the control rats at the end of the experiment.

One cup will protect about 40 pounds of grain. For optimum coverage, mix grain and dust in small quantities. It's also a good idea to wear a mask while mixing: breathing the dust can cause lung irritation.

One warning: don't buy the kind sold by swimming pool manufacturers. It has been heat-treated and crystallized for use as a water-filtering compound, making it inedible. You can, however, use this kind of diatomaceous earth as a garden pesticide.

TD

Source: Diatect Corporation, c/o Gordon Dill, 410 E. 48th Street, Holland, MI 49423.

Where to get grains

Buying grain

"Grown with care and respect," says the small handlettered sign on the barrel of cornmeal at Beautiful Day Trading Company in College Park, Maryland. "The field was not burned, high heat wasn't used in drying, no molinate was used to kill grass; no Sevin or Bux was used to kill bugs. No chemical residues."

If that information makes you wonder about what's omitted from supermarket labels, then you've found one good reason to look for alternate sources of grain and flours.

Co-ops offer both lower prices and a far broader selection of grains and flours. These products aren't necessarily organic, as funky wooden barrels or hefty grain scoops may suggest. There's the opportunity (or the requirement) of trading labor in return for prices close to wholesale.

A *buyers' co-op* can be no more than a group of friends pooling their dollars to buy in bulk at wholesale prices.

The prices aren't quite as good at *natural food stores,* but they are a source of hard-to-find grains and flours.

Warehouses may sell retail, in large quantity—say, cases of milk or 50-pound bags of flour.

The Yellow Pages may list health food or natural food stores and co-ops. Shopkeepers can likely direct you to other shops, as well as any warehouses. Buying co-ops (and even some storefronts) may be practically invisible since they don't advertise. The basic guide to co-ops, co-op warehouses, and co-op information sources around the country is the *Food Co-Op Directory* ($3.00 from The Food Co-op Directory, 106 Girard, SE, Albuquerque, NM 87106). The directory lists several regional organizations that may be able to lead you to sources in your area:

You have to dig in with a scoop, and risk paper bag blow-out, but co-op shopping offers a good variety of fresh flours.

SW Fed Communication Exchange, Box 890, Tempe, AR 85281.
Earthwork, 3410 19th, San Francisco, CA 94110.

Nutrition Information Center, 239 SE 13th Avenue, Portland, OR 97214.
Chicago Area Co-op Info., Box 2559, Chicago, IL 60690.
Maine Federation of Co-ops, Box 107, Hallowell, ME 04347.
Consumers Co-op Alliance, 7404 Woodward Avenue, Detroit, MI 48202.
Buffalo Co-op Community Council, 169 Leroy Avenue, Buffalo, NY 14214.
Cleveland Co-op Federation, 1223 W. Sixth Street, Cleveland, OH 44113.

Delaware Valley Coalition of Co-ops, 201 N. 36th Street, Philadelphia, PA 19104.
Community Warehouse, 2010 Kendall Street, NE, Washington, DC 20002.
D.A.C.E., 1401 S. Fifth, Minneapolis, MN 55454.

If you can't find a co-op and want to organize one, consult *Food Co-ops: An Alternative to Shopping in Supermarkets* (William Ronco, Bea-

SOURCES FOR GRAINS, SEEDS, NUTS, AND BEANS

It's tough to outdo the best pizzaria in town, especially with a standard oven, whole wheat flour, and moderate use of oil, but it's fun trying. In time you may come to prefer whole wheat pizza and a lighter topping.

con Press, $3.95); *NonProfit Food Stores: A Resource Manual* ($3., from Strongforce, 2121 Decatur Place, NW, Washington, DC 20008); or *Co-op Stores and Buying Clubs: Operations Manual* (free from Community Services Administration/Consumer Action and Co-op Programs, 1200 19th Street, NW, Washington, DC 20506).

Co-ops can intimidate the shopper accustomed to the streamlined impersonality of supermarkets. You are often entrusted with the responsibility of measuring your own bulk produce (into the bags you're encouraged to bring from home).

Typically, co-ops buy their grain in 50- or 100-pound bags at wholesale cost. They mark it up anywhere from 10 to 35 percent to cover labor and costs. The savings are considerable. For example, wholesale wheat bran costs 13 cents per pound in a 25-pound bag at the co-op warehouse in Washington, DC. Co-ops mark it up to about 18 cents. By contrast, a local "wholesale" discount grocery was selling bran for 35 cents in bulk (53 cents packaged) and a regular supermarket for a spectacular $1.38, packaged.

Mail-order companies offer a greater variety of grains and flours than the supermarket. Prices vary (compare 32¢/pound and 46¢/pound for organic whole wheat flour from two different firms), so it's best to

(Continued on page 82)

Arizona

The Grover Company, 2111 S. Industrial Park Avenue, Tempe, AZ 85282. Buckwheat, corn, rice, rye, triticale, wheat. *Mail order.*

Magic Mill Centre, 6246 N. 43rd Avenue, Glendale, AZ 85301. Stocks over 40 varieties of grains, legumes, seeds and raw nuts. *Local.*

California

Moores' Flour Mill, 1605 Shasta Street, Redding, CA 96001. Most whole grains, and all whole grain, nut and bean flours. *Local and mail order.*

Moores' Flour Mill, 955 East Avenue, Chico, CA 95926. Most whole grains, and all whole grain, nut and bean flours. *Local.*

Erewhon Los Angeles, Inc., P.O. Box 58064, Vernon, CA 90058.

Colorado

Son-Shine Whole Grains of Wiggins, S.R. #2, 6547 County Road H, Wiggins, CO 80654.
Wheat flour, cornmeal, rye flour, whole grain wheat and corn. *Serves northeastern Colorado and delivers to Denver area.*

John N. Strahan, 2000 57th Lane, Boone, CO 81025.
Barley, rye, wheat. *Local. Will consider mail order if customer pays shipping costs. Minimum order, 10 pounds.*

Illinois

Allen Larson, *Pine Hill Farm,* 21543 Capron Road, Capron, IL 61012.
Wheat, soybeans. *Local.*

Rip and Ruth Sparks, *Red Top Farm,* Route 1, Ipava, IL 61441.
Wheat. *Local. Minimum order, 60 pounds.*

Indiana

Doyle Robinson, Route 2, Box 169, Fremont, IN 46737.
Buckwheat, oats, wheat. *Local.*

Iowa

Paul's Grains and Vegetables, Route 1, Box 76, Laurel, IA 50141.
Barley, buckwheat, corn, millet, oats, rice, rye, wheat, sorghum molasses, flour from each grain, soybeans, yellow and white popcorn. *Mail order and custom-made orders.*

Kansas

Loyd Fight, Route 3, Box 493, Leavenworth, KS 66048.
Cornmeal, wheat, soybeans. *Local.*

Massachusetts

Brean and Circus, Inc., *The Natural Food Supermarket,* 1524 VFW Parkway, West Roxbury, MA 02132.
also at
392-396 Harvard Street, Brookline, MA 02146.
Natural Foods Supermarket carries all whole grains in bulk and packaged forms.

East West Journal, 17 Station Street, Brookline, MA 02146.
Buckwheat, cornmeal, millet, rice, and others. *Mail order. Orders under $10 add $1 service charge. Please send for sample catalog.*

(Continued on page 82)

SOURCES FOR GRAINS, SEEDS, NUTS, AND BEANS
(CONTINUED)

Massachusetts (continued)

Erewhon, Inc.,   3 East Street, Cambridge, MA 02141.
Barley, buckwheat flour, cornmeal, millet, soyflour, and others. *Operates 3 retail stores in Boston. Mail order. Minimum order, $250.*

Michigan

Homestead Flour,   911 W. Camden Road, Montgomery, MI 49255.
Stone-ground soft whole wheat flour, yellow cornmeal, rye flour, buckwheat flour, soybeans. Will do custom milling. *Local and mail order.*

Tom Vreeland,   5861 Geddes Road, Ypsilanti, MI 48197.
Yellow corn, oats, wheat, soybeans. *Local and mail order.*

Minnesota

Dan Lundy,   Millstead Farm,   RFD#2, Rushford, MN 55971.
Barley, buckwheat, millet, oats, wheat. Different varieties of beans and seeds for sprouting. *Local and mail order.*

Valley View Farms,   Lester Frohrip,   Morgan, MN 56266.
Whole yellow corn, soybeans, wheat in tonnage lots wholesale. *Local.*

Missouri

Edwards Mill,   School of the Ozarks, Point Lookout, MO 65726.
Buckwheat flour, cornmeal, medium rye flour, whole wheat flour. *Local and mail order. Wholesale accounts welcomed.*

Hodgson Mill Enterprises, Inc.,   P.O. Box 126, Gainsville, MO 65655.
Buckwheat flour, cornmeal, rye flour, whole wheat flour. *Local and mail order.*

New York

Tom MacDonald,   R. D.#1, Trumansburg, NY 14886.
Rye and wheat, sold in 50-pound bags. Soybeans, kidney beans, black turtle and adzuki beans sold in 25-pound bags. *Local.*

New Hope Mills,   R. R.#2, Moravia, NY 13118.
Whole wheat flour, buckwheat grits, rolled oats, rye and cornmeal, soy flour. *Local and mail order.*

North Carolina

Old Mill of Guilford,   Box 623, Route 1, Oak Ridge, NC 27310.
Barley flour, buckwheat flour, yellow and white cornmeal, millet flour, rye flour, and whole wheat flour, yellow and white grits. *Local and mail order.*

Ohio

Daisy Hill Farm,   545 State Route 98, Caledonia, OH 43311.
Whole wheat flour. *Local*

L & M Organic Farm,   17725 Rheinecke-Schipper Road, Botkins, OH 45306.
Corn, oats, wheat. *Local.*

Oklahoma

Carl L. Barnes,   Route L, Box 32, Turpin, OK 73950.
Whole corn and cornmeal, sorghum, wheat and wheat flour, beans and peas in season. *Local. 20 pound minimum.*

(Continued from page 81)
check a few suppliers before placing your order.

Prices for organically grown grains may be slightly higher (compare 63¢/pound for organically grown short grain brown rice versus 59¢/pound for non-organic). But that's not always true, especially if you check prices for organic products at a co-op against non-organic prices elsewhere.

If you have the space to store grains and want to buy in large amounts, the best route is finding your way to a warehouse outlet by asking the nearest co-op or health food store where they get their grain. (Co-ops are more likely to deal with local sources.) In Washington, D.C., that would lead you to the Community Warehouse, a huge non-profit wholesaler open for public sales two days a week. Buying clubs come here, but so can the fairly rare individual who wants to buy 25-, 50-, or 100-pound bags of flour, grain, and other goods. Bagged grains and flours are stacked on the floor in aisles. You pick out what you need on a cart and fill out your own invoice.

Sara Ebenreck

Wheat in the garden?

The idea sounded preposterous. Wheat was something farmers grew in huge fields. Why were we planting a plot of it just 15 feet wide and 100 feet long?

First of all, my wife and I felt we could hardly brag that our two-acre mini-farm was an honest attempt at raising all our own food if we didn't grow wheat. From wheat comes bread—right? How can you subsist on your own land without your own flour for the staff of life? In a slightly more practical vein, the plot I wanted to grow wheat on had been in vegetables for several years and needed a rest by rotating clover or some other nitrogen-enriching legume on it that could be plowed under for green manure. Wheat as a nurse crop can be grown right along with clover, so even if the grain didn't amount to much, the effort wouldn't be wasted.

Our experiment with garden wheat turned out to be our most suc-

Gene Logsdon at work.

cessful venture yet in raising our own food. Not only did we discover a good rotation for vegetables, but the whole grain flour from our wheat is providing us (at very low cost) with some of the tastiest dishes ever to come out of our kitchen. Here's how we did it.

We bought our wheat seed from a farm supply store since it is not generally available from seed catalogs. There are five distinct types of wheat and many varieties of each type—hard red winter and hard red spring, both used mainly for bread; soft red winter used for cake dough and similar pastries; white wheat used for bread or pastries; and durum, used for spaghetti and macaroni. Each type is grown in more or less specific regions, though the areas sometimes overlap. In our locality, soft red winter wheat is usually grown, so that's what we planted. If you want to try wheat, my advice is to grow what has proven itself in your area. The same advice holds for the variety planted. And in any event, try to get certified seed that's germination-tested and clean of weed seeds.

Winter wheat is so named because you plant it in the fall. It grows a while before cold weather, dies back a little, then resumes growth in the spring. Wheat growers follow the

(Continued on page 84)

SOURCES FOR GRAINS, SEEDS, NUTS, AND BEANS

Oregon

Bob Cooperrider,   Route 1, Box 308, Sheridan, OR 97378.
Buckwheat, rye, soft wheat, garbanzo beans, red clover and vegetable seeds for sprouting. *Local and will ship mail order throughout the Northwest.*

Butte Creek Mill,   402 Royal Avenue N., Eagle Point, OR 97524.
Barley flour, buckwheat flour, cornmeal, millet flour, oat flour, rice flour, rye flour, whole wheat flour. Many different varieties of cereals and beans. *Local and mail order.*

Moore's Flour Mill,   4001 S.E. Roethe Road, Milwaukie, OR 97222.
All grain, nut and bean flours, and most whole grains. *Local.*

Pennsylvania

Armadillo Downe,   550 Prescott Avenue at Olive Street, Scranton, PA 18510.
Barley, buckwheat, millet, oats, rice, rye, wheat. *Store sells locally, but will accept mail orders.*

Mrs. Glenn Ford,   R. D.#1, Carlton, PA 16311.
Cornmeal, oats. *Local.*

Great Valley Mills,   101 S. West End Boulevard, Quakertown, PA 18951.
Buckwheat, cornmeal, oats, wheat. *Local and mail order.*

Paul Hartz,   Route 1, Box 86, Morgantown, PA 19543.
Whole wheat flour. *Local.*

Donald and Rebecca Kretschmann,   R. D.#1, Box 50, Ziegler Road, Rochester, PA 15074.
Rye, wheat, soybeans, mung and black beans. *Local.*

Walnut Acres,   Penns Creek, PA 17862.
Barley, buckwheat, cornmeal, millet, oats, rice, rye, triticale, wheat. *Mail order.*

South Dakota

Dennis W. Schaefer,   R. R.#2, Box 56, Tripp, SD 57376.
Barley, millet, wheat, open-pollinated corn. Prefers to sell in bulk. *Local.*

Texas

Lamb's Grist Mill,   Route 1, Box 66, Hillsboro, TX 76645.
Stone-ground cornmeal. *Sells to warehouses and distributors in Texas, Arkansas, New Mexico, Oklahoma, Louisiana. Will ship mail order.*

Vermont

Vermont Country Store,   Weston, VT 05161.
Buckwheat, cornmeal, oat flour, rye flour, wheat. *Mail order. Minimum order, $3.*

Virginia

Fangorn Organic Farm,   Route 3, Box 141B, Rocky Mount, VA 24151.
Corn, cornmeal, rye and rye flour, wheat and wheat flour, soybeans and soy flour. *Local and mail order. Minimum order, 25 pounds.*

Canada

Herb Eldridge,   R. R.#1, Ethel, Ontario, Canada N0G 1T0.
Barley, rye, wheat. *Local.*

EARS

On a still night a few years ago seven plant scientists, equipped with wire recorders, microphones, and wind gauges, went deep into a 100-acre Wisconsin cornfield. Their purpose: to *hear* corn grow. Perhaps some of the seven were skeptical when they went in. All seven, coming out, said they believed they had heard it.

The old corn farmers used to say they could hear corn growing, in July and August, in the hot humid windless Midwestern nights. If you think that is a myth, choose such a night and go listen. Go into a great cornfield—not a little Eastern hillside patch but one of the vast fields of the corn belt, so deep that for a quarter-mile in every direction corn is surrounding you, standing higher than your head. You will have placed yourself, then, in the center of a distinct and special world made up of earth, air, corn, and nothing else. There is no wind: in the motionless and brooding night there is no reason for any sound. But listen and you will detect a whisper, a faint crackling, coming from nowhere and everywhere at once. This ghost of a sound could be nothing but the minute stretching of a billion corn leaves, stalks, husks, kernels . . . . It is the sound of growth, a sound perhaps heard by insects as a roar of creation.

From page xiii in *Native Inheritance: The Story of Corn in America* by Howard T. Walden 2d. Reprinted by permission of Harper & Row, Publishers, Inc.

(Continued from page 83)
rule of planting after the danger of Hessian fly has passed—usually after September 15 in the north and central regions where winter wheat is grown. I manured and rotary-tilled my plot into a fine seedbed in the middle of September and sowed the wheat on the 21st.

Wheat is usually planted at a rate of about five or six pecks per acre. I sowed my seed by hand, simply broadcasting it as evenly as possible on top of the ground, a handful at a time, with a sweeping motion of my arm. Then, using a hand cultivator, I went over the ground with a harrow-like action that covered most of the wheat seeds with about an inch of dirt. Then I limed the plot.

By the middle of October, my plot was bright green with new wheat. In fact the wheat grew so well I could allow the chickens to feed on it and get some green stuff in them before winter—just as farmers sometimes graze their wheat in the fall. (Don't let chickens eat very much though. The wheat needs plenty of leaves to develop a good strong root system.)

By April, the wheat had come to life again. In May, it began to joint (send up stalks) and headed out beautifully in June. By July 4th, our crop was almost ripe. On a regular farm, wheat is usually harvested when it is dead ripe, directly from the stalk, with a combine. However, on our little patch, we intended to revert to 19th century agriculture and thresh our grain by hand. That method entails cutting the wheat before the grain is completely ripe, binding the stalks into sheaves, and then arranging the bundles into shocks where the grain remains until it is ripe.

To tell if the wheat is ready to cut, pull a few heads and shell out the grain in the palms of your hands with a rubbing motion. The wheat should come free from its husks fairly easily, but not too easily. Blow the chaff out of your cupped palm and chew a few grains. They should feel hard when biting into them. If the grain is still milky, wait a few more days. If the grains are very hard and shatter out of the husks very easily, you've waited too long. The stalks should be nearly all yellow, with only a few green streaks remaining. When wheat is dead ripe, there is no more

green in the stalk.

To harvest grain by hand the right way, you need a grain cradle—a scythe equipped with three or four long wooden tines arranged about six inches apart above the scythe blade. You see them often in antique shops. When you swing the cradle, the cut stalks of wheat gather against the wooden tines. The cut stalks then fall in a neat little pile to the left of the swath you are cutting as you complete your stroke. These little piles are then easily tied into bundles.

It takes practice to develop the proper rhythm for cradling. But if you've ever done any scything, you can catch on to it in a hurry. The trick of scything is to cut a rather narrow strip, letting the scythe blade slice through the standing stalks at a sidewise, 45-degree angle. Don't try to whack off the stalks with the blade at right angles to the stalks. The blade should be very sharp and always held parallel to the ground. Don't let the point dip down to catch in the ground.

Unfortunately, my grain cradle was 500 miles away, stored in my father's barn. Instead, I contented myself with an ordinary scythe, which lets the wheat stalks fall where they are cut. This meant that we had to gather the stalks into bundles, which could become a laborious job on a bigger area.

We tied the stalks into bundles with ordinary binder twine, available from farm stores. A bundle should measure about eight inches in diameter at the tie. The bundles are then set up in shocks.

To build a shock, I grab a bundle in each hand, sock the butt ends firmly down on the ground and then lean the tops against each other. Two more bundles are set the same way on either side of the first ones. Sometimes your beginning shocks will fall over until you get the knack of it. With the first 4 bundles in place, you can then stand about 6 or 8 more evenly around them. When you have a fairly sturdy shock of 12 bundles or so, you can tie a piece of twine around the whole shock to make it stand more solidly.

After my ten shocks dried in the hot July sun for five days, I put the bundles in an airy part of the barn to finish ripening where no rain would fall on them. If you intend to store grain in a barn in this way, put them in a mouse-proof place or keep a few cats around. We follow the latter method and have three barn cats in perpetual residence.

Having wheat bundles in the barn is not having flour in the kitchen. Our next venture was threshing. Grain was threshed for centuries by beating the straw with flails. So I made a flail following a drawing from an old book. We spread a clean white sheet on the sidewalk, laid the bundles of grain on top and proceeded to whale the daylights out of them. The grain fell easily onto the sheet.

But my son made a discovery the oldtimers had no chance to find. Unsuccessful in persuading Dad and Mom to let him use the flail (we were having too much fun), he went into the house and got his toy plastic ball bat. It worked better than the flail. The plastic was firm enough to knock the wheat from its husks, yet flexible enough not to crack the wheat grains. And it was light and easy to swing.

Once the straw bundles were relieved of their grain and removed, there still remained lots of chaff with the wheat. Traditionally, grain was winnowed simply by pouring it slowly into a bucket in a brisk wind. The breeze blew away the light husks and the heavier grain fell directly to the bucket. The wind, however, wouldn't cooperate with us. Instead, we brought our big electric fan outside to generate a winnowing wind. It

worked like a charm. We repeated the process until the grain was clean—or nearly so. A few grains did not hull out completely, but we found that to be no drawback in the flour.

We did not want to treat our wheat with chemicals to keep weevil out, so we borrowed an idea from Walnut Acres. We stored the grain in our refrigerator, where it keeps very well. Weevil don't like the cold.

Gene Logsdon

SMALL-SCALE GRAIN RAISING

Gene Logsdon. 1977.
Rodale Press, 33 E. Minor Street,
Emmaus, PA 18049. 305 p. softcover
$4.95, hardcover $8.95.

Gene Logsdon writes for Rodale Press from his farm in central Ohio. He admits that the "very small garden is no place for grains," but encourages those with more ambitious plots to sow a patch.

Figuring space requirements

You don't need much space to raise at least some grains. A normal yield of wheat grown organically would be about 40 bushels to the acre. So you'd need only 1/40th of an acre to produce a bushel. That would be a plot of ground 10 feet wide by about 109 feet long. A really good wheat grower with a little luck could get a bushel from a plot half that size. Wheat yields have been recorded as high as 80 bushels per acre and even higher.

But using the same kind of average calculations as above, here's the amount of space you'd need to grow a bushel of the following grains:

Field corn:	10 feet by 50 feet
Sweet corn:	10 feet by 80 feet
Popcorn:	10 feet by 80 feet (for the larger-eared varieties. I don't know per-acre yields for the small varieties, like strawberry popcorn.)
Oats:	10 feet by 62 feet
Barley:	10 feet by 87 feet
Rye:	10 feet by 145 feet
Buckwheat:	10 feet by 130 feet
Grain sorghum:	10 feet by 60 feet
Wheat:	10 feet by 109 feet

Don't hold me too tightly to these figures. They're estimates to give you an idea of how big the playing field is. Weather, fertility, variety, and know-how could alter the figures. All I'm trying to show really is that nine bushels of assorted grains might be raised on 1/6 of an acre and provide you with the major portion of your diet.

Beans
Rediscovering a neglected food

The time is ripe for beans to come out of the cupboard and take their rightful place on our tables. The centuries, after all, haven't tarnished the virtues of this versatile food. They were ideal for their enormous role in early agriculture because they could be easily grown, easily dried, and easily stored. They still can be. Because most beans have a rather mild taste, they mix well with other foods in casseroles, soups, and salads.

Ancients relied on beans because they had little meat. We need beans because we've fallen into the habit of eating too much meat—and eggs, and milk. High in fiber and protein, low in fat, beans are literally what the doctor ordered as an antidote for the overly rich American diet of the late 20th century. What a pleasant coincidence that they taste so good.

Poor man's meat

In the beginning, according to the Pequot Indian legend, man had only meat to eat. Beans were the gift of a ghostly sky-maiden. In the legend of another tribe, a benevolent crow brought the first bean and grain of corn, one in each ear.

Anthropologists think it more likely that man discovered beans early in his history by the earthly process of trial and error, but it's easy to see why he considered them heaven-sent. As a forager, he ate wild beans, and when he first began to grow his own food, back in the Stone Age some ten thousand years ago, beans were among his first crops.

Today, the affluent world has all but forgotten the bean. In the United States, beans can be found in a bowl of chili, sitting on the counter of an all-night diner, or in the cold pork and beans that plop cheerlessly out of a can. Edible, but hardly a dish for company or good times. A dish, rather, for lean times, and for those nights when good eating isn't worth the effort. Humble food, cheap food: "poor man's meat."

Actually, there's more than a little truth to the epithet "poor man's meat." It's not much of an exaggeration to call man's struggle to feed himself well down through the ages a quest for protein, the essential ingredient for growth and tissue repair, and man has looked to beans with their 20 to 40 percent protein.

The protein in beans, let it be admitted, is no match for the protein in meat, eggs, or dairy foods. Those are "high quality" protein, containing the essential amino acids we need in roughly the same proportions in which we need them. Beans have all the essential amino acids too, but some are in short supply.

Fortunately, there are other vegetable foods whose amino acids are "unbalanced" in a way that *complements* beans. Put beans and grains together, for example, and you have a fairly complete protein. Man apparently made this valuable nutritional discovery early. Primitive peoples grew grains and beans together (like the Indians of Latin America, who plant beans in their corn fields, allowing the vines to climb the cornstalks). And they stayed together at the table, down through the ages, as is evident in traditional dishes like beans and rice, beans and cornbread, refried beans and corn tortillas, and baked beans and brown bread.

When there was a choice, however, most of our ancestors chose meat over vegetables. Generally expensive or in short supply, meat on the table has long been a sign of status and wealth. In ancient Rome, beans already had "plebeian" connotations, and in the Middle Ages, it was a sturdy food for the peasants to eat, while their betters nourished themselves with fowl and roast beast. Today, we still talk of a good provider "bringing home the bacon"—not the beans.

In modern America, though, overabundance of meat on the menu is likely to be more of a problem than undersupply, which is one reason many people are giving beans a second look. In meat, eggs, and dairy, protein usually comes with an unwelcome partner: fat. Too much animal fat has been implicated in heart disease, and too much fat, it seems, raises the risk of cancer. Beans provide protein without the fat—most beans have one-tenth the fat, gram for gram, of club steak, for instance. It's an arrangement that a health-conscious person should not ignore.

Beans also supply in abundance something that meat lacks utterly: fiber, the indigestible residue that much modern research has found vital for health. The diseases linked to a fiber shortfall sound like an honor roll of modern maladies: bowel cancer, diabetes, gallstones, colitis, hemorrhoids, and varicose veins. What about heart disease? A low-fiber diet is associated with high levels of cholesterol in the blood.

In the typical American diet, fiber is conspicuous by its absence. The fibrous husk or bran of whole grains is scrupulously removed to produce refined products like white bread. Meat, the rich man's beans, is not just poor in fiber; it's quite bankrupt.

These bean-eaters are gaining other nutritional benefits, too: healthy doses of the minerals iron and magnesium, and a sampling of the B vitamins. Large lima beans and pink beans, for example, supply more thiamine than beef liver or wheat germ, more pyridoxine (vitamin B_6) than yeast or wheat germ, and more folacin than liver or yeast. They are also high in niacin.

If a burgeoning respect for beans sends you out in search of them, you shouldn't have to look very far. Despite their cosmopolitan reputation, beans have American roots that go down deep; they were filling bellies on this continent several millenia before Columbus ever set foot here. When Captain John Smith sat down to dinner with the Indian Chief Powhatan, one principal course of their festive meal was beans, which Smith pronounced "as sweete as filbert kernells." In the Massachusetts Bay Colony, beans satisfied not only hunger, but the thirst for democracy. Balloting was done with black and white "Indian beanes" (a custom that goes back to the ancient democracies of Greece and Rome). One nutritional lesson that the New England colonists gained from the Indians was a dish of corn and beans that is still popular today; its name, succotash, comes from the Indian *m'sikwatash*.

Some idea of their varied roles in American life can be glimpsed in the names of the countless varieties of beans that have evolved through the years. There are *navy* beans and *soldier* beans, *Jacob's cattle* beans and *wild goose* beans, *case knife* and *lazy wife*. Each one, surely, has a tale to tell.

While bean dishes may have gradually dropped in status, they have never lost their place in traditional American kitchens. Today, as fifty or a hundred years ago, pea beans are baking in many a New England oven, while savory pots of pinto beans are simmering atop numberless Southern stoves. And in the dining room of the United States Senate, hungry legislators recoup their energy after a hard day of lawmaking with bowls of its specialty—bean soup.

Carl Sherman

Beans, foreign and domestic

Beans have been more popular with poorer peoples who, in general, live closer to the equator than those in carnivorous countries. So, to find inspired ways of making complete, nutritious meals of beans, we look to southern Asia, the Middle East, Africa, and tropical America. (An interesting statistic: the protein portion of legumes in the tropical American diet is 20 times the portion in the diet of the temperate world.) Nutritional folk wisdom is reflected in the recipes that follow. There is a valuable legacy in the spicing, too: the unobtrusive taste of beans invites the herbs, resins, peppers, oils, and essences.

Note that in each of the following recipes the incomplete protein of the bean is complemented with another food, either grain, dairy, or seeds.

Mexican beans

Mexicans traditionally eat rice with beans, a combination that increases the usable protein content up to 40 percent. Cornbread complements the beans in chili. Cornmeal or flour tortillas make a nutritious partner with a filling of refried beans (frijoles refritos); the product is known as a burrito.

Supermarkets near Hispanic neighborhoods often carry a wide variety of beans.

Hummus and whole wheat pita give complete protein. The two also happen to complement each other's taste. They make a good appetizer or party food—light and yet kept interesting by the garlic and lemon juice.

Frijoles Refritos

Refried beans are served at breakfast with eggs, as a side dish, and as the filling for burritos.

2   cups (1 pound) dry, pink pinto beans
4   cups water
⅔   cup tomatoes, pureed
2   tablespoons olive oil
1   cup onion, coarsely chopped
1½   teaspoons garlic, minced
1   green chili pepper, finely chopped
     (⅓ cup)

Soak beans overnight, then drain. Place beans and water in a large pot. Bring to a boil, cover, and simmer for 20 minutes.

Remove lid, stir in pureed tomatoes and simmer uncovered, stirring occasionally, until the beans have absorbed most of the cooking liquid (about 1 hour).

Heat the olive oil in a large skillet. Saute the onion, garlic, and chili pepper until soft. Gradually add the beans, mashing them in the pan with a potato masher. Stir frequently to prevent scorching. When all of the beans have been added to the skillet, fry to your taste.

Yield: 4 to 6 servings

Burritos

A burrito is a tortilla filled with mashed, fried beans.

6   tortillas (page 30)
     frijoles refritos (yield of above recipe)
1   tomato, chopped
2   cups cheddar cheese, shredded
1   cup lettuce, shredded
     red chili sauce (optional)

Preheat oven to 300° F.

Fold or roll tortillas about bean filling, and top with tomato, cheese, lettuce, and if you wish, a spicy tomato sauce, with onion, garlic, and chili powder or anchos chilis to taste.

Bake at 300° F. until warmed through (about 15 minutes) or less if the beans were hot to begin with.

Yield: 6 burritos

Middle Eastern beans

A Middle Eastern solution to protein complementarity combines humus, a chickpea dip, and flat bread (pita, page 30).

Hummus

2   cups cooked chickpeas (or 1 cup dried)
½ to 1 cup tahini (sesame butter)
2   cloves garlic, crushed
6 to 8 tablespoons lemon juice
     a couple dashes of hot pepper or cayenne, to taste
     olive oil (optional)
     tamari (optional)

Soak 1 cup dried chickpeas in water overnight. Drain. Add fresh water and cook until tender — about 1½ hours. Add more water if necessary. Either mash the chickpeas with a fork, or puree them in a blender or processor using a little of the cooking water to make things easier for the motor. Add the tahini, garlic, and lemon juice and mix until smooth. Spread on a plate and garnish with hot pepper or cayenne, or a drizzling of olive oil and a couple shots of tamari. Serve as a dip with warm pita bread, Indian chapatis, or wheat tortillas.

Yield: 2 cups

Felafel

Sauteed felafels are quite meatlike in personality.

2   cups cooked chickpeas (or 1 cup dried)
⅓   cup water
     crumbs from 1 slice bread
1   tablespoon flour
2   cloves garlic, finely minced
1   egg, lightly beaten
2   tablespoons parsley, chopped
¼   teaspoon ground cumin
¼   teaspoon basil
¼   teaspoon marjoram
¼   teaspoon (or less) cayenne pepper
1   tablespoon tahini
     flour for coating
5   tablespoons oil for frying

Mash or grind the cooked chickpeas (you can put them through a blender with the ⅓ cup water). Combine the chickpeas and all remaining ingredients except the flour for coating and the oil. The mixture will be soft. Form it into 1-inch balls (or patties), coat with the flour, and saute until crisp in the oil. Then turn and saute on the other side. Serve in whole wheat pita garnished with tomatoes, lettuce, olives, and more tahini. Or, serve with brown rice and salad.

Yield: 8 servings

Indian beans

In India, the country that leads the world in legume consumption, beans are combined with complementary foods in a rich variety of ways. Dal, a souplike lentil puree, is eaten with rice or breads such as chapatis. (See page 30.)

You can use the standard drab lentil, or shop around for a brilliant orange-red variety that cooks up quicker, being smaller in size. Sadly, these legume extroverts turn a bland shade when cooked.

Simple Dal

1   cup lentils
4   cups water
1   teaspoon curry powder or ¼ teaspoon each of cayenne, coriander powder, and turmeric powder
½   teaspoon crushed fresh ginger

Rinse the lentils and look over for foreign particles. Add 4 cups boiling water and cook covered for 15 to 25 minutes, or until soft. Stir in spices.

Yield: 5 cups

A pancake for the adventurous is the Indian dosa. To us, it is a pancake in name only, even though the batter is cooked according to our conventions. The dosa is distinguished by its fermented grain-and-bean batter—or slurry, to be more exact—as the rice and mung beans are soaked and then pulverized wet, in a traditional version of the mortar and pestle or in a blender. Herbs, chilis, and spices add excitement to the bland, pleasant-smelling batter.

The variation that follows was described for us by the Ramakrishnan family of Madras, India. They eat dosas frequently for breakfast, refrigerating leftover batter for up to 3 days.

Dosa

1½   cups dry rice
⅔   cup dry green or black mung beans

Wash rice and beans and remove any foreign matter. Soak for 3 to 6 hours in separate containers, rice in 1½ cups water and beans in ⅔ cup water. Then grind the mung beans finely and the rice less so, using a traditional mortar and pestle, blender, or processor.

The two are combined and allowed to sit overnight in a warm place so that naturally occurring yeasts can turn them into a light, fluffy batter.

The prepared batter can be spiced up with minced onion, curry leaves, or fresh coriander leaves. Pour onto an oiled, heavy pan in spoonfuls, as you would pancake batter. Dosas should be thin, more like crepes than pancakes; add water to the batter if necessary. Cover the pan, and you won't have to flip the dosa to cook the top side, suggests Mrs. Ramakrishnan.

Yield: serves 3 or 4

The Madras Woodlands Restaurant, New York City outpost of a Brahman group of restaurants, serves dosas rolled about a filling of potato or apple. A first-time visitor here is apt to be amazed that the light, fragile dosas contain no eggs or dairy products. Dosas can be served with chutney or dal.

The overnight fermentation of grain and legume batter enhances nutrition, thanks to naturally occurring microorganisms. The batter rises as did the first bread doughs—without the addition of yeast or baking soda. Thiamine, riboflavin, and niacin all increase over a six-hour period, according to research done at the University of Baroda, India. This same study found that a more nutritious (and nearly as tasty) batter could be made with soybeans standing in for the traditional black mung bean. The researchers were trying to find a way to slip the soybean into the Indian diet—a difficult task, as Indians and Sri Lankans disdain the beany taste. In the Far East, soybeans are popular in the form of tofu, tempeh, tamari, and miso, but preparation of these foods appears long and cumbersome to the Indians and Sri Lankans.

Another traditional use of fermented batter is idli, fluffy steamed cakes often served at breakfast. The variation we describe here is that of Mrs. Ramakrishnan, a botany professor. She says the batter is similar to that for dosa, except for a coarser texture and a larger proportion of rice.

Idli

  2 cups dry rice
2/3 cup dry green or black mung beans, or halved soybeans
2 2/3 cups water

Wash the rice and beans, removing any foreign matter. Soak separately for 3 to 6 hours, rice in 2 cups water and beans in 2/3 cup water.

In a mortar, blender, or processor, grind the rice coarsely and the dal finely. Then combine and set aside in a warm spot to rise overnight. Next day, the batter should have risen somewhat. Indians cook it in a special steamer—simply a stack of 3 or more metal tiers with stamped, perforated indentations to hold the batter. To encourage the cooked idlis to come free, the cups are either oiled or lined with cheesecloth. You can improvise with small cloth-lined bowls, egg poachers, strainers, or colanders. Traditional idlis are 3 or 4 inches in diameter, but there's no reason you couldn't make one large, stomach-filling cake.

You'll also be breaking with tradition, harmlessly, if you use brown rice instead of snowy white. White rice will make a fluffier, more subtly flavored idli.

Steam for 10 to 15 minutes, or longer for larger cakes; they're ready when a toothpick passed through the center comes out clean.

Idlis are eaten with chutney or dal. Cold, leftover idlis can be buttered and warmed in a toaster-oven for a snack.

Yield: roughly 8 servings

Three unfamiliar foods that are easy to like: round idlis, large crepelike dosas, and a dish of sambar condiment. (Photo courtesy New York's Food and Life Sciences Quarterly.*)*

Across a bit of monsoon-tossed water from India is the island country of Sri Lanka (or Ceylon, on out-of-date maps and old tea boxes). The Sri Lankans have taken the dosa and made a hopper out of it—a crepelike cake of rice flour, flavored with a bit of coconut milk. The Western tongue is given a challenge by dishes of sambar or curry served as customary accompaniment. A dish of grated coconut serves the purpose of yogurt on the Indian table: to extinguish the tongue's spicy fire between bites.

Our recipe takes the liberty of adding mung beans or soybeans for the sake of optimum nutrition.

Brigit Phillips, the wife of the vicar at Nuwara Eliya, high in the hills of Sri Lanka, suggests that a sticking pan can be remedied by scrambling an egg before cooking hoppers.

Hoppers aren't easy to make. They involve planning ahead to soak the ingredients, and even Sri Lankans of very limited means will buy ready-made hoppers. Westerners have no choice but to make their own, but luckily we don't have to pulverize our grain and beans with the equipment at the vicarage: an enormous stone pestle shaped like a teardrop is rolled about the pocket of a still larger stone. This kitchen appliance must weigh more than a refrigerator, and is found throughout Sri Lanka and India—even on the kitchen cars of long-distance passenger trains.

Each tier of this idli steamer holds four of the cakes.

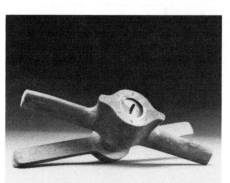

Hand-held presses are used in India and Sri Lanka to extrude noodles of rice flour. The noodles are either steamed or deep fried. Different templates are available for a variety of noodle shapes and sizes.

(Continued on page 90)

(Continued from page 89)

Adzuki bean

A variety of soybean, not well known in the United States but extremely popular in Japan.

Appearance: tiny deep red, almost maroon, beans.

Uses: since often bitter if pressure cooked, they're tastiest when soaked overnight and then boiled; good in casseroles, soups, and baked goods.

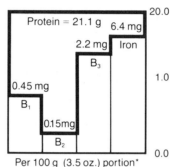

Protein = 21.1 g

B₁ 0.45 mg
B₂ 0.15 mg
B₃ 2.2 mg
Iron 6.4 mg

Per 100 g (3.5 oz.) portion*

Hoppers

```
2    cups dried rice
⅔    cup dry mung beans or halved
       soybeans
     pinch bicarbonate of soda
     fresh grated coconut
```

The batter is prepared as for dosa, but the grains are soaked in coconut water, and must be blended smoother, as the Sri Lankan hoppers are thinner, more delicate.

Coconut water is made by soaking a handful of grated coconut in a cup of water for an hour or so, then hand-squeezing to express the coconut-flavored water.

The batter is poured in the center of a curved cast-iron pan, shaped much like a wok. A standard flat-bottomed iron skillet will do, but it produces hoppers that lack a characteristic round pillow at the center of the otherwise paper-thin cake.

Sambar

Mrs. Ramakrishnan makes this lentil dal fresh daily, as it doesn't store well and flavor is especially important: sambar is used to add interest to the daily staples of idli, rice, dosa, and chapati.

```
1    cup lentils
6    cups water
1    small onion, finely chopped
½    teaspoon black mustard seeds
½    teaspoon curry powder, or more to
       taste
1    cup assorted vegetables, chopped
       (cauliflower, tomato, spinach, as
       available)
1    tablespoon dried tamarind soaked in
       ¼ cup hot water (or juice of ½
       lemon and 1 teaspoon honey)
3    teaspoons oil
```

Wash lentils and inspect for foreign matter. Boil in the water for 15 to 25 minutes, or until soft. In the meantime, fry the onion and mustard seeds in oil, then add the curry powder and saute the chopped vegetables only until still crisp. Combine with the cooked lentils and tamarind juice or lemon juice and honey, and remove from heat.

Yield: 4 to 6 servings

■

Black bean

Closely related to turtle beans, black beans are a major source of food in Mexico, the Caribbean, and the American Southwest.

Appearance: small, dark oval.

Uses: primarily in soup and casseroles; the Chinese ferment the cooked beans. Feijoada is a Brazilian black bean soup; it's best not to pressure cook black beans as their skins fall off easily and may clog cooker's valve.

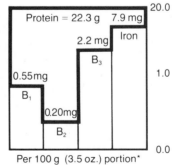

Protein = 22.3 g

B₁ 0.55 mg
B₂ 0.20 mg
B₃ 2.2 mg
Iron 7.9 mg

Per 100 g (3.5 oz.) portion*

Although most Indian foods can be prepared with the tools in your kitchen, something is lost thereby, just as recipes are diminished when their exotic ingredients are left out.

There's no sense in investing in exotic ingredients and equipment unless a foreign cuisine truly sweeps you up and changes your eating

*B₁ is thiamine, B₂ is riboflavin, B₃ is niacin. Sources: *Composition of Foods, USDA Handbook No. 8; Food Composition Table for Use in East Asia* (HEW and FAO, 1978).

habits. Should your Indian repertoire grow beyond the standard raisin-and-peanut curry, we'll describe a few traditional utensils. These are available at some Indian shops in larger cities.

The *karhai* is a deep bowl-shaped pan of iron, similar to the Japanese wok. Sizes range from five inches to two or three feet in diameter, the larger karhais intended for commercial use. This pan is used for both deep-frying and sauteing in a minimum of oil.

The *tava* is a slightly concave disc of heavy iron, measuring about nine or ten inches in diameter and about ½ inch thick. It is placed directly on the stove burner or hot coals and is used to cook unyeasted breads such as chapati (with the convex side up) and paratha in oil held by the concave side. Spices and nuts are roasted or fried in oil on the concave side. Use a large cast-iron skillet as a substitute.

Spices and mustard seeds are also fried in a *karcha,* a heavy iron spoon. A small cast-iron skillet will work well.

Indians grind whole dry spices in a tool similar to our mortar and pestle. Fresh herbs are ground between a grooved stone base, or *sil,* and similarly grooved roller, or *batta.* In the Western kitchen, a grinder or electric blender will reduce dry and fresh ingredients, but this is a nuisance with small quantities.

The clay oven known as a *tandoor* was introduced to the Indian subcontinent by the Moghuls. The traditional tandoor is shaped like an urn and imbedded in earth to hold heat. It is fired by wood, and when baking temperature is reached, the fire is raked out and small breads are slapped on the inside to bake. The resulting flavor is said to be beyond the ken of modern gas and electric ovens.

A smaller, portable form of the tandoor is the *mitti ka bartan,* which means literally "earthen pot." This round pot with lid can be easily thrown by amateur potters. It is fired without glaze. The pot is placed on hot coals, and coals can be placed upon the lid to achieve an even heat. Meat cooked in clay takes on a subtle flavor unique to cooking in unglazed pottery.

Shahnaz Mehta

Black-eyed pea

Also known as cowpeas and black-eyed beans.

Appearance: oval and creamy, white, marked with a distinctive small eye.

Uses: blackeyes cook quickly and become tender in less than an hour; popular in southern United States, used in Hoppin' John; said to bring good luck if eaten with rice on New Year's Day (a bit of folklore based on sound nutritional practice).

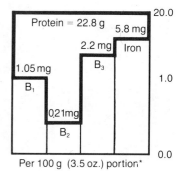

Protein = 22.8 g

B_1 1.05 mg · B_2 0.21 mg · B_3 2.2 mg · Iron 5.8 mg

Per 100 g (3.5 oz.) portion*

Chickpea

Entire cookbooks have been written about the versatile chickpea, or garbanzo, popular along the Mediterranean, through the Middle East to India.

Appearance: roundish and pale.

Uses: bring a nutlike flavor and a crunch to salads; in the Middle East they're boiled and pureed with garlic and tahini to make hummus or ground with wheat to make felafel; Spanish cuisine features them with tomato sauce. With roasting and grinding, the beans are transformed into a coffee substitute. Always a good source of protein, calcium, iron, potassium, and B vitamins, they were cultivated in the Near East as early as 5400 B.C. according to archeology reports.

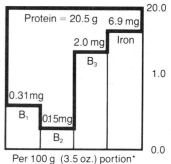

Protein = 20.5 g

B_1 0.31 mg · B_2 0.15 mg · B_3 2.0 mg · Iron 6.9 mg

Per 100 g (3.5 oz.) portion*

Fava bean

Also known as broad beans. Can be purchased either dried or young and fresh.

Appearance: large and flat, of various shades.

Uses: resemble lima beans in appearance and method of preparation; often substituted for chickpeas in Arabic recipes.

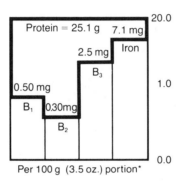

Protein = 25.1 g   7.1 mg   20.0
2.5 mg   Iron
B₃
0.50 mg   1.0
B₁   0.30mg
B₂
Per 100 g (3.5 oz.) portion*   0.0

Kidney bean

The chili bean first grown in Central and South America.

Appearance: light and dark red varieties, shaped like kidneys.

Uses: all-purpose beans, for soups, salads, casseroles, and chili; in Mexico, simmered with onions and green chili peppers in large earthenware pots called ollas; in Brazil, mashed and fried into cakes; red beans often found canned, and needn't be soaked and cooked for long.

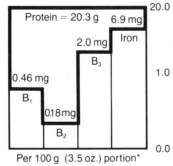

Protein = 20.3 g   6.9 mg   20.0
2.0 mg   Iron
B₃
0.46 mg   1.0
B₁   0.18mg
B₂
Per 100 g (3.5 oz.) portion*   0.0

Yankee beans

This New England dish, served with whole grain bread, makes a solid meal—the United States' best-known contribution to international bean cuisine.

Baked Beans

Baked bean recipes seem to have only two things in common: beans and dry mustard. At one time or another, recipe writers have pulled down everything from the shelves in service of this dish, even though beans will do quite well all by themselves, judging by a recent discovery at the Rodale Press Test Kitchens. Surprisingly, soybeans simmered for several hours take on a new personality, becoming sweet and rich enough to be mistaken for molasses-saturated baked beans. They even turn a nice brown all by themselves.

A basic recipe calls for soaked beans (navy, pinto, white, kidney), onions (either minced or whole), molasses (between ½ tablespoon and ½ cup per pound of dry beans), and dried mustard. Common additions are tomatoes (paste, canned, fresh, ketchup) and any of several flavorings (curry, basil, cloves, apple, vinegar, ginger). Oil is optional. So is a bean pot—any oven-proof covered dish will do. Cooking times vary greatly from cook to cook, ranging from an hour to all day long.

The recipe we print here is ready to accept your embellishment.

2   pounds dried beans
3   medium onions
1   cup dark molasses
3   teaspoons dry mustard

Wash and pick over beans. Soak overnight in water to cover. Discard soaking water the next day, and then cover again with water. Bring to boil, then lower heat and simmer gently until beans are tender (about 1½ hours). Drain, reserving the liquid.

In a 4-quart pot add the 3 whole onions, combine the reserved liquid with the molasses and mustard into pot. Add beans. If there is not enough liquid, add more water to come to within 1 inch of the top of the pot.

Cover and bake in oven at 350°F. Bake 2 hours covered and 2 hours uncovered. Add boiling water if necessary to keep beans moist.

Yield: 12 servings

*B₁ is thiamine, B₂ is riboflavin, B₃ is niacin. Sources: Composition of Foods, USDA Handbook No. 8; Food Composition Table for Use in East Asia (HEW and FAO, 1978).

Senate Navy Bean Soup

This version of the famous soup of Congress was developed for us by JoAnn Benedick of the Rodale Test Kitchens.

    1   pound dried navy beans, soaked
          overnight and drained
   13   cups chicken stock
    1   bay leaf
    1   tablespoon tamari
   ½   teaspoon dried thyme
   ½   pound potatoes, peeled and cut into
          2-inch cubes
   ¼   cup milk
    2   tablespoons olive oil
  1¼   cups celery, finely chopped
  1¼   cups onion, finely chopped
    1   tablespoon garlic, minced
   ¼   cup parsley, finely chopped
   ½   teaspoon oregano

Combine soaked beans, 12 cups of the chicken stock, bay leaf, tamari, and thyme in a large soup pot. Bring to a boil, reduce heat, cover, and simmer until beans are tender, about 30 to 40 minutes.

Meanwhile, put the cubed potatoes in a small saucepan, cover with water and bring to a boil. Simmer until potatoes are tender. Drain potatoes and put them through a food mill or ricer. Heat the milk and beat it into the potatoes, using a wooden spoon.

Heat the oil in a skillet and saute the celery, onion, garlic, and parsley until soft. Add the sauteed vegetables, potatoes, and oregano to the beans. Add the remaining cup of stock. Simmer soup, covered, for 1 more hour, and stir now and then.

Remove 1 cup of the beans, mash them well, return them to the soup, and cook until thickened.

Yield: 6 to 12 servings

Peanut Butter, Smooth or Chunky

Use roasted peanuts, either purchased that way or roasted in the oven for 20 minutes or more (another 10 minutes if in the shell) at 300° F. Roll them about to prevent burning. Grind them in a food processor fitted with its metal blade. After 1 or 2 minutes you will have a crumbly, dry peanut butter. Three or 4 minutes produce a smooth peanut butter because extended processing generates enough heat to extract oil from the peanuts.

For chunky peanut butter, add a few peanuts to creamy peanut butter and then grind about 15 seconds, or until the size of the chunks is to your liking.

A meat grinder or a blender will also make peanut butter. Mild-flavored oil can be added to make the blender easier to spin, but do not add more than ¼ cup oil for 2 cups of peanuts or you'll end up with tanning lotion.

Lentil

The United States was slow to discover this quick-cooking legume, but it is now the leading exporter.

Appearance: small discs of green, brown, and reddish-orange.

Uses: absorb flavors readily and work well with spices and flavorings. Combined with brown rice by the Egyptians to make kushari; vegetarians of the United States appreciate the full-bodied character of lentil stew; a prolific sprouter (cooked sprouts have a nutlike flavor; raw sprouts are somewhat peppery and are best combined with milder sprouts).

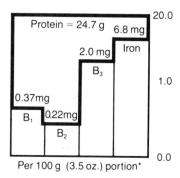

Protein = 24.7 g

B_1 0.37mg · B_2 0.22mg · B_3 2.0 mg · Iron 6.8 mg

20.0 / 1.0 / 0.0

Per 100 g (3.5 oz.) portion*

Lima bean

Sometimes called butter beans.

Appearance: pale, broad flat beans, in a variety of sizes (most common are the large lima and the baby lima); available fresh, frozen, dried, or, in Europe, canned.

Uses: baby limas excellent as plain vegetables, or in casseroles or soups; large limas best in casseroles or soups, particularly in succotash.

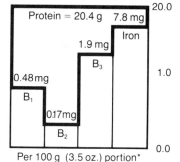

Protein = 20.4 g

B_1 0.48mg · B_2 0.17mg · B_3 1.9 mg · Iron 7.8 mg

20.0 / 1.0 / 0.0

Per 100 g (3.5 oz.) portion*

Mung bean

If you've ever ordered chop suey, you've probably eaten mung beans, in sprouted form; a protein-rich staple in India and the Orient, served as porridge, pancakes, and noodles; innocent of producing gas; used by some cultures to end fasts because of high digestibility.

Appearance: pale, drab green and lozenge shaped; Indian varieties, known as grams, include a black mung bean.

Uses: excellent sprouts, eaten raw or lightly stir-fried; combined with rice by Indians and Sri Lankans to make cakes known as idli, and stewed into bland or spicy dals; more expensive than most other beans, but they can be grown in backyard gardens north to Maine: a 25-foot row in that state yielded 2½ pounds of mung beans or close to $2.10 worth.

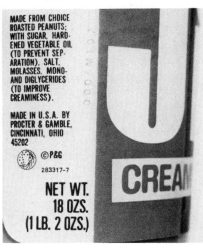

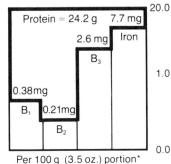

Protein = 24.2 g — 7.7 mg Iron
2.6 mg B₃
0.38 mg B₁
0.21 mg B₂
Per 100 g (3.5 oz.) portion*

Food scientists have lavished their attention on humble peanut butter.

Communal beans

On The Farm, we live on soybeans. They supply us with the protein part of our diet, taking the place of meat, fish, eggs, milk, and dairy products. We are complete vegetarians and don't eat any of those foods. We are growing 150 acres of soybeans this year to feed our community of 1,200 folks.

We also grow grains, fresh garden vegetables, white potatoes, sweet potatoes, and fruit trees. We freeze and can summer produce for the winter. So we have a basic diet we can grow ourselves in abundance right here in the South.

We eat whole cooked soybeans, plain or in a sauce, roast them into soy nuts, and make soyburgers, soy loaf, and soy sausage. We have a soy dairy which makes milk, cheese, yogurt, ice cream, and milk shakes, all from soybeans. We are experimenting with processes that turn soybeans into foods with other flavors and textures. Most cooking oil and

Peanut

Bean disguised in a nutlike shell.

Appearance: known well to circus audiences, elephants, and owners of beige plastic Mr. Peanut pencils.

Uses: raw peanuts are soft and taste similar to raw peas; commercial peanut butter often perverted with sugar, coloring, stabilizers, preservatives, and extra oil (peanut butter recipe, page 93); peanut butter is surprisingly at home in vegetable burgers, loaves, sauces, and oatmeal, try the nuts in Indian cooking, salads, and granola.

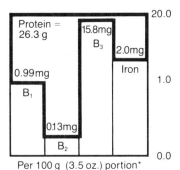

Protein = 26.3 g
15.8 mg B₃
2.0 mg Iron
0.99 mg B₁
0.13 mg B₂
Per 100 g (3.5 oz.) portion*

*B₁ is thiamine, B₂ is riboflavin, B₃ is niacin. Sources: *Composition of Foods, USDA Handbook No. 8; Food Composition Table for Use in East Asia* (HEW and FAO, 1978).

margarine is made from soybean oil, and from it we make soy mayonnaise, salad dressing, whipped cream, and dessert frostings. In winter we can sprout the beans to make a delicious fresh vegetable. Tastewise, we are entirely satisfied with our diet and find that we do not miss the foods that we no longer eat.

Buy your soybeans at a farmer's supply store. Get a one bushel (60 pounds) sack of seed grade beans. (A bushel is enough to last an average family of four well over a month.) They will be uniform, clean, high quality beans with good germination in case you want to sprout them. It will be much cheaper to buy beans this way than in other stores. If you buy your soybeans from a feed or seed store, be sure that they have not been treated with mercury or any other poisonous chemical. Mercury poisoning can be fatal or cause permanent central nervous system damage.

Store your sack of beans tightly closed in a cool, dry place. Always hand sort beans before cooking because even the best seed cleaner occasionally passes a soybean-sized rock.

Our recipes are based on the commercial oilseed type of soybean, which is the major type grown in this country. There are also vegetable-type soybeans, which are larger, cook quicker, and are generally more expensive.

From the introduction to *Yay Soybeans* (1974), a cookbooklet by The Farm, Summertown, TN 38483.

Pinto bean

Best known in Mexican bean dishes.

Appearance: a distinguished, dappled, salmon color.

Uses: in most recipes can be interchanged with red and kidney beans; pinto flakes cook up to round tender beans in about 40 minutes.

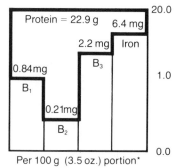

Protein = 22.9 g. Per 100 g (3.5 oz.) portion*

Soybean

Ranks first among legumes in nutrition and versatility.

Appearance: round, of various shades.

Uses: soymilk, tofu, tempeh, tamari, and miso are the best known of many Oriental foods based on the soybean; beans sold as defatted and fatted flour, granules, grits, flakes; raw soybeans can be roasted or boiled; flakes require no presoaking and cook quicker; both yellow and black varieties good for sprouting and will be easier to digest and more nutritious if stir-fried briefly. See elsewhere this chapter for sections on tofu, tempeh, and other soyfoods.

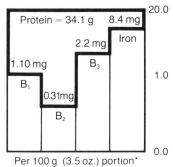

Protein = 34.1 g. Per 100 g (3.5 oz.) portion*

BEAN CUISINE, A CULINARY GUIDE FOR THE ECOGOURMET

Beverly White. 1977.
Beacon Press, 25 Beacon Street,
Boston, MA 02108. 142 p. hardcover
$7.95.

Everything about *Bean Cuisine* is modest: its small size (roughly 5½ by 8 inches), circumscribed subject, straightforward recipes, and simple yet exquisite drawings by Julie Maas, illustrator of the two volumes of *The Vegetarian Epicure*. The bean dishes are inspired; we highly recommend the book.

RY

I am confident that one response to our earth's crises will be the emergence in American culture of a new breed: the bean cook, highly skilled in the art of preparing beans in combination with grains, nuts, and vegetables to complete their protein value.

Mexican Kidney Bean Salad, No. 1

1   9-ounce can tomato sauce
¼   cup chili sauce
1   teaspoon hot mustard
2   tablespoons onion, grated
2   tablespoons horseradish
½   teaspoon chili powder
1   teaspoon dried basil
1   clove garlic, crushed
¼   cup vinegar
¼   teaspoon salt
¼   teaspoon Tabasco
2   tablespoons olive oil
2   cups cooked kidney beans
        (available canned)

Combine all ingredients except oil and beans and simmer for 10 minutes. Blend with the oil and add the beans.

Marinate for 2 hours or more. Serve cold on lettuce with avocado slices and cornsticks.

Yield: 4 servings

Split pea

Of pea soup fame.

Uses: Indian dals, pea soup.

Appearance: Yellow or green hemispheres, of uniform size.

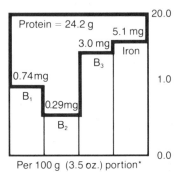

Per 100 g (3.5 oz.) portion*

White bean

This category includes great northern, small white, navy, marrow, and pea beans.

Appearance: white, kidney shaped.

Uses: vegetable soups and fish chowders; hold their shape well under long, slow cooking, and consequently make outstanding baked beans.

Profiles by Carol Munson and Krissa Strauss

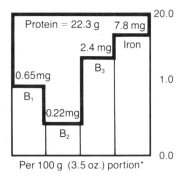

Per 100 g (3.5 oz.) portion*

*B_1 is thiamine, B_2 is riboflavin, B_3 is niacin. Sources: *Composition of Foods, USDA Handbook No. 8;* Food Composition Table for Use in East Asia (HEW and FAO, 1978).

Sources

Buying them

Compared with meat, cheese, and vegetables, the dry beans and peas at the local supermarket seem a bargain. These dry goods aren't perishable, and there's a low loss rate and consequently less mark-up tacked on. But supermarket dry beans and peas are likely to come in small, one-pound packets, and usually aren't labeled as organically grown. Supermarkets also may not offer a great selection (stores in Hispanic areas may be better stocked.)

For variety, try co-ops, natural food stores, co-op warehouses, or mail order.

Sources

Casa Moneo Catalog, Moneo and Son, Inc., 210 W. 14th Street, New York, NY 10011.

This catalog lists 40 pages of staples and curiosities from Spain, Mexico, and South America. The selection of beans is impressive: pink, white, red, black, pinto, chili, mexe, red kidney, white kidney, chickpeas, pigeon peas, and more. They sell paella pans in 17 sizes, tortilla presses, and comals. Casa Moneo requires a minimum order of $25.

Vermont Bean Seed Co., Garden Lane, Bomoseen, VT 05732.

Vermont Bean Seed offers "the largest bean and pea seed selection available anywhere in the world."

Johnny's Selected Seeds, Albion, ME 04910.

Johnny's offers varieties suited to a cool climate.

The displays may not be pretty, but co-ops are a cheap source of beans.

Threshing them by hand

Bean-fancier John Withee of Lynnfield, Massachusetts, has devised a threshing system that is easy on both beans and beaner. He beats the dry plants in a burlap bag sewed up especially for threshing. The finished bag is a cone-shaped tube about 5 feet long, with the top opening forming a circle about 30 inches in diameter. The bag is tapered to a 15-inch diameter at the bottom, which is closed by puckering and tying with an attached light rope. The top has two ropes sewn to the reinforced rim, opposite, each about 4 feet long.

A rough wood hoop, 30 inches in diameter, holds the bag open for loading. The dry bean plants are easily stuffed into the bag to crushing capacity. The wood hoop is then removed and the loaded threshing bag raised to a good height to be beaten with a smooth hardwood stick, a method that is less traumatic for the beans than beating them against a hard surface. All of the beans and fine chaff fall to the narrow bottom.

(Continued on page 98)

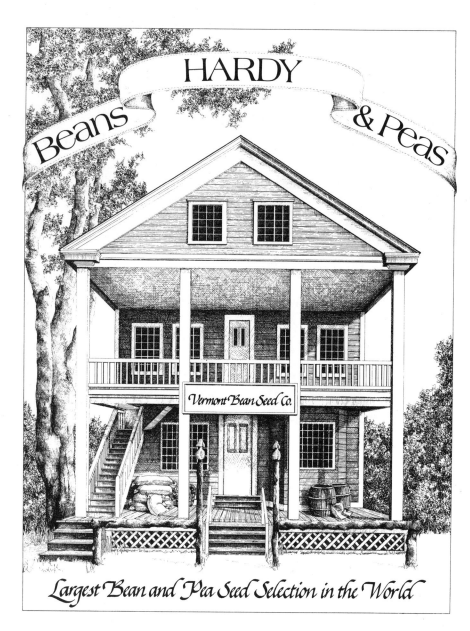

Beans HARDY & Peas

Vermont Bean Seed Co.

Largest Bean and Pea Seed Selection in the World

LEATHER BRITCHES MAKE GOOD EATING

"Leather britches" is the traditional name given to dried, long, green pole beans.

To dry, you thread a large needle with heavy thread, knotted at the end, and string the freshly picked beans just below the blossom end, as though you were making a necklace. Do enough for a meal for your family on each string.

Hang them to dry in a warm place—over a wood stove or in your attic. They can be left hanging all winter or stored in large glass jars. Make sure that no moisture can gather in the jars.

To cook, soak the britches over-night; the next day, boil them for three hours. About an hour before mealtime, add a piece of bacon or ham. Continue boiling for another hour until there is just enough liquid left to barely cover them in the pot. Don't drain, but serve them as they are.

Doris E. Stebbins

"Get familiar with those beans. Go ahead, stick your hand in there all the way up to the wrist and let those cool bean beads give you a little massage."

Phil Levy, advising first-time visitors to natural food stores, in "Natural Shopper Shock," *East West Journal*.

Most Sri Lankans don't like it, some don't mind it, while a few may even like it.

Soyanews, a Sri Lankan periodical devoted to promoting the soybean, commenting on the unfortunate beany flavor of that legume.

(Continued from page 97)

By controlling that opening with the rope tie, a stream of beans is released and, with a strong wind blowing, most of the chaff is blown away as the beans fall. The rough beanstalks and pods are shaken out through the top and another threshing is started. "Truly a Yankee idea," boasts Withee.

Once threshed, the bean seeds must be protected from damage by weevils. These tiny legless grubs eat their way into the fresh beans after hatching from eggs laid on the pods by the adult weevils. They remain in the seed, covered by the skin which hardens over them. To prevent the weevils from hatching out and decimating the seed beans, Withee spreads the beans out and lets them air-dry for a week or two, then puts them in the freezer for three days. Once this is done, the beans need only be kept in dry bags or bottles to retain viability for three or four years.

Steve Smyser

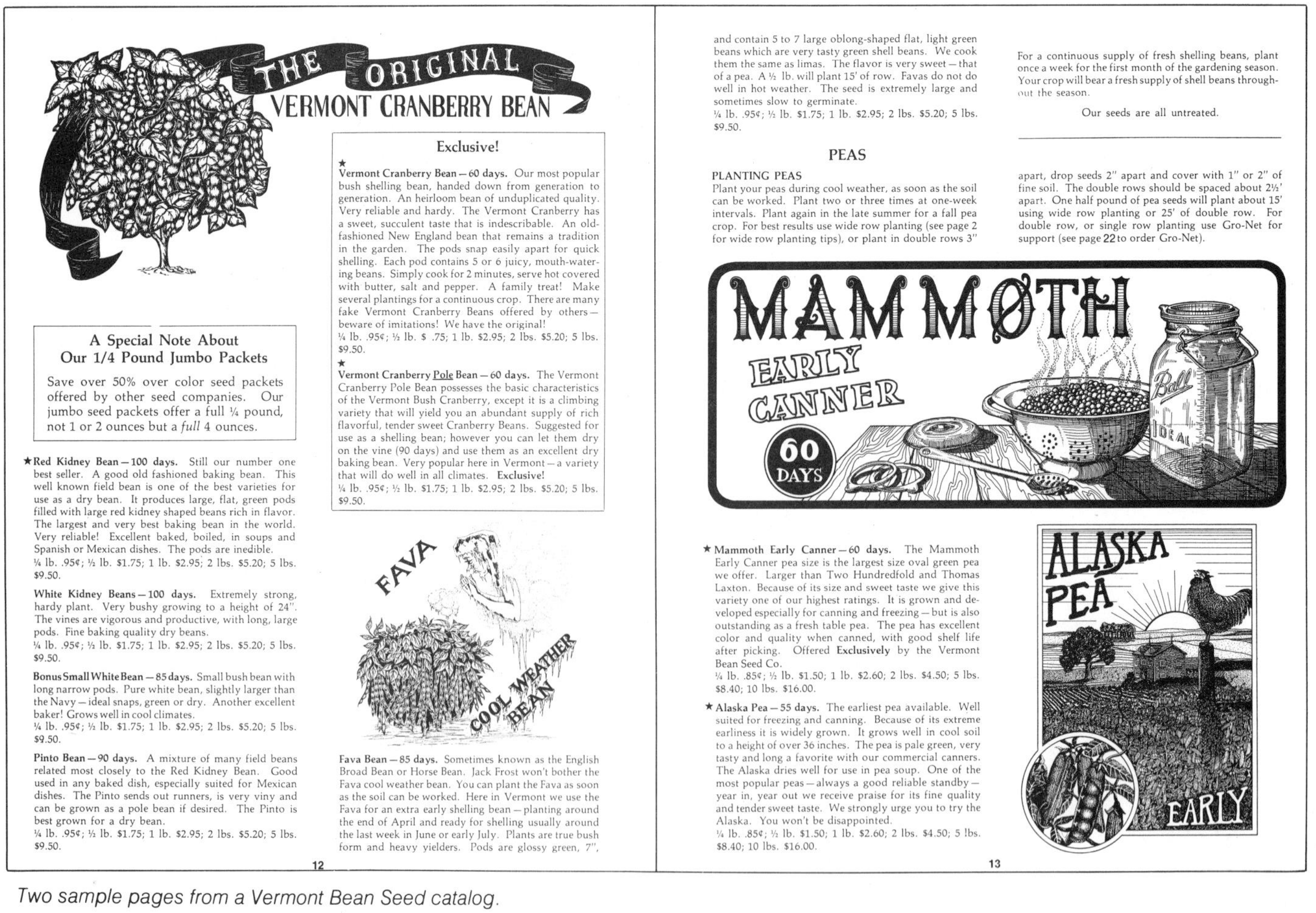

THE ORIGINAL VERMONT CRANBERRY BEAN

Exclusive!

★ **Vermont Cranberry Bean – 60 days.** Our most popular bush shelling bean, handed down from generation to generation. An heirloom bean of unduplicated quality. Very reliable and hardy. The Vermont Cranberry has a sweet, succulent taste that is indescribable. An old-fashioned New England bean that remains a tradition in the garden. The pods snap easily apart for quick shelling. Each pod contains 5 or 6 juicy, mouth-watering beans. Simply cook for 2 minutes, serve hot covered with butter, salt and pepper. A family treat! Make several plantings for a continuous crop. There are many fake Vermont Cranberry Beans offered by others – beware of imitations! We have the original!
¼ lb. .95¢; ½ lb. $ .75; 1 lb. $2.95; 2 lbs. $5.20; 5 lbs. $9.50.

★ **Vermont Cranberry Pole Bean – 60 days.** The Vermont Cranberry Pole Bean possesses the basic characteristics of the Vermont Bush Cranberry, except it is a climbing variety that will yield you an abundant supply of rich flavorful, tender sweet Cranberry Beans. Suggested for use as a shelling bean; however you can let them dry on the vine (90 days) and use them as an excellent dry baking bean. Very popular here in Vermont – a variety that will do well in all climates. **Exclusive!**
¼ lb. .95¢; ½ lb. $1.75; 1 lb. $2.95; 2 lbs. $5.20; 5 lbs. $9.50.

A Special Note About Our 1/4 Pound Jumbo Packets

Save over 50% over color seed packets offered by other seed companies. Our jumbo seed packets offer a full ¼ pound, not 1 or 2 ounces but a *full* 4 ounces.

★ **Red Kidney Bean – 100 days.** Still our number one best seller. A good old fashioned baking bean. This well known field bean is one of the best varieties for use as a dry bean. It produces large, flat, green pods filled with large red kidney shaped beans rich in flavor. The largest and very best baking bean in the world. Very reliable! Excellent baked, boiled, in soups and Spanish or Mexican dishes. The pods are inedible.
¼ lb. .95¢; ½ lb. $1.75; 1 lb. $2.95; 2 lbs. $5.20; 5 lbs. $9.50.

White Kidney Beans – 100 days. Extremely strong, hardy plant. Very bushy growing to a height of 24". The vines are vigorous and productive, with long, large pods. Fine baking quality dry beans.
¼ lb. .95¢; ½ lb. $1.75; 1 lb. $2.95; 2 lbs. $5.20; 5 lbs. $9.50.

Bonus Small White Bean – 85 days. Small bush bean with long narrow pods. Pure white bean, slightly larger than the Navy – ideal snaps, green or dry. Another excellent baker! Grows well in cool climates.
¼ lb. .95¢; ½ lb. $1.75; 1 lb. $2.95; 2 lbs. $5.20; 5 lbs. $9.50.

Pinto Bean – 90 days. A mixture of many field beans related most closely to the Red Kidney Bean. Good used in any baked dish, especially suited for Mexican dishes. The Pinto sends out runners, is very viny and can be grown as a pole bean if desired. The Pinto is best grown for a dry bean.
¼ lb. .95¢; ½ lb. $1.75; 1 lb. $2.95; 2 lbs. $5.20; 5 lbs. $9.50.

Fava Bean – 85 days. Sometimes known as the English Broad Bean or Horse Bean. Jack Frost won't bother the Fava cool weather bean. You can plant the Fava as soon as the soil can be worked. Here in Vermont we use the Fava for an extra early shelling bean – planting around the end of April and ready for shelling usually around the last week in June or early July. Plants are true bush form and heavy yielders. Pods are glossy green, 7",

and contain 5 to 7 large oblong-shaped flat, light green beans which are very tasty green shell beans. We cook them the same as limas. The flavor is very sweet – that of a pea. A ½ lb. will plant 15' of row. Favas do not do well in hot weather. The seed is extremely large and sometimes slow to germinate.
¼ lb. .95¢; ½ lb. $1.75; 1 lb. $2.95; 2 lbs. $5.20; 5 lbs. $9.50.

For a continuous supply of fresh shelling beans, plant once a week for the first month of the gardening season. Your crop will bear a fresh supply of shell beans throughout the season.

Our seeds are all untreated.

PEAS

PLANTING PEAS

Plant your peas during cool weather, as soon as the soil can be worked. Plant two or three times at one-week intervals. Plant again in the late summer for a fall pea crop. For best results use wide row planting (see page 2 for wide row planting tips), or plant in double rows 3" apart, drop seeds 2" apart and cover with 1" or 2" of fine soil. The double rows should be spaced about 2½' apart. One half pound of pea seeds will plant about 15' using wide row planting or 25' of double row. For double row, or single row planting use Gro-Net for support (see page 22 to order Gro-Net).

MAMMOTH EARLY CANNER — 60 DAYS

★ **Mammoth Early Canner – 60 days.** The Mammoth Early Canner pea size is the largest size oval green pea we offer. Larger than Two Hundredfold and Thomas Laxton. Because of its size and sweet taste we give this variety one of our highest ratings. It is grown and developed especially for canning and freezing – but is also outstanding as a fresh table pea. The pea has excellent color and quality when canned, with good shelf life after picking. Offered **Exclusively** by the Vermont Bean Seed Co.
¼ lb. .95¢; ½ lb. $1.50; 1 lb. $2.60; 2 lbs. $4.50; 5 lbs. $8.40; 10 lbs. $16.00.

★ **Alaska Pea – 55 days.** The earliest pea available. Well suited for freezing and canning. Because of its extreme earliness it is widely grown. It grows well in cool soil to a height of over 36 inches. The pea is pale green, very tasty and long a favorite with our commercial canners. The Alaska dries well for use in pea soup. One of the most popular peas – always a good reliable standby – year in, year out we receive praise for its fine quality and tender sweet taste. We strongly urge you to try the Alaska. You won't be disappointed.
¼ lb. .85¢; ½ lb. $1.50; 1 lb. $2.60; 2 lbs. $4.50; 5 lbs. $8.40; 10 lbs. $16.00.

12 / 13

Two sample pages from a Vermont Bean Seed catalog.

The soybeaning of the American kitchen

The movement has begun to get soybeans out of the feed troughs and onto the dinner plates of America. Beef, hogs, and poultry just don't need to eat so richly. The United States is awash with soybeans (136 billion pounds harvested annually). But instead of savoring foods made directly from soybeans, we convert them into meat and dairy products.

Meanwhile, East Asians, who have lived with the soybean for milennia, and who are not as affluent as we, have a simple, economical, and healthful answer: simple, nutritious, inexpensive foods such as tofu, tempeh, soymilk, miso, and many others less known to the West.

Americans rediscover an abundant natural resource

Discovered and domesticated in China around 2500 B.C., the soybean didn't arrive at our shores until the 1850s. In 1922, American farmers raised 4 million bushels of soybeans for food oil, but up until World War II, the United States was a net importer of soybeans from Europe. With the war, which shut off the convenient European supplies, the United States switched its soybean production into high gear, and by 1979, 70.5 million acres produced 2.27 billion bushels—the largest soybean crop ever. The success of soybeans is understandable. They are notable both for their high levels of protein (40 percent) and oil (20 percent), and for their ability to produce more protein per acre than any other crop. One estimation projected that an acre of soybeans will provide enough protein to sustain a moderately active man for 2,224 days, as compared with 877 days for an acre of wheat and 354 days for an acre of corn.

Of this vast quantity of soybeans, produced mainly in the Midwestern states (Illinois, Iowa, Indiana, Ohio), the vast majority goes for livestock meal and cooking oil, with a small portion finding industrial applications. Of the 2 or 3 percent employed directly as human food, much is converted to various high-technology soybean foods, such as soy protein isolates, concentrates, spun protein fibers, textured vegetable proteins, soy powders, and vitamin-enriched soymilks. These laboratory babies are disguised as ingredients in other foods, meat extenders and analogs, and special weaning foods and dietary substitutes. (Former President Nixon reportedly remarked glibly that he had never seen a soybean.) It is not surprising that so few Americans have ever seen their nation's number-one cash crop.

But work goes on quietly to turn Americans on to soyfoods. The USDA's post-harvest research group in Peoria, Illinois, and various colleges (including Cornell, Texas A & M, and University of Illinois) have been engaged in what proved to be pioneering work in the development (or, in a sense, the domestication) of the traditional Eastern soyfoods—tofu, tempeh, miso, soymilk—into commercially viable and publicly acceptable products.

But it took a major book and its zealous author to begin the real colonization of America by Eastern soyfoods. Zen student William

(Continued on page 100)

This tofu US of A was made up for the jacket of Rodale Press's The Tofu Cookbook, *but the food is so anonymous-looking that the point was lost. No doubt tofu would catch on quicker in the West if it looked a bit more like pizza or french fries.*

The soybean makes up for its homeliness in nutritional virtue.

THE OATS, PEAS, BEANS AND BARLEY COOKBOOK
Edyth Young Cottrell. 1974.
Woodbridge Press, Santa Barbara, CA
93111. 271 p. paperback $4.95

This is a nutrition-conscious cookbook of the old school—most recipes are the product of kitchen puttering rather than directly inspired by foreign peasant cuisines (in the manner of *The Vegetarian Epicure* volumes or *The Book of Tofu*). Consequently, this is a good source for those of a more cautious nature. The recipe names tell the story: Soy-Oat Patties with Tomato Sauce, Beans with Bread, Graddaughter's Favorite Patties, Creamed Peas. One unconventional influence is the author's affiliation as research nutritionist at Loma Linda University, run by the Seventh-Day Adventists. The Loma Linda people have worked out meat analogs that are either tantalizing or disquieting in their verisimilitude, depending on the beholder. The author tells how to make gluten, a refined wheat product, and suggests using this pliant substance to form steaks and burgers.

RY

Parsnip Patties

- 1　large parsnip (2 cups shredded)
- ½　cup soaked garbanzos
- ¼　cup water
- ½　teaspoon salt
- 1　tablespoon oil

Shred raw parsnips on medium-coarse shredder. Blend raw soaked garbanzos in water until fine or grind using fine blade in grinder. Combine all ingredients.

Drop from tablespoon on to lightly oiled skilled (about 325° F.). Cover. Let cook 5 minutes, or until nicely browned. Turn. Cover. Reduce heat to 300° F. and let cook 15 minutes.

Note: May be spread thin for crisp texture or made into a regular patty.

Yield: 6 servings

(Continued from page 99)
Shurtleff studied tofu and miso production in Japan for several years in the early 1970s, then returned to the United States and with his wife, Akiko Aoyagi, published *The Book of Tofu* in 1975, an event that was to put tofu on the culinary map and eventually launch a new industry to the continent. Since then, they have published basic review books on miso and tempeh, in addition to detailed technical manuals for the commercial production of tofu, tempeh, and miso. And today, besides lecturing and maintaining a soyfoods resource center, they are preparing a comprehensive overview of the many soyfoods.

Back in 1975 there were virtually no Caucasian-operated tofu shops in America. Few people, besides the Oriental populations con-

Tofu and other soy products are often available at hard-to-find Oriental markets. Check the Yellow Pages.

centrated in the larger urban centers, had either heard of tofu or knew how to cook with it. There were 53 tofu shops at that time, spread out across the country. By September 1978, this figure had swelled to 112, of which 62 were new Caucasian companies. By the spring of 1980, the number had jumped to nearly 170. In addition, two miso companies have opened, and there are more than 20 tempeh shops in production. Industry sales growth hovers at an annual 30 percent, while some companies can point to a staggering sevenfold sales increase over three years. Yearly sales of tofu are estimated at $33 million (average retail price: about $1 per pound), approximately 13.5 million pounds of soybeans are used and, including all soyfoods (but not the high technology ones mentioned above), Americans are spending something like $45 million a year now for these once-unknown Eastern soyfoods.

Why did these obscure foods catch on with such a vengeance? First, the natural foods movement had evolved to the point that people were ready for an alternative vegetarian source of protein, and soyfoods were unbeatable. Coupled with this was the growing recognition of the world protein shortage and America's culpability in creating it through over-consumption of protein staples. Soyfoods eliminate the depressing inefficiency of livestock protein conversion. A final condition for a soyfoods boom was supplied by people with an active desire for a low-technology, socially valuable, foodcraft livelihood.

The first market for tofu and tempeh was natural foods stores, co-ops, and vegetarian restaurants. Then private and certain chain supermarkets began to carry packaged tofu, and now institutions—college food services, nursing homes, hospitals—are looking seriously at the merits of soyfoods. Like yogurt, which began 20 years ago as an exotic, ethnic specialty food, tofu is expected to graduate from the sandlots and stride into the major leagues. Privately, yogurt industry spokespeople predict that soyfoods sales could easily soar to 30 times that of yogurt in the years to come.

Home-made soyfoods are best

Just like bread, tofu and tempeh are best savored when fresh and, being perishable foods, they lose their peak flavor fast. Shops may not store them properly, further reducing shelf life. Home-made tofu can be rinsed daily, and either tofu or tempeh can be frozen immediately after its manufacture. Commercial tofu and tempeh are often a week old before being purchased. In part this is because soyfood producers are still few and far between, but with the continued boom, soyfood markets will shrink from national to regional, or even local—much like the neighborhood bakeries of a few years ago, and similar to Japan's 38,000 neighborhood tofu shops.

Until a tofu shop moves next door, and a tempeh shop is established across the street, consider making your own from scratch. Not only will the foods from your kitchen be fresh, but you'll save money as well. A pound of tofu, selling for an average of $1, can be made from between 15 and 20 cents worth of beans.

Further, tofu and tempeh are satisfying to make, and the necessary equipment is simple.

Before proceeding with the recipes, we must warn picky eaters that soyfoods are tough to hate, thanks to the vastly different forms they take.

Soymilk can be the base of many foods—frozen soymilk desserts, such as ice cream, sherbets, yogurt, cream cheese, cottage cheese, sour cream, kefir, buttermilk, and other cultured drinks. Soymilk should keep in the refrigerator as long as dairy milk.

Tofu can be whipped in a blender to make mayonnaise, dips, and creamy salad dressings; it stands in for cheese in lasagna, cheesecake, and sandwiches; pan-fry cubes of tofu with spices for a meaty, chewy texture; deep-fry tofu, and it takes on another form; freeze it, and the curious result is spongy, chewy and riddled with little air holes.

Tempeh is not quite as versatile as tofu, but its texture is amazing—a world away from its humble beany origin. Pan-fry or deep-fry it, then use

(Continued on page 102)

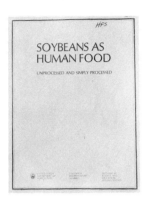

SOYBEANS AS HUMAN FOOD—UNPROCESSED AND SIMPLY PROCESSED

Utilization Research Report #5, Science and Education Administration of the USDA, Stock No. 001–000-03957-1. 1979.

Superintendent of Documents, U.S. Government Printing Office, Washington, DC 20402. 54 p. paperback $2.40.

Six soy experts describe traditional soybean food uses throughout the world, and adapt the procedures, equipment, and methods of serving to American ways.

TS

Dried soybean seeds, when roasted and finely ground like dry coffee beans, make an excellent substitute for coffee because the color and flavor are similar to coffee.

To prepare soybean coffee, clean the dried beans of dust and other foreign materials and roast them in a pot or frying pan as is done with coffee beans. Roasting usually requires from 15 to 20 minutes, depending upon the moisture content of the beans, the color of the roast desired (light, medium, or dark), and the skill of the operator. After roasting, the beans are ground finely in a grinder. The coffee is then ready for boiling. Although the protein content of soybean coffee has not been reported, it does contain soluble soybean protein.

Chien chang is formed by pressing thin layers of bean curd between sheets of cloth. Chien chang, which looks like a piece of cloth, can be cut into thin strips and used as noodles in soups. It can also be stewed with meat. Sometimes it is wrapped around ground meat and then cooked.

THE BOOK OF TOFU
William Shurtleff and Akiko Aoyagi.
1975.
Autumn Press, 25 Dwight Street,
Brookline, MA 02146. 334 p.
paperback (8½ by 11 inches) $6.95.

1979. Ballantine Books, 201 E. 50th
Street, New York, NY 10022. 433 p.
paperback (4¼ by 7 inches) $2.95.

This is the book that launched the soyfoods revolution in America. It masterfully guides the reader from the basic ingredients and an awareness of nutrition and world hunger through recipes (500 in all) that use the various forms of soybeans throughout the tofu-making process (puree, curds, whey, okara, tofu, soymilk, yuba). Particularly inspiring are the two sections, Japanese Farmhouse Tofu and The Traditional Tofu Shop (found in the original, longer Autumn Press edition). Here you can read of true craftsmanship inherent in Japanese tofu-making. The Ballantine edition is revised, condensed, and shrunken for brisk selling on supermarket shelves.

This and following soyfood book reviews by Richard Leviton, except as noted.

For the true tofu master, practice is a living reality, giving energy and an ungraspable, deep meaning to daily work. To watch such a master at work is a rare and beautiful experience. His every gesture seems to emerge from a deep, still center. Grace and economy of movement give a feeling of dance to even the most mundane of his actions. A sense of rhythm, alertness, and precision shows the result of years of patient training and untiring striving for excellence. A panoramic awareness allows the master to be wholly concentrated on the one activity before him while being simultaneously attentive to the full field of activity around him.

(Continued from page 101)
in stews, sauces, pizza, as a burger, and in salads. If you happen to make too much tempeh, you can freeze it for months. Unlike tofu, it defrosts in much the same condition it went into the freezer.

Richard Leviton

Tofu

This recipe yields about 1 pound of tofu—fine for a first try, or for one-person households, but time-consuming considering the modest-sized brick you get for your effort. Most will want to move up to larger batches once the method is mastered.

Tofu
- 1 cup dry soybeans
- 11 cups water, approximately
- 1¼ teaspoons nigari, Epsom salts, or calcium sulfate (solidifier)

Soak soybeans in 4 cups water overnight, well covered to allow for expansion. In the morning, drain the beans. Liquify them for 2 minutes in a blender, using 1 cup of soaked beans to 1½ cups of water. (An alternative but less preferable method is to grind the beans in a meat grinder with a fine screen.)

If you used a blender, combine the liquified beans with 4½ cups water. If you used a hand grinder, add the beans to 6 cups of water. Bring this mixture to a slow rolling boil in a heavy pot, and stir frequently to minimize sticking. After it has reached the boiling point (when the foam suddenly rises, threatening to overflow, which it often does before you can catch it), lower the temperature and simmer the soy mixture for 10 to 15 minutes. This is very important, as uncooked soybeans contain a factor known as trypsin inhibitor, which will prevent the body from assimilating the protein.

Strain the mixture through a coarsely woven cloth or sack to extract the soymilk. (To make a straining bag, sew a sack approximately 18 by 20 inches out

The beans are soaked.

Soaked beans are turned into pulp by a blender or processor, then cooked.

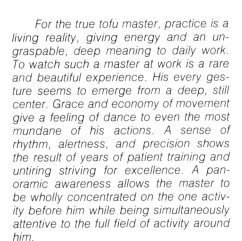

A jar saves the hands from the heat of pressing the last drops of soymilk.

After simmering, the soymilk is ready to be separated with nigari or an acid.

of unbleached muslin, a dishtowel, or any other similar material of medium-to-loose weave. Be sure the seams are well sewn, as this sack will see a lot of twisting.)

Arrange the sack over a colander, jelly cone, or a wooden rack, making sure it is over a pan or bowl to catch the liquid that will run out. Ladle the hot soybean mixture into the sack, and strain as much of the milk as possible out of the bag, leaving a residue called okara. Place the sack in a container of cold water, and stir the water through the okara. Again strain out the moisture into the bowl holding the liquid obtained from the first pressing. Then add 2 cups of water to the okara and again strain this through the bag, twisting, squeezing, and pressing it to remove all of the water into the bowl. What you now have gathered is soymilk. It may be used as is, or you may go on to make tofu out of it. The okara may be used for other recipes.

Tofu is curdled from soymilk in much the same way that cheese is curdled from dairy milk. In cheese-making, rennet is commonly used, but in tofu-making, the most common solidifiers are calcium sulfate, nigari (the traditional salt, evaporated from sea water) and magnesium sulfate. All are perfectly natural ingredients that pose no threat to your system. The easiest to obtain is probably magnesium sulfate, which is nothing more than Epsom salts. Tofu can also be made using acids such as lemon juice or vinegar, but are disappointing, and you will be much better off obtaining a salt solidifier. The traditional Japanese solidifier, nigari, can be purchased from The Learning Tree, P.O. Box 76, Bodega, CA 94922, or Farm Foods, 156 Drakes Lane, Summertown, TN 38483.

To prepare the solidifier for use, dissolve 1¼ teaspoons nigari, calcium sulfate, or Epsom salts in 1 cup water. Add a third of the solution to the soymilk which has been reheated to the boiling point, and stir well. Wait for the milk to settle and sprinkle another third over the top, stirring gently. Cover and let stand for 7 to 8 minutes, then gently sprinkle the remaining third of the solution into the milk. Again let the soymilk stand, this time about 4 minutes, then gently stir the top 2 or 3 inches of the mixture while the curdled milk gathers together into soft curds. The yellow, clear liquid left between the curds is the whey.

Some sort of mold is necessary to allow the whey to run off while forming the curds. You can use any type of mold, as long as it allows you to apply pressure at the top and lets liquids run off. Options include a cloth-lined colander or coffee can with holes punched in the sides and bottom.

Position the mold over a bowl or pan, and line it with settling cloths. Gently ladle the curds into it. The whey will drain out holes in the sides and bottoms, leaving the curds behind. Save this whey, as it is a good soup stock (high in protein and the B-complex). When all the curds are in the mold, fold the cloths over the top and place a flat object over the top of the cloth-wrapped curds. A weight of about 3 pounds is all you need. The whey will continue to drip out under the pressure; after 30 or 40 minutes, invert the mold under a sinkful of cold water to cool the tofu. When it has cooled well, unwrap. Tofu may be stored under water in the refrigerator for a week or slightly more if you change the water occasionally.

Juel Andersen and Cathy Bauer

THE WHAT TO DO WITH TOFU COOKBOOKLET
Eleanor Rovira, editor. 1979.
The Grow-cery, 6526 Lansdowne
Avenue, Philadelphia, PA 19151. 31 p.
paperback $2.

Thirty-nine vegetarian recipes printed in red ink on beige paper that get you and your tofu safely from breakfast to dinner.

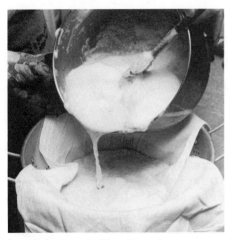

Soymilk is separated from the pulp.

Curds are spooned into a cloth-lined settling box.

The heavier the weight on the press, the firmer the tofu. Soft curds are good in soup, firm curds are needed for stir-frying.

Scrambled Tofu

    3   cloves garlic
    3   small onions
    2   green peppers
   10   large mushrooms, sliced
    3   tablespoons safflower oil
    6   cakes tofu (about 24 ounces)
    1 to 2 teaspoons turmeric
        cayenne pepper to taste
        tamari soy sauce to taste

Saute garlic, onions, peppers, and mushrooms in oil. Add tofu and mash in pan while frying. Saute tofu 15 to 20 minutes or until done (water will boil off). Add turmeric until the color of tofu looks like real scrambled eggs. Add cayenne pepper and tamari to taste. Serve with a grain dish and other vegetables for a complete meal.

Yield: 6 servings

TOFU GOES WEST
Gary Landgrebe. 1978.
Fresh Press, 774 Allen Court, Palo Alto, CA 94303. 114 p. paperback $4.95.

An engaging primer for the vegetarian who would both prepare tofu with dairy and eggs and combine tofu with grains. Landgrebe also offers main dishes that use frozen tofu, such as enchiladas, stuffed peppers, chowders, and lasagna. On the cover, tofu descends from the clouds over a desert Western landscape of two cowboys on horseback—a metaphor for tofu's new frontier.

Cinnamon Oat Muffins

1   cup whole wheat flour
1   cup rolled oats
¼   cup raisins (optional)
8   ounces tofu
2   eggs
¼   cup oil
½   cup honey
1 to 1½ teaspoons cinnamon
½   teaspoon salt
2   teaspoons baking powder
¼   teaspoon baking soda

Mix the first 3 ingredients well in a large bowl. Mix remaining ingredients in a blender until smooth. Pour the blended ingredients into the flour mixture. Stir well. Fill oiled muffin tins 2/3 full. Bake at 425° F. for 12 to 15 minutes.

Yield: 12 muffins

A skillet will do for stir-frying, but a wok is best for keeping vegetables crisp.

THE TOFU COOKBOOK
Cathy Bauer and Juel Andersen. 1979.
Rodale Press, 33 E. Minor Street, Emmaus, PA 18049. 188 p. paperback $8.95.

Tastefully printed in a pleasant format with handsome woodcuts. Bauer and Andersen make an ambitious effort to Americanize tofu, arguing that "tofu deserves a chance to make it in the American kitchen." Their book is written with the aim of "making tofu an everyday staple in your kitchen," and addresses itself to the "skeptical reader," to "all those who have tried tofu—once."

Tofu Quiche

2   tablespoons cornstarch
½   cup cold soymilk or dairy milk
       (salt—optional)
1 or 2 eggs
½   cup cottage cheese
½   cup yogurt
1   cup tofu
       pepper, to taste
       dash of freshly ground nutmeg
       dash of Worcestershire sauce
Your options:
½ to 1 cup cheddar, Swiss, blue, or
       combination of these cheeses,
       grated
1   medium onion, chopped and sauteed
       in 1 tablespoon oil or butter
½   pound fresh mushrooms, chopped or
       sliced
1   cup´any cooked vegetable, well
       drained
½   cup cut-up shrimp or any flaked fish,
       with 1 tablespoon dry sherry
       herbs, to taste

A quiche can be served as a full dinner accompanied by a salad and stout bread and perhaps preceded by a soup. Our quiche has few eggs and lots of tofu, and the proportions can be varied at will.

Use the quiche pie crust that follows. Preheat oven to 350° F.

Combine cornstarch, milk, and salt in a blender and process until smooth. Add the remaining ingredients and process until very smooth.

Stir in your choice from the cheese,

vegetable, and shrimp options. Pour into the baked quiche shell, and bake for about 35 to 40 minutes, until a knife inserted into the middle comes out clean. If the quiche is not brown, place under the broiler for a few minutes, or until nicely browned. Let stand for about 5 minutes before cutting.

Yield: 6 to 8 servings

Pie Shell for Quiche

1   cup crumbs
½   cup okara
       (salt—optional)
¼   cup Parmesan or other dry cheese,
       grated
¼   teaspoon paprika
3   tablespoons melted butter, margarine,
       or oil

Note that this is not a crust for sweet dessert pies. Preheat oven to 325° F.

Mix the dry ingredients well. Blend butter, margarine, or oil into the above mixture.

Pat into a quiche pan or a pie pan. Quiche pans have straight sides so try to press the crumbs up the sides of the pan. (A pie pan does just as well.) Bake for 10 to 20 minutes, or until the crust is just beginning to brown. Fill with quiche mixture, and bake according to recipe specifications.

Yield: 1 crust

A basic principle of soycrafting is to look on the soybean as if it were the udder of a tiny cow. In fact, the soybean has often been called "the cow of China," and for good reason. The oil, protein, and some of the minerals that are in regular milk are packaged neatly inside each soybean. All that's missing is water, and some heat to emulsify the mixture.

Soybeans are hydrophilic, which means they accept water readily.

Soak them in water overnight and each pound of beans will become about 2½ pounds of tiny udders, waiting to be warmed up and squeezed so they can give milk.

Robert Rodale. *Prevention.*

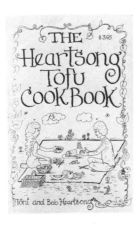

THE HEARTSONG TOFU COOKBOOK

Toni and Bob Heartsong. 1978 (revised edition).
Banyan Books, P.O. Box 431160, Miami, FL 33143. 92 p. paperback $3.95.

The hand-lettered text and amateurish line drawings of curlicues, flowers, and sunny faces give one the impression this book was put together by a family one sunny Saturday morning in May. The recipes, which utilize whole grains and basic staples, are praiseworthy for their creative avoidance of dairy and animal ingredients throughout the various soups, salads, desserts, dressings, sauces, and main dishes.

Pumpkin Tofu Pie

3¼ cups mashed pumpkin
1 cup soymilk
1 8-ounce cube of tofu
¼ cup arrowroot
¼ cup cashew butter
½ cup date sugar
1 tablespoon vanilla
1 tablespoon cinnamon
1½ tablespoons molasses
½ teaspoon salt

Mix ingredients in a blender. Pour into a pie shell and bake at 425° F. for 15 minutes, then at 250° F. for 1¼ hours. It will be firm when cooled.

Soysage

"Soysage" is the name coined by The Farm, a Tennessee commune, to describe its okara-based food. Okara is the crushed bean that is a by-product of making tofu. It is moderately high in protein, but often discarded or used as animal feed. "Soysage" is also the brand name of a spiced loaf sold by North Country Soyfoods, Inc., of Bethlehem, New Hampshire, under the Cloud Mountain label. The company recommends pan frying soysage for use in grilled sandwiches, with eggs and pancakes, in chili, or pizza, with sauerkraut, and in spaghetti sauce. Soysage, prepared by this recipe, combines grain protein (in the wheat flour and wheat germ) with the bean protein of okara.

2½ cups whole wheat flour
¼ cup wheat germ
¼ cup nutritional yeast
   recommended spices: sage, fennel seed, celery seed, dry mustard, garlic, cayenne pepper, allspice, cumin, oregano, cinnamon, nutmeg, caraway, curry
2½ cups okara
⅓ cup vegetable oil
1 tablespoon tamari
4 teaspoons honey
1 tablespoon vinegar
⅓ cup soymilk

Soysage can be shaped into a loaf and sliced for frying.

Combine whole wheat flour, wheat germ, yeast, and desired spices together. Add okara and mix well. Combine oil, tamari, honey, vinegar, and soymilk. Add this to the okara mixture and mix together with your hands.

Form the soysage into a loaf and place in an oiled loaf pan; or, fill an oiled tin can for a round loaf. Cover with aluminum foil. Steam on a rack over boiling water for about 45 minutes. Cool and remove from pan. Store in the refrigerator or freezer.

Prepare tempeh by cutting off slices or cubes and frying.

THE GREAT AMERICAN TOFU COOKBOOK

Patricia Gaddis McGruter. 1979. Autumn Press, 25 Dwight Street, Brookline, MA 02146. 124 p. paperback $6.95.

The Great American Tofu Cookbook is designed to place tofu prominently on the dinner plate of America, whether the diner be a meat lover, health enthusiast, penny-pincher, or vegetarian. Tofu's blandness and simplicity are given as its prime assets.

Tofu Stroganoff

1 tablespoon butter
3 tablespoons cold-pressed soy oil
2 pounds firm tofu, drained, pressed and cut into slices ½ inch thick and 2 inches long
   dash of shoyu (natural soy sauce)
½ cup whole green onions, thinly sliced
½ pound mushrooms, thinly sliced
1 tablespoon whole wheat or unbleached white flour
½ cup white wine
½ cup sour cream
1 tablespoon heavy tomato puree
1 teaspoon sea salt or to taste
   freshly ground black pepper to taste
2 tablespoons fresh parsley, finely chopped

In a heavy 12-inch skillet, heat the butter and 2 tablespoons of the oil over medium heat until butter foams; add the tofu. Sprinkle with shoyu and fry the tofu until lightly browned on both sides. With a slotted spoon, transfer to a plate. Pour the remaining oil into the skillet; add the onions and mushrooms and saute, stirring often, for 3 to 4 minutes, or until lightly browned. Gradually mix in the flour, wine, and sour cream, stirring constantly. Blend in the tomato puree, salt, and pepper. Add the fried tofu, coating thoroughly with the sauce. Cover and simmer for 2 to 3 minutes, or until heated through. Sprinkle with parsley and serve with noodles or brown rice.

Yield: 4 to 6 servings

THE SOYSAGE COOKBOOK
Maxine H. Cloud. 1979.
Cloud Mountain, Inc., RFD 1, Box
118, Hyde Park, VT 05655. 33 p.
paperback $3.95.

A charming book from the New England countryside that features 30 recipes for using soysage (the sausagelike patties made from okara, or soy pulp left from tofu-making) in sandwiches and main dishes, including burgers, omelets, spaghetti, pies, and vegetable casseroles. This slim book is of spiral-bound thick paper and features appealing illustrations on nearly every page. It's no doubt the only cookbook available devoted exclusively to soysage.

Dear Cooks: After we first tasted soysage, we knew it would become a real staple in our diet. It tastes great, it is a good source of protein, and it is versatile. To make an analogy—soysage can be to vegetarians as hamburger is to meat-eaters. It can be cooked quickly for meals-in-a-hurry, or it can be turned into a true delight for the most discriminating and skeptical of palates.

Soysage with Veggies in Cream Sauce

6 ounces soysage, broken into small pieces
½ onion, chopped
1 large carrot, sliced
1 stalk celery, sliced
1 medium to large potato, cubed heaping handful of fresh mushrooms, sliced
⅓ cup sweet peas (approximately) any other veggies you have lying around—green beans, turnips, parsnips, kohlrabi

In a sauce pan cook all the vegetables together in water until fork tender. Drain liquid into 2-cup measuring cup. Put veggies and soysage in a casserole.

Cream Sauce

Melt 4 tablespoons butter in a sauce pan. Add ¼ cup whole wheat flour mix to form a roux. Add enough milk to vegetable liquid to make 2 cups. Add to roux. Stir and cook over medium heat until thick and bubbly.

Pour over vegetables. Add sea salt, ½ teaspoon garlic powder, ½ teaspoon onion powder, and ½ teaspoon cayenne. Mix thoroughly. Bake at 350° F. for 20 to 30 minutes.

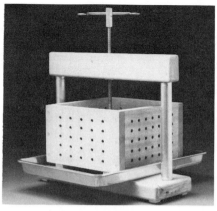

A larger press, suited to families who really appreciate tofu.

This handsome hand-thrown tofu press is marketed by CIS, 2813 Arizona NE, Albuquerque, NM 87110. The press, pressing cloth, and instructions are $20., or $27.50 with Rodale Press's The Tofu Cookbook.

NUTRITIONAL VALUES OF TOFU, MEAT, AND CHEESE
(4 oz. or ½ c. serving)

	Calories	Protein (g.)	Fat (g.)	Cholesterol (mg.)
Tempeh (fresh)	179	22	9	0
Tofu (medium)	82	8.8	4.9	0
Tofu (firm)	100	12	6	0
Ground beef (lean, 10% fat)	202	23.4	11.3	106
Chicken (light meat, without skin)	168	35.6	3.9	90
Cottage cheese	120	15.6	4.8	22

Sources: William Shurtleff and Akiko Aoyagi, *The Book of Tempeh* (Harper & Row, 1979). Shurtleff and Aoyagi, *The Book of Tofu* (Autumn Press, 1975). *Nutritive Value of American Foods,* Agricultural Handbook #456. "Cholesterol Content of Foods," *Journal of the American Dietetic Association.*

A small (one-pound) tofu press, marketed by The Learning Tree, P.O. Box 76, Bodega, CA 94922.

Vic Sussman made this large tofu press to handle the appetite of his vegetarian family. The Sussmans are featured in the Grains and Juicing chapters.

Before rushing your first batch of tofu into a recipe, sample it fresh and still warm with a dash of tamari and perhaps sliced scallions.

Tempeh

Centuries ago, the Indonesians developed a way to turn rock-hard, protein-bound, unappetizing soybeans into tempeh. A friendly mold is encouraged to grow on the beans, and a microbiological wonder takes place. Fried tempeh not only tastes better than you'd think it had any right to, but it is more nutritious as well. As described below by Keith H. Steinkraus of the New York Department of Food Sciences and Technology, enzymes from the mold make the proteins, lipids, and other components more accessible to us.

■

Everyone knows the great contribution that cheese, pickles, pickled fish, sauerkraut, fermented sausages, yogurts, fine wines and beers, and sourdough and yeast breads have made to the diet of the Western world. Few Americans, however, know the flavor and aroma of Eastern fermented foods, such as genuine soy sauce, miso, tempeh, idli, and

dosa. Much of the soy sauce sold today is produced by hydrolyzing (digesting) soybeans with hydrochloric acid. This acid splits the proteins and lipids and other soybean components, yielding, when neutralized, a salty, tasty liquid that imparts a meatlike flavor to bland foods. However, the genuine, fermented product, known as shoyu, has an even richer flavor, enhanced by aromatic esters, acids, and alcohol, all of which are largely absent in the acid-hydrolyzed product. To produce shoyu through fermentation, soybeans are soaked, steam cooked, coated with ground roasted wheat, inoculated with a special mold (Aspergillus oryzae), and incubated until the beans are over-grown by the mold. Then they are covered with an approximately equal volume of 20 percent salt brine. Over the next 6 to 12 months enzymes from the mold hydrolyze the protein, liberate peptides and amino acids, split the fats to free

fatty acids, and generally solubilize much of the soybean and wheat. At the same time, a Lactobacillus and a yeast are usually present; they produce acid and some alcohol, respectively. The acids combine with the alcohol to produce esters which create a desirable aroma. By filtering the digested soybean-wheat mixture to remove insolubles, fermented soy sauce is obtained. Once one becomes accustomed to the genuine, fermented product, it is difficult to be satisfied with the artificial, acid-hydrolyzed product.

The invention of shoyu thousands of years ago by the Chinese was of great significance because it added, to their generally bland and relatively low-protein rice diet, meatlike flavor and missing amino acids.

A product closely related to shoyu is Japanese miso. It is a high-protein shoyu-flavored paste made from soybean with either rice or bar-

(Continued on page 108)

A SIMPLER TEMPEH

Perhaps novice tempeh-makers would do best to begin with a type of tempeh that eliminates the most difficult step of cracking and hulling the soybeans. We have found that an excellent product can be made from canned chickpeas, also known as garbanzo beans. To make this easy tempeh, begin by draining and rinsing the peas, then cutting them into small bits. Dry the pieces in the oven for 20 minutes, at 250° F. Cut the beans again, sifting the fine meal away from the granular bits. Then, inoculate the beans as you would soybeans.

Researchers who tried the chickpea tempeh found the texture firm and the flavor after cooking to be very agreeable. Each portion contains 18.8 percent protein, 6 percent fat, and 2.3 percent fiber. Tempeh made from fava beans contains 31.2 percent protein, 1.3 percent fat, and 1.7 percent fiber.

Once you've tried elementary tempeh, become comfortable with the incubation procedure, and experimented with tempeh's many culinary uses, advance to preparing soybean tempeh. Each serving is 38 percent protein, more than twice the protein of chickpea tempeh.

Carol Keough

Tempeh Cookies

⅔ cup peanut tempeh, chopped
1 cup wheat flour
1 teaspoon cinnamon
1 tablespoon molasses
1 tablespoon honey
¼ teaspoon vanilla extract
3 tablespoons butter, melted
¼ cup yogurt
1 egg
¼ cup raisins
¼ cup coconut

Combine tempeh, flour, and cinnamon and mix well. Combine molasses, honey, vanilla, butter, yogurt, and egg and blend well.

Mix dry and wet mixtures together well and stir in raisins and coconut. Mix thoroughly.

Drop by the teaspoon onto a greased cookie sheet. Bake at 350° F. for 20 minutes.

Yield: 30 cookies

(Continued from page 107)

ley. Soybeans are soaked, steamed thoroughly, and mixed with rice that has been over-grown previously with the shoyu mold. Again, mold enzymes hydrolyze the proteins, lipids, and other components, producing the soy sauce flavor. The salt concentration is lower, and miso fermentation can be completed in as short a time as one week. Such miso must be consumed rather quickly, since its keeping quality is limited. When miso is made with a higher salt content, it can be kept for a longer period of time. Most Japanese eat miso in the form of a soup for breakfast nearly every day. As meat prices rise and per capita consumption decreases, Americans might find it desirable to add some miso to their diets as a good source of flavorful protein.

Keith H. Steinkraus. Courtesy *New York's Food and Life Sciences Quarterly.*

Tempeh

2½ cups whole dry yellow soybeans, washed and drained
4½ quarts water (18 cups)
1½ tablespoons white vinegar
1 teaspoon tempeh starter

Combine soybeans and 8 cups of water in a large pot, and bring to a boil. Cover, remove from heat, and let stand for 10 to 16 hours. Or, simmer covered for 20 minutes, then remove from heat and let stand for 2 hours.

Drain the beans and rinse well. Cover the beans with water and rub them between your hands to remove the hulls. This will also split the beans. As the hulls rise to the surface of the water, skim them off and discard. You will have to drain the beans and add more water several times to clean the beans of their hulls. A few hulls remaining on the beans will not matter, but the majority should be removed. It is important to remove the hulls because they prevent the growth of the tempeh mold.

The dry soy beans are first ground.

Soaked or canned chickpeas are cut into small bits.

Either bean is then inoculated by sprinkling with starter culture.

When finished, the tempeh should look like the pan on the right—firm and covered with a uniform layer of white. If the tempeh looks like the pan on the left, return it for further incubation.

Allow the beans to cool to room temperature, then pat with a towel to remove excess moisture. Combine the dehulled beans, 10 cups of water, and the vinegar in a large pot. The vinegar lowers the pH of the beans, ensuring better mold growth. Boil, uncovered, for 45 minutes. Drain the beans well in a colander.

Transfer the beans to a baking tray, lined with toweling. Spread them in an even layer, and allow to stand for 20 to 30 minutes until cool.

Transfer the beans to a large mixing bowl or another baking tray. Sprinkle tempeh starter with a salt shaker evenly over beans and mix for about 2 minutes with a large spoon to distribute starter evenly. Tempeh starter can be purchased from: Farm Foods, 156 Drakes Lane, Summertown, TN 38483: and Microcultures, Drawer A, Bayville, NJ 08721.

Spread the beans ½ inch deep in a shallow pan and cover with foil. Punch several very small holes in the foil at ½-inch intervals so excess moisture can escape. Place the trays in the incubator, switch on the heater or light, and incubate the tempeh for 24 to 30 hours at about 85° to 100° F.

After that time, the tempeh cakes should be solidly bound with white mold, possibly streaked with gray or black. Tempeh that has turned completely black or smells strongly of ammonia should be discarded. A thin slice of good tempeh will feel firm and flexible and will not crumble easily. If the tempeh isn't quite ready, the white growth will not be full and the bean pieces will be very loosely held together and will crumble easily. If part of the cake seems finished, cut off that section and return the rest for further incubation. Portions not intended for immediate cooking can be stored in the refrigerator or frozen for later use.

Tempeh becomes a lot less mysterious when you pop a few slices into a skillet and fry it up as you would meat.

Constructing the tempeh incubator

Materials
1   Styrofoam picnic basket 16 to 18 inches by 10 to 12 inches by 12 to 14 inches*
1   ceramic light socket
1   7½-watt clear bulb
1   6-foot extension cord
    electrical tape

Procedure
1)   With a pencil, punch a hole in the middle of the cover of the Styrofoam picnic basket.
2)   Thread the two leads of the ceramic socket through the hole so that the socket can be attached to the underside of the cover.
3)   Tape the socket on both sides of the cover.
4)   Splice the leads from the socket to the extension cord.
5)   Screw the bulb into the socket.

A simple tempeh incubator can be made of a foam ice chest with a low-wattage bulb for heat.

Preparing the pan

Materials
aluminum foil pan 8¾ inches by 5¾ inches by 1⅛ inches (available in most hardware and department stores)
aluminum foil sheet to cover the pan
bottle caps or small lids

Procedure
1)   Perforate the entire bottom of the pan. Space the holes ¼ inch apart.
2)   Perforate enough aluminum foil to cover the top of the pan.
3)   The covered pan of inoculated beans should be set on the bottle caps inside the incubator. This allows plenty of air to the tempeh on all sides.

*Be sure to use the correct size Styrofoam box to maintain the optimum temperature (between 92 and 96° F.).

THE BOOK OF TEMPEH

William Shurtleff and Akiko Aoyagi.
1979.
Harper & Row, 10 E. 53rd Street,
New York, NY 10022. 245 p.
hardcover $16.95.

Tempeh, an Indonesian fermented soyfood, is brought to a new home and the American cuisine by soyfoods' best friend, Bill Shurtleff. We are served Tempeh Chops with Applesauce, Batter-Fried Tempeh Crisps, Breaded Tempeh Cutlets, Delectably Crunchy Tempeh Condiment, and scores of other tantalizing American and Indonesian tempeh dishes. This is a rich and informative book, replete with world food use polemics, scientific appendices, tempting recipes, fascinating cultural information, nutritional studies, processing details and instructions, and delightful, vivid illustrations by Akiko Aoyagi.

Masters of their own culture and guardians of their cuisine, the people of Yogya also have a great university and a traditional market that exceeds in size and color even the fabled markets of West Africa. Here was tempeh in the greatest abundance and variety we had seen, each small cake wrapped in shiny green pieces of banana leaf. Each morning we went to the marketplace to bargain (in Indonesian) for our breakfast fruits—fresh papayas, mangosteens, salaks, durian, and tiny bananas. The laughing women in their colorful batik sarongs, the delightful melange of earthy aromas, the great crimson mounds of fresh chilies, and the grains and beans in bulging handwoven baskets were a feast for the senses.

Already tempeh has established a firm foothold in North America. At least 22 tempeh shops are in operation in the United States and two in Canada, and more than half of these are run by non-Indonesian Americans—the first time in Western history that tempeh has been produced by indigenous craftsmen for local consumption.

Tempeh Cabbage Rolls

This recipe is from the Glass Onion restaurant, Lincoln, Nebraska.

1	clove garlic, minced
4 to 6	green onions, chopped
8	ounces (2½ cups) tempeh, thawed and diced or crumbled
½	cup celery, chopped
½	cup parsley, chopped
1½	tablespoons basil
1	tablespoon dill
2 or 3	dashes Tabasco
2½	cups cooked brown rice
	steamed cabbage leaves
2	cups tomato sauce
1	tablespoon olive oil (sprinkled over rolls)

Saute garlic and onions in vegetable oil. Add tempeh and brown. Stir in celery, parsley, basil, and dill. Add 2 or 3 dashes Tabasco to taste.

Stir in the cooked rice and remove from heat. Spoon the mixture into cabbage leaves and roll. Put into an oiled baking dish, then pour on tomato sauce and sprinkle with olive oil. Bake at 350°F. for 20 to 30 minutes or until cabbage leaves are tender.

Yield: 6 to 8 servings

Tempeh can be incubated without supplemental heat in a plastic bag, perforated with small holes.

A more elaborate incubator of wood, made by Rodale Press.

Tempeh Chili

3 to 4	tablespoons oil
1	large onion, chopped
2	green peppers, chopped
1	large can tomatoes, or 4 cups fresh tomatoes, coarsely chopped
5	teaspoons chili powder
1	pound pinto bean tempeh
1¼	cups tomato juice
	cayenne, to taste

Saute chopped onion in oil. Add green pepper, canned tomatoes, chili powder, and crumbled tempeh. Thin with tomato juice as desired. Add cayenne to taste, and simmer 20 to 30 minutes. Serve with grated Monterey Jack or cheddar on top, or as is.

Yield: 6 to 8 servings

Soyfoods glossary

Agé. Also called deep-fried tofu cutlets, these are thinly sliced strips of tofu that, after deep-frying, puff up to make pouches.

Curds. The cloudlike coagulated soy protein solids that are pressed into tofu. Sweet and delicious, curds are traditionally savored with a dash of tamari.

Hamanatto. A traditional Japanese fermented condiment, made from whole soybeans inoculated with a mold. Also called savory soy chunks or nuggets, the beans look like dark raisins and are served as salty appetizers.

Kinako. Roasted soy flour, ground from dry-roasted soybeans. With its nutty flavor, kinako is used as a condiment or mixed with oil or butter to form a nut paste.

Miso. A fermented seasoning made over a period ranging from two months to three years, depending on the variety of miso, from soybeans, grain (either rice or barley), salt, water, and mold starter Miso has the consistency of peanut butter and is used as a condiment, soup starter, or spread.

Natto. Fermented whole soybeans, inoculated with a bacterium that produces a sticky surface. Natto is used as a topping for rice dishes, simmered in soups, and served in salads.

Nigari. The mineral-rich salt remain-

(Continued on page 112)

Pinto bean tempeh: Lincoln, Nebraska

Pinto bean tempeh has a more familiar taste than the soybean variety, especially when used in bean soup, chili, and casseroles. Many who have eaten this chili at the snack bar of the Open Harvest co-op in Lincoln, Nebraska, have had difficulty believing no meat was used, presumably because of the hearty flavor. Pinto bean tempeh also tastes great fried, perhaps even better than soy tempeh.

You can try making tempeh from other beans, as well, using the technique below. Cooking time and proportion of water to beans can be adjusted through trial and error.

First crack 18 cups of pinto beans in a grist mill. A grist mill is not normal kitchen equipment, but can come in handy. I got mine through the Sears farm catalog, and I power it with a 3-horsepower electric motor on 220 volts. You may wish to try steel burrs in a hand mill. Or you can cook the beans till semi-soft, then put them through the mill. Or cook until quite soft and process in the blender. Whatever kind of mill you use, be sure no flat beans pass through without getting cracked, as they will not "temp."

Winnow the beans to remove loose hulls by pouring them between two large pans in a gentle, steady breeze, or in front of a large window fan.

Boil 21 cups of water in a large, thick cooking pot that will spread the heat and prevent scorching. A big pressure canner (without the lid) will work fine.

When the water is in a full rolling boil, pour in the beans. Boil for eight minutes, stirring constantly. Hulls that rise to the rim should be removed with a spoon. Remaining hulls do not support the growth themselves. That disadvantage is well compensated,

Gale Randall at his mill.

however, since the hulls provide much of the familiar taste that makes pinto bean tempeh so versatile in cooking. Hulls also provide air spaces that help the growth of rhizopus.

Cook the beans until they are as dry as possible without burning. Spread them in a tray, allow to cool, and inoculate as for soy tempeh.

Pack the beans in perforated containers and incubate at 90° F. for 24 hours. To retard taste degradation during storage, blanch the tempeh thoroughly in boiling water 5 or 10 minutes. Smaller pieces take less time. Use a slotted utensil to remove the tempeh. Package and freeze.

The tempeh will not be as firm as that made from soybeans, but for use in casseroles and soups, its ability to be sliced is not important. Slice pinto tempeh a bit thicker—at least ¼ inch—and use a thin, serrated knife.

Gale Randall

Gale packs tempeh into casings so that the final product can be sliced and fried in the manner of sausage.

THE RODALE TEST KITCHENS— AMERICANIZING SOYFOODS

Rodale Press wants to popularize soyfoods. They reason that soybeans have no cholesterol, are low in fat, can be easily digested, and are thought to promote health in other, newly discovered ways. This all appeals to the head, but soybeans need a tad of work before they become appealing to the palate.

Tom Ney, Rodale Press's Director of Test Kitchens and Food Services, samples a cup of experimental soy coffee

The Rodale Test Kitchen staff is preparing dozens of the basic soyfoods to find the best processing methods, the best soybean varieties, and the best recipes for each item. The kitchen experimentation includes sprouting; grinding meal and flour; roasting; making milk products, tofu, and tempeh; and finding uses for okara and whey. In addition, Test Kitchen technicians have developed a practical tempeh incubator that can be built at home for less than $10, and a soymilk extraction press for less than $25—both made expressly for efficient home kitchen use. Hopefully, the result of all this work will be a wealth of soyfood recipes to broadcast in the Press's magazines and books.

Meanwhile, agronomists at the Rodale Organic Gardening and Farming Research Center near Maxatawny, Pennsylvania, are developing small soybeans for sprouting, and work with varieties from around the world. When the work at the experimental farm is completed, the seeds will be turned over to seed houses for stock increases and popular distribution.

(Continued from page 110)
ing after seawater has been evaporated and the sodium chloride (table salt) removed. Nigari is traditionally used to coagulate soymilk to form tofu. Calcium sulfate (gypsum) is also used.

Okara. Literally, "honorable hull." The fibrous, insoluble by-product from making soymilk and tofu, left behind after the soymilk has been extracted from the ground soybean puree. This pulp is used to feed livestock, in baking, in making tempeh, and as a base for soysage, a food developed by The Farm commune in Tennessee.

Shoyu. A naturally fermented soy sauce, made over a period of a year from soybeans and cracked, roasted wheat in equal portions, water, salt, and mold starter. Commonly confused with tamari.

Soy coffee. A hearty, aromatic beverage brewed from roasted, cracked soybeans. It contains no caffeine.

Soyfoods. A new generic term that refers to the panoply of lightly processed soybean foods made for immediate human consumption, including tofu, tempeh, miso, soymilk, and all their by-products.

Soymilk. The rich, milklike water extract obtained from soybeans that have been soaked in water, ground into puree, and cooked. Soymilk is used to make tofu, and is a good beverage in its own right.

Soy nuts. Peanutty deep-fried soybeans that are soaked in water before frying.

Soysage. An okara-based food developed by The Farm commune in Tennessee.

Soy sauce. In the West, this term has come to mean the inferior-quality sauce made by the rapid, chemically induced fermentation of soybeans, salt, and water, usually with sweeteners and colorings added. Soy sauce is also a generic term that includes tamari and shoyu.

Sufu. Known also as bean cake cheese, this soft, cheeselike food is made by inoculating cubes of tofu with a fungus, followed by brining and aging.

Tamari. Fermented soy sauce, made from soybeans, salt, water, and inoculated grain (koji) but no wheat. Miso-tamari is the delicious liquid that accumulates at the top of miso during its aging process.

Tempeh. A traditional fermented food of Indonesia, made from tender-

(Continued on page 115)

BEANS

. . . My beans, the length of whose rows, added together, was seven miles already planted, were impatient to be hoed, for the earliest had grown considerably before the latest were in the ground; indeed they were not easily to be put off. What was the meaning of this so steady and self-respecting, this small Herculean labor, I knew not. I came to love my rows, my beans, though so many more than I wanted. They attached me to the earth, and so I got strength . . . why should I raise them? Only Heaven knows. This was my curious labor all summer,—to make this portion of the earth's surface, which had yielded only cinquefoil, blackberries, johnswort, and the like, before, sweet wild fruits and pleasant flowers, produce instead this pulse. What shall I learn of beans or beans of me? I cherish them, I hoe them, early and late I have an eye to them; and this is my day's work. It is a fine broad leaf to look on. My auxiliaries are the dews and rains which water this dry soil, and what fertility is in the soil itself, which for the most part is lean and effete. My enemies are worms, cool days, and most of all woodchucks. The last have nibbled for me a quarter of an acre clean. But what right had I to oust johnswort and the rest, and break up

their ancient herb garden? Soon, however, the remaining beans will be too tough for them, and go forward to meet new foes.

Those summer days which some of my comtemporaries devoted to the fine arts in Boston or Rome, and others to comtemplation in India, and others to trade in London or New York, I thus, with the other farmers of New England, devoted to husbandry. No that I wanted beans to eat, for I am by nature a Pythagorean, so far as beans are concerned, whether they mean porridge or voting, and exchanged them for rice; but, perchance, as some must work in fields if only for the sake of tropes and expression, to serve a parable-maker one day. It was on the whole a rare amusement, which, continued too long, might have become a dissipation. Though I gave them no manure, and did not hoe them all once, I hoed them unusually well as far as I went, and was paid for it in the end, "there being in truth," as Evelyn says, "no compost or laetation whatsoever comparable to this continual motion, repastination, and turning of the mould with the spade." "The earth," he adds elsewhere, "especially if fresh, has a certain magnetism in it, by which it attracts the salt, power, or virtue (call it

either) which gives it life, and is the logic of all the labor and stir we keep about it, to sustain us; all dungings and other sordid temperings being but the vicars succedaneous to this improvement." Moreover, this being one of those "worn-out and exhausted lay fields which enjoy their sabbath," had perchance, as Sir Kenelm Digby thinks likely, attracted "vital spirits" from the air. I harvested twelve bushels of beans.

Plant the common small white bush bean about the first of June, in rows three feet by eighteen inches apart, being careful to select fresh round and unmixed seed. First look out for worms, and supply vacancies by planting anew. Then look out for woodchucks, if it is an exposed place, for they will nibble off the earliest tender leaves almost clean as they go; and again, when the young tendrils make their appearance, they have notice of it, and will shear them off with both buds and young pods, sitting erect like a squirrel. But above all harvest as early as possible, if you would escape frosts and have a fair and saleable crop; you may save much loss by this means.

Henry David Thoreau. *Walden.*

THE SOYBEAN BOOK

Phyllis Hobson. 1978.
Garden Way Publishing, Charlotte,
VT 05445. 172 p. paperback $5.95.

This versatile guidebook takes you from seed to kitchen to dinner plate and back to the garden again. Author Hobson starts by explaining the growing and harvesting of soybeans in the garden, then tells how to make tofu, soymilk, tempeh, sprouts, yogurt, and other soyfoods. She gives 275 recipes for breads, soups, salads, candies, cookies, puddings, and stews, and returns to outdoors to show us how to use soybeans as animal feed and soil improvements. Useful, refreshing, and thorough, *The Soybean Book* is a reliable beginner's handbook for working with soybeans.

Inoculation

Seed may be inoculated with soil taken from another part of the garden or a field in which soybeans (any variety) have been grown successfully. Or you may use commercial inoculant powder, which is sold by seed dealers. Be sure to buy inoculant made specifically for soybeans. Soybeans cannot be inoculated with bacteria from any other legume.

To coat the soybeans with the bacteria-rich soil or commercial powder, first moisten the seed with water in which sugar or molasses has been dissolved. One-half cup of sugar dissolved in one quart warm water will moisten up to one bushel of seed. Spread the seed out and dust it with the commercial inoculant listed on the package or with one gallon of finely sifted soil. Mix well by hand or with a hoe until the seeds are well coated.

THE FARM VEGETARIAN COOKBOOK

Louise Hagler, editor. 1978 (revised edition).
The Book Publishing Co., 156 Drakes Lane, Summertown, TN 38483. 223 p. paperback $4.95.

Nearly two-thirds of the text is devoted to preparing and cooking with tofu, soymilk, tempeh, and okara. Instructions for home production are accompanied by clear, explanatory photographs. This is a vegetarian cookbook, using no meat, fish, poultry, eggs, or animal products in its wide range of recipes: soybean burritos, tempeh, lettuce and tomato burgers, honey banana ice bean, deep-fried tofu with parsley garnish, tofu "grilled cheese" sandwiches, fried tempeh strips, rich soymilk drinks.

Banana Smoothie

1   cup soymilk
1   cup ice
3   medium-size bananas
2   tablespoons honey
    dash of salt

Combine all ingredients in a blender and blend until smooth. Serve immediately.

Banana Smoothie #2

A good smoothie can be made with frozen bananas instead of ice. Peel and freeze bananas in a plastic container or plastic bag. Start blending 2 cups soymilk, 2 tablespoons honey, and dash of salt in blender and add about 4 small frozen bananas. Drop in one banana at a time. The soymilk will get thick and frothy and should be served immediately. More honey may be added if you want it sweeter.

FIRST ENCOUNTERS WITH SOYFOODS, FONDLY RECALLED

About ten years ago, when I had just become a vegetarian, my roommate brought home a few cans of bean curd, as the label called it, imported from Japan. For breakfast, we scrambled the bean curd, or tofu, with chickpeas and onions, and topped it all with ketchup. I had no idea what tofu was, or how it was made, but I made certain that when it came my turn to shop, I bought enough cans of the fascinating white substance—watery, firm, vaguely chewy, bland, light, altogether surprising—to last us a week.

Three years later, on a visit to a basement tofu shop in Boston's Chinatown, I saw how this food is made. The dimly lit room held a fascinating scene of running water, steam, foam, barrels of plump, yellow soybeans, stone grinders, shining cauldrons, and workers crowded by all the equipment. Standing at the doorway, I breathed in the quiet hum, aroma, and whirr of tofu-making. One man leaned over a barrel of soymilk, a cigarette with a long, faltering ash dangling from his mouth. I bought a pound of fresh, unpressed curds and a big cake of tofu. I remember an unforgettable light sweetness, mildly reminiscent of chicken.

Finally, Bill Shurtleff published his milestone book on tofu and I was ready to make tofu in my kitchen. The first aroma was heady—a grassy, beany, even paintlike smell, something like linseed oil—as the wet beans were ground into puree by a flour mill. As the puree began to simmer on the stove, the smell became the delicate chickeny aroma of the Oriental tofu shop. I experienced the magical transformation of hot soymilk into cloudlike white curds, triggered by adding the curdling agent, nigari. Use too much, and you have a bitter, dense curd, while just the right amount will yield a bouquet of little cloud puffs of curds and an evanescent taste that never makes it to the store shelf. The flavor of just-made tofu fades within hours after it is made, and can be tasted at its best as unpressed curds with a dash of tamari. Next best is the warm, trembling cake, before it is stored in water.

Richard Leviton

SOYFOODS MAGAZINE

Soycrafters Association of North America, Sunrise Farm, Heath Road, Colrain, MA 01340. Quarterly, 64 p. $15 yearly.

Soyfoods covers production, marketing, and research for the soyfoods industry. The magazine works to articulate and encourage the growing awareness of soyfoods in the American marketplace and to draw together the strands of the world's emerging soybean network: soyfoods producers, researchers, food distributors, and the consumer.

Soy Deli Update

The Yellow Bean, one of five soyfoods delicatessens nationwide, is situated in a neighborhood Timothy Huang describes as "borderline between a rich suburb, Grove Point, and the East Side of Detroit, the black ghetto." As for clientele, "We get everybody—Orientals, blacks, rich, poor, a good cross section of what there is." While the deli is not currently supporting itself, grossing about $600 for a six-day week, the Huangs expect the introduction of hot consumable items to greatly improve the picture. Carol Huang placed an advertisement at a nearby high school for a course on vegetarian soybean cooking and nutrition and received six pages of interested registrants. Timothy plans to expand the Yellow Bean's wholesale produce line because, in some instances, their customers include restaurants that have deli bars that could handle bulk soy mayonnaise and tofu salads. The Huangs also expect to retail tempeh burgers and are considering collaborating on a tempeh operation in their basement with a tempeh-maker from northern Michigan.

FUTURE FOOD: ALTERNATE PROTEIN FOR THE YEAR 2000

Barbara Ford. 1978. William Morrow & Company, 105 Madison Avenue, New York, NY 10016. 290 p. paperback $4.50, hardcover $10.95.

In the context of worrying about how the Earth's billions will be fed, the author uncovers many curious things about the diet of the affluent American. Of particular interest are the chapters on beans and peanuts.

RY

Every country has its own legume favorites. In India, where beans and peas are called dal or dhal, the most popular beans are several kinds known as grams, the pigeon pea, the chickpea, and the lentil. The cowpea is the favorite in Africa, along with the peanut. Central America's most popular bean is the black bean, while kidney beans lead the list in Mexico. The soybean, as might be expected, is the number one bean in Japan, followed by the mung bean. In Europe, people eat lentils, black-eyed peas, peas, and broad beans, among other species. Beans are often eaten whole, the way we eat them, but in various countries they are also served mashed, ground to a powder, fermented, or extracted into a drink or watery solid product like tofu.

Our own bean consumption in the United States is only about six pounds per year (7.4 grams, or 0.27 ounces, per day), a figure which has been slowly falling for many years. Beans are a much bigger food item in developing countries and even in some highly developed nations such as Japan. The Japanese eat about 64 grams (2.38 ounces) a day. Since the average-size bean weighs about a gram, that means 60 beans spread out over three meals a day. Indians eat about the same amount. Some people eat many more. In a world survey

of bean-eating habits, one Indian family was recorded as eating more than 420 grams (14.7 ounces) of beans per day per person. The Chiga tribe of Uganda ate about 400 grams of beans per day per person. In most countries, there is a definite relationship between family income and bean eating. As income rises, people eat less beans. But in South America, everyone eats lots of beans—rich, poor, and middle class alike.

Why don't Americans eat more beans? Bean growers who would like to sell more beans, and nutritionists who would like to see Americans getting more of their protein from cheap and efficient plant protein sources, often ponder the question. Part of the answer seems to be that we are a rich people, comparatively speaking, and bean consumption, as indicated above, tends to fall with rising income. In a 1958 survey of bean eating in the United States, people with the lowest incomes in both urban and rural areas ate four times as many beans as people with the highest incomes. For the rich, or aspiring rich, meat becomes a sort of status symbol, attractive not only for its good taste and protein value, but for its ability to evoke envy and make its consumers feel as though they are enjoying the better things of life. Silly? Of course. But our steak restaurants, full of executives on expense accounts and people celebrating special events, are built on creating such psychic satisfactions.

FLATULENCE FROM DIFFERENT LEGUMES

	Ratio*
Phaseolus vulgaris	
California small white	11.1
Pinto	10.6
Kidney	11.4
Phaseolus lunatus	
Lima, Ventura	4.6
Lima, Fordhook	1.3
Phaseolus mungo	
Mung	5.5
Glycine max	
Soya, Lee, or yellow	3.8
Arachis hypogaea	
Peanut	1.2
Pisum sativum	
Pea, dry	5.3
Pea, green	2.6
Bland test meal	1.0

*Ratio of flatulence from test meal of 1000 g. (dry weight) for a three-hour period measured from four to seven hours after ingestion as compared to a bland test meal.
Source: Murphy, 1972.

(Continued from page 112)
cooked soybeans inoculated with the *Rhizopus* mold. Tempeh is high in protein, the highest vegetarian source of vitamin B$_{12}$, and is delicious when sauteed or fried.

Tempeh starter. A mold, *Rhizopus Oligosporus,* that initiates the fermentation process in soybeans for tempeh-making.

Tofu. Also known as bean curd and soy cheese. Tofu is a firm custardlike cheese made from soybeans, water, and nigari. It was invented in China over 2,000 years ago, to become a staple of East Asia, especially Japan.

Whey. The watery residue in tofu-making that is left behind when the soymilk is coagulated into curds. It's a clear, yellowish liquid with a sweet taste, useful as an alkaline cleaning agent, or as a soup base.

Yuba. Also called bean curd skin or

In Indonesia, tempeh is fermented in large leaves. (Photo courtesy New York's Food and Life Sciences Quarterly.)

tofu robes, yuba is the thin skin that forms on the surface of simmering soymilk (similar to the skin that forms on simmering dairy milk). Yuba is skimmed off and hung in sheets to dry; it is then used in cooking as rolls or pourches for vegetables, or added to soups.

Richard Leviton

THE ECLECTIC DIET

It may come as a surprise to readers of funky peasant-cuisine cookbooks—those paperbacks printed in brown ink on whole-wheat-colored paper—that not all peasant food systems are healthful. Baking, gardening, and cheese-making aren't practiced in the quest for holistic health, but out of necessity. This clears up the apparent contradiction of, for example, Yugoslavian families who both make yogurt daily and grow, dry, and roll their own intensely powerful cigarettes.

We of the affluent West are fortunate in having the freedom of means and thought to select the best systems of a culture, to put together our own cuisine.

RY

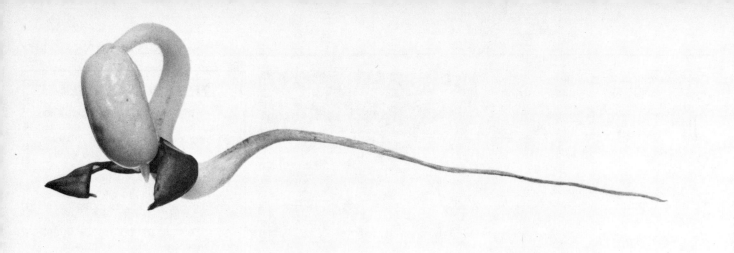

Sprouting
Your garden in a jar

Many of us have eaten bean sprouts for years, perhaps without knowing it, in dishes at Chinese restaurants. Only recently have Westerners learned what the Chinese have known for centuries: that sprouts are tasty, nutritious, and simple to grow.

Even so, sprouting hasn't exactly taken the Western world by storm. Sprouts are gaining recognition, but in the public mind they are still a curiosity, not a basic food that can be produced on top of the refrigerator or in a desk drawer at the office.

What is a sprout?

A sprout is a seed that has started to grow, or germinate. In its dormant state, a seed holds a tiny embryo packed in a hard, dry container of starch. When the seed gets the right temperature, a little water, and a little oxygen, the embryo begins to grow.

All growing things are miracles, but sprouts are special. They grow so rapidly (ready to eat in from two to five days) that you can virtually watch them change. The act of a tiny seed becoming a full-grown plant is played before your eyes.

As the seed sprouts, it feeds itself on the starch surrounding the embryo, and so it can grow for several days without nutrients from the soil. Even so, the sprout's own nutritional value increases. Both vitamin and protein content rise significantly.

Mung beans and alfalfa sprouts are most popular, but dozens of

seeds can be turned into tiny edible plants. Sprouted wheat tastes nutty and sweet. Radish sprouts have a crisp, tangy taste.

Probably the only seeds that shouldn't be sprouted are tomato and potato seeds. The seeds of these members of the nightshade family are poisonous in large amounts.

By trying a variety of sprouts, you'll learn that they are not only a garnish, but basic food—a novel form of vegetables. Add them to soups, breads, and main dishes, or eat them alone as salads or on sandwiches.

In her *Beansprout Book,* Gay Courter tells the story of a family of seven in Utah that survived a full winter—six long months—by eating nothing but sprouts. The family had apparently fallen on hard times, and were left with no alternatives. During this period, they spent a total of

$52.50 on food, and suffered no colds or other sicknesses. While an all-sprout diet may be extreme, this story demonstrates their value and versatility.

Nutritional blessings

Taste is not the only virtue of sprouts. Soybean sprouts have a vitamin C content far higher than unsprouted beans. Wheat and millet seeds contain 2.6 milligrams of vitamin C per 100 grams, but when sprouted they contain 15.0 and 23.8 milligrams of vitamin C, respectively.

During World War I, D. Cyrus French conducted a curious experiment on troops suffering from scurvy. He gave four ounces of lemon juice per day to one group, and four ounces of sprouted beans to another. After one month, just over half the soldiers on lemon juice were free from scurvy symptoms while 70 percent of the sprout eaters got well.

Sprouted oats have 13 times as much riboflavin (B_2) as their dormant form; sprouted wheat offers 4 times as much folic acid. A Yale University study on grains, peas, and beans showed that sprouting resulted in significant increases in almost all B

vitamins, with the amounts in some cases tripling or quadrupling.

Studies at the USDA lab in Madison, Wisconsin, concluded that sprouts contained 1.3 to 2.1 times as much protein as corresponding seeds. At the same time, they are low in carbohydrates because the growing sprout uses up the starch in the seed. Consequently, a cup of mung bean or alfalfa sprouts has only about 16 calories.

Sprouts are tasty additions to almost any food: soups, salads, sandwiches, and casseroles.

You can grow sprouts anywhere it's warm, including a desk drawer at work.

Save with sprouts

Home-grown sprouts are cheap. A pound of mung beans retailing for about $1.50 will produce about eight pounds of sprouts. Calculating that each pound will feed at least four people, each serving costs less than five cents. Soybeans, one of the most nutritious beans to sprout, can usually be purchased for less than $1 per pound. Some seeds—such as mustard, alfalfa, and radish—may cost considerably more, especially in small quantities. But keep in mind that a pound of seed renders eight pounds of highly nutritious food.

Taking our economic pitch a step further, sprouting also conserves the fuel needed to work and harvest fields, process food, haul it to market, package and store it. You save on your travel to the supermarket and back, as well.

Sprouts are less work than growing vegetables out back. There's no ground to dig, no weeds to pull, no pests to fight. You simply sow the seeds, water them a little, then sit back and let Mother Nature do the work. They'll grow year round, no matter what the weather.

Most sprouts are tasty raw. Those few that do require cooking take less time to prepare than their unsprouted cousins. They can be quickly boiled or stir-fried, and served alone or in combination with other vegetables or meats.

Seeds store easily, too. They are small, and don't need to be frozen or preserved to keep them from spoiling. One small cupboard can hold enough assorted seeds to feed your family for many months.

TD

SPROUTS VS. CANCER

Recent research suggests that sprouts may help prevent cancer. Scientists at the University of Texas System Cancer Center in Houston showed that the action of a potent cancer-causing agent on test bacteria was greatly inhibited by an extract of sprouted wheatgrass, mung beans, or lentils.

The scientists believe the active factor is chlorophyll, which may interfere with the enzymes that activate certain carcinogens. It is not clear whether sprout chlorophyll does this better than any other kind of chlorophyll, however.

The potential value of this discovery is also unclear. The researchers are proceeding with tests on mice and emphasize that "we have no evidence as yet that this material will prevent cancer in animals or humans. This will require a great deal more work over a period of at least two or three years."

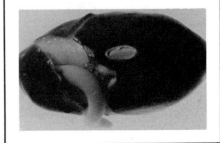

How to sprout

Sprouting is simple. The only things you need are seeds, a jar, a piece of cheesecloth and a rubber band, and clean water. Alfalfa is a good beginner's seed. It grows easily, and makes a crisp and tasty sprout for salads, sandwiches, or just snacking a pinch at a time.

You can buy the seeds in most health food stores or from a wide variety of mail-order houses. Use untreated seeds that are especially for sprouting. Other seeds are likely to contain chemicals, pesticides, or dyes which can make them poisonous as sprouts.

To begin sprouting, soak the seeds in warm water. Alfalfa does not require soaking, but soaking up to three hours will speed germination.

After soaking the seeds, pour off the water. (It has a mild flavor and some nutrients; you can use it in cooking or to water plants.) Rinse the seeds and put them in the sprouting jar. Stretch the cloth over the jar mouth and hold it in place with the rubber band. The cloth keeps sprouts from tumbling out during rinsing. Most seeds should be rinsed at least two times a day to provide water and to wash away the by-products of growth that encourage spoilage. Simply fill the jar with water, swish it around briskly, then drain it thoroughly.

A good temperature for sprouting alfalfa is a steady 70° F., out of

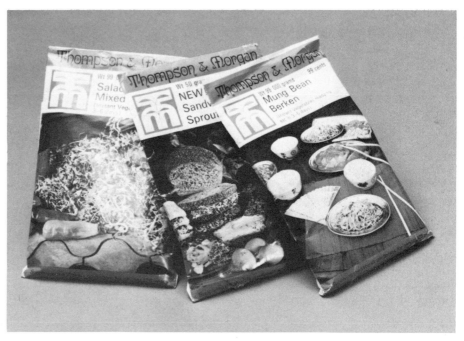

Sprouting seeds are often premixed for flavorful combinations.

direct sunlight. The sprouts can be harvested in three to five days.

Sprouts grown in the dark may benefit from a few hours of indirect sunlight before harvesting. They'll be able to produce additional chlorophyll, increasing their nutritional value and coloring them a pleasant green. Store harvested sprouts in the refrigerator to keep them crisp.

To begin the sprouting process, soak the seeds in a jar of warm water for several hours.

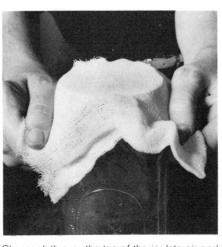

Cheesecloth over the top of the jar lets air and water through but keeps the seeds in.

The sprouts should be rinsed at least twice a day, but they'll thrive on more rinsings.

Variations on the technique

You don't even need a jar to sprout. A bowl, pie tin, sink strainer, or bean pot will do. If you use a bowl or pot, rinse the sprouts in a strainer, and keep them moist by putting a wet towel over the top of the container.

Virtually any container will work, but avoid those of wood or metal. Wood absorbs moisture, and may grow mold or mildew. Some metals may give sprouts a bad taste. Glazed pottery containers should be food-safe; some glazes incorporate lead, a highly toxic element that's easily absorbed by the body.

There is considerable disagreement over whether sprouts should be grown in darkness or light. Most sprout growers favor darkness, arguing that plant roots—which sprouts basically are—grow in darkness. Others contend that too much light will make the sprouts tough or give them a bitter taste from too much chlorophyll.

Light does have an effect on certain nutrients. Recent research at the University of Puget Sound indicates that sprouts grown in indirect light have more enzymes, vitamins, and protein than sprouts grown in the dark. (An exception to this was mung bean sprouts.)

Studies on soybeans and alfalfa sprouts showed that sprouts grown in light had considerably more vitamin C than those grown in darkness. Just the reverse is true, however, of riboflavin (vitamin B_2); even though con-

(Continued on page 120)

After rinsing, drain the sprouts thoroughly. Standing water in the jar can cause mold.

Alfalfa

The Spanish word alfalfa evolved from the Arabic al-facfacah, meaning "the best sort of fodder." The plant is used mostly for animal forage, but alfalfa sprouts make good fodder for human beings, too. The tiny golden seeds turn into long, thin sprouts with a delicate nutty sweetness. The seeds are a good source of vitamin K and many minerals.

Sprouts will be an inch long or more in three to five days. Given light a day before harvesting, the sprouts will turn an attrac-tive green, the work of chlorophyll. When the two tiny leaves turn bright green, the alfalfa crop is at its peak of flavor and nutrition.

They add light sweetness to sandwiches, salads, soups, and egg dishes. Alfalfa sprouts also may replace nuts in baking. Sprouts harvested at 24 hours (before the moisture content is too great) are a fresh addition to bread dough, higher in protein than wheat flour.

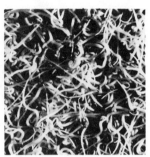

Almond

Like beans and seeds, nuts are plants-to-be and therefore will sprout. But we had no luck with almonds, even though many sprouting books blithely place them in the "what to sprout" list. The best we got, even after several trials, was a tiny white nodule on the tip of one nut.

If you'd like to have a go at it, here are the usual instructions. Presoak almonds overnight, then spread them on a damp, double-thick layer of paper towels in a glass or stainless steel tray. Cover loosely with a second layer of damp towels with enough room for air to circulate freely. Water the almonds by sprinkling them lightly, and resoak the top towels when they dry out. In three to five days, the sprouts are supposed to be ¼ inch long, and crisp and crunchy. If you succeed, the almond sprouts may be used in recipes calling for unsprouted almonds. Almonds are superior to most other nuts in protein content, and are rich in calcium and potassium.

Good luck.

(Continued from page 119)

Amaranth

Amaranth was a major grain crop of the Aztec Empire, but since then this plant with its beautiful showy flowers has been largely ignored. The protein-rich seeds, because of their composition of amino acids, come closer to perfect nutritional quality than any other grain.

The tiny seeds, no bigger than a pinhead, burst into ¼-inch sprouts in two days. The sprouts are a good source of calcium.

Sprinkle amaranth sprouts on green salads and soups. To add protein and flavor to baked goods, use young, small sprouts that won't introduce much water to the recipe.

tinually synthesized by the sprout, it is destroyed by light.

Most commercial mung bean producers favor darkness for aesthetic reasons: customers prefer white sprouts to green. In fact, mung sprouts are often bleached with sodium hypochlorite to make them still whiter (and to discourage mold growth, even though the chemical destroys some of the nutritive value).

Another disagreement among sprouters centers on how long seeds or beans should be soaked. Many books recommend simply soaking overnight. That is fine for mung beans and other harder beans, but most seeds need only two or three hours of soaking. Curiously, the length of time seeds or beans are soaked does not affect the quality of the sprouts or the time required to reach maturity. Soaking for too long, however, can cause seeds to lose some water-soluble nutrients.

Research at the USDA in Philadelphia indicates that boiling for very short periods (about 20 seconds) may eliminate the need for soaking and produce higher germination percentages. But it is questionable whether this is worth the extra effort, as nutrients may be lost in the process.

Some growers claim that sprouts should be rinsed as often as possible, up to five or six times daily. Others say two rinses are sufficient. In tests here at Rodale Press, we found that the number of rinses consistently affected neither yield nor taste.

In general, tap water is acceptable for soaking and rinsing. If your water is heavily chlorinated, and you are soaking the beans or seeds for a long period, you might want to use spring or distilled water.

What's the optimal temperature? A good range for most sprouts is 70° to 75° F., but cool-weather plants—wheat, rye, radishes, peas—will sprout well at temperatures as low as 50° F. In fact, one study showed that winter wheat sprouted at 41° to 46° F. had more than twice the vitamin C than if sprouted at 65° to 72° F. As a general rule, the higher the temperature, the faster the sprout will grow. But higher temperatures (above 80°

1 tablespoon =

SPROUTING TO REDUCE COOKING TIMES

Sprouting not only enhances the nutritional value of beans, it also cuts down on the time required to cook them. This chart compares cooking times for soaked beans with beans that have been sprouted three days.

Beans	Dry volume (c.)	Cooked volume (c.)	Boiling time (min.) Soaked beans*	Boiling time (min.) Sprouted beans
Baby limas	1	3	60	15
Black beans	1	2½	45	20
Black-eyed peas	1	2½	30	3
Chickpeas	1	2⅔	30	15
Great northern	1	2	60	45
Kidney	1	2½	90	30–40
Lentils	1	3	10	3
Mung beans	1	2⅔	7	2–3
Pinto beans	1	2½	30	15
Soybeans	1	2½	120	3

*The times given are for cooking beans uncovered in rapidly boiling water. With some beans, however, rapid boiling breaks the shell. If you want a whole bean, longer, slower cooking may be necessary.

F.) will also favor mold and spoilage.

The best time for harvest depends on your taste buds. Beans and grains can be harvested within 24 hours, or as soon as the sprout appears. Other seeds may take as long as six or seven days to reach maturity. If the seeds are sprouted too long, you will know it. They lose their flavor and become bitter. Also, vitamins and other nutrients may be lost if the sprouts get too old.

TD

Chia

Chia comes from a sage plant. The small silvery brown seeds are used by some cultures to make beverages. The pungent sprouts are rich in potassium, copper, calcium, phosphorus, magnesium, and iron.

Chia seeds become gelatinous when wet, and will mildew if they remain too damp, so they should not be presoaked. To sprout, sprinkle them onto water in a glass dish, using equal amounts of water and seed. Sprinkle with a little more water when the seeds become dry. They will be about ¼ inch long and ready for use in one or two days. At that length they can be ground in a blender and added to breads and pancakes. After four days the 1-inch plants are ready to become a part of dips and spreads, sandwiches, salads, and soups.

TROUBLESHOOTING CHART

Seeds don't sprout, or sprout poorly

- old or faulty seeds
- temperature too cold or too hot
- seeds not soaked enough to soften outer hull
- not enough oxygen getting to seeds
- seeds too crowded in container

Sprouts rot, mold, or mildew

- too much water in container; drain well, place dry towel at bottom after each rinse
- temperature too hot
- insufficient rinsing
- sprouts too crowded
- failure to remove seeds that don't sprout

Dry, withered, or tough sprouts

- too much direct sunlight
- not enough water

Sprouts taste bad

- sprouts harvested too early, or too late
- too much sunlight
- sprouts exposed to toxic fumes
- untrained taste buds

Fenugreek

Fenugreek seeds make aromatic tea sometimes used as a gargle for sore throat. Fenugreek is also often recommended for sprouting. But every time we tried, the seeds got slimy—and just sat there. Apparently some people have had success, though; several sprout books claim the fenugreek seeds sprout in three to five days, and may be harvested from one to three inches.

The sprouts are extremely rich in iron. If you can get them to grow, you'll discover their spicy bitter taste, said to be appropriate in curries and rice dishes.

From 2 table-spoons alfalfa seeds to 2½ cups sprouts in 4 days, for 22 ¢

Lentil

Lentil plants have been cultivated since prehistoric times for their edible seeds. Sprouted lentils can be used just like a vegetable, their robust taste replacing celery and green peppers in stews, soups, and casseroles. Lentils also are good tossed raw into salads, or steamed and served as a vegetable, seasoned with butter and spices.

Lentils offer substantial amounts of vitamin B, iron, and phosphorus.

1 tablespoon

Mung bean

Say "bean sprout" to someone, and chances are he'll think of mung beans. They're among the easiest sprouts to grow.

Mung beans make big thick sprouts—1½ to 2 inches long—in three to five days. They contain vitamins A, C, and E, choline and niacin from the B complex, and potassium, calcium, phosphorus, and iron.

They combine well with chicken, fish, and brown rice. Steamed and buttered, mung bean sprouts may be served as a vegetable. Their green crunch is also good raw, either in salads or as a snack.

1 tablespoon

- moist, but not wet
- at least 70°F.
- rinsed at least twice a day

Serious sprouting: Emmaus, Pennsylvania

When Sara Bell started cooking at an inner-city day care center in Allentown, Pennsylvania, she wanted to prepare all natural foods. The people in charge of the center were receptive to her suggestions, so every day Sara made natural foods meals for about a hundred children. Sprouts were one of her staples.

Sara sprouts only at home now, but her method is the same whether she's sprouting for 2 or 102. She uses the basic sprouter: a quart jar with a cheesecloth cover. She tries to rinse the seeds three times a day if possible, but a busy schedule often dictates only two daily dousings.

To use sprouts regularly, you have to keep track of when each batch will be ready. Sara gets around this potential problem by making weekly meal plans, a habit acquired at the day care center.

Sara uses tap water for rinsing, although she drinks only bottled water. "I'd like to use bottled water to rinse my sprouts, too," she says, "but the cost makes it impractical." Because the water used to soak and rinse sprouts picks up vitamins and minerals, Sara uses it for soup base or watering her plants.

Sara doesn't toast the sprouts as some people do, figuring that the less you mess around with them, the less you destroy their nutritional value. A favorite dish is a simple cooked grain with a sprout salad on top. She also likes to serve a main dish with several side dishes around it ("like constructing a taco"). Fresh sprouts usually make a good addition to those meals.

Sara uses a variety of sprouts in salads. "Sunflower seeds and lentils are great," she says. "They sprout almost immediately. You can soak a jarful overnight, rinse them the next morning, and by evening the sprouts will be ready to eat."

CB

SPROUTING VS. SOCIAL DIS-EASE

Fans of sprouting claim that beans cause less flatulence if sprouted. This isn't strictly true: it's not the sprouting that helps, but the first soak and subsequent rinsings.

According to Dr. Joseph Rackis of the USDA, flatulence results when undigested complex sugars in the lower intestine are acted upon by naturally occurring bacteria.

These complex sugars (called trisaccharides) are water soluble, and soaking and frequent rinsing wash them away.

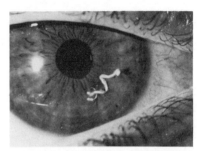

EYEBALL SPROUTING

It's hard to stop a determined seed.

A seven-year-old South African boy came home from school one day with a swollen left eye. There was no obvious explanation for the problem. He saw the family doctor, who prescribed an eye ointment. After a few days, the swelling went down and the whole incident was forgotten.

About 16 months later, the boy again had eye trouble: occasional bouts of redness, itchiness, rubbing, and shutting of his left eye when watching television. After the problems had persisted for about 10 weeks, he was taken to an eye specialist.

The examination showed a small scar on the cornea, left from an entrance wound. A closer look at the eye with a slit lamp revealed a seed embedded in the iris, with a spiral-shaped sprout growing into the aqueous humor. The seed itself was impaled on an eyelash, which probably penetrated the eye at the same time as the seed.

Shortly afterward, doctors at the Groote Schuur Hospital in Cape Town surgically removed the sprout from the boy's eye. There was no permanent damage to his vision.

QUICK REFERENCE CHART

	Rinses/ day	Harvest sprout length (in.)	Sprout time (days)	Approximate yield	
Alfalfa	2	1–2	3–5	3 T.= 4 c.	Easy to sprout. Pleasant, light taste.
Almonds	2–3	¼	3–5	1 c. = 1½ c.	Similar to unsprouted nuts. Crunchy, nutty flavor.
Amaranth	3	¼	2–3	3 T. = 1 c.	Mild taste. Sprouts smell like corn silk.
Anise	6	1	2	3 T. = 1 c.	A strong, anisey flavor. Good if used sparingly.
Barley	2–3	sprout is length of seed	3–4	½ c. = 1 c.	A chewy texture and pleasant taste. Not sweet. Toasting enhances flavor.
Beans (all kinds except those listed individually in chart)	3–4	1	3–5	1 c. = 4 c.	Taste like unsprouted beans. For tender sprouts, limit germination time to 3 days.
Buckwheat	1	¼–½	2–3	1 c. = 3 c.	Simple to sprout. Buy raw, hulled groats for sprouting.
Chia	1	¼–1	1–4	2 T. = 3–4 c.	Hard to sprout, because of their tendency to become gelatinous. Sprinkling rather than thorough rinsing can help prevent this problem. Their strong flavor adds zip to any dish.
Clover (red)	2	1–2	3–5	1½ T.= 4 c.	Similar to alfalfa sprouts.
Corn	2–3	½	2–3	1 c. = 2 c.	Sweet corn taste, with chewy texture. Difficult to find untreated kernels for sprouting.
Cress	2	1–1½	3–5	1 T. = 1½ c.	A gelatinous seed. A strong, peppery taste.
Fenugreek	1–2	1–3	3–5	¼ c. = 1 c.	Spicy taste, good in curry dishes. Bitter if sprouted too long.
Flax	2–3	1–2, greened	4	2 T. = 1½–2 c.	Tend to become gelatinous when wet. Sprinkle rather than rinse. Sprouts have a mild flavor.
Garbanzos	4	½	3	1 c. = 3 c.	A raw bean flavor. Best lightly cooked.
Lentils	2–4	¼–1	3	1 c. = 6 c.	Chewy bean texture. Can be eaten raw or steamed lightly.
Millet	2–3	¼	3–4	1 c. = 2 c.	Similar to barley sprouts.

	Rinses/ day	Harvest sprout length (in.)	Sprout time (days)	Approximate yield	
Mung beans	3–4	1½–2	3–5	1 c. = 4–5 c.	Easy to sprout. Popular in oriental dishes. Sprouts begin to lose their crispness after 4 days of storage.
Mustard	2	1–1½	3–4	2 T. = 3 c.	Spicy, tangy taste, not unlike fresh English mustard.
Oats	1	lead sprout is length of seed	3–4	1 c. = 2 c.	Only unhulled oats will sprout. Water sparingly; too much water makes sprouts sour.
Peas	2–3	sprout is length of seed	3	1½ c. = 2 c.	Taste like fresh peas. Best when steamed lightly.
Pumpkin	2–3	¼	3	1 c. = 2 c.	Hulled seeds make best sprouts. Light toasting improves flavor.
Radish	2	⅛–2	2–6	1 T. = 1 c.	Sprouts taste just like the vegetable.
Rice	2–3	sprout is length of seed	3–4	1 c. = 2½ c.	Similar to other sprouted grains. Only whole-grain brown rice will sprout.
Rye	2–3	sprout is length of seed	3–4	1 c. = 3½ c.	Easy to sprout. Very sweet taste, with crunchy texture.
Sesame	4	sprout is length of seed	3	1 c. = 1½ c.	Only unhulled seeds will sprout. Delicious flavor when young; sprouts over 1/16″ turn bitter.
Soybeans	4–6	1–2	4–6	1 c. = 4–5 c.	Difficult to sprout because they ferment easily. Need frequent, thorough rinses. Should be cooked before eating for optimum protein availability.
Sunflower	2	sprout no longer than seed	1–3	½ c. = 1½ c.	Good snacks, especially if lightly roasted. Become bitter if grown too long.
Triticale	2–3	sprout is length of seed	2–3	1 c. = 2 c.	Similar to wheat sprouts.
Vegetable seeds	2	1–2	3–5	1 T. = 1–2 c.	Usually easy to sprout. Best eaten raw.
Wheat	2–3	sprout is length of seed	2–4	1 c. = 3½–4 c.	Simple to sprout. Very sweet taste.

Unconventional twists

Harvey Lisle's fermented sprout cereal

If you're looking for a new way to prepare sprouts, you might try fermenting them. Harvey Lisle of Norwalk, Ohio, does. He sprouts wheat, rye, and millet, then grinds them up and lets them ferment. He eats the results as a cereal.

Anita Hirsch, Rodale's home economist, followed Mr. Lisle's directions except for the last step: she couldn't stomach his concoction. "All the benefits he insists come from eating this oatmeallike mixture would not persuade me to eat it," she explained with a grimace.

Mr. Lisle realizes his cereal isn't for everyone. He wrote us, "After smelling your fermented grains you will exclaim that they smell spoiled and all of your work has been for naught. But just remember all fermented foods smell strongly. Think of sauerkraut or sour milk or sour pickles. . . . If you don't like the strong smell of fermented foods, then this dish is not for you."

On the other hand, if you do like fermented foods, here's what to do. Use equal amounts of the three grains (⅔ cup of each is the recommended amount), sprout for two to three days, grind them together, pack loosely in two bowls for maximum exposure to air, and let them ferment for two days. During fermentation, the protein quality of the sprouts improves, and they warm up to 90 or 100° F. (Wrapping the bowls in a towel will help keep them happily warm.) Mr. Lisle likes to eat his cereal with raisins, bananas, yogurt, and blackstrap molasses.

Dried sprouts

Sprouted grains develop a sweet flavor after drying in an oven at temperatures over 170° F. Buckwheat sprouts are light and crunchy, and can be used whole. Wheat berry sprouts are a bit harder, and you may prefer them coarsely chopped. Rye sprouts are hard to chew; try grinding

them. Long-grain brown rice and flax sprouts are also excellent dried.

The drying temperature must exceed 170° F. for the starch in the sprouts to develop into sugar. A good drying temperature is 250° F.; at this temperature most sprouts will dry thoroughly in 45 minutes to one hour. Before storing the sprouts, be sure they are completely dry or they'll get moldy.

Dried Sprout Cereal

1½ cups dried wheat berry sprouts
¾ cup dried buckwheat sprouts
¾ cup oat flakes, toasted
¾ cup triticale flakes, toasted
¼ cup dried rye sprouts, ground

Mix ingredients together well. Serve plain or with milk.

Yield: 4 cups

For a change in taste and texture, try drying sprouts in your oven.

Chinese soil sprouting

If you're an old hand at sprouting, you might want to try your green thumb on growing sprouts in soil, a method used by the Chinese for thousands of years. Why bother messing around with soil, when sprouts will grow on air and water? One reason is that soil sprouts need less tending: rather than rinsing twice or more a day, you simply sprinkle as needed. Sprouts grown in soil pick up extra nutrients, as well, and according to our laboratory tests they may have double the vitamin C. The Chinese, earth sprouters for centuries, report even higher levels of this vitamin.

They make bean sprouts inside a wooden box partly buried in the earth. According to a Chinese technical publication, "The bottom is first covered with a layer of fine soil and over it are spread beans with a thickness not surpassing two beans. The pit is again covered with a layer of fine soil. No watering is needed"—the soil naturally retains the moisture needed to promote sprouting, so once the seeds are soaked and the earth wet down, you can go away for a few days and come back for the harvest.

(Continued on page 128)

We made our bottomless sprouting box of ¾-inch stock. Slats hold the cover in place. Work the frame into fine, loose soil to a depth of three inches.

Radish seed

The radish seed, a small brown teardrop, sprouts from ⅛ inch to more than 2 inches long over a period of from two to six days. Its peppery bite gets hotter with length.

Radish sprouts taste just like the vegetable, so they complement all kinds of salads: tuna and seafood, chicken, and mixed greens. They also are good when sprouted in combination with alfalfa.

Radish sprouts contain a relatively high level of potassium.

Soybean

Because soybeans are a larger bean, they should be presoaked for 12 hours or more. They are not easy to sprout, tending to ferment unless rinsed well and often—at least four times a day. Harvest the sprouts after four to six days.

Soybean sprouts have a raw bean taste, but steaming or stir-frying them for a few minutes will bring out a rich, nutty flavor and make their protein more digestible. They are an excellent meat substitute for chop suey dishes, soups, stews, and casseroles, and blend well with hearty meats like beef and pork. Try them as a salad vegetable, too.

Sunflower seed

The familiar black-and-white seeds make good sprouts. Sunflower sprouts must be harvested quickly, however, before they develop a strong flavor that burns the back of the throat. The root should be no longer than the seed. Sprouts will reach that length in 24 to 72 hours.

Some sprouters recommend that only unhulled sunflower seeds be used for sprouting, but raw hulled seeds work as well. The sprouts contain significant amounts of a whole list of minerals.

Use sunflower sprouts in salads, and mix them with dates, fruits, nuts, and raisins for snacks. Combine sunflower sprouts in a blender with honey and water for a breakfast drink.

Wheat

Wheat sprouts get sweeter as they lengthen. Some may be harvested in as little as two days, but it generally takes four days for the sprout to reach optimum size—just double its original length.

As wheat sprouts grow, their vitamin C content increases dramatically. They are also distinguished by plentiful thiamine, niacin, and vitamin E.

Knead sprouts into bread dough just before shaping the loaves. Use them ground in casseroles and meat loaf, and as a coating for chicken or fish. If dried in a warm oven, wheat sprouts turn crunchy and can replace nuts in baked goods; or, enjoy their sweetness simply by munching them.

Many enthusiastic claims have been made for a drink of liquified wheatgrass grown in compost. See pages 117 and 130.

(Continued from page 127)

We tried the earth-box technique at our Organic Gardening and Farming Research Center, and found that sprouting comes easily. Heat and moisture can lead to mold, but this can be prevented by keeping the cover of the sprouting box open to improve ventilation.

You don't even need a wooden box to sprout in soil. The Chinese also use a "sheltered ground pit"— simply a hole with a cover that serves to provide shade and keep in some of the moisture.

We sprouted successfully in vegetable flats and windowsills located in greenhouses, cold frames, and our homes. The seeds included large and small beans, grains, and herbs. Viktoras Kulvinskas, author of books on natural foods, sprouts in a layer of compost held by a cookie sheet. The sheets can be placed in simple wooden racks.

We harvested our yields using two methods. We cut off the larger sprouts (such as sunflower) close to the soil surface with a scissors or sharp knife. This method sacrifices some of the sprout but the yield is still substantial. With alfalfa and other smaller sprouts, we simply plucked the sprouts from the soil. Mixing some sand (about 20 percent) with the fine top layer of soil keeps dirt from sticking to the sprouts.

And the taste? Opinion among members of the *Home Food Systems* staff varied. Generally, we'd say that soil sprouts are more flavorful than regular sprouts—similar in taste to certain leafy salad greens.

—TS

Sprouts for breakfast

Cereal for a Bowl or Glass

¼ cup wheat sprouts
¼ cup wheat germ
2 tablespoons sesame seeds
1 banana, sliced, or berries in season
½ cup milk or yogurt

Mix everything gently in your cereal bowl. Or, put all the above in a blender, and whiz till smooth for a breakfast meal of another texture.

Yield: 1 serving
Jane Kinderlehrer

Scatter seeds evenly over the soil surface, then cover lightly with soil and put the lid in place to conserve heat. You may have to wet down the soil if dry; otherwise, just rely on natural moisture. The earthbox works just as well indoors, if you take the trouble to add a ¾-inch bottom. Fill the box with three inches of loose soil and sow your seeds.

COLD-WEATHER SPROUTING

Sprouting need not stop in cold weather, even if you like to save energy by keeping your house considerably below 70° F. Most homes have warm places—the top of a refrigerator or water heater, a radiator, a south-facing window— where sprouts can grow comfortably.

Another good sprouting location is your oven—if you have a gas stove. The pilot light will maintain a fairly even temperature. If it's too warm in the oven, you can probably adjust the pilot light (within a limited range), or you can put something in the door to keep it open a crack.

The biggest danger with growing sprouts in the oven is forgetting them. Left for a couple of days without rinsing, the sprouts will dry up or rot. Or you may preheat the oven to bake a casserole, and open the door to find roasted sprouts. So if you're the forgetful type, leave yourself some reminder of the incubating sprouts.

If none of these sites is suitable, you can easily make a "warm box"—a light bulb in an enclosed

In a cool house, the top of the refrigerator may provide a warm spot for sprouting.

box or Styrofoam chest, of the sort used to make tempeh (see page 109). You can control the temperature in the box by changing to a light bulb of a different wattage, or by varying the amount of air you let into the box.

Another cold-weather option is to try sprouting at work. Most work places are kept at a temperature well within the proper range for sprouting. Seeds don't take up much space, and you can sprout in a jar on your desk, or in a desk or file drawer.

TD

Hippocrates Health Institute: let raw foods be your medicine

Dr. Ann calls them her "human angels."

They appear in various forms: Ed, the cook, whose specialty is raw spaghetti and meatballs (the spaghetti made from the stems of buckwheat sprouts, the meatballs made from cauliflower); Loretta, whose hypoglycemia was so severe when she came to the Hippocrates Health Institute that she couldn't climb a flight of stairs without resting, but now runs the Institute (all six, elevatorless stories of it) when Dr. Ann is on tour; Steve, the Institute's public relations director, who closes his correspondence not with a mere "Sincerely," but with, "May peace

and health bless your life."

Dr. Ann is Ann Wigmore, D.D., Ph.D., N.D., founder and director of the Hippocrates Health Institute in Boston, Massachusetts. The Institute bears Hippocrates' name because it was he who said, "Let food be your medicine." Dr. Ann has slightly modified that approach in treating the thousands of patients who have come to the Institute since 1963. She says, "Let *raw, living* foods be your medicine."

"It is an irrefutable fact that the raw, live food diet is an immediate and radical body healing method for all disease without exception," she says. And she means it.

Those who come to the Hippocrates Health Institute, no matter what their ailment—cancer, heart disease, arthritis, allergies, hypoglycemia, diabetes, ulcers—are treated with the same raw food diet. And the most important item in that diet is a food which Dr. Ann believes is the most purifying and nutritious in existence: wheatgrass.

Every patient who comes to the Institute plants wheat berries in shallow, soil-filled trays. In a week the wheatgrass is tall enough to be harvested and juiced. Only a hand juicer is used; Dr. Ann believes electric juicing destroys the enzymes which are crucial to proper digestion and therefore to perfect health. The juice ready, the patient either drinks it or gives himself a wheatgrass "implant"—an enema. Sound strange?

The cancer patients who claim to have been cured by wheatgrass therapy don't think so. Wheatgrass is 70 percent chlorophyll, which some research suggests may inhibit cancer cell activity. It also, says Dr. Ann, is rich in laetrile (which some people claim is an anti-cancer substance), and is high in protein and minerals. These factors, she says, improve the digestion, cleanse the body of toxins and build up a store of nutrients in the bloodstream, three steps which she believes are the key to healing not only cancer but all disease.

Wheatgrass is only a part of the Institute's healing program. The patients (or "guests," as they are called) also prepare much of the rest of their therapeutic diet. They grow indoor greens from buckwheat and sunflower seeds. They prepare their own sprouts, rinsing them daily. And they make "rejuvelac"—a liquid concoction Dr. Ann claims is a rich source of enzymes—by simply soaking wheat berries in distilled water for a day or so and then drinking the water. They also eat fruits, vegetables, seeds, grains, and edible weeds.

The guests drink cocktails like the "Bloody Mary," made from tomato juice, alfalfa sprouts, celery, garlic, tamari, horseradish, lemon juice, and freshly ground black pepper. They have sherbet or cookies for dessert—the sherbet is mixed fruits blended with apple juice and frozen; the cookies are mashed dried fruit mixed with banana, spread on grated coconut and dried. They eat sun-baked bread (made from sprouted grains) and seed cheese (made from rejuvelac and sunflower seeds blended to a creamy consistency and fermented). There are dips and soups, sauces and salads, all raw and many with sprouts as a main ingredient.

The Institute also offers nourishment for the soul. Ann Wigmore is a Doctor of Naturopathic Medicine and a Doctor of Divinity; the Institute is a research institution, a live-in hospital, and a temple. She isn't your typical cleric, however. Her religion, "Rising Sun Christianity," is best described as a combination of Christian mysticism, astrology, yoga, macrobiotic philosophy, pyramid power, and occultism. In the Institute's most comprehensive publication, *Naturama Living Textbook,* subjects include "Prosperity Can Be Yours," "The Secret of Life," "Reincarnation," and "Divine Law."

Echoing Christ's words, Dr. Ann says that the body is the "temple of the soul." Only in a healthy body, she believes, can each person rise to know his Creator and serve mankind.

Dr. Ann's own journey to health began in war-torn Europe, where she stayed alive—and healthy—eating only roots, grass, and rough bread made from rye flour and straw. But when she arrived in the United States during her teens (she's now in her 70's), she glutted herself on processed food. Her health soon began to fail.

"Because the physicians and drugstores seemed powerless to help me," she says, "I instinctively turned to God for guidance. I began to study my Bible, and I recall, as though it were only yesterday, when these simple words came to me: 'Become a minister and build my temples.'"

The temples, she would realize years later when her Divinity degree was almost earned, were human bodies. And the building material was grass.

Through intuition and months of experiments she selected wheatgrass, and the first person she healed with it was herself: "My shattered health experienced miraculous recovery." Today, healthy enough to need only four hours sleep, she has opened new institutes in California and India (the former Prime Minister of India, Moraji Desai, wrote the foreword to one of her books) and is continually expanding her healing mission.

"We must become living examples to help others achieve happiness and well-being," she says. "Only by doing so do we become co-creators of Our Heavenly Father who intended for us to be in perfect health. Let us not settle for less."

Bill Gottlieb

Wonder water sprouts

If you want to grow wonderful sprouts, you might try soaking and rinsing your seeds in so-called "wonder water," degassed water that reportedly stimulates all sorts of biological activity.

Fred Ponting of Red Bluff, California, sprouted soybeans, mung beans, and alfalfa seeds using both tap water and wonder water. He reported that after four days the wonder water soybean sprouts "were plumper, more succulent, and had a cleaner appearance" than the tap water soybeans. Comparing yields, he estimated that the wonder water beans weighed 15 percent more.

He obtained similar results with alfalfa. After three days, "the wonder water sprouts were ahead at least 20 percent by bulk due to longer, plumper, more thrifty sprouts, and from better germination." His mung beans weren't as fond of wonder water, sprouting just as well in tap water.

To make your own wonder water, bring bottled or tap water to a rolling boil for five minutes. This removes all the dissolved gases. To prevent regassing, store the water in airtight containers. ∎

Elaborations on the Mason jar

There are lots of sprouters available, and they all work. That's less a tribute to the manufacturers than testimony to the amazing vitality of seeds, nuts, and beans. Given a little water and warmth, they'll sprout in a nondescript bowl or jar. After all, nature didn't intend them to have an elaborate support system of domes and meshes.

In fact, none of the commercial sprouters we tested surpassed the Mason jar for productivity, price, or ease of harvesting and cleaning. Some were more attractive than a plain jar, and others offered such features as separate sprouting areas or semi-automatic rinsing, but none grew better sprouts.

That doesn't mean the Mason jar is for everyone. If you're going to have your sprouter on continual display in the kitchen, for example, you might want a more comely model. Or

if you're gone from home a lot, you might choose one that can be ignored for longer periods. But if it's just good sprouts you're after, start with a jar.

Our tests focused mainly on rinsing and cleaning. Since most sprouts need to be rinsed twice a day—and some thrive on more rinsings—we checked ease of rinsing, how much water was required, and how many sprouts were washed away during rinsing. Then we judged the sprouters on ease of cleaning.

With a multi-level sprouter, you can sprout different seeds or beans at the same time.

THE BEANSPROUT BOOK

Gay Courter. 1973.
Simon and Schuster, Rockefeller
Center, 630 Fifth Avenue, New York,
NY 10020. 96 p. hardcover $2.95

This book includes sprouting basics, a myriad of recipes, directions for sprouting large quantities, and methods for sprouting in strange places, including a knapsack.

Mary Jurinko

The portable knapsack sprouting method

You will need the following materials:

1) *Seeds for sprouting—mung, lentil, rye, wheat, and soy work best.*
2) *A heavy plastic bag.*
3) *A second lightweight plastic bag.*
4) *A few paper towels or a clean rag.*
5) *Twister seals or rubber bands.*

Directions:

1) *Soak the beans or seeds in the heavy bag. Seal the bag with twister to keep the water in.*
2) *Take the lighter bag and punch small holes over about ⅓ of the bag's lower surface, one inch apart.*
3) *Pour the soaked beans into the sieved bag and shake, or hang the bag on a tree limb until all the water is drained out.*
4) *Soak the towels or cloth in water, squeeze out gently and tuck around the beans in the sieved bag.*
5) *Place the sieved bag containing the beans and moist toweling into the heavy bag. Twist the tops of the two bags together and fasten with tie or rubber band.*
6) *Tuck the two bags into your knapsack.*
7) *To rinse, remove the inner sieved bag and flush with water for a few minutes. Drain completely. Then repeat steps four and five.*
8) *Repeat rinsing and draining twice a day until you harvest your crop.*

Beale's Famous Seed Sprouter

$9.50

This unusual sprouter is a clay bowl with a matching lid. Soaked seeds rest in the bowl, which sits on a plate holding about ½ inch of water. The lid is placed on the bowl. Water absorbed through the sides of the bowl keeps the sprouts moist.

Harvesting and cleaning were easy for us. You can rinse sprouts in the bowl, but a sieve or strainer is necessary to keep small sprouts from going down the drain.

Beale's Famous Sprouter, Box 323, Fort Washington, PA 19034.

Biosta/Biosnacky

$14.50 (green), $16.50 (clear)

The clear plastic Biosnacky (and its green twin, called the Biosta) work on a siphoning principle. The three bowls are filled with seeds—a different kind on each level, if you choose—and water is poured into the top bowl, to slowly siphon through the seed bowls into a drainage bowl on the bottom. The seeds get just enough moisture, and little rinsing is required.

You pay for this simplicity with a lower yield, because regular rinsing removes waste and increases the percentage of germination. The siphoning tubes had a tendency to become air locked and clogged with seeds. Also, small seeds often stuck in the grooves on the bottom of the sprouting bowls, and had to be removed with a toothpick or brush.

Miracle Exclusives, Inc., 3 Elm Street, Locust Valley, NY 11560.

Kitchen Garden

$7.95

The Kitchen Garden is a round white plastic tray with square drain holes in the bottom. It comes with two removable partitions to make two or four compartments for different kinds of sprouts. Sprouts are rinsed in the tray, and the water drains out the bottom.

The drain holes are too large, however, and we found that small seeds such as alfalfa and millet could wash through. The holes allow very rapid drainage, so it took a considerable amount of water for rinsing, and the sprouts were never sufficiently flooded.

Kitchen Garden Sprouter, 2465 S. Main Street, Salt Lake City, UT 84115.

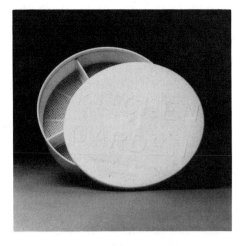

Little Green Acre

$18.95

This three-piece plastic sprouter looks like a beach ball. A transparent red hemisphere with a center cone sits on an identical transparent purple hemisphere. A clear, circular sprouting tray rests between them. The seeds are scattered on the tray, and the growing sprouts hang through the rectangular drain holes.

The Little Green Acre works only with large seeds and beans. Smaller seeds, such as millet, alfalfa, and chia, fell through the sprouting tray or stuck in the drain holes. Larger sprouts often stuck in the holes, making harvesting and cleaning a slow process. Sprouts could be rinsed in the tray, but with the large holes it required a lot of water, and the sprouts were never well flooded.

Little Green Acre Sales, Inc., 760 NW. 57th Street, Fort Lauderdale, FL 33309.

Redwood Sprouter

$12.50 (small), $17.10 (large)

The sides are redwood, the top and sprouting tray are plastic. The tray is covered with drain holes, so the sprouts can be rinsed in the sprouter. Because the redwood sides retain moisture, the sprouts grow in a humid atmosphere.

Rinsing was easy and the sprouter drained well, although the first rinse flushed out some alfalfa seeds. The redwood sides unfold for cleaning. A brush is required to scrub out the tray.

The Redwood Sprouter Company, 2506 Southland Drive, Austin, TX 78704.

Seedsmith

$12.50

The Seedsmith consists of three round growing trays and four covers. Each of the opaque plastic trays has two rows of small drainage holes around the outside. The rinse water drains slowly, so a small amount of water can flood and rinse the sprouts properly.

With three different trays, three kinds of sprouts can be grown simultaneously. The sprouts were easily harvested and cleaning was uncomplicated.

Kienholz Products, Inc., 4400 Loch Alpine West, Ann Arbor, MI 48103.

THE COMPLETE SPROUTING COOKBOOK

Karen Cross Whyte. 1973.
Troubador Press, 385 Fremont Street,
San Francisco, CA 94105. 120 p.
paperback $4.95.

This is a good, simple sprouting manual. Besides explaining four easy methods of sprouting, the author gives concise information on the history of sprouting, why and how seeds sprout, nutritional benefits of sprouting, and instructions for storing sprouts. There is a wide range of recipes which uses only natural ingredients.

Ms. Whyte gives specific directions for sprouting 25 different seeds or beans. Each item has its own page with interesting tidbits about its background, medicinal uses, and nutritional data.

Mary Jurinko

Minestrone

1	cup red bean sprouts
1	cup white or navy bean sprouts
1	cup lentil sprouts
1	cup pea sprouts
1	potato, diced
2	carrots, diced
½	cup cabbage, chopped
1	cup spinach or chard, chopped
1	tablespoon parsley
2	small zucchini, chopped
1	cup whole wheat spaghetti
1	quart water
¼	teaspoon sweet basil
2	cloves garlic, chopped
	dash of cloves
¼	cup olive oil

In a large pot, combine sprouts, vegetables, spaghetti, and water. Cook until tender, about 15 to 20 minutes. Make a paste of basil, garlic, a dash of cloves and olive oil. Add to soup. Serve with Parmesan cheese.

Yield: 6 servings

SPROUTING FOR ALL SEASONS

Bertha B. Larimore. 1975.
Horizon Publishers, P.O. Box 490, 55
East 300 South, Bountiful, UT 84010.
139 p. paperback $3.95.

A comprehensive guide to sprouting. More than half of the book is devoted to recipes, some of them rather strange perversions of the sprout, including soy nut brittle, a dip of sprouts and bacon drippings, and sprout fudge. The book is redeemed by the author's imagination.

Mary Jurinko

Instead of purchasing canned or packaged baby foods, sprouts may be used for preparing a variety of nutritious dishes. They taste delicious, contain no preservatives, and cost but a fraction of the cost of prepared foods. The dried sprout powder or granules can be kept on hand to make feedings as needed, or several servings can be prepared and frozen in small portions for later use. Many sprout and meat or fruit combinations can be used, depending what is on hand or what has been cooked for the family.

Instant Sprout Cereal

1 cup wheat sprouts, dried
1 cup oat sprouts, dried
1 cup rice sprouts, dried
1 cup soybean sprouts, dried

Grind the dried sprouts in the blender until they are very fine. Store in airtight jar and use as needed. For serving, stir milk or formula into desired amount of sprout cereal. A raw or lightly cooked egg yolk can also be added if desired. Makes 2 cups of cereal powder.

Any combination of dried sprouts can be used to make the instant cereal. The powder can also be stirred into mashed or pureed vegetables for more nutrition.

Sprout Garden

$4.99

This sprouter works like a wide-mouth jar. It's made of clear plastic and comes with a meshed snap-on top. The mesh, however, was a bit large, and on the first rinse we lost some alfalfa seeds. Soaking and rinsing are done in the sprouter, and cleaning was simple.

Vitality Farms, Inc., Box 7049, Fruitvale Station, Oakland, CA 94601.

Sprout-Ease Tube Sprouter

$7.95

This clear plastic sprouter works like a Mason jar, but it is open at both ends and comes with several threaded lids—two solid and three meshed. A small-gauge mesh works well for both small and large seeds, but larger meshes are included.

Soaking and rinsing are done right in the sprouter, and cleaning proved to be simple and quick. With the open-end design, you can easily push mature sprouts out of the tube in perfect shape. The instructions include six recipes using sprout "slices" cut from a "loaf" of compacted sprouts.

Bima Industries, Inc., P.O. Box 88007, Tukwila Branch, Seattle, WA 98188.

Wide-mouth jar

$.50 (Including lid and cheesecloth)

A wide-mouth quart jar, a rubber band, and a piece of cheesecloth make a fine sprouter. Or for a more permanent alternative to cheesecloth, you can buy a plastic or stainless steel screen and cut it to fit inside the jar's screw ring.

To rinse the sprouts, you just fill the jar with water through the cloth or screen, shake it, and drain. It takes little water and little time. Cleaning is equally simple, and the wide mouth makes harvesting easier for pudgy hands.

The Sprout Monster

Rodale engineers have been among those trying to design the perfect sprouter. A large, early wooden model was nicknamed "The Monster" by the Fitness House Kitchen staff. A more recent creation is the Food Generator, a curious self-rinsing device fabricated from stainless steel and plastic.

This sprouter is arranged in tiers with a tray for water collection at the bottom, a four-compartment sprouting tray, and an automatic rinser at the top. The automatic rinser consists of a tray with a hole and a wick, and another tray covered with small holes. The top tray is filled with water, which flows through the opening (the flow rate is controlled by the wick) into the perforated tray. From there, it drips down into the sprouting compartments, keeping the sprouts moist and rinsing them at the same time. The water then collects in the reservoir at the bottom.

In our tests, the Food Generator grew fine sprouts without any manual rinsing. But there were problems with both harvesting and cleaning. No further experiments are planned.

Juicing
Drink your garden

Fresh from the garden or orchard, fruits and vegetables are so perfect that it seems a sacrilege to do anything but sit down and eat them. But if you're up for a treat that runs *au naturel* munching a close second—sit down and drink them.

Compared to the produce whence they come, fresh juices are both better and not quite as good: not quite as good, because removing juice from the pulp means losing some natural goodness, including a fraction of the vitamins and minerals, and all of the fiber; better, because juices are the concentrated form of fruits and vegetables, which makes them well worth having around, whether your interest is nutrition or food storage.

When those baskets of tomatoes and beets come pouring in from your garden all at once, you can distill the pounds of vegetables into their essence, delicious drinks that will resurrect the taste of summer long after the garden plot has been covered with snow. You can't find a simpler, safer way to put up a bountiful harvest in limited space.

Even if your garden is scaled down to your immediate needs, producing a more modest flow of fruits and vegetables, you might do well to make part of the harvest drinkable. Fresh juices have their own important contribution to make to good health. When we think about nutrition, we're likely to pay a lot more attention to eating than to drinking. But we need some three quarts of water per day to stay in the best of health—and many of us, doctors say, fall short. When we're sick, or hot, or active, we may need a lot more.

Finding naturally good drinks may be harder than finding wholesome foods. Pure water is just about the perfect fluid, but most of us get it inadvertently polluted and advertently chemically treated. And, it's just human nature to want a different drink from time to time. That's where the trouble begins. Soft drinks, the ubiquitous alternative, are crammed with artificial flavorings and colorings, sugar, or artificial sweeteners said to cause cancer. A cold beer is fine at the right moment, but a steady diet can hardly be called a healthy habit. While commercial fruit and vegetable juices are better, they are often adulterated with salt, sugar, or chemical preservatives.

Pour yourself a glass of freshly squeezed juice. This glass of juice contains most of the vitamins and minerals that were in the original fruit or vegetable, including substantial amounts of vitamin C, significant

When your garden just won't quit, juicing may solve the problem of over-abundance.

traces of the B vitamins, and such vital minerals as potassium, magnesium, and calcium. They are here in a concentrated form—you'd have to eat two oranges to get the vitamin C of one eight-ounce glass of orange juice.

For children, juices can repair gaps in haphazard eating habits. The same kid who flees from a salad may come running for a cold glass of tomato juice. Slip some spinach into your blender when you make that tomato juice for a pleasant tang, a boost of nutrients, and a welcome that spinach never got at the table.

Through a splendid example of nature's wisdom at work, fresh juices are most readily available just when we need them most. On a hot summer afternoon when you'd rather drink than eat, a tall glass of vegetable juice allows you to indulge yourself, with little nutritional sacrifice. You need more water, minerals, and vitamins on that summer's day to replace what you are losing through perspiration. Juices supply them well. (Runners and other vigorous exercisers take note: you can sweat out as much as two quarts of water in an hour of a hot day, plus the sodium, potassium, magnesium, and other minerals that are dissolved in it. Juices are valuable replacement fluids.)

From a nutritional standpoint, there's no comparison between homemade juices and the store-bought kind. A commercial manufacturer will pick fruits and vegetables at his convenience, not necessarily at the peak of the vitamin content. He may allow days to elapse before juicing his harvest, as its nutrient level rapidly deteriorates. While the juice sits on its warehouse shelf on its way to your supermarket, its food value declines still further. Your own juice, from fruits and vegetables that were picked at their peak and squeezed within hours, starts out way ahead.

As part of the preservation process, commercial juices are heated: this destroys enzymes and reduces vitamin C. For better storage properties, some manufacturers wash juices to remove water-soluble minerals—the same minerals that made them worthwhile in the first place. When commercial juices are clarified to keep better and look more appealing, they may lose much vita-

min A. One series of tests of commercial tomato juices found that, on average, they lost one-third of their vitamins A and C in the processing. You'll find that juices you can at home fall short of fresh juices in flavor and nutritional value just as your canned vegetables are no match for garden vegetables.

If fresh juices sound good to you, you can probably start tasting their pleasures with equipment already at hand. The ordinary kitchen blender will competently juice soft-fleshed fruits and vegetables such as strawberries and tomatoes. Citrus fruits can be juiced by hand with a simple citrus reamer. If you get serious about juices, consider investing in a specialized machine. In the following pages several are discussed.

Where do the best juices come from? Citrus fruits, berries, apples, and grapes are favorite fruits for juicing; among the vegetables, tomatoes, beets, and carrots are good bets. But there are no rules and regulations. The spirit of adventure well may lead you to blends that no one ever told you about. Blend in yogurt for a rich juice drink; add spring water for a light refresher. Will a complete salad—tomatoes, spinach, pepper, celery, cucumber, sprouts, and a hint of garlic—make a delightful juice, or something to be forgotten quickly?

It might be fun to find out.

Carl Sherman

Cheap fruit for juicing

If you don't grow your own fruits and vegetables, or if you want juices from foods not in your garden, there are alternatives to buying at supermarket prices.

At a pick-your-own farm you can climb ladders into peach trees or forage among cucumber plants to gather your own fresh produce. Picking on a summer day is its own pleasure, and blackberries at 50 cents a quart are a good buy compared with blackberry juice at $2 a quart at the health food store.

Your county agricultural extension agent should have a list of where you go, along with estimates of ripening times. Call ahead to be sure that fields or trees are really ready for picking. Ask the farmer if you need to bring containers. Usually the grower will furnish quarts or bushels for picking and measuring, but you'll likely have to pour them into your own bags or boxes. Ask the farmer what's been sprayed on the plants or trees for insect or disease control so you'll know what kind of washing to do.

A second source of fresh produce in quantity is the farmers' market, an increasingly popular alternative to the supermarket. Try bargaining with the seller for large quantities. Some sellers may be willing to let you visit their fields and

(Continued on page 138)

(Continued from page 137)

orchards to pick at lower prices. Look at the market as a place to meet people who grow the food.

Road-side stands can be another route to fresh local produce. Again, bargaining may get you a lower price for large quantities.

Food co-ops could have contacts with local farmers for in-season vegetables and fruits. Out-of-season produce probably comes to the co-op through the usual wholesale marketing outlets (flown or trucked in to a central warehouse) and is already marked up by middlemen. Still, out-of-season apples at a co-op sold for 17 cents each, for example, compared to a supermarket price of 29 cents each. (For more information on co-ops and how to find them, see the last section of the first chapter, Grains.)

Another option is to do what co-ops do—buy produce at wholesale markets yourself. That's not easy. Buying at produce wholesale warehouses means getting there at 4:00 A.M. (produce bought daily has to be on supermarket shelves ready to sell by 8:00 A.M.) and bargaining with shrewd dealers who can spot and exploit a novice. But if you're willing to shop carefully and compare prices, you can buy cases of produce at considerable savings. State USDA offices have information on wholesale produce markets, and can mail you weekly wholesale price guides.

Most stores dump produce that's overripe, bruised, or spotted—that's the reason for the relatively high mark-up on it. Such fruit can often be picked up cheaply at co-ops and supermarkets. In a grocery, ask the produce manager about a deal and arrange a pick-up time.

Sara Ebenreck

What to juice?

Almost any fruit or vegetable can be juiced. With some of the softer fruits, like tomatoes, oranges, and grapes, you can make juice with your bare hands—or feet. Most vegetables and harder fruits require a juicer or a press.

The only food we tested that couldn't be transformed into juice was bananas. They produced a rather runny paste, which could not truly be called liquid. But bananas are a great addition to almost any other fruit juice, and make for a hearty mixed drink.

Juice therapy

Fruit and vegetable juices, according to tradition and folk medicine, can prevent and cure a host of ailments ranging from ulcer to cancer. Is there any basis for these beliefs?

On Cape Cod in Massachusetts, where cranberries are grown, cranberry juice has long been highly regarded as a remedy for urinary infections. Now that cranberry juice is available just about everywhere, its use as an over-the-counter treatment for these painful, often resistant infections has become widespread.

Apparently, cranberry juice works. When one physician gave 60 patients 16 ounces of cranberry juice daily for three weeks, all reported some improvement in their urinary problems, and more than one-third were cured. Another doctor reported substantial relief for a group of women with chronic infections, including one 66-year-old who had suffered for five years with pyelonephritis (a severe infection that carries the threat of permanent kidney damage). The juice succeeded where powerful drugs had failed.

Cranberry juice also seems to protect the urinary tract by preventing the formation of kidney stones. It has been used successfully in this way with bedridden patients, whose risk of developing these painful stones is particularly high.

What cranberry juice does, apparently, is make the urine more acidic. This can kill bacteria and prevent calcium from coming out of solution and forming stones. Hippuric acid, which the body manufactures from the acids in cranberry juice, has proven to exert a particularly powerful germicidal effect.

Although scientific support is not nearly so strong, there may be some basis for belief in the power of other fruit juices to prevent infections. In the laboratory, grape juice and apple juice have demonstrated an ability to render viruses inactive. Do they do the same in the body? This depends on exactly what happens when the juices are digested, and the answer is not clear. Studies have suggested that people who drink orange juice to prevent colds are in fact giving their natural defenses a valuable boost. Here, the vitamin C and bioflavonoids of orange juice may be responsible.

Where fruits have a special therapeutic effect, their juice may offer a convenient way to gain the benefit. Many people have found relief from the joint pains of gout by eating cherries (why this works is not fully understood). Cherry juice, a concentrated form of cherries, so to speak, can provide the same relief.

Physicians have used bromelains, enzymes that occur naturally in pineapple, to reduce swelling and inflammation throughout the body. Blood clots, infections, and the painful swelling that follows injury and surgery have responded well to bromelains, which apparently dissolve the proteins that keep fluids trapped in damaged tissue to cause inflammation and pain. It has been suggested that drinking pineapple juice—the original source of bromelain—may provide the same kind of help when the inflammation is not too severe.

Carl Sherman

MAKE YOUR JUICER YOUR DRUGSTORE

L. Newman, N.D. 1970.
Benedict Lust Publications, Box 404,
New York, NY 10016. 192 p.
paperback $1.95.

Laura Newman starts her book with the premise that man does not die—he kills himself. If we would but take proper care of our bodies, we could—theoretically—live indefinitely. So she talks first about what we ought to eat and drink, and why.

The bulk of *Make Your Juicer Your Drugstore* details different types of physical problems and tells how to alleviate them with juice. The reason juice can cure is that "there is, after all only ONE DISEASE—deficient drainage!" The particular disease you get depends on where the poisons that accumulate in your body decide to settle. According to Dr. Newman, the proper kind of juice will build up weakened cells that contain poisons, and assist the body in eliminating these poisons.

FRESH VEGETABLE AND FRUIT JUICES (formerly RAW VEGETABLE JUICES)

N. W. Walker, D.Sc. 1936.
Norwalk Press, 2218 E. Magnolia,
Phoenix, AZ 85034. 165 p. paperback
$3.95.

Dr. Walker's juice book is the most comprehensive and specific of those we reviewed. It contains an alphabetical listing of various juices (mostly vegetable) and their relative benefits. Then there is a
(Continued on page 140)

Apple

Apples are the most popular temperate-zone fruit for juicing. Different varieties produce juices differing in flavor and aroma. Cider is simply unpasteurized apple juice that air got to: air browns the clear, amber juice almost immediately. Cider, left to its own devices (that is, untampered by chemical preservatives, pasteurization, or refrigeration), is a happy medium for alcohol—producing yeast, and soon becomes "hard" cider. Commercial cider won't turn hard if the tiny print says preservative has been added. Some states require that cider not sold within a short period of pressing must be so adulterated, to protect both the consumer and the state's right to tax liquor.

Certain yeast, often borne by fruit flies, may turn cider to vinegar. While this is anathema to winemakers, fruit vinegars are aromatic, strong, and stable. You later may want to filter and dilute your homemade vinegar.

Left out over a very cold night in a plastic jug, hard cider will separate out into ice and apple jack. Just pour the still-liquid portion from the frozen mass. Hard cider

left to mature and clear in a bottle or keg in time will become apple wine, the unsophisticated (but wholesome) country cousin to grape wine.

Apple juice loses much of its charm if pasteurized. To preserve the taste of fall, it's best to freeze cider.

Cider is boiled down with apples to make apple butter. Apple cider alone reduces to a nicely tart jam that's nothing like bland supermarket apple jelly.

Blackberry

Early American settlers discovered blackberries made excellent wine. They ascribed various medicinal qualities to blackberries, and country doctors were sometimes judged by the quality of their blackberry tonic—a mixture of blackberry wine and spices fortified with brandy—that they prescribed for almost any ailment. In some areas of North America, blackberry tonic is still available as an over-the-counter remedy for diarrhea. Whatever their curative value, blackberries are high in vitamin A and potassium. Crush the berries and simmer them—without any extra water—until you get a juicy mixture. Squeeze the mixture through a jelly bag and strain with a damp cheesecloth.

Cabbage

There's a reason cabbage juice isn't overly popular: it doesn't taste good. But this plebian vegetable has its merits. It's been successful both in treating ulcers and preventing them. Back in the 1930s, Garnett Cheney, M.D., discovered that cabbage juice seemed to block ulcers in guinea pigs. After the war, he tried it on humans, and reported that of 13 ulcer patients treated with fresh cabbage juice, 11 had no ulcers after nine days (California Medicine, January 1949). Later experiments confirmed these findings, as did a 1963 Hungarian study.

Cabbage juice never caught on as standard ulcer treatment, perhaps because it sounded like a joke. If you have a nervous stomach, however, you'll like what it does for you. Cabbage juice can't be made without a juicer, but if you don't have one, just eat raw cabbage for the same effect.

But you don't have to have ulcers to enjoy cabbage juice. If the flavor doesn't appeal to you, try chilling it or adding—in whatever amount necessary—celery, pineapple, tomato, or citrus juice.

Carrot

If only one-tenth of the claims made for carrot juice were true, it would still have to be considered a miracle liquid. Depending on whom you read—or believe—carrot juice can prevent infection of the eyes, nose, throat, and lungs; cure intestinal and liver diseases; dissolve ulcers and cancers; and beautify the skin and hair. If you drink enough, it can also give your eyes and body an orangish cast.

Amidst this dubious fanfare, the best-kept secret about carrot juice may be its taste. Fresh juice has a sweet, full-bodied flavor, and it's packed with vitamin A and minerals.

To make carrot juice you need a juicer. Fresh, firm, dark yellow carrots make the best juice; older carrots tend to be dried out. One pound of carrots makes approximately one cup of juice.

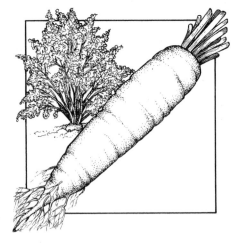

(Continued from page 139)

table of 87 juice formulas, followed by a directory of 161 various ailments, ranging from apoplexy to nymphomania to varicose veins. After each ailment, the formulas, for proper treatment are given by number.

If, for example, you suffer from emphysema, use formula 87: 6 ounces of carrot juice, 4 ounces of parsnip, 4 ounces of potato, and 2 ounces of watercress. If bedwetting is the problem, use formula 30: 10 ounces of carrot juice, 3 ounces of beet, and 3 ounces of cucumber.

Prior to the list of ailments and formulas, the book warns that "It is not legal to diagnose and prescribe anything whatever in case of illness except by a doctor licensed to do so. The following AILMENTS and corresponding FORMULAS are listed here as a guide for the HEALING PROFESSION and are given for their general information. While they are based on past experience, they are not intended to be used as prescriptions."

LIVE FOOD JUICES
H. E. Kirschner, M.D. 1957.
H. E. Kirschner, M.D., 748 Crescent Drive, Monrovia, CA 91016. 120 p. paperback $1.75.

Live Food Juices surveys the field of juice therapy, and focuses on maintaining health. It has chapters on the various vitamins and minerals needed by the body, as well as on a few of Dr. Kirschner's favorite juices: watercress, parsley, alfalfa, celery, onion, and predictably, carrot.

The most interesting parts of this book are the case histories. Along with the healing of the mysterious Mrs. X (see photos next page) and Mr. B, there are tales of juice cures for leukemia, arthritis, and kidney disease.

BEVERAGES FOR YOUR HEALTH AND WELL-BEING
Ruth Adams and Frank Murray. 1976. Larchmont Books, 6 E. 43rd Street, New York, NY 10017. 286 p. paperback $1.75.

Beverages covers the whole field of drinks—good and bad. It not only discusses the benefits of fruit and vegetable juices, carob, and milk, but also makes a case against coffee, tea, cocoa, soft drinks, and alcoholic beverages. There is a large section on water, stressing the need for drinking lots of it and attacking the practice of fluoridation. It is generally well researched and has a bibliography.

Cranberry

The tart, sour flavor of the berries makes for a tangy juice. To mellow the taste, try adding honey and some whole cloves or cinnamon sticks. Or combine with apple, pineapple, or citrus juice.

Cranberries aren't sorted by weight or color, but by their bounce. Commercial sorters give the berries seven chances to bounce over four-inch-high hurdles. The best berries hop over the hurdles, while the softer, lower-quality berries are unable to make the jump.

Cranberry juice has been cited by doctors for its value in discouraging urinary tract infection and kidney stones.

Dandelion

Most folks consider dandelions a pest, good for nothing but blowing on the fluffy seed clusters. Wild-food epicures know better, however; the roots and leaves make fine food and drink—high in vitamins A and C, calcium, and potassium—and wine can be made from its flowers.

To make dandelion tea, fall harvest is best, when the roots are full of stored nutrients. You'll need a trowel or shovel to unearth them, since they go quite deep.

Scrub the roots clean and dry them on a cookie sheet for four hours in a 300°F. oven. (Thoroughly dry roots will be dark brown in the center when you snap them in two.) Grind the dried roots in a blender or coffee grinder, and mix with hot water, using one level teaspoon of ground dandelion root per cup of water.

AN ALLEY-HOPPING ORCHARDIST

Recently, biking along an alley of my hometown in north-central Ohio, I spied an immaculately kept rose garden, in the center of which a home-owner was cutting down a peach tree killed by the bad winter of 1978. I sympathized with him, having just cut down a peach and a nectarine myself. The gardener pointed to a corner of his yard, to a peach tree glowing with life and heavy with fruit. "Now if all our peach trees were like that one, we wouldn't have to worry," he said. "That tree is extremely hardy and almost always bears fruit—big peaches, too."

Quite matter-of-factly he told me about the tree—as if it were common knowledge. "It's a seedling of Lemon Free," he said, "and it comes true from seed. I've given seeds to friends around town, and they have peaches regularly.

As if that were not discovery enough, the gardener (who prefers to remain anonymous because there is no way he can supply readers with seed) took me to his brother's place next door and showed me a cherry tree that "must be nearly 70 years old."

"The tree's been there as long as we remember," he continued. "The cherries are the sour type, but they get almost as big as plums! We've grown seedlings from this tree too. They are very hardy but their cherries are normal size." It's not hard to guess where my next peach and cherry trees are going to come from.

Gene Logsdon

Testifying to the rejuvenative wonders of carrot juice, here is Mrs. X at 65, 94, and a normal 135 pounds. The metamorphosis was effected by drinking more than 2 tons of carrot juice—a gallon a day for 18 months.

From 1 pound of carrots to an 8-ounce glass of juice in 5 minutes.

Grape

People have been drinking grape juice for millenia, often with considerable gusto: letting the juice ferment into wine was a prime method of storage. But in 1869, a temperate dentist named Thomas Welch bottled the grape without the alcohol. He pressed some juice from his Concord grapes, capped it in bottles, then boiled the full bottles in a pot of water. When he checked a few months later, the juice was still sweet, and he began marketing it as Dr. Welch's Unfermented Wine.

You can still get Welch's grape juice, but it won't taste any better than what you can make at home. Most commercial grape juices made in this country are pressed, which means the grapes are heated prior to being processed. This better preserves the rich purple color, but destroys some vitamins.

To make grape juice, wash, stem, and crush the grapes, then

squeeze them through a jelly bag and let the juice stand overnight in the refrigerator. The next day, pour off the clear juice without mixing the strong-tasting sediment—primarily tartaric acid crystals. (If the sediment carries over, just re-strain your juice before drinking it.)

Orange

Prior to 1930, orange juice was a rare breakfast treat for people outside Florida and California. Then, with newly perfected techniques of extracting and freezing concentrate, orange juice became the number-one fruit juice, passing first grape juice and then apple cider in the hearts of the public.

For the best juice, look for firm, tree-ripened fruit with no soft spots. And don't squeeze your oranges too hard; the rind contains an oil that will turn your juice bitter.

COFFEE, TEA, AND CAFFEINE

According to legend, coffee was discovered around A.D. 850 by an Arab goatherd named Kaldi. This observant herdsman supposedly noticed that goats who fed on the berries of the evergreen coffee plant sometimes frolicked all night long. Puzzled by these antics, he picked some of the berries and brewed himself a drink from them. He experienced a sense of exhilaration shortly after imbibing, and soon shared his discovery with the world. Coffee quickly became popular, not least because it helped people endure long religious services.

Tea was introduced to Europe from China in the early 1600s, and soon became England's favorite hot drink. The American colonists first preferred tea, but switched to coffee after the Boston Tea Party, which they had staged as a protest against heavy taxes on tea.

Caffeine is a natural ingredient of both coffee beans and tea leaves. It is also a powerful stimulant that acts on the central nervous system. It peps you up and makes you feel more energetic. When your central nervous system is excited by caffeine you feel fine. But as the effects wear off, you get a let-down feeling—drowsiness, tiredness, lethargy, and fatigue followed by depression. So more coffee or tea is necessary to restore your energy. This constant cycle of stimulation can be disastrous to your health.

But you don't have to choose between good health and a flavorful morning brew. There are many varieties of tasty herb teas available. Or you can simply heat some apple or tomato juice with your favorite spices.

THE JUICER COOKBOOK

Jackie Kramer with A. R. Addkison. 1978.
Cornerstone Library, Inc., 1230 Avenue of the Americas, New York, NY 10020. 112 p. paperback $5.95.

The best over-all introduction to juicing we've found yet. It has a brief section of nutritional information, discusses the different types of juicers, and talks about how to use juices in cooking.

The majority of the book is devoted to recipes: beverages, desserts, vegetables, main dishes, soups, and salads and dressings. The recipes, while they focus on natural fruit and vegetable juices, do use considerable amounts of sugar, salt, and white flour. This recipe, however, is all natural.

Go-Get-Em Cocktail

If this doesn't work, nothing will!

1	cup carrot juice
½	cup celery juice
¼	cup parsley juice
¼	cup spinach juice
¼	cup watercress juice

THE SHRUB: A CURIOUS OLD-TIME DRINK

A novel beverage concentrate can be made of berries and vinegar. To one quart of cider vinegar, add three quarts of ripe berries. Let this stand for a day, strain, and add up to one pound of mild-flavored honey. Simmer for 30 minutes, and store in refrigerator when cool. Make the shrub by adding two tablespoons of concentrate to a cold glass of water.

Rhubarb

If you're tired of strawberry-rhubarb pie, you might want to try imbibing some of your rhubarb crop. The leaves are poisonous, but the tender stalks make fine juice. Just don't drink rhubarb juice straight—all that oxalic acid will pucker you up good. Use it to liven up both vegetable and fruit juices.

Cut the rhubarb into short pieces, and place it in a pan of water—just one quart of water for each five pounds of rhubarb. Bring it to a boil, press through a jelly bag, strain (damp cheesecloth works well), and sweeten to taste with honey.

Tomato

Tomatoes are the number-one home garden plant in this country. That represents a real turnaround: a hundred years ago most Americans thought tomatoes were poisonous. (Indeed, the leaves, stems, and seed sprouts *are* toxic.) One reason for the comeback—in addition to their taste—may be that throughout much of their history, tomatoes were believed to be aphrodisiacs.

Homemade tomato juice is a hearty, nutritious beverage, and a good source of vitamins A and C, which commercial heating, handling, and homogenizing greatly reduce. It can be drunk straight, mixed with other juices or spices to make a cocktail, used as a soup base, or cooked down for sauce.

To make juice, just run quartered, cooked tomatoes through a food mill or sieve. It is essential, however, that every spot or blemish be removed, and that suspicious-looking fruits be taste-tested. One tomato with an off-flavor can ruin the whole batch.

Tomatoes are the basic ingredient of gazpacho, a cold Spanish soup ideal for summertime.

Gazpacho

3	cups ripe tomatoes, peeled and chopped (about 6 medium tomatoes)
½	cup onion, finely chopped
1	cucumber, diced, peeled, and seeded

½	cup green pepper, finely chopped
1	small clove garlic, minced
3	tablespoons parsley, chopped
2	tablespoons chives, chopped
2	cups tomato juice
1/3	cup red wine vinegar
¼	cup olive oil
¼	teaspoon cayenne pepper
	thin slices of cucumber for garnish

To loosen tomato skins for peeling, stick a fork into the tomatoes and plunge into boiling water for a few seconds. Peel immediately.

In a large bowl, combine the tomatoes, onion, cucumber, green pepper, garlic, parsley, and chives. Add tomato juice, wine vinegar, olive oil, and cayenne; mix together. Taste and correct seasoning.

Cover bowl and place in the refrigerator for at least 2 hours to chill and blend flavors. Garnish with thin slices of cucumber.

Yield: approximately 6 cups

Making juice concentrate by freezing—rather than boiling—preserves more of the nutrients.

BACK PORCH CONCENTRATE

One of the simplest methods of putting up juice is to freeze it in concentrated form. First, pasteurize the juice. Next, cool it in the refrigerator in a closed container. Keep exposure to air at a minimum. The greater the surface touching the air, the greater the vitamin C loss. When the juice is cool, pour it slowly into a bottle or jug with a narrow neck. Do not fill the bottle more than three-fourths full. Put the container into the freezer (or out on the porch in weather well below freezing), upside down. As the juice freezes, it will expand into the empty space.

When the juice is thoroughly frozen, open the bottle and place it upside down in a glass pitcher in the refrigerator. The concentrated juice essence will thaw and run into the pitcher, leaving nearly colorless ice behind in the bottle. Taste the drippings frequently. As soon as they are no longer sweet, remove the bottle and allow the ice to drain out. Twice more, return the concentrate to the bottle, and repeat freezing, thawing, and dripping. Freeze the remaining concentrate in a small container. Reconstitute by adding three parts water to one part concentrate.

Combining juices

If solitary flavors don't excite you, try mixing them. Make your own V-3, or -4, or -5 juice. Here's a cocktail that uses either garden or refrigerator leftovers.

Lush Green Drink

    1 handful Swiss chard
    1 medium tomato
    2 celery stalks with leaves
    ½ medium zucchini
   15 green beans

Yield: 2 servings

Kale Cocktail

Soft, leafy vegetables like spinach, chard, kale, or parsley will go through the juicer much more easily if twisted into a strand.

Kale is a nutritious vegetable, and persistent: it will survive well into winter, and cold weather actually improves its flavor. Kale juice, by itself, has a strong and unappealing taste. By combining kale with other vegetables, you can enjoy the benefit of its plentiful iron, vitamin C, and calcium. It doesn't wilt easily and keeps well in the refrigerator.

    8 medium kale leaves with stalks
  6 to 8 large lettuce leaves
  8 to 10 Swiss chard leaves
    ½ green pepper
    1 medium-large tomato

Yield: 1 or 2 servings

Mixed Vegetable Juice

A nice way to enhance the flavor of your juice combinations is to add different herbs or spices. But it's best to start small. Garlic and parsley, for example, are quite pungent. Mix small amounts in a base of carrot or tomato juice.

    3 medium tomatoes
    ¼ lemon, peeled
    1 celery stalk with leaves
    2 parsley sprigs
    ¼ onion
    ¼ green pepper
       celery seed
       freshly ground pepper

Yield: 1 or 2 servings

If these combinations are heartier than you prefer, lighten them up with spring water or a naturally carbonated water like Perrier or Saratoga. Or use larger amounts of thinner juices, such as celery or cabbage. Experiment to find out what proportions suit you.

Fruit juices combine well too. To make banana-orange, just put a cup of orange juice into your blender with a banana. You'll get a drink with the consistency of a milk shake, that's just as sweet but a lot better for you.

You may need a juicer for certain fruits, but when it comes to mixing juices together, or adding bananas, ice, or powdered milk, a blender works best. To thicken your drinks, you can add milk powder or yogurt. Here's an all-purpose recipe for fruit shakes.

Fruit Shake

    ¾ cup water
    6 teaspoons dry milk powder
fruit: 1 peach, 1 orange, ½ banana, ½
       cup diced cantaloupe, or ½ cup
       berries
    2 ice cubes

Combine water, milk powder, and fruit in blender. Then feed the ice cubes into the blender one at a time while the machine is running and process them until the ice is liquified.

Yield: 1 serving

Fran Wilson

Garden Way.　　　　　*Little Giant.*　　　　　*Hunger Mountain.*

Home cider presses

There was once a time when almost every American farm had a little cider press. It likely had a capacity of one bushel, and could make about two gallons of cider per pressing.

The little presses were as alike from farm to farm as the Model-T out in the barn would be a generation later. They were mass produced, and sold by mail. If you wanted one, you looked in your Sears or Montgomery Ward catalog. In 1897, you could order a Little Giant from Sears for $7.95, or the Senior Improved Cider Mill, weighing 410 pounds, for $20.50.

By 1930, the presses had disappeared from mail-order catalogs. Farmers were going to the movies on Sunday afternoons, or out for a spin. Pasteurized apple juice had joined the list of American food-processing triumphs. It had no particular flavor, but it would keep almost indefinitely. Sweet cider, by contrast, keeps somewhere between three days and a week, depending on the condition of the apples you made it from. (It then has another few days of being hard cider, an apotheosis that pasteurized apple juice never even dreamed of.)

There is a happy ending, though. A few years ago, home cider presses entered a modest revival.

Now, in 1979, there are two companies making quite different models: Hunger Mountain Crafts in Worcester, Vermont, and the much larger Garden Way in Charlotte, Vermont. Once again it is easy to make cider on Sunday afternoons.

Which press, if either, should you get? *Consumer Reports* is never likely to tell you. But maybe I can. I love making tests. I decided to compare cider presses.

Since I already owned a somewhat battered, early Hunger Mountain, my first act was to sell it and buy a new press from each company. Then I called a neighbor who has an old Sears Little Giant that she got at a farm auction ten years ago. She

agreed to bring *that* over. So one Sunday afternoon there were three small cider presses set up in a row in front of our barn. Beside each was a two-person crew and a couple of boxes of windfalls from my orchard— mostly McIntoshes, but some Duchesses mixed in. Cider is generally better if you use two or more varieties of apples.

On the left was the Garden Way. It's the biggest press of the three, and handsome in a slightly crude way. The frame is yellow pine, the hopper a high-quality hardwood plywood, the pressing tub rock maple. It comes with a nylon pressing cloth that exactly fits the tub. It is billed

(Continued on page 146)

(Continued from page 145)

to handle one bushel of apples, and to make "up to three" gallons of cider per pressing. It costs $189 at the factory in Charlotte. Delivery charges raise the price gradually to a maximum of $220 in California.

In the middle was the Little Giant. Stained and rusty from 80 years of hard work, it is not handsome at all. I doubt if it was even in 1900. What it looks is ingenious, with its cast-iron hopper and heavy flywheel. It's the kind of machine you expect to see patent numbers and "Pat. Pending" marks all over. The Connecticut Yankee could have invented it personally, before he went off to King Arthur's court. It, too, is a one-bushel mill.

On the right was the Hunger Mountain press. It's tiny. Capacity: one-third of a bushel. Claimed output: three quarts to a gallon per pressing. It is also much the best looking of the three presses, being made entirely of rock maple. (Except, of course, for the hardware, which is mostly steel.) It is so handsome that I find myself hoping it can really compete against its larger rivals.*

All crews are now ready. Two pairs of children are working the new presses. Alice and Don Lacey, who own it, are handling the Little Giant. He's our local doctor; she runs their farm.

At exactly 3:07 P.M. the timer (my wife) gives the signal. All crews begin to make pomace, though not all in the same way. One of the Garden Way crew is turning the wheel that operates the grinder, while the other feeds apples into the hopper. Ditto with the Hunger Mountain, except that here it's a crank handle that Amy Perrin is turning. But over at the Little Giant, Alice and Don are chopping apples in an enormous old chopping bowl they have brought with them, using an archaic apple chopper for the purpose. They have learned over the last ten years that their Little Giant is not comfortable with whole apples, but will grind nicely with quarters and eighths.

3:10—The first action comes from the Hunger Mountain. The pomace it makes looks practically like applesauce. After only three min-

utes, a little cider has begun dripping out of the press, even though less than one-fourth of the apples have been ground.

3:11—The Garden Way crew is having trouble. This is still early in cider season, and the apples are quite hard. The wheel is a light one. Jon Roberts, age nine, who's turning it, finds that whenever his partner puts more apples in the hopper, he loses all his momentum. The wheel does not, as it should, serve as a flywheel. He finds it hard to get started again.

3:14—Jon says he's had it, and turns the grinding over to Elisabeth Perrin, 17. She's enough stronger so that jam-ups now almost cease—though she, too, complains that the wheel lacks momentum. But the pomace is really pouring through. Alice and Don are still chopping.

3:15—Trouble on Hunger Mountain. There is now so much fine pomace in the grinder that it will no longer drop through into the pressing tub. Amy tries reversing direction; it helps hardly at all. Tracy Roberts pushes hard on top of the hopper with the wooden hopper plate. That helps some, but makes the crank much harder to turn.

3:17—My wife starts cutting apples in half for the Hunger Mountain press. She uses a common kitchen knife. That helps a lot, though the pomace *still* won't go through easily. Pomace continues to pour through the Garden Way, though there are no drips of cider yet. Over at the Little Giant, Alice is still chopping, but Don has begun to grind. The old press puts chopped apples through with ease.

3:22—The Garden Way has finished grinding a bushel of apples. It took exactly 15 minutes. The crew fit on the pressing plate, and begin to turn down the press.

3:24—At last! A stream of cider is gushing out from the Garden Way. Elisabeth and Jon look extremely pleased with themselves.

3:25—The Hunger Mountain has finished grinding its one-third of a bushel. It took 18 minutes. Amy and Tracy fit on their pressing plate.

3:27—We begin to bottle at both presses. Since this is a test, we're putting it all in quarts for more accurate measurement. Alice and Don are still chopping and grinding.

3:31—Finished at both presses. The Hunger Mountain made just over three quarts of cider in 21 minutes. The Garden Way made six quarts in 24 minutes. Twice as much cider, but it used three times as many apples. The pomace is noticeably coarser.

3:41—Alice and Don run their last chopped apple through the grinder and start pressing.

3:48—Alice and Don finish pressing. They made 5½ quarts in 41 minutes. None of the yields seem what you'd call impressive. But neither are any of the crews complaining. The cider is delicious.

That was the first of six tests we ran at intervals during cider season, and the only one that included the old Sears press. The other five tests did not produce any startling new results. The times gradually went down and the output gradually went up, but the performance ratios remained nearly constant.

The conclusion seems clear. Of the two presses currently available, the Garden Way is preferable for anyone who has access to free apples, or even to very cheap ones. It produces less cider from a given quantity of apples, but is enough easier to operate to more than make up. The one seriously unsatisfactory feature is the light wheel. Until Garden Way starts supplying a much heavier wheel with its presses, the buyer would do well to look for a replacement. Any good junkyard should offer a choice of cast-iron wheels—or if you happen to know the right collector of decorator items, there is an even simpler way. Quietly remove the heavy wheel from his old Sears press, being careful not to disturb the jonquils in the pressing tub. Replace it with your light one. This is quite decorative, and you're actually doing him a favor.

So much for people with free apples. Anyone who has to buy apples at retail prices would fare better with the Hunger Mountain. It yields almost a gallon more of cider per bushel of apples—and if you are paying $4 or $5 for that bushel, much more if you are buying plastic sacks of apples in a store, the extra production time will be worth it.

In a strictly economic sense, to be sure, neither press is "worth it." Even with free apples, you would have to make 100 gallons of cider

*As of May 1980, the Hunger Mountain is not in production. *Editor.*

just to pay for the press—and then ten more to pay for the 110 containers. By then you would be so far behind in paying for your time that you might never catch up. But anyone who views life in these terms is probably drinking pasteurized apple juice anyway, and should keep right on.

Noel Perrin
Reprinted by permission from *Blair & Ketchum's Country Journal.* Copyright © 1979 Country Journal Publishing Company, Inc.

Sources for cider presses

Cumberland General Store, Route 3, Crossville, TN 38555.
Cumberland's fat, fun $3 catalog features the Garden Way press and carries other grinders and presses, the most expensive of them being electric.

To store the squeezings, oak and basswood kegs are available in sizes from 1½ gallons to 30 gallons, priced from $19.70 up to $59.90. Used whiskey barrels (42 gallons) go for $18; you may be able to pick these up locally by consulting the Yellow Pages.

Day Equipment Corp., 1402 E. Monroe, Goshen, IN 46526.
Day is primarily commercial; their smallest model goes for $600.

Garden Way, Charlotte, VT 05445.
In addition to the press mentioned in the above article, Garden Way's catalog carries mills, dehydrators, sprouters, baking paraphernalia, canning equipment, and more.

Good Nature Products, Inc., Olean Road, P.O. Box 233, East Aurora, NY 14052.

Last fall in Potomac, Maryland, writer Vic Sussman made 70 gallons of cider with a Good Nature shredder and press. It's the medium-priced of three models, and he bought it for about $1,000—a considerable sum, considering that the Sussmans own no apple trees. But Vic says that a wealth of unsprayed apples is to be had from unused trees for the asking. His press can handle 2½ bushels at a time, and in one day cranked out 26 gallons (with a lot of muscle power at the crank).

Happy Valley Ranch, P.O. Box 9152, Yakima, WA 98909.
The Homesteader single-tub press, with grinder, sells for $249. A double-tub model is $369. Kits are available. ∎

Juicers

The taste of fresh, homemade juice sells a lot of juicers. But buying a juicer is a major expenditure, and the choices are both many and wide.

There are three basic types of juicers: the reamer or squeezer, which juices only citrus fruits; the centrifuge, handling vegetables and some fruits; and the steamer, good with both vegetables and non-citrus fruits. The consistency of juice varies from one type to another: a reamer or squeezer makes clear juice with some pulp; a centrifuge's product is usually creamy or foamy; and a steamer's is always clear.

Some people feel that juicers aren't worth the investment because a lot of produce yields relatively little juice. For example, one pound of carrots yields approximately one cup of juice. But there are advantages to this reduction. Those with an excess of fresh garden produce find it a blessing. And this concentration means that a glass of juice can contain many more nutrients than a regular serving of the same food.

As you read through our evaluations, note that not all juicers can be used with all produce.

Reamers

If you have an abundance of oranges, lemons, limes, and grapefruits, a reamer is right for you. A reamer is little but a grooved cone over which a halved fruit turns under pressure. (It's important to halve the center and not from end to end.) Some manual reamers have cranks to turn the cone. Electric models do the spinning for you. The juice is fairly clear, with some pulp.

The most familiar reamer is the $2 glass model found in many kitchens. But reamer prices range on up to $40, $80, and even $150. Some are made of hard plastic, some of

(Continued on page 148)

(Continued from page 147)

stainless steel. The more expensive ones are intended for restaurant use but occasionally creep into homes because they're so fast and powerful. Beware of those made of cheap, flimsy plastic—more likely than not, you'll need another soon.

Most reamers are easy to clean, although some cannot go into a dishwasher. If you are hurrying off to work, they can be quickly disassembled, leaving the parts to soak in warm, sudsy water. Electric models are usually just wiped clean because the base that houses the motor cannot be submerged in water. However, the cone can be soaked.

A valuable feature to look for is a cone that can ream different-sized fruits, from limes to grapefruits.

Squeezers

We coined our own name for this category of juicer. These machines do not rub any of the pulp from citrus fruits; rather, their principal action is squeezing. They generally produce less juice than reamers, and the juice is clearer.

There are two styles of squeezers. The most familiar is the tall, heavy commercial model that squeezes a cone against the fruit half. It works on all citrus fruits except grapefruits, which are too large. A plunger, attached to a lever arm, presses against the fruit, squeezing the juice out.

The other kind of squeezer juicer pushes the fruit together from the

This centrifuge basket holds the pulp and must be emptied periodically by hand. The result: more work, and more juice.

sides. A catch basin collects the juice, and must be emptied after every fruit half is squeezed.

Centrifuge juicers

If you plan on juicing more vegetables and hard fruits than citrus fruit, consider a centrifuge juicer. This type is favored for vegetable juices, but prices range from $80 to $300. The centrifuge works by grinding the produce at high speed on a whirling metal disc with tiny blades. The whirling produces a centrifugal action which throws off the juice. Pulp is trapped on a strainer.

On some models, an automatic pulp ejector removes the pulp from

the inside of the strainer and into a separate compartment, eliminating the need to stop the machine for cleaning. All of the automatic-pulp-ejector models we tested were offensively noisy, and though they produced juice faster, they did not get out all the available juice. That's important. Since a whole pound of produce gives only a cup of juice at best, you'll want to squeeze out every possible drop.

Centrifuges work best with hard fruits and vegetables, such as apples and carrots. You can feed carrots and celery into the centrifuge without removing the leaves. Radishes, beets, onions, and parsley also juice well in a centrifuge. Softer foods, including cucumbers, tomatoes, pears, and grapes, won't give up all of their available juice to a centrifuge; very soft or over-ripe fruits just turn into mush.

Some centrifuges can handle citrus fruits adequately. They extract less juice than reamers and omit fiber in the form of pulp. But all the activity of a centrifuge produces a very pleasant, foamy orange or grapefruit juice. The foam really makes it special. Still, citrus juice made with a reamer is usually sweeter and pulpier because the reamer doesn't get any of the bitter skin.

Cleanup is not something to look forward to with centrifuge juicers. Even a few seconds of juicing will make a mess.

Manufacturers' literature repeatedly warns about proper cleaning techniques. Some grades of plastic

This centrifuge basket automatically throws off the pulp, so the juicer can run for a long period without being emptied.

This juicer top has many corners and crannies where pulp can collect, so cleaning is time-consuming and tedious.

cannot withstand hot water and will stain from juice if not cleaned immediately. Stains can be removed with a mild bleach solution.

Juicers without automatic pulp ejectors have fewer parts to clean and there are few, if any, nooks and crannies to negotiate. Still, there's work to be done. Pulp has to be scraped from inside the centrifuge basket and dumped. This process can be speeded if there is a liner inside the strainer basket: when the liner is pulled out, it releases compacted pulp from the basket. The small holes in the strainer need the help of a brush to come clean. All other parts require a sudsy rinse and the base (which cannot be submerged) will come clean with a damp cloth.

Centrifuges with automatic pulp ejectors are harder to clean because their molded plastic bodies have many tight corners that are hard to reach even with the help of a brush. Immediate rinsing helps, but it's clear that those who engineered the machine didn't have cleanup in mind while at the drawing board.

Steamers

If you have a bushel of apples just waiting to be eaten, but not enough people to do the job, a steam juicer will turn them into clear juice with a minimum of effort. Steamers retail from $35 to $80, depending on size and construction materials. Stainless steel, enamel, and aluminum models are available. It's easy to boil a steamer dry, so spending more for one with a reinforced bottom might be a good idea. For durability, it is wise to choose a model that has fruit and juice pans made from stainless steel.

A steam juicer consists of three pans, one on top of another. A shallow-bottomed pan sits directly on the stove burner and is partially filled with water. The next pan, shaped somewhat like an angel food cake pan, allows steam to rise through a cone-shaped opening in its middle while catching juice that drips from the perforated fruit basket above. This basket holds the sliced fruits or vegetables. Steam from the bottom pan rises to draw juice. A rubber tube drains the pan, and a clamp on

(Continued on page 150)

Reamers

Acme

$29.95. Juices 6 oranges in 6 min. 20 sec.

This reamer is a white plastic attachment that fits onto the $189.95 Acme centrifuge juicer. It locks into place easily, and doesn't engage until you press the fruit half down on the cone. It produced more juice than any of its competitors, and the powerful reamer never faltered. But the comparatively high speed of the reamer made the orange difficult to hang onto—especially at the end when we were pressing hard to get every last drop of juice.

The Acme reamer handled all citrus fruits, and produced smooth, pulpy juice. The action of the reamer shredded the pulp somewhat, so that it was not at all obtrusive. A few seed bits did find their way into the juice.

Cleanup was time-consuming. It took the longest of all the reamers—close to 5 minutes. A brush for cleaning the strainer would probably speed things up.

Acme, Tenth & Lowther Streets, Lemoyne, PA 17043

Krups Pressa

$40. Juices 6 oranges in 4 min. 52 sec.

The electric Krups Pressa is easy to use: you press down on the reaming cone to start it spinning.

The Krups has plenty of power, and performed well on grapefruits, oranges, lemons, and limes. Grapefruits were too large to fit tightly on the reamer, but they can be juiced by pressing the fruit against the cone one section at a time. Juice volume was moderately high. Juice from the Pressa had some small seed bits in it, but it was nicely pulpy and had a good taste.

Cleanup time was under 2 minutes.

Robert Krups North America, Allendale Industrial Park, 3 Pearl Court, Allendale, NJ 07632.

Lillian Vernon

$4.98. Juices 6 oranges in 17 min. 25 sec.

This small, plastic juicer is operated by a crank on the side. A lot of pressure is needed to extract any juice; our tester had a sore arm at the end of the six-orange test. The sieve did not drain well and left a wet pulp, and we spent extra time trying to press more juice out of it with a spatula. The sieve was small and had to be cleaned after every two oranges.

Juice from the Lillian Vernon was clear; it had the least pulp of all we tested. Some seed bits did get through but these were easily swallowed. The volume of juice produced was the lowest for all the reamers—a third less than the most productive we tested. The Lillian Vernon had mixed success with grapefruits, but was able to get juice from oranges, lemons, and limes.

Cleanup took 1 minute.

Lillian Vernon, Box LV, Dept. OXGE, 610 S. Fulton Avenue, Mt. Vernon, NY 10550.

The several parts of a steam juicer.

(Continued from page 149)

the tube keeps you in control of the draining. Your only responsibility in this process is to periodically check to make sure the water hasn't completely evaporated from the bottom pan.

Many steamers now come in the nine-liter size (slightly more than two gallons). This size will make about seven quarts of juice from a basketful of high-yield produce such as apricots or tomatoes. Grapes and cherries will yield about four quarts of juice, and low-yield fruits such as apples and peaches, only two quarts or so.

The whole process takes from 1½ to 3 hours. The best thing about steamers is that the fruit can be used whole; cores, peel, seeds, and even stems can be tossed in the basket. When the pulp has been juiced, the remainder can be put through a strainer and dried to make fruit leather.

Juice that comes from a steamer can be canned or frozen immediately, since it has gone through a pasteurization process during heating. But remember that this process

Livingood Citropress

$12.95. Juices 6 oranges in 7 min.

This stainless steel juicer has an unusual design. A barbed lid hangs on to the fruit while a crank on top turns the fruit. (With most other reamers, the *cone* turns.) Success doesn't come easy until you read the directions; then, the unit does just what it claims to do.

The reamer is made of rigid wire bent into a cone. This did a fine job of separating the juice from the fruit, but there was nothing to filter out seeds and larger hunks of pulp—so, the volume of juice was quite high but the quality was low. It was full of seeds and pulp. Seeds float to the top in time, and can be skimmed off the top. Most juicers don't require this step.

The Citropress is lightweight and not intended for heavy use. We sometimes had difficulty getting the fruit properly centered. This reamer was not comfortable to use for a long period of time. After several oranges, our arm muscles began to tire

rapidly. Grapefruit fans will rue the fact that these fruit are too large to fit inside.

Cleaning time is just ½ minute.

Livingood Enterprises, 2716 Hubbard Lane, Eureka, CA 95501.

destroys some of the vitamins and all the enzymes.

In addition to making juice, a steamer will steam meats, vegetables, breads, and puddings. See the piece on the Mehu-Maija (below) for a recitation of all that a steamer can be.

DB

Testing juicers

It's not hard to record times, weights, and dimensions—these are objective measurements no one can argue with. But when it came to rating the aesthetic qualities of juice, we had difficulties. Should juice ideally be clear and free of pulp? Or rich with fiber (and implicitly healthier and less wasteful)? Our taste testers found they had fairly similar standards for juice. From reamers and squeezers, they liked citrus juices with a moderate amount of pulp. Very clear juice wasn't as pleasant as juice with some texture to it. Clear juice tended to be sweeter—almost too sweet. But there was a point at which pulp was too thick or too lumpy to swallow.

In general, our taste testers liked a creamy juice from the centrifuges. They didn't like small food fibers in the juice. They preferred some body, but clear juice was favored over thick or pulpy juice, and bits of leaves or chunks of food were not acceptable. They also concurred that a moderate amount of foam—up to ¼ inch on top—was pleasant, but more than that was too much of a good thing.

Juice didn't differ from one steamer to the next, so judging quality wasn't a problem. With apples, however, there remains the question of how long to run a batch: steamers continue to extract color from the skins long after the apple flavor is gone.

DB

Mrs. Hibbard and the Mehu-Maija

I call it my magic machine, although it has no moving parts and nothing to break down, come unscrewed, or fall apart. The Mehu-Maija (meh-hoo my-yah) extracts

(Continued on page 152)

Rival Juice-O-Matic

$29.95. Juices 6 oranges in 7 min. 5 sec.

We found this plastic juicer easy to operate with all citrus fruits. It has an on–off switch, rather than relying on pressure to engage the reaming cone. The Rival was powerful enough to keep running no matter how much pressure we applied. The large sieving basket strained out hunks of pulp and seeds and kept the machine from clogging during the six-orange test. The Rival was heavy and balanced enough to keep it stable during juicing.

The Rival extracted a good volume of juice. The juice itself was nicely pulpy but rather thick and it contained many tiny seed bits.

Cleanup took almost 4 minutes.

Rival Manufacturing Co., P.O. Box 19517, Kansas City, MO 64141.

Waring

$14.99. Juices 6 oranges in 11 min.

This electric reamer, made of white, yellow, and clear plastic, has a fine, clean-cut look. Unfortunately, it worked slowly and lacked power. Although it can handle any citrus fruit, it struggled with them all, as our six-orange test suggests. The strainer for pulp and seeds was small and had to be cleaned out after juicing three oranges. The pulp from the strainer was so wet we took some extra time to force it through the strainer with a spatula. When juicing, the machine walked and wobbled a bit.

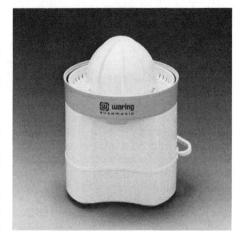

The juice from the Waring was nicely pulpy but there were many tiny seed bits in it. The volume of juice was low for a reamer.

Cleanup took under 2 minutes.

Waring Products Div., Route 44, New Hartford, CT 06057.

Squeezers

Cumberland

$13.94. Juices 6 oranges in 5 min. 35 sec.

While this aluminum juicer looks like the Foodco, it worked better for us. Resting on four legs and a wide base, the machine doesn't have to be held steady with one hand. The Cumberland has an arm connecting the base to the pressing area, providing a good source of leverage. You can get all the juice out in one squeezing.

The Cumberland's reservoir fills up quickly with juice and must be emptied after each orange. The pulp has to be cleared away after three oranges.

Juice from the Cumberland had no seeds, but orange juice had lumps of pulp and a slightly bitter taste. (Lemons and limes did not produce lumpy juice like oranges did.) It ranked in the middle of the three squeezers for volume produced.

The pressing area was not large enough for grapefruit but it did accommodate

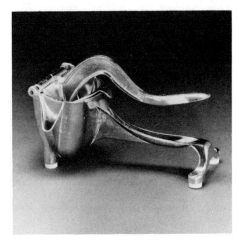

other citrus fruits.

Cleanup took a quick ½ minute.

Cumberland General Store, Route 3, Crossville, TN 38555.

Foodco

$23.95. Juices 6 oranges in 10 min. 15 sec.

This cast-aluminum hand juicer employs a plunger to squeeze fruit halves from the sides. A toothed slot sieves out seeds and pulp from the juice.

During our tests, this slot clogged periodically and had to be cleared with a probing finger. After one squeezing, oranges had to be shifted and squeezed again to get all the juice.

The small base allowed the unit to tip when we were pressing the plunger and there was no way of applying leverage to the handle for hard squeezes.

The Foodco will not accommodate grapefruit. Of all the reamers and squeezers we tested, this model produced the least juice. Its juice was clear and had very little pulp, but it tasted somewhat bitter.

Cleanup (including time to clean up the

table) took 2½ minutes.

Foodco Appliance Corp., 405 E. Marion Street, Waterloo, IN 46793.

(Continued from page 151)

juice by steam that rises through the food being processed and then drips into a self-contained juice collecting pan. The whole rig costs less than $50 and I saved that much the first year by not buying stuff to drink, not to mention other products I made. (See evaluation, page 158.)

In one year, working together, we painlessly, simply, and cheaply bottled 375 fifths of juice. We made countless quarts of soup stock and pounds of tomato puree and spaghetti sauce. We made fruit butters: apple, peach, pear, plum, grape, cherry, currant, and strawberry. We disguised dozens of summer squash I couldn't bear to waste. We made the juices for wine and jelly for our Christmas presents. We made fruit leathers. We even fed the compost with the leavings.

The real beauty of this machine is that I don't have to spend hours peeling, paring, coring, sorting out bits of leaves, and stirring. I do strawberries and currants and elderberries stems and all. I wear gloves and strip gooseberries right off the canes. Grapes go in by the bunch. None of our stuff is sprayed so I don't even have to wash them.

Many's the time I've slid summer squash in with whatever fruit was being extracted—as much as 25 percent by volume. It makes me feel virtuous not to waste even a lowly zucchini, and nobody has ever noticed the difference. You can only give away so many squash. Juices with squash make jelly with no trouble at all.

After the first few bottles of juice, I'm left with a mess of pulp. I make a second extraction simply by doing the whole works over again. A look at the steamed-out pulp may make you think that's foolish, but it isn't. After all, the bottles are free, and they store neatly in liquor store boxes which we stack clear to the ceiling on the north side of our house for insulation. We cover the windows right over (this is on the inside of the house, of course.)

I'm writing this in bed at 10:30 on a cold and frosty morning. The kids are all grown with kids of their own. I'm just getting used to not feeling guilty if I'm not up at 6 o'clock rushing around. I have pleasant memories of the years spent caring for a full house, canning and juicing, with kids and animals underfoot, and the stove always too small for enormous pots.

When I juiced, if I'd just stuck to juicing, the stove would have been large enough. But I used the pulp, and that's when I ran out of room. A sensible person would add the pulp to the compost pile, or feed the pigs and chickens. Not I. Housewifery is deadly dull if you don't use your creative abilities. That's a flat statement of fact. Buying stuff at the store and collecting waste cans, waste paper, foil, and green stamps is a big bore. So I'd be juicing one batch and making something out of pulp in another—something like fruit leathers, made from sieved pulp and honey, and spread on oiled cookie sheets to dry. Creative cooks can even see the possibilities in vegetable pulps. When I extracted potato juice, we had mashed potatoes mixed with hot milk and butter. Other times I used the potatoes in shepherd's pie, ham casseroles, potato soup, stew, pancakes, or bread dough.

My Mehu-Maija does more than just make juice and pulp. I use it to steam onions to loosen the skins. It's a quick way to process a big batch of onions for canning, freezing, or making soups and relishes. You don't cry when you skin steamed onions and the skins loosen in minutes.

Shelling peas is a lot of work when you're shelling to freeze quarts and quarts. I steam them in the pod until the color changes to bright green, then plunge them in ice cold water, drain, and shell. Presto! They shell like greased lightning and need not be further processed. Simply shell and freeze.

This juicer will steam clams, lobster, and fish, or old-fashioned cloth-wrapped puddings, breads, and grains. I use it to steam meaty bones and fowl racks. And, of course, I save the extracted juice for soup and gravies.

Its great virtue is that I needn't watch a boiling pot to keep it from sticking. It's a time-saver. It allows me wintertime leisure to make other things from its extractions. It's a joy. I couldn't live without my Mehu-Maija.

Betsy Hibbard

Hamilton Beach

$169.90. Juices 6 oranges in 5 min. 10 sec.

This plastic-and-stainless steel juicer is common in restaurants, and the design seems intended for the harried waitress. The plunging top swivels out of the way to allow room to get at the squeezed rind. A cup on a spring-loaded arm sits ready to catch the juice, and can be quickly removed for dumping. The cup swings out of the way to allow filling a glass. It is an extremely heavy unit which stays in place without clamps—or even the help of a second hand.

The juice from this squeezer was quite clear of pulp and very sweet. It produced a moderate amount of juice—which is understandable as there was so little pulp.

The expensive Hamilton Beach is a restaurant-quality machine, but our great expectations for it weren't always met. Cranking the pressing arm down wasn't a smooth procedure. Also, we found it was tricky to remove the fruit rind. Grapefruits were too big for the Hamilton Beach, but it worked well on oranges, lemons, and limes.

Cleanup involves washing three removable parts and wiping down the whole unit—a 1½-minute procedure. Because of its heft, you would most likely want to find a permanent spot for this juicer on your counter.

Hamilton Beach, P.O. Box 1158, Washington, NC 27889.

Centrifuge juicers

Acme Juicerator

$149.95 (plastic), $189.95 (stainless steel)

The Acme is one of the most popular centrifuges. The housing is made from cast aluminum and stainless steel, while the parts that contact the food come in either stainless steel or plastic. Citrus fruit can be juiced with an optional reamer attachment.

The Acme is a quiet, solid machine that runs virtually without vibration. It consistently extracted more juice from various foods than the other centrifuges did. The juice was quite clear, and contained little pulp. The motor never struggled, even when large carrots were fed rapidly into the machine.

Cleanup was not difficult. If you use disposable filters, it's fairly easy to get out the pulp, although cleaning the centrifuge basket does require a brush. Because the machine has no automatic pulp ejector, juicing large quantities necessitates stopping periodically for cleaning.

The Acme does a fine job with firm fruits and vegetables, and the centrifuger can also handle citrus fruits. It comes with a ten-year guarantee.

Acme, Tenth & Lowther Streets, Lemoyne, PA 17043

Preserving juices

Fruit and vegetable juices can be preserved as easily and successfully as any other food in the home. There is only one problem: how can you retain the fresh, uncooked taste while heating the juice enough to reduce or eliminate enzymes and microorganisms that can spoil its flavor and wholesomeness?

Because just-picked flavor is essential to preserved juice, freezing is the best preservation method. Thawed juices taste uncooked and retain the lion's share of vitamins. Canning is a satisfactory alternative (and some juices, such as tomato or vegetable cocktail, actually taste better cooked). Juices also can be stored as syrups or in a stable fermented state, like wine and hard cider.

Any method of processing takes its toll on vitamins. Some are retained nearly intact (the B vitamins) while the vitamins A and C are reduced by one-third on average. While the main culprit is heat, the length of heating is also a factor. The rule of thumb for cooking juice: briefer is better. Heat the product just enough to inactivate enzymes and to safely wipe out microorganisms.

Tomato juice can lose as much as one-half of its vitamin C in just the time it takes for a cook to find a stray canner lid. Cool juices as quickly as possible by submerging storage containers into a series of hot-to-cool water baths. (See how in the instructions for citrus juice, below.) Use only utensils made of stainless or enameled steel. Other metals may react with the acid in juice, to pollute, color, and flavor your drink. In addition, copper destroys vitamin C.

Pasteurizing and freezing

To pasteurize, first fill the bottom of a double boiler with water, and bring to a rolling boil. Put juice into the top pan, and heat to 190°F. Use a candy or jelly thermometer to assure accuracy. As soon as the juice reaches 190°F., remove the upper pan from the double boiler and pour the juice into containers for freezing or canning. If you are pasteurizing large amounts, you'll have to move quickly to insure the juice does not drop below 185°F. before reaching the storage container. If the juice does cool, return it to the top of the double boiler and reheat to 190°F.

To *freeze* juice, pasteurize (if di-

A freezing jar is wider at the top than at the bottom so that contents can be removed easily when only partially thawed.

rected to do so by the instructions below) and pour into glass jars or freezer containers, leaving 1-inch headspace for expansion. Seal the container and freeze immediately.

Note that the following procedures vary somewhat. The high temperatures of a long boiling water bath can ruin the flavor of many fruit juices; for this reason, some juices are poured boiling hot into sterilized jars and sealed without further processing. Others are processed in a *hot water* bath (the water simmers steadily at 185 to 190°F., but not boiling). A few must be processed in a *boiling water* bath (for specific information on this procedure, see the chapter on canning).

Berry juice. Wash, crush, simmer until soft. Add water only if necessary to prevent scorching. Strain through a jelly bag for several hours, until the pulp releases no more liquid.

Freezing: Pasteurize. Pour hot into freezer containers.

Canning: Pasteurize. Pour hot into hot jars, leaving a ¼-inch headspace. Process pints and quarts for 30 minutes in a 190°F. *hot water* bath.

Apricot/peach nectar. Not really a juice, this beverage is preserved as a puree to which ice water is added before serving. Wash, drain, pit, and measure fruit. Add 1 cup boiling water to each cup fruit. Cook until fruit is soft, then puree. Add honey, if desired, and 1 tablespoon lemon juice to each quart. Pasteurize.

Freezing: Pour into containers and freeze.

Canning: Leave a ¼-inch headspace. Process half-pints and

pints for 15 minutes and quarts for 20 minutes in a *boiling water* bath.

Citrus juice. Freezing: Pasteurize. Pour juice into *glass* freezer jars, cool, and freeze immediately. Store up to 3 or 4 months.

Canning: Pasteurize. Working quickly, pour juice into hot, sterile canning jars, leaving a ¼-inch headspace. Cap the jars and turn upside down for 3 to 4 minutes to sterilize the caps. To retain vitamin C, cool the jars quickly. To avoid breaking the glass, first place the jars in a large kettle or canning pot filled with water at 120°F. After 5 minutes, remove one-third of the water and replace it with cold water. Repeat 3 times at 5-minute intervals. Finally, run cold water into the kettle for 5 minutes, or until the juice is cooled. Store in a cool, dark place.

Cranberry juice. For every cup of berries, use 1 cup of water. Boil in a pan until the berries burst. Strain through a jelly bag, then return to boil for 1 minute. Add honey if desired.

Freezing: Pour boiling juice into jars and cool. Freeze immediately.

Canning: Pour boiling juice into sterilized pint or quart jars, leaving a ⅛-inch headspace. Seal jars. No further processing is needed.

Grape juice. Place grapes in a pot and cover with boiling water. Simmer until the fruit is very soft. Strain through a jelly bag or several thicknesses of cheesecloth. For clear juice, refrigerate for 1 or 2 days and strain once more. If desired, heat to a simmer and add honey to taste.

Freezing: Pour hot juice into sterilized jars, cool, then freeze.

Canning: Pour hot juice into jars, leaving a ¼-inch headspace. Process pints and quarts in a 190°F. *hot water* bath for 30 minutes.

Tomato juice. Cut up tomatoes and simmer until soft. Press through a sieve or food mill. Add spices as you wish.

Freezing: Pour juice into sterilized jars, cool, and freeze.

Canning: Reheat juice to just below boiling and pour hot into sterilized jars, leaving a ¼-inch headspace. Process pints 10 minutes and quarts 15 minutes in a *boiling water* bath.

Carol Keough

Krups Biomaster

$80.

The handsome, white plastic Biomaster comes with an aluminum strainer basket and stainless steel cutting blades. It worked quietly and produced a good volume of juice.

The Biomaster made the clearest juice of all the centrifuges, and rated high with our taste testers. It did a fine job with firm fruits and vegetables, but the manufacturer does not recommend juicing citrus fruits. Indeed, when we started feeding it oranges, the machine began to wobble rather violently.

In order to catch all the juice from the Biomaster, you need to hold a cup right up to the spout. This is because the cooling fan blows air past the drain area, causing the juice to blow out the spout.

Because there is no automatic pulp ejector, the basket needs to be emptied frequently. Otherwise, the machine will begin to shake and walk across the counter. The motor is less powerful than some of the other centrifuges, and carrots must be fed slowly to prevent the motor from straining. The Biomaster should not be run for more than five minutes at a time.

Cleanup was rather simple, especially when the liner was used in the strainer basket. The white plastic housing, however, needs to be cleaned immediately or it will stain.

The Krups Biomaster works well with most firm fruits and vegetables. It has a one-year guarantee.

Robert Krups North America, Allendale Industrial Park, 3 Pearl Court, Allendale, NJ 07632.

Miracle CE 11

$99.95.

This machine has a white plastic body, a clear plastic top, and stainless steel blades. It produced a good volume of juice, but the juice was very pulpy. As with other pulp-ejector models, the motor was quite loud, and we found it difficult to hear anything else while it was running.

The CE 11 did a fair job with citrus fruits, but the motor had trouble juicing harder vegetables. Carrots often kicked sideways and slowed the motor almost to a stop, and then were forced through without being fully ground.

Cleaning was tedious and time-consuming. There are many sharp corners that are hard to reach, and the separating basket must be cleaned with a brush. Juice leaked onto the motor housing directly below the cutting blades, and this could cause mechanical problems.

The CE 11 comes with a brief set of instructions and a one-year limited guarantee.

Miracle Exclusives, Inc., 3 Elm Street, Locust Valley, NY 11560.

THE IMPORTANCE OF BEING FRESH

Make juice from fresh produce only, never from obviously over-ripe or rotting produce, or the flavor and nutritional value will suffer. Fresh, raw juices taste best drunk immediately. For storage, they should be tightly sealed and refrigerated, but ideally for no longer than 48 hours. If you must keep juice longer than two days, freeze it.

Carrot juice offers a graphic illustration of the importance of freshness. Carrots produce rich, creamy, orange-colored juice. It's almost sad to drink it. But just 20 to 30 *seconds* after carrot juice has come from the juicer it begins to turn brown. Oxidation is taking place, and the nutrients are beginning to deteriorate. Even with proper storage, the fresher the juice, the better.

THE CHOP-RITE COMPANY: STILL THE SAME

The Chop-Rite Company has been around a long time. I know because I frequently see their products for sale at farm auctions here in Pennsylvania Dutch country. They're always well worn and have acquired plenty of character, but they still work. Few auction-goers realize that these sturdy hand-cranked machines can be bought new, from Chop-Rite of Pottstown, Pennsylvania. The products haven't changed much over the years, so replacement parts are available even for the antiques.

Chop-Rite's products include hand-cranked food and meat choppers (with sausage stuffing attachments), a cherry pitter, a juice extractor, and a combination sausage stuffer and lard or fruit press.

Someday I'd like to visit the Chop-Rite folks; reading through their booklets is like stepping back into the late 1800s.

At one auction three years ago I watched as a combination sausage stuffer and press was swept up for just $9. (New, with all the attachments, the machine costs nearly $200.) The woman who bought it wasn't sure if she would use it as a planter or a lamp base.

DB

Miracle XM 10

$129.95.

Unlike the other centrifuges we tested, the XM 10 has a blender attachment. It resembles its little brother, the CE 11, in appearance, with similar plastic construction, stainless steel blades, and lots of noise.

The motor in the XM 10 struggled a bit with carrots, but otherwise ran smoothly. It produced less juice than most of the other centrifuges, but the texture was very good. It handled citrus fruits adequately.

Cleaning was not easy, although no juice leaked onto the housing below the blades. Considerable poking, scraping, and brushing were required to get the machine clean.

Overall, the XM 10 performed well, and the blender (which we did not test thoroughly) might be a plus to some buyers. The brief instructions come in both Italian and English, and the machine has a one-year guarantee.

Miracle Exclusives, Inc., 3 Elm Street, Locust Valley, NY 11560.

Miracle Ultra-matic

$299.50.

The Ultra-matic struck us as an attractive, serious-looking juicer, with its stainless steel body and blades. It has a plastic top and pulp holder. The powerful motor ran smoothly, even when faced with heavy loads.

The motor was almost too powerful, in fact. It flung pulp out of the machine before all the juice was extracted, explaining the low output. Juicier fruits, such as oranges, were sometimes thrown back up the feed hole, so a hand or plunger needed to be held over it. And like all the juicers with automatic pulp ejectors, the Ultra-matic was noisy.

Cleaning took time, but it was easier than with the other pulp ejectors. A plastic pulp catcher sits under the discharge chute, and allows for quick dumping. You must be sure the catcher is aligned properly, however, or you'll have a minor mess.

Rather than simply resting on the housing, the cutter blades screw on and off with a large plastic key—a nuisance, especially when your hands are covered with pulp.

The Ultra-matic is well made, but we found its performance does not meet the same standard. It comes with a one-year guarantee on parts and any defects, and a five-year guarantee on the blade.

Miracle Exclusives, Inc., 3 Elm Street, Locust Valley, NY 11560.

Oster Automatic Pulp Ejector

$86.

The compact, businesslike Oster performed well, even with heavy loads, despite its small motor. Construction is all plastic, except for the stainless steel blades.

The Oster placed near the top in quantity of juice produced, but near the bottom in quality. The juice was very pulpy, sometimes even lumpy. The strainer let through cabbage pieces the size of a dime, for example.

Operation was fairly quiet for a juicer with an automatic pulp ejector, and the ejector itself worked well. This was the only unit with a juice catcher. The catcher did its job as long as it was aligned perfectly in the rack. If not, some juice spilled.

Cleaning is difficult. There are many nooks and corners for food pulp to hide in, and these must be scrubbed with a brush. The cleaning directions are explicit: do not use hot water; do not submerge the base; do clean the plastic immediately or it may stain.

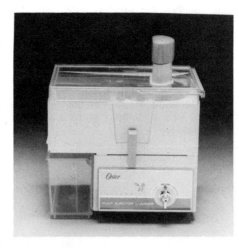

The Oster will juice firm fruits and vegetables, as well as citrus fruits. It comes with a booklet of instructions and recipes and carries a one-year limited guarantee.

Oster, 5055 N. Lydell Avenue, Milwaukee, WI 53217.

STOCKING UP
Carol Hupping Stoner, editor. 1977. Rodale Press, 33 E. Minor Street, Emmaus, PA 18049. 532 p. hardcover $13.95.

For three years we have preserved the fruits and vegetables of our summer harvest and enjoyed them all through the year, supplementing with little, if any, commercial produce. *Stocking Up,* by the editors of *Organic Gardening* magazine, is the only book we've used to become veterans in home food preservation.

This complete guide gives detailed instructions for preserving almost everything—fruits, vegetables, dairy products, nuts, grains, and seeds—without chemicals or sugar.

Heather Stoneback

Steamer juicers

Ashdown House Steamer

$69.95 (stainless steel, satin finish), $76.50 (stainless steel, brushed finish), $95.50 (brushed finish, aluminum-clad bottom), $105.50 (highly polished, aluminum-clad bottom).

Any part of the Ashdown House Steamer that touches food is made of heavy stainless steel, except the rubber tube that drains the juice. The customer is offered pots with different finishes, and either an aluminum-clad or stainless steel bottom. It's somewhat tall (17¼ inches) and has 2-inch handles, so you need lots of shelf room.

Ashdown House, 612 E. Pheasant Way, Bountiful, UT 84010.

Green Drink with Tomatoes

3 tomatoes
1 large cucumber
4 large sprigs of watercress, both
    leaves and stalks
4 spinach stalks, and leaves
1 cup cold water
1 tablespoon kelp
6 basil leaves
4 sprigs of dill
1 cup yogurt

This green drink is best if you make it just after you pick the ingredients. Wash and cut up the ingredients and put them a few at a time into a blender that already has the cup of water in it. Start with the tomatoes and greens, and last of all add the yogurt. Only blend long enough to liquify.

This is a very cooling and refreshing drink. You can also include sprigs of wild plants such as lamb's-quarters, winter cress, plantain, and sorrel.

If you wish to can or freeze this drink, don't add the yogurt before doing so. Rather, add it at time of serving.

To can, bring juice to a boil and pour into sterilized jars, leaving a 1-inch headspace. Process pints 20 minutes and quarts 30 minutes in a pressure canner at 10 pounds pressure.

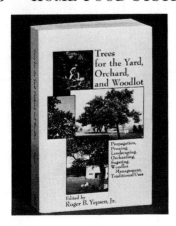

TREES FOR THE YARD, ORCHARD, AND WOODLOT.
Roger B. Yepsen, Jr., editor. 1976. Rodale Press, 33 E. Minor Street, Emmaus, PA 18049. 305 p. hardcover $8.95, paperback $6.95.

The writers of this book obviously like trees, and speak of them in warm, almost human tones. "After addressing a huge fallen oak with a handsaw it's hard not to come away impressed. Trees are noble even in repose, lying on the forest floor like beached serpents."

The book, as the title implies, covers a broad range of subjects. Selection and care of trees, prevention and treatment of tree diseases, and use of wild trees are all treated in depth. Then there's a chapter entitled "Forgotten Uses for 120 Trees: Food, Furniture, Dyes, Drink, Medication, and More."

The chapter "The Home Fruit Orchard" gives you about everything you need except the seedlings. (And if you want to start your trees from seeds, that's covered in Chapter Four, "Propagating, Grafting, and Pruning.")

TD

Fruit trees offer an advantage not generally found in the typical shade tree—you can buy fruit trees according to the desired mature size. Such sized trees are the result of grafting top wood from popular varieties onto roots that allow the tree to grow only so tall. If you only have space for a nine-foot tree, your nurseryman should be able to offer apple, pear, and peach trees to suit your need.

The smallest trees, called dwarfs, mature at a height of six to ten feet, and you can easily pick the fruit just by standing on the ground. Many popular varieties of fruit have been developed as dwarf trees and, contrary to mistaken opinions of some people, only the tree itself is small. The fruit is in every way the variety named; a Red Delicious apple from a dwarf is identical to one from a full-sized tree.

Mehu-Maija Steamer Juicer

$74.90 (stainless steel), $44.90 (aluminum).

The 9-liter Mehu-Maija measures 13 inches in both height and diameter (with 2½-inch handles), and it should fit into a conventional cupboard.

Mehu-Maija also offers rubber caps to fit old soda or wine bottles and gallon jugs. After sterilizing the bottles, just fill them with hot juice from the drain tube and cap immediately.

Osmo O. Heila, R.D. 1, Podunk Road, Trumansburg, NY 14886.

Defying classification

Chop-Rite Health Fountain

$54.80.

The Chop-Rite Health Fountain No. 27 Juice Extractor is the only muscle-powered juicer we know of. It looks something like a cast-iron meat grinder with a long snout, and it behaves like one too.

The snout houses an auger which grinds fruits, vegetables, and even leaves into a pulp. A large adjustable screw at the end of the snout puts pressure on the pulp inside the juicer before allowing it to exit around the screw. That squeezes the juice out. Beneath the auger is a small screen, through which the juice drips into a catch basin. More pressure produces more juice but it also makes cranking harder; too much pressure may make the auger stop feeding.

We had the best luck with firm vegetables, producing clear, sweet juices with no pulp. With the pressure screw properly adjusted, the Health Fountain outproduced most of the centrifuges (though this took two runs through the juicer). The Chop-Rite also proved strong at extracting juice from leafy foods such as Swiss chard, parsley, and wheatgrass, a job on which other juicers fell short. But apples and tomatoes turned to mush, which made the first run slow and a second impossible.

The Health Fountain is slow—it took an average of 15 minutes to extract 8 to 10 ounces of juice from a pound of carrots (it would only take 2 to 3 minutes with a centrifuge). Add the time it takes to wash and cut vegetables (standard with the centrifuges) and you would have a 20- to 25-minute project ahead of you. But it's a workable alternative to a high-speed, noisy electrical centrifuge.

Chop-Rite Manufacturing Co., Box 294, 857 Cross Street, Pottstown, PA 19464.

Vita Mix

$325.

It's difficult to classify the Vita Mix. It looks like a blender with a stainless steel container, but is sold as a juicer, grain mill, dough kneader, and ice cream maker.

There's nothing mystical about the Vita Mix's design that allows it to perform all of these functions. It is simply a heavy-duty, high-powered machine with multiple speeds and reverse. The motor has two to three times as much muscle as an ordinary blender, and this gives the four stainless steel blades the power to chop through ice or tough grains. Reverse permits thorough masticating, throwing food back into the blades for further grinding; models without reverse just throw food up against the sides of the blender.

Some juices, like carrot, came out a thick, mushy puree. That motivated us to look at the Vita Mix's juice press attachment, an optional device used after the mixing process for straining very pulpy juices. It consists of a straining bag that fits between two pressing blocks. The bag is inserted into the container and a screw press slips onto the top. Tightening the screw pushes the top block against the juice-filled bag to produce a strained, clear juice.

To make carrot juice, we washed one pound of carrots, broke them into small pieces, and half filled the container. The other half we filled with ice cubes. Then the "Action Dome"—a clear plastic dome with a hole for stirring contents—was latched in place on top. Using the forward and reverse action as directed, mixing took 12 minutes, including a stop to scrape the sides with a spatula. (Without reverse, we would have had to scrape every minute or so.) Setting up the press took only a minute, and pressing took 3 minutes. We tried to cut down on the ice to get a stronger flavor, but there wasn't enough liquid and we ended up with chopped carrots.

The Vita Mix did a fair job of cleaning itself; we just filled it with water and turned it on. A small amount of scrubbing finished the job.

So our glass of strained carrot juice took 15 minutes to prepare and almost 5 minutes to clean up after. Frothy, thick orange juice took 10 minutes plus another 2½ for cleanup. Tomato, also foamy, was the quickest of all—3½ minutes to juice and 2 minutes for cleanup. (All these times could probably be less with experience.)

Ice cream is made by mixing milk or

cream with frozen fruit. We used strawberries—once with milk and once with cream. After 2½ minutes of grinding, we had ready-to-eat ice cream.

The Vita Mix did an acceptable job of grinding wheat into bread flour. The flour was not as fine as that from a grain mill, but as long as you're not trying to make light pie crusts or pastries with it, you should be satisfied.

The Vita Mix generated more heat during grinding than most mills, but the flour temperature never exceeded 180°F., the point at which proteins in the flour become denatured. In fact, the heat generated during the grinding process is an advantage. With the flour already warmed, dough can be mixed immediately in the Vita Mix container, at a temperature that's ideal for yeast. You just let the mixer run a few minutes. We were skeptical about the Vita Mix's kneading ability but suspicions were laid to rest when we took the bread from the oven.

After 25 minutes for rising and 45 minutes for baking, the bread was done—just 1½ hours after starting with the whole wheat berries. Both texture and flavor were just right. Bread from slightly coarse flour like this tends to be crumbly, and is best thickly sliced.

Should you buy a Vita Mix? That depends on what you're looking for. It doesn't do any job as well as a machine designed for a single function. But if you're making juice, bread, and ice cream at home, it could spare you the expense and clutter of two, three, or even four machines.

Vita Mix Corp., 8615 Usher Road, Cleveland, OH 44138.

Drying
Taking the water out

Drying is the most ancient method of food preservation. Nature used it first, sun-drying grains, beans, nuts, and other seeds so they can wait out droughts, or extreme heat or cold, until conditions are right to grow.

Man must have learned from nature how to dry food. He may have noticed that the wheat grains he gathered kept throughout the winter. Or he may have found dried fruits under a tree, curiously changed from their fresh form but still edible and apparently nutritious. Certainly the principle is a simple one.

People in the Near East sun-dried meat after tenderizing it by beating the pieces with rocks. They dried fruit by burying it in hot sand. Northern tribes learned that cold, dry air and a strong wind dried food almost as quickly as would an arid environment.

South American Indians still use a crude version of freeze-drying that was developed in pre-Columbian times. They freeze potatoes outside at night, then trample them the next day as they thaw to squeeze out the juice. When these stomped potatoes are air dried, they keep for months.

For centuries, drying was about the only way to keep food from harvest to harvest, or from coast to coast if you were exploring new worlds. The Pilgrims survived the Mayflower crossing by dining on "salt horse" (as they called dried beef) and dried fish; pemmican—an Indian invention consisting of dried meat pounded into a paste with fat and berries—helped win the West.

With the advent of home canning and then freezing, drying declined in popularity, even though today there are hundreds of commercially dried products, good and not so good, that remind us of this method's utility: powdered milk, beans and nuts, yeast, pasta, soup mixes, instant coffee, raisins, potato flakes, herbs.

Why dry?

Drying foods is a beautifully simple cycle: the same sun that made the food grow preserves it; the water that brought it to life is removed to lengthen its life.

Simply beautiful your food isn't, however, when you've finished dehydrating it. Your storage shelves turn into a witch's den of bizarre shriveled things sitting around in the dark. Fixings for your soup pot look like fish food; string beans become worms flattened on the road after rain.

But add water to a handful of wrinkled BB pellets and you have a bowlful of green peas. You don't even have to reconstitute dried fruit to enjoy it. Throw your dried onion or parsley into a soup, and enjoy their flavors at a fraction of supermarket cost.

Cheesecloth protects drying food from curious insects.

·Drying isn't just for when your freezer's jammed, or you can't stand any more canning. With many foods, it keeps the taste and nutrition as well as any other preservation method, often with less energy and a lot less storage space.

The equipment needed for drying can be as simple as sunlight itself. If your climate is hot and dry, all you need are some trays. Heavy humidity can be overcome with a simple frame and glass cover. An attic that's getting free heat anyway is ideal for drying herbs.

Preparation is minimal for drying. With many foods, you just wash, slice, and place in a dryer. Some may also need blanching.

Serving dried foods is equally simple. Most dried meats and fruits can be eaten straight from the jar (or bag). Dried vegetables can be reconstituted either by soaking them in cold water for several hours, or by placing them in boiling water for a few minutes.

The initial preparation cost for dried foods (with an electric dryer) is about equal to that for canned foods, and long-term storage is much cheaper than freezing. You can also save cash by making large, in-season purchases and then drying what you can't eat fresh.

Finally, drying can introduce you to new, even unheard-of foods: zucchini chips, a healthful alternative to potato chips; dried bananas, a super-sweet but wholesome dessert; tomato leather, a portable spaghetti sauce; and turkey jerky, a high-protein, no-salt snack.

Every supermarket stocks a host of dried foods—most of them so familiar we seldom think of them as dried.

What happens to the water?

The combined agents of warmth, low humidity, and moving air, will dry anything—your clothes, your hair, or your food. Nature likes things to be balanced, so to equalize a moisture difference between your food and the air, dry air accepts some of the water from the food. If the surrounding air is too humid, the food won't dry. But if there's an air flow to carry away the moisture-laden air and bring in dry air, drying will continue.

Although heat pops into mind

Some foods, like these peppers, look good enough when dried to string up and hang on the wall.

with the thought of drying, low humidity and air movement are even more important. Clothes will dry outside in the dead of winter with snow on the ground as long as the air is dry (although warm air holds more moisture than cold air). For the necessary air movement, most electric dryers have fans, and non-electric dryers are designed to provide an adequate air flow.

A proper balance among these three factors is also essential. If the temperature is too low, or the humidity too high, the food will not dry quickly enough, and may spoil. If the temperature is too high, you may get case hardening: the outside of the food dries too quickly and hardens, locking moisture inside that will later cause problems.

The two chief enemies of food preservation are enzymes and microorganisms. Drying slows these reactions; heat over 140°F. halts them.

Microorganisms such as yeast, mold, and bacteria thrive when the moisture is high and hasten the decomposition of food. Temperatures over 140° F. inactivate most microorganisms, but removing the water in food stops them cold.

The growth of spoilers also depends on the acidity of the food being dried. High-acid foods like fruit are naturally resistant to microorganisms. They don't need to be blanched before being dried, and keep nicely with about 80 percent of their moisture removed. Most vegetables, being lower in acid, are usually blanched prior to drying, and about 90 percent of their water must be removed for safe storage.

Diana Branch and Pat Pleatman

The method

Fruits and vegetables

Start with top-quality produce. Drying will preserve much of the original goodness, but it won't improve your food. Wash the fruits or vegetables thoroughly, preferably in cold water since it preserves freshness best. Soaking is not recommended; some vitamins will be dissolved, and the food will absorb moisture, making drying more difficult.

There are several opinions about the best pretreatment for drying. Most authorities recommend blanching vegetables. This stops the enzyme activity, sets the color, and speeds drying by softening the plant walls so water can escape more easily. (At the same time, it destroys some of the water-soluble vitamins and minerals, as well as the enzymes.) The need for blanching is directly related to how long you want to store the vegetables. If you don't blanch, their color will be poorer (though we found that rehydration restored much of the lost color) and enzyme reactions will slowly continue, eventually resulting in loss of flavor, color, and quality. For storage of less than six months, you really don't need to blanch; for longer storage, it's recommended. You can blanch vegetables by placing them in boiling water or steaming them until they're heated through. Steam blanching is preferable because it minimizes the loss of water-soluble vitamins and minerals. Just loosely pack the vegetables into a sieve, colander, or wire basket. Then suspend this container in a large pot holding two or so inches of boiling water, cover tightly, and steam for the recommended time. (See table.)

With fruit, blanching or sulfuring is usually advocated to prevent discoloration and deterioration. While sulfuring does help preserve the natural color of fruit, and reduces the loss of vitamins A and C, it destroys the B vitamins and can give fruit a disagreeable taste. We see no need for adding sulfur to your foods. If you're concerned about maintaining the color, you can blanch fruit or soak cut fruit in a solution of ascorbic acid (vitamin C) for about five minutes. (Use one tablespoon per quart of

soaking water.) But for storage up to six months, no pretreatment is necessary.

Some fruits, including cherries, plums, grapes, and figs, have a wax-like coating that slows drying. In the past, these coatings were often "checked," or removed, by soaking the fruit in lye. There are two easier—and less dangerous—ways to get around this water barrier: make a few nicks in the skin with a sharp knife so moisture can escape more easily; or, loosen and remove the skin by dipping the fruits in boiling water and then plunging them into cold water.

There's no ideal drying temperature: anything between 95 and 145°F. will work for most foods. We prefer a medium-range temperature of about 120°F. At that level, food dries fairly fast but there's little loss of enzymes to overheating. Higher temperatures bring quicker results, but most enzymes are destroyed at 140°F. Unless you have a thermostat, the temperature in the dryer will tend to rise slowly as the food dries. So if you're drying at the upper limits of temperature, check periodically to make sure you're not cooking your food.

Food can be cut into various shapes for drying: halves, cubes, slices, strips. The thinner and smaller the pieces, the quicker they'll dry. The only rule is that they should be as uniform as possible: if pieces vary much in size, they will dry at different rates, and you'll end up with an inef-

The more uniform the food pieces, the more evenly they will dry.

ficient dryer and some overdone (or underdone) food.

After a few loads, you'll know where any hot spots are, and be able to shift food accordingly. If your food has a skin side, start by putting that side down on the tray. By the time you turn it, the exposed side will be dry enough to have lost its stickiness.

How dry is dried?

It's hard for a beginner to judge when food is properly dried. There are three basic ways to determine dryness: 1) by timing the drying period with the dryer set at a particular temperature; 2) by weighing the food before and after drying; and 3) by knowing the different characteristics of dried foods.

The complicating factor is that drying is not done in a static environment. If you're drying on a humid day, the air will receive less moisture from your food, and the food may absorb a bit of that humidity as well. The size of the food pieces, how full you pack the dehydrator, the outside air temperature—these are just some of the variables that affect drying.

As a result, it's impossible to give an exact time for drying particular foods. One excellent drying chart we have (for electric dehydrators set at 140°F.) gives ranges of 8 to 14 hours for drying green beans, 3 to 9 hours for onions, and 10 to 18 hours for tomatoes. There's a lot of room for uncertainty in such figures.

The most precise—as well as the most difficult—method for determining dryness is to weigh the food before and after drying. First find the water content of the food. (This may be the hardest thing of all. One sure source is *Nutritive Value of American Foods,* Agriculture Handbook 456, a massive paperback published by the USDA.) For our example, sweet cherries, the figure is 80 percent. Next, weigh your food when it's prepared for drying. Let's assume we're drying 10 pounds of cherries. Determine how much total moisture is in the food. In this case, 10 pounds x 80 percent equals 8 pounds. Then figure

(Continued on page 164)

THE SPECIFICS
Vegetables

	Preparation	Steam blanching time (min.)*	Characteristics when dry
Asparagus	Use only tender, young stalks, or the top 3 inches of older stalks.	3 to 5	tough to brittle
Beans, green	Slice lengthwise or crosswise in short pieces, holding knife diagonally.	2 to 2½	leathery, brittle
Beets	Remove tops and roots. Slice thinly, or cut into ¼-inch cubes or shoestring strips.	6 to 8	brittle
Broccoli	Use stalks with bright-green heads only. Trim, then quarter stalks lengthwise.	3 to 3½	crisp
Brussels sprouts	Cut in half lengthwise through stem.	6 to 7	tough to brittle
Cabbage	Quarter, core, and slice thinly.	2½ to 3	brittle
Carrots	Cut in thin slices or strips. Peel if desired.	3 to 3½	tough to brittle
Cauliflower	Remove flowerets from core. Split stems so flowerets are less than 1 inch thick.	4 to 5	crisp
Celery	Trim base, cut stalks into small slices or strips. Leaves may be dried along with stalks.	2	very brittle
Corn, cob	Husk, remove silk, trim.	2 to 2½	brittle
Corn, cut	Husk, remove silk, cut without including any cob. (If blanching, cut after blanching.)	2 to 2½	dry, brittle
Eggplant	Peel, cut into ½-inch cubes or slices.	3½	leathery
Mushrooms	Trim tough stems. Leave small ones whole; slice larger ones lengthwise.	none	leathery

*The purpose of blanching is to heat food through and kill the enzymes. You may need to adjust these recommended times, since they can vary depending on the condition of the food and the size and number of food pieces.

(Continued on page 164)

FIREPLACE DRYING

We have been fascinated by the sight of trays of sliced fruit drying in front of a fireplace. At one time this was an extremely common way of preserving foods for the winter. Now, although it is no longer a necessity, some mountain people continue the habit. Mrs. Grover Bradley, who had both a churn and several trays of sliced apples warming beside her fireplace when we last visited with her, said, "We had to eat things like that. They wadn't no other way to live. We dried everything."

APPLES—Apples are either sliced up into thin slivers, or cored and sliced into rings. One woman claimed that with a peeler, she could core, peel, and slice a bushel of apples in 54 minutes.

The rings were strung on a broomstick or a pole: slices were spread out on boards. Then they were set out in the sun or in front of the fireplace, depending on the weather, until the slices were brown and rubbery. This usually took two to three days. Some people say that they brought the fruit in at night to protect it from the dampness. Others simply covered the fruit with canvas at night. While drying, it was turned over frequently so that it would dry evenly.

Peaches were dried just like apples. Small berries, such as blackberries, were simply spread out on boards to dry and were not sliced.

Excerpt from *The Foxfire Book* edited by Eliot Wigginton. Copyright © 1968, 1969, 1970, 1971, 1972 by The Foxfire Fund, Inc. Reprinted by permission of Doubleday & Company, Inc.

THE SPECIFICS
Vegetables (continued)

	Preparation	Steam blanching time (min.)*	Characteristics when dry
Okra	Use only young pods. Trim, slice crosswise about ¼ inch thick.	none	very brittle
Onions	Trim, remove outer paper shells. Slice thinly or dice.	none	brittle
Parsley	Separate clusters. Discard long or tough stems.	none	brittle, hard
Parsnips	Trim and peel if desired. Cut into ¼-inch slices or ⅜-inch dice.	3 to 5	tough to brittle
Peas	Shell.	3	wrinkled
Peppers, green	Trim, slice into thin strips.	none	leathery to brittle
Peppers, hot	Leave whole or cut into 1-inch dice.	none	shriveled, crisp
Potatoes	Slice, grate, or dice, depending on how you plan to use them. Peeling is optional.	6 to 8	brittle
Pumpkin	Cut in half, remove stems and seeds. Slice into small strips, peel.	2 to 3	tough to brittle
Rhubarb	Cut into thin strips.	3	brittle
Rutabagas	Trim, peel if desired. Cut into ½-inch slices or dice.	3 to 5	tough to brittle
Spinach, other greens	Trim, cut coarsely into strips.	2 to 2½	crisp
Squash, summer	Slice into thin strips. Peel if desired.	2½ to 3	leathery to brittle
Sweet potatoes	Cut into thin strips or ⅜-inch dice.	2 to 3	tough to brittle
Tomatoes	Remove skins by dipping in boiling water, then cold water. Cut into ¾-inch slices.	none	crisp
Turnips	Trim, peel if desired. Cut into ¼-inch slices or ⅜-inch dice.	3 to 5	tough to brittle
Zucchini	Cut into ⅜-inch slices for chips, or grate.	2 to 3	brittle

*The purpose of blanching is to heat food through and kill the enzymes. You may need to adjust these recommended times, since they can vary depending on the condition of the food and the size and number of food pieces.

(Continued from page 162)

out how much moisture should be removed. For fruits, that amount is usually 80 percent. So, 8 pounds of water times 80 percent equals 6.4 pounds, the amount of water that must be evaporated. Subtract this figure from the original weight to determine what the dry food should weigh. In our case, 10 pounds minus 6.4 pounds equals 3.6 pounds, the proper weight of our fully dry cherries. This all explains why our chart states the characteristics of different properly dried foods. This is inexact, but once you get the hang of it, you can judge dryness under any circumstances. At first, keep records of drying times so that, if you make a mistake, it can easily be corrected the next time. Whenever you're in doubt, dry a bit more; it's better to overdry than underdry. And let the food cool before testing.

Fruit leather

Fruit leather has been around a long time. But the product in the supermarket likely is a far cry from the kind that has been made for hundreds of years by the Himalayan Hunzas. They pound apricot pulp until it's smooth and then dry it in the sun on a flat surface; our modern version is made with frozen fruit, sugar and corn syrup, and artificial color.

That's reason enough to make your own. All you need to do is puree your fruit into a smooth, thick liquid, then dry it.

Fruit leather gets its name from

the texture—not the taste—of the product. As the puree dries, it takes on a shiny, leatherlike appearance and becomes pliable enough to be rolled up for storage (although it may be a bit sticky and need to be rolled with wax paper). If dried too much, the leather will become brittle and break, and you'll have to use it as pieces. (These pieces taste great in cereals and desserts.)

To prevent sticking, the puree should be dried on oiled trays or trays lined with plastic wrap. Dry it using any of the usual methods, or simply cover the tray and put it in a warm place in the house. The leather will dry in six to eight hours in an electric dehydrator, or in one to two weeks at room temperature. It is done when the center is no longer sticky.

When the leather is removed from the dryer, it should be placed on a cake rack for a few hours to make sure it is dry on both sides. Then, when it's completely dry, dust with cornstarch or arrowroot powder before rolling or stacking for storage.

With most leathers, you'll find that you don't need to add any sweetener; drying concentrates the sweet flavor. If you want a very sweet product, try adding some honey or apple juice.

Apple Leather

12 to 15 apples
1½ cups apple cider
    cinnamon, cloves, or nutmeg
       (to taste)
    honey (optional)

Peel and core apples, cut in pieces, then run through a food mill or blender. Place the ground apples and juice in a large, heavy pot and add 1 cup of cider. Place the pot over low heat and bring the apples to a boil. Add more cider if needed to keep the apples from sticking to the bottom of the pot. When the mixture is boiling well, add the spices and, if the apples are especially tart, a few table-spoons of honey.

When the mixture has the same consistency as thick apple butter, remove from heat and spread the pulp about ¼ inch thick on trays.

Yield: 3 sheets of leather, about 14″ by 18″

(Recipe adapted from Stocking Up; *review on page 157.) Other fruits especially good for making leather include apricots, berries with seeds, cherries, peaches and nectarines, pears, pineapples, and strawberries.*

THE SPECIFICS
Fruits

	Preparation	Steam blanching time (min.)*	Characteristics when dry
Apples	Pare, core, and cut into thin slices or rings. Don't peel unless the apples have been heavily sprayed.	5	soft, pliable, no moist area in center when cut
Apricots	Pit, then halve or cut in slices.	3 to 4	pliable
Bananas	Peel. Slice lengthwise or crosswise in about ¼-inch thicknesses.	none	pliable to crisp
Berries	Halve strawberries, leave others whole. If necessary, crack skins by nicking with a knife.	none	leathery
Cherries	Pit and remove stems. Crack skins with a knife, or cut in half.	none	pliable
Figs	Remove skins. Cut into halves or quarters.	none	leathery
Grapes	Remove stems. Crack skins with a knife.	none	raisinlike texture
Peaches and nectarines	Pit, then halve or quarter. Peel if desired.	8	same as for apples
Pears	Halve and core. Peeling recommended.	6	same as for apples
Plums	Pit. Can be halved or left whole. If left whole, nick skin with a knife.	none	leathery

*The purpose of blanching is to heat food through and kill the enzymes. You may need to adjust these recommended times, since they can vary depending on the condition of the food and the size and number of food pieces.

Meat

The most common form of dried meat is thin strips known as jerky. If you're drying your own meat, the leaner the better. And unless your climate is hot and dry, it's best not to sun dry meat, because the risk of spoilage or contamination is high.

With beef, the best cuts are flank, rump, brisket, or round. First trim all the fat, then cut the meat with the grain into narrow strips about ¼ inch thick. Partially freezing the meat before cutting will make slicing easier, or you can ask the butcher to do it for you. With game, any of the meat can be used for jerky, but the leaner loin, flank, or round cuts are best. (Also, to prevent disease, it's a good idea to freeze game for a couple months at 0° F. This will kill any microorganisms.)

Raw pork and raw poultry don't make good jerky. Drying temperatures aren't high enough to kill the harmful bacteria in pork, particularly

(Continued on page 167)

Give dried vegetables a chance

Most people are familiar with dried fruits, herbs, and meat. But their only close encounter with dried vegetables has probably been on a camping trip, when they ate freeze-dried stew because they were famished, not because it was tasty. Dried vegetables have two bad raps: they look ugly and taste bad.

The appearance problem is unavoidable. Dried vegetables *are* wrinkled, faded, shrunken. They look old. In a society that places high value on youth and beauty, it's no wonder that they are scorned. But like people who have grown old gracefully, dried vegetables are not without their worth.

A common mistake is expecting these foods to equal their fresh counterparts. They can't. (Neither can frozen or canned foods, for that matter.) Dried vegetables are not necessarily ugly, however; they just look strange to someone who's used to seeing them fresh. And when you add a little water, they perk up amazingly well.

The test-tasters at work.

As for taste, that's partly an approach problem, too. We took three vegetables—peas, corn, and broccoli—and asked 13 people to compare the taste of canned, dried, and frozen versions. We prepared them simply, with no sauces or accompaniments to mask their merits (or demerits).

In general, frozen vegetables received the highest marks for taste and appearance, but the dried vegetables were rated as highly as the canned ones. We didn't come up with any revelations, but we did learn two things from our experience and the comments of the tasters.

1) While dried vegetables won't always measure up to other preserved vegetables in head-to-head competition, in certain dishes—peas in cream sauce, baked broccoli with cheese—dried vegetables would work as well as frozen ones. (Why? Precisely because they're drier and resist getting soggy.) In other dishes, such as corn casserole, dried vegetables can provide the desired taste and texture better than canned or frozen ones.

2) Dried vegetables, however closely they resemble their fresh ancestors, are really a new food. If you treat them like fresh or frozen vegetables, you'll be disappointed. But if you creatively explore some of the possibilities of dried vegetables, you'll get interesting results. You'll

likely find that some dried vegetables just don't suit you. Others, however, will surprise you with their versatility and charm, and find a permanent place in your yearly cycle of food preservation.

Speaking of versatility:

- *Use dried vegetables in soups, stews, or with sauces, where their flavor can be enhanced by the savory surroundings.*
- *Powder them in a blender and mix with dried milk for your own instant soup. (Experiment with proportions to find what suits you.)*
- *Make vegetable chips—zucchini is especially good—by drying thin slices. For extra flavor, sprinkle with herbs or spices before drying.*

Winter Corn Pudding*

¾ cup dried corn
3   cups boiling water
2   eggs, slightly beaten
2   tablespoons butter, melted and
        slightly cooled
2   cups light cream
2   tablespoons onion, chopped
2   teaspoons honey
⅛ teaspoon pepper

Rehydrate corn by adding to boiling water and allowing to stand for 20 minutes. Simmer corn until tender (approximately 1 hour). Drain off excess water and save for soup or gravy.

Preheat oven to 325°F. and grease a 1-quart casserole. In a large bowl, combine corn, eggs, melted butter, light cream, onion, honey, and pepper. Pour into the greased casserole and bake for 35 minutes or until a knife inserted in the center comes out clean.

Yield: 4 servings

*From *Home Drying of Foods*, by Ruth N. Klippstein and Katherine J. T. Humphrey, an extension publication of the Division of Nutritional Sciences, Cornell University, Ithaca, NY 14853. See review, page 178.

A test kitchen employee tries to find words to describe a rehydrated sample.

(Continued from page 165)

the parasite that causes trichinosis. You can make jerky from fully cooked ham, but its safe storage life at room temperature is only about two weeks. The flavor and texture of raw poultry meat is unpalatable, but cooked turkey or chicken breast will make tasty jerky.

To make jerky, first parboil the strips of meat for 15 to 30 seconds—long enough for them to change color. Then you can simply dry the strips or marinate them before drying. Most commercially dried meats are heavily salted and contain sodium nitrate or sodium nitrite, chemicals that increase the storage time and help preserve color. But they are both unnecessary for safe drying, and may have detrimental side effects.

Marinate the meat, if you wish, with this mixture (enough for two pounds).

Marinade

1   teaspoon ground black pepper
8   tablespoons tamari
16 cloves garlic, crushed
4   teaspoons champagne vinegar
1   teaspoon fresh chives, minced

Mix the seasonings with just enough water to cover the sliced meat, and soak for 8 to 12 hours in the refrigerator. Then drain off the liquid, pat the meat dry, and place it in the dryer.

Once in the dryer, the meat should be occasionally blotted with paper towels to remove beads of oil that will appear. It's wise to place aluminum foil below the trays to catch the drippings. The jerky is done when a cooled piece will crack but not break. It should be stored in an airtight jar or sealed plastic bag, or frozen for long-term storage.

Fish

Any lean fish can be dried, as long as you're quick. Fish is the most perishable of foods, and it must be preserved or eaten soon after being caught to prevent deterioration.

First clean, scale, and dress the fish. Small fish may be dried whole or cut in half; larger fish should be cut in strips or small pieces. As with meat, the fish can be dried plain or marinated first.

When you think the fish is dry, squeeze it to check for moisture. If the flesh is firm and springs back when pressure is released leaving no finger imprint, it's dry. The individual

Drying Meat

Trim all excess fat. Cut the meat with the grain into strips about ¼ inch thick.

Parboil the strips of meat until they change color. This usually takes only 15 to 30 seconds.

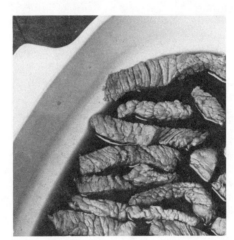

Marinate the meat for 8 to 12 hours in an herb-and-spice mixture, then pat the pieces dry and place them in a dryer.

Dry the meat at 140°F. for 4 to 10 hours. When it's dry, the jerky should bend into sharp corners but not break.

pieces should be wrapped in moisture-proof paper and stored in a lidded container. The dried fish should be used within three months.

TD

Herbs and spices

Harvest leafy herbs shortly before they blossom, when the aromatic oil content is high. Just snip off the top growth—about six inches of stem—with scissors.

Dry herbs out of direct sunlight, since sun darkens the leaves unnecessarily. Herbs were traditionally dried in the warm, dry air of attics or above the fireplace. Electric dryers should be set fairly low—not above 105° F.

The foliage of large-leaved herbs

can be removed from the stems and dried in trays. Other herbs should be tied in small bundles and dried with the leaves pointing down, so the oil in the stems will flow into the leaves.

To protect herbs from light and dust, dry them in a paper bag with holes to allow for circulation. Just put in a herb bunch, gather the bag together around the stems, and close it with a string or rubber band.

Seed herbs should be harvested when the pods or heads have darkened but before they begin to shatter. The seed parts can be dried on trays, then separated easily by rubbing between the palms of your hands. Or place the whole plant upside down inside a paper bag, and the seeds will drop into the bottom of

(Continued on page 168)

(Continued from page 167)

the bag when they're dry.

With roots, a fall or winter harvest is best, as the plant is then full of stored food. Dig up the entire plant carefully, harvest a few of the more tender roots, then replant it. (Don't be greedy, or the plant won't survive.) Scrub them with a brush, then dry them on trays. For quicker drying, cut thick roots into slices or strips.

When the herbs are properly dry, the leaves should crumble easily. Roots should snap in two, and seeds should shatter readily. To be doubly safe, dry seeds for a day or two after they're separated from the pods.

Herbs keep better and retain more essential oil if they are kept

Since heat rises, the rafters of a warm, dry room are a fine place to dry herbs.

whole, to be crushed just before using. Likewise, seeds should be ground only when needed, since they deteriorate rapidly after their shell is broken. (Don't forget that dried herbs are much more concentrated than fresh ones: one tablespoon of fresh herbs equals ½ teaspoon of dried herbs or ¼ teaspoon of dried, powdered herbs.) Place dried herbs in tightly sealed jars, and if you later find moisture inside the glass or under the lid, remove the contents and spread them for further drying.

Dried herbs should not be stored in cardboard or paper containers because these containers will absorb oil and leave herbs tasteless.

While herbs are usually stored separately, you can experiment with storing them in combinations. During storage, the flavors and aromas will blend, making exotic mixtures that will liven up your cooking. See "Tongue Foods," page 246, for suggested herb blends.

Storing dried food

Normal drying temperatures are not sufficient to destroy insect eggs or other contaminants, and you may want to sterilize dried food by freezing or heating.

Store dried foods in glass, and you won't be likely to forget about them.

To sterilize by freezing, place the food in airtight bags in a freezer set below 0°F. for a minimum of 48 hours. To pasteurize your food, spread it loosely on trays, then heat it in a 175° F. oven for 15 minutes, or a 160° F. oven for 30 minutes. Freezing is preferable, since it destroys fewer nutrients than heating.

Even with the best dryers, some finished pieces will have more moisture than others. For the longest shelf life, the food can be "conditioned," or

(Continued on page 170)

Sun-dried preserves: Texarkana, Texas

If you like fruit preserves but don't enjoy the work that goes into making them the old-fashioned way, you can make them like Joshua Tree does. He uses the sun's heat to evaporate the excess moisture, and adds just enough of a light-colored, mild-flavored clover honey to give his preserves the correct syrupiness.

His evaporator is nothing more than a cold frame. During the summer, when it's not being used for starting plants or hardening them off, Josh fills it full of shallow glass pans, and creates sunshine preserves.

First he lines the dirt floor of the cold frame with a section of black plastic stapled to the inside edges of the structure. Over the plastic he lays a flooring of 4 x 8 x 16-inch concrete partition blocks, then he places the partially cooked preserves on these blocks. Finally, Josh covers the entire top of the cold frame with a piece of

clear plastic pulled taut and held in place by a few stones along the outside of the frame. A narrow gap is left along the top edge to allow moisture to escape.

Processing begins with paring and slicing large fruits such as peaches, pears, and quinces into ½-inch sections, or with halving smaller fruits such as berries, apricots, and cherries.

Next, combine five cups of the fruit with one cup (or less) of mild honey and from two tablespoons of lemon juice for small fruits to ¼ cup for larger fruits. Bring this mixture slowly to a boil in a saucepan, stirring constantly, then turn the heat up high and boil vigorously for 5 minutes without stirring. Let cool 30 minutes in the uncovered saucepan.

Pour the preserves gently into shallow glass pans to a depth of ½ inch. The pieces of fruit may have to be carefully spooned to prevent

breakage. Place pans in the cold frame–evaporator and stir occasionally until fruits become plump and the syrup is rich and thick.

Be patient. This evaporation process will take anywhere from 2 hours to around 12 hours, depending upon the climate, the time of year, and the amount of available sunshine.

What if you should leave them in too long? No matter. You'll have fruit leather. This confection can be frozen in jars or a paper sack until wanted.

Your preserves may now be canned or frozen. They will keep in the refrigerator about a month. To freeze the preserves, it is only necessary to allow the fruit and syrup to cool, then spoon into small freezing containers. Or, can the warm preserves by pouring them from the evaporator directly into sterilized canning jars, then processing in a boiling water bath for ten minutes.

Drying: Hopkinton, New Hampshire

When it comes to self-reliance, consider Derek and Patty Owen and their three children. They grow most of their vegetables, have several fruit trees, and forage for wild berries and other foods. They have two milk cows, a heifer, a steer for meat, several pigs, chickens, two geese, a goat, some rabbits, sheep, and a couple of bee hives. For fun they ride their horses (the Owens grow most of the hay, and barter vegetables for the rest).

With all this food on their hands, they've got much to preserve. Patty fills their root cellar, cans, makes butter and cheese, and dries. "I prefer drying because it takes less energy and less space than canning," Patty says. She does most of her drying in the sun, or in the oven of her wood stove. She dries herbs in the shade, since direct sun blackens them.

Patty's favorite dried vegetables are squash, kale, dandelions, green and hot peppers, and corn. She also dries herbs: peppermint, spearmint, tarragon, sage. And the bush cherries she forages are dried into fruit leather, as are strawberries and tomato paste.

It's only been recently, however, that Patty has done a lot of drying. "I was raised on a farm, but we didn't do any drying," she says. "A few years ago I started reading things about drying in books and in *Organic Gardening* magazine, and it made a lot of sense. So gradually we started doing it more and more."

Although Patty enjoys eating dried vegetables as snacks, most of them go into soups, which are flavored with her dried herbs. The mint herbs are used in tea, and the corn makes chowder or casseroles.

While the Owens enjoy their self-sufficient life, Patty points out that "it's not a simple life. With animals, we can't just pick up and go on vacation, or take a few days off. But we like it, and it's especially good for the kids, giving them a chance to learn different skills and responsibilities."

FREEZE-DRYING: AS YET, NOT A HOME FOOD PROCESS

One of the more recent commercial developments in food drying is freeze-drying. It's a great way to preserve food. Unfortunately, it's a complicated process that's impractical for home food dryers. The food is placed on trays and frozen. Next, it's placed in a vacuum chamber, where the air is pumped out and heat is applied. The heat melts the ice in the food, but because of the vacuum it sublimes, or changes directly into water vapor, and the vapor is pumped out of the chamber. The whole process takes about ten hours.

What makes freeze-drying superior to other forms of drying is that it gets food very dry—the moisture can be reduced below 2 percent—so they can be safely stored for longer periods. More nutrients are preserved because the food is frozen rather than heated, and by the time it is warmed above the freezing point, most of the water has been removed.

JERKY

Dried meat is called jerky, not because it was pulled or shaken, but from a corruption of the Peruvian word *charqui* (shar-key), meaning strips of dried beef. The original South American version was unsalted.

RODALE DRYER

After experimenting with several different models, the Rodale Research and Development Department has designed a solar dryer that can dry 15 pounds of food in two sunny days, and also has an electric back-up in case of rain or clouds.

The Rodale dryer can be set up against a table or with a stand, and it folds up for compact storage. Materials to construct the dryer cost about $90 to $100. A book with complete plans, directions on how the dryer works, and information on how and what to dry, will be available after May 1, 1981. Write to: Rodale Press, Plans Department, 33 E. Minor Street, Emmaus, PA 18049.

Patty Owen.

(Continued from page 168)

processed, to distribute the moisture evenly so the chance of molding or spoiling is minimized.

To condition dried food, pack it loosely in a closed container. Each day, shake the container to separate the pieces. As time passes, the excess water from some of the pieces will be absorbed by the drier ones. Also check regularly for signs of water condensation. If you see any, the food should be dried more thoroughly. After a week, the moisture content of the food should be equalized.

The best storage spot is a dry, cool, dark place. Suitable containers include airtight glass jars, polyethylene plastic freezer bags, or metal containers with tight-fitting tops. Metal containers should be lined with plastic or paper bags so that the metal won't give the food a bad taste. It's best to pack food in small quantities, so if one package spoils, the problem won't spread far. If your storage place gets too humid, you may want to store the food in your refrigerator until things dry out a bit.

All dried foods deteriorate with time. Dried carrots, onions, and cabbages have a shelf life of six months or less. Most other dried vegetables will last up to a year, though the longer they're stored, the less tasty they'll be. The same is true for dried fruit.

Dried food should be checked periodically for mold, spoilage, or other problems. If you find the food is getting too moist, unpack it and dry it out some more. If you discover bugs or worms, don't be in a hurry to throw your food out. Remove the bugs, spread the food on a tray, and put it in a 300° F. oven for 20 to 30 minutes. This will destroy any vermin and sterilize the food as well.

Rehydrating

To bring dried food back to life, simply reverse the drying process by adding water, a step known as "refreshing." Soak or cook the food until it looks good to you.

Dried food can be soaked in cold water, hot water, or boiling water. The warmer the water, the shorter the soaking time. No sugar or salt should be added to the water, since both substances cause food to absorb less water.

We recommend soaking dried vegetables in cold water. This saves the energy that heating water requires, and the long soaking time tenderizes the vegetables. If you drop them directly into boiling water, refreshing takes less time, but the vegetables will be tougher. Cold soaking times range anywhere from ½ to 1½ hours for root or stem vegetables to overnight for dried peas or beans. Once the vegetables are refreshed, beware of overcooking them: rehydrated foods cook more quickly than fresh ones.

Fruits generally refresh more slowly than vegetables, but they don't have the same problem with toughness. Cold soaking can take up to eight hours, but you can revive fruit in minutes with boiling water, as this recipe proves.

Stewed Fruit

1   cup any dried fruit
1   cup boiling water
2   pinches cinnamon
1   pinch ginger
1   pinch nutmeg
    yogurt and honey (optional)

Pour the boiling water over the dried fruit. Let stand 5 minutes. Sprinkle on cinnamon, ginger, and nutmeg. Stir in yogurt and honey, if desired.

Yield: 1 serving

VEGETABLES
Cups water added to 1 cup dried food

Asparagus	2¼
Beans, green	2½
Beans, lima	2½
Beets	2¾
Cabbage	3
Carrots	2¼
Corn	2¼
Okra	3
Onions	2
Peas, green	2½
Pumpkin	3
Spinach, other greens	1
Squash	1¾
Sweet potatoes	1½

FRUITS
Cups water added to 1 cup dried food

Apples	1½
Peaches	2
Pears	1¾

Source: *Home Drying of Foods*, by Ruth N. Klippstein and Katherine J. T. Humphrey, an extension publication of the Division of Nutritional Sciences, Cornell University, Ithaca, NY 14853.

Thermos cooking

One good use for dried vegetables is to make soup in a thermos. With this recipe, you just fill your thermos in the morning, and have hot soup at lunchtime.

Some dried vegetables that are excellent in this soup are tomatoes, peas, onions, broccoli, zucchini, and celery leaves. We also tried brown rice, lentils, and split peas, but they came out undercooked.

Tom Ney

Vegetable Grain Soup

1   tablespoon each of 5 or 6 dried
        vegetables
1   tablespoon buckwheat groats, bulgur
        wheat or barley
1   tablespoon broken whole wheat
        spaghetti or other small pasta
¼   teaspoon dried parsley
¼   teaspoon dried sweet basil
    garlic powder to taste
    ground pepper to taste
¼   teaspoon tamari
2   cups boiling hot vegetable, chicken,
        or beef broth

Place the vegetables, grains, pasta, parsley, sweet basil, garlic, pepper, and tamari in a full-size thermos bottle. Bring broth to a rolling boil and pour over dry ingredients. Quickly cover the thermos and close securely.

Allow the dry ingredients to steep in the hot broth for 3 to 4 hours. Open thermos, pour and serve.

Yield: 1 to 2 servings

A little water can transform a pile of wrinkled pellets into a tasty serving of peas.

Dehydrators

You can get started drying by placing your first batch of food on an old, but clean, window screen set out in the sun. That's about what our grandparents did; they often spread food on a flat, woven, basketlike tray and left it in the sun to dry. Loose weaving allowed for some ventilation, but the food still had to be turned to ensure even drying.

Other early dryers were designed for use on woodburning stoves, and they are becoming popular again. One type had wooden trays stacked on a metal bottom that sat on the stovetop. Warm air trapped by the base would rise to carry away moisture by natural convection. Another antique design, now sold as the Bumble Bee dehydrator, is a large, flat, metal drying tray warmed by a reservoir of water. Simply set the dryer on a stove, warm the water in the reservoir, and distribute the food over the tray to dry.

But by far the most common—and convenient—way to dry food today is with an electric dehydrator. Many of these have modern designs which closely resemble microwave ovens: handsome metal cabinets with see-through plastic doors and operating controls alongside. Such aesthetics are not always functional, however. A see-through door means you can check the drying process without opening the door and losing some heat, but a control panel on the front just makes the dryer unnecessarily wider. The following information is your guide through the forest of designs, options, and technical features available in electric dehydrators.

Construction

The best dryer cabinets are durable, washable, and insulated so they can't burn you. Improperly treated wood might pick up moisture (and odors) and then warp, but we didn't observe this.

Drying shelves or trays should be made of screening or other grid material so air can pass through the shelves and around the food. Shelves should also have enough strength to support the weight of a full load of heavy fruits. Nylon screening, stretched within an aluminum frame, can sag or pull loose from its frame when supporting a load of wet produce. Some trays are reinforced for extra strength and have a removable

(Continued on page 172)

Antique stovetop dryer.

Bee Beyer's Food Dryer

$199.95. Drying space, 11.6 sq. ft. Cost to dry 1 lb. of peas, 2.1¢.

Bee Beyer's Food Dryer has the modern, microwave look: an attractive gold-colored metal cabinet with a dark translucent door and colorful front-mounted controls which turn on the unit, regulate the thermostat, and adjust the amount of ventilation.

The eight shelves are made of nylon window screen fit into an aluminum frame by a rubber strip. The shelves are large, and when they're full the screen has a tendency to stretch or break out of its channel if not handled carefully.

A unique option with this dehydrator (as far as we know) is stainless steel drying trays. This is the best material to use for parts contacting food, but it's expensive and very difficult to find in screen form.

The Bee Beyer's Dryer took a little longer than average to dry peas. The drying pattern was similar to other dryers; more drying occurred near the heating element.

Rotating the trays once or twice during the drying period increased the efficiency.

Bee Beyer's Food Dryer, 1154 Roberto Lane, Los Angeles, CA 90024.

Dri Best Food Dehydrator

$229. Drying space, 14.7 sq. ft. Cost to dry 1 lb. of peas, 2.1¢.

The Dri Best is a serious processing tool. It's big and is made entirely of unpainted, textured aluminum.

A powerful fan mounted in the bottom of the unit creates a vertical air flow. With just one or two tray rotations (no more than with horizontal flow units) we got a fairly even drying pattern.

During our trial with peas, the desired 130° F. was reached by the second hour but the temperature throughout the trial ranged from 119 to 152° F., an excessive swing that can be blamed on the Dri Best thermostat.

The 12 trays are made of a heavy-gauge steel with an enamellike baked acrylic finish. One-half-inch holes provide ventilation. On top of each shelf is an excellent medium-fine nylon mesh which makes removing food and cleaning easier.

Laughlin Enterprises, P.O. Box 9045, 2784 East 3000 South, Salt Lake City, UT 84109.

(Continued from page 171)

screen or mesh for easy access to the dried food. On some models, the mesh is not fine enough to keep dried corn or herb flakes from falling through.

The capacity of a dryer is the total surface area of its trays. Each square foot of shelf space will hold 1.2 to 2 pounds of produce—a little more than one quart of peas or four medium-size apples. A large capacity is important when the bounty is plentiful and must be dried quickly to prevent spoilage.

Different dryer manufacturers make their trays with different mesh sizes. A large mesh may allow herb flakes or other small food pieces to fall through.

Air flow patterns

The pattern and velocity of air movement in a dehydrator affects the rate and evenness of drying. For maximum efficiency, every piece of food (on all its sides) should come in contact with dry air that can carry its moisture away. Proper air movement also evenly distributes heat so that there are no hot or cold spots.

A dryer may have some air flow without a fan because of natural convection—warm air rises—but a fan will increase the speed of drying considerably. Dryers without fans tend to have hot spots, so that food may cook or burn if not carefully watched. In our opinion, a strong fan is as important as a reliable heating

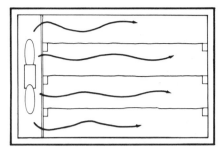

A horizontal air flow dryer.

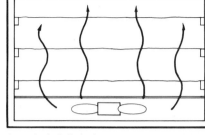

A vertical air flow dryer.

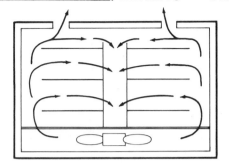

A converging air flow dryer.

element. Dryers with fans should also have some method of filtering dust and dirt from the air before it enters the dehydrator.

There are three kinds of air flow patterns in dryers with fans: horizontal, vertical, and converging. A dryer employing a horizontal air flow has the fan and heater mounted either on the side or back of the unit, and hot air is blown across the drying trays. Advertisements claim that horizontal air flow provides the most even drying pattern—one requiring no tray rotation. We found definite patterns, however; more drying took place near the heat source and fan, and food on the middle shelves tended to dry faster than that on the top and bottom shelves. With horizontal air flow, the warm air is saturated with moisture before it gets halfway across the trays and is then less effective in drying the remaining food.

A dryer with the heating element on the bottom has a vertical air flow whether it has a fan or not, since hot air rises naturally. This system has a reputation for uneven heating, with temperatures highest near the heat source. But our comparison of a horizontal air flow dryer and a comparably sized vertical air flow dryer—both with fans—revealed an equal need for tray rotation. By rotating, you can both shorten the drying time by as much as two hours and prevent overdrying.

A converging air flow pattern is found in round, columnar dryers. Hot air is forced up along the sides of the dryer and across the trays. The air then converges at the top and exits back down through the middle. Our analysis revealed this pattern to be the most efficient; drying was even and tray rotation wasn't necessary. Converging air flow dryers were also the most quiet.

One so-called energy-saving

feature found on several dryers is recirculating air flow. Dryers with this extra have adjustable vent openings that can be closed to keep the hot air in. This means the heater can stay off longer, and thus (theoretically) use less energy. Our tests showed, however, that a full dryer with recirculating air flow used about the same amount of electricity to dry food as one without this feature. Further, the recirculating mode should be used only at the end of the drying cycle when the amount of residual moisture is small. If used when food still has a high water content, this mode will

lengthen drying time and actually cost more.

Heat sources

The least-expensive electric dehydrator you can buy uses an electric light bulb as a heat source, but it bleaches the color out of food.

The two most common heating elements found in electric dryers are nichrome wire coils (often used in space heaters) and insulated sheath tubes (as in most electric ovens). Nichrome coils (nichrome is an alloy of nickel and chromium) are the most

(Continued on page 174)

David Sellers and Diana Branch, the director of product testing at Rodale, measure a dryer's heat patterns with a Datalogger machine.

Equi-Flow (Nature's Way Food Dehydrator)

$169.95. Drying space, 12.6 sq. ft. Cost to dry 1 lb. of peas, 2.1¢.

The Nature's Way Food Dehydrator is more commonly known as the Equi-Flow. Equi-Flow is the term the makers use to describe its air flow pattern.

The Equi-Flow is taller and deeper than it is wide to make good use of counter space. The plastic cabinet is covered with a wood grain vinyl, and has a clear Plexiglas door, hinged at the bottom. As with all other large-capacity horizontal air flow models, the drying efficiency of the Equi-Flow improved with a minor amount of tray rotation. An on–off switch and thermostat are located on the back.

The ten shelves are constructed of a heavy, clear, rigid plastic. Vent holes in the shelves are suitable for most applications but flakes from dried herbs will fall through. (To solve the problem, just crush the herb *after* it has dried.)

There's one thing peculiar about the Nature's Way shelves: although they are square, they fit into the dryer only when

four little tabs rising from the corners are positioned to catch on the shelf rack above. So, the shelves fit in only two ways—as with a rectangular shelf—instead of all four.

B and J Industries, Inc., 514 State Street, Marysville, WA 98270.

Excalibur

$149. Drying space, 14.3 sq. ft.

The Excalibur has a brown steel cabinet with a see-through plastic door and controls up front. The door is not hinged, but attached magnetically for easy removal.

The nine shelves have a rigid plastic frame which supports a foldable coarse plastic mesh (which helps when dumping a tray full of food). The mesh works well with most foods, but it will let herb flakes filter through.

We didn't have much of a chance to test the Excalibur. Prior to our trials, we turned on all machines for two months, 8 to 12 hours a day, and the Excalibur's thermostat failed, allowing the internal temperature to exceed 200°F. We had a meltdown. The plastic shelves and plastic lining softened just enough to intertwine in the middle of the dryer—never to do any drying again. Excalibur tells us they now have a safety shutoff switch in their cur-

rent models, but we haven't had an opportunity to test the newest version.

Excalibur Products, 6083 Power Inn Road, Sacramento, CA 95824.

(Continued from page 173)

popular, since they're very efficient heaters. But they tend to produce too much heat and, when used with a thermostat, are constantly switched on and off. Insulated tubes are less-efficient heaters, but they can supply heat at a low, continuous rate, resulting in greater durability and longer service life (although we had no problems with any of the heaters during our 1,000 hours of dryer operation).

The heating element should be enclosed with protective screening. An insulated or double-walled cabinet is ideal: the outside of the cabinet won't get hot enough to burn. UL (Underwriters Laboratories) approval indicates the electrical components have met independent testing laboratory standards, or are being evaluated if the product is new.

The recommended drying temperatures are roughly 100° F. for herbs, 120 to 130° F. for fruits and vegetables, and 145° F. for meats. A good dryer should have a temperature dial that's adjustable over a wide range (80 to 150° F.). At the lower temperatures you can make yogurt, raise dough, incubate tempeh, or hatch chicken eggs.

A thermostat is no frill. It keeps heat production under control so there's an even temperature, no chance of burning food, and security against a fire. During our testing, one thermostat malfunctioned, allowing the dryer temperature to flare up over 200°F. The plastic shelves and interior melted into one big blob. A high temperature shut-off switch—found on many dryers—would have prevented this problem.

Speed and efficiency

As a general rule, the more wattage a dryer has, the faster it will dry food. Even though a powerful unit will use more electricity, its speed may mean the total energy consumption will be equal to—or less than—a low-powered unit. However, if the drying space is very large, even a powerful unit will dry food slowly.

The evenness of drying determines the dryer's efficiency. The dehydrator must be run until the last pieces are dry. Rotating the trays will improve evenness and the speed of drying, and thus increase efficiency.

Our criteria

- Temperature differences within the dryer which could cause uneven drying.
- Evenness of drying and identification of the drying patterns.
- Accuracy of the thermostat.
- Drying time of the dryers when loaded to capacity.
- Energy consumed in drying a full load.
- Quality of the dried food.

For our test we chose peas. Peas are small enough to slip through dehydrator trays if the mesh is large. They are uniformly sized and wrinkled as they dry, enabling us to closely observe the drying pattern. Peas do not require peeling or slicing, eliminating the variable of preparation techniques. And they are a moderately moist vegetable, requiring the standard time needed for drying most vegetables—one day.

We spread the peas evenly over the trays, with a small space for air flow around each pea (about 1¼ pounds of peas takes up one square foot of tray), and turned on the dryer.

To monitor the heat distribution within the machine, we placed temperature probes from our Datalogger—an electronic thermometer and recorder—at three locations in the loaded dryers. Temperatures were recorded every 20 minutes. We could soon pinpoint the dryer's hotspots. To then identify the drying patterns, we noted the appearance of the peas and weighed individual shelves hourly. The peas were considered dry when they had lost 80 percent of their original weight. Our drying times and costs are reported for dryers in which the shelves were rotated twice during the drying cycle—a procedure insuring best results.

With a watt-hour meter we measured how much power the dryer was using and determined the total electric consumption. A few more calculations and we had the cost to dry one pound of peas. We chose to use this number for operating cost comparisons, rather than the cost to dry one full load of peas, because of the great range in dryer capacity. While the comparison numbers are small, over a long period of time the cost difference can be significant.

THE PEA TEST

	Drying space (sq. ft.)	Drying time (hr.)	Cost to dry 1 lb. peas (cents)
Bee Beyer's	11.6	9	2.15
Dri Best	14.7	10	2.1
Equi-Flow	12.6	9.5	2.1
Excalibur	14.3	—	—
Harvest Maid III Preserver	3.8	8	3.25
Harvest Maid FD-300	10.9	10	1.9
Hulware	14.6	7	2.1
Jack's	3.6	11	1.4
Little Harvey	12.7	8.5	2.75
Nutri-Flow	16.2	12	1.3
Ronco	2.2	15 +	2.9
Sun Pantry (oven)	4.7	8	5
Waring	3.4	6	1.4

Harvest Maid III Preserver

$109.95, extra shelves, $15.95 for two, plus $3.95 for each plastic mesh liner. Drying space, 3.8 sq. ft. Cost to dry 1 lb. of peas, 3.25¢.

The Harvest Maid Preserver is a small, round dehydrator made of white plastic. It comes with four drying shelves, but as your drying needs grow, so does the dryer. Up to six more shelves may be purchased separately and stacked on top of the first four.

The shelves are made of rigid plastic with a large mesh. Removable plastic screening rests on top to keep food from falling through. To check how the food is drying, you must unstack the shelves.

The converging air flow is controlled by an on-off switch and variable-temperature thermostat. The heater and fan were adequate for the four standard shelves but we didn't test with optional shelves. The unit runs quietly.

Alternative Pioneering Systems, 109 Portland Avenue South, Minneapolis, MN 55401.

DRY IT, YOU'LL LIKE IT

Gen MacManiman. 1974.
Living Foods Dehydrators, P.O. Box
546, Fall City, WA 98024. 74 p.
paperback $3.95.

Although this little book is published by a dryer manufacturing company (Living Foods Dehydrator), it avoids being a marketing pitch. In fact, *Dry It* encourages the first-timer to try drying over the radiator or in other warm places around the house. It also suggests such resourceful tricks as washing large batches of herb roots in the washing machine. All along, this optimistic book reassures the reader with "don't worry." Its friendly, casual approach is reinforced by its design: all 74 pages are written by hand.

Dry It has a representative, but not exhaustive, fruit and vegetable survey, a brief herb chapter, and a short treatment for meat, fish, and fowl. There is also a substantial chapter on grains and seeds—a rarity for books on drying—with directions for making wafers, flatbreads, and chips. About half of the book is devoted to recipes.

The final chapter has plans for building an electric dryer, and the instructions are detailed so clearly that even an unskilled worker can follow them. The heater must be purchased assembled, however.

Banananut Freezecreme

4   ounces dried bananas soaked in
      enough water to make 1½ cups of
      bananas and liquid
¼   cup oil
1   teaspoon vanilla
¼   cup honey
1   cup walnuts

Blend soaked bananas and add oil, vanilla, and honey–blend in thoroughly. Add walnuts and blend briefly (so nuts will still be chunky). Serve as is, or soft-freeze for a most surprising, delicious, creamy, marshmallowy dessert!

Harvest Maid Home Food Dehydrator FD-300

$199.95. Drying space, 10.9 sq. ft. Cost to dry 1 lb. of peas, 1.91¢.

The Harvest Maid FD-300 is a deluxe cabinet model with double-wall insulation, a decorative walnut vinyl veneer, and an air intake filter. The thermostat and on-off switches are located on the front. The tinted see-through plastic door opens to the side.

The eight plastic shelves are covered with flexible plastic mesh that allowed some small particles of food to fall through.

This Harvest Maid has a horizontal air flow and a recirculating mode. According to our tests, energy-saving claims made in the instruction booklet for the recirculating air flow are suspect. With an empty dehydrator, the temperature could be maintained with less energy, but a filled dehydrator used as much energy as other dryers.

Alternative Pioneering Systems, Inc., 109 Portland Avenue South, Minneapolis, MN 55410

Hulware

$79.95. Drying space, 5.5 sq. ft. Cost to dry 1 lb. of peas, 2.1¢.

The Hulware dehydrator is small, taking up just a little more than one square foot. There is a walnut vinyl veneer over the formed plastic body and a see-through door.

This little guy has a horizontal air flow and a powerful fan, but even with its small shelves there was an uneven drying pattern. One or two tray rotations are recommended for vegetables. The thermostat ranges from 60 to 170°F., but the numbered markings don't indicate the temperature of the setting. You need a thermometer to know what the setting is.

The Hulware's seven trays are not quite square, and it's hard to tell which side is which when loading them. The shelves have nylon window screening which sometimes can pull loose under stress, although we didn't have this problem.

Hulbert Warehousing, 1921 Parkwood Drive, Olympia, WA 98501.

Jack's Food Dehydrator

$50. Drying space, 3.6 sq. ft. Cost to dry 1 lb. of peas, 1.4¢.

Jack has built the basic economy dehydrator. There's no fan, no thermostat, no switches, just a heating element in the bottom of its unpainted aluminum body. It is shipped disassembled in a thin box, but can be assembled in just 15 minutes with a wrench and screwdriver.

The heating element was different from the other dryers, so we gave Jack a call to chat about it. He told us it was his own patented design, specially insulated and designed with a relatively low power output to generate a slow-but-constant source of heat that (he claims) eliminates the need for a thermostat. Our experiences didn't bear him out. If the box gets too hot you can lower the temperature by opening the doors. A thermometer is necessary for this dehydrator.

The four drying trays are ventilated by small holes spaced about an inch apart. Nylon window screening rests on the shelves to keep food from falling through the holes. We had serious problems during our pea test with uneven drying, and had to continually stir the food so that

each piece could have a turn over a ventilation hole and so that some pieces wouldn't overdry. The temperature can be controlled by adjusting the drawerlike trays, but this procedure required constant attention.

Wheeler Enterprises, Inc., P.O. Box 78007, Skyway Branch, 7855 South 114th Street, Seattle, WA 98178.

FOOD DRYING AT HOME THE NATURAL WAY
Bee Beyer, 1976.
J.P. Tarcher, Inc., 9110 Sunset Boulevard, Los Angeles, CA 90069.
193 p. paperback $4.95.

This book is part information and part tract; in addition to food drying information, it has many pages extolling the advantages of natural living. It presents the most basic methods for drying food—no preblanching, soaking, or sulfuring.

The author offers drying specifics for a wide selection of fruits and vegetables, recommending only one method: the electric "preassembled home food dehydrator." If you want to dry food with an oven or in the sun, she suggests you'll likely fail.

Little Harvey Dehydrator

$169.95. Drying space, 12.7 sq. ft. Cost to dry 1 lb. of peas, 2.75¢.

There's nothing terribly fancy about the Little Harvey Dehydrator, but it gets the job done. It has a cream-colored metal cabinet and a tinted, see-through plastic door on the front which opens from the top. Little Harvey is deeper than it is wide, so it takes a minimum of counter space. The thermostat and on–off switch are located on the back and control a horizontal air flow.

The eight shelves have aluminum frames and fine nylon screening. They withstood many heavy-duty jobs drying fruits, leathers, and soybean mush, and never stretched. There are no reinforcements underneath the screening.

The thermostat controls a wide temperature range, but markings are only every 30°F. We recommend you use a thermometer. Shelves were rotated twice to insure even drying.

Little Harvey Dehydrator, P.O. Box 15481, 3272 South West Temple, Salt Lake City, UT 84115.

Fresh Corn Chips

Corn is one of our most nutritious foods. We destroy much of this by turning it to starch when we cook it. Here's a delicious corn chip that is never cooked.

2 cups (3 or 4 ears) cut fresh corn (about 1⅓ cups pureed)
¼ teaspoon dried powdered onion, if desired
¼ teaspoon dried powdered garlic, if desired
3 tablespoons green pepper, chopped
3 tablespoons tomato, chopped, including peel and seeds.

Puree corn, onion, and garlic in blender. Prepare one tray. Spread plastic wrap lengthwise over tray, and tape each corner to tray with transparent or masking tape. Pour corn mixture into a rectangular shape on lined tray. Using rubber scraper, spread ¼-inch thick making rectangle about 9 by 13 inches. Sprinkle with green pepper and tomato. Dehydrate 8 to 10 hours or overnight, until crispy and crinkled. Corn mixture may be broken into chips and served in a bowl as a snack.

Yield: 1 or more cups of chips

HOME DRYING OF FOODS
Ruth N. Klippstein and Katherine J. T. Humphrey. 1977.
Information Bulletin 120, Mailing Room, Building 7, Research Park, Cornell University, Ithaca, NY 14853. 14 p. paperback $ .50.

This booklet will get you started drying. It gives clear, detailed instructions for drying in an oven or electric food dehydrator, with enough trouble-shooting tips to keep you out of danger (although we recommend you ignore the instructions on sulfuring fruit.)

Don't be fooled by the small size: *Home Drying of Foods* has a yield table for 16 fruits and vegetables, a preparation chart for many more, and a rehydration table. It covers leathers, jerky, and herbs and includes 12 recipes. A bibliography is included so you can progress. And it only costs 50 cents.

Sweet Tomato Leather

Small cherry tomatoes or varieties with high solid content are best for leathers. Wash thoroughly and remove stems and blemishes. Puree in a blender. Begin with a few wedges of tomato to obtain juice, then add more tomatoes to the desired amount. The addition of a lemon wedge and 1 tablespoon of honey per cup of puree makes a delightfully sweet leather.

We stored all we could outside, and when we ran out of room we put the rest in the barn.

Attributed to Abraham Lincoln, when describing a bumper crop of hay in Illinois.

Nutri-Flow Food Dehydrator

$149. Drying space, 16.2 sq. ft. Cost to dry 1 lb. of peas, 1.3¢.

The Nutri-Flow Food Dehydrator has a boxy metal cabinet with a vinyl wood veneer finish. The narrow, deep design makes good use of counter space. The clear plastic door slides up and down in a groove.

The 12 rigid plastic shelves have a coarse mesh that is fine enough for most foods and will work for herbs if you flake them after drying.

There's lots of drying space here; almost more than the Nutri-Flow can handle. With a full load it's pretty slow going, as we discovered in our pea test. The drying pattern was typical—most drying took place near the heater—so trays need to be rotated occasionally.

Nutri-Flow, 5201 S.W. Westgate Drive, Suite 102, Portland, OR 97221.

Ronco Electric Food Dehydrator

$24.99. Drying space, 2.16 sq. ft. Cost to dry 1 lb. of peas, 2.9¢.

The secret to the $25 Ronco is a light bulb: 100 watts of heat with no fan.

This round unit has an airway up the middle through which the light-bulb-heated air rises by natural convection. The Ronco is made entirely of white plastic and occupies less than one square foot of counter space.

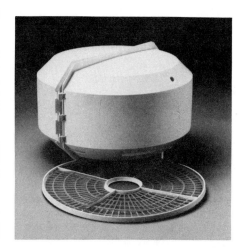

The ventilation holes in the three plastic shelves are quite large—fresh peas rolled through unhindered. We cut screening to fit over the shelves to solve that problem.

The dryer is hinged in the back and divides in half down the middle for loading and unloading. At first, closing the dryer with three full shelves of peas was a two-person job. The difficulty was in juggling three shelves so they would slide into their respective notches while closing the two sides. It wasn't easy, but with practice we finally learned to do it single-handed. We dried the peas for over 15 hours, rotating the shelves and stirring the peas every few hours. Those in the center of the dryer and on the bottom shelf dried fastest. By the end of the day many peas had not dried; rather they had soured and bleached from the light.

Ronco recommends replacing the 100-watt bulb with a 25-watt bulb for making yogurt. We did not test this function.

Ronco Teleproducts, Inc., 1200 Arthur Avenue, Elk Grove Village, IL 60007.

Sun Pantry Oven Fruit Dryer

$25.95. Drying space, 4.7 sq. ft. Cost to dry 1 lb. of peas, 5¢ (in oven).

The Sun Pantry Dryer is just a wooden rack with four window screen trays. But it could be just the dryer to get you started drying, especially if you don't want to spend a bundle for an appliance because you're not quite sure how much use it will get. The Sun Pantry can be used out in the sun (with cheesecloth draped over it to protect food from insects), set in a warm attic, or placed in a warm oven.

We tested the Sun Pantry in the oven, and found that the peas dried just as quickly as in electric dehydrators. We averaged about 8 hours with the oven on low and the door opened about 6 inches (the temperature averaged about 150°F.). Drying was quickest on the top and bottom trays and around the tray edges, so we rotated the trays and stirred the peas.

Ambit Enterprises, Inc., P.O. Box 1302, 14137 Garfield Avenue, Paramount, CA 90723.

HOW TO DRY FOODS
Deanna DeLong. 1979.
H.P. Books, Box 5367, Tucson, AZ 85703. 160 p. paperback $5.95.

This book includes virtually everything you need to know to dry almost any food common to the North American diet, presented in a straight-forward, accessible manner. It makes dried foods look so pretty and suggests so many uses for them that it's inspirational as well as informational. While it's inclusive, *How to Dry Foods* is also succinct. All the process basics, including equipment, are covered in two pages. It includes tips for every drying method, an excellent check list for buying a dehydrator, and plans for building your own electric or solar model.

Our only caveats: The book recommends sulfuring fruits, and extolls the curing and preservative benefits of sodium nitrite, saying only that this chemical is being investigated by the FDA and should be used carefully.

Waring Food Dehydrator

$75. Drying space, 3.43 sq. ft. Cost to dry 1 lb. of peas, 1.44¢.

The Waring Food Dehydrator is an attractive little unit (it weighs only 6 pounds) with a brown translucent dome covering five gold plastic trays. It takes up just over one square foot of space. The Waring has no adjustable temperature control; the thermostat has been preset for 140°F. at the factory.

The Waring—with its converging air flow—was the fastest of the dryers we tested to dry peas, and had the lowest cost for those with fans.

Waring Products Division, Dynamics Corporation of America, P.O. Box 207, New Hartford, CT 06057.

Easy Indian Chutney

Sweet and spicy flavor enhances curry, meat, or rice dishes.

¼ cup boiling water
1 cup chopped dried apricots, peaches, pears, or pineapple
¼ cup raisins or dried currants
3 tablespoons honey
1 tablespoon red wine vinegar
¼ teaspoon ginger, ground
¼ teaspoon coriander, ground
dash cayenne pepper
dash cloves, ground
⅓ cup cashews, chopped

In a medium saucepan, pour boiling water over dried fruit. Cover and bring to a boil. Remove from heat and let stand 10 minutes or until water is absorbed. Add raisins or currants to fruit mixture. In a small bowl, combine honey, vinegar, ginger, coriander, cayenne pepper, and cloves. Stir to mix well. Stir into dried fruit mixture. Add cashews; stir. Store in refrigerator up to 2 weeks.

Yield: 1 to 1½ cups chutney

COMPARING THE DRYERS

Dryer	Cost	Weight (lb.)	Dimensions W x D x H (in.)	Capacity		Air flow pattern	Heat source
				Shelves	Drying area (sq. ft.)		
Bee Beyer's	$229.99	33	23½ x 15½ x 12	8	11.6	horizontal (with re-circulating option)	nichrome coil
Dri Best	229.00	26	17⅝ x 16 x 24	12	14.7	vertical	nichrome coil
Equi-Flow	169.95	29	15 x 18⅜ x 15¼	10	12.6	horizontal	nichrome coil
Excalibur	149.00	28	22 x 17¼ x 12	9	14.3	horizontal	nichrome coil
Harvest Maid III Preserver	109.95	10	15 round x 9½	4	3.8	converging	nichrome coil
Harvest Maid FD-300	199.95	24	21½ x 17½ x 10½	8	10.9	horizontal (with re-circulating option)	nichrome coil
Hulware	79.95	8	13⅝ x 12½ x 11⅞	7	5.5	horizontal	insulated sheath tube
Jack's	50.00	7	12½ x 12¼ x 16	4	3.6	natural con-vection	specially insulated nichrome coil
Little Harvey	169.95	35	17 x 24 x 13	8	12.7	horizontal	ceramic cone with nichrome coil
Nutri-Flow	149.00	25	16 x 22 x 14	12	16.2	horizontal	nichrome coil
Ronco	24.99	2	14 round x 9	3	2.2	natural con-vection	light bulb
Sun Pantry	25.95	2.5	17 x 13 x 7	4	4.7	natural con-vection	sun, oven
Waring	75.00	6	13¼ dia. x 10¾	5	3.4	converging	insulated sheath tube

Wattage	Air intake filter	Fan	On-off switch	Protected heat element	High-temperature cutoff	Grounded plug	Thermostat
not available	−	+	+	+	+	+	100-140° F.
1,000	+	+	+	+	−	+	120-170° F.
1,050	−	+	+	+	+	+	low-medium-high
1,050	+	+	+	+	+	+	85-145° F.
525	−	+	+	+	+	−	95-160° F.
600	+	+	+	+	+	−	90-145° F.
530	−	+	+	+	−	+	60-140° F.
165	−	−	−	−	−	−	none
1,200	−	+	+	+	+	+	95-155° F.
not available	−	+	+	+	+	−	90-150° F.
100	−	−	−	−	−	−	none
none	−	−	−	none	−	−	none
275	−	+	−	+	+	−	140°F.

Solar drying

For thousands of years, heat from the sun has been used to dry food. The energy is free. But nature is a capricious servant at best. One day may be hot and clear—perfect for drying—while the next day (or two or three) may be rainy. Your food could get off to a great start, and then spoil if you are without an oven or electric dryer for backup.

Bob Flower, a researcher at Rodale's Organic Gardening and Farming Research Center, spent several months studying the possibilities and problems of solar drying. He recently shared his findings—and ideas—with us.

To sun dry your food, you don't have to move to Arizona. You can successfully solar dry in any area of the United States, if you plan with care.*

Three clear, sunny days in a row will provide an ideal drying period, and such weather is likely for many areas of the country only in late summer or early fall. Two such days will probably be enough to dry most foods; with only one, you'll be in trouble. So watch the weather reports, and whenever possible adjust your harvests for sunny periods.

The best drying weather for the northeastern United States conveniently comes in September and October. These are not the hottest months, but they have the most usable sunlight, along with light winds and low rainfall. This is the time when nature ripens and dries most crops, and gardeners can intensify this process in a solar dryer.

Along with heat from the sun, relative humidity is another key factor. It is a direct indication of the moisture-absorbing capacity of the air; the lower the relative humidity, the greater its capacity to absorb moisture from food. One surprising thing we learned was that if air coming into a solar dryer can be heated to 100° F. or higher, its moisture absorbing capacity increases so tremendously that neither the temperature nor relative humidity of the incoming air has a major effect on the

drying rate. So if you've got bright sunlight, you can probably dry your food even if the humidity is high.

Imagine, for example, a hot, humid, late-summer day with a temperature of 90°F. and a relative humidity of 80 percent. Not a great day for drying anything. But if your solar dryer can raise the temperature of the incoming air 10° to 100°F., the relative humidity will fall below 60 percent—a tolerable level for drying

many foods. At 110°F. the humidity will be less than 50 percent, and at 120° F., it will be about 35 percent.

Drying cycles

There are two periods in the food drying cycle. During the first period (roughly equivalent to the first day of drying), when the moisture content in the food is high and fairly evenly distributed, a high air flow will remove

Bob Flower discovered that the Rodale Solar Growing Frame, which is designed to grow vegetables year-round, could also be used as a solar dryer.

*The specifics given here are from our tests in eastern Pennsylvania and will vary for different parts of the country.

more moisture than a high temperature, so adjust your dryer accordingly. (This is where a simple solar device has an advantage over some electric dehydrators—you can vary both the temperature and the air flow.)

During the second (or finishing) period, you try to remove the small amount of remaining water from the center of the food through the already dry outer layer. To accomplish this, high air temperature is more effective than high air flow, since the food must be heated to cause this moisture to move. By limiting the air flow through your dryer, the temperature will automatically rise.

The object of solar drying is to remove as much moisture as possible during the first two days, when the flavor and vitality of the food are at their peak. Fruits and vegetables continue to ripen for several days after picking, so the ideal is to finish the drying process just as ripening is completed.

However, unless you've had two days of strong sunlight in a row, you should probably dry the food for a third day to be sure it's fully dry. If you had a sunny first day, the food can survive a cloudy second day if the third day is relatively bright. If the food must be left in the dryer longer than three days, the trays should be loosely covered with a layer of white muslin or other light cloth to prevent loss of color and taste. After three days in the sun, food quality will begin to deteriorate.

We found that it's not necessary to seal up the food at night (though it should be covered to exclude insects and morning dew). Some drying will take place during darkness. If it rains, however, grab the food and run indoors with it.

Solar dryers

There are all kinds of solar dryers, from cardboard boxes lined with aluminum foil to elaborate devices with solar powered fans. They all try to do the same thing—improve the drying efficiency of the sun.

The concept is simple. No matter how warm the sun is, if you hold a small amount of air in a relatively confined space, that air will get warmer. The less air movement

(Continued on page 184)

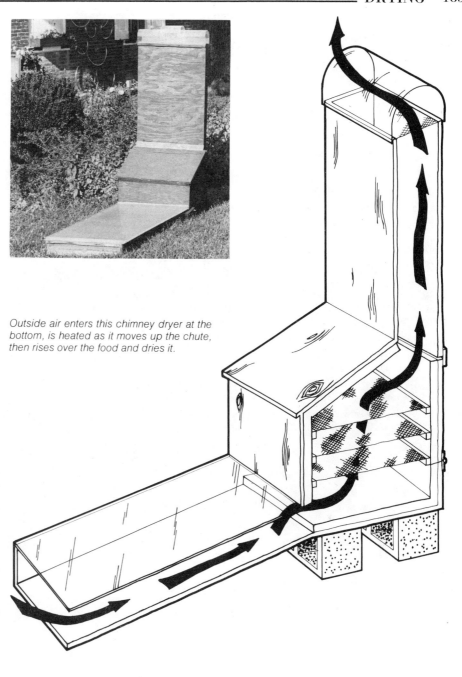

Outside air enters this chimney dryer at the bottom, is heated as it moves up the chute, then rises over the food and dries it.

THE GREEN BEAN TEST

Dryer	Estimated relative drying rate (1.0 is the fastest)	Estimated time needed to dry beans
Electric	1.0	15 hours at 115° F.
Cold frame	0.90	2 sunny days
Chimney	0.66	3 sunny days
Barrel	0.34	3 sunny days
Brace hot box	0.25	4 sunny days
Open screen	0.17	4 sunny days

From 5 pounds of fresh pears to 1½ pounds of dried pears in 10 hours for 13¢.

(Continued from page 183)

through the space, the hotter it will be. Dryers are designed to both amplify the sun's heat and provide an air flow for efficient drying.

Bob Flower's research at the Organic Gardening and Farming Research Center included testing solar dryers to see how they performed. Among the dryers he tested were: 1) a chimney-type dryer; 2) a cold frame designed by Rodale Press for growing food in winter, slightly modified for drying; 3) a barrel dryer; 4) a box dryer designed by the Brace Research Institute of McGill University, Montreal, Quebec, Canada; and 5) a screen tray set in the open sun. For a comparison, he used an Equi-Flow electric dryer. Tests were done with green beans and green peppers.

The fastest dryer, as the chart shows, was the electric one. But the tests showed that a well-designed solar dryer can virtually equal the performance of an electric dryer. And while the drying rates varied considerably, most of the dryers got the job done. Similar results were obtained in tests with green peppers; all dryers (except the open screen) dried them fully by the end of the third day.

The conclusion: If you can adjust your drying periods to nature's moods, you shouldn't have any problem drying your food.

(Continued on page 186)

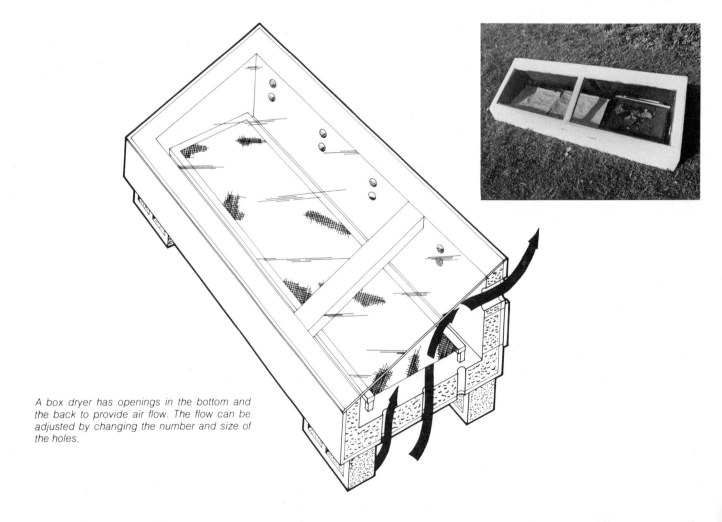

A box dryer has openings in the bottom and the back to provide air flow. The flow can be adjusted by changing the number and size of the holes.

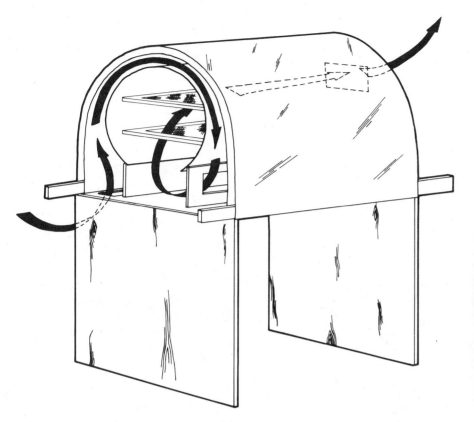

In this barrel dryer, the air enters through a slot in the bottom and circulates around the food before exiting.

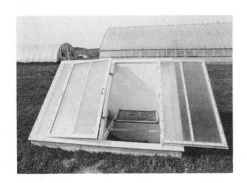

Since the Rodale Solar Growing Frame is not designed as a dryer, air movement occurs only when the door is open. Cool air comes in the bottom and heated air goes out the top.

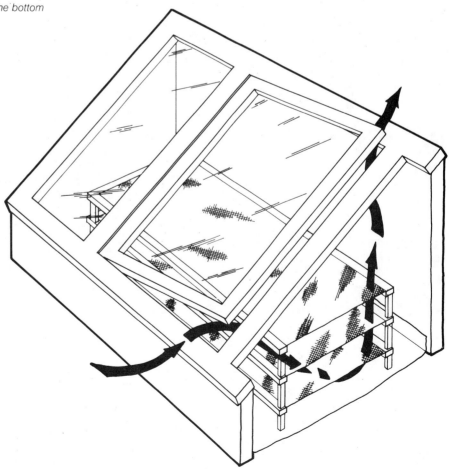

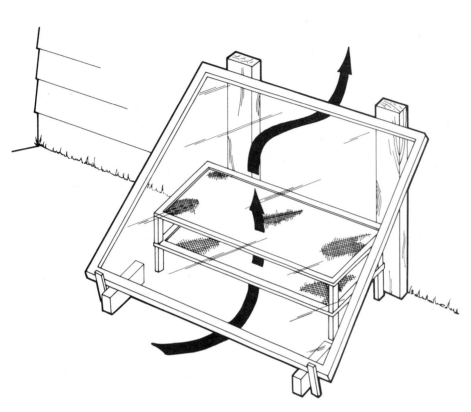

You can make a solar dryer from some wood and screen trays, a storm window, and a little scrap lumber.

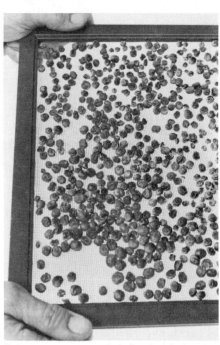

If the temperature and air flow inside the dryer vary too much, the food will dry unevenly.

(Continued from page 184)

Building a solar dryer

If you'd like to build a dryer, you can buy plans (see below) or design your own. In hot, dry climates, a tray or two are all you'll need; in other areas, some refinements will be necessary. The following design, suggested by Bob Flower, has several options, and it's intended to use equipment and materials you may already have around the house. You can take this idea and adapt it to your situation.

You can build basic trays by making a wooden frame, then fastening nylon or fiberglass screen on the bottom. One option is to size them so they'll fit into your oven, in case the sun doesn't always cooperate. At least one of the trays should have legs that provide about an eight-inch clearance between the ground and the bottom of the tray, to allow for air circulation. It's smart to screw the legs on so they can be removed easily. You can build all the trays with legs, or make them stacking, or both.

For solar drying, place the trays beside a south-facing wall, then lean storm windows against the wall above the trays. (If you don't have storm windows or something similar, you can rig up a frame and use clear plastic.) A block of wood can be put

The simplest method of drying: just place your food on a tray in the sun.

between the wall and the top of the window to leave a gap for the rising hot air to escape through. As it flows out, it will pull in new air.

Depending on how hot the sun is, the windows may be all you need. If they don't get the temperature high enough, tape or tack clear plastic sides between the window and the wall. To insure an air flow, leave an opening at the bottom. Adjust the size of the opening to control both the air flow and the temperature.

The Lightning Tree
P.O. Box 1837
Sante Fe, NM 87501

This press publishes the book Homegrown Sundwellings *by Peter van Dresser, a long-time inventor and experimenter in the area of solar energy. It includes plans for several chimney-type solar dryers, as well as information on solar home design and solar greenhouses. The book costs $5.95.*

Brace Research Institute
Macdonald College of McGill
    University
Ste. Anne de Bellevue
Quebec, Canada H9X 1C0

The Brace Research Institute develops equipment and techniques for making dry lands more agriculturally productive. They sell plans for a wide variety of solar equipment, including dryers, stills, steam cookers, and water heaters. Their plans for a cabinet solar dryer – the Brace Hot Box – are $1.50, plus a $1.50 handling charge per order. They'll also send you a complete list of their publications on request.

Solar Survival
Box 275
Harrisville, NH 03450

Solar Survival's plans and instructions are for a 20-pound barrel dryer similar to the one we tested. The cost is $10.75, postpaid.

Solstice Designs, Inc.
Box 2043
Evergreen, CO 80439

Solstice's plans package tells you how to build four types of solar dryers: a family food dryer (similar to the Brace Hot Box), a collapsible dryer, a window box dryer, and a cardboard box dryer. Cost for all four plans is $16.

Rodale Press
Plans Department
33 E. Minor Street
Emmaus, PA 18049

We discovered that our solar grow frame can be modified slightly to make a fine dryer. Simply line the bottom with a sheet of black polyethylene plastic, place the trays inside the grow frame, and prop the door open slightly to provide ventilation. Cost of the Solar Growing Frame book (with plans) is $14.95.

(Continued on page 189)

A bracelet of milk: Kangan, India

In parts of India, milk is actually stored on a string.

It is first boiled until quite thick (about four hours), then made into thin patties and set out in the sun for two days or so. The resulting discs are perforated, placed on a string like beads, and hung up until needed. The discs are reconstituted by frying and then boiling in water until they reach an edible, somewhat cheeselike consistency. They are eaten with rice and chapatis.

Home Food Systems picked up this novel process from Ghulam Mohd Wani of Kangan, a small Kashmiri village. He had just finished building a low wood-fired kitchen stove of clay and stone—it was smooth and still malleable to the touch, as attractive as a piece of sculpture.

G. M., as he is called, is almost forcefully hospitable, and we were directed to take seats on the floor of the living room. There a lunch of cold rice, fresh yogurt, dal, and a samovar of spiced green tea was served by G. M.'s young wife, a woman attractive in a sulky, small-town way. The couple watched with polite curiosity as we did our best to deliver the food to our mouths without utensils, as is the custom.

The S/A Bumble Bee

The makers of this stovetop dryer call it the Bumble Bee because it doesn't look like it will work, just as the bumble bee looks too heavy to fly. But work it does. Water in the reservoir below the drying tray tempers the heat from a stove (or any other primary heat source) and evenly distributes it over the stainless steel tray. Heat transfer to the food is direct, so drying time is fairly short. Using an electric range, we dried four pounds of peas in about 5 hours using 15¢ worth of energy.

The best application of the Bumble Bee is on top of a wood stove. The heat is almost free, and slow but steady. A bonus is that the water bath humidifies the house as steam rises from the dryer.

Constant heat isn't a must. The dryer can be set aside for an hour or so once the water starts steaming if you need to use the stove. However, full of water it weighs 40 pounds, so you may want a partner for toting it around.

Write S/A Distributors, 700 E. Water Street, Suite 730, Midtown Plaza, Syracuse, NY 13210.

There are some indications—very slight in the case of human beings—that man is not completely devoid of the capacity to select a nutritionally balanced diet . . . in the ages during which men have been dependent upon a system of food provision in which there was division of labor, and men—and/or women—collected or produced food, which was then prepared for the family by the wife, or the husband—so that the child had no experience of foraging for itself, and half or more of the adults of one sex usually ate food prepared by someone else, the viability of any food pattern has come to rest upon an intricate pattern of traditional behavior, rather than upon the instinctual choices of each individual.

Margaret Mead in *Centennial*, collected papers presented at the Centennial Celebration, Washington, D.C., 1948.

USDA ON TV DINNERS

TV dinners are costing the American consumer more in dollars and providing him less in nutrition, according to a recent study by the Consumer and Food Economics Institute of the USDA. The commercially prepared, frozen entree does save time and effort in meal preparation, but it costs nearly twice as much as identical portions of a home-cooked entree containing the same ingredients. In this study, conducted in the Washington, D.C., area in the fall of 1978, the cost of individually purchased ingredients was totaled, and then the cost per serving of the home-prepared dinner was estimated; finally, the cost of the commercially prepared equivalent was compared. The figures showed that not only did the TV dinner cost more; it was also less nutritious!

Prices varied according to the meat content of individual dinners. For example, convenience-form lasagna cost 127 percent more per serving than an equivalent serving of kitchen-prepared lasagna. However, chicken pie frozen in individual portions cost only 9 percent more—but, the commercially prepared chicken contained less meat.

Frozen, commercially prepared entrees have other drawbacks besides cost, however. Consumers may prefer more meat, poultry, fish, cheese, or other ingredients than a serving of the convenience item contains. In addition, persons on therapeutic diets—such as low-fat, -sugar, or -salt—have less control if a commercial product is used because they usually cannot determine the presence and amounts of such substances in the product from the container label.

Family Economics Review (USDA), as quoted in *Journal of the American Dietetic Association.*

(Continued from page 187)

Oven drying

If you'd like to try drying food, but don't want to start by buying an electric dehydrator, try your oven. It's not terribly energy efficient, but it works.

You can use the oven racks as trays, but cover them with cheesecloth or nylon mesh. Cookie sheets will work also, but because they're solid the food will dry more slowly and must be turned to dry evenly. Or you can make trays with a wooden frame and nylon screen to fit your oven.

Since heat rises, all the heat should come from the bottom of the oven if possible. If heat is coming from the top, place the food well down in the oven and put some kind of shield (a cookie sheet or piece of aluminum foil) on the top shelf.

The biggest problem is maintaining the desired temperature. Because air circulation is essential, you need to dry with the oven door slightly open. The temperature should be set no higher than 145° F. If

you can't set the control that low, put it on the lowest possible temperature and adjust the door opening to regulate the heat. Either way, you should have an oven thermometer so you can monitor the actual temperature inside the oven.

Because an oven can overheat quickly, check your food every half-hour or so. The heat may also be uneven, in which case you'll need to move the trays around and stir the food. Be especially careful near the end of the drying time, since a little too much heat can scorch food and ruin it.

FINE-TUNING YOUR APPETITE

By simplifying the diet for a time, we can fine-tune our appetites so that taste will exert its beneficial, guiding influence over what foods we choose to eat. One becomes highly aware of salt overdoses, cloying desserts, intoxicating beverages, the heaviness of meats and oils and high-fat cheese, the sludge of overcooked vegetables, the bogginess of overprocessed starch. To the person with a simple, sane diet, departures from common sense at the table make themselves known by internal rumblings, a dry tongue, upset bowels, a greasy face, sluggishness, an overstimulated appetite, a reawakened craving for sweets.

How to simplify? For a few days, try planning your meals around a basic food—rice, or chapatis, or soyfoods—and experiment with recipes from third-world cultures.

RY

Canning

Putting a lid on it

Home canning has fallen on hard times. It used to be that nearly everyone who had a garden put something up for the cold months. Now, with home freezers, the year-round availability of many kinds of fresh produce, and the variety of canned foods in the supermarket, fewer people are canning.

Canning *seems* outdated as well. The work conjures up visions of grandmother spending endless hot hours over a stove to insure the family had plenty to eat through the winter. Today, the glass jars and metal lids, boiling kettles, hot pads, and cooking racks seem a little primitive and even frightening. Canning truly is not as simple as dropping a bag of vegetables in the freezer. But this is no obituary.

Home canning is not dead

It might surprise you that 40 percent of American families can one thing or another. They've learned that it's more than a low-cost, energy-efficient way to keep foods. There's a unique, basic sense of satisfaction that comes with those finished jars.

Canning rates just behind freezing when it comes to flavor and nutrition, but it wins in long-term storage. There's no space problem, either, and the possible products are almost limitless.

Home canning is a whole-family activity. With an assembly line of responsibilities, everyone can help, from picking the fruits and vegetables to washing, pitting, slicing, stirring, filling jars, watching the pot and the time, checking the seals, labeling jars, cleaning up, and of course,

taste-testing along the way.

By canning your own food, you control how it's prepared and what goes into it. You're not forced to eat foods with additives and pesticides. Even when your home-grown supplies are exhausted, you can seek out local organic growers through newspaper ads, and co-op members might be able to steer you toward a pesticide-free patch.

Perhaps you've never thought of home canning as a means of eliminating waste, saving energy, or preserving and renewing natural resources, but it does contribute favorably to the earth's ecology in several ways: canning reduces the need to purchase off-season items that are transported many miles, using fuel for travel and storage; home-canned foods require no refrigeration; they don't involve the wasteful packaging of store-bought items; and trimming

from preserved produce can be recycled through the compost pile.

The benefits of canning become even more obvious when the freezer is overflowing: you can go on canning until your basement bulges.

Susan Hercek

Birth of the can

The year 1795 was not a good one for France. Napoleon's far-flung battles had drained the country's manpower, a revolution was in progress, and the country was at war with several of its neighbors. Provisioning the armed forces became a serious problem; soldiers and sailors living on putrid meat and stale bread did not make outstanding fighters. So the five-man Directory, France's governing board, offered a prize of 12,000 francs to anyone who could invent a successful method of long-term food storage.

Francois Appert—the eventual winner—was a confectioner with no technical training or experience. He was, however, a careful observer, and in his cooking he noticed that food that had been heated and sealed in airtight containers tended not to spoil. He didn't know why: Pasteur's discovery of microbes was still some 50 years away.

For his work, Appert used glass bottles, since they were "most impermeable to air." The ordinary narrow-neck bottles were inconvenient to pack, and broke easily, so he ordered special bottles with openings as large as four inches. He sealed them with custom-made cork, which held most of the time. Before boiling the bottles, he put them in individual sacks—both for protection and for easy removal in case of breakage.

Everything was by trial and error, so the work was slow. But by 1804, Appert had succeeded in preserving a wide variety of food. A contemporary report on three-month-old provisions "prepared according to the process of Citizen Appert" stated that

the beef was "very edible" despite a broth that was "weak." Green peas and beans had "all the freshness and the agreeable savor of freshly picked vegetables."

In 1810 Appert published a treatise on his method of food preservation, which he modestly called Appertizing. The book, *The Art of Preserving All Kinds of Animal and Vegetable Substances for Several Years,* sets forth the four basic steps to successful Appertizing: 1) enclosing in bottles the substance to be preserved; 2) corking the bottles

carefully, "for it is chiefly on the corking that the success of the process depends"; 3) immersing the bottles in a *boiling water bath;* and 4) withdrawing the bottles from the bath at the end of the prescribed period. These steps—with minor refinements—are still current canning procedures.

While Appert was reaping his prize, other experimenters were trying different approaches. Peter Durand, an Englishman, received patents in 1810 for both glass and metal containers that could be used in

(Continued on page 192)

> I composed this jelly, according to the prescription of a physician, of calves feet and lights, red cabbage, carrots, turnips, onions, and leeks, taking a sufficient quantity of each. A quarter of an hour before I took this jelly from the fire, I added some sugar-candy with some Senegal gum. I strained it as soon as it was made. After it was cold it was put in bottles, which were corked, tied, wrapped up in bags, and put in the water-bath, which was kept boiling one quarter of an hour, and this jelly was preserved and remained as good as it was the day on which it was made.
>
> Francois Appert. *The Art of Preserving All Kinds of Animal and Vegetable Substances for Several Years* (New York: D. Longworth, 1812).

A meat-canning factory in 19th century France. The meat was prepared in the room shown above, then passed through the wall to the room shown below, where the cans were filled and sealed.

(Continued from page 191)

canning. The handmade metal containers were crude and bulky, but stronger and more durable than glass. When the English had success with these primitive tin cans, Appert tried them. The French were poor tin artisans, however, and he ended up using wrought iron with a coating of tin plate.

Early canners used a hole-in-cap method of sealing (which is still used with canned milk). After the can was formed, and the ends were hammered on, a hole was cut in one of the ends. The can was filled through the hole, then closed by soldering on a small disc of tin plate. The cans were tested in a closed room heated to at least 90° F. Defective cans would bulge.

This tedious method—with variations—was used by most commercial canners throughout the 19th century. Since filling the cans through a small hole was inconvenient, a method of soldering on the entire end was soon developed. Other improvements followed: coatings to minimize the effects of corrosion on the tin and machines to solder the seams and ends. Pasteur's work in the 1860s showed why food spoiled, and provided a scientific basis for canning procedures.

Because soldering was so important, different canning companies had secret mixtures they guarded carefully. There was continual controversy over whether the end of the can should be soldered on the inside or the outside. Some countries required outside soldering, though one authority ridiculed this as being based on a "whimsical belief that the danger from lead poisoning is lessened by this maneuver." This whimsy was probably sound, however, since solder is an alloy of tin and lead.

In 1900 the open top or "sanitary" can appeared. The breakthrough was lining the can ends so a hermetic seal could be produced without soldering, despite irregularities in the thickness of the tin or wear on the seaming rolls. These cans were sealed by crimping the edges of the top and bottom over the can ends. This quickly became standard practice, and remains so today.

TD

Why can?

Picture a shelf in your pantry lined with rainbow jars of home-canned fruits and vegetables. You'd have the staples, of course—bright red stewed tomatoes, amber applesauce, ruby beets sliced with a fluted cutter. Perhaps a few jars of chunky dilled carrots and some hot-pickled green beans.

Then there are all kinds of specialty foods you can make in your kitchen. How about gooseberry conserve, kippered cherries, peach chutney, quince preserves, plum butter, corn relish, tomato mincemeat? None of these gourmet-quality foods requires any special expertise—just good fresh food (preferably home-grown) and an ability to follow directions step by step.

It's fun, too, to turn out a mixed pickle that is more than the sum of its parts. At the tag-end of the gardening season when you have a single head of cauliflower, a handful of carrots, a quart of tiny onions, a smattering of snap beans, two cukes and not enough limas to make a meal—you can combine them all to make garden relish.

Before you know it, you may have a specialty—a pickle your friends rave about or a fruit butter you do especially well. A teen-age boy we know is widely respected for his super cucumber dill pickles. He got so good at it, in fact, that he decided to take over the growing of the cukes too. An aunt of mine has traditionally given a basketful of her home-canned peaches, pears, tomato juice, and pickles as gifts to her favorite people for special occasions.

Many foods come out better canned than frozen. Canned snap beans, I've found, have more flavor than frozen ones. Applesauce also tastes better canned than frozen. Canned tomatoes are a homey staple, necessary for all manner of winter soups and casseroles. And while tomatoes freeze perfectly well, for me freezer space is too scarce and expensive to fill with tomatoes when the canned ones taste so good.

Then there are all the pickles, relishes, jams, and butters that are best preserved in jars. Watermelon pickle is a specialty of the house here—one for which I take a lot of teasing for my eagerness to pounce on the rinds as soon as people are through eating the melon flesh. Frozen fruits are good, but canned fruits—especially peaches, pears, cherries, and apricots—are easier to use because you needn't plan ahead and thaw them first. And even though I keep a good supply of beets in the

root cellar, I also can some pickled beets in summer for instant use on busy winter days.

Peas, broccoli, and corn are best frozen, I think, and I never attempt to can them. When berries are canned in syrup the seeds become quite pronounced, so I prefer to freeze berries or make canned juice or jam out of them. Cabbage, potatoes, parsnips, escarole, celery, and other good winter-keepers do well in the root cellar without any processing at all, and since I don't go around looking for extra work, that is where I keep them.

When you start to explore the art of home canning, you'll hear that you should use only jars that have been made especially for canning. I will only say that in the 14 years I've been canning in a big way, I've used countless pint and quart mayonnaise jars fitted with regulation narrow- and wide-mouthed lids, and they've sealed and held up as well as the

heavier canning jars. Some folks think that you must add sugar to fruits when canning, but that isn't true either. Water-packed fruit is a healthful alternative to sickly sweet sugar syrups. I use a light-flavored honey (such as clover) in canning, and I sweeten foods sparingly. The family can always add more fresh honey at the table.

It's been 23 years since I filled my first canning jars—with jam made from wild blackberries my husband and I picked the first summer of our marriage. As my repertoire has grown, so has my satisfaction in this process that preserves the goodness of the food and at the same time lets you admire your handiwork and keep easy tabs on your store of provisions. Begin with the food you like best, or have in greatest abundance. But don't wade through poison ivy, as we did that first year, to get to it!

Nancy Bubel

Be cautious: stay safe

When the vegetables of your labor reach their peak of nutrition and flavor, you will want to preserve them as lovingly as you grew them. But if you're careless about canning procedures, all your efforts—in the garden and in the kitchen—won't be worth a hill of moldy beans.

A forewarning

Begin by thoroughly washing your produce to cut down on contamination by bacteria. Then examine the canning jars for nicks, cracks, and sharp edges. Do not use any that are imperfect. Wash the jars and lids in warm, soapy water and rinse. Place the lids in a pot of hot water, bring to a boil, and simmer. Fill a dishpan with *hot* water and stand the jars in it until they are ready to fill. Keeping them hot will prevent them from cracking when they are filled with very hot food. Never put boiling liquid into a cool jar.

Fill the jars, pressing out air bubbles. Leave the amount of head space recommended in your recipe. This head space is important because some foods swell as they are processed—especially starchy

foods. Wipe each jar's lip and top with a clean, damp cloth or paper towel so that no food particles remain to interfere with sealing.

Jars with vacuum lids are recommended because even a novice can easily check to see if they are completely sealed. Mayonnaise jars or other jars from commercially canned produce aren't as strong because their glass is too thin and often not highly tempered.

Once filled, the jars are covered with vacuum lids, held in place by screw-on bands. At this point the food is ready to be processed.

For a boiling water bath canner, a special metal rack is needed to raise the jars above the bottom of the pot, allowing boiling water to surround them entirely. A rack also separates jars so they do not jostle

(Continued on page 194)

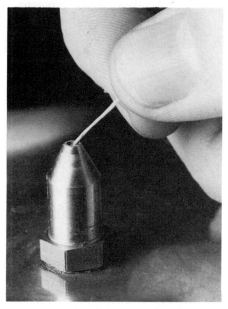

For a pressure canner to work properly, the petcock must be cleaned regularly.

BOTULISM

Botulism is caused by the bacteria *Clostridium botulinum,* a normally benign organism found in water and soil throughout the environment. Once inside the airless milieu of a canning jar, however, the spores divide and produce toxin.

Despite the danger, the prudent canner has little to worry about as long as the directions for canning are followed carefully. Botulism spores are quite heat resistant, but they can be destroyed.

Because acid retards the growth of *C. botulinum,* high-acid foods can be safely canned in a boiling water bath where the temperature reaches only 212° F. For low-acid foods, a pressure canner is necessary, since the food must reach a temperature of 240° F.

Other precautions are required as well. Make sure that you adhere to the recommended processing time, that you get an airtight seal, and that all your equipment is clean and functions properly. For extra safety, boil your food for 15 minutes in an open pan before serving. (This can devastate many of your lovingly canned products, but it may be tolerable with green beans, corn, spinach, peppers, and asparagus, the foods most commonly involved in botulism poisoning.)

In rare cases, botulism can appear in even the most perfectly canned foods. If there is any abnormality in your product—bulges, foam, mold, cloudy contents, malodor—throw it out where no person or animal can get to it. Don't sample questionable items: one trillionth part of a gram of pure *C. botulinum* toxin is enough to kill.

Botulism poisoning hits from 12 to 36 hours after the toxin is ingested, and usually lasts three to six days. The first symptoms may be severe nausea and vomiting, but in many cases these do not appear at all. More-certain signs include double vision, speech and swallowing difficulty, and progressive paralysis of the heart and respiratory system.

In the past 50 years, there have been more than 700 deaths from botulism in this country. If you suspect poisoning, call a doctor immediately; there is an effective antitoxin. Despite its high fatality rate (over 50 percent), botulism is not lethal if it's caught in time.

—TD

(Continued from page 193)

and crack during boiling. Be sure the jars are covered with one to two inches of boiling water so the heat also penetrates the top of the food.

Pressure canners are equipped with either a dial gauge or a weighted gauge that indicates pounds of pressure. Gauges must be checked periodically for accuracy. (Your county extension agent or the canner manufacturer can tell you where to have the gauge tested.) If the dial is off by as much as five pounds per square inch, it should be replaced. If the variation is less than five pounds, adjust the reading to compensate for the variation.

Clean the petcock and safety valve by running a thread through their openings. Clogged vents can cause a pressure canner to explode. Place jars in a metal holding rack, and process according to recipe directions. Allow ten minutes for cooling before removing jars from the canner. To avoid breakage, do not place them on a cold surface or in a draft.

Use vacuum lids (or rubber rings) only once. They are not sturdy enough to insure a good seal more than one time.

Some foods may be canned by alternative procedures. Small jars (pint and half-pint) can be processed in a regular pressure cooker. Because these pots heat and cool more quickly than canners, add 20 minutes at ten pounds pressure to the processing time called for in the recipe.

Carol Keough

Sealed or not?

The biggest fear of home canners is that their jars will somehow fail to seal. At best, this means reprocessing; at worst, the produce may spoil and result in food poisoning.

The basic causes of aborted seals are: 1) underprocessing; 2) using damaged equipment, such as jars with nicked or cracked rims, or warped screw bands; 3) leaving too little headroom so the jars overflow; and 4) catching food between the lid and jar rim. These problems can all be avoided by careful attention to proper canning procedures.

There are several ways to check for a proper seal. With screw rings and flat caps, push on the center of the cap. If it's already down or stays

down, the jar is sealed. If it moves up and down, the jar isn't sealed. You can also tap on the jar with a spoon: a clear ringing sound indicates a good seal; a dull sound means danger.

To test the zinc caps of bail jars, tip the jars upside down and check for leakage or air bubbles. A better test for glass jars is to release the bail wires and lift the jar by the glass lid. With a good seal, the lid will be firmly attached. (But don't lift it more than an inch or two in case the seal isn't good.)

To make things even easier you can now buy lids which have a bubble or button in the center. This pulls down under a vacuum or pops if the seal isn't perfect.

If you catch unsealed jars within 24 hours, the food can be reprocessed, or frozen, or refrigerated and eaten within a few days. For reprocessing, however, the entire canning procedure must be repeated, with additional loss of quality and nutrition. If you discover a bad seal more than a day after canning, the food should be thrown away.

—TD

What happened?

Even the most careful canners sometimes go wrong. Here are some common canning problems and their probable causes.

Bacteria. These pesky creatures cause such blights as putrefaction, salmonellosis, and botulism poisoning. Sustained high temperatures are necessary to destroy both the active bacteria and their dormant spores, making it essential to use only approved canning methods and follow processing directions carefully. (If you observed the directions scrupulously and still had problems, have your pressure canner checked.)

Crystals. The yellow crystals that may occur on canned green vegetables are glucoside. They're natural and perfectly harmless. The tartaric acid crystals that turn up in grape products should be removed. (Let them settle to the bottom, then pour off the juice and fruit.) They won't hurt you but they taste terrible.

Discoloration. Some loss of color in food processing is normal, but canned food occasionally darkens considerably or develops a gray tinge. This is usually caused by a chemical reaction between food and the minerals in hard water or metal utensils. Use deionized water and avoid copper or iron equipment.

Sometimes fruit darkens after you remove it from the jar. This is caused by enzymes in the food, and it indicates that your processing time was too brief to destroy them.

To help maintain the color of fruit, soak it for 20 minutes in a solution of two tablespoons lemon juice or vinegar to one gallon of water.

Rinse before processing.

Flat sour. This is a common term for spoilage caused by a bacteria that gives food a sour, acid taste. It is most common in starchy vegetables, such as corn. This particular bacteria thrives at temperatures between 130 and 150° F., and it grows when food sits too long between canning steps.

Floating fruit. The fruit is lighter than the syrup, and sometimes sinks only reluctantly. Any fruit that stays above the liquid level won't spoil, but will probably discolor. To help prevent

floating, heat the fruit first and pack it very tightly.

Yeasts and mold. These scourges grow and flourish even in high-acid foods, but they are killed by heat much easier than spore-forming bacteria. Yeast causes fermentation, forming bubbles that break the seal and ruin the contents. Mold can be found in virtually any canned food. The causes of yeasts and mold are underprocessing (insufficient temperature or heating for too short a time) and improperly sealed jars.

Canning equipment

Successful canning requires little equipment. All you need is a canner, along with jars and lids, but a few other small items will make the job much easier. With good care, everything but the lids will last for years, and can be reused over and over.

Canners

A *boiling water bath canner* is nothing more than a pot with a wire rack in it to hold the jars and allow water to circulate under and around them. The most common are of black enamel with white spots, but any large pot will do. The most important requirement is that it be at least 11 inches deep, so there is plenty of room for the water to boil around the standard 7½-inch-tall jars.

Pressure canners work like pressure cookers. The pot has a tight-fitting lid with a pressure regulator.

When water is heated in the pan, it turns to steam. Pressure builds up, and the temperature rises. At 10 pounds of pressure the temperature is 240° F.; at 15 pounds, 250° F. All canners have safety devices that will blow out if the pressure gets too high.

There are two types of pressure canners: those with a dial gauge to show the pressure visually, and those with weights that make a noise when

the required pressure is reached. Canners with weight controls usually have three settings—for 5, 10, and 15 pounds pressure.

Neither of these canners will hold the desired pressure, and you may have to practice regulating the heat. Just before the proper pressure is reached, the heat should be turned down. Then, minor adjustments may

(Continued on page 196)

Boiling water bath canner.

The pressure canner on the right has a gauge; the one on the left uses weights.

(Continued from page 195)

be necessary to keep the temperature steady; uneven pressure can force liquid from the jars.

Other useful items

A *jar lifter* makes it easy to lift jars in and out of boiling water, or move hot jars around without burning your hands or wetting a potholder. Most hardware stores stock these.

To fill jars without leaving bits of food or liquid on the rim, a *wide-mouth funnel* comes in handy. Glass or plastic is preferable to metal, which may interact with high-acid foods.

A *colander* or *sieve* is useful for draining fruits and vegetables, straining juices for jelly, and dipping produce in boiling water.

A *blancher* is a perforated basket that sits in a large, flat-bottomed pan. The food in the basket can be easily lowered into boiling water, dipped in cold water, and then drained. This makes blanching or wilting leafy vegetables simple.

Filling jars with hot liquid can be done easily with a *ladle* or a *glass measuring cup.* Putting hot pieces of food into jars is best done with a *long-handled, slotted spoon,* which also works well for stirring cooking foods.

If your canner does not come with a wire rack, you'll need one to keep the jars off the bottom of the

canner and let the water circulate freely around them. A rack for holding jars while they cool is also useful. A cake rack will work, as will a couple of towels.

It is also nice to have a few large *measuring cups,* some good sharp knives, a wire basket for dipping fruits and vegetables into boiling water, and a *food mill* or *press.* These last items can help with such preparation tasks as slicing, grinding, and squeezing out juice.

Jars

Jars and lids are the most basic canning items. Jars come in several sizes; the most common are pints and quarts. Canning with quart jars is more energy efficient than using pints, but depending on the size of your family and the food being canned, it may make sense to use small jars at times. Half-gallon jars are available, but they have some drawbacks. Most boiling water bath canners aren't deep enough for half-gallons, and processing times

may be hard to find. Also, heat may not penetrate well into the center of densely packed foods, resulting in dangerously inadequate processing.

Standard-size jars are available in wide-mouth versions; these are easier to fill but cost more. Wide-mouth jars are suited for freezing because you can remove the food before it is completely thawed.

For years, there has been a running dispute about whether it is safe to reuse commercial jars for canning. Most experts advise against them; many home canners have used them for years with success.

We believe it is best not to reuse commercial jars. They are not intended for repeated use or for the extremes of heat and pressure that are part of home canning. These lightweight jars therefore break more easily than canning jars, and waste valuable produce.

If you plan to reuse commercial jars anyway, be extra careful. Use regular canning lids, and be sure they match the jar. Check the jars for nicks and cracks, and use them only

---

CANNING ECONOMICS

For low-cost food preservation, don't dry or freeze your produce—can it. That's the conclusion of three Pennsylvania State University food scientists who compared the cost of drying, freezing, and canning for peaches and sweet corn.

The canning tests were done with both quarts and pints. The investigators found that with many acid foods, including peaches and cherries, there was no difference in cost between raw pack and hot pack canning, since the raw pack foods required longer processing times.

For information on freezing, the testers used a 1975 Cornell University study. All figures are based on an energy cost of 4 cents per kilowatt hour.

These few items will simplify your canning.

PRESERVATION COSTS PER POUND

	Convection drying	Forced-air drying	Oven drying	Freezing	Canning pints	Canning quarts
Peaches	3.9¢	2.7¢	19¢	9.8¢	1.6¢	1.4¢
Corn	3.6¢	2.5¢	11.5¢	9.8¢	2¢	1.7¢

in a boiling water bath canner with high-acid foods. To reduce breakage, be sure they are hot before you fill them with hot food, or put them in a boiling water bath. When removing them from heat, be very gentle, and avoid drafts or quick changes in temperature. By exercising great caution, you may have success. Better yet, use regular canning jars and save the mayonnaise jars for jams and jellies.

Lids

Since the great lid shortage of a few years ago, the market has greatly expanded. More than 20 brands of lids are now available. Many look alike, but they don't all work the same way, so follow the manufacturer's directions closely.

There are three basic types of lids. The most common are *two-piece lids,* made up of a screw band and a dome lid. A rubber compound on the inside of the lid makes the seal. After the jar is sealed, the band can be removed and stored where it won't rust. The lids should not be reused.

Porcelain-lined zinc caps are still used in many places. They seal with a removable rubber ring, and can be used repeatedly if boiled for 15 minutes before reusing. The rubber rings are not reusable.

The third type, *bail-wire clamp jars,* has a glass lid held in place with a clamp. A rubber ring is placed under the lid to make the seal. These jars must be clamped tightly after processing; when instructions say "complete seal if necessary," they are referring to this type of jar.

When buying lids, check the inner sealing ring to see that it is uniform in thickness and covers the entire area of the lid that will come into contact with the jar. A good lid often has a series of concentric circles at different levels, which allows the vacuum to pull down the middle part of the lid, forcing the outer ring into good contact with the jar.

A good lid has the thick, uniform inner coating necessary to prevent quick corrosion. A small dimple in the middle of the lid is another desirable feature. These dimples give a loud clicking sound when a proper seal is made, and make it easier for you to see that the lid is pulled down—another sign of a good seal.

Reusable, non-corroding screw rings and canning lids of plastic are fairly new on the market. Although the companies making plastic lids have independent testing agency reports claiming good, safe performance, engineers at Cornell University do not believe plastic can ever perform satisfactorily for canning purposes, according to Ruth Klippstein, Associate Professor of Nutritional Sciences. This fear stems from plastic's inability to completely hold its shape under the constant stress of a vacuum.

If you are in doubt about the quality of your canning lids, and the manufacturer lists its name and mailing address on the package, you can write for test results; or ask your local

HOW MANY JARS WILL YOU NEED?

Raw produce	Measure and weight (lbs.)*	Pounds (approx.) needed for 1 qt. jar
Fruits		
Apples	1 bu. (48)	2½–3
Apples (for sauce)	1 bu. (48)	2½–3½
Apricots	1 lug (22)	2–2½
Berries	24 qt. crate	1½–3
Cherries	1 bu. (56)	2–2½
	1 lug (22)	
Peaches	1 bu. (48)	2–3
	1 lug (22)	
Pears	1 bu. (50)	2–3
	1 box (35)	
Plums	1 bu. (56)	1½–2½
	1 lug (24)	
Tomatoes	1 bu. (53)	2½–3½
	1 lug (30)	
Tomatoes (for juice)	1 bu. (53)	3–3½
Vegetables		
Beans, green or wax	1 bu. (30)	1½–2½
Beans, lima (in pods)	1 bu. (32)	3–5
Beets (without tops)	1 bu. (52)	2–3½
Carrots (without tops)	1 bu. (50)	2–3
Corn, sweet (in husks)	1 bu. (35)	3–6
Okra	1 bu. (26)	1½
Peas, green (in pods)	1 bu. (30)	3–6
Spinach and other greens	1 bu. (18)	2–6
Squash, summer	1 bu. (40)	2–4
Sweet potatoes	1 bu. (50)	2–3

*The standard weight of a bushel, lug, or box is not the same in all states.

(Continued on page 198)

(Continued from page 197)

county extension agent for any available test results on that brand.

If lids aren't available in your area, you can order them by mail. The following brands have been tested (for sealing ability *only*) in the Rodale Test Kitchens and worked well.

Klik-it lids are one-piece, reusable lids that should last for several seasons; they make a loud popping sound to let you know a good seal has been made. M & B Sales, 6080 N. Whipple Street, Chicago, IL 60659.

Torque-Rite makes both metal replacement lids and bands. Edison Packing Corp., P.O. Box 150, Edison, NJ 08817.

Poly-Vac lids have been made in New Zealand for 35 years, and are now being sold in this country. Poly Commodity Corp., 25th floor, 1350 Avenue of the Americas, New York, NY 10019.

How to can

Many people are afraid to can. They've heard more talk about botulism than about the benefits of canning, and they know that it can be a hot, laborious job. Canning just doesn't seem worth the hassle.

Canning takes time, and at first glance it may seem complicated as well. But canning is nothing more than a series of simple, logical steps, and when you've finished them you'll have a unique and satisfying product.

Canning and acid

Apples may be safely canned in a boiling water bath; string beans require the higher temperatures of a pressure canner. The reason behind this is that the acidity in apples has an inhibiting effect on harmful microorganisms (including botulism spores).

Most of the microbes that carry disease or spoil food grow well in a low-acid environment. Because these creatures are particularly heat resistant, low-acid foods must be processed in a pressure canner where the temperature reaches 240° F. foods include fruits, pickles, and vegetables, meat and fish, and soups.

More-acidic foods can be safely canned in a boiling water bath canner where the temperature reaches only 212° F. Besides apples, these foods include other fruits, pickles, and tomatoes.

Boiling water bath canning: high-acid foods

The same basic procedures used to can tomatoes will apply to any high-acid foods. (See page 200.)

Unless otherwise noted, food in the jars should be packed to ½ inch

of the top, then covered with hot juice or syrup leaving a ½-inch headspace. For raw pack, have water in canner hot but not boiling; for hot pack, have water boiling. When adding boiling water to the canner to cover jars, don't pour it directly on them. Count processing time when water in canner comes to a rolling boil. For processing times, check the chart.

Apples. Hot pack only. Pare, core, cut into pieces. To keep from darkening, place in water with 2 tablespoons vinegar per gallon. Drain, then boil 5 minutes in syrup, juice, or water before packing.

Applesauce. Hot pack only. Make applesauce, then pack hot to ¼ inch of top.

Apricots. Wash, remove skins and pits, soak in anti-darkening solution (see under Apples), and drain. Raw pack: Put prepared fruit in jars. Hot pack: Heat fruit through in hot juice or syrup.

Beets (pickled). Hot pack only. Cut off tops, leaving 1 inch of stem and root. Wash and boil until tender. Remove skins and slice. Make pickling syrup (2 cups vinegar to 1 cup honey) and heat to boiling. Cover packed beets with boiling syrup.

Berries (except strawberries). Wash berries and drain.
Raw pack: Best for raspberries,

blackberries, boysenberries, dewberries, and loganberries. Shake berries down gently when filling jars.
Hot pack: Best for blueberries, cranberries, currants, elderberries, gooseberries, and huckleberries. Add ¼ cup honey for each quart fruit, bring to boil in pan, then pack.

Cherries. Wash and remove pits. (Sweet cherries need not be pitted, but if the skins aren't pricked they may burst during processing.)
Raw pack: Shake cherries down gently.
Hot pack: Add ¼ cup honey to each quart fruit (and a little water to unpitted cherries), and bring to boil before packing.

Figs. Hot pack only. Wash ripe figs, but do not remove skins or stems. Cover with boiling water and simmer for 5 minutes. Pack hot figs and cover with boiling syrup or juice. Add 2 teaspoons lemon juice to pints and 4 teaspoons to quarts.

Grapes. Wash and stem seedless grapes.
Raw pack: Pack tightly, while being careful not to crush fruit.
Hot pack: Bring to a boil in syrup or juice, then pack tightly.

Grapefruits, oranges, tangerines. Raw pack only. Remove fruit segments, peeling away the white membrane that causes bitter flavor. Seed with care.

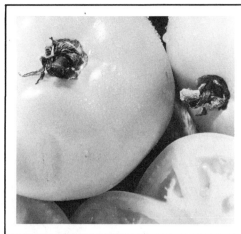

Six quarts of tomatoes, from garden to jar, in 1½ hours at 14¢ a quart. (Includes 45¢ for tomato seeds, 7¢ for electricity and 6¢ for each lid.)

Peaches. See directions for apricots.

Pears. Peel, cut in halves, and core. Follow directions and timetables for apricots, either raw pack or hot pack.

Pineapple. Peel, core, and cut into uniformly sized chunks.
Raw pack: Put prepared fruit in jars.
Hot pack: Simmer fruit in syrup or fruit juice for about 10 minutes.

Plums, Italian prunes. Wash. To can whole, prick skins. Halve freestone varieties and pit.
Raw pack: Put prepared fruit in jars.
Hot pack: Heat to boiling in syrup or juice. If fruit is very juicy, you can heat it with honey, adding no liquid.

Rhubarb. Hot pack only. Wash and cut into ½-inch pieces. Add ¼ cup honey to each quart rhubarb and let stand to draw out juice. Bring to boil.

Strawberries. Hot pack only. Wash and hull berries. Using ¼ to ½ cup honey for each quart of berries, spread berries one layer deep in pans and drizzle honey over them. Cover and let stand at room temperature 2 to 4 hours. Place in saucepan and simmer in their own juice for 5 minutes, stirring to prevent sticking. Pack without crushing and add extra syrup if berries don't produce enough juice of their own.

Tomatoes. Use only perfect, ripe tomatoes. Scald just long enough to loosen skins; plunge into cold water. Drain, peel, and core.
Raw pack: Leave tomatoes whole or cut in halves or quarters. Pack tomatoes, pressing gently to fill spaces.
Hot pack: Quarter peeled tomatoes.

Bring to boil and cover with extra liquid if tomatoes have not made enough juice of their own.

Steam pressure canning: low-acid foods

The basic procedures of steam pressure canning are similar to those for boiling water bath canning. The major differences lie with the pressure canner itself.

1) Put two or three inches of hot water in the bottom of the canner and start it boiling.

2) Wash the canning jars in hot soapy water, rinse, and put them in hot water until needed. Pour boiling water over the lids and set them aside.

3) Prepare vegetables for canning as usual.

4) Pack the raw or hot vegetables into jars, leaving the recommended headspace. Pour in enough boiling water to cover them. Slice

(Continued on page 201)

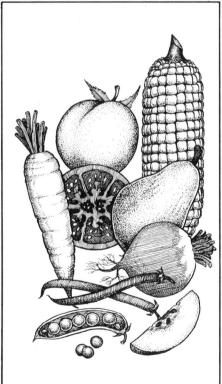
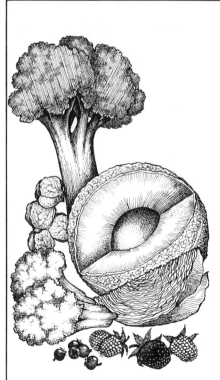

While the foods on the left are suited to canning, those on the right are best frozen or placed in cold storage.

Boiling water bath canning, step by step

Examine the jars to be sure they have no nicks or cracks. Wash them in hot soapy water, rinse, and put in hot water until needed. Pour boiling water over the lids and set them aside. (If you're using rubber rings, don't pour boiling water over them. Wash with soapy water and rinse thoroughly.) Fill a saucepan half full of water, and bring it to a boil; this will later be used for blanching.

Select enough tomatoes to fill up your canner (2½ to 3 pounds will fill up a quart jar). Wash them carefully, trim out the core, and cut out any soft or bad spots.

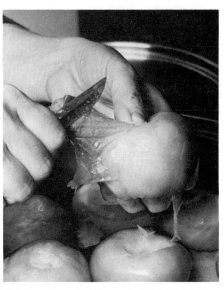

Put some tomatoes in a wire basket and lower them into boiling water in the saucepan. Hold them in the water until the skins start to crack (about 30 seconds). Then dip them in cold water. The skins should slip off easily. (If you don't have a wire basket, you can dip the tomatoes with a large dish towel.)

Remove the air from the jars by running a spatula or plastic knife along the side of the jar. Metal utensils can chip or otherwise damage the jar.

Wipe the tops and threads of the jars with a clean towel or cloth and screw the top on firmly. Food caught between the lid and the rim could prevent proper sealing.

Place the closed jars upright on the rack, lower the rack into the canner, put the lid on, and bring the water to a boil. The water should be 2 inches over the top of the jars. Begin to count time as soon as the water begins to boil, and process quarts for 45 minutes (pints for 35 minutes). Check the canner periodically, and add more boiling water as needed to keep the jars covered.

Pack the whole tomatoes (large ones may be cut into quarters) into jars to within ¹/₂ inch of the top. Using a plastic or wooden spoon, gently press the tomatoes to release the juice and fill up the empty spaces. If there is not enough juice to fill the jars to within ¹/₂ inch of the top, add more hot water or juice.

When the time is up, remove the jars from the canner. If you aren't using sealing caps, complete the seals as soon as the jars are removed from the kettle. Cool the jars—on racks if possible—making sure they are far enough apart so air can circulate freely around them. Do not place jars in a draft or cold place, or cover them while cooling.

When jars are cool, check the seals by pressing on the center of the lid. If it is down already or stays down when pressed, the seal is good. Label the jars carefully, and store them in a cool, dark place.

(Continued from page 199)
through tightly packed greens so the heat can reach the center of the jar more easily.

5) Wipe the tops and threads of the jars and screw the lids on tightly.

6) Set the closed jars on a rack in the canner so that steam can circulate around them freely. Also use a metal rack between layers of jars.

7) Fasten the canner cover securely so that no steam escapes except at the open petcock or weighted gauge opening. Bring the water in the canner back to a boil.

8) Allow steam to escape from the opening for ten minutes so all the air is driven out of the canner. Then close the petcock or put on the weighted gauge and let the pressure rise to the desired level.

9) Start counting time as soon as the desired pressure is reached, and process for the required time. Keep the pressure as uniform as possible by regulating the heat.

10) At the end of the processing time, gently remove the canner from the heat.

11) Let the canner stand until the pressure returns to zero. After a minute or two, slowly open the petcock or remove the weighted gauge.

12) Unfasten the canner cover, tilting the far side up so the steam escapes away from you. As you remove the jars from the canner, complete the seal if the jars are not the self-sealing type. Set the jars upright on a rack, far enough apart so air can circulate around all of them.

These step-by-step procedures should be modified with the following particulars. Unless otherwise noted, pack food to ½ inch of the jar top, then cover with hot juice or water to leave a ½-inch headspace. For processing times, check the accompanying chart (page 210).

Artichokes. Hot pack only. Wash and trim (so chokes can fit in a wide-mouth jar). Cook 5 minutes in a solution of ¾ cup vinegar in 1 gallon water. Discard solution. Cover packed chokes with liquid made from ¾ cup lemon juice per gallon of water. (For easy removal and to prevent chokes from falling apart, tie a string firmly around the petals.)

Asparagus. Wash, then trim off scales and tough ends and wash again. Cut in 1-inch pieces.
Raw pack: Pack as tightly as possible without crushing.
Hot pack: Boil prepared asparagus for 2 or 3 minutes, then pack loosely.

Beans (dry, with tomato or molasses sauce). Hot pack only. Sort and wash. Boil 2 minutes, remove from heat, and soak 1 hour. Heat again to boiling and drain, saving liquid for sauce. Fill jars ¾ full with hot beans, then fill to ½ inch of top with hot tomato or molasses sauce.

Beans (fresh, lima). Shell and wash beans.
Raw pack: Pack small beans loosely to 1 inch of top of jar for pints and 1½ inches for quarts. For large beans fill to ¾ inch of top for pints and 1¼ inches for quarts. Cover with boiling water.
Hot pack: Bring beans to boil. Pack loosely to 1 inch of top of jar. Cover with boiling water, leaving 1 inch at top.

Beans (snap or green). Wash beans. Trim ends and cut into 1-inch pieces.
Raw pack: Pack tightly.
Hot pack: Boil for 5 minutes, then pack loosely.

(Continued on page 202)

(Continued from page 201)

Beets. Hot pack only. Cut off tops, leaving a 1-inch stem and root, and wash. Boil until skins slip easily. Skin, trim, cut, and pack into jars. (For pickled beets, see table of recommended processing times [page 210].)

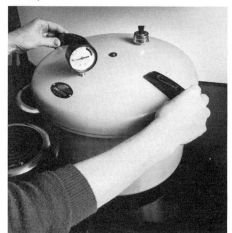

The top of a pressure canner locks in place to form an airtight seal.

pH VALUE OF VARIOUS FOODS

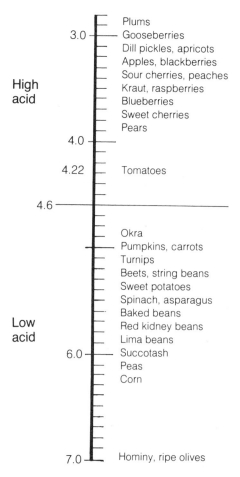

High acid

3.0	Plums
	Gooseberries
	Dill pickles, apricots
	Apples, blackberries
	Sour cherries, peaches
	Kraut, raspberries
	Blueberries
	Sweet cherries
	Pears
4.0	
4.22	Tomatoes
4.6	
	Okra
	Pumpkins, carrots
	Turnips
	Beets, string beans
	Sweet potatoes
	Spinach, asparagus
	Baked beans
	Red kidney beans
	Lima beans
6.0	Succotash
	Peas
	Corn
7.0	Hominy, ripe olives

Low acid

Broccoli and brussels sprouts. Hot pack only. Cut off woody, tough stems and old leaves and yellowing blossoms. Soak in cold, salted water (about 1 tablespoon salt to each quart water) for 10 to 15 minutes to drive out bugs. Rinse well. Cut into 2-inch pieces. Boil 3 minutes, then drain, reserving liquid. Pack tightly and cover with boiling liquid, leaving 1 inch at top.
(Note: Not recommended for canning because the processing intensifies the strong flavor and discolors the vegetable. Much better frozen or pickled.)

Cabbage. Hot pack only. Clean, cut up in small wedges, and process as for broccoli. (Not recommended for canning. Fresh cabbage is much better kept in cold storage.)

Carrots. Wash and scrape carrots. Slice, dice, or leave whole.
Raw pack: Pack tightly to 1 inch of top.
Hot pack: Bring carrots to boil, then pack to ½ inch of top. Cover with hot cooking liquid and water.

Cauliflower. Prepare and process like broccoli. (Not recommended for canning. Much better frozen.)

Celery. Hot pack only. Wash and trim off tough leaves and woody bottoms. Cut into 1-inch pieces. Boil 3 minutes. Drain, reserving liquid. Pack jars and cover with hot liquid, leaving 1 inch at top.

Corn (cream style). Husk corn and remove silk. Wash. Cut corn from cob at about center of kernel and scrape cob. (Process corn in pint jars only, since time required for quarts darkens corn.)
Raw pack: Pack corn loosely to 1 inch of top.
Hot pack: Add 1 pint boiling water to each quart of corn. Heat to boiling. Pack hot corn to 1 inch of top.

Corn (whole kernel). Husk corn and remove silk. Wash. Cut from cob at about 2/3 the depth of the kernel. Pack loosely to 1 inch of top of jar with mixture of corn and liquid.

Eggplant. Hot pack only. Wash and peel, then slice or cube. To draw out bitter juice, line a colander with eggplant, sprinkle with salt, put another layer of eggplant over that, then salt, and so on. Let stand 1 hour.

Then press eggplant against sides of colander before rinsing well and removing. Boil in fresh water for 5 minutes. Drain, reserving liquid. Pack in jars and cover with hot liquid, leaving 1-inch headspace.

Mushrooms. Hot pack only. Use only tender, young mushrooms. Wash thoroughly and trim off tough stalks. Cut in slices or leave small caps whole. Steam for 4 minutes. Pack into hot jars and cover with boiling water, leaving 1-inch headspace.

Okra. Hot pack only. Use only young, tender pods. Wash and trim stems. Leave whole or cut into 1-inch slices. Boil 1 minute, then drain, reserving liquid. Pack in jars and cover with hot liquid, leaving 1 inch at top.

Onions (small, white). Hot pack only. Choose onions of uniform size, about 1 inch in diameter. Peel, trim off roots and stalks, and wash. (If you push the hole in the onion end downward through the middle with a sharp object like a clean finishing nail, the centers will cook with less chance of shucking off outer layers.) Boil gently for 5 minutes. Pack loosely and cover with boiling liquid.

Peas (green). Shell and wash peas.
Raw pack: Pack loosely to 1 inch of top, and cover with boiling water.
Hot pack: Bring to a boil before packing loosely to 1 inch of top. Cover with boiling water.

Peppers (bell). Hot pack only. Remove stem, core, seeds, and inner white membrane. Remove skins by plunging peppers in boiling water for a few minutes, then running them under cold water, and finally removing split skins with a sharp knife or potato peeler. Slice peppers or flatten whole halves and pack carefully in layers. Cover with boiling water. Add ½ teaspoon of lemon juice or 1 tablespoon of vinegar per pint. (This is necessary to increase the acidity and make the peppers safe for canning.) Process at 5 pounds pressure, as a higher pressure injures both flavor and texture.

Potatoes (white). Hot pack only. Wash and peel.
Hot pack (cubed): Cut into ½-inch cubes. Cook 2 minutes in boiling water. Pack and cover with boiling water to 1 inch of top.

Hot pack (whole): Use potatoes 1 to 2½ inches in diameter. Cook in boiling water 10 minutes. Pack and cover with hot cooking liquid or water to 1 inch of top.

Pumpkin or winter squash. Hot pack only. Wash, remove seeds, and pare. Cut into 1-inch cubes.
Hot pack (cubed): Add just enough water to cover cubes. Bring to boil. Pack hot in jars.
Hot pack (strained): Steam cubes until tender (about 25 minutes). Put through food mill or strainer. Simmer until heated. Pack hot in jars.

Soybeans. Hot pack only. Shell beans. Bring to boil, then drain, reserving liquid. Pack beans loosely and cover with hot liquid, leaving 1-inch headspace.

Spinach (and other greens). Hot pack only. Pick over and wash thoroughly. Cut out tough stems and midribs. Place about 2½ pounds of spinach in cheesecloth bag and steam 10 minutes or until well wilted. Pack loosely.

Sweet potatoes. Wash, then boil or steam 20 to 30 minutes to facilitate slipping off skins. Cut into uniform pieces.
Dry pack: Pack hot potato pieces lightly, pressing gently to fill air spaces, to within 1 inch of jar top. Do not add liquid.
Wet pack: Pack hot potato pieces to within 1 inch of jar top and cover with either boiling water or medium syrup to within 1 inch of top.

Turnips, parsnips, rutabagas. Hot pack only. Wash and peel, cube, or slice. Boil 3 minutes and drain, reserving liquid. Pack into jars and cover with hot liquid, leaving 1-inch headspace. (Not recommended for canning; much better kept in cold storage.)

Zucchini, yellow squash. Wash and slice; don't peel unless squash is large and the skin is tough.
Cut into ½-inch slices, and halve or quarter larger slices.
Raw pack: Pack tightly in jars and cover with boiling water, leaving 1 inch at top.
Hot pack: Bring slices to boil. Drain, reserving liquid. Pack loosely and cover with hot liquid, leaving ½-inch headspace.

Full steam ahead with your pressure canner

Do you have an expensive pressure canner that gathers dust between canning seasons? Some people will not use canners for regular cooking because of questions about aluminum migration into food, but this doesn't rule out using it as a steamer, since food does not come in contact with the metal.

When the pressure canner is used as a steamer, as in the five jobs described below, the petcock is left open. Attune your ear to the regular sound of the steam, lowering the heat if it hisses violently, raising it if the sound stops altogether. Observe the safety rules recommended in your pressure canner booklet. Never attempt to remove the lid if the overpressure plug is up. Always remove the lid *away* from you to avoid a hot steam bath.

To cook or reheat one-dish meals. Fill a heat-proof casserole dish with your favorite stew, loaf, or custard. Cover the dish and place it on a rack in the canner. (If it's not covered, the water that condenses on the canner lid may get into the food.) Add two cups of water to the bottom of the canner, put on the lid, and build up the steam. With the petcock open, steam for 15 to 20 minutes, depending on casserole size and whether the food is to be cooked or simply heated through. This method takes less fuel than over-baking, and the food will stay hot and moist for a long time, as on a restaurant steam table—a real bonus when mealtimes are stretched out to accommodate different schedules.

To make bread. Steamed breads mix up in fewer steps than regular yeast bread. They freeze well, too. Follow any recipe for steamed brown bread, date bread, or plum pudding. Fill oiled cans or molds with batter and cover with foil or waxed paper. Place cans on the rack, and add enough water to reach halfway up the cans. Secure the lid, build up steam, then reduce the heat so that steam escapes steadily but not frantically for one hour. Let cool 15 minutes before removing bread from cans.

(Continued on page 204)

PRESSURE CAN HIGH-ACID FOODS EASILY

If you want to save energy and nutrients by canning all your produce in a pressure canner, there is a simple method for high-acid foods normally canned in a boiling water bath. Processing them at *five pounds pressure for ten minutes* will destroy any yeast, mold, or bacteria that might thrive in a high-acid environment.

LOW-ACID TOMATOES

Tomatoes fall close to the dividing line between low- and high-acid foods. Because they are slightly on the high-acid side, they have generally been canned in a boiling water bath.

This was challenged in 1974 when the U.S. Center for Disease Control attributed two cases of botulism to home-canned "low-acid tomatoes." About the same time, ads for low-acid varieties of tomatoes appeared in some seed catalogs. People began to wonder if canning tomatoes in a boiling water bath was safe.

To clear up the matter, researchers at the U.S. Department of Agriculture and the University of Minnesota tested the acidity of 100 different varieties of tomatoes. They found that none of the "low-acid" tomatoes would support dangerous bacterial life. In fact, many of the so-called low-acid types were more acidic than average; apparently a higher sugar content masked the sourness.

The investigators did discover four varieties of truly low-acid tomatoes: Garden State, Ace, 55 VF, and Cal Ace. These should only be processed in a pressure canner.

This study also found that as tomatoes ripen, they become less acidic. Consequently, if overripe tomatoes are canned, there is a possibility that harmful bacteria could grow. But any firm, ripe, red tomatoes can be safely processed in a boiling water bath.

If this talk has made you suspicious of your tomatoes, there's a way to be doubly safe. Adding 4 teaspoons of lemon juice or 2 tablespoons of vinegar or ½ teaspoon of citric acid per quart of tomatoes will make them sufficiently acidic.

Steam canner.

THE STEAM CANNER

Canning is much easier and quicker at our house since I've been using a steam canner (not to be confused with a steam *pressure* canner). I need only two quarts of water to process seven jars of food—less than one-fourth the amount of water used in the large kettle of boiling water when I followed the cold pack or boiling water bath method. This saves fuel and time, and keeps the kitchen cooler.

The filled jars are positioned on a perforated rack set over a shallow pan of water, then covered with a vented domed lid. The steam that accumulates under the lid penetrates the jars as effectively as hot water, and much more neatly. Jars come out sparkling, with no mineral deposits from the water as they often do when boiled. It is a relief not to have water boiling over on my stove—a by-product of canning that is not only messy, but can shorten the life of the stove-top control in certain electric stoves. The steam canner is perfect for fruits, pickles, tomatoes, and preserves. It is not a pressure canner and should not be used to put up meats and low-acid vegetables.

Nancy Bubel

MASON JARS

Canning jars are often called Mason jars. They got the name from John L. Mason, who made the first practical canning jar back in 1858. His jar company is no longer in existence and the term Mason now means any canning jar.

(Continued from page 203)

To blanch vegetables for the freezer. Place vegetables in an enamel colander on the rack with one cup of water in bottom of canner. Secure lid and steam for 2 to 3 minutes. Upon opening canner, check with tongs to see that beans and peas are uniformly bright in color and that greens are lightly wilted throughout. Pour vegetables into a basin and cool quickly by tossing with ice cubes. Drain thoroughly before packaging.

To process high-acid fruits and tomatoes. Fill canning jars and follow the regular procedure for canning low-acid vegetables. After the air is cleared from the canner, process the fruit or tomatoes at five pounds pressure for ten minutes. This method requires less fuel than the regular water bath and is handy at the peak of the harvest season when your enamel canner is already filled with another batch of jars.

To tenderize tough poultry or meat. Put the meat in a heat-resistant dish, placed on the rack with one cup of water in the cooker. Secure the lid, and steam for 15 or 20 minutes. Transfer food to a stainless-steel or cast-iron pot and continue to cook slowly, adding vinegar for a calcium-rich broth (the vinegar works on bones to release calcium). Add grains, legumes, and other vegetables as desired for soup or stew.

Another good use of a pressure canner is to cook beans. If you have a stainless-steel cooker, allow three or four cups of water for every cup of beans and cook under pressure (15 pounds) for 25 to 30 minutes. Don't fill the pot more than ⅔ full of water, however. If cooking soybeans, add

one or two tablespoons vinegar to the water to prevent the beans from foaming. If you have an aluminum canner, place the water and beans in a stainless-steel bowl, and cover tightly with foil. Put the pot in two inches of water in the canner and proceed as above. This way, there is no danger of aluminum migration or frothing of the beans.

Regular use of your steam pressure canner makes your investment in this piece of equipment pay off throughout the year. Just as importantly, it will give you greater confidence when it comes to the exacting work of canning low-acid vegetables and meat.

Judy Hinds

What does canning do to nutrients?

As soon as food is picked, it begins to lose nutritional value. Some nutrients are harmed by air and water, while others are sensitive to heat and light. The goal of all methods of food preservation is to minimize the losses by careful and efficient handling.

Canning involves high temperatures to destroy molds and bacteria, and heat and water destroy or carry off minerals and vitamins. The trick is to destroy the organisms without destroying the nutrients.

Experiments have shown that the faster raw foods are cooked, the fewer nutrients they lose. Therefore, a steam pressure canner, with its shorter processing time, preserves more goodness than a boiling water bath canner, for both high- and low-acid foods. (A pressure canner also takes less energy.) And if you use the liquid in which the food is processed, the water-soluble vitamins and minerals are not really lost.

You can also preserve nutrients by careful trimming. Don't cut off any more of a fruit or vegetable than necessary. Broccoli leaves have more vitamin A than the stalks or buds, and the cabbage core is high in vitamin C. In other cases, try leaving on fruit and vegetable skins. Some will wrinkle and others darken, but they contain valuable nutrients and fiber, and the extra nourishment will look good on you.

WATER NEEDED TO PROCESS IN PRESSURE CANNERS*

Size of canner (qts.)	Amount of water (qts.)
4	1
6	1½
8	1½
16	2
21	2

*Increase the amount of water by 1 pint if the canner is not filled with jars.

High-pressure canning: low-acid foods

Home canning is normally done at 10 pounds per square inch. Two food scientists at the University of Minnesota, however, discovered that foods could be processed even faster using 15 pounds of pressure rather than 10. The higher temperature destroys botulism toxins quicker. (For exact processing times at 15 pounds pressure, check the chart of recommended processing, page 210.)

Storing jars or cans in a cool, dark place minimizes the loss of heat- and light-sensitive nutrients. For example, a year's vitamin C loss can increase from 10 percent at 65° F. to 25 percent at 80° F.

When you're ready to serve your canned foods, pressure cook them if possible to preserve their nutrients. If you must boil, use as little water as possible, and save the remainder for soup stock or gravy.

Pickles without brine

Pickles are one of the most popular canned foods—from traditional cucumber dills to green tomato or watermelon pickles. They are virtually always made with salt, but we have discovered (and taste tests have confirmed) that delicious pickles can be made without this traditional ingredient. So gather your dill heads and cukes; here are some recipes to try.

Susan Asanovic

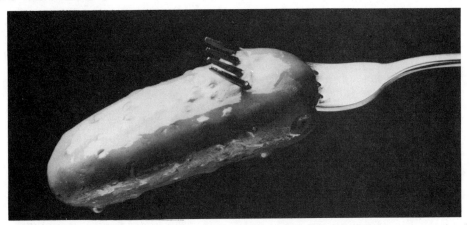

Spicy Deli Dills

Take 20 to 25 medium cucumbers. Place in water in the refrigerator for 12 hours. Drain, and cut into spears. Put 8 sterilized quart jars in the water bath canner to heat while you prepare the vinegar mixture.

16 cloves garlic, peeled
8   heads of fresh dill
2   dry red hot peppers
8   grape leaves (optional)
8   slices horseradish root, peeled (optional)
2   quarts vinegar
2   quarts water

Bring the above ingredients to a boil and simmer while you pack the cucumbers into the jars. (The grape leaves—in case you're wondering—will help keep the pickles crisp.) Pour the hot mixture over the pickles, making sure the liquid covers them completely, and leave ½-inch headspace. Cap jars and submerge them in boiling water. Process for 10 minutes from the time they are placed in the boiling water. Remove promptly and cool. Allow them to mellow a few weeks before tasting. Since you're not likely to come out exactly even to a given number of jars, there will be samples to nibble immediately.

Yield: 6 to 8 quarts

Icicle Pickles

Adding honey or maple syrup to counteract vinegar's bite will yield a mellow pickle, but you'll have to overlook honey's tendency to darken. The subtle flavor will certainly please you.

20 medium cucumbers (about 4 inches long)
6   small onions, quartered
6   small stalks of celery
1   tablespoon mustard seed
1   quart vinegar
1   cup water
1¼ cups light honey or fancy-quality maple syrup

Soak cucumbers in ice water for 8 to 12 hours. Drain and cut lengthwise into quarters. Begin to heat the jars in the boiling water bath canner. When the water is boiling remove the jars and pack each one with a piece of celery, an onion, and the cucumber spears. Bring the vinegar, water, and honey to a boil with the mustard seed. Pour this marinade over the vegetables, filling jars to ½ inch from the top, as with all pickle recipes. Adjust caps and place the jars in boiling water. The water should cover them by 2 inches. Process for 10 minutes and remove immediately to cool.

Yield: about 6 pints

Dilled Fresh Zucchini

When the cucumbers fail and zucchini takes over, many gardeners have tried pickling the bounty. They've found that young squash makes crisper, firmer pickles than cucumbers.

6   pounds young, firm zucchini
2   cups celery, thinly sliced
2   cups onion, chopped
½   cup light honey (optional)
4   tablespoons dill seed
4   cups apple cider vinegar
6   cloves garlic, scalded

Peel, seed (optional), and slice zucchini lengthwise into strips about 4 inches long and ¼-inch thick. Mix all vegetables together in a large bowl and cover with ice cubes. Let stand at room temperature for about 3 hours. Drain. Start heating the water in the boiling water bath canner. Combine dill seed, vinegar, and honey, if used, in a saucepan and bring to a boil. Add the vegetables and reheat to boiling. Pack into hot pint jars adding a clove of garlic per jar. Make sure the vegetables are completely covered, and leave ½-inch headspace. Cap jars and process 15 minutes in a boiling water bath. Remove immediately and cool.

Yield: about 6 pints (seeded), about 8 pints (unseeded)

(Continued on page 206)

(Continued from page 205)

Green Tomato Relish

Green tomatoes are often abundant after early frosts. They make such a delicious pickle, however, that you will make them even without a frost.

7	pounds green tomatoes
6	medium onions
1	tablespoon each: allspice, whole cloves, mustard seed, black peppercorns, celery seed (except for severely restricted diets)
½	lemon, sliced
2	sweet red peppers
1	cup light honey (omit to save calories, or if you enjoy a sharp pickle)
3	cups vinegar

Wash and trim tomatoes and peppers. Slice about ¼ inch thick. Peel onions and slice the same way. Place on top of ice in a bowl and leave several hours in the refrigerator, covered with plastic wrap. Begin to heat water in boiling water bath canner. Tie the spices in a cheesecloth bag and place in a kettle with the vinegar and honey. Bring to a boil, then add the lemon and the vegetables. Simmer about 5 minutes or until just tender but crisp. Be very careful not to overcook or you will have a mushy pickle. Remove spice bag, pack vegetables into hot pint canning jars, and cover with the boiling liquid, leaving ½-inch headspace. Cap jars. Process in boiling water for 10 minutes, then remove immediately.

Yield: about 6 to 8 pints

Meat canning

With the rise of the home freezer, the popularity of home meat canning has declined. Yet canning remains an excellent way to keep meat, and while it takes more work than freezing, it's cheaper in the long run.

Because meat is high in protein and has little acid, it is especially susceptible to spoilage. Only top-quality meat should be used, and it is best to can it as soon as the animal heat is gone. If there is a delay, the meat should be frozen at 0° F. until canning time. You can process frozen meat, but it is difficult to work with, and extra time is required to cook it sufficiently. Thawing is best done gradually at low temperatures (from 32 to 40° F.).

To protect against bacteria, cleanliness is even more crucial than

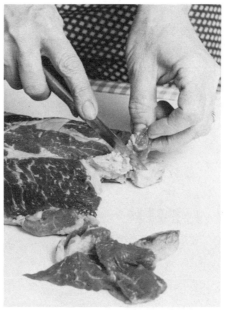

Bone the meat, trim off the fat, and cut into jar-sized pieces.

godliness. Scrub all surfaces and equipment with soapy water, and use disinfectant on the cutting board and any other wooden utensils. After the canning is done, clean these surfaces again to insure that no pieces of food are left behind to encourage bacteria.

Salt is often added to canned meat, but only for taste—it is not necessary as a preservative, and you can omit salt safely. Do not use a salt substitute, as the high temperatures of processing sometimes result in a disagreeable taste.

Meat can be packed either precooked or raw. Either way, it should be packed loosely, so it will heat through more easily and won't overflow when processed. The size of the pieces can vary according to the meat.

To prepare the meat, trim off the excess fat and debone it if possible; this saves space and makes packing

CHILLING TIMES FOR FRESHLY KILLED ANIMALS (HRS.)

Chicken, duck, goose, gamebirds	6–12
Turkey	8–14
Goat, lamb	18–24
Mutton, pork, veal, venison	24–36
Beef, moose	36–48

Cook the meat until only a slight pink color remains.

easier. (Too much fat may cause a strong flavor, and can eat away the rubber seal.) Cut tender meats into pieces with the grain of the meat running lengthwise. Tough pieces of meat can be ground or cut into chunks for stew, and bony pieces can make broth or stock.

Poultry is best cut into container-sized pieces. Remove the bones from meaty parts such as the breast, separate the thighs from the drumsticks, and keep the giblets to process separately.

Before processing, many canners heat the meat to a temperature of 170° F. to exhaust all the air in it. This is not necessary to get a vacuum seal, but it helps insure a good seal. With precooked meat, the frying, boiling, or cooking should be sufficient. To remove air from raw pack meat, the open, filled containers can be put in a pan of boiling water. (The water level should be about two inches from the top of the containers.) Heat the meat slowly to 170° F., or for 75 minutes if you don't have a meat thermometer.

Canned meats, like other canned goods, should be kept in a cool, dry place. Sunlight or warm temperatures will lower their quality; freezing may damage the seal so spoilage can begin.

Don't give meat that may have gone bad a taste test. Instead, boil it for 20 minutes in a covered pot to destroy any dangerous toxins that

Pack the hot meat into clean jars.

Cover the meat with boiling liquid, leaving a 1-inch headspace.

Process the meat in a pressure canner for the recommended time.

might be present. In general, boiling is the best way to find out if meat is safe to eat, because heat intensifies the characteristic odor of spoiled meat. So if your meat develops a bad smell after cooking, throw it out.

Before opening jars, check them for signs of spoilage: bulging jar lids or rings, gas bubbles, and leaks are all danger signs, and such jars should be disposed of safely. Don't worry, however, if the meat has darkened or if the metal jar lid is discolored. Meat often changes color after canning, and sulfur in some meats may discolor in contact with metal.

Pressure canning meat

Review procedures for using a pressure canner. Unless otherwise noted, meat should be packed to 1 inch of the top of the jar, then covered with hot broth or water to the same level. Refer to the chart of recommended processing times (page 210).

Cut-up beef, veal, lamb, pork. Use tender cuts for canning in strips. Making sure that grain of meat runs lengthwise, cut meat the length of the jar. Cut less-tender cuts into cubes for stewing or soups.

Raw pack: Pack cold, raw meat loosely.

Hot pack: Precook meat in skillet or

(Continued on page 209)

YIELD OF CANNED MEAT FROM FRESH

Cut of meat	Pounds of meat per jar	
	pts.	qts.
Beef: Round	1½–1¾	3–3½
Rump	2½–2¾	5–5½
Pork loin	2½–2¾	5–5½
Chicken: Canned with bone	1¾–2	3½–4¼
Canned without bone	2¾–3	5½–6¼

BOILING WATER CANNING

Altitude (ft.)	Add this much canning time (min.):	
	20 min. or less	over 20 min.
1,000	1	2
2,000	2	4
3,000	3	6
4,000	4	8
5,000	5	10
6,000	6	12
7,000	7	14
8,000	8	16
9,000	9	18
10,000	10	20

Canning:
Breinigsville, Pennsylvania

Sarah Henninger didn't need a bumper crop of tomatoes to persuade her to start canning. "It just came naturally," she says. Her mother had always canned for their family of 14. Then when Sarah and her husband moved onto the Pennsylvania dairy farm that had belonged to his parents, Sarah took over her mother-in-law's huge garden.

Canning was the logical way to save money and stock up on all their home-grown foods. As the family grew (the Henningers have seven children), Sarah did more and more canning. "I'd see that what I put up last year didn't go around, so I'd do more this year to make it reach."

Canning hundreds of quarts of vegetables, fruits, and jams each year is a lot of work. First, the garden must be cared for. After her husband plows the land, it's all Sarah's project. "I'm the planter, the weeder, the hoer, and the picker," she says. But she enjoys working in the garden, being in the afternoon sun, and getting exercise all at the same time.

Once the vegetables start to ripen, the canning process swings into operation. In the Henninger household, canning goes on nearly every day except Sundays from July through September. When all four daughters were at home, they helped Sarah with an assembly line of washing, cutting, and preparing the food. Now, however, the three older girls have jobs of their own, and Sarah does most of the canning by herself.

Sarah makes tomato sauces, ketchup, red beets, corn relish, chow chow, and six kinds of pickles from her garden vegetables. Fruit trees on their property provide enough fruit to make canned pears, peaches, applesauce, apple butter, and several kinds of jams. A friend gives Sarah the apricots for apricot jam, and another friend gives her enough grapes to make all her own grape juice.

Everything Sarah cans is home-grown—"otherwise it wouldn't pay." Her canning expenses are minimal. To her fresh fruits and vegetables she adds only sugar and water, or vinegar and spices.

Because friends bring her all their empty jars, Sarah doesn't invest money in canning jars. She hasn't bought any for the past six or eight years. The only things she must replace are the flats for the lids.

During the months when she is canning, the Henningers' electric bill may be higher because of all the cooking, but as Sarah says, "you can afford to pay a little extra to the electric company when you can your own things." Sarah estimates it costs her about 30 cents to make a jar of jam that would cost $1.50 or more to buy. And hers has no chemical additives. "I feel good when I look at everything I've canned and know there are no preservatives or additives in there. When I look at my product, I think it's worth it."

In that respect, canning suddenly took on extra importance for Sarah five years ago when her sister-in-law died of cancer in middle age.

That experience jolted Sarah into thinking more about protecting her family's health. "I have seven children. I tried to find some answers."

She began to read articles about healthful living and learned a lot about staying away from chemicals in food.

Now, almost all the food she serves her family is grown and prepared naturally, with no additives that can build up in the body day after day. Sarah wholeheartedly believes in natural foods—"the things God provides, without synthetics." She uses whole wheat flour instead of bleached white flour, and raw honey and brown sugar instead of white sugar.

"I learned in Sunday school that your body is your own little house to take care of. You are responsible for what you put in it. And I believe in that saying, 'You are what you eat.'"

CB

Sarah Henninger.

(Continued from page 207)

saucepan in just enough water to keep meat from scorching. Stir occasionally so all pieces heat evenly. Cook until medium done. Pack hot meat loosely.

Ground meat. Raw pack: Pack raw ground meat into jars.

Hot pack: If you are making patties, press meat into patties that will fit easily into jar. Meat can also be left loose. Precook patties in a slow oven or over low heat in a skillet until medium done. Pour off all fat; do not use any in canning. Pack patties or loose meat in jars.

Poultry, small game, rabbits. Sort meat into meaty and bony parts and can separately.

Raw pack (with bone): Bone the breast and cut other parts with bone in, into jar-size pieces. Trim off excess fat. Pack poultry loosely by placing thighs and drumsticks on the outside and breasts in the center of the jar. Raw pack (without bone): Remove bones from meaty pieces, but not skin. Pack loosely.

Hot pack (with bone): Bone the breast and cut bony parts into jar-size pieces. Trim off excess fat. Heat pieces in pan with broth or water. Stir occasionally until meat is medium done, when only slight pink color remains. Pack meat loosely, with thighs and drumsticks on the outside of jar and boned pieces in the center.

Hot pack (without bone): Remove bones, but leave skin on meaty parts. Cook meat as above, and pack loosely in jars.

Giblets. Hot pack only. Separate hearts and gizzards from livers and cook and can separately to avoid blending of flavors. Cook giblets in a pan with water or broth until medium done. Stir occasionally so meat heats evenly. Pack hot giblets into jars.

Liver. Hot pack only. Wash, remove skins and membranes. Slice into container-sized pieces. Drop into boiling water for 5 minutes. Follow hot pack directions for cut-up meat.

Heart. Wash and remove thick connective tissue. Cut into container-size strips or 1 inch cubes. Cook in water at a slow boil until medium done. Follow hot pack directions for cut-up meat.

ADJUSTING PRESSURE CANNER POUNDAGE

Altitude (ft.)	Poundage:		
	if 5 lbs.	*if 10 lbs.*	*if 15 lbs.*
1,000	5½	10½	15½
2,000	6	11	16
3,000	6½	11½	16½
4,000	7	12	17
5,000	7½	12½	17½
6,000	8	13	18
7,000	8½	13½	18½
8,000	9	14	19
9,000	9½	14½	19½
10,000	10	15	20

Tin can canning

The term "canning" is really a misnomer for most home food storers. Home use of tin cans has become increasingly rare since the Depression, when extension agents traveled from town to town with the necessary equipment.

The chief virtue of cans is that they're less destructible than jars. To use them, however, you need all the equipment required for glass jar canning plus a can sealer, a relatively costly item. Nevertheless, we were intrigued by the idea, so we wrote to the Ives-Way Company and asked them to send us one of their can sealers for an informal test.

Mastering the machine

When it arrived in several pieces, the sealer looked somewhat complicated and cumbersome. But once we got it together, it turned out to be a relatively simple and solid piece of machinery. The sealer is quite versatile, and can handle cans ranging in size from #1 to #3, as well as cans of varying heights.

The most difficult part was finding a can supplier who would sell in small quantities (a list of suppliers follows). Once we got the cans, it took only two attempts and one minor adjustment to get a feel for operating the sealer. We were amazed at how easy it was to seal the cans. Our submersion tests showed that they had a fine, airtight seal.

We'd heard reports that you could take a used can, trim the end

(Continued on page 210)

RECOMMENDED PROCESSING TIMES (IN MINS.)

(If process time is not shown, procedure is not recommended.)

	Pack method	Boiling water canner		Pressure canner 5 psi.		10 psi.		15 psi.	
		pts.	qts.	pts.	qts.	pts.	qts.	pts.	qts.
Fruits, tomatoes, and tomato products									
Apples	Hot	15	20	10	10	—	—	1	1
Apples (for sauce)	Hot	15	15	10	10	—	—	1	1
Apricots—see Peaches									
Berries, except strawberries	Raw or hot	15	15	—	—	—	—	—	—
Cherries, sweet or sour	Raw	20	25	10	10	—	—	1	1
Cherries, sweet or sour	Hot	15	15	10	10	—	—	1	1
Figs	Hot	85	90	10	10	—	—	—	—
Grapes	Raw or hot	15	20	10	10	—	—	1	1
Grapefruits, oranges, tangerines	Hot	10	10	—	—	—	—	—	—
Peaches	Raw	25	30	10	10	—	—	—	—
Peaches	Hot	20	25	10	10	—	—	—	—
Pears	Hot	20	25	10	10	—	—	—	—
Pineapples	Raw	30	30	10	10	—	—	—	—
Pineapples	Hot	20	20	10	10	—	—	—	—
Plums	Raw or hot	20	25	10	10	—	—	—	—
Rhubarb	Hot	15	15	10	10	—	—	1	1
Strawberries	Hot	10	15	—	—	—	—	—	—
Tomatoes, whole or quartered	Raw or hot	35	45	—	—	10	10	1	1
Tomato sauce (no other vegetables, thin)	Hot	35	35	10	10	—	—	1	1
Tomato sauce (no other vegetables, thick)	Hot	20	20	10	10	—	—	1	1
Tomato sauce with vegetables (half-pints or pints)	Hot	35	—	10	—	—	—	1	1
Tomato sauce with meat	Hot	—	—	—	—	60	75	50	50
Vegetables									
Artichokes	Hot	—	—	—	—	25	30	—	—
Asparagus	Raw or hot	—	—	—	—	25	30	15	15
Beans and peas, dried	Hot	—	—	—	—	75	90	50	50
Beans, fresh lima	Raw or hot	—	—	—	—	40	50	30	30
Beans, snap	Raw or hot	—	—	—	—	20	25	15	15
Beets	Hot	—	—	—	—	30	35	15	15
Broccoli and brussels sprouts	Hot	—	—	—	—	25	30	—	—
Cabbage	Hot	—	—	—	—	25	30	—	—

(Continued from page 209)
with tin snips, and recycle it. Those reports are true—sort of. After some effort, we got an airtight seal on a trimmed can. But it's not a very practical process, taking considerable time to get an even trim and to flare out the can top so a new lid would fit. Then, after a couple of trial trims, we were left with a rather small can.

Because the cans are expensive and cannot be reused easily, we do not recommend tin can canning unless the use of jars might be impractical (if the food needed to be shipped, for example). In such cases, it is a simple, dependable option.

Canning with metal cans is similar to using glass jars, except for three differences.

1) Pack the cans *full* of vegetables or fruit, then cover them with hot water or syrup.

2) To remove air from the cans prior to sealing, the food must be heated to at least 170° F. For hot pack foods, normal preparation should be enough. For raw pack, place the open, filled cans in a pot with boiling water about 2 inches below the can tops. Cover the pot and bring the water back to boiling. Boil the cans until the food reaches 170° F. (about 10 minutes). Remove the cans from the water one at a time. Replace any liquid spilled from the cans by filling them with boiling water. Place a clean lid on each and seal at once.

3) After the cans have been processed, they should be cooled in cold water, changing the water often

| | Pack method | Boiling water canner | | pts. | Pressure canner | | | | |
| | | | | | 5 psi. | 10 psi. | | 15 psi. | |
		pts.	qts.		qts.	pts.	qts.	pts.	qts.
Vegetables (Continued)									
Carrots	Raw or hot	—	—	—	—	25	30	15	15
Cauliflower—see Broccoli									
Celery	Hot	—	—	—	—	30	35	—	—
Corn, cream style	Raw	—	—	—	—	95	—	—	—
Corn, cream style	Hot	—	—	—	—	85	—	—	—
Corn, whole kernel	Raw or hot	—	—	—	—	55	85	—	—
Eggplant	Hot	—	—	—	—	30	40	—	—
Mushrooms (half-pints or pints)	Hot	—	—	—	—	45	—	40	—
Okra	Hot	—	—	—	—	20	40	—	—
Onions, small, white	Hot	—	—	—	—	25	30	—	—
Peas, edible pod	Raw or hot	—	—	—	—	20	25	15	15
Peas, fresh green	Raw or hot	—	—	—	—	40	40	30	30
Peppers (bell)	Hot	—	—	—	—	50	60	—	—
Potatoes, cubed	Hot	—	—	—	—	35	40	20	20
Potatoes, whole	Hot	—	—	—	—	30	40	20	20
Pumpkins, cubed	Hot	—	—	—	—	55	90	20	20
Pumpkins, strained	Hot	—	—	—	—	65	80	—	—
Soybeans	Hot	—	—	—	—	55	65	—	—
Spinach, other greens	Hot	—	—	—	—	70	90	35	35
Sweet potatoes	Dry	—	—	—	—	65	90	—	—
Sweet potatoes	Wet	—	—	—	—	55	90	—	—
Turnips, parsnips, rutabagas	Hot	—	—	—	—	25	30	—	—
Soup (all types)	Hot	—	—	—	—	60	75	50	50
Zucchini, yellow squash	Raw	—	—	—	—	25	30	—	—
Zucchini, yellow squash	Hot	—	—	—	—	30	40	—	—
Chicken, meats									
Chicken or rabbit (with bones)	Raw or hot	—	—	—	—	65	75	30	30
Chicken or rabbit (without bones)	Raw or hot	—	—	—	—	75	90	50	50
Meat—ground, chopped, strips, cubes, chunks (beef, veal, lamb, pork, venison, bear)	Raw or hot	—	—	—	—	75	90	50	50
Pickled products									
Pickled beets	Hot	30	30	—	—	—	—	—	—

enough to cool them quickly. Remove the cans from the cooling water while still slightly warm so they can air dry.

Check with your local county extension agent for processing times.

TD

Can sealer: Ives-Way Products, Inc., 820 Saratoga Lane, Buffalo Grove, IL 60090. Tin cans: Cumberland General Store, Route 3, Box 479, Crossville, TN 38555. Embarcadero Home Cannery, 2026 Livingston Street, Oakland, CA 94606. Freund Can Co., 155 W. 84th Street, Chicago, IL 60620. May and Company, Inc., 100 Grand Avenue, Brooklyn, NY 11205. McKay's Hardware, 1068 Fireweed Lane, Anchorage, AK 99503. Wells Can Company Ltd., 3434 Lougheed Highway, Vancouver, BC V5M 2A4, Canada.

Canning glossary

Boiling water bath canning. The simplest and most popular canning method, in which the filled, capped jars are completely immersed in rapidly boiling water for a specified period of time. This is used primarily with high-acid foods such as fruits, tomatoes, and pickled vegetables.

Headroom, headspace. The space between the top of canned food and the jar lid. Follow headspace directions carefully, because while most raw foods contract on heating, a few—like corn and peas—expand. Too little headroom can cause overflow; too much may result in an improper seal.

Hot pack. Canning precooked foods. This usually permits more compact packing, especially with greens, and cuts down on cooking time. Hot pack is best for foods that tend to discolor, but because of the extra heating it does cause an additional loss of nutrients with some foods.

Hot water bath. Recommended only for sweet or acid fruit juices. The jars are simmered at 180 to 190° F., thus pasteurizing the liquid. Hot water bath canning preserves the taste and the nutrients better than a boiling water bath, but it is unsafe for canning most foods.

Open-kettle canning. An outmoded and dangerous canning procedure in

(Continued on page 212)

(Continued from page 211)
which the cooked food is simply packed into sterilized jars and sealed. A vacuum seal is formed (hopefully) by steam from the hot food condensing inside the jars as they cool. This method is recommended only for jams and jellies containing sugar.

Oven canning. Another obsolete and risky canning method, in which the filled jars are heated inside an oven. Oven canning is perilous for two reasons: the temperatures reached inside the jars are often insufficient to kill botulism spores, and the jars can explode if they seal while the liquid inside is still vaporizing.

Pressure canning. Canning inside a pressurized kettle so that temperatures higher than 212° F. (the boiling point of water at sea level) can be reached inside the jars. This is used with low-acid foods—primarily vegetables—because sustained temperatures of 240°F. or more are necessary to kill any botulism spores.

Raw pack/cold pack. Packing uncooked fruit or vegetables into canning jars. Hot liquid—either water or syrup—is then added before the jars are processed.

Vacuum seal. An airtight bond formed between the canning jar and the lid which protects the food from contamination. As the jars boil during the canning process, some of the liquid inside turns to steam and displaces the air at the top, which vents out under the lid. Then when the jar cools, the steam condenses back into liquid, forming a vacuum which holds the lid tightly in place.

Cherry and plum pitters

Cherry pitters are among the most ingenious mechanisms ever invented. They range in complexity from simple hand-held squeezers to clever (and more expensive) devices that can pit 80 cherries a minute. Prices vary as much as the technology.

The simple, one-at-a-time, hand-held units work fine for recipes that require only a few pounds of cherries at a time. They'll do if you have a small tree. They are faster than the old method of digging the pit out with a hairpin or pushing them through with the head of a pencil that has had its eraser removed. And they do a neater job than squeezing the pit out between your fingers.

The larger pitters come in handy if you are canning a lot of cherries or plums, and they will pay for themselves in a season if you have just one good-sized tree. Not only do they speed things along—they're actually a treat to use.

Chop Rite Cherry Stoner

$16.64. Processes 1 lb. cherries in 2 min. (plus stemming and cleaning pits).

The Chop Rite has a traditional design, and is made of a rough metal casting with a hot-dipped tin coating. The unit is riveted together and cannot be taken apart, but the parts are loose fitting and easy to clean. A screwclamp is built into the unit; it worked fine but would likely mar a soft tabletop.

Stemmed cherries are placed in a feeding trough with the left hand as the right hand turns a small crank connected to the pitting plate. The plate rubs the cherry open and drops it out the bottom as the pits exit down a chute in front. Though no pits clung to the cherries, about 30 percent of the cherry mass went with the pits— enough to warrant picking through the pits by hand and pulling away the fruit.

Cherries have to be forced into the pitting chamber while you're cranking, and this presents a clear danger to fingers. Both cranking and feeding require some forcefulness, and a handle with a larger grip would have been more comfortable.

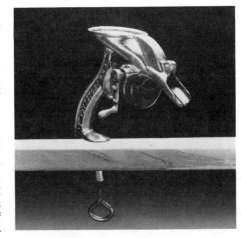

Ads for the stoner said it pits cherries without hurting the fruit. But it did not produce whole cherries; instead, it turned out masticated fruit which would be fine for a pie but would not make an attractive cherries jubilee.

Cumberland General Store, Route 3, Crossville, TN 38555.

Inca Kirschensteiner

$20.50. Processes 1 lb. cherries in 1 min. 15 sec.

The plastic body of this unit has an attractive, modern look, and is made of quality materials. The base container into which the pits are ejected is made of clear gray plastic so you can see when the pits come out and when to empty the container.

The plunger has four stainless steel blades designed to cut the cherry cleanly around the pit. This makes a clean path for the reamer to follow without mutilating the cherry. The plunger assembly and plastic cradle are removable for easy cleaning and storage. The pitter comes with a removable clamp which holds it secure but may leave marks on a soft surface.

To operate the Inca, you place stemmed cherries in the feeding tray. One at a time they roll into place on the plastic cradle, where the plunger blades go through the cherry and pop the pit out. The pit falls into a clear container which forms the base of the unit. The plunger picks up the cherry and drops it down a chute when the blades slip back behind a plastic

shield. Then whole pitted cherries roll right into your bowl.

The plunger has to be hit swiftly and hard to get the pit through the rubber cradle. In our 1-pound time trial, all 73 cherries were pitted but 5 pits went into the bowl with the whole cherries.

Garret Wade Co., 302 Fifth Avenue, New York, NY 10001.

RECYCLING CANS WITH PEDAL POWER

Tin can canning is a good long-term method. The biggest drawback is the initial high cost, both for materials and equipment.

Joe Thompson, the supervisor of a Mormon Church cannery in Mesa, Arizona, has found a way around this problem: recycling with home-made equipment.

Instead of paying for new cans, the Mesa cannery buys large, used #10 cans from the local school system for a half-cent each. (New cans would cost from 30 to 50 cents, depending on the quantity purchased.) Then, they use a pedal powered device to cut the top off and reflange it so a new lid will fit on.

Thompson made the cutter/reflanger because he couldn't find a commercial version that would work on #10 cans. Because it's pedal powered, the speed can be easily controlled, lessening the danger of a cut from the sharp edges of the cans or the metal trimmings.

Once the cans are cut and reflanged, they can be resealed with new lids. If the cans are pitted, cannery workers pack the grain or dry milk in a plastic bag first. Before sealing the cans, they drive out the oxygen by filling them with carbon dioxide, and this helps prevent spoilage. Wheat stored in these cans will last ten years under good storage conditions, and dry milk up to three years. Recycled cans are not recommended for liquid foods.

RICH MAN, POOR MAN

It is curious to note how a country's cuisine is modified by the very rich and the very poor. At one extreme, *Home Food Systems* happened upon a Himalayan family boiling a dinner of rice and an herb in its lean-to of pine boughs; they could not afford the traditional accompaniment of curds or dal or curry. At Bombay's genteel Hotel President, we noted, the accompaniment evolves into rich, cloying main dishes, while rice or chapatis are reduced to a side order.

Lobster Canning

In many cases the wonder is not that the canned product is so bad, but that it is as good as it really is. Many of the factories are mere hovels with inadequate appliances for ordinary cleanliness, and under the best conditions it is to be remembered that the quantity and kind of offal connected with the process is admirably suited for the growth of putrefactive microorganisms. The factories are seated upon the shore with stages leading into deep water for the accommodation of boats, or the buildings themselves are at the end of a stage connecting with the shore . . .

For the employees, men and women, the business is not unpleasant. The work is not laborious, the pay is good, and on stormy days and wet they have good leisure to indulge their propensities, which sometimes unfortunately run in undesirable channels.
The food is abundant and good, if not very delicate, nor the cooking of it over dainty.

From "Discoloration in Canned Lobsters" by Sir Andrew MacPhail, supplement to the 29th Annual Report (1897) of the Canadian Department of Marine and Fisheries.

We eat what we can, and what we can't, we can.

Abraham Lincoln, describing what becomes of a good harvest in Illinois.

Kenberry

$.98 Processes 1 lb. cherries in 5 min. 40 sec. (plus stemming).

This simple device is made of formed and plated steel. Your thumb fits through a ring to pump the plunger. There is no spring action. Single cherries fit on a cradle with a hole in it, and the plunger meets the pit and forces it through the other side.

Unless the cherry sits upright in its cradle with its bottom carefully centered, there's a good chance the pitter will miss the pit. A second shot will get it but some mutilation of the cherry results. This inexpensive pitter works, but slowly.

Kenberry Div. Etamco Ind., Montgomery Street, Belleville, NJ 07109.

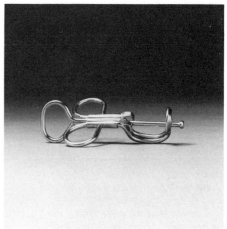

Kernex Cherry Pitter

$7. Processes 1 lb. cherries in 4 min. 40 sec. (plus stemming).

This small cast-aluminum hand pitter holds cherries—placed by hand one at a time—in a round cradle. The cradle has holes to allow the pit and plunger to pass through. If time is taken to position the stem properly, it also will be discharged through an opening.

A firm squeeze of the handle sends a barbed, star-shaped concave reamer into the cherry and pushes the pit on through, leaving a clean hole. The pit is ejected rather forcefully, so you may want to use this pitter outside.

The cherry is supposed to be lifted by the reamer and knocked away as the spring propels the reamer back behind a barricade. However, the reamer doesn't always pick up the cherry, so it must be cleared from the cradle with a twist of the wrist. Sometimes the pit clings to the cherry, requiring the operator to stop and separate it. Cleaning involves only a simple, soapy rinse.

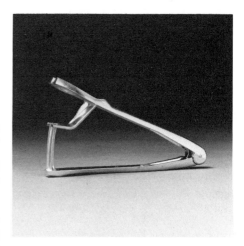

The Kernex is a handy tool. The pitting time is fast enough, and can be improved by a third if the stems are removed ahead of time.

Creative House, 190 W. Ashland Street, Doylestown, PA 18901.

Steinex Combi

$21. Processes 1 lb. cherries in 3 min. (plus stemming), 1 plum in 10 sec.

This pitter works for both cherries and plums. Four stainless steel blades pit plums and cut them into quarters, and a plastic reamer with four barbs pierces cherries and forces the pit out. (These barbs are showing signs of wear with only our moderate amount of use.) The unit has a rugged aluminum frame, and four legs with rubber feet to make it conveniently freestanding. The rubber feet are not glued in place, however, and one dropped off several times. The plunger handle slips on and off for easy storage.

One by one, either a cherry or a plum is placed on a cradle with a hole in it for the pit to pass through. The reamer automatically ejects the cherry, but often misfired in our tests.

The Steinex Combi worked best if the plunger was hit briskly. Sometimes the pit hung on to the cherries. Since both pits and cherries fall into the same pan, the best operating method is to stand the pitter inside a large cake pan to collect pits and cherries. This slows things down, however, and makes for a messy job.

With plums, the slicer cuts all the way through the plum on one side as the fruit

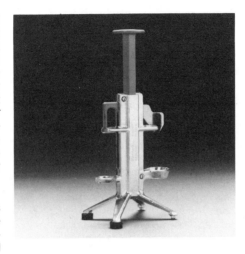

sits upright in its cradle. If the fruit is nicely ripened, the pit will exit out the bottom of the plum, taking along just a bit of fruit—maybe a half a bite's worth. The other three blades simply score the plum into sections but do not cut all the way through. If you want quarters, the plum has to be pulled open by hand and cut with a knife. The whole operation works best with slender plums.

Creative House, 190 W. Ashland Street, Doylestown, PA 18901.

Steinex Plum Pitter

$5.50. Processes 1 plum in 10 sec.

This pitter has a sturdy aluminum frame and four blades for pitting plums, cutting them open, and scoring them into quarters. A bent metal spring pops the tool open after each use. The blade points are very sharp but the blades are not.

A single plum is set upright on a round cradle with a hole in it. The four-bladed reamer pierces the plum, wedges the pit between the blades, and forces it out. The plum is cut all the way through on one side and scored in three other places for cutting into quarters. The plum must be removed from the pitter by hand. The pit falls through the bottom with only a small amount of fruit left on it.

Creative House, 190 W. Ashland Street, Doylestown, PA 18901.

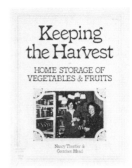

KEEPING THE HARVEST: Home Storage of Vegetables and Fruits

Nancy Thurber and Gretchen Mead. 1976.
Garden Way Publishing, Charlotte, VT 05445. 202 p. paperback $5.95.

Keeping the Harvest is a terrific book. It provides the beginner with inspiration, instruction, and reassurance, and yet is comprehensive enough for the experienced canner. The information is organized and notated for quick reference. And it is one of the few canning books we saw with photos clear enough to be instructive.

Keeping the Harvest is far more than a canning book. Its freezing section also speaks with the voice of experience. It warns, among other things, that you should find a source for large supplies of dry ice *before* the electricity is cut off for 48 hours.

Keeping the Harvest has a large format that allows for delightful layout and illustrations to complement the book's sweet and helpful tone. It provides diverting asides to amuse the casual reader. People who think that home food preservation is an obscure discipline should read this book and learn otherwise.

Judy Rittenhouse

Cinnamon Pears

14 pounds pears
1 teaspoon stick cinnamon
1 teaspoon whole cloves
1 quart cider vinegar
6 pounds honey

Peel pears. Put spices in cheesecloth bag. Heat honey and vinegar to boiling, adding spice bag and pears. Cook until pears are tender. Remove pears and spices. Boil syrup until thick. Pack pears into hot clean jars, leaving ½-inch headspace. Cover with hot syrup, leaving ½-inch headspace. Adjust lids. Process in boiling water bath; pints for 20 minutes and quarts for 20 minutes. Remove jars. Complete seals if necessary.

Makes 6 quarts.

Westmark Kernomat Cherry Pitter

$25. Processes 1 lb. cherries in 1 min. 36 sec. (plus stemming).

The Kernomat's frame is made of aluminum casting, and the plunger, reamers, feeding tray, and pit receptacles are plastic. The pit receptacles snap into place (and occasionally popped out during use). The rubber suction foot works well.

This two-at-a-time pitter was fast and fun to use. Just load the feed tray with stemmed cherries and you're ready to pit. One hand hits the plunger as the other helps funnel cherries to either side of the tray where they automatically drop into place under the reamer.

The reamer/plunger penetrates the cherry, drives the pit out through the bottom and into a chamber, and picks up the cherry as it returns to its original position. The cherry then slips off the plunger as the plunger springs back into the body of the unit. It falls on to a flipper and rolls into a waiting dish or pan.

Clean-up was easy. The rubber pads through which the pits are forced presented the greatest cleaning challenge, but we found that with the help of a pen or pencil the rubber could be slipped through the side and removed for cleaning. Small pieces of cherry get caught inside the unit and should be cleaned out before you store it for the season.

Cherries did not always roll into place on both cradles, so sometimes only one cherry was pitted at a time. If you do a bushel of cherries a year, this unit is probably worth the cost.

Creative House, 190 W. Ashland Street, Doylestown, PA 18901.

Apple peelers

Apple peelers work. If you've got a bushel or two that needs peeling, these machines can turn a slow, tedious job into a relatively quick, painless task. But before rushing out to buy one, you may have a basic question: Why peel apples in the first place?

Store-bought apples that aren't organically grown may be waxed, in addition to containing traces of pesticide, and peeling strips away this unhealthful layer.

With organic apples, peeling is really a matter of preference. Apple skins are a source of fiber, but they have little nutritive value. Most people don't like peels in their applesauce, and if you dry apples without peeling them, the skin develops a tough, plasticlike quality.

Our four test models range in price from $12.95 to $80. Three of them do more than just peel: two core and slice and a third even knocks the apple off when the peeling cycle is complete.

For the peelers to hold them properly, apples must be fresh and *crisp*. A soft apple will break free from the holding fork at the core so that only the core spins, not the whole apple. Soft apple skins tend to clog the blade. So, peelers are definitely an August-through-October tool. We did our testing at the end of November and had difficulty finding crisp apples.

Apples don't have to be perfectly round or all the same size to work on a peeler. In one test, we changed from Winesap to Delicious to McIntosh and never had to make any adjustments. There wasn't much waste. In fact, when adjusted properly, peelers will take off just the skin and produce a thin peel five to seven feet long. They beat any attempts I'd ever

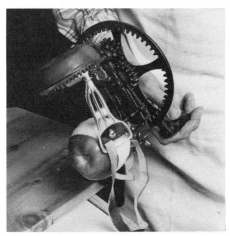

This peeler has two blades: one for the sides of the apple and one to get the ends.

made in competition with my sister for the longest peel on pie baking day. The long curly peels ended up gracing the Christmas tree.

Apples aren't the only thing these machines will peel. The Pefra worked on oranges and lemons, and made some progress on cooperative potatoes. We tried all the peelers on potatoes but had only marginal success. Don't buy a peeler just for potatoes; if you already have one it will help with big potato jobs, but you'll still have to finish the job by hand.

—DB

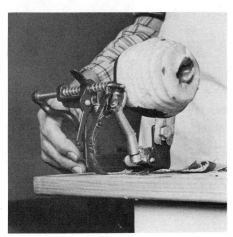

The cutting blade on the bottom slices the apple as it is peeled.

Pickers

During my many hours of blueberry picking I used to dream of ways to make things go faster. Could I hang a basket around my neck so I could pick with both hands and hold

(Continued on page 220)

Johnny Apple Peeler

$13.95. Peels 6 apples in 5 min. Apples only. Peels, cores, and slices.

This machine is designed to peel, slice, and core crisp apples. It relies on manual skill more than any of the other peelers. One hand cranks as the other guides the peeling blade. (All other peelers guide the blade mechanically, though the White Mountain requires extra guidance by hand.) The blade pivots in a ball-and-socket arrangement, giving the operator full flexibility of direction—and more possibility for error.

After the peel is removed, a panel of eight serrated blades can be lifted against the apple to cut it into slices as the operator continues cranking (the faster the better). When that's done, a coring blade can be pushed into the rotating apple to cut the slices off the core. Any of these functions can be done independently of the others; however, coring must be done last, because the peeler can't rotate a coreless apple.

Setting up the Johnny Apple Peeler is tricky. Only a firm, non-porous surface will cooperate with the suction base. Since we were using a wooden workbench, we had to bring in a piece of glass for the peeler to grip. Still, we had problems getting it to adhere all the time.

As long as a firm apple was used, the slicing and coring features worked well. But the peeling procedure was awkward, requiring practice and some manual skill, and performance was spotty. This unit removed a considerable portion of apple along with the peel, and made many skips. It worked best on round apples.

Because of the square blade, the machine wasn't able to remove much of the peel at

either end. The blade is not adjustable, so it's not very versatile for different-sized apples. On small apples the blade rode high on the apple, making it difficult to get it to dig in. Sometimes the ball base of the blade arm popped out of its socket.

Mounting an apple on the long, four-edged plastic spear was sometimes awkward. The tendency is to push the apple on with the palm of your hand, but that leaves your hand in line to be speared. Instead, the instructions recommend using two hands to mount the apple—one hand on each side. This is both safer and easier. Apples with crooked or off-center cores do not align properly, and won't peel well.

Cleanup is easy; you simply rinse the plastic and stainless steel parts.

Leisure Technology, Inc., 5440 W. 125th Street, Savage, MN 55378.

"Your honor, I will prove to the satisfaction of this court that whenever Little Red brought groceries to her grandmother, they were the kind of package goods heavily loaded with chemical additives, homogenized with insecticides, and fungicides, alkaloids, propiolates, polyoxyethylene monostearates, prokylene oxides, methyl bromides and assorted diacetate compounds, benzoate of soda and 2,4-D."

"No wonder they tasted like a pig's behind," says Grandma. "Yech!"

Testimony of a wolf in J. I. Rodale's play "The Hairy Falsetto."

CONVENIENCE

Consumers have played a role in the denaturing of food through their headlong pursuit of convenience. Convenience is an indelible part of modern life, but we pay a dear price for it, whether the currency be petroleum, integrity of the landscape, or food quality. Convenient foods are rarely the best foods. Foods become convenient only when someone else processes them for you. Processing is inversely proportional to food value: the more work you do to a food, the less flavor, beauty, and nutrition you get out of it.

—RY

Hudson Food Preservation Center: struggling toward success

Community canning centers are a great idea: people working together, sharing ideas and experience, making the best use of energy and equipment. But they seldom work out. The centers invariably have financial problems; if they charge enough to cover costs, they price themselves out of the market.

The Hudson Food Preservation Center in Hudson, New York, is trying to overcome this problem. It opened in 1977 with funds from a national church organization. But it soon became obvious to the director, Tom Brenner (photo at left), that the center could not always be dependent on outside funding.

Brenner decided to turn the center into a dual operation: a place where community residents could can for a minimal cost *and* a commercial canning business. Such a program would mean additional funds for the cannery, and avoid having the equipment sit idle on weekdays or during the winter and spring.

It was an elusive goal. The center had enough equipment for local groups to can, but not enough for a commercial operation. By finding used machinery, Brenner was able to buy equipment worth more than $100,000 for about $3,500. During 1978, the Center canned nearly 6,000 quarts of produce for local residents, and nearly 12,000 quarts of applesauce, pear sauce, apple butter, and pear butter for sale (including versions without sugar). The products were priced competitively, and they had no trouble selling out.

The commercial operation began to thrive in 1979, and by early fall was showing a small profit. The center was still having trouble attracting local residents, however. Then, in November, the center lost most of its funding, and had to close down temporarily. Brenner is now lining up new sources of money, and planning to open again in the summer. He remains confident that the center can be self-supporting within three years.

So the work goes on. These pictures of the Hudson Food Preservation Center show some of the joys of community canning.

Pefra Peeling Machine, Type 105

$80. Peels 6 apples in 2 min. 35 sec. Peels apples, oranges, and lemons. Peels only.

This machine, imported from Italy, is the Cadillac of apple peelers: you pay a high price, but you get top quality. It peels apples, oranges and lemons very well, but it does not core or slice.

The Pefra mounts onto a table or counter edge with two clamps which are removable for compact storage. Setup is simple, though the clamps open only to a 1¹¹/₁₆-inch width. The apple mounts on two forks. One delivers the rotating force to the apple, and the other holds the opposite end in place. Securing the fruit at both ends was a feature unique to the Pefra, and gave it more success with softer apples than the other three units.

Gearing between the crank and the turning fork speeds the revolution of the apple. Peeling is quick, slick, and easy. The sharpness and contour of the U-shaped blade is responsible for its superb performance. A "stop-nose" controls the depth of the cut at both the front and the edges of the blade. This control can be adjusted by a thumb screw. The U-shape enables the blade to keep in contact with the fruit at either end as well as throughout the

middle. As the handle turns, the apple rotates and the blade moves from one side to the other.

Cleanup can be done easily with a damp cloth. The only required maintenance is an occasional oiling.

A. W. Austad, 9853 Alpaca, S. El Monte, CA 91733.

Reading Apple Parer

$19.95 East coast, $20.95 West Coast. Peels 6 apples in 2 min. Best with apples, but can peel oranges. Peels only.

The Reading peels apples; it does no slicing or coring. It has two blades, one for the main body of the apple which will conform to almost any shape and one to peel in close to the stem end. As the crank turns, the apple rotates on a fork while a blade travels around the apple on another gear.

It's important to start the peeler at the designated starting place, because when the blade reaches the end it trips a lever which pushes the apple off the fork. (In our tests, the apple came completely off only about one-third of the time.)

This machine, in production since 1864, was the fastest peeler we tested. It didn't skip at all. The blade is razor sharp and adjustable, allowing a very thin peel to be taken off in one long piece.

Like most other peelers, the Reading requires a very crisp apple or the fork will

break the core away from the apple and lose traction. Cleaning was simple—just a quick rinse and wipe.

Sterling G. Withers, P.O. Box 2081, Sinking Spring, PA 19608.

(Continued from page 217)

the bucket close to the bush? If I had wide cuffed sleeves, could I just drop berries down my sleeve and dump them into a bucket all at once? It wasn't the picking that frustrated me, it was the long trip between the bush and basket. Holding the basket next to the bush helped, but that left the basket hand unproductive.

There are now a variety of pickers available. None of them can do the work for you, but they can make it easier, quicker, and more enjoyable.

DB

This shoulder hamper is meant for commercial use, but it could come in handy for any serious fruit picking. The basket holds 24 quarts and has a six-foot-long nylon strap. It can be ordered from an orchard supply manufacturer, and costs about $20.

Pole fruit pickers

Pole pickers are useful alternatives to unsteady step ladders and house ladders that scrape fragile branches. A wire or cloth basket is attached to the end of a long wooden or aluminum pole (the basket usually is sold without the pole). Open wires catch the fruit, sometimes bending across the top so fruit can be tugged in a number of directions. The fruit then falls into the basket. Some units provide padding to avoid bruising the fruit; if not, you can add a little of your own.

Most pickers have small baskets, so that the pole is constantly raised and lowered for emptying. But filling a larger basket could damage the fruit, and a pound or two of fruit at the end of a ten-pound, 14-foot pole can be heavy.

With prices ranging from $2 to $7 for a basket, a pole picker may seem like a modest investment. Wait until you price a pole. Shopping locally, we found a variety of handles, wooden and aluminum, from $7.50 to $20 depending on length. Notice the weight of wood as compared to aluminum. (See table of poles for pickers, page 222.)

Cumberland Picker, $3.89: Cumberland General Store, Route 3, Crossville, TN 38555. Garden Way Padded Picker, $6.95: Garden Way, Charlotte, VT 05445.

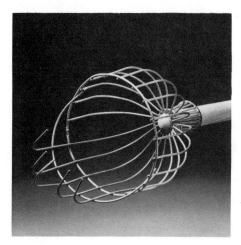

White Mountain Apple Peeler

$18.75. Peels 6 apples in 5 minutes. Apples only. Peels, cores, and slices.

This device does more than peel; it can peel, core, and slice all in one step. A single fork holds the apple and connects it with the hand-cranked drive. The paring blade is stationary but spring loaded to hold it against the apple.

The White Mountain works only with fresh, crisp apples. Coring, slicing, and peeling put a great deal of drag on the apple, and a soft one will break at the core. The blade can be adjusted for different-sized apples by loosening a screw and manually moving it.

Slicing and coring is done by a single L-shaped blade. As the crank is turned, the blade cuts a continuous ⅛-inch-thick spiral of apple, severing it from the core at the same time. To make separate slices, the spiral-cut whole apple must be cut in half. It will then fall into ⅛-inch half slices.

This machine peels comparatively slowly, but since it slices and cores at the same time, this is no disadvantage. If the apple is fresh, few skips occur.

Cleaning is fairly easy, and for best results the paring knife should be removed first. The unit should be dried thoroughly after rinsing to prevent rust.

White Mountain Freezer, Inc., Lincoln Avenue Extension, Winchendon, MA 01475.

Quicker Picker

$4.95 (1¼-quart size), $5.95 (3-quart), $8.95 (10-quart).

The Quicker Picker is a nylon pocket with a rigid rim that slips over your hand as a catching mitt for fruits, vegetables, and berries. It hangs from the back of your hand, leaving fingers free to be very selective about what they pick. Since there's no aiming involved in hitting the pocket, you can use one on each hand. The berries fall right into the bag as long as your hands are working in a vertical position.

To prevent berries from escaping between the pocket and your hand, a leather reinforcement at the palm cups itself for a form fit. You can pick one berry at a time or slide your hand over a whole cluster to release a dozen or two at once.

The vertical positioning is just right for blueberries, cherries, crab apples, currants, blackberries, grapes, and gooseberries. In the garden, the Quicker Picker can help with pole beans, limas, peas, and staked cherry tomatoes, but low bush beans might be a problem because you can't get underneath them.

For larger fruits, there's a Quicker Picker

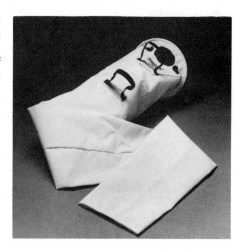

with a long bag and an open end to deliver apples, peaches, and plums from your picking hand to your free hand near the basket. With an open end, there's no pile-up to damage the fruit and no all-at-once dumping. Your free hand can catch each piece of fruit after its trip down the chute and carefully place it in the basket.

Garden Way, Charlotte, VT 05445.

EASY GRAPE JUICE

You can make grape juice without going to the trouble of stewing and straining a kettleful of the fruit. This quick, easy, and neat method will give you juice that retains all the delicate overtones of grape flavor normally lost in overcooking. The flavor is fuller when the processed jars stand for a week or so before serving.

Wash grapes and remove from stems. Put one cup of grapes and ¼ cup honey into each clean quart jar. Pour in boiling water to ¼ inch from the top of the jar.

Apply seals and screw on rings. Process jars in boiling water bath for 10 minutes, then allow to cool on a rack away from drafts. Inspect seals and refrigerate any unsealed jars to keep them until the full flavor develops.

To serve, simply pour juice off through a strainer and discard grapes and seeds.

POLES FOR PICKERS

Item	Source	Length (ft.)	Construction material	Weight (lbs.)	Cost (Aug. 1979)
Tree trimming extension	Local hardware	Adjustable 7–14	Wood	5–6	$14.95
Extension trimmer	Local hardware	7 each, possible to add more	Wood	10–15	$7.49
Straight extension	Local hardware	16, no adjustment or addition	Wood	13–16	$.79/ft.
Telescopic pool net handle	Swimming pool supply	8–16	Aluminum	1–3	$16.00
Electrical conduit	Electrical supply	10–20	Aluminum	5–7	$20.00

Pureers and strainers

For gardeners and home food storers who like apples and tomatoes, pureeing is one of the very first processing chores. We looked at three straining tools: the Foley Food Mill; a cone-shaped strainer with a wooden pestle; and a Squeezo Strainer. Our tests are prefaced with our remarks on doing these tasks by hand.

DB

By hand

Tomatoes really aren't difficult to process by hand. Blanching in boiling water loosens the skins so that they peel right off, and there is only the stem end to be removed, not a whole core. (Forget about getting the seeds out by hand, however.)

To make our pot of tomato puree, we placed five pounds of tomatoes in a soup pot and covered them with boiling water. After 10 minutes the skins cracked and were ready to peel. Most peels came off easily, and in another 10 minutes we had them peeled, cut into wedges, and mashed with a potato masher. In 7 minutes the mixture was simmering nicely and getting juicy. We mashed it again and in another 15 minutes, once more. After 30 minutes of cooking, the puree was finished. (Total time was 72 minutes.) More cooking would have been necessary for a thick sauce or paste.

To extract some of the water and reduce cooking time, you can freeze tomatoes whole before pureeing. As they thaw, some water will separate and can be dumped off. It will be slightly pink but there's no real loss of flavor. A small amount of nutrients may go down the drain.

Apples were another story. It took 25 minutes just to peel five pounds of apples. Coring took another 5. But then making the sauce was just a matter of simmering for 15 or 20 minutes (for a total processing time of 50 minutes).

Pumpkin was a toughie. Peeling the thick skin was a chore—and a little dangerous when we used a knife. After peeling, we cut the pumpkin into squares, boiled them until they were soft, and mashed them with a potato masher. The pumpkin was somewhat stringy and fibrous but still ready for the pie shell.

To avoid peeling, you can put the whole pumpkin in the oven to bake, then just scrape the innards out. This is fairly easy, but the pumpkin must bake a couple of hours and that energy costs money.

With a Squeezo Strainer, puree comes out one end and skins, seeds, and cores go out the other.

Foley Food Mill

$9.90 (1¾-quart model), $13.21 (3½-quart model).

Our tester, Diana Branch, bought a Foley Food Mill at a yard sale a few years ago. She was attracted to its simple design and guessed she might need such a tool someday. It's shaped like a two-quart saucepan and has small holes in the bottom. A single handle with a horizontal cranking action turns a spring-loaded metal plate which forces the food through the holes. A spring-loaded wire underneath scrapes the bottom clean. A hand nut on the bottom lets it all come apart easily for cleaning—a three-minute job.

The Foley Food Mill did a good job of straining cooked vegetables, making baby food, mashing bananas for banana bread, and separating cooked pumpkin from its shell. But it didn't do quite as well at straining items with skins. For instance, tomato innards, including seeds, were separated from their skins but the skins coated the bottom of the pan, covering the holes and making a quick cleaning necessary. Simply scraping them free and tipping the pot was all that was necessary to get started again.

The same was true for apples. Apple

seeds and cores were left behind in the pan and had to be dumped every two or three loads. Even with the skin problem, we were able to strain 5 pounds of cooked apples in 3½ minutes—almost ten times faster than by hand. Straining does require precooking until apples are soft—about 20 minutes.

Cumberland General Store, Route 3, Crossville, TN 38555.

HOME CANNING OF FRUITS AND VEGETABLES (Home and Garden Bulletin No. 8)
Consumer and Food Economics Institute, Agricultural Research Service, USDA. 1976.
Superintendent of Documents, U.S. Government Printing Office, Washington, DC 20402. 30 p. booklet $.45.

HOME CANNING OF MEAT AND POULTRY (Home and Garden Bulletin No. 106)
Human Nutrition Research Division, Agricultural Research Service, USDA. 1966.
Superintendent of Documents, U.S. Government Printing Office, Washington, DC 20402. 24 p. booklet $.15.

These booklets are bargains, especially if you like the tone of USDA films on home economics: "As soon as you take jar from canner, screw cap down tight to complete seal" or, "Organisms that cause food spoilage—molds, yeasts and bacteria—are always present in the air, water and soil."

Fruits and Vegetables includes a good sample of specific directions for fruits and vegetables. Beginners should first read it straight through: occasionally, vital information, such as adding 20 minutes additional processing time when canning in a pressure saucepan instead of a large canner, is separate from the timing chart.

The second booklet obviously isn't intended as a speedy reference to glance at when you are elbow-deep in feathers. Its information is broken up, so that a beginner should make a step-by-step outline from the booklet before going to work. But *Meat and Poultry* does make a good introduction.

Note: There is a $1 minimum charge for each mail order, perhaps to encourage larger orders for Christmas giving.

Judy Rittenhouse

Kitchen Craft Strainer

$15.26.

This simple, effective tool sits about a foot high in its own wire stand and is covered with small holes. As with most models, this one is aluminum, but stainless steel strainers are made. A large wooden pestle comes with it, made just right to fit snugly in the bottom of the cone and squeeze the puree through. Strainers may be hard to find; the best sources are specialty catalogs and well-equipped hardware stores.

Using cooked fruit, it took just 2 minutes to strain 5 pounds of apples (and 3 minutes to clean the strainer). Skins didn't clog up the works. The tomato puree was good, but there were tiny bits of seeds in it which were broken when forced through the straining holes. The strainer also worked well with pumpkin. If you cook the pumpkin in sections with the skin on, you don't need to peel it, and straining helps

rid the puree of fibers.

Cumberland General Store, Route 3, Crossville, TN 38555.

THE BALL BLUE BOOK

The Ball Corporation. 1974.
Ball Corporation, Muncie, IN 47302.
112 p. booklet $1.

The Ball Blue Book is a comprehensive instruction and recipe resource. Its canning and freezing information is accessibly arranged in a small size that lends itself to quick reference as well as inspirational perusal.

It includes many good charts and lots of recipes to help the reader survive canning and enjoy its rewards. The charts provide easy reference on heat tolerances, suggest needed garden space, and estimate the number of jars you'll need. Included are specific directions for a wide sampling of fruits (all the way to loquats). Vegetables are very well represented, as are meats (including rabbit and cracklings). Some canning books try to discourage the home canner from treating seafood; *The Blue Book* tells you how to do it.

The recipes range from meat sauces to mincemeat to Chicken a la King (to can!). There are seven pages of pickle recipes plus chutney. Unfortunately, the book omits meat and vegetable stock methods and seasonings, which are vital to great meats and soups. That would appear to be its only oversight.

The book contains a wide variety of other useful material. It tells how to can homemade baby food. It has a comprehensive troubleshooting check list. It tells how to freeze foods in Ball products.

A silly marketing tone occasionally intrudes to promote the product and treat the reader like a six-year-old. Except when it pictures molds, yeasts, and bacteria like Halloween monsters, this tone can be ignored.

The Ball Blue Book is a bargain for a dollar. It is the sort of guide you pull out of a kitchen drawer as you hold the tomato blancher with the other hand. It covers about every legitimate use you might find for a Ball jar except for draining engine oil or taking a specimen to the doctor.

Judy Rittenhouse

Squeezo Strainer

$29.90, extra screens $5.50 each.

When your needs exceed the capacity of the Foley Food Mill or pestle strainer, you're ready for a Squeezo. With this device, you can go through a bushel or two before tiring.

The Squeezo (and its almost identical twin, the Victorio strainer) uses an auger to crush food and force it through a cone mesh. A large hopper directs food into the auger. A plunger assists the flow. Puree squeezes through the screen onto a chute and oozes into your bowl. Skins, seeds, cores, and stems exit from the end of the auger into another bowl. All this happens under your own power via a hand crank.

It was novelty to watch the Squeezo perform the first few times. Cooked quartered apples or uncooked quartered tomatoes went in the top, puree rolled out the front, and waste was routed out the end. Sometimes a piece of skin or a seed would sneak into the puree, and sometimes the puree would run over the back end if not nudged down the chute, but otherwise it worked quite well. We strained 10 pounds of apples in 4 minutes. Cranking wasn't hard and the machine was so safe a child could help. The hardest work was the plunging.

Puree from the Squeezo was excellent: clear and nearly free from all tomato seeds and seed bits. And there was no stopping for scraping out the skins; it operated continuously until the job was done.

Cleanup was time-consuming. The Squeezo just was not worth getting dirty for small jobs. Eight of the nine parts will usually rinse clean with the help of a brush, but the strainer requires a brush and patience—especially since the brush flings soapy water at you from the strainer holes.

Reassembling all the parts can be tricky. We invariably forgot that the chute goes on before the strainer gets bolted in place, and had to take it apart again. And if you haven't watched how the parts came apart, you may be left with a piece in your hand when it's all back together.

The Squeezo also comes with a fine screen for straining berries (seeds and all) and a coarser screen for pumpkin. We did not test these. A motor driven model is available as well.

Squeezo Strainer: Garden Way, Charlotte, VT 05445.

Victorio Strainer: W. Atlee Burpee Co., 300 Park Avenue, Warminster, PA 18974.

PACKING IN HOT SYRUP

It isn't necessary to sink canned fruit in syrup. Sweetening helps fruit hold its shape, color, and flavor, but it is not needed to prevent spoilage. Unsweetened fruit can be processed just as sweetened fruit.

If you wish to use sweet syrup, try packing your fruit in its own juice or that of other juices. Or excellent syrup can be made by blending two cups of honey with four cups of very hot water. Use a light-flavored honey so the honey taste will not overpower the flavor of the fruit. To prevent the fruit from darkening, pack it in hot syrup as soon as it is peeled and sliced.

BEHOLD THE HOUSEHOLD ECONOMY

A home flour mill generates a return on capital greater than that achieved by Pillsbury or General Mills if it grinds bread flour for a family of four. Home yogurt makers, peanut butter grinders, and other simple machines generate returns in excess of 30 percent a year. The simple fact is that it makes good economic sense for any family with a small plot of land to produce and process as much of its own food as possible.

Scott Burns. *Organic Gardening.*

PUTTING FOOD BY

Ruth Hertzberg, Beatrice Vaughan, and Janet Greene. 1973.
Stephen Greene Press, Brattleboro, VT 05301. 360 p. hardcover $8.95. Bantam Books, 666 Fifth Avenue, New York, NY 10019. paperback $2.95.

In addition to covering the more common preservation methods, this book draws on Vermont pioneer tradition to include introductory methods for rendering lard, waterglassing eggs, and making sausage and cheese.

The canning section is a somewhat haphazardly arranged, oddly edited compendium of how-to. Lacking a quick-reference canning guide, first-timers will have to wade through pages of passages such as: "And though the occasional homemaker . . . may still get away with doing green beans and corn and meats and suchlike in the old wash boiler. . . ." I've been hoping to meet an occasional homemaker ever since I read that.

Putting Food By approaches canning seriously, starting with biological dangers—as most canning guides do—and listing no less than 13 *essential* items of equipment for the canning enterprise. It is thorough even to listing warnings, limitations and techniques for "maverick canning methods" such as cold-water canning and making hominy grits from corn with lye.

Judy Rittenhouse

Canning nut meats

All nuts are rather fatty, and it's this fat that turns rancid and spoils the meats (even nuts in the shell can spoil after a while). The two ways to delay or prevent rancidity are to can the nut meats or freeze them.

General handling. Pressure Canning or Boiling Water Bath. Use dry *hot* pack only (nut meats are oven-dried before canning). Use dry, sterilized jars no larger than pints, and with self-sealing lids.

Spread a shallow layer of nut meats in baking pans, and bake in a very slow oven—not more than 275° F.—watching the nuts and stirring once in a while, until they are dry but not browned: they must not scorch. Keep hot for packing.

Hot pack only. In self-sealing jars. For pints (or ½ pints), fill dry, sterilized jars, leaving ½-inch headspace; adjust lids. Pressure-process at 5 pounds (228° F.)—10 minutes for ½ pints or pints. OR process in a boiling water bath—but with the water level well below the tops of the jars—for 20 minutes.

Freezing and cold storage
Taking the heat out

Just after the turn of this century, a group of explorers traveling in eastern Siberia found the remains of a woolly mammoth buried in ice near the Beresovka River. The massive mammal was virtually intact, with clotted blood in its chest and unswallowed food in its mouth. The value of the discovery was not lost on the expedition's dog team, which eagerly feasted on the carcass.

In addition to its paleontologic worth, the discovery demonstrated the remarkable preservative powers of cold. Meat, quick-frozen some 30 millennia earlier, was still edible, as if a cave-wife had placed it in her freezer only a few weeks previously.

People have used ice to keep food cool whenever it's been available. The first home delivery of ice in America was in 1802, and most of our great-grandparents were familiar with some kind of icebox. Ice was harvested from ponds and streams in the winter, then stored for summer use. In a typical northern city in 1855, you could have 15 pounds of ice delivered for $2 per month in the summer. (Your refrigerator now costs $5 to $10 per month to operate.)

The principles of mechanical cooling were known through the last half of the 19th century, but there was little demand for a motor-driven icebox. Than, in 1890, an unusually mild winter caused a severe ice famine, and the idea became more attractive. Nathaniel B. Wales developed the first electric refrigerator in 1914 and marketed it under the name "Kelvinator," after William T. Kelvin, the British physicist who developed a scale for measuring extremely cold temperatures. Freezers followed shortly afterward.

The cold that preserved a woolly mammoth for thousands of years keeps much of our food stored safely today. Refrigerators are America's most-used appliances and are found in 99.8 percent of our homes. Over 45 percent of American families have freezers, and many have discovered for themselves the benefits of home-frozen foods.

How does cooling work?

Cold is the enemy of most living things. Low temperatures retard chemical and enzymatic actions, and slow the growth and spread of microorganisms which cause food to deteriorate. The lower the temperature, the slower these actions. And if it gets cold enough, everything stops; at absolute zero—minus 460°F.—even air freezes.

A few foods, including pumpkins, squash, and sweet potatoes, will keep for considerable periods at temperatures as high as 60°F. Most foods prefer temperatures from 32 to

40°F., the recommended range for root cellars and refrigerators. At these temperatures, root vegetables, cabbages, apples, and pears will keep for many months; other vegetables, fruits, and dairy products will stay fresh for a week or two. At usual home freezer temperatures—around 0°F.—most foods will keep safely for six months to a year.

Cold preserves vitamins and minerals better than any other storage method. Freezing itself does not destroy nutrients; the only losses occur during preparation and thawing. So the more rapidly food is frozen, the higher its nutritive value.

Freezing is also kind to taste and quality. Foods that are frozen at their peak retain much of that excellence when thawed. Frozen foods require minimal processing, and once thawed, are easy to prepare.

This doesn't mean, however, that freezing is best for all foods. Some fruits and vegetables, such as lettuce, tomatoes, and green peppers, turn mushy when they're thawed. Other foods that can be frozen, such as apricots or peaches, are particularly suited to drying or canning. And since freezer space is limited, you'll want to experiment with all kinds of storage methods to see which ones best suit your needs.

The only drawback to freezing is its cost—about 15 cents per pound of food when you include the cost of a freezer and the energy to run it. That's the biggest advantage of cold storage. With a root cellar, springhouse, or simple earth mound, you can preserve a whole range of foods for months without spending a penny for energy.

And the initial investment, in both work and money, will likely be minimal.

TD

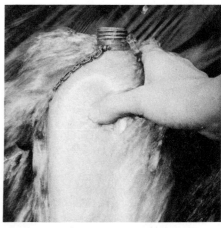

A basic kind of evaporative cooling.

Refrigerators

In the beginning was the icebox—a small insulated wooden cabinet that was kept cool by a large block of melting ice. Small boys were given the chore of emptying the drip pan into which the ice melted. When you needed more ice, you put an "Ice" sign in your front window. The iceman, making his rounds in a truck, would deliver a new block.

With the coming of the electric refrigerator in 1914, the traditional block of ice was simply replaced with the cooling coils of an evaporator unit. The unit was connected to a compressor and condenser, placed either beneath the unit or in the basement.

In the early days you could buy a new refrigerator or have your local dealer convert your trusty icebox by installing a refrigeration unit. Now you can choose among a dazzling array of refrigerator sizes, styles, and colors, and opt for such extras as ice-makers, water dispensers, egg trays, glass shelves, and butter keepers. But refrigerators still work the same way.

Sealed inside the coils of your refrigerator's cooling system is a liquid refrigerant (usually Freon 12), pumped through the system by a compressor. As the liquid refrigerant passes through the coils of the evaporator unit, it absorbs warmth from inside the refrigerator. This heats the liquid, causing it to evaporate. These vapors are then circulated back to the coils of the condenser, which are outside the refrigerator. Here, under pressure, the refrigerant is cooled by the air and converted back to a liquid. The liquid is then returned to the evaporator coils inside the refrigerator and the cycle repeats itself.

As refrigerators have grown in size, they've required more electricity to operate. Refrigerators now account for about 20 percent of all residential electricity used in the United States. Much of this energy could be saved if Americans would trade in their big refrigerators for small ones—just as they are now switching to smaller cars.

But the large refrigerator has become part of the American way of life. A family's weekly shopping now can be done in one trip to the supermarket. Americans now buy food to store it, not to eat it right away. This is in sharp contrast to the way of life in Europe, where most people shop daily for their food and refrigerators have remained quite small.

In recent years manufacturers have developed more energy-efficient refrigerators by modifying the compressors, increasing the surface area of condensers, and switching from fiberglass to polyurethane foam. If the nation's 70 million old refrigerators were replaced with new, more efficient models,

(Continued on page 228)

LOVINS ON REFRIGERATORS

Q. You often talk about the inefficiency of household appliances. Most people don't realize this down-home energy glut. Can you give us an example?

A. Let's take your refrigerator as a text. Around the end of World War II, refrigerators had efficient motors that were mounted on top. Now they have inefficient motors mounted underneath. The heat comes up where food is, and your refrigerator probably spends half its effort taking away the heat of its own motor. Then the manufacturers began skimping on the insulation so the heat comes right in through the walls. Because of that, and because it is so badly designed that when you open the door the cold air falls out, it frosts up inside, so you have probably several hundred watts of electric heaters to keep it from frosting. Then you have more heaters around the door to keep the gasket from sticking because they haven't bothered to use a Teflon coating. The radiator on the back is probably pressed right into the thin insulation to help the heat get back inside, and then the refrigerator is probably installed next to a stove or dishwasher to heat it up some more. When you get through fixing up all these points, you have a refrigerator that does the same job with about one-sixth as much electricity. You recover its small extra capital costs in only a few years.

The Federal Trade Commission is just starting to require that labels disclosing estimated annual costs be placed on all major appliances. But the consumer is still at the mercy of the manufacturer. The funny thing is if there was a really efficient refrigerator on the market that was priced right, it would probably sell like crazy. Have you heard any rumors about manufacturers cashing in on this potential goldmine? I got a letter from an engineering student in Santa Barbara who built one just for fun and it is four times as efficient as the best refrigerator on the market. . . . I don't really know for sure how to move the manufacturers. One way is to spread the rumor that the Japanese are doing it.

Amory Lovins. New Roots Magazine.

(Continued from page 227)

an estimated $1.1 billion a year in energy costs would be saved. If the entire nation switched to smaller, European-style refrigerators, the savings could be twice that much.

Many people have the habit of storing almost everything edible—cereal, fruit—in their refrigerators. Bread, of course, can be kept in a bread box, and fresh food is always better than stored food. If you keep this in mind when planning your family's meals, you'll be taking the first step toward making the best use of your refrigerator.

When buying a refrigerator, try to buy the smallest model that will meet your needs. You'll also save money in the long run if you buy one of the more energy-efficient models. You may have to pay an additional $50 to $100, but you'll make that up in savings on electric bills in 3½ to 5 years, and most refrigerators last 10 to 14 years.

The Association of Home Appliance Manufacturers (AHAM) tests refrigerators for energy usage. If your local dealer doesn't have a copy of the AHAM directory to efficiency ratings of each refrigerator model, you can order one for 50 cents directly from AHAM, 20 N. Wacker Drive, Chicago, IL 60606.

You may save electricity by buying a manual-defrost model instead of a frostless refrigerator, but not very much. Energy consumption increases significantly as soon as frost starts to build up in a manual refrigerator, and you use extra energy after defrosting when you have to recool both the food and the refrigerator.

Some refrigerators now come with an energy-saving switch. On hot, humid days you can switch from "normal power" to "extra power" to stop condensation and maintain the cool temperature inside the refrigerator. These energy-saving switches will save most if you live in a dry climate where you can run on normal power most of the time.

Helping your refrigerator keep cool

No matter what type of refrigerator you have, you can make it more efficient. It should not be placed near any heat source, such as a stove or heat register, or in direct sunlight. The

cooler your refrigerator's location, the less energy will be required to keep your food cool.

Periodically check the rubber gasket around your refrigerator door for wear. To do this, close the door on a dollar bill. You should feel definite resistance when pulling it out. If the bill pulls out easily, replace the gasket. Make this check at several points. If the door is sealed at one point, but not another, it may not be aligned quite properly.

To align the door, loosen the top and bottom hinge screws. Twist the door gently to attain proper alignment, then tighten the screws.

Your refrigerator should be at least four inches out from the wall so that air can circulate freely around the condenser, which is attached to the back of most models. In some models, the condenser is under the refrigerator and is ventilated by a fan. Make sure the fan is in working order or the condenser will be hampered.

Dust can lower your refrigerator's efficiency by collecting on the condenser coils and the compressor housing and make the unit work harder than necessary. At least twice a year you should unplug your refrigerator, pull it away from the wall, and vacuum these parts. But be careful; the coils are easily damaged and fixing them is a major repair job.

There's no perfect inside temperature for a refrigerator. Fresh meats and most fruits and vegetables keep best in a near-freezing chill. But butter won't spread if it's too cold, and some vegetables keep better at temperatures of 40 to 45°F. A good compromise is 37°F., with the freezer compartment set at 0°F.

Temperatures vary within a refrigerator. For instance, door shelves usually are 5 to 7°F. warmer than the rest of the storage compartment, and may be as much as 10 to 15° warmer—too warm for long-term storage. To test the temperature ranges in your refrigerator, put a thermometer in a glass of water and leave it overnight at the point you want to test. This way you can find out exactly what the temperature is in different parts of your refrigerator, and arrange your food accordingly.

Most refrigerators have a vegetable crisper—one or two metal drawers beneath a glass shelf. Temperatures in such drawers usually are about 5°F. warmer than the rest of the refrigerator, and are good for green beans, cucumbers, eggplants, melons, and squash.

Other fruits and vegetables such as apples, peaches, pears, lettuce, tomatoes, carrots, celery, and radishes do better at cooler temperatures. They should be stored in the main part of your refrigerator, not in the vegetable crisper.

Most refrigerator doors either have removable egg trays or indented egg shelves, but eggs will keep better if they are stored on a shelf in the coldest part of the refrigerator section. Butter and cheese, however, do well stored in their compartments; the covers keep them at 40 to 45°F., which makes for easier spreading and cutting.

But the door on your cheese compartment won't keep strong cheese odors from migrating to the rest of the refrigerator. Such cheeses should be securely wrapped or kept in airtight containers.

Objectionable odors inside your refrigerator usually can be eliminated by washing the storage cabinet with warm water and baking soda. If that doesn't work, place a container of activated charcoal or a box of baking soda in your refrigerator. You can buy activated charcoal at an aquarium supply store.

Bill Keough

Freezers

Freezing is one of the simplest methods of food storage, and it does a fine job of preserving taste and nutrition. But it's also the most expensive, energy-intensive way of storing food. So if you're going to freeze food, shop for the most efficient freezer.

A typical 16-cubic-foot freezer costs $375 to buy and about $4.25 each month to run. Figuring on a 15-year lifespan and 500 pounds of storage space, the cost to freeze 1 pound of food for a year is more than 15 cents. What's more difficult to calculate is the energy cost of producing that freezer. But if you grow your own food or purchase locally grown produce in season, you're avoiding the energy costs of intensive chemical agriculture, processing, packaging, transporting, distributing, and marketing those foods.

Freezers come in two types: an "upright" has a door on the front like a refrigerator; a "chest" is a waist-high box with a lid that hinges up. Uprights use less floor space and the food is easier to reach. But some of the space goes to waste since food has to be packed so it won't fall out the front. Because it is tall, temperature tends to stratify in an upright. A freezer thermometer will help you keep track of the warmest and coldest spots for temporary and long-term storage.

When the door of an upright is opened, cold, dry air spills out and warm, more-humid air takes its place. Humidity condenses and freezes on the cold innards of the freezer, eventually building up as frost. Since cold air stays put in a chest freezer, not as much humid air can get inside and less frost builds up. Self-

(Continued on page 230)

A classic icebox. Since cold air sinks, the ice compartment was on the top and the shelves below held the food.

(Continued from page 229)

defrosters are available with uprights, but they use 40 to 100 percent more energy than comparable manual defrost models, and they tend to dry out food that's not well wrapped.

A chest freezer uses more floor space than an upright and literally keeps you on your toes to find what you're after. It's easy to bury a small package as you pull a big one up from the bottom. Chests keep a fairly even temperature, though they're a bit colder around the sides and on the bottom. You can take advantage of that by leaving long-term storage items on the bottom and placing ice cream on top, where it's easy to reach and soft enough for scooping.

Though the experts insist that there's no significant difference in energy efficiency between upright and chest models, the U.S. Department of Energy has run some tests in homes which show chest freezers are 18 percent less expensive to run than comparable upright models. But the difference may come from how a freezer is used rather than how it's designed.

Because they don't require much floor space, an upright freezer is more likely to be located in a heated part of the house, where it has to work harder to keep cool. Since it's closer to household traffic, it may be opened more frequently too. A chest will likely be relegated to a spot in the garage or basement, where more space is available and it's less likely to be opened by hungry passers-by.

Deciding between an upright and a chest model isn't nearly as important as finding one with good insulation, an efficient compressor, an

effective door seal that works as an insulator as well, and a unit with manual defrost. To give an example of what efficient designing can do, Whirlpool just came out with a high-efficiency, no-frost freezer which they claim uses only 80 kilowatt hours per month—a 33 percent energy saving simply by improving the insulation and using a better seal. Other improvements are sure to be forthcoming, so compare energy ratings carefully before you buy.

Buying a freezer

If you're serious about energy conservation, you may have thought of cutting back on freezing this year. If you can cut back to where you need only the space your refrigerator-freezer provides, that'll be a savings. But if you try to save by buying a tiny freezer, your efforts may be all for naught.

Small freezers cost less to run and therefore use less electricity than large freezers, but per package they consume far more energy. Using the 1977 AHAM Directory, we calculated the energy cost per square foot of freezing space per year (at the present national average of 5 cents per kilowatt hour). We found that you pay about $34 to run the first five cubic feet, and about $8 for each additional five cubic feet. With a larger freezer, you are sharing the cost of those first five cubic feet with more and more food space, so the cost per package steadily decreases.

Larger units also save energy because they hold the cold better and work more efficiently. Heat can get into the freezer only through its sides, and with larger freezers the amount of space inside the freezer increases greatly with only a small addition to the outer surface area.

Chest freezers generally cost less than uprights. The larger they are (up to 25 cubic feet) the less they cost per cubic foot to buy. Best buys are mostly in the sizes between 15 and 20 cubic feet. No-frost models are much more expensive—both to buy and operate.

If you're in the market for a freezer, you might try finding someone willing to share with you. That way you can cut the buying price in half, get a more efficient model, reduce your per-package cost, and

A smaller freezer is cheaper to buy, but a larger one operates more efficiently. The 5-cubic-foot freezer on the left uses about $34 worth of electricity per year; the 15-cubic-foot model on the right uses about $50 worth.

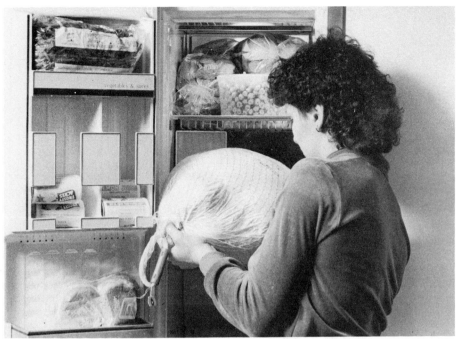

When you're buying a refrigerator-freezer, consider what kinds of foods you want to freeze. Side-by-sides usually have more freezing space, but there may not be room for large items.

avoid the energy costs of producing a second freezer.

Before you buy, send for the AHAM *Directory of Certified Refrigerators and Freezers* (Association of Home Appliance Manufacturers, 20 N. Wacker Drive, Chicago, IL 60606) and do some serious energy comparisons. When you decide which models might be best for you, check around for a reputable dealer in your area with a good service record (although freezers are a low-maintenance appliance). A few kilowatts may not seem like much, but over 15 years the savings can really add up.

Parking your freezer

Since room air is used to cool the condenser, the cooler the location, the less the freezer will have to run. However, problems can arise when the ambient temperature goes below freezing, so it's best to ask about manufacturers' recommendations for locating the particular unit you're about to buy. Ironically, freezers are made to operate at temperatures up to 110°F., but they aren't designed for low temperatures. One dealer threatened to void a warranty on his freezer if it was kept in an unheated garage. That isn't standard procedure, but it is worth asking about.

If possible, keep your freezer out of humid areas to reduce the accumulation of frost. Frost acts as an insulator, keeping heat in the freezer from being absorbed by the condenser coils. It also robs you of valuable freezer space. Don't worry about condensation on the outside of the unit; that's a natural result of warm, moist air touching a cool surface, and it doesn't affect the operation of the freezer.

Some freezers have an energy saver switch that is to be turned on in humid weather. It operates a little heater around the door seal that evaporates condensation. It's a nice convenience if you don't like droplets on your appliance, but it only saves energy when it's turned off. Many freezers have no switch, so that the heater operates all the time.

DB

Defrosting

If you don't have a frost-free freezer (and we recommend that you don't since they use more energy and tend to dry out food), you'll need to defrost occasionally. Most manufacturers suggest you shouldn't let more than ¼ inch of frost build up, but

(Continued on page 234)

WHAT TO DO IF YOUR FREEZER STOPS

Freezers stop for two reasons: mechanical failure and power outage. If yours quits, keep the freezer closed and try to estimate how long it will be off. If it will only be a few hours until the electricity comes on again or a repairman can visit, you shouldn't have any problem. With the door closed, food will usually stay frozen in a fully loaded freezer for two days, or for one day in a unit with less than half a load.

If you can't get the freezer working again in that time, you have four options: 1) take the food to a freezer-locker plant; 2) cool the freezer with dry ice; 3) can or dry the food; or 4) invite the neighbors over for a gigantic feast.

The simplest option is the first one. If you can get locker space, wrap the food in newspaper or blankets, and rush it to the plant.

If you favor the second option, use about 25 pounds per 10 cubic feet of freezer space. Place heavy cardboard on top of the food containers, then lay the dry ice on top of that. (Handle it with gloves; it can burn you.) If you can pack in the ice soon after a failure, it will keep a half-full freezer below 32° F for two to three days, and a full freezer for three to four days. To increase efficiency, you can cover the freezer with blankets. Just make sure you don't cover the motor vent openings if there is a chance that the electricity may come on suddenly. The motor could overheat and burn out.

The third option—canning or drying—is workable, but it takes a lot of work and inevitably causes a further decline in food quality. The last option is the most fun, but then the food is not there to help you through a tough winter.

The best protection for freezer stoppage is to be prepared. Know the location of a freezer-locker plant or a local source of dry ice. Choose a well-insulated freezer, so it will hold the cold for a maximum period. And if you think a power failure is imminent, set the freezer lower than usual so it can build up some additional cold.

VEGETABLES FOR THE FREEZER

Artichokes. Pick small plants. Remove outer leaves, cut off bottom stem, and trim bud. Steam blanch 8 to 10 minutes.

Asparagus. Select young, firm stalks. Remove scales, and sort for size to speed packing. Steam blanch 2 to 4 minutes.

Beans, lima. Pick bright-green, slightly rounded pods. Shell. Steam blanch 2 to 4 minutes.

Beans, snap. Pick before seeds mature. Snip ends, sort for size, cut if desired. Steam blanch 3 to 4 minutes.

Beets. Harvest young, mild-flavored plants. Trim tops and cook until tender. Skin and cut if desired.

Broccoli. Choose compact, dark-green heads. Peel, trim, and cut into fairly small pieces. Soak in salt water ½ hour to drive out worms; rinse. Steam blanch 5 minutes.

Brussels sprouts. Use green buds only. Remove outer leaves and trim. Steam blanch 3 to 5 minutes.

Cabbage. Trim tough outer leaves. Cut in wedges or shred. Steam blanch wedges 4 minutes and shredded cabbage 3 minutes.

Carrots. Harvest young, tender roots. Trim and peel. Use small carrots whole, cut others into cubes or slices. Steam blanch 4 to 5 minutes.

Cauliflower. Pick solid, well-formed heads. Break into flowerets, peel and split stems. Soak in salt water ½ hour to drive out worms; rinse. Steam blanch 4 minutes.

Celery. Select crisp stalks. Trim, cut across the rib into 1-inch pieces. Steam blanch 4 minutes. Freeze leaves separately and use in soups and stews.

Corn, on cob. Harvest when fully ripe. (The kernels should be milky when tested with a fingernail.) Husk and remove silk. Steam blanch 6 to 8 minutes.

Corn, kernel. Proceed as above. After corn is blanched, remove kernels from cob with a sharp knife or corn cutter.

Eggplant. Pick firm, medium-size plants with tender seeds. Peel, cut into slices or cubes. To preserve natural color, soak for 5 minutes in a solution of 1 tablespoon lemon juice to 1 quart water. Steam blanch 4 minutes, then resoak in lemon juice solution before freezing.

Greens. See *Spinach*.

Herbs. Harvest just before plants blossom. Trim and rinse in cold water. Steam blanch 1 minute.

Kohlrabi. Choose tender, mild-flavored plants. Trim, peel if desired, and cut into slices or dice. Steam blanch 1 to 2 minutes.

Mushrooms. Pick firm, small to medium-size plants. Trim stalks, cut large mushrooms into pieces. Soak for 5 minutes in a solution of 1 teaspoon lemon juice and 1 pint of water to prevent discoloration. Steam blanch 3 to 5 minutes.

Okra. Pick young, tender pods. Trim stems, but do not cut into pods. Steam blanch 5 minutes.

Onions. Pick fully mature, crisp plants. Trim and chop. No blanching needed.

Parsnips. Dig smooth, firm roots before they become woody. Trim tops, peel, and cut into slices or dice. Steam blanch 3 minutes.

Peas. Harvest when bright-green and still slightly immature. Shell. Steam blanch 1½ minutes.

Peppers, green. Pick fully ripe fruits with thick, glossy skin. Halve, remove seeds and pulp, slice or dice. Blanching is not necessary; if desired, steam blanch 2 minutes.

Pumpkin, winter squash. Harvest fully mature, hard-shelled fruits. Cut in half and scoop out seeds. Bake in 350 to 400° F. oven until tender. Scoop out pulp and mash.

VEGETABLES FOR THE FREEZER
(CONTINUED)

Rutabagas, turnips. Pick young, tender roots. Trim tops, peel, cut into strips or small cubes. Steam blanch 2½ minutes.

Spinach, other greens. Harvest small, tender leaves. Remove thick stems and large midribs. Steam blanch 3 minutes. Stir while blanching to prevent leaves from matting together.

Squash, summer. Pick small fruits with tender rinds. Trim ends, peel if desired, cut into slices. Steam blanch 1 minute. For winter squash, see pumpkin.

FRUITS FOR THE FREEZER

Apples. Pick firm, ripe, bruise-free fruit. Peel, core, and slice. Pack dry or in honey syrup with 2 tablespoons lemon juice per quart to prevent browning. Applesauce may be frozen with no additional processing.

Apricots. Choose ripe, yellow, firm fruits. Halve, remove pits, and peel. (Skins can be loosened by dipping fruit in boiling water for 15 seconds or so.) Pack in light syrup.

Blackberries. Pick plump, firm berries. Remove leaves and debris. Pack dry or drizzle small amounts of honey over berries.

Blueberries. Select ripe, firm berries. Remove stems and leaves. Steam blanch 1 minute to prevent skins from toughening. Pack dry or in light syrup.

Cantaloupe. Select firm, vine-ripened fruit. Halve, remove seeds, cut flesh into cubes or balls. Pack in light syrup. Best if served before fully thawed.

Cherries. Pick firm, tree-ripened fruits. Stem and pit. Pack in syrup.

Cranberries. Choose deep red, glossy-skinned berries. Stem. Pack dry, or make into sauce before freezing.

Currants. Pick fully ripe, bright-red fruit. Stem. Pack dry or add honey to taste.

Figs. Select tender, soft, tree-ripened fruits. Cut off stems, peel, leave whole or slice. Pack in thin syrup.

Gooseberries. Pick fully mature berries. Remove stems and blossom ends. Pack dry or in syrup.

Grapefruits. Choose mature, tree-ripened fruit. Peel, section, remove all white membrane. Pack dry or add honey to taste.

Grapes. Pick firm grapes with full color. Stem. Leave seedless grapes whole; halve seeded grapes and remove seeds. Pack dry or in thin syrup.

Oranges. Select firm, heavy fruit. Peel, remove sections from heavy membrane. Pack dry or add honey to taste.

Peaches. Use tree-ripened, fully mature fruit. Peel without blanching, halve or slice, remove pits and stems. Pack in syrup with lemon juice or ascorbic acid added.

Persimmons. Pick soft, red, fully ripe fruit. Slice or puree. Add honey to taste, and 2 tablespoons lemon juice per pint of fruit to prevent discoloration.

Pineapple. Freeze only fruit that ripened on plant. Peel, trim, core, and slice or dice fruit. Pack in own juice or thin honey syrup.

Plums, prunes. Select tree-ripened, firm, sweet fruits. Halve and pit. Pack in syrup.

Raspberries. Pick sweet, plump berries. Remove stems. Pack dry or sprinkle honey over to taste.

Rhubarb. Pick crisp, tender, red stalks; early spring varieties freeze best. Remove leaves, trim tough ends, cut stalks into 1-inch lengths. Pack dry or in syrup.

Strawberries. Choose vine-ripened, bright-red fruit. Hull, cut in half or leave whole. Pack dry, sweeten to taste with honey, or use a thin honey syrup.

HOME FREEZING OF FRUITS AND VEGETABLES (USDA Home and Garden Bulletin No. 10) 1976.
Superintendent of Documents, U.S. Government Printing Office, Washington, DC 20402. 48 p. booklet $.50.

This small book presents specific freezing information, handy tips, and thawing advice in a succinct, almost rapid manner. The graphics are unimaginative (aside from the cover), and some important points, like the warning against using galvanized ware with fruit, are buried and easy to miss. Terms are occasionally unspecific, as when it refers to "wide" or "narrow" or "tall" containers, and there's a heavy emphasis on sugar. But it's illustrated, and indexed, and it covers the whole subject without undo chatter.

Thawing fruit

For best color and flavor, leave fruit in the sealed container to thaw. Serve as soon as thawed; a few ice crystals in the fruit improve the texture for eating raw.

Allow 6 to 8 hours on a refrigerator shelf for thawing a one-pound package of fruit packed in syrup. Allow 2 to 4 hours for thawing a package of the same size at room temperature—½ to 1 hour for thawing in a pan of cool water.

Fruit packed with honey thaws slightly faster than that packed in syrup. Both honey and syrup packs thaw faster than unsweetened packs.

Thaw only as much as you need at one time. If you have leftover thawed fruit it will keep better if you cook it. Cooked fruit keeps in the refrigerator a few days.

(Continued from page 231)

research has shown little loss of efficiency at that point. When you reach a ⅜- to ½-inch build up, it's definitely time to defrost.

Many people like to defrost in the middle of summer, when they've finished most of last year's food and haven't started freezing this year's crop. Those in cold climates prefer the dead of winter because they can simply take the food outside in the interim. But frost doesn't always cooperate, and you may need to do your defrosting under less favorable circumstances.

The day before defrosting, turn your freezer control down so the food will be colder and keep longer outside the unit. Unless it's going to some place just as cold, wrap the displaced food in newspapers or blankets—anything to insulate it from the heat.

To speed defrosting, you can use warm water or blunt instruments to help remove the ice. But be careful; it is easy to pierce the side of a freezer and puncture a Freon tube. When the frost is gone, turn the freezer on and reload the food. Since it should still be frozen, it will help the unit cool more quickly.

Freezing food

Fruits and vegetables for freezing should be harvested in the morning, when they're freshest. Slightly immature, tender produce is better than fully mature food; overripe or damaged produce should not be frozen. Freezing in small quantities not only speeds things up, but also means you can pick food at its peak. If you need to harvest large amounts, refrigerate the produce until you are ready to go to work.

All produce should be washed carefully in cold water and trimmed. Vegetables need to be blanched to inactivate the enzymes which can change flavor, color, and texture. In addition, blanched frozen vegetables retain more of their vitamin C.

You can blanch vegetables by immersing them briefly in boiling water or by steaming. Steaming is preferable because it retains more nutrients. Just place the vegetables in a sieve, colander, or wire-mesh holder, and set them over boiling water in a closed pot for the recommended time. (See chart, above.) After blanching, cool the vegetables quickly by plunging them into cold water, then drain and pack.

Fruits don't need to be blanched, but most people add some sweetener before freezing them. This isn't necessary for safe storage, but with many fruits it does help preserve the flavor and color. If you prefer adding some sweetener, we recommend honey rather than white sugar, since it will add some food value to your frozen fruit. Pick a light, mild-flavored variety that won't affect the fruit's taste.

There are several ways to pack fruit. You can simply dry pack it—with no additions. Raspberries, blueberries, and cranberries are especially suited to this. For the traditional "sugar pack," just drizzle honey over the fruit. This method is recommended for juicy fruits or those that don't darken. Whole fruits and those that tend to turn dark are usually packed in syrup. When you use honey instead of sugar for packing

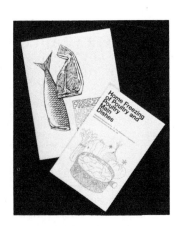

HOME FREEZING OF POULTRY AND POULTRY MAIN DISHES (USDA Information Bulletin No. 371) 1975. 29 p. booklet $.50. FREEZING MEAT AND FISH IN THE HOME (USDA Home and Garden Bulletin No. 93) 1973. 23 p. booklet $.45. *Superintendent of Documents, U.S. Government Printing Office, Washington, DC 20402.*

Much of Meat and Fish is photos and diagrams for carving beef, pork, and lamb carcasses and for cleaning or pan dressing fish. Yield charts, wrapping illustrations, and a storage chart round out the booklet. But it leaves one with the hunch that carving up flesh will take some practice if it's to be as neat as the hands in the photo make it appear.

The instructions for poultry are more comprehensive, with charts for storage times and thawing, and 33 recipes for poultry entrees. The recipe selection is composed mainly of American standards such as chicken croquettes, turkey pie, chicken creole, and cordon bleu. These far-sighted recipes are for 24 servings each, broken into four frozen batches of 6 servings apiece. Very handy!

While both booklets talk about carving and disjointing, they fail to mention one factor: the essential importance of a sharp knife. As Juila Child says: "Without a sharp knife, you can't do nothin'."

Turkey Chop Suey

1½	quarts turkey broth
2	cups bean sprout liquid
1	cup cornstarch
1	cup soy sauce
1½	quarts celery, cut in 1-inch strips
1	quart onions, sliced
4	pounds bean sprouts
1	pound water chestnuts, drained sliced
2	quarts (about 2½ pounds) turkey, cooked, diced
⅛	teaspoon pepper

Bring broth and bean sprout liquid to a boil.

Mix cornstarch with soy sauce and stir into boiling liquid.

Cook until thickened, about 5 minutes.

Add vegetables, turkey, and seasonings to sauce.

Divide mixture into four parts.

To serve without freezing, simmer for about 20 minutes or until celery is of desired doneness.

Serve over rice.

To freeze, pour parts to be frozen into freezer containers. Each part will require a container or combination of containers with a 1½-quart capacity.

Cool for 30 minutes at room temperature. Seal, label, and freeze immediately.

To heat frozen food, remove food from freezer containers. Place in heavy pan and reheat over low heat, stirring occasionally until hot, about 1 hour. Serve over rice.

Calories per serving: about 155 without rice.

Yield: 24 servings, about 1 cup each

fruit, cut the amount by half. For syrup pack, add three cups of hot water to one cup of honey for a nice light syrup; an equal mix of two cups each makes a medium syrup. Be sure the syrup is cool before you add it to the fruit.

To prevent peaches, pears, apricots, and sweet cherries from turning brown during freezing, you can add ascorbic acid (vitamin C) or lemon juice. Add 1,500 milligrams of ascorbic acid or one to two tablespoons of lemon juice per cup of syrup (which should be enough for one quart of fruit).

Food can be frozen in almost any container that can be sealed tightly.

Containers

You can freeze in any container that will keep air out and moisture in. Best are glass or plastic containers, with tight-fitting tops, and sealed plastic bags, since they can be made airtight. Because foods wrapped in butcher paper or waxed paper are exposed to some incoming air, they lose moisture and will not keep well for long periods. Items like meat that have an irregular shape should first be wrapped carefully in aluminum foil or polyethylene plastic wrap, then put in a plastic bag with the air pressed out and sealed. To make the best use of space, freeze only in rectangular containers, not round ones.

Since water expands as it freezes, you should always leave some headspace in containers. Unless you're using narrow-neck bottles, an inch of space per quart should be sufficient. If you freeze in small, thin packages, your food will freeze and thaw more quickly.

A full freezer operates more efficiently than one that's half full, and it will stay cold longer if the electricity goes off. Some people fill plastic milk jugs or similar containers with water and stick them in empty spots to help hold the cold. But since this freezing requires extra energy, you'll save most if you freeze the water outdoors in winter.

To avoid overworking your freezer, don't add more than 10 percent of its volume at one time, and then give it a chance to catch up. You may want to turn it colder for a day or so until the temperature stabilizes at 0°F. Food will obviously freeze at higher temperatures, but above zero its storage life will be greatly reduced.

Before you pack your food in the freezer, it should be labeled with the name of the product, freezing date, and other relevant information such as the number of servings. Freezer tape or masking tape makes nice labels, or you can write on the containers with a special pen.

Once frozen foods are thawed, they deteriorate more rapidly than fresh foods. If food is to be cooked, don't wait for it to thaw—just heat it up. When food needs to be thawed, if possible do it in the refrigerator rather than in hot water or at room temperature, since this minimizes deterioration. Many frozen fruits taste best only partially thawed, when they're still crisp.

Foods that are only partially thawed or that have been thawed but are in good condition may be safely refrozen. The quality will suffer somewhat, and they shouldn't be stored for long periods, but they'll be safe to eat. Most commercial food packagers recommend against refreezing not because it's dangerous, but because the food won't taste as good.

Freezing temperatures kill many microorganisms and inactivate most others. But since these pests are active in temperature ranges from 40 to 120°F., problems can develop during freezing and thawing. Keep foods in

(Continued on page 236)

FREEZING COMBINATION MAIN DISHES (USDA Home and Garden Bulletin No. 40) 1976.
Superintendent of Documents, U.S. Government Printing Office, Washington, DC 20402. 22 p. booklet $.50.

The title of this booklet is misleading. It's about the thrifty practice of cooking ahead and freezing. (If more people did this, TV dinners wouldn't be big business.) It provides inspiration and information for getting started in this time-saving approach to home cooking. The booklet presents a broad sample of entrees which freeze well.

The single most valuable idea in the booklet is this tip for anyone who has tied up all the baking dishes in the freezer. Line the dish with freezer wrap, pour in the prepared food, wrap and seal it, freeze it, then remove it from the casserole. Obviously it will stack up like a block once it's frozen, and the dish is freed. When you need the food, unwrap it, put it back into the dish, and bake.

FROZEN WALRUS

We are indebted to an Arctic explorer for the following Eskimo recipe for a frozen dinner: "Kill and eviscerate a medium-sized walrus. Net several flocks of small migrating birds and remove only one small wing feather from each wing. Store birds whole in interior of walrus. Sew up walrus and freeze. Then two years or so later, find the cache if you can, notify clan of a feast, partially thaw walrus. Slice and serve." Simplicity itself.

Reprinted from *Joy of Cooking*, Irma S. Rombauer and Marion Rombauer Becker, with permission of the Bobbs-Merrill Co., Inc. See review, page 465.

(Continued from page 235)

this temperature range for as short a time as possible.

Unless food is carefully wrapped, it will dry out in a freezer. Even a tiny hole can lead to moisture loss. This is known as freezer burn, and any food afflicted with it must be thrown out, since it cannot be rehydrated.

Most meat and poultry can be frozen for short periods in the containers they are purchased in. If you freeze home-grown meat, it should be aged for a week or so at temperatures just above freezing. It's best to pack meat in the form you plan to use it, with waxed paper between slices, patties, or pieces. Fish can be neatly frozen in water in milk cartons. Meat may be cooked frozen or thawed first; if thawed first, it will cook and brown more evenly. Poultry and fish should be thawed before being cooked.

Butter and hard or semihard cheeses, if tightly wrapped, can be safely frozen for periods of six months or more. Milk can be frozen for several months, cream for two to three months. However, the butterfat in cream will separate out, so previously frozen cream won't whip well. It may be used in cooking or for ice cream. If eggs are frozen in their shell, they'll crack and break. But you can crack them and freeze whites and yolks either separately or mixed together. These frozen dairy products should be thawed in the refrigerator in their unopened containers. Egg whites will whip to a larger volume if they are brought to room temperature before whipping.

While most foods freeze well, you may have a problem with some casseroles. At freezing temperatures, many herbs and spices lose their flavor, pepper turns bitter, and starchy food such as potatoes and pasta tend to get soggy. So it's best to freeze casseroles unspiced, then add the seasonings and starches later.

TD

Non-electric cooling

Using outside air. The simplest way to keep food cool is to use cold outside air. With a little ingenuity you can make your garden harvest last several months longer just by placing it in bags, boxes, or an extra garbage can, stashing it in a cold spot. Keep in mind that outside air cooling is only as reliable as the weather.

Almost every house has at least one suitable space for cold storage: a tool shed, unheated hallway or spare room, cellar, enclosed porch, or outside window well. If you live in an apartment, use the balcony or fire escape. Under the front porch is ideal. The attic is great for pumpkins and squash, which like it warmer. An unheated garage will also work, if you make sure the food is protected from toxic car fumes by careful wrapping.

The goal of cold storage is to keep food cool but not frozen, so if the temperature of your storage area is too low, you'll need to protect the food. A Styrofoam picnic cooler is a fairly cheap way to keep small quantities of food; for larger amounts you can insulate a box or barrel with fiberglass batts.

If you have a yard, and your average winter temperatures don't go below 30°F., try earth mounds. (These are also called clamps.) Place food—with fruits and vegetables in separate mounds—on the ground, and cover it with straw and earth for insulation. (See illustration.) Mark where you put your mounds; snow can be deceiving!

If you already have a garden, you can simply leave root crops right in the ground. Under a layer of mulch, carrots, parsnips, radishes, and kale can keep right through frost and snow. Trenches will protect cabbage and celery several months, and covering even delicate crops like tomatoes will keep them around a few weeks longer.

Evaporative cooling. When you feel a a chill after a swim or a shower, you're experiencing evaporative cool-

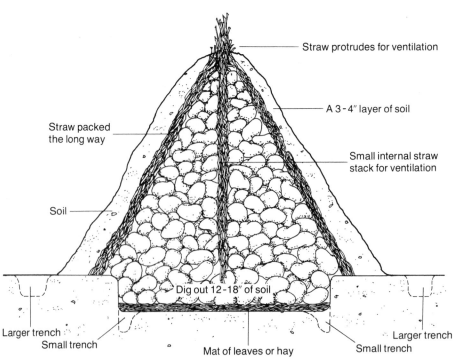

Straw protrudes for ventilation

A 3 - 4″ layer of soil

Small internal straw stack for ventilation

Straw packed the long way

Soil

Dig out 12-18″ of soil

Larger trench
Small trench

Mat of leaves or hay

Larger trench
Small trench

An earth mound, or clamp is a good place to store root vegetables. Dig a small hole, pile in the vegetables, then cover them with a layer of straw and a layer of packed earth, leaving space for ventilation at the top. A small trench around the vegetables and a larger trench around the entire mound will insure adequate drainage.

ing. In a dry climate, you can put this process to work for you.

You can make a simple cooler from an orange crate or box frame. First, make shelves of chicken wire or open wooden slats to provide air circulation around the food. Then cover the box completely with burlap, allowing a flap to put the food inside and space on top to slide a pan of water between the box and the burlap. (See illustration.)

When the burlap is placed in water, capillary action soaks up the water and spreads it over the whole surface of the burlap. The water then slowly evaporates, cooling the box. Placing the box in a breeze (or using a fan) will speed evaporation, and thus keep the box cooler.

You can protect the food from animals and insects by covering the box with screen or cloth.

Ice. Another economical way to keep food cool is to do what our grandparents did—use an icebox. If you buy an antique one, you'll probably pay a good price. But if you make one, and harvest your own ice, the cost will be minimal.

The only energy an icebox uses is yours, but it can use quite a bit. First you've got to make the box—an insulated wooden chest with a space for ice and a drain hole for melting water. Then you've got to get the ice. (If you live in a warm climate, go on to the next section.)

To harvest ice you need a week of 0°F. weather, a few tools, and a simple shed for storage. A hard day's work at a pond can supply all the ice you'll need for a year—80 to 100 blocks. If you have a well or stream on your land, you don't even need to trek to the pond. You can divert the water right into a shed with a block form, freeze a block at a time, and separate each block with newspaper.

To harvest pond ice, wait until the ice forms to the depth of the blocks you need for your icebox, allowing for some melting during storage. (While you're waiting, keep snow off the ice because it slows up the freezing process.) Take a chisel and mark off the squares. With a large hand saw or chain saw, cut most of the way through the ice, then tap it with a chisel. It should break off with clean, even edges.

Using sawdust for insulation, a three-sided roofless storage shed will keep ice through the summer with minimal melting. Pile old sawdust two feet thick on the bottom, sides, and top—the older the sawdust, the better (the moisture in fresh sawdust makes it a poor insulator). You can also build a double-walled shed and line it with fiberglass or Styrofoam. Bury it in a hillside to make use of the earth's moderate temperature.

Spring water. If you have a spring-fed stream on your land, you may be letting a cooling freebie float right by. Just as the best grocers keep their greens sprinkled with a fine mist of cold water, a spring can revive your endive and crisp your cukes. If it's handy to the house, you can cool your drinks in it and store butter and milk.

A spring never freezes up, but the water is always icy cold and should keep vegetables and fruits crisp and fresh as long as the refrigerator can. And since it's both more humid and more airy than the fridge, it won't dry out foods or mix odors.

There are several ways to cool with spring water. You can build a bottomless box with a hinged top and open shelves, and set it right over the spring water. Or you can build a box with holes in the lower part of the sides and let the water flow through. Food in containers can sit right in the water, while things like meat and butter stay on shelves above it. Use non-rusting nails for whatever structure you build.

For more permanent storage, there is the traditional springhouse, with an earthen floor to retain moisture and troughs of varying depths for setting milk cans, butter, and other foods right in the water. Old

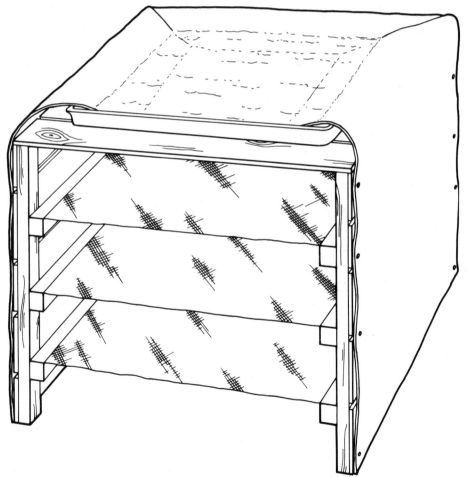

This evaporative cooler is made from orange crates and burlap, with shelves made from screen with wood frames. The burlap covering gradually soaks up the water, and as the water evaporates, it cools the food inside the crates.

(Continued on page 238)

(Continued from page 237)
springhouses are almost always of stone, but concrete block or well-insulated brick walls will retain the cool in the hot months and allow storage through most of the winter without freezing.

Underground: the cold cellar, and holes in the ground. A root cellar is by far the best method for long-term cold storage of vegetables and fruits. And if you think it means eating potatoes, beets, and turnips till you're sick of them, you may be missing the real joys of the root cellar. You can have cantaloupe for Halloween, tomatoes in a Christmas Day salad, and carrots right into May.

Underground storage provides the best protection from fluctuating temperatures because of the earth's moderating effect on climate extremes. It's a practical method where average winter temperatures are 30° F., or so, a little less reliable in

The best cold-storage foods

- *Apples*
- *Beets*
- *Cabbage*
- *Carrots*
- *Garlic*
- *Horseradish*
- *Jerusalem artichokes*
- *Onions*
- *Parsnips*
- *Pears*
- *Potatoes*
- *Rutabagas*
- *Squash*
- *Turnips*

A buried barrel is a good storage place for hardy root vegetables.

warmer climes. Where the ground freezes hard, it's the only reliable method for maintaining a steady, above-freezing temperature.

Compared to other cooling methods, the underground cellar initially requires the most time, labor, and money. But once you've invested, you can put up each year's harvest for no cost and little effort.

The typical root cellar traditionally is either in the basement or near the house, underground, unheated, with an earthen floor to keep the room humid. If you have an older home, you may be able to use the basement as a root cellar with virtually no changes.

But even a heated, finished basement can be partitioned off to provide cold storage space. An area eight by ten feet can store 60 bushels of produce. The space must be insulated from the basement's heat and needs a window or two for ventilation. Humidity is kept high by a damp gravel or dirt floor.

Store the produce in bins, crates, cardboard boxes, baskets, even buckets. To insulate and ventilate the produce, you can pack it in leaves, hay, straw, mulch, sand, burlap, or sawdust. You should also have a thermometer—because it's best to keep a fairly constant temperature of 32 to 40°F.

The cold water running through a springhouse keeps the air cool. Some foods, such as milk, can be stored in containers placed directly in the water.

Even better than your basement is an outside root cellar. It's less affected by the outside temperatures, and isn't bothered by any heat from the rest of the house. It does require more work, because of the excavation involved. But there are shortcuts like digging into a bank, or combining the job with a new porch or shed.

Cheaper, but somewhat less effective, are underground pits, barrels, trash cans, drain tiles, or discarded refrigerators and lockers. Even a simple wooden box can be built and buried. These are an improvement over above-ground storage because they use the constant, moderating temperature of the earth below the soil surface. But unlike a root cellar, these areas cannot be controlled for ventilation, humidity, and temperature.

Pat Donahue

Produce in a root cellar can be kept in a variety of containers, but they should all be raised off the floor to allow air circulation.

Storing foods

Timing is important. Soft foods like tomatoes and peppers, which will keep for a short time in cool storage, should be gathered before frost can nip them. Hard-shelled foods like squash and pumpkin should be protected from heavy frost. Let them mature on the vine as long as possible, or until their rinds are too hard to dent with your fingernail. Immature vegetables don't keep well. Potatoes are ready to dig when their tops die down. Sweet potatoes should be dug when frost kills the vines. Leafy vegetables can stay out through moderate frosts, but must be picked before a killing frost. Unprotected root vegetables may be left in the ground until killing frost, but should be dug before the ground freezes hard. Kale, parsnips, and salsify remain in good condition all winter right in the garden row if protected by a blanket of mulch. In fact, the flavor of each of these three vegetables is improved by frost. Onions and garlic should be pulled when their tops have bent over at the neck.

Dig the vegetables carefully, from the side of the row, so the shovel won't nick so many of them. Cleaning isn't necessary for vegetables you plan to store. Just brush off excess dirt. Select only your best, well-developed produce to store. Any vegetables that have been damaged

in digging or partly eaten by insects should be used promptly. They won't keep long.

Clip off all the leafy tops immediately after digging. Cut off carrot and turnip tops right at the root shoulder. Leave a one-inch stub of leaf stem on beets so they don't bleed. Snip off onion and garlic leaves to a length of one inch before spreading them in the sun to cure. Root tips should not be broken off because skin breaks invite spoilage. Leave stems on squash and pumpkins. If any do break off accidentally, use those vegetables first. After cleaning and clipping, cool the vegetables before putting them away.

Curing is important for some storage vegetables. White potatoes keep best if allowed to develop a tough skin at 60 to 75°F. for 7 to 10 days. Sweet potatoes also need to be cured by spending 10 to 14 days in a warm (80 to 85°F.), rather humid place before they are put away. Squash and pumpkins keep better if exposed to sun and air for 10 to 14 days after picking to dry and harden their skins. Onions and garlic should cure in the sun for 5 to 7 days. Cabbage, endive, celery, carrots, parsnips, leeks, turnips, rutabagas,

Building a root cellar against a hillside will mean less digging for more space.

beets, and most other root vegetables do not require a curing period before storage.

Vegetables and fruits can be divided into several groups according to the storage conditions in which they last the longest. In practice, many people simply divide their vegetables into two storage groups: those that keep well in a dry place, and those that store best in moist air. The requirements given here are ideals; most of these foods will keep at slightly higher temperatures or different humidities, but for a shorter time. Foods not listed keep only for very short periods.

VEGETABLES FOR STORAGE

Beets. Pick mature vegetables before first killing frost. Pack in layers of damp sawdust or sand. Requirements: 32 to 40°F., 90 to 95 percent humidity. Life: 4 to 6 months.

Brussels sprouts. Harvest after a few mild frosts. Store in perforated plastic bags. Requirements: 32 to 40° F., 90 to 95 percent humidity. Life: 3 to 5 weeks.

Cabbage. Pick firm, solid heads, and trim off loose outer leaves. Place heads on shelves, string them up by the roots, or wrap in newspaper and store in boxes or bins. (Beware: cabbage emits a strong odor during storage.) Requirements: 32 to 40°F., 80 to 90 percent humidity. Life: 2 to 4 months.

Cabbage, Chinese. Store only mature, solid heads, picked before a severe frost. Store as with regular cabbage, or pack roots in damp soil or sand. Requirements: 32 to 40°F., 90 to 95 percent humidity. Life: 2 to 4 months.

Carrots. Dig before any hard freezes. Store in cartons of sawdust, sand, or leaves. Requirements: 32 to 40°F., 90 to 95 percent humidity. Life: 6 to 8 months.

Celery. Store in trenches in the garden, or place roots in boxes of sand or soil. Requirements: 32 to 40°F., 90 to 95 percent humidity. Life: 1 to 2 months.

Endive, escarole. Pick after moderate fall frosts. Store with roots in soil or sand. Requirements: 32 to 40°F., 80 to 90 percent humidity. Life: 2 to 3 months.

Garlic. After curing, store in paper bags, or braid into strings and hang from rafters. Requirements: 35 to 40°F., 60 to 70 percent humidity. Life: 6 to 8 months.

Horseradish. Dig anytime in fall. Store in damp sand or sawdust. Requirements: 32 to 40°F., 90 to 95 percent humidity. Life: 4 to 6 months.

Jerusalem artichokes. Mulch in garden or store in plastic bags or damp sand. Requirements: 32 to 40°F., 90 to 95 percent humidity. Life: 1 to 2 months.

Kohlrabi. Remove leaves and roots. Store in damp sand or sawdust. Requirements: 32 to 40°F., 90 to 95 percent humidity. Life: 2 to 3 months.

Onions. After curing, remove tops and store in bins or string bags, or braid and hang from rafters. Requirements: 35 to 40°F., 60 to 70 percent humidity. Life: 4 to 6 months.

Parsnips. Mulch in ground or store in damp sawdust, sand, or leaves. Requirements: 32 to 40°F., 90 to 95 percent humidity. Life: 4 to 6 months.

Potatoes, sweet. After curing, wrap individually in newspaper and pack in baskets, or pack loosely (make sure potatoes aren't touching) in cartons of sawdust. Requirements: 50 to 60°F., 60 to 70 percent humidity. Life: 3 to 5 months.

Potatoes, white. Late potatoes store best. After curing, pack in baskets or boxes. Don't store near apples, which give off a gas that promotes sprouting. Requirements: 32 to 40°F., 80 to 90 percent humidity. Life: 4 to 6 months.

Pumpkins. After curing, place on shelves or in boxes. Requirements: 50 to 60°F., 60 to 70 percent humidity. Life: 4 to 6 months.

Rutabagas. Dig after a few mild frosts. Store in damp sand, sawdust, or moss. Requirements: 32 to 40°F., 90 to 95 percent humidity. Life: 2 to 4 months.

Salsify. Mulch in garden, or store like parsnips. Requirements: 32 to 40°F., 90 to 95 percent humidity. Life: 1 to 2 months.

Squash, winter. After curing, pack on shelves or in boxes. Requirements: 50 to 60°F., 60 to 70 percent humidity. Life: 4 to 6 months.

Tomatoes, green. Pick mature green fruits before frost. They will gradually ripen at 55 to 70°F. Requirements: 50 to 60°F., 60 to 70 percent humidity. Life: 4 to 6 weeks.

Turnips. Harvest before a heavy freeze. Store like carrots. Requirements: 32 to 40°F., 90 to 95 percent humidity. Life: 2 to 4 months.

(Continued on page 242)

Containers

A wide variety of containers can be used for cold storage. The most common are wooden fruit boxes or barrels. These are good for foods that need to be packed in sand, sawdust, or leaves. Simply layer the packing materials and produce alternately, finishing with two or more inches of packing at the top. When stacking boxes, place furring strips between them, the floor, and other tiers to permit good air circulation.

Metal tins can be used open-topped in place of barrels or boxes. Make sure they're galvanized or you may have trouble with rust.

Orange crates or mesh bags are excellent for storing onions because they allow maximum air circulation. For storing large quantities of potatoes or other root crops, you can build wooden bins. These should stand a few inches off the floor to allow for air circulation underneath.

Conditions

The most important thing you can do to prolong the storage life of food is keep it cold. Temperatures just a few degrees above those recommended can drastically reduce the keeping time of produce. In most areas, you'll need to adjust the ventilators almost daily to maintain the desired temperature.

Since vegetables and fruits are mostly water, if they're not kept moist during storage they'll shrivel up. There are three ways to keep food humid: wrap or cover it; keep a moist environment in the storage area; or put water directly on it.

Wrapping food in plastic or burying it in soil, sand, sawdust, leaves, or a similar material will minimize moisture loss. The advantage to this method is that the atmosphere of the storage room can then be less humid, and a greater variety of food can be safely stored.

You can add moisture to the air in your storage room by placing large

(Continued on page 242)

A low-cost root cellar: Salt Lake City, Utah

A few years ago Ingo Haidenthaller built a 5 by 7-foot root cellar. It took him about 20 hours and less than $30. Here's his basic plan.

First, dig a hole (Ingo's measured 5 by 7 feet and was 4 feet deep). Short-handle spades are best for this job. You'll also need to dig a small area to serve as an entrance later on.

It is wise to build a ladder first, before digging is completed, to make getting in and out of the hole easier. Ingo used 2 x 2s for the sides and 1 x 2s for the rungs. Nailing the ladder together is not recommended, since the rungs may split. Two-inch wood screws, with pilot holes drilled, will work better.

To construct the framework, use six 2 x 4 studs, cut to dimensions as shown and pointed at one end. To preserve the wood, soak the lower ends in old motor oil for about 30 minutes. The studs must be at least six inches in the ground to provide sturdy support.

After the studs are set and leveled, the braces and rafters should be nailed in place, as shown. Ingo made the roof at an angle to let rainwater run off.

The frame must be sturdy and self-supporting. Shelves along the walls will add to the strength of the frame. If lining boards are desired, they can be added at this point. A layer of tar paper on the outside will keep moisture out.

The entrance frame and the doors are next. Ingo hung a regular door that opens to the inside, then added a trap door to cover the entrance hole, which prevents snow from falling in and provides safety.

For proper air circulation, cut a 3 by 12-inch opening in the lower part of the door and nail a grill over it. A vent pipe in the roof can be added later.

The best roofing, according to Ingo, is ¾-inch plywood. It is strong enough to support the dirt cover. A low-grade plywood will do the job. Do not save money by using particle board. The moisture will cause it to sag, and the roof may cave in.

Painting the outside of the roof will seal the wood sufficiently.

Insulating the top is the final step. If it is properly done, the inside temperature will be pretty constant throughout the year. Ingo also used some old trash bags as additional insulation. If the moisture level gets too high, add more vents.

Before he covered his structure with dirt, Ingo spread a two-inch layer of straw over the top, and then covered the straw with six to eight inches of dirt from the original hole excavation. The sides were bermed up carefully, with flat stones from his rock garden. They prevent the rain from washing away the dirt.

Ingo's root celler was built in a corner of the yard and is barely visible. He recommends planting grass or foliage to make the dirt mound a bit more attractive. A final suggestion: spring or early summer is a good time to start the project, so the root cellar can dry out during the summer months.

Ingo Haidenthaller dug a simple root cellar at the rear of his yard.

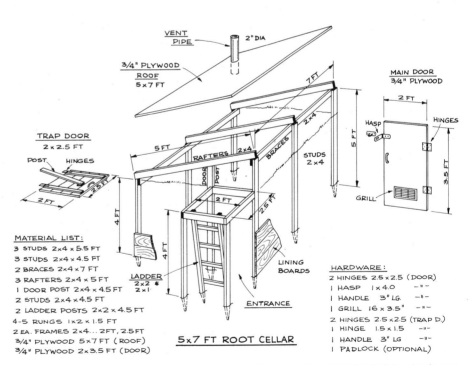

MATERIAL LIST:
3 STUDS 2×4 × 5.5 FT
3 STUDS 2×4 × 4.5 FT
2 BRACES 2×4 × 7 FT
3 RAFTERS 2×4 × 5 FT
1 DOOR POST 2×4 × 4.5 FT
2 STUDS 2×4 × 4.5 FT
2 LADDER POSTS 2×2 × 4.5 FT
4-5 RUNGS 1×2 × 1.5 FT
2 EA. FRAMES 2×4... 2FT, 2.5FT
¾" PLYWOOD 5×7 FT (ROOF)
¾" PLYWOOD 2×3.5 FT (DOOR)

HARDWARE:
2 HINGES 2.5 × 2.5 (DOOR)
1 HASP 1×4.0   —"—
1 HANDLE 3" LG. —"—
1 GRILL 16 × 3.5" —"—
2 HINGES 2.5 × 2.5 (TRAP D.)
1 HINGE 1.5 × 1.5 —"—
1 HANDLE 3" LG   —"—
1 PADLOCK (OPTIONAL)

This skeleton shows the positions and dimensions of the materials needed to build a simple root cellar.

(Continued from page 240)

pans of water around the fresh air intake, by covering the floor with wet materials such as straw or sawdust, or by sprinkling water on the floor. If you use the sprinkle method, you should put a layer of gravel two to three inches thick on the floor. This will give you a much greater surface area, and speed evaporation.

Storing carrots in sand keeps them moist and fresh.

Adding moisture directly to food works best if the food is packed in burlap or cloth. The packing material will retain moisture and keep foods humid, but there will be no standing water.

Some spoilage of stored food is inevitable. This can be minimized by storing only the best food, handling it carefully, keeping the storage area as clean as possible, and checking your food regularly. While this task may seem like a chore, it can give you a weekly chance to gloat over your hoard. The only way to stop rot, mold, and decay is to remove the infected foods. And if you catch signs of decay early, you can cut out the bad parts and still use the food.

Occasionally some of your food may freeze, either because it was left out too long or because your storage area wasn't insulated well enough. Don't throw frozen foods out too quickly. Carrots, beets, onions, cabbage, and some other root vegetables can be used after they've thawed. But once frozen and thawed, they spoil readily, so use them as soon as possible.

■

Rediscovering the cooling cabinet

Many of the old houses in southern California had floor-to-ceiling food cooler cabinets built into their kitchens. The cabinets worked on a simple principle: in a climate where nights were cool, it was possible to store night air beneath the house and use it during the day to cool food in the cabinet. Even on days when the temperature outdoors rose to 90°F., the inside of a cooler cabinet usually stayed around 65°.

Originally these coolers were all families had to keep their butter and milk cool. Later they became adjuncts to the household ice box: highly perishable items went on ice, less perishable foods went into the cabinet. As refrigerators became common, builders stopped installing cooler cabinets in kitchens. Many cabinets already installed were ripped out during remodeling. The lowly food cooler became a victim of kitchen progress.

Now, however, an energy-conscious generation too young to recall the days of iceboxes is rediscovering the virtues of the curious cabinets they often find intact in their old houses. And I recently discovered in a posh new architect-designed house the biggest cooling cabinet I've ever seen! Signs of these times.

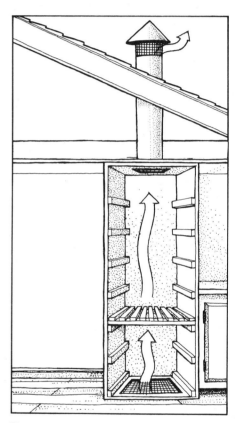

The cooling cabinet works by natural convection. Cool air from under the house is continually drawn up toward the warmer roof.

Externally, a cooling cabinet looks like any other floor-to-ceiling kitchen cabinet. It may have a single door, or—preferably—a tier of doors. The tier makes it possible to root around in one part of the cabinet without allowing warm room air to dilute all the cool air.

Inside, however, this cabinet is different. First, the shelves are not solid—they are made from wood slats (or sometimes from ⅛-inch mesh hardware cloth stretched on a wood frame) so that air can circulate

FRUITS FOR STORAGE
(CONTINUED)

Apples. Store only tree-ripened fruit, and leave stems on. Keep in shallow baskets or boxes, since fruits bruise easily. Requirements: 32 to 40°F., 80 to 90 percent humidity. Life: 4 to 6 months.

Cantaloupes. Pick just before maturity. (Fully ripe cantaloupe will keep only 1 to 2 weeks.) Store on shelves or in boxes or baskets. Requirements: 40 to 50°F., 85 to 90 percent humidity. Life: 4 to 6 weeks.

Grapes. Store only fall ripening grapes. Keep bunches in trays or baskets one layer deep, or hang in bunches. Requirements: 32 to 40°F., 80 to 90 percent humidity. Life: 1 to 2 months.

Oranges, grapefruits. Keep in boxes or baskets. Sort periodically and remove soft fruits. Requirements: 32 to 40°F., 80 to 90 percent humidity. Life: 1 to 2 months.

Pears. Pick mature—but not fully ripe—pears. Wrap individually and store in shallow layers in boxes or baskets. Requirements: 32 to 40°F., 80 to 90 percent humidity. Life: 2 to 3 months.

through the cabinet. At both the top and the bottom of the cabinet there are openings: at the top to a roof vent, and at the bottom to the enclosed crawl space beneath the house. These openings are screened to keep out insects and mice.

The crawl space acts as a reservoir of cool air. The air stays cool because it's never hit by sunlight and is in contact with the cool, bare soil. Each night the supply of cool air is replenished. The roof vent acts as an escape valve for the warmest air. With the cabinet doors shut, a continuous gentle flow of air rises from the cool crawl space to the warmer roof.

The floor vent covers almost the entire cooler bottom. This large vent allows for the free movement of cool air into the cabinet. The screening in the floor must be sturdy—small mesh hardware cloth is good—for the principal unwanted entrants from below are mice.

The lowest shelf sits several inches above the floor vent. Sturdy, simple shelves can be made from 1½ by ½-inch slats nailed to 1 by 2-inch strips. The slats should be spaced ⅜ to ½ inch apart. The shelves, in turn, are set atop furring strips nailed along the cabinet's sides. That way the shelves may be removed easily for cleaning.

The top vent is smaller than the bottom. A six-inch diameter or six-inch square vent is big enough for most cabinets. This vent is screened to bar insect entry. Since leaves, and sometimes birds, get caught on top of the screen, it should be removable from the inside of the cabinet. The vent connects with a standard sheet-metal duct of the same size as the vent opening. The duct runs through the ceiling and attic and exits to the roof. A standard sheet-metal hood covers the top of the duct to keep out rain.

The cabinet may be whatever size is convenient. Mine is built into the end of the kitchen counter, so it's about 22 inches deep inside, the same as the rest of the counter cabinets. It's 18 inches wide, and has six shelves, spaced about 15 inches apart. It holds as much food as a 17-cubic-foot refrigerator. The front has three doors—one the same height as the under-counter cabinets, one the same height as the above-

counter cabinets, and one to fill the space in between.

The cabinet doors should fit snugly. If they don't, the kitchen may be quite drafty when the wind blows. In that case, a bit of weather stripping around the doors solves the problem.

Unrestricted air flow is necessary for the cooler to function. The cabinet must not be packed so full of food that the flow is blocked. The cabinet will be slightly cooler near the floor than the ceiling. I typically find a 2 to 3°F. difference between top and

bottom on a warm afternoon. Surprisingly, there is only a small temperature fluctuation between day and night.

Clearly, a cooler cabinet cannot entirely replace a typical family's refrigerator. Rather, it is an adjunct to refrigeration—a place to store things that need to be kept cool, but not cold: things like fruit, onions, potatoes, whole grains, flour, and fresh eggs. Much of what we store in refrigerators really doesn't need to be there.

Richard Schmidt

ROOT CELLARING
Mike and Nancy Bubel. 1979. Rodale Press, 33 E. Minor Street, Emmaus, PA 18049. 297 p. hardcover $12.95.

This entire book is about root cellars. You probably won't sit down and read it from cover to cover, but if you've got a question—any question—about cold storage of fruits and vegetables, you'll find the answer here.

Root Cellaring tells you what to plant for winter storage, when and how to harvest it, how to store it, and what to store it in. Then there are recipes for the food you've stored.

This book has detailed plans for all kinds of root cellars—from buried trucks to conventional basements to window boxes. You can decide which model suits your situation, and build it. And you meet a wide range of people who have built their own cold storage areas.

TD

Double-purpose basement stairs

Proof that ingenuity and common sense can find good storage spots for vegetables in almost any setting can be found in the basement of Eleanor and Elbert Kohler's snug Cape Cod house. The Kohlers grow many vegetables in the garden which occupies the entire side yard of their small lot. They've lived in this house, on a quiet street in a large town, for over 40 years—time enough to develop a system! In their 80s now, they still enjoy working together in the garden and preparing their produce for winter use. On the early April day when we visited
(Continued on page 244)

Using a little ingenuity, you can turn some unused space—like under the basement stairs—into a cold storage area.

STORING VEGETABLES AND FRUITS IN BASEMENTS, CELLARS, OUTBUILDINGS AND PITS (USDA Home and Garden Bulletin No. 119) 1978.
Superintendent of Documents, U.S. Government Printing Office, Washington, DC 20402. 18 p. booklet $.50.

This direct and simple booklet provides a good overview of cold storage plans and structures. Its recommendations range from a barrel covered with straw to an elaborate underground cellar which doubles as a nuclear radiation shelter. It is specific about types of materials and amounts of insulation to use in various storage structures, providing details without being intimidating.

To store vegetables and fruits over winter (long-term storage) in a basement that has a furnace, you will need to partition off a room and insulate it.

Build the room on the north or east side of the basement, if practicable, and do not have heating ducts or pipes running through it.

You need at least one window for cooling and ventilating the room. Two or more windows are desirable, particularly if the room is divided for separate storage of fruits and vegetables. Shade the windows in a way that will prevent light from entering the room.

Equip the room with shelves and removable slatted flooring. These keep vegetable and fruit containers off the floor and help circulation of air. The flooring also lets you use water or wet materials (such as dampened sawdust) on the floor to raise the humidity in the room.

Store vegetables and fruits in wood crates or boxes rather than in bins.

(Continued from page 243)
them, Mr. Kohler had part of his garden dug and raked and had already planted a wide row of onion seeds in a specially prepared raised bed. And he still had potatoes and onions in good condition in his basement storage drawers.

The drawer storage system the Kohlers have worked out has proven to be a convenient and easily managed kind of small-scale root cellar. They don't need great quantities of vegetables. There are just the two of them—they are small people and they eat small meals. But they like to be independent and well prepared.

The Kohler's vegetable storage bins are long drawers built into the steps leading from their basement to the back yard. Mr. Kohler made the drawers out of scraps of one-inch wood. This beautifully simple plan allows them to retrieve vegetables at their convenience regardless of outdoor weather. At the same time, the natural refrigeration of cold air descending into the cellar cools the storage bins enough to keep root vegetables in good condition. On bitter cold nights, Mr. Kohler opens a vent in the inside basement door to permit warm air from the heated basement to keep the temperature of the steps above freezing.

The outside door and the ceiling and walls of the cellar entrance are insulated with one-inch Styrofoam covered with Masonite panels. A thermometer is attached to the outer surface of the inside cellar door.

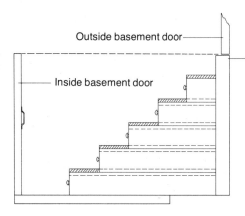

Outside basement door

Inside basement door

As the sketch indicates, the drawers form the step risers. They are supported underneath by two 2 × 4s, which run the entire length of the drawers.

The basement entrance is on the north side of the house, so the area is cool enough in October for early fall storage of root vegetables. In addition, there is no heated house area directly above the storage bins; the outside entryway extends out from the back of the house. A small shedlike structure attached to the house protects the area from weather.

We can't imagine a better use for this ordinarily wasted space. Retrofitting, you see, need not involve a large outlay of time and money; small corners like this, used to their fullest capacity, can provide a surprising amount of handy storage.

RAW VEGETABLES

One of my first experiences with raw vegetables was with peas. It is ridiculous to cook them. They are so easy to take raw, and anyone will agree that in cooking, something is lost. When you eat them raw you get everything. Don't overlook the fact that peas are seeds. They are living things and will grow if placed in the soil. When you cook them, that living quality is destroyed. It is ironic to see people eating cooked peas, followed by the taking of vitamin pills. The vitamins are in the peas. Cooking destroys part of them.

Another terrible blunder is to cook carrots. It is absolutely unforgivable. Cooking not only dissipates valuable nutritional qualities but it vitiates the taste of this wonderful vegetable. Then sugar is added to give it flavor. My next encounter was cauliflower and when you raise it without chemical fertilizers and eat it raw you get a glimpse into a new world of taste. When you cook it, you then usually add a cream sauce of some kind to bring back the taste which you sold down the river. And so on it goes—cabbage, beets, kohlrabi, turnips. These vegetables lend themselves wonderfully to be eaten raw.

I would say my greatest surprise came when I experimentally tasted some raw sweet corn just off the stalk. It was so sweet and full of milk that it was a most delightful gustatory experience. That is the way I have been eating corn since. Why waste the time cooking it, when you can get it fresh off the plant? Older corn usually loses its taste when eaten raw.

J. I. Rodale

FUTURE FOOD

Colin Tudge. 1980.
Harmony Books, New York. 205 p.
paperback $9.95.

Future Food is a thoughtful, provocative book that tells how we can *all* eat well through the next century. The ideas in it are workable, rational, and quite mundane—they can be put into practice by suppertime tonight.

"By growing or buying good things and cooking them well, you, and your neighbor, can effect the small but collectively crucial changes that could take the human race, and its fellow creatures, safely through the twenty-first century. People often ask me, when I proselytize in public places: 'But what can I do?' Cook, is the answer. Cook with knowledge. Cook and evangelize."

Mr. Tudge divides the edible substances of the world into three categories. Foods of the First Kind are the protein-supplying staples—beans, grains, potatoes. Foods of the Second Kind—vegetables, fruits, and meat—"abet and enhance" the staples and provide vitamins, minerals, and necessary fat. Foods of the Third Kind include everything else: mushrooms, nuts, spices, fish, etc. By changing our agricultural and culinary habits to give each category its proper due, Tudge believes we can insure future abundance. The many recipes show clearly that sensible eating can also be tasty and exciting.

TD

Rational agriculture could be so dramatically productive that it could restructure entire countries; and once farming had concentrated upon the problem of producing human food, it would then be freed for a whole range of other functions. Britain, a middle-sized country with a high population, illustrates the point. It has about 55 million people, and its agriculture (which British farmers claim is "the most efficient in the world") supplies them with about half their food.

Yet one study shows that if Britons were all vegetarian, and the agriculture were adjusted accordingly, Britain could support 150 million people without imports. Livestock does not necessarily reduce total productivity if it is slotted in as described above, so Britain could presumably support at least 150 million, even if people were not vegetarian but had a diet something like that of peasant China. Another estimate suggests that if all Britons cultivated this blessed plot as assiduously as some gardeners do already, then it could support 250 million people. These figures are based on carefully gathered data; but even if they are 50 per cent wrong, it is obvious that Britain, which traditionally has been one of the world's most avaricious food importers, could theoretically support its population unaided—twice, or several times over.

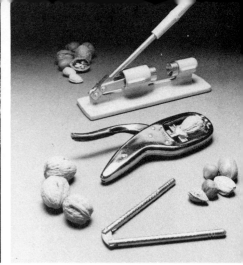

Tongue foods

The embellishment can wreck the meal

Beware of commercial tongue foods: those little bottles and cans of toppings, condiments, and other flavorings that are high in salt, sugar, and additives. The most boorish of tongues can still recognize sugar and salt, and supermarket foods cater to this low denominator. Additives are there in full force to imitate freshness; the long list of ingredients on a bottle of creamy salad dressing testifies to the food chemists' attempts to suggest freshness. That's a tall order for anyone less than a god. Freshness is as hard to fake as spontaneity.

This chapter will suggest how to make your own tongue foods from scratch. The following section talks about herbs—growing them, preparing them, and making herbal sauces, vinegars, salt substitutes, and more. Beginning on page 253, the topic is condiments, including such favorites as ketchup, mustard, relish, mayo, chutney, and horseradish. This is followed by short pieces on carob (page 257), oils and extracts (page 258), cider and the mother (page 259), crackers (page 260), and healthful jams and jellies (page 260). We go on to address the sticky subject of sweeteners—sticky because of controversy over whether or not natural sweeteners are a great deal better than refined sugars (page 262); this section offers an introduction to beekeeping and maple sugaring, two traditional activities that yield sweeteners with at least a trace of nutrients. Penultimately, a section treats the matter of beverages, likely the weakest part of the diet (page 280). Finally, nuts and seeds deserve a few words (page 282), and we take a look at several nutcrackers.

It should be said that heavy-handed use of even a pure, wholesome condiment can be too much of a good thing. Good fresh foods don't need the crutch of heavy seasoning.

RY

Herbs

North American cuisine borrows freely from all around the world, as our spice racks testify. While we have a great variety of herbs and spices at hand, we are dilettantes: too many choices can be debilitating.

Help is to be had from the recent spate of herbals on the market (a few are reviewed at the end of this section) and from the herb farms popping up in the United States and Canada. (Note our list of mail-order sources.)

You can make do with the supermarket's selection of herbs, but this is a costly source once you learn to use them often. Certain herbs are very easy to grow and seem to thrive on neglect (mints, basil, parsley). Others are choosier about their surroundings. See the Drying chapter

246

for ways of storing herbs with their essential oils intact. In this chapter, a section tells how to store an herb's personality in an essence (page 258) or vinegar (page 259).

Culinary herb receipts

It's no longer a well-kept secret that food can be enlivened with a smidgen of herbs. Most every serious cook with a green thumb has discovered how easy it is to grow savory herbs for the soup pot or souffle dish. What about fancy blends of herbs such as Fines Herbes, Bouquet Garni, or Italian seasoning? They are usually purchased rather than homemade, but you can whip up a batch or two of your own special combinations—the ingredients are right there in the herb garden.

Gather herbs for drying on a dry day. Cut perennials and biennials only half-way to the ground, and annuals to the ground. For prime foliage and flavor, the harvest should be

Fresh dill weed has a flavor that the jar-bound herb only hints at.

Basil is easy to grow, and when fresh, has a wonderful clovelike aroma.

made just before the herbs begin to bloom. Mulched herbs will likely be clean and require no washing. Dirt-splashed herbs should be hosed, then shaken hard to remove as much water as possible.

Tie bunches together tightly with strong string or a rubber band. Hang upside down in a well-ventilated, dark, dry place for one to two weeks. To reduce evaporation of volatile oils and subsequent flavor loss, always dry herbs in darkness (an attic or spare room is ideal). Lacking a dark room, bunches may be placed in a brown paper sack and hung in the garage or kitchen. To dry leaves and small stemmed plants which can't be tied in bunches, place them in thin layers in a shallow box or on newspapers (to provide darkness and dust protection cover with another layer of newspaper). Cut seed heads when they are mature, but before they begin to shatter and fall. Dry in shallow layers in boxes in a dark and dry place. When dry, shake out the seeds.

As a test for dryness, herbs should crumble into small pieces between your fingers. Holding the stem in one hand, strip off the leaves with the other, starting at the base of the stem. Store crumbled herbs in tightly covered glass jars. Herbs stored in plastic or paper will quickly lose flavor; stored in metal, they may

develop a metallic taste. For retaining flavor and color, jars of herbs are best stored away from sources of extreme heat and light.

Les Fines Herbes. Les Fines Herbes traditionally is a crumbled blend of two parts each of chervil, chives, and parsley to one part tarragon. As all four of these herbs either change flavor or lose flavor when dried, the combination is best used either fresh or frozen. To freeze Fines Herbes merely chop fine, mix the blend, and freeze in tiny amounts convenient for a recipe. If herbs are washed first, be sure to dry them as much as possible between layers of paper towels before freezing.

A versatile blend of Fines Herbes that dries nicely is equal parts summer savory, basil, sweet marjoram, and thyme. Add a touch of this blend to all types of egg and meat dishes.

To add new life to the same old soup, all green vegetables, and potato dishes, give this Fines Herbes a try: blend equal parts parsley (or chervil), thyme, basil, oregano, and lovage (or celery leaf).

Make your own combinations from any of the following: chervil, parsley, thyme, savory, basil, sweet marjoram, oregano, celery (or lovage), sage, rosemary, tarragon, and chives.

(Continued on page 248)

HERBS ARE FOR EVERYONE

Bonnie Fisher. 1976.
Hickory Hollow Press, Peterstown,
WV 24963. 51 p. paperback $3.

This self-published book covers planting and propagation and uses of herbs, all on few pages.

Herb Oils

Extracting oil from an herb takes a little patience and lots of fresh flowers or herbs. To start, fill a ceramic crock with rose petals, lavender blossoms, lilac blossoms, mint leaves, rosemary, thyme, basil, or another herb of your choice. Gently press the petals or herbs down and slowly pour in rain water or spring water until covered. Place the crock outside in full sun, bringing inside only if rain should threaten. In about one week an oily scum should appear on the surface. This is the precious oil you've been waiting for. Soak up the oil carefully with a small piece of cotton and squeeze the oil into a small container. Before storing the herb oils in tiny sterilized colored glass bottles make sure no water is left in the oil. Tie a piece of cheesecloth over the bottle for several days to allow any remaining water to evaporate. Close the bottle with a tight-fitting lid.

Singing woad, the oft-tred-upon savant of low places, offers musical, rather than culinary, talents. Indeed, this herb has no flavor or aroma whatsoever. The reclusive and habitually morose Pinhead tribe of boreal Athabasca is fond of decocting this plant, finding cheer in the bird-pitched warbling and whistling that issue from the pot.

Orson Fowler.
Bog Whimsey (1861).

(Continued from page 247)
Bouquet Garni. Whereas Fines Herbes and other blends are made with crushed herbs that are left in the food, Bouquet Garni is a tied bunch of fresh or dried sprigs (or "bouquet") of herbs immersed in a soup or stew and removed before serving. Old English recipe books call these "faggots of herbs." The usual Bouquet Garni consists of three sprigs parsley, one sprig thyme, and one bay leaf (and sometimes a small piece of orange peel). The traditional Bouquet Garni imparts a mild touch of flavoring and is a good starting point for those uninitiated to the many flavors of herb cookery. To dry this blend for future use, string a needle through a fresh or dried bay leaf, add the parsley and thyme, tie securely, and hang till dry. Leave a long end on the string so it can be tied to the pot handle.

I prefer this more pungent Bouquet Garni: two sprigs of both parsley and thyme (or lemon thyme), one sprig of both rosemary and basil, and one bay leaf. This combination retains its flavor well when dried and is an all-purpose Bouquet Garni. For chicken dishes use a sprig or two of tarragon in place of the rosemary (or for fish cookery make it a sprig of fennel).

Italian herbs. Italian herbs belong in everything that speaks of Italy—spaghetti sauce, pasta sauce, lasagna, and pizza. Just a touch sparks

A good selection of herbs can be grown indoors. They're right at hand, and available year round.

up macaroni and cheese, omelets, tomato dishes, tomato juice, soups, and meat dishes. To three parts of both oregano and thyme add two parts rosemary and one part each sweet marjoram, sage, and basil. For variety, substitute summer savory for the basil or sage.

Other herb mixes. Concoct some "European herbs" by mixing three parts of both chervil and basil, two parts sweet marjoram, and one part lemon balm. Use this to enhance omelets, souffles, quiches, cheese spreads and dips, cheese dishes, green vegetables, potatoes, green salads, and poultry. If you like cumin, try this stronger version: two parts of both chervil, basil, and sweet marjoram to one part of both lemon balm

The attic is a good place to dry quantities of herbs.

and crushed cumin seed. This mix is especially tasty with green beans, legume dishes, and bean soups.

Dill and tarragon have an affinity for fish. Here is a mix of fish herbs that adds a distinguishing touch to baked or boiled fish, fish chowders and soups, fish pates, tuna or seafood salad, and newburgs. Stir together three parts tarragon, two parts dill weed, and one part paprika. If desired, include two parts sesame seeds (they give a nice crunch to the blend). Another pleasing addition is a small amount of either peppermint or celery seed.

Jazz up the salad bowl with either of these recipes. Salad Herbs #1: four parts tarragon; two parts each poppy seed, basil, and thyme; one part of both paprika and garlic powder. Salad Herbs #2: three parts sesame seeds, two parts of both tarragon and thyme, one part celery seeds. Use these salad herbs with all types of greens (raw salads or cooked potherbs) or include in homemade oil and vinegar salad

dressings (use one teaspoon for each cup dressing). For herb mayonnaise stir one tablespoon into a cup of homemade mayonnaise.

Poultry herbs or seasonings usually contain a combination of savory, sweet marjoram, basil, thyme, parsley, lovage (or celery greens), sage, lemon peel, oregano, and paprika. This first mix is especially nice on fried chicken, as paprika aids in browning and gives a nice color to the chicken. For Poultry Herbs #1, combine and blend equal parts sage, paprika, thyme, and oregano. Poultry Herbs #2: 2 parts each summer savory, basil, thyme, and lovage (or celery greens); one part each parsley, sage, and sweet marjoram. Both blends suit roast chicken, duck, or goose; fried or baked chicken; chicken noodle soup; chicken or turkey salad; and creamed turkey or chicken. Give them a try with vegetables or stuffed eggs. Concoct your own poultry bake coatings by mixing two tablespoons of either Poultry Herbs with one-half cup cornmeal or flour (this amount will coat one fryer).

If you grow those potent little chili peppers, you are already on the way to a homemade chili powder so popular in Mexican dishes such as chili con carne and tamales, as well as curries that suggest India. Chili peppers are healthful in small amounts as they aid in digestion and are loaded with vitamins A and C. As some varieties are exceptionally hot, proceed with caution. To make chili powder, lightly roast whole dried chili peppers (the seeds add flavor and bite) in a skillet on medium heat until they are dark red to light brown (do not burn). Or roast in the oven on cookie sheets until slightly browned. (I sometimes omit the roasting and proceed as below.) Grind small amounts in an electric blender on high speed (a small jar works best) for 30 seconds at a time until powdered. The finished powder is fiery hot. The typical American palate prefers a toned-down blend; combine three parts powdered chilis with one part each oregano, garlic powder, and cumin seed (powdered in a blender). For an even milder blend add additional cumin and oregano. Some commercial brands also use powdered allspice, cloves, or coriander to help tone the fire. Homemade

(Continued on page 250)

A MODERN HERBAL
Mrs. M. Grieve. 1971.
Dover Publications, Inc., 180 Varick Street, New York, NY 10014. Two volumes, $5 each.

This is a two-volume reprint of an herbal originally published in 1931. Unlike car parts and tube radios, herbals aren't ever victims of obsolescence. In fact, most newly published herbals owe an enormous debt to past volumes.

The illustrations are indifferently reproduced, but Mrs. Grieve offers an excellent and detailed description of the uses to which flora have been put. Her herbal is fun to read through at random, as the following excerpts suggest.

To carry a raw potato in the pocket was an old-fashioned remedy against rheumatism that modern research has proved to have a scientific basis. Ladies in former times had special bags or pockets made in their dresses in which to carry one or more small raw potatoes for the purpose of avoiding rheumatism if predisposed thereto. Successful experiments in the treatment of rheumatism and gout have in the last few years been made with preparations of raw potato juice.

Let your nutts be green as not to have any shell; then run a kniting pin two ways through them; then put them into as much ordinary vinegar as will cover them, and let them stand 30 days, shifintg them every too days in ffresh vinegar; then ginger and black peper of each ounce, rochambole two ounces slised, a handfull of bay leaves; put all togeather cold; then wrap up every wall nutt singly in a vine leaf, and put them in putt them into [sic] the ffolloing pickel: for 200 of walnutts take two gallans of the best whit vineager, a pint of the best mustard seed, fore ounces of horse radish, with six lemons sliced with rin(d)s on, cloves and mace half an ounce, a stone jar, and put the pickel on them, and cork them close up; and they will be ffitt for use in three months, and keep too years.

SOURCES FOR HERBS

California

Green Valley Seeds, 11565 E. Zayante Road, Felton, CA 95018.

China Herb Company, 428 Soledad, Salinas, CA 93901.

Nature's Herb Company, 281 Ellis Street, San Francisco, CA 94102.
Herbs and spices; *catalog, 25¢.*

Taylor's Garden, Inc., 1535 Lone Oak Road, Vista, CA 92083.

Colorado

Mini Mountain, Box 91, Indian Hills, CO 80454.

Connecticut

Caprilands Herb Farm, Silver Street, N. Coventry, CT 06230.
Catalog, 10¢.

Georgia

Creekwood Herb Craft, Box 15323, Atlanta, GA 30333.
Catalog, $1.

Indiana

Indiana Botanic Gardens, Inc., P.O. Box 5, Hammond, IN 46325.
Wide variety of botanicals and herbal products; *catalog, 25¢.*

Kentucky

Rutland Herbs, Box 583, Georgetown, KY 40324.
Catalog, $1.

Maine

Johnny's Selected Seeds,* Organic Seed and Crop Research, Albion, ME 04910.
Catalog, 50¢.

Massachusetts

Borchelt Herb Gardens, 474 Carriage Shop Road, E. Falmouth, MA 02536.
Catalog, 10¢.

Michigan

Misty Morning Farm,* Richard and Margaret Dykstra, 2220 W. Sisson Road, Hastings, MI 49058.
Herb plants and dried herbs.

New Hampshire

Tom Woods, P.O. Box 64, Jefferson, NH 03583.

New Jersey

Well-Sweep Herb Farm,* 317 Mt. Bethel Road, Port Murray, NJ 07865.
Price lists for seeds, plants, and products, 35¢.

New York

Herbst Brothers Seedsmen, Inc., 1000 N. Main Street, Brewster, NY 10509.

Shash Georgi, 7233 Lower East Hill Road, Colden, NY 14033.

*Sources which list organically grown herbs.

chili powders are a brigher red than commercial brands and also tend to be hotter, so use sparingly.

Homemade pickling spices are no further away than the herb patch with a little assist from the spices in your herb cabinet. Most pickling spices contain 8 to 12 of the following in varying amounts: bay leaf, dill weed or seed, mustard seed, chili peppers, celery seeds, ginger, garlic, tumeric, black peppercorns, allspice, cloves, mace, coriander, cinnamon, and cardamom. Try creating your own mixture or use this recipe: ½ cup coriander seed, ½ cup bay leaf (broken in ½-inch pieces), ½ cup mustard seed, ¼ cup chili peppers, ¼ cup dill seed, ¼ cup whole allspice, one tablespoon black peppercorns, one tablespoon whole cloves, and one tablespoon ginger root (grated and dried). This will make about one pint—the ginger may be left out, if desired.

Bonnie Fisher

Perennial herbs are suited to formal gardens.

Herbs in a small place

If I had to choose between growing herbs or vegetables, I would take herbs. One can always buy some kind of vegetable—if you are lucky, sometimes even a vegetable as fresh as garden-grown—but I know of no market that regularly sells fresh herbs. To cook properly, we need fresh herbs every day. They should be located near the kitchen if they're not to be much more difficult to pick than to select from a spice rack.

Even though I have always lived in apartments, I've never been without a little garden. The herbs grow in their own pots, cans, and containers of various kinds. I've concentrated on perennials because they are the chief ingredients of so many of our dishes: two kinds of thyme, tarragon, marjoram, oregano, chives, five kinds of mint, winter savory, and two kinds of bay leaf, the Turkish and the Pacific coast bays.

Keeping herbs in containers now seems the only way that is convenient. We can move them right outside the kitchen door when their season for prime use arrives. My wife and I once visited an English estate which had marvelous gardens. Down near the end of the extensive gardens was a secluded enclosure devoted to herbs—which seemed to be a quarter of a mile from the house. When servants are available, the herb garden can be located a good hike from the kitchen as at this 17th century Tudor estate; but when you and I have to go after our own, I think a short distance is more convenient for cooking. Also, having herbs ready at hand means that the busy cook will be more agreeable to using them.

To grow any perennial in containers, watch the watering, feeding, and root development. Container growing requires more water because the plants dry out quickly. The root zone covers the inside of the container where water flows through. A vigorous herb like oregano will fill a medium-sized container in two years. It can be pushed into a third year, but it may seriously decline during the summer of the fourth year. (Plants such as geraniums, on the other hand do better when pot-bound, and can be kept in a container for years.)

When the garden hasn't room for the odd plant of basil or a spot of borage, then we put it in a pot or a can or a box. In past years we have had containers of garlic, parsley, chives, Chinese leeks, Japanese shiso, and Mexican tea outside our kitchen door. Although some of these may be exotic if now downright useless to the average cook, I cannot imagine the cook who can do without some of the basic and common herbs. I consider a pot of chives near at hand as important as a pot for boiling potatoes.

John Meeker

Sources for herbs

If you can't grow your own dill weed for pickling, mint for tea, or basil for spaghetti sauce, you'll want to look for alternatives to supermarket herbs in tiny bottles. Your savings can be impressive: compare $1.76 a

(Continued on page 252)

New York (continued)

Aphrodisia, 28 Carmine Street, New York, NY 10014.
Herbs, spices, oils; *catalog, 25¢.*

Caswell-Massey Co., Ltd., 320 W. 13th Street, New York, NY 10014.
Catalog, free.

Wide World of Herbs, Ltd., P.O. Box 266, Rouses Point, NY 12979.
Botanicals.

Oregon

Nichols Garden Nursery,* 1190 N. Pacific Highway, Albany, OR 97321.
Herbs and other plants, seeds, botanical products; *catalog 15¢.*

Casa Yerba,* Star Route 2, Box 21, Days Creek, OR 97429.
Seeds and plants.

The Dutch Mill Herb Farm, Larry and Barbara Remington, Route 2, Box 190, Forest Grove, OR 97116.
Herb plants and products: culinary, aromatic, medicinal, landscaping; *they do not ship.*

Herbs N. Honey Nursery, c/o Mrs. Chester Fisher, Route 2, Box 205, Monmouth, OR 97361.
Catalog, $1.

Pennsylvania

Haussmann's Pharmacy, 534-536 W. Girard Avenue, Philadelphia, PA 19123.
Unusual botanicals; *catalog available.*

Penn Herb Company, 603 N. Second Street, Philadelphia, PA 19123.
Dried herbs, seeds, herb products.

W. Atlee Burpee Company, 300 Park Avenue, Warminster, PA 18974.
Seeds and plants.

Rhode Island

Greene Herb Gardens,* Greene, RI 02872.
Herbs, herb plants, seeds.

Meadowbrook Herb Garden,* Wyoming, RI 02898.
Seeds, plants, herbal products; *catalog 50¢.*

Texas

Hilltop Herb Farm, Box 866, Cleveland, TX 77327
Herbs, herb plants, seeds, unusual botanicals; *catalog, 30¢, and stamped, self-addressed envelope.*

Yankee Peddler Herb Farm, Highway 36 North, Brenham, TX 77833
Herb plants, seeds, teas, botanicals, and books; also many Colonial-day favorites and hard-to-find varieties.

Utah

Old Fashioned Herb Company, P.O. Box 1000-G, Springfield, UT 84663.

Washington

Cedarbrook Herb Farm,* Don and Karman McReynolds, 986 Sequim Avenue South, Sequim, WA 98382.
Kitchen, rockery, and tea herbs; *they do not ship; April through October only.*

(Continued on page 252)

THE RODALE HERB BOOK

William H. Hylton, editor. 1974.
Rodale Press, Inc., 33 E. Minor
Street, Emmaus, PA 18049. 653 p.
hardcover $13.95.

Unlike many publishers, Rodale Press writes many of its own books, especially massive hardcovers that seek to be the last word on a subject. We at the Press refer to them as our "Big Books," as opposed to our slimmer, more topical volumes.

The Rodale Herb Book is a staff-edited Big Book. It has seen 19 printings to date, and covers its subject well enough that the Press has yet to follow its success with a splashier or more specialized effort.

This is a good candidate for the one-herbal library.

Freezing is a very simple way to store the culinary herbs for winter use. Gather the herbs at the specific times previously mentioned for drying, wash them if necessary, shake dry, and then place in plastic boxes or bags, properly labeled. Place these immediately in the freezer. The herbs can be chopped or left whole.

The herbs can also be blanched before freezing, although it is not necessary.

Do not defrost the frozen herbs before using. If the recipe calls for minced herbs, it is easier to chop them while they are still frozen, since they break apart so readily. Chives, sorrel, parsley, dill, oregano, sweet marjoram, lovage, tarragon, and mint leaves freeze well.

Another method to freeze herbs is to place the chopped herbs into ice cube trays filled with water. After freezing, place the cubes in a plastic bag, label, and store in the freezer. When needed just pop an ice cube with herbs into soup, stew or casserole.

(Continued from page 251)
pound for bulk rosemary at a co-op to $11.52 a pound (in half-ounce bottles) in the supermarket. You can order basil through the mail for around $5 a pound, in bulk.

Because herb teas and novel spices are enjoying a burst of popularity, many supermarkets or kitchen specialty stores carry wrapped packages holding one or two ounces at half or two-thirds the price of buying in small bottles. But if you're making your own herb teas, or need lots of basil and oregano for canning your own tomato sauces, hunt for a place that will sell to you by the pound.

Your Yellow Pages will lead you to herb shops (sometimes under herbs, sometimes under natural or organic foods, or just "foods"). Many natural health foods stores and storefront co-ops sell by the pound. (See the grains chapter for description of these stores and how to find them.)

County agricultural extension agents may know of local herb growers. We list here several growers who will sell by mail.

Sara Ebenreck

Kicking the salt habit

Do it gradually. If you've been using a teaspoon of salt to a pot of potatoes, rice, kasha, or whatever, cut down to a half-teaspoon for a week or two. Use a touch of ginger, paprika, and grated lemon peel. I now combine these flavor enhancers and keep them handy in a shaker at the stove. As I decreased the salt, I increased the spices until I was using no salt at all at the stove. Remember that a taste for salt is learned. It can also be unlearned.

But don't we need some sodium? Yes, a tiny bit. Between half a gram and two grams a day is all the body requires, and you get that naturally in foods. Water, milk, meat, vegetables, and fruits all contain natural sodium. There is a little danger of not getting enough.

In the United States, we get 10 to 15 times as much sodium as we need, and the consequences are disastrous. Health professionals believe that the high salt content of the American diet may be an important factor in the development of hypertension, which affects about 35 million Americans and contributes to several hundred thousand deaths from heart attacks and strokes every year.

The usual medical treatment for high blood pressure is diuresis—administration of a drug which pulls sodium and fluids from the tissues. Anyone who's been on a diuretic will tell you that his sleep is frequently disturbed by an emergency call from his kidneys.

Doesn't it make far more sense to eliminate salt and not suffer from high blood pressure in the first place?

SOURCES FOR HERBS
(CONTINUED)

West Virginia

Enfluerage, P.O. Box 416, Athens, WV 24712.

Hickory Hollow,* Route 1, Box 52, Peterstown, WV 24963.
  Herbal products; *catalog, 25¢.*

Wisconsin

Herbarium, Inc., Route 2, Box 620, Kenosha, WI 53140.
  Botanical drugs and spices.

Canada

Wide World of Herbs, Ltd., 11 Sainte Catherine St. East, Montreal, PQ H2X IK3.
  Botanicals.

Otto Richter & Sons Ltd., Goodwood, ON L0C 1A0.
  Herb seeds and books; *catalog, 75¢.*

*Sources which list organically grown herbs.

In his book *Sea of Life* (David McKay Co., 1969), Dr. William D. Snively reports a study of a large group of Americans questioned about the amount of salt they normally use. Out of 100 who reported that they never salted their food at the table, only one had high blood pressure; among 100 who added salt even before tasting, 10 had high blood pressure. The doctors who worked on the study came to the conclusion that the most common form of high blood pressure or hypertension will not develop unless salt intake is excessive.

Why does excessive salt cause your pressure to go up? There is an old saying, "Water goes where the salt is." Because sodium clings to water, an excess of salt in the cellular fluid carries extra liquid. This extra liquid increases the volume of the plasma. In order to distribute the expanded blood, Dr. Snively explains, the heart must create additional pressure and therefore pump harder.

Excessive salt intake may also trigger migraine headaches. In one study, 10 out of 12 patients with proven migraine had lower incidences of recurrent headaches after restricting salt intake from snack foods on an empty stomach (*Minnesota Medicine*, Vol. 59, No. 4, 1976).

Let's start with breakfast. Are you accustomed to putting salt on your eggs? Try a combination of dill, oregano, and chopped chives. Grind them up to a fine powder and put them in the salt shaker.

Do you like hot cereal? How can you cook oatmeal or wheat cereal without a touch of salt? Here's how. With a touch of cinnamon, a speck of ginger, and a bit of lemon peel. Experiment with these and other spices to find the combination that pleases you best. Keep that combination in a shaker near the stove.

At Fitness House, the Rodale Press dining room, we simply eliminated salt from bread and cake recipes without making any substitutions. We got hardly any flak. One or two die-hard salters remarked that the bread tasted flat. Personally, I find our no-knead whole wheat bread more delicious in the saltless state.

The subtle flavors of the wheat manage to get through. (See the recipe on page 25.)

Fran Wilson, chef-manager of the Rodale dining room, uses a bit of ginger on carrots and in cream of carrot soup, and a light dusting of nutmeg on potatoes. For a zesty meat loaf, she relies on chopped onions, tomatoes, and chopped celery, including the tops. Nobody misses the salt.

Here are some more tips for making your food tasty without salt:

- Vegetable and meat soups: add a little vinegar, wine, or lemon juice.
- Chicken soup: use ginger, one clove, bay leaf, a few peppercorns.
- Cream soups: a touch of cinnamon and nutmeg.
- Cucumbers: marinate in tarragon vinegar.
- Asparagus: sprinkle tips and stalks with nutmeg before serving.

Jane Kinderlehrer.
Prevention Magazine

Condiments

Here follow recipes and thoughts on popular condiments and processed seasonings. You may or may not save money, but the point here, as throughout the book, is that foods processed at home taste better and are unharmed by the imaginations of food scientists.

Horseradish

You can find jars of commercial horseradish that are adulterated with nothing worse than a bit of salt and perhaps beet juice for a rich magenta hue. So, unless you happen to be a true horseradish fan, this recipe may not have much appeal.

To further deflate one's incentive, hand-grating the fresh root can sting the eyes and cause the nose to run. If you are undeterred by this warning, procure a fresh root from the market, peel it, then grate it either by hand or feed chunks to a blender or processor.

Add vinegar until you reach the desired texture (wine vinegar will color the mixture a pale pink). Refrigerate in a tightly closed jar. Try not to make more than you'll use in a month's time, as horseradish browns and loses flavor even if kept cool.

For a meat condiment, good in meat sandwiches, combine grated horseradish with homemade mayonnaise. Store and use with the same precautions you would for mayonnaise.

Horseradish Sauce

⅓ cup white vinegar
⅔ cup horseradish root, grated
⅓ cup mayonnaise
1 tablespoon vegetable oil

Blend vinegar and horseradish root together at high speed. Add mayonnaise and blend well. Add oil and continue blending at high speed until fairly smooth.
Yield: 1 cup

(Continued on page 254)

A homemade curry powder can be made of cumin, turmeric, coriander, red pepper, cinnamon, cardamom, and other spices as you please: there is no set formula.

(Continued from page 253)

Chutney

3   mangoes, near ripe but still a bit firm
1   teaspoon vegetable oil
¼   cup cider vinegar
¼   cup water
⅓   cup gold raisins
3   tablespoons Barbados molasses
    juice of 1 orange (⅓ cup)
4   teaspoons ginger root, minced
½   teaspoon garlic, minced
1   1-inch stick of cinnamon
2   small dried red chili peppers
½   teaspoon ground coriander
¼   teaspoon ground cardamom
½   teaspoon turmeric
2   cups apples, chopped in ½-inch
        cubes

Slice the flesh away from the mango in pieces slightly larger than bite-size. Combine all remaining ingredients save the apples in a saucepan. Cook over medium heat, stirring frequently until the mango is very soft and the mixture somewhat thickened.

Stir in the apples and cook until just tender. Remove the cinnamon stick and chilis.

Yield: 3 cups

Ketchup

2   quarts tomatoes, quartered (5
        pounds)
⅓   cup onion, chopped
¼   cup celery, chopped
1   teaspoon garlic clove, chopped
⅓   cup champagne vinegar
1   tablespoon honey
2   teaspoons unsulfured molasses

Bouquet Garni: (tie in cheesecloth)
½   bay leaf
⅛   teaspoon celery seeds
1   small dried red chili pepper
        (about 1 inch long)
¾   teaspoon mustard seeds

Puree the tomatoes in a blender. Then puree the onion, celery, and garlic, adding some of the tomato puree as liquid to work the blender.

Combine the tomato puree, onion, celery, and garlic in a stainless steel or enameled pot and bring to a boil. Stir in the vinegar, honey, and molasses. Add the Bouquet Garni and cook over medium heat, stirring frequently, for 30 minutes. Remove bouquet garni.

Continue to cook until the ketchup becomes somewhat thickened.

Pass the ketchup through a food mill and then transfer ½ cup to a blender. Blend at high speed, ½ cup at a time, until smooth.

Yield: About 1½ cups

Ingredients for a good chutney.

German Mustard

¼   cup brown mustard seed, freshly
        ground (grind to personal
        preference—the finer the grind,
        the smoother the mustard)
5   tablespoons dry mustard
½   cup hot water
¾   cup champagne vinegar
2   tablespoons cold water
2   large slices onion (approximately 1
        ounce)
2   teaspoons honey
1   teaspoon unsulfured molasses
2   cloves garlic, peeled and halved
¼   teaspoon dill seed
¼   teaspoon ground cinnamon
¼   teaspoon ground allspice
¼   teaspoon dried tarragon, crumbled
⅛   teaspoon ground cloves

Soak the ground mustard seed and dry mustard in the hot water and ¼ cup of the vinegar at room temperature for at least 3 hours.

Combine the remaining vinegar, cold water, onion, honey, molasses, garlic, dill seed, cinnamon, allspice, tarragon, and cloves in a small saucepan. Bring to a boil, boil for 1 minute, remove from the heat, cover, and let stand for 1 hour.

Transfer the soaked mustard mixture to a blender. Strain the spice infusion into the mustard mixture, pressing the spices against the sides of the strainer to extract all the flavor. Process until the mixture is the consistency of a coarse puree.

Pour the mixture into the top of a double boiler, set over simmering water, and cook until thickened, about 20 to 25 minutes. (The mustard will thicken a bit more when chilled.)

Remove from the heat, and pour into a jar. Let cool uncovered and then put a lid on and store in the refrigerator.

Yield: 1 cup

Hand-beaten Mayonnaise

For best results, have all ingredients at room temperature.

2   egg yolks, lightly beaten
2   tablespoons lemon juice or vinegar
½   teaspoon dry mustard
1⅓ cups vegetable oil
2   teaspoons boiling water (optional)

Warm a glass or stainless steel bowl and a wire whisk in hot water. Dry thoroughly. Add the egg yolks to the bowl with 1 tablespoon of the lemon juice or vinegar and the dry mustard. Beat to mix well. Continue beating, constantly, as you add the oil, one drop at a time. Be sure the yolks are absorbing the oil; this may require you to stop adding the oil and just beat the yolks for a few seconds.

After about ⅓ cup of the oil has been incorporated into the yolks, the remaining oil can be added by the tablespoon, beating well after each addition of oil. When the mayonnaise is thick and stiff, beat in the remaining lemon juice or vinegar to thin it out. Then continue to beat in the remaining oil. To keep the mayonnaise from breaking, 2 teaspoons of boiling water can be blended in at this point.

Store in a covered glass jar in the refrigerator.

Yield: 1¼ cups

Yellow Mustard

4   tablespoons dry mustard
4   tablespoons hot water
3   tablespoons white vinegar
⅛   teaspoon garlic powder
    pinch of dried tarragon
¼   teaspoon unsulfured molasses

Soak dry mustard in hot water and 1 tablespoon of vinegar for at least 2 hours. Combine the remaining vinegar, garlic, and tarragon in a separate bowl and let stand for ½ hour.

Strain the tarragon from the second vinegar mixture and add the liquid to the mustard mixture. Stir in the molasses. Pour the mustard into the top of a double boiler, set over simmering water. Cook until thickened, about 15 minutes. (The mustard thickens further when chilled.)

Remove from the heat and pour into a jar. Let cool uncovered, and then put on a lid and store in the refrigerator.

Yield: ½ cup

Homemade Onion Powder

Peel off the skin of one large onion and thinly slice. Separate the slices and dry in the oven or dehydrator. When the onion slices are dried, pound or grind them to a smooth powder as for chili powder. Store in a shaker jar.

Yield: 3 tablespoons powder

Blender Mayonnaise

For best results, have all ingredients at room temperature.

2   egg yolks, lightly beaten
2   tablespoons lemon juice or vinegar
½   teaspoon dry mustard
1⅓ cups vegetable oil
2   teaspoons boiling water (optional)

Warm the blender bowl in hot water and dry thoroughly. Combine the yolks, lemon juice or vinegar, and mustard in the blender bowl. Blend at medium speed for about 1 minute.

Gradually add the oil, a few drops at a time, until ⅓ cup of oil has been incorporated into the yolks. At this point the oil can be added 1 tablespoon at a time until all the oil is used.

To insure against the mayonnaise breaking, 2 teaspoons of boiling water can be blended in at this point.

Store in a covered glass jar in the refrigerator.

Yield: 1½ cups

Homemade Chili Powder

Wash the chili peppers, discard stems, and dry in the oven or dehydrator. Smaller peppers are usually dried whole. If the peppers are cut, make sure to keep your hands away from your face because the pepper will burn your eyes. Chili peppers can also be dried by stringing. Run a needle and thread through the thickest part of the stem. Hang them outdoors or in a sunny window to dry. They shrink and darken considerably, and will be leathery when dry. Although dried peppers can be kept in storage containers, they are best left hanging in a dry place.

Grind in a mill, blender, or mortar, and store in a standard spice tin or bottle.

Chili Sauce

4   red chili peppers, split and seeded
2   cups champagne vinegar, boiling
2   whole garlic cloves
1   cup tomatoes, chopped
2   small dried chili peppers

Combine the chili peppers, vinegar, and garlic cloves in a pint jar. Cover and let sit for at least 48 hours. Combine the garlic and the chilis with the chopped tomatoes in a blender and liquify.

Transfer contents of blender to a double boiler. Add 2 small dried chili peppers. Cook for 30 minutes over medium-high heat. Strain and discard solids. Refrigerate.

Yield: ¼ cup

Homemade Garlic Powder

Separate the cloves of one garlic head and remove skins. Slice the garlic or leave whole and dry in the oven or food dryer. When cloves or slices are completely dry, pound or grind them as for chili powder. Store in a shaker jar. Note that homemade powder may be stronger than commercial.

Yield: 4 teaspoons powder

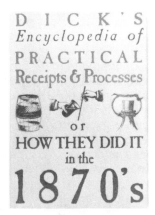

DICK'S ENCYCLOPEDIA OF PRACTICAL RECEIPTS AND PROCESSES
William B. Dick. 1975.
Funk & Wagnalls, Inc., 666 Fifth Avenue, New York, NY 10019. 607 p. hardcover $9.95.

The jacket copy of this reprint calls *Dick's Encyclopedia* "a sort of 'Whole Earth' catalog of the 1870s," but inside the modern-day publisher has seen fit to print a "special note" warning that "this book is intended solely as an historical record. . ." Indeed, a reader taking every receipt literally, if anyone could be so naive, would be a dangerous person. Receipt 5695: "If a person falls in a fit, let him remain on the ground, provided his *face be pale;* . . . but if the *face be red or dark colored* raise him on his seat, throw cold water on his head immediately. . . . " Such adventures go on page after page. "5812. Simple cure for earache. Take a common tobacco-pipe, place a wad of cotton in the bowl, drop upon it 8 to 10 drops of chloroform, and cover with another wad of cotton; place the stem in the affected ear, then blow into the bowl, and in many cases the pain will cease almost immediately."

Between the laughable and quaint items are forgotten lessons waiting for a return to a more self-reliant time.

RY

Tomato Paste

5   pounds of tomatoes, preferably Italian plum

Quarter tomatoes and puree in a blender. Simmer over low heat in a large flat saucepan, stirring frequently, until all the liquid has cooked off.

Yield: about 1½ cups
(depending on the fleshiness of the tomatoes)

(Continued on page 256)

To Make Good Cider Vinegar.

Take 10 gallons of apple juice fresh from the press, and suffer it to ferment fully, which may be in about 2 weeks, or sooner if the weather is warm; and then add 8 gallons like juice, new, for producing a second fermentation; in 2 weeks more add another like new quantity, for producing a third fermentation. This third fermentation is material. Now stop the bunghole with an empty bottle, with the neck downward, and expose it to the sun for some time. When the vinegar is come, draw off ½ into a vinegar cask, and set it in a cool place above ground, for use when clear. With the other half in the first cask, proceed to make more vinegar in the same way. Thus one cask is to make in, the other to use from. When making the vinegar, let there be a moderate degree of heat, and free access of external air. The process is hastened by adding to the cider, when you have it, a quantity of the mother of vinegar, as it is called—a whitish, ropy coagulum, of a mucilaginous appearance, which is formed in the vinegar and acts as a ferment. The strength of vinegar depends on the amount of sugar or starchy matter to be ultimately converted into acetic acid.

Walnut Catsup.

Take young, tender walnuts, prick them in several places, bruise them with a wooden billet, and place in a jar with sufficient water to cover them, adding a handful of salt for every 25 walnuts; stir them twice a day for 14 days; then drain off the liquor into a saucepan. Cover the walnuts with boiling vinegar, crush to a pulp and strain through a cullender into the liquor in the saucepan. Add, for every 2 quarts, 2 ounces each black pepper and ginger, 1 ounce each cloves and nutmeg pounded fine, a pinch of cayenne, a shallot minced fine, and a thimbleful of celery seed tied in a muslin bag. Boil all together for an hour, and when cold, bottle. In the above manner an excellent catsup may be made from butternuts.

(Continued from page 255)

Basic Tomato Sauce

5   ripe, fresh tomatoes, chopped coarsely
2   tablespoons olive oil
2   cloves garlic, minced
1¼ cup onion, chopped
1   bay leaf
1   teaspoon basil
1/6 teaspoon celery seed
    pinch of thyme
3   sprigs parsley, chopped finely
    oregano to taste

Puree the tomatoes in a blender. In a large saucepan, heat the olive oil and saute the garlic and onions. Add the pureed tomatoes, bay leaf, basil, celery seed, and thyme. Cook over low heat for at least 45 minutes, the longer the better.

Add the parsley and cook for 2 or 3 minutes. Season to taste with oregano and cayenne.

Yield: 2 cups

Blueberry Syrup

2   cups blueberries (fresh or frozen)
1¼ cups water
1   tablespoon arrowroot
3   tablespoons maple syrup

Combine blueberries and water in a saucepan and bring to a boil. Lower heat and simmer for 5 minutes. Pour the hot blueberry mixture through a strainer into a bowl or saucepan. Press out as much juice as possible with a wooden spoon. Put the blueberry juice back on the stove and add arrowroot and maple syrup. Continue heating and stirring on medium heat until the mixture thickens slightly. (The blueberry pulp can be added back into the syrup to make a blueberry sauce. This utilizes the entire berry. Or you can put the pulp in a blender to break up the skin, and then add it to the syrup.)

Yield: 1¼ cups syrup, or 1½ cups sauce

Fruit Syrup

Fresh fruit syrup is nothing more than a puree or thickened juice. You don't need to add sweetener – take applesauce and apple butter as examples – but this general recipe calls for honey.

2   pounds fruit or berries
½   cup butter
1   cup honey

Remove pits and skins of larger fruit. Puree the fruit through strainer or in blender. Melt the butter and add pureed fruit. Add honey and dilute with a little water or juice. A ½ teaspoon of cinnamon may be added to peach, apricot, and nectarine syrups. If too sweet, add lemon juice.

Yield: approximately 1 pint

Curious truth about the peppercorn

Of all the basic spices for the home the peppercorn is the nation's favorite. And to my mind this is curious, in that most recipes call for minute quantities—a pinch or a dash, for example—of black or white pepper. And this predilection for the spice is no recent phenomenon, not by a long shot.

When in the fifth century Rome was besieged by Goths part of the ransom demanded from the city was 3,000 pounds of peppercorns. A thousand years ago, peppercorns were exacted from outsiders for the privilege of trading in London.

Although there are many fin becs and feinschmeckers (including some of the world's great chefs) who deem white pepper to be far superior to black pepper, Americans by and large prefer the black. Or at least they choose and use black pepper indiscriminately because that's what they are accustomed to.

It has been observed that the British, Germans, and Scandinavians—en masse—prefer the white. So do the Chinese. One reason that chefs are inclined to use white pepper—in addition to its flavor—is that the specks or fine grinds of the pepper are not noticeable to the eye when added to white wine sauces, cream sauces, and so on.

There are, by the way, a variety of "grinds" of black pepper and, theoretically, each serves a different culinary purpose. The most common and best-known grind is known as "table grind" or "normal grind," which is what you find on supermarket shelves. Other grinds include "cracked" and "coarse," either of which may be massaged into the meat for steak au poivre before cooking, and "fine grind," or pulverized, which is to be added to sauces. White peppercorns customarily are finely ground or pulverized. Or they are used whole as for pickling, marinades, and so on.

Finely ground white pepper is what gives the "hotness" to the Chinese soup known as hot and sour.

It is an interesting point that most spice manuals are quick to point out that producers of ground pepper are more open to blending their product with impurities, such as powder, than

any other spice. Which is one reason why many a gourmet prefers to buy his peppercorns whole and grind his own, using a peppermill (or, in the case of cracked pepper for steak au poivre, with a mallet or the bottom of a flat, clean heavy skillet). Obviously, too, freshly ground pepper has more pungency than that which has been packaged for even a short time.

Incidentally, both black and white peppercorns are one and the same except for age and treatment. Black peppercorns are those which are picked before they are fully ripe. They are allowed to dry in the sun during which period they turn black, or a blackish brown.

The peppercorns for white pepper are picked after maturity. At that point they have a reddish cast. The peppercorns are packed or stacked in considerable quantity and tend to ferment. They are then washed well and dried. In Tom Stobart's excellent book, *Herbs, Spices and Flavorings* (Penquin Books, 1977), he points out that "In those days, white pepper may also have been made by rubbing off the outer layer of black pepper by machinery, though this is strictly not white pepper but decorticated pepper."

Pure white pepper is not as pungent or as aromatic as black pepper.

Those who prefer and have a certain connoisseurship of black pepper often aver that Tellicherry pepper is the finest of all. Although Tellicherry is a coastal city in India, the name is applied generally to all first-grade, "bold," large peppercorns grown on the entire Malabar coast of India. As someone has said, that isn't exactly earth-shaking news but it should be of interest to devotees of peppercorns. Look at your black peppercorns. If they are large and goodlooking, they are what is called in the spice trade "bold."

Within recent years, one of the great innovations in the field of food has been the use of green peppercorns in the kitchen—duck with green peppercorns, green peppercorn butter, steak with green peppercorns, and so on. These are the soft, green, unripe peppercorns taken from the vine and packaged. They were introduced only about a decade or so ago when a method was finally developed in the Malagasy Republic to process and ship them. They came about almost simultaneously with the much touted nouvelle cuisine of France.

One point of great interest to me: although steak au poivre has become one of America's and the world's favorite dishes, it does not exist in the works of Georges Auguste Escoffier. All of which is to imply that it is a culinary invention of the past 40-odd years. Escoffier died in 1935.

If you want to know the chemical appeal of pepper, it is said to be an alkaloid called piperine that stimulates the taste buds, makes saliva and gastric juices flow and, therefore, benefits digestion.

Some dishes are immeasurably improved with a last-minute (which is to say at-table) grinding of peppercorns—preferably black (the color is bolder and more appealing)—

from a pepper mill. Among them I would include most smoked fish (salmon, trout, sturgeon); fish in cream or white wine sauces; many pates and terrines; and almost all cream soups. In the kitchen, I would recommend —almost invariably—freshly ground or preground white pepper for cooking.

Craig Claiborne © 1979 by The New York Times Company. Reprinted by permission.

Carob

Carob is a sweet powder ground from the long, leathery pods of a tree indigenous to the Mediterranean. Donkeys carry burlap bagfuls down to the roadside to be trucked away. The pods can be eaten right off the tree for a chewy snack—all but the rock-hard, lentil-shaped seeds. In this form, carob doesn't have much of a resemblance to the food it often stands in for—chocolate. The similarity is brought out only through processing.

Carob is more nutritious and less fattening than chocolate. For years, commercial cocoa manufacturers have used it as a cocoa extender because it is cheaper but has a somewhat similar taste and texture. Carob has only 51 calories to the ounce compared to cocoa's 84 (and 151 for sweet chocolate). And yet it is almost 50 percent sweeter, allowing you to cut all or most of the sweetener from your chocolate and cocoa recipes.

Unlike cocoa, carob has no caffeine or theobromine, another stimulant. And people who are allergic to chocolate can safely enjoy carob dishes. What's more, carob—unlike chocolate—contains no oxalic acid to interfere with the body's ability to assimilate calcium.

Carob has a long and interesting history. It is believed the fruit of this "chocolate tree" was used to feed Mohammed's armies. This fruit also sustained John the Baptist during his sojourn and meditation in the wilderness (Mark 1:6), and provided for the Biblical prodigal son (Luke 15:16) who was hungry and without money. As recently as World War II, troops on the island of Malta and in Greece survived the German occupation by eating carob.

The carob powder or flour found in supermarkets and health food stores is produced by drying, grinding, and roasting the pods. Various types of sugar, chocolate, and cocoa may be mixed in.

Because carob powder is sweet, it can be used instead of sweetener in bread and pastry products— including bread, waffles, cakes, pie, pancakes, cereals (hot or cold) crepes, and muffins. Of course, carob will make these foods chocolaty brown in color and give them a somewhat chocolaty flavor. If this is undesirable, you can try mixing various proportions of carob and honey to find the mixture that suits you best.

If you purchase pure carob powder in bulk, you'll likely find it slightly coarser than usual, and it can be used as a flour substitute. For best results, the carob powder is mixed approximately half and half with another flour, such as wheat, corn, or rice flour.

Western readers who have a source of the whole pods may want to try making their own powder. You can also buy pods from Italian specialty shops at Easter and the mail order source mentioned below. The first step is to remove the hard seeds, which are notorious for gumming up

grinders (see recipe). One method is to place the washed pods in a pressure cooker at about 15 pounds pressure for 20 minutes. When cool and dry, they can be split open easily to remove the seeds. The soft pods can be cut into small sections and powdered in the blender.

Another technique is to break open pods with a pair of pliers, remove the seeds, and dry the pods slowly in the oven at a very low heat. Then, the deseeded pods are ground in a stone grinder, at a coarse setting since carob has a tendency to foul up the grinder. The coarsely ground pods are dried for a few hours more and then re-ground at a finer setting.

Source of pods. Survival Shop, P.O. Box 42216, Los Angeles, CA 90042.

Carob powder

Commercial carob powder is usually marketed as a chocolate substitute, but when we ground our own, the product wasn't even reminiscent of chocolate. This isn't to say that our rough grind was less flavorful—in fact, we preferred our powder to the talcum-fine storebought carob, which was more chocolaty but less sweet and even a

(Continued on page 258)

(Continued from page 257)

trifle bitter. We concluded that our powder was better suited as a sweetener than a mock chocolate.

We wondered why our carob was so different from commercial carob, besides the fineness of the grind. The label on a box of carob gave a clue: the powder had been toasted and enhanced with natural flavors. So, we tried toasting our powder. Some of us found the results more chocolaty—like cocoa, one said—but some only could say it had gained a bitter edge. As for the added natural flavors, vanilla is a likely one.

The Rodale Test Kitchen found that homemade carob powder cannot replace the commercial in recipes, as it absorbs liquid differently and has an altogether different flavor. Our tester Linda Gilbert says the homemade powder was successful

as a sweetener in its untoasted form, but that you should be cautious in using it, as additional liquid may be required.

While commercial powder can be mixed with an equal amount of water to produce a smooth, dark, chocolaty syrup, our powder absorbed twice the amount of water and still didn't resemble a syrup.

Basic Carob Syrup

1   cup carob powder
1   cup water

This basic syrup is used in our natural recipes in place of bitter or unsweetened chocolate. If you wish to convert one of your own recipes which calls for melted, semi-sweet chocolate, use the above recipe, adding ¼ cup honey and (if you wish) 2 tablespoons butter. The same procedure is recommended.

Carob Powder

½   pound carob pods
2   cups boiling water

Wash the carob pods well, and soak them in boiling water for 15 minutes. Drain well and let air-dry for about 30 minutes.

Break the pods into 2-inch pieces with your hands. Grind the broken pieces, a small quantity at a time, in the blender until broken well. Pick out all of the brown seeds and discard—they are rock hard and bitter.

Spread the partially ground carob and carob pieces out flat to dry. When dry, grind the carob to a powder.

If toasted carob is desired, preheat the oven to 300°F., spread powder out on a cookie sheet, and toast for 7 minutes.

Yield: about 1 cup

Essential oils and extracts

The preparation of plant-based oils and extracts passed from kitchen to factory long ago. Old herbal and receipt books give ways of coaxing the flavor and aroma out of flowers, herbs, and berries and fruits. Many fine sources, such as *Dick's Encyclopedia* (reviewed in the Condiments section) and Mrs. M. Grieve's *A Modern Herbal* (see Herbs), have been reprinted. Tucked between the quaint nostrums and downright deadly recipes can be found forgotten wisdom on making one's own stovetop preparations. Cooking oils are another matter. It often takes special equipment to express the oil from even the most lubricious of seeds and nuts.

Flavorful oils, extracts, and essences

Relatively simple methods can be used to prepare flavorful oils from herbs, including infusion or decoction, whereby the plant is bruised and then either steeped for some time or simmered in water or alcohol. The oil can be separated from the other liquid by distillation; water and alcohol, having lower boiling points, will vaporize first, leaving the oil behind. Or, you can put berries or herbs in a bottle of grain alcohol for at least a week and then pour off the liquid into a shallow vessel to evaporate. The oil will be left behind.

Peppermint oil can be made at home with a still. Fresh leaves are

minced and soaked in water. This infusion is then gently heated to drive off the volatile oil in steam. The product is a dilute preparation, called peppermint water in old books. In time the oil will rise to the top and, in large batches, can be separated off.

Lemon oil is traditionally collected in a simple, if tedious, fashion. Empty lemon halves are turned inside out, and the oily rind (now the inside of the bowl) is wiped with a sponge. The sponge, when saturated, is squeezed to release the oil. Orange oil can be prepared the same way.

A method for making vanilla extract follows, but the most direct way to use the beans is to finely grind them and add the powder as you would any seasoning. A venerable

text on commercial baking apologizes for this method: "in light-colored cakes and ices the appearance of what look not unlike particles of snuff scattered throughout the substance is unsightly." In today's snuff-scarce world, the vanilla flecks have come into fashion as a sign of natural flavoring in ice cream.

RY

Vanilla from the bean

The vanilla bean is the seed pod of an orchid. To get from the original seed pod to the finished bean is quite a long, tedious process. The beans are handpicked, spread out in the sun to dry, wrapped in blankets to ferment, and otherwise babied and pampered. Thus the high cost of vanilla. Vanilla beans are used more in Europe than in America. In this country the flavor is extracted by alcohol and is sold as pure vanilla extract.

The seeds in the long, thin bean seem to contain most of the flavor. They are edible, whereas the tough outer part of the bean can be used but then must be discarded.

To flavor a hot liquid, one splits the bean and soaks a piece of this split bean in the liquid. To flavor other items, such as cakes and cookies, one must split the bean, scrape out the seeds, and put these scrapings into the batter. Try using one-quarter or one-half of a bean for your first trial batch of cookies.

The remaining seedless husk can be used in baked custards, rice puddings, and so on.

Homemade Vanilla Flavoring

1 or 2 vanilla beans
½ cup boiling water
1 teaspoon liquid soy lecithin
2 tablespoons honey
2 tablespoons cold-pressed oil

Split the beans lengthwise, cut up into small pieces, and add boiling water. Let stand 6 to 8 hours in a tightly sealed container so flavor does not escape into the air.

After soaking, pour the bean and water into a blender and liquify. Strain through a small strainer and press to get out all the juice. Return the juice to the blender and add lecithin, honey, and oil. Blend for a few seconds. Pour into a jar with a tight lid. Keep in the refrigerator. This vanilla keeps well, as the lecithin and honey act as natural preservatives. Shake or stir well before using.

Courtesy Walnut Acres
Foods-By-Mail, Penns Creek, PA
17862.

Cider and the mother

The best cider is made from an assortment of apples. The fruit need not be perfect, but it should be firm and ripe. For large quantities of cider, you will need a fruit press or the services of a local cider mill; small quantities can be processed by hand.

Cut the apples and run them through a blender, food chopper, or processor, or crush them with a rolling pin on a chopping block. Put this crushed pulp (pomace) into a clean muslin sack—an old but clean pillowcase will work—and squeeze out all the juice possible. (For variety, try letting the pomace stand for a day exposed to air before pressing it. Some experts claim this improves the flavor of the cider.) Catch the juice in a glass or enamel container. Aluminum or unglazed metal will discolor the juice and affect its taste.

To store, pour the juice into glass bottles and refrigerate. The cider will stay sweet for about one week, and then begin to turn hard.

Once you've got apple cider, making vinegar is easy. You just let naturally occurring organisms turn the drink into vinegar. Strain the cider through a cloth to remove the sediment, and pour it into a crock, a wooden container, or a dark-colored glass jar. Leave ample headspace—about ¼ of the container—to accommodate the working liquid. Cover the container with a tea towel or a triple layer of cheesecloth to let the air in but keep insects and dust out. If the vinegar doesn't happen spontaneously, you can help it along by adding a lot of good wine—the cheaper jug brands contain preservatives. Commercial vinegars don't carry the vital spark, either, as they too have been sanitized.

Store the brew in a cool, dark place such as an unheated basement or garage. After four months, remove the cover and taste the vinegar. If it is strong enough, strain it through a triple layer of cheesecloth and store in sealed bottles. Let weak cider continue working. You may want to dilute strong vinegar with a little water.

(Continued on page 260)

Place sprigs of your favorite herb in vinegar; use flavored vinegar with oil for salad dressing.

(Continued from page 259)
The layer that forms on top of working vinegar is called the "mother." When the vinegar is strong enough to use, strain off the mother and save it to get the next batch of vinegar off to a quick start.

If you have no luck culturing vinegar, you might consider spending $49.95 for a half-gallon vinegar barrel full of live vinegar. You simply add wine as you use the vinegar. This system is available from Franjoh Cellars, P.O. Box 7462, Stockton, CA 95207. Wine and the People (907 University Avenue, Berkeley, CA 94710) sells a culture for $2.55.

Crackers

Commercial crackers are expensive and often tell a sad tale in their ingredient list. Try these baked wafers, and feel free to play with the recipe to suit your taste.

Cheese Wafers

For best results, be sure the cheese is finely grated and the nuts well-ground. If the cheese isn't fine enough, your crackers will be more lacy and fragile, and will brown quicker.

¼   pound butter
½   pound sharp cheddar cheese, finely grated
¾   cup whole wheat pastry flour
⅓   cup pecans, ground
¼   teaspoon cayenne pepper

Cream butter and cheese together. Mix in other ingredients and knead together well.

Form dough into a log about 1½ inches thick and about 11 inches long. Wrap well and refrigerate until firm. With a very sharp knife, cut log into thin slices (about ⅛-inch thick). Arrange on a foil-lined baking sheet.

For heftier crackers, simply form dough into small balls and flatten with a fork.

Bake at 375°F. for 8 to 12 minutes, until lightly golden. Watch them carefully —they can burn easily.

Yield: about 4 dozen

It's not easy to reproduce commercial crackers in the kitchen. Someday someone will write a cracker cookbook that'll tell how the big bakeries make those light, fluffy nothings.

Until then, you can experiment with ingredients and procedures— and you shouldn't have any trouble making a more healthful cracker.

Many cracker recipes call for lots of oil, but excellent "flat breads" imported from Scandinavia contain no oil whatsoever. Flat bread is a good term to keep in mind as you bake crackers. Try to turn your favorite bread recipe into a flat bread recipe. Just keep in mind that crackers need a very short stay in the oven.

Jams and jellies

In the struggle to make jam or jelly with honey instead of sugar, many people end up with jars of thin, syrupy products that lack firmness. Commercial pectin helps. But for careful label readers, there may be a disquieting feature about those products—they usually contain sugar. Take heart. You can make pectin—often called apple jelly stock—ahead of time and preserve it for later use, if you enjoy making combination jellies or blending them with other fruits in season when fresh apples are not available. Or, if you are lucky enough to have the fresh apples at the same time as the other fruits, you can use the pectin immediately.

Homemade pectin

Use the jelly stock, or pectin, with any fruits that are low in pectin, or lack it entirely. These are the ones that do not jell readily without some help from apple pectin. Fruits such as pears, peaches, and cherries, along with berries such as strawberries, blueberries, elderberries, mulberries, and raspberries will all ap-preciate that extra boost of pectin from the apples.

What kind of apples should you use to make pectin? Apple thinnings—those small, immature green apples sold in the early summertime—are rich in both acid and pectin. They will make good jelly stock and give a snappy tartness to the product. However, if you prize the clarity of the jelly product, be warned that such apples will not produce as clear and transparent a jelly as pectin made from fully mature apples. If you happen to have a bumper crop of apples, you will like the idea of using some of the surplus to make apple stock. It represents one more good use for that large supply. You can use imperfect fruit, even with insect damage, bird pecking, bruises, or cuts from dropping from the trees. Merely cut away the imperfect sections, and use the sound parts.

Wash the apples carefully, trim, and cut pieces into thin slices. Measure one pint of water for every pound of apples. Place the slices in a kettle, and pour the measured water over them. Cover the kettle and boil for 15 minutes.

Strain off the free-running juice through one thickness of old-fashioned cheesecloth, without attempting to squeeze the pulp. Return the pulp to the kettle, and add the same measure of water again. This time, cook the mixture at a lower temperature for 15 minutes. Allow it to stand for 10 minutes, then strain the second batch of juice through one thickness of cheesecloth. Again, do not attempt to squeeze the pulp. Allow it to cool

Sugarless Amaranth-Strawberry Jam

Faye Martin, who developed this recipe, has served as consultant to Rodale's kitchens. She says this jam is a nutritious replacement for standard jam. "Spread on whole wheat bread, it makes a nutritious breakfast for people in a hurry, or a snack for growing children."

Like many great discoveries, amaranth's gelatinous quality was found quite by accident. "I was cooking the grain to determine what kind of hot cereal it would make, when I noticed its sheen, and the way it held together in the saucepan. It looked like a good quality jam."

Faye combined it with fruit, so that the product would have a lively, natural sweetness. The resultant spread needs no sugar or added pectin.

Almost any fruit can be combined with the amaranth grain, but the recipe we offer here is Faye's personal favorite.

2 cups water
1 cup raw amaranth seeds
2 cups fresh strawberries, washed, hulled, and chopped
2 tablespoons honey (optional)
1/3 lemon, chopped very fine with rind included

In a small saucepan, bring water to a boil. Add amaranth, strawberries, honey, and lemon. Stir until the mixture comes to the boiling point again, then lower heat and cook slowly for about 45 minutes, until mixture is thick and amaranth grains are tender.

Yield: 2 cups spread

enough so that you can handle it. Squeeze out the remaining juice, and combine all that you have. There should be about one quart of juice for every pound of apples you used.

You can use this stock immediately for blending with other fruit juices to make jelly or jam, or you can preserve it for future use. If you wish to can the stock, heat it to the boiling point and pour it immediately into hot, sterilized canning jars. Seal, and invert the jars to cool.

If you prefer to freeze the stock, allow it to cool, and then pour into freezer containers. Allow one inch of space at the top for expansion.

What proportion of pectin to other ingredients do you use in making jelly? To make a simple honey jelly, for each 2½ cups of honey and ½ cup of water, use a pint of homemade pectin (i.e., 2 cups). Proportions would vary somewhat using fruit juice, depending on the fruit or fruits used. The more pectin used (up to a point), the thicker the jelly. However, the more pectin used, the weaker the flavor of the other fruit.

(Continued on page 262)

PECTIN TEST

Before you seal your jars, here is a simple test to determine if you really have made pectin. Remove about one teaspoon of the juice you have made and put into a cup. If this liquid is not cool, put cup into the refrigerator a few minutes. Then add one teaspoon of grain alcohol or rubbing alcohol, stir gently, and then let stand three to five minutes. Pour this mixture into a saucer. If lumps of gelatin form, it is moderately high in pectin. If nothing happens, or if there are only a few lumps of gelatin, you may need to boil your liquid pectin extract again to further concentrate the pectin. Don't boil the extract too rapidly because too high heat or even heating too long can degrade pectin. Remember *not to taste* the mixture tested with rubbing alcohol because it is poisonous. Throw it out, and rinse all your utensils thoroughly.

Before testing for the pectin content of any fruit, it must be cooked for 5 to 10 minutes and then cooled.

Grape Jelly with Homemade Pectin

2½ cups unsweetened grape juice
½ cup honey (mild-flavored)
2 cups homemade pectin

Combine ingredients, bring mixture to a boil, and boil vigorously for 10 minutes. Stir the jelly to keep it from foaming over the top, but do not turn down the heat. Pour jelly immediately into sterilized, hot jelly jars and seal. Do not move the jars until they are cool and partially set. Refrigerate to complete the gel.

Beatrice Trum Hunter

FRUIT COMBINATIONS THAT REALLY GEL

Fruits that make good jelly are high in both pectin and acid. However, other fruits can be used if the pectin or acid content is adjusted by the cook. For example, blueberries are high in pectin but low in acid. By adding lemon juice or any fruit high in acid (like strawberries), you will get a good fruit jelling combination. If a fruit is low in pectin, you can add your homemade pectin extract.

The following listings show where fruits stand in the pectin and acid ratings.

Fruits high in both pectin and acid. Sour apples, lemons, limes, cranberries, red currants, sour oranges, grapefruit, sour plums, Concord grapes, wild grapes, and unripe quinces.

Fruits high in pectin, low in acid. Sweet apples, sweet oranges, blueberries, clingstone plums, sweet cherries, ripe quinces, crab apples, and tangerines.

Fruits high in acid, low in pectin. Fresh or dried apricots, rhubarb, sour cherries, ripe, and unripe strawberries, and pineapple.

Fruits low in both acid and pectin. Bananas, nectarines, pears, raspberries, peaches, and all overripe fruits.

Combine different fruits to achieve high-pectin, high-acid combinations. For instance, you can add rhubarb (high acid) to ripe quinces (high pectin) to make a satisfactory jelly. Likewise, you can combine peaches with strawberries and homemade pectin for a good product. Use your imagination to create one-of-a-kind preserves.

(Continued from page 261)

Fruit butter

To prepare a fruit butter, put quartered fruit, unpeeled for extra flavor, in a large roasting pan (not aluminum); add ½ to 1 cup water, depending on the amount of natural juice in the fruit; cover; set the oven for 300°F.; and grab a book or a broom while the fruit cooks. Stir occasionally so that it cooks evenly. When fruit is soft, follow your own recipe for the type of fruit butter you are preparing. Measure ingredients back into clean roasting pan and mix well. Return to a 300°F. oven (Do not use the bottom shelf, as it is too close to the heating unit.) Cook butter until it holds its shape on a cool saucer and no juice sneaks out the edges. Put into hot, sterilized jars. Adjust sterilized caps and leave at room temperature to cool. Make sure each is sealed before storing.

Irene Piveral

Sweeteners

Most natural foods books spurn sugar, while talking up honey and maple syrup. Honey and maple syrup *are* natural foods, by some definitions, but beneath the delicate flavor and the fine traditions of hive and maple is that controversial food, sugar. How does sugar act on the body? What nutritional advantage do the so-called natural sweeteners have?

A nation sweet on sugar

Like a dictator whose every command is heeded by the governed, the sweet tooth is sovereign over the American palate. Scarcely a single ice cream parlor or an aromatic bakery can be encountered these days without the sweet tooth making sly, but powerful, commands to drop in and sample two scoops piled high on a cone, or some warm jelly donuts, or a thick, flaky napoleon.

American food manufacturers have taken full advantage of this national craving. Why should white sugar be in desserts alone? is the industry's thinking. Why shouldn't it be in, say, hot dogs, or blue cheese dressing or even salt for that matter? Sugar has become so popular an ingredient in the food supply as to be almost ubiquitous, as the accompanying chart shows.

While naturally occurring sugars make up only 6 percent of the calories in our diet, the percentage of calories from added sugar is three times that amount. Annually, we consume close to 95 pounds of table sugar. This is more than twice what it was 100 years ago. Also greatly on the rise is the use of other sweeteners, such as corn sugar (dextrose), corn syrup, and honey. In 1978, we ate more than triple the amount of corn syrup than in 1960. Adding up all the sugars the average American eats, we consume about 135 pounds of sugar per year.

Most naturally occurring sugars do not present a problem because we eat them in foods which also have protein, minerals, vitamins, and fiber. But, as the U.S. Senate Committee on Nutrition said, the main problem with added sugar is "the danger in displacing complex carbohydrates, which are rich in nutrients." Simple sugar furnishes nothing but calories to our diet. These "empty" calories either replace nutritious foods or increase the number of calories we take in, which may lead to obesity. Obesity, in turn, increases the risk of heart disease, diabetes, and other health problems. Moreover, sweeteners, especially when eaten in sticky foods between meals, cause tooth decay, a problem that costs Americans $3 billion annually.

From an article in *Nutrition Action*, a monthly publication of the Center for Science in the Public Interest. *Nutrition Action* is available from CSPI, 1755 S Street, NW, Washington, DC 20009, for $10 a year.

The rap against sugar

Prompted by a jump in sugar prices several years ago, the *New York Times* financial page ran a story on alternative sweeteners. One soft-drink executive was telling the writer about differences in the way the alternatives tasted in soda pop when he suddenly stopped himself. "But really," he asked, "who can tell? If you ice sodas or even beer sufficiently, no one can tell the difference, much less what they are drinking. What Americans want is something wet, cold, and sweet—that's all."

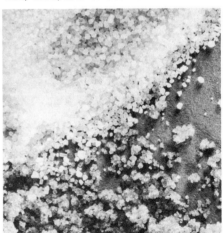

Is there such a thing as a healthful sugar?

Now, you may not want to be told exactly what you want, but this guy does have statistics behind him. In 1977, the American diet included 137.8 pounds of sweeteners for every man, woman, and child in the country. That figure represented a 21 percent rise from the 114.2 pounds consumed per capita in 1961. Sweeteners (both added and naturally occurring) now comprise nearly one-fifth of America's calorie intake.

(Continued on page 264)

The Home Food Systems *staff asked Anita Hirsch of the Rodale Test Kitchens for suggestions on replacing sugar with other sweeteners. Here's her reply.*

MEMORANDUM   DATE: April 16, 1980

TO: Jean Polak

FROM: Anita Hirsch

RE: Substituting Other Sweeteners for Sugar in Recipes

Generally, each cookbook suggests a different level for substituting sweeteners for sugar.

The [Allentown, Pennsylvania] *Morning Call* advises replacing sugar with up to one cup of honey, "but don't try to go beyond that amount or you'll be courting disaster." It suggests reducing the liquid by ¼ cup for each cup of honey and lowering the oven temperature by 25° F.

Some cookbooks suggest complicated conversions for every cup of sugar:

1½ cups molasses or maple syrup minus 3 tablespoons liquid used in the recipe.

3/4 cup honey minus 3 tablespoons liquid used in the recipe

1 cup honey and reduce liquid by ¼ cup, and in baked goods add ½ teaspoon baking soda for every cup of honey and lower temperature 25° F.

And then there are some books that advise not only using 3/4 cup of honey for each cup but also decreasing the liquid by ¼ cup; but if there is no liquid, you must add ¼ cup of flour.

The Joy of Cooking says to use 7/8 cup of honey for each cup of sugar in baking cakes and cookies; but in making bread and rolls, use 1 cup of honey for 1 cup of sugar and reduce the liquid by 3 tablespoons for each cup of honey substituted.

All this gets very complicated! I have found that in substituting honey for sugar, the amount of honey used is half that of sugar. I never change the temperature nor have I ever added baking soda. On a rare occasion I have decreased the liquid when a batter or dough would appear to contain too much liquid because of substituting honey.

Generally, baked products are heavier when sweetened with honey (as they are when made with whole wheat flour).

Sugar is not healthful and this includes honey and the other sweeteners, so cutting down the amounts of any sweeteners in a recipe makes sense.

The choice of sweetener also depends on the flavor you wish to have in the final product. Clover honey is the most mild and dark buckwheat honey is strong-flavored. A strong-flavored honey will overpower a recipe. Buckwheat honey will hide the taste of a delicate tea.

Memorandum to Jean Polak
Page 2

The degree of sweetness of sweeteners varies, and there are no standards. Even clover honeys vary in sweetness, depending on time of the year, brand, or other factors.

The functions of sweetener in a cake are to color the crust, sweeten the cake, increase the volume, and increase the tenderness. If a cake has too little sweetener, it can lack sweetness, be tough in texture (this also depends on the butter content because butter also contributes to tenderness), turn out low in volume, and have a pale crust. If there is too much sweetener, the cake may fall, there will be excessive browning of the surface, and the crust will be sticky and appear crystalline.

The following recipe for gingerbread was prepared eight times, using a different sweetener for the molasses each time.

    2 cups Barbados molasses
    3/4 cup butter
    1 teaspoon baking soda
    1 teaspoon nutmeg
    ½ teaspoon cinnamon
    3 cups whole wheat flour
    1 cup buttermilk or sour milk
    1 egg beaten lightly
    1 teaspoon baking powder

Bring molasses and butter to a boil in a big pot. Remove from heat and add baking soda and spices. Add flour and milk in alternate dollups. Then add the egg and baking powder. A 12 by 16 by 2-inch jelly roll pan should be oiled and heated ahead of time. Pour the mixture over the warm pan and bake 15 minutes at 350° F.

The sweeteners used were:

    2 cups maple syrup
    2 cups sugar
    4 cups sugar
    2 cups sorghum
    2 cups corn and barley malt syrup (very thick syrup, a natural food brand)
    2 cups Barbados molasses
    2 cups honey
    2 cups blackstrap molasses

The results were tested for appearance, volume, flavor, and crumb texture.

The maple syrup sample tasted sweet but bland. It rose well, had good texture, but was very pale in color.

The 2 cup sugar sample rose well, tasted sweet and sugary, had good texture with a spotted caramel top. The volume was high on the sides and low in the middle.

The recipe with 4 cups sugar had poor volume, a very granular sugary texture, and was browned as a result of too much sugar.

Memorandum to Jean Polak
Page 3

The honey sample had a very moist texture, sweet honey taste, and good color.

The barley and corn syrup sample was very moist with good color, mild flavor, and volume high on sides and low in middle.

The blackstrap sample had a moist texture, good volume, good strong molasses flavor, and very dark appearance because of the color of the molasses.

The sorghum sample had acceptable volume (lower than all above except the high-sugar recipe), good taste, good color, good appearance, and kept its shape when cut. The volume was equal throughout--other samples sank in the middle.

The Barbados molasses sample had good color, was very moist (and therefore difficult to cut), tasted good, sank in the middle, with the volume about the same as sorghum.

Members of the Home Food Systems staff were invited to taste these gingerbread samples and each had his own preferences. I prefer the taste of the sorghum or blackstrap samples.

Combinations of these liquid sweeteners could be used to achieve different tastes. A combination of blackstrap and honey or molasses would result in a milder tasting product and still give some of the mineral benefits of black-strap molasses.

There are also taste changes from product to product. One molasses may taste stronger than another. Sorghums differ in taste.

Another factor to consider when using various sweeteners is cost. Right now (April, 1980), the costs are:

	Quantity	Retail cost
	16 oz.	1.89
	16 oz.	1.23
Carob syrup (clear)	16 oz.	1.69
Corn and barley malt syrup	16 oz.	1.53
Rice syrup	15 oz.	2.10
Sorghum	16 oz.	3.62
Molasses (Barbados)	15 oz.	1.59
Maple syrup	16 oz.	1.20
Blackstrap molasses		
Honey		

In conclusion, I believe that when substituting honey or any other liquid sweetener for sugar, first start with half the amount stated in the recipe. After you taste the product, you may find you would prefer a different sweetener, more or less sweetener, or a combination of sweeteners. You would, of course, be guided by the product you are making; for example, you wouldn't use all blackstrap molasses as a sweetener in custard.

Anita Hirsch

AH/ch

(Continued from page 262)

The largest single user of sweeteners is the beverage industry which, in the process of satisfying the American craving for something wet, cold, and sweet, accounts for about 23 percent of the sweeteners used in this country. You have to understand that the American sugar addiction is not a matter of people heaping tablespoons of sugar on their salads, mashed potatoes, and roast beef. Big business, the food processing industry, feeds our habit. At the turn of the century, about two-thirds of the sugar consumed in this country was used in the home, and one-third was used by the food processors. Today, the figures are reversed.

By far the largest bulk of the sugar we eat comes not from the sugar bowl, but from the overprocessed, overadvertised products we take off the supermarket shelves. Sugar, in one form or another, is found in soy sauce, chicken pot pies, ketchup, crackers, peanut butter, instant soup, bouillon cubes, mayonnaise, spaghetti sauce, even dog food. One salad dressing recently tested by the *Consumers Union* contained more than three times as much sugar as Coca Cola by weight. Even if you *try* to avoid eating sugar, the food processing industry makes it difficult to do so.

Clearly, there is something very basic about the human craving for sweets. The sweeter a formula is, the harder a newborn infant will suck at the bottle. Even a five-month-old fetus suspended in its mother's womb will increase its swallowing rate when saccharin is injected into the amniotic fluid.

But if the basic desire is present in all of us, that does not mean that the desire blossoms into addiction all by itself. Scientists in England recently surveyed a group of 94 mothers on their dietary habits and the diets of their infants. The babies received sweets an average of 4.3 times a day. Seventy-seven percent of the meals and snacks the babies received contained sugar. About a third of the mothers were even adding sugar to their babies' milk.

There was a definite relationship between the way the mothers ate and the way their babies ate. Of the babies eating high amounts of sugar, 61 percent had mothers with a high sugar consumption, and only 13 percent had mothers eating low amounts of sugar. Only 17 percent of the mothers who did not drink coffee with sugar put sugar in their babies' milk, while 83 percent of the mothers who took their coffee with sugar also spiked their children's milk with the stuff.

Whether those children will grow up to eat as much sugar as their mothers is anyone's guess, but they certainly were given ample opportunity to pick up the habit. "Even in these early months," the British scientists concluded, "sugar is given the connotation of 'nice' and 'good' by the way that adults use sweet foods and drinks as comforters, gifts, bribes, and rewards. Once the baby has learned to associate sugar with 'good' during early childhood it may be difficult for health educators and parents to alter habits for the older child."

The truth of the matter is that sugar is by no means "nice" or "good." The sugar industry vigorously contests every charge that their product endangers human health, but even they are hard pressed to produce evidence against some of the accusations.

It's firmly established, for example, that sugar causes tooth decay. Sugar-eating bacteria in the mouth produce a thick, sticky substance called glucan. Glucan clings to the teeth, speeding the buildup of deposits on the teeth. Plaque holds the bacteria against the teeth, and the bacteria feeding on the sugar produces a number of different acids which eat away at the tooth. The result is a mouthful of cavities.

Sugar is also a major contributor to America's weight problem. Sugar in and of itself does not make you fat. It's when you eat calories in excess of your body's needs that fat begins to build up. But sugar-sweetened foods like ice cream, pies, and cake are often excessively rich in calories from sources other than sugar, and we've already seen how a little sugar makes you want to eat more of it.

In addition, there are indications that sugar may, through some unknown mechanism, cause a bigger buildup of fat than an equal amount of other carbohydrates. USDA scientists have discovered that rats fed table sugar get fatter than rats fed starch, even if both feedings contain comparable amounts of calories.

The scientists at the USDA compared the effects of sugar and starch on people as well as rats. They found that the same amount of sugar consistently produces higher levels of sugar and insulin in the blood. Insulin is the hormone which regulates the levels of sugar in the blood. When insulin fails to do its job, blood sugar rises dangerously and diabetes results. The onset of diabetes in adults is preceded by a rise of insulin in the blood, exactly the response in people fed sugar rather than starch. Sugar also affected the levels of several fats in the blood. High levels of the blood fat cholesterol, and a group of fats called triglycerides, are associated with an increased risk of heart disease. Again, the USDA study found that people eating sugar rather than starch had significantly higher levels of both cholesterol and triglycerides in their blood.

Other ailments have been linked to sugar. A British survey of patients suffering from Crohn's disease, an inflammatory ailment of the intestine, recently revealed that the patients' intake of refined sugar was considerably higher than normal. Sugar has even been linked in laboratory studies to a possible breakdown of the body's basic defenses against infection in general.

There is plenty of evidence that can be used to deny the possible links of sugar with various diseases. We simply don't know enough to prove these connections beyond a shadow of a doubt. But no one can deny the fact that sweeteners making up one-fifth of the American diet are displacing foods that are much more essential for human health.

Sugar is a carbohydrate. There are various kinds of sugars, one of which is table sugar, or sucrose. The others, lactose, glucose, fructose, maltose, are, like sucrose, simple carbohydrates. These sugars enter the body fast and are used fast. The complex carbohydrates found in fruit, vegetables, rice, and grain are broken down more slowly. The foods containing complex carbohydrates contain other essential nutrients that are totally absent in the sugars, nutrients like fiber, vitamins, and minerals.

Seventy years ago the various

(Continued on page 266)

ADDED SUGAR IN PROCESSED FOODS

Food	Serving size	Calories per serving	Tsp. sugar in serving	Sugar, % weight
Beverages				
frozen concentrate grape juice	6 oz.	100	1.0	2.0
Coke	12 oz.	144	9.2	10.0
Welch's grape juice drink	6 oz.	110	3.9	9.3
Kool Aid (sugar sweetened flavors)	6 oz.	68	4.3	10.0
Bright & Early frozen concentrate imitation OJ	6 oz.	100	5.4	13.0
Cereals				
General Mills Cheerios	1¼ c.	110	0.2	3.6
GM Wheaties	1 c.	110	0.7	11.0
GM Total	1 c.	110	0.7	11.0
GM Kix	1½ c.	110	0.5	7.1
GM Lucky Charms	1 c.	110	2.7	39.3
GM Nature Valley Granola Cinnamon & Raisin	⅓ c.	130	1.7	25.0
Post Alphabits	1 c.	110	2.7	39.0
Post Raisin Bran	½ c.	90	2.2	32.0
Ralston Purina Cookie Crisp (chocolate chip)	1 c.	120	3.0	46.0
Kellogg's Fruit Loops	1 c.	110	3.5	50.0
Kellogg's Sugar Pops	1 c.	110	3.2	46.0
Kellogg's Special K	1¼ c.	110	0.5	7.1
Kellogg's Corn Flakes	1 c.	110	0.5	7.0
Kellogg's Raisin Bran	¾ c.	110	0.7	10.7
Kellogg's All Bran	⅓ c.	60	1.0	14.3
Kellogg's Apple Jacks	1 c.	110	4.0	57.0
Kellogg's Sugar Frosted Flakes	⅔ c.	110	2.7	39.0
Kellogg's Country Morning	⅓ c.	130	1.5	21.0
Kellogg's Rice Krispies	1 c.	110	0.7	11.0
Kellogg's Sugar Smacks	¾ c.	110	4.0	57.0
GM Boo-Berry	1 c.	110	3.2	46.0
GM Cocoa Puffs	1 c.	110	2.7	39.0
GM Count Chocula	1 c.	110	3.2	46.0
GM Golden Grahams	1 c.	110	2.7	39.0
GM Crazy Cow	1 c.	110	3.0	43.0
Nabisco 100% Bran	½ c.	70	1.5	21.4
Instant Quaker Oatmeal with cinnamon and spice	1⅝ oz.	176	4.0	35.2
Quaker Captain Crunch	¾ c.	110	3.0	43.0
Quaker Life	⅔ c.	105	1.2	18.0
Quaker 100% Natural	¼ c.	140	1.5	21.0
Shredded Wheat	1 biscuit	90	—	—
Condiments				
blue cheese salad dressing	1 tbsp.	75	0.2	6.7
french salad dressing	1 tbsp.	65	0.7	18.8
Hellman's Spin Blend salad dressing	1 tbsp.	60	0.7	18.8
italian dressing	1 tbsp.	85	0.2	6.7
cranberry sauce	½ c.	203	11.7	35.1
catsup	1 tbsp.	16	0.6	17.0
Protein				
bacon (Oscar Mayer)	2 slices	70	0.0	1.7
beef bologna (OM)	2 slices	150	0.3	3.0
Canadian style bacon (OM)	2 slices	80	0.0	0.4
canned ham (OM)	3 oz.	105	0.1	0.4
luncheon meat (OM)	2 slices	190	0.3	3.0

From *Nutrition Action*.

(Continued on page 266)

(Continued from page 264)

sugars made up only 30 percent of our total carbohydrate intake. Now these sugars make up half of our total carbohydrate diet. The more sugar we eat, the less complex carbohydrates we seem to take in, and the harder it is to get enough of the vitamins, minerals, and fiber we need.

Michael Jacobson, a microbiologist and Director of the consumer organization Center for Science in the Public Interest, recently summed the problem up nicely. "When you consume 20 percent of your calories from sugar, that means you have to get all of your nutrients from 80 percent of your food. It's unlikely that you're going to get them."

For this reason the various sugar substitutes, used by the food processing industry because they are becoming cheaper than table sugar, and used by health food advocates because they are supposedly better for you, are really not much better than sucrose itself. The sugar industry tried to diffuse the alarm over high sugar consumption by pointing out that the amount of table sugar consumed by Americans has not changed since 1925. That's true, as far as table sugar goes, but the use of corn syrup and corn sugar (dextrose) has more than tripled since 1915. These are the kinds of sweeteners that are slipped to us by the food processors, and they are just as useless nutritionally as table sugar.

Health food advocates who substitute honey or blackstrap molasses for sucrose are at least getting *something* for their money. Honey is largely made up of glucose and fruc-tose, two simple sugars, but it does contain a few nutrients, like potassium. However, an adult would have to eat almost 300 tablespoons of honey each day to meet the government's Recommended Dietary Allowance for the nutrient. Blackstrap molasses does better than that nutritionally, but still has a high ratio of calories to nutrients.

Then there are the unrelated chemical substitutes for sugar, including cyclamates, saccharin, and xylitol. Not only are these chemicals nutritionally worthless, but a number of them are suspected cancer-causers. You don't have to go too far through the list of sugars and sugar substitutes before you realize that *none* of the alternatives shines brighter than the rest.

John Yates

ADDED SUGAR IN PROCESSED FOODS

(CONTINUED)

Food	Serving size	Calories per serving	Tsp. sugar in serving	Sugar, % weight
Protein (continued)				
peanut butter	2 tbsp.	182	0.3	4.0
pork sausage	3 links	195	0.3	2.0
hard salami	6 slices	210	0.1	1.0
Spam	3 oz.	264	0.8	4.0
weiners (OM)	one	140	0.3	3.0
vanilla ice cream, hard	½ c.	135	3.1	19.0
vanilla ice milk	½ c.	112	3.4	15.0
yogurt, lowfat flavored	8 oz.	194	4.1	7.0
yogurt, fruit	8 oz.	231	7.5	13.0
Snacks				
applesauce	½ c.	116	4.3	14.0
Columbo frozen yogurt (whole milk)	8½ oz.	138	5.3	22.0
canned pineapple, heavy syrup	½ c.	90	3.1	10.0
chocolate pudding	4 oz.	161	3.9	12.0
Cool Whip	1 tbsp.	13	0.2	23.0
Dannon frozen yogurt (vanilla)	½ c.	90	2.8	11.0
Gino's vanilla milkshake	12.1 oz.	310	7.5	9.0
graham cracker	2 crackers	55	0.9	25.0
Hershey's milk chocolate	1.2 oz.	185	4.4	51.0
Hunt's vanilla Snackpack	5 oz.	190	4.4	13.0
Jello, cherry	½ c.	80	4.5	15.0
Kellogg's brown sugar cinnamon Poptarts	1 tart	210	3.8	31.0
Vegetables				
beets, pickled (Del Monte)	½ c.	57	2.1	9.9
sweet peas (canned)	½ c.	75	0.9	4.5
tomato sauce	4 oz.	30	0.7	2.4
white refined sugar	1 tsp.	15	—	—
	1 tbsp.	46	—	—

From *Nutrition Action.*

A guide to sugars

Sucrose. This is ordinary table sugar. Sucrose consists of equal parts of the simple sugars glucose and fructose chemically joined together. Table sugar is refined from sugar cane or sugar beets. Syrup extracted from the beets or cane is boiled in steam evaporators, sugar is crystallized from the syrup in vacuum pans, and the sugar crystals are separated in a centrifuge. The remaining syrup, molasses, is treated several more times, and more and more sucrose is removed from it at each step. Further treatment of the raw sugar produces the 99.9-percent-pure sucrose we put in our sugar bowls.

Glucose. Glucose is also known as blood sugar. It is one of the most important substances found in the human body, though it is also present in nature in fruits and vegetables. Many of our basic foods are eventually broken down into glucose, which is burned by the body for energy and plays a key role in a number of other metabolic processes.

Glucose is what the sugar industry is really talking about when they say that "sugar" is the vital fuel that keeps our children running and jumping. Table sugar is converted to glucose all right, but so are the sugars and starches found in vegetables and starchy foods like potatoes. Those foods, as we have seen, deliver a solid package of a variety of nutrients, not just a quick shot of energy.

Fructose. Fructose, or fruit sugar, commonly found in fruits, berries, and honey, has been suggested as a healthful alternative to sucrose, particularly for diabetics. Unlike sucrose, fructose does not need insulin to be broken down for use by the body, and therefore, the argument goes, diabetics with insulin problems would be able to tolerate this sugar. Fructose is also about 50 percent sweeter than sucrose, and as a result a person would theoretically need fewer calories of fructose to achieve the same level of sweetness.

Fructose, however, has a number of drawbacks. It can cause diarrhea. Like sucrose, it produces increases in the levels of triglycerides and other fats associated with heart disease. The peculiar way that fructose is broken down in the body has been shown to disrupt the normal metabolic action of the liver. And in the end, fructose, like sucrose, remains nothing but empty calories.

Dextrose. Also known as corn sugar or grape sugar, this sweetener is not as sweet as sucrose but has the same food value. It is produced from cornstarch and, along with the other corn sweeteners, is used extensively by food processors. The jump in U.S. sweetener consumption in the last two decades has been caused primarily by the jump in the use of corn sweeteners in processed foods.

Corn syrup. This is the liquified form of dextrose, also used extensively in the processing of foods. The existence of the various forms of sugar additives enables the processors to play a shell game with the sugar content of their products. When they list a product's ingredients on the label, the largest component of the produce is always listed first, but almost never will you find "total sugar" listed as an ingredient. Instead, sugar content is broken down into categories and strung out through the list— sucrose, dextrose, corn syrup, honey—so there's no way of telling how much sugar you're really getting.

High-fructose corn syrup. HFCS is fructose isomerized from dextrose. It may eventually capture a good half of the food processing market. Japanese scientists perfected the process, which gives processors the high sweetness of fructose and the low cost of corn sugar. Many soft drinks now contain HFCS, and Smuckers, the jellies and jams people, use the sweetener exclusively in their products.

Sorghum syrup. Sorghum syrup is a sweetener derived from the starch of grain sorghum, just as corn syrup is derived from cornstarch.

Maple syrup. Maple syrup is made by concentrating the sap of the sugar maple tree, collected in New England in the early spring. The syrup is basically the sap of the tree with the water removed. It consists largely of sucrose.

Honey. Honey is the result of the action of enzymes contained in flower nectar, or in the bodies of the bees that collect the nectar. Fermentation of the nectar takes place under the influence of those enzymes, and the bees evaporate excess water from the product, resulting in the syrup we call honey. Honey contains a combination of several different sugars, mainly fructose and glucose. Between 75 and 80 percent of honey is sugar of one kind or another, so this is hardly a health food.

However, there are some nutrients present. Honey contains modest amounts of potassium, calcium, phosphorus, B vitamins, and vitamin C. It is about twice as sweet as table sugar, which is an advantage if you're counting calories.

Raw sugar. The first sugar crystals separated from cane syrup in the refining process are called raw sugar. In this form, raw sugar is banned from sale in the United States, since it contains impurities like dirt, molds, bacteria, lint, waxes, and insect parts. Sanitized, it is sometimes sold in natural foods stores as turbinado sugar, but this product is still 96 percent sucrose. Raw sugar is not much different in nutritional quality than table sugar.

Brown sugar. Supposedly a less refined product than table sugar, brown sugar in actuality is nothing more than white sugar sprayed with a small amount of molasses to give it a darker color. Nutritional differences between brown and white sugar are so tiny as to be meaningless.

Molasses. Standard molasses bought in a supermarket is not a byproduct of sugar refining. It is the specially manufactured, full juice of the sugar cane, clarified and concentrated to a thicker consistency. Blackstrap molasses, on the other hand, is the final by-product of the refining process, and is much more nutritious. It is cane syrup from which it is no longer profitable to extract any sugar. It contains about half the sugar in the original cane juice, but retains all the nutrients.

Blackstrap molasses contains small amounts of the vitamins thiamine, riboflavin, niacin, and pantothenate. Its mineral content is more impressive. There is almost as much calcium in a tablespoon of blackstrap molasses as there is in a half cup of

(Continued on page 268)

(Continued from page 267)

milk. The same tablespoon contains twice as much iron as a cup of raw spinach, and more potassium than a banana or two oranges. Blackstrap also contains the trace minerals magnesium and chromium.

Unfortunately, the taste of blackstrap molasses can seem pretty powerful to the uninitiated, so its use might better be confined to cooking at first. The taste may remind you that blackstrap molasses is the one sweetener with any substantial nutritional value.

John Yates

Honey

Any beekeeper worth his pollen can tell you the *bees* are the ones who do all the work, and the one or two stings you get a year are worth the harvest. You can keep bees in Manhattan or in the Mohave as long as there are fields or flowers within a few miles. And two hives will produce at least seven gallons a season in return for about 16 hours of work. Your fruits and vegetable garden will appreciate the bees' pollinating efforts, too.

Costs for starting beekeeping will run from $150 to over $300, depending on whether your hives and equipment are new or used, and whether you buy new or already established colonies.

With an established colony, you can get honey in the first season. Don't get stung trying to construct the hives yourself. Hives come in a standard size, and your workshop model may not be interchangeable with other equipment.

You'll need a hive tool, a veil for protection, a smoker for subduing the bees while you work the hives, and an uncapping knife for processing the honey. The extractor is a major investment, but necessary to getting the honey from the frames, unless you can share one with other beekeepers.

Make your own honey and you avoid the pasteurization, chemicals, and sugars that pervert some commercial honeys.

∎

The Food and Drug Administration has more than a passing interest in honey as a packaged food prod-

uct, and there are some (you'll excuse the expression) sticky problems with the sweet stuff that the FDA has to contend with. For one thing, some drugs are used in raising bees, since the species is subject to a variety of illnesses. For another, a form of botulism is alleged to be associated with honey and has been linked to the sudden infant death syndrome. And finally, because honey is expensive and high-fructose corn syrup, or invert sugar syrup, is cheap but looks and tastes similar to honey, some adulterated honey finds its way onto the market.

Two drugs used to treat bees are Fumidil-B (fumagillin) and Terramycin (oxytetracycline). Fumidil-B is used to prevent nosema disease, caused by a parasite that attacks the digestive tracts of the insects. Terramycin is another preventive drug, used as an aid in the control of American and European foulbrood disease, a bacteria-borne malady that gets into the hive and destroys young bees. The FDA is concerned about drug residues in honey.

Clostridium botulinum is the offending organism believed to be responsible for some cases of the sudden infant death syndrome. Reporting in the June 1979 *American Journal of Diseases of Children,* a California research group said that infants less than a year old should not be fed honey because of the danger of transmitting "infant botulism."

Studying a number of children who had become ill from infant botulism, the researchers discovered that almost half had been fed honey prior to the onset of the illness. Testing honey samples, they found up to 10 percent contain viable *C. botulinum* spores. In a California portion of the study, no spores were found in any of the several hundred food items tested other than honey, according to Drs. Richard O. Johnson, Susan A. Clay, and Stephen S. Arnon.

Typical of the honey adulteration cases was one earlier this year in Bowling Green, Kentucky. That state's Department of Human Resources was suspicious but not certain of some samples it collected of honey being sold in Bowling Green. The state agency notified the FDA, which provided more complete testing of the samples and ascertained

that the honey had indeed been watered down—or, if you will, "syruped down." The product, packed by Anthony's Syrup Co. of Philadelphia, Mississippi, was subsequently seized by a deputy U.S. marshall. Taken were 467 quarts and 83 pints valued at $846.50.

The reason for the adulteration is simple economics. In mid-1979 honey was being sold by the producers at about 50 cents per pound. On the other hand, a high-fructose corn syrup, which acts much like honey in some lab tests, was selling in bulk for about 10 cents a pound. It doesn't take much more than fourth grade arithmetic to figure out that substituting a 10-cent product for a 50-cent product can lead to many trips to the bank.

Adulteration may range from less than 10 percent to more than 70 percent, although the usual adulterated mixture is believed to be 20 to 35 percent invert sugar syrup. The extent of the problem is indicated by the detention of a total of 1.3 million pounds of imported honey in 1976 because it didn't meet Mother Nature's (and the FDA's) exacting standards. Recently, more sophisticated laboratory techniques had to be developed to detect the adulterated honey. That's because the cheaters have been using a high fructose corn syrup that, in mixture, would pass for pure honey under some older testing methods.

In addition to the FDA efforts to keep honey pure, the honey industry does its own self-policing, often relying on Dr. J. W. White, a retired USDA honey expert who has set up his own business, Honeytech, Inc., in Navasota, Texas, to test honey samples. White also gets referrals from honey packers, dealers, and users who want to be sure they're getting the real thing.

Dr. White says most of the defiled honey is sold to the commercial food processing market. However, for individual consumers, he advises them to watch out for honey that is thin, that is selling below the regular market price, and that has a mild or faint flavor of honey.

Consumers who get a batch they suspect should notify the FDA.

Roger W. Miller.
FDA Consumer.

Beekeeping:
New Tripoli, Pennsylvania

Eighty-year-old Clarence Yahn has kept bees almost all his life, and admits he still hasn't mastered everything there is to know about bees. But he certainly is a storehouse of information about his favorite insects.

Clarence has been watching honeybees live and work since 1908, when his father brought home a hive. That was the modest beginning of the Yahns' experience with beekeeping. The elder Mr. Yahn was a poultry farmer, and decided to add bees as a sideline. Within a few years he had 100 hives.

In time, Clarence Yahn established his own retail poultry farm, and delivered eggs directly to his customers' homes. He gave up bees for a few years when he was newly married and getting settled. Bees were just too fascinating, though, to give up forever. Pretty soon his hives were buzzing as his father's had. Even now, more than 70 years after his first encounter with honeybees, Clarence keeps 20 hives. With about 60,000 bees living in each one, that means he and his wife have the company of more than one million bees on their Pennsylvania farm.

The hives are set up differently from the standard backyard apiary. Clarence has them inside an old chicken coop, lined up against a wall and open to the outside so the bees come and go as freely as if the hives were free-standing. But during the winter the building shelters the bees from the worst of the snow and wind.

Each super—one story of a bee "apartment building"—weighs about 60 pounds when full of honey, and lifting them (gracefully so as not to upset the bees) is no easy job. When Clarence had a heart attack eight years ago his wife told him he'd finally have to abandon such strenuous work. Then he remembered that the chicken coop had a track in the ceiling that once had been used to move hay. He rigged it up with a block and tackle that he could slide along the row of hives. Now the system is electrified so he can lift a heavy super merely by pulling a string.

Bees actually require little day-to-day care. "Bees are self-sufficient,"

Clarence Yahn, beekeeper.

Clarence says. "They prefer not to be disturbed too much." He says you can tell a new beekeeper—they're always out checking their bees. He explains with admiration for the industrious honeybee that it takes 20,000 bees to carry the nectar needed to make one pound of honey. They visit two million blossoms for that same pound, and in the process fly 36,000 miles or more—which is like flying one-sixth of the way to the moon.

Just like people, bees have distinct temperaments. When he offered to open a hive for our inspection, he selected one holding quiet bees he knew would not intimidate us—even

though we were protected with traditional beekeepers' straw hats and veils. And before he opened the hive Clarence used his smoker to calm the bees. A smoker looks like a tin can with a spout in the front and a bellows in the back. A beekeeper lights a fire in the can with something that will make a smoky fire—Clarence uses dried-up, rotten wood. Pumping the bellows forces smoke out the spout and into cracks in the hive, scaring the bees so that they fill up with honey in preparation for escape. The point is, a bee full of honey theoretically won't sting.

Without coaching from Clarence,

(Continued on page 270)

(Continued from page 269)

the bees know what jobs need to be done, and which bees should do what. For about the first four weeks of their lives, bees spend time in the hive cleaning cells, feeding the larvae, making wax, and building comb. Some bees must guard the brood. Older bees venture out of the hive on test flights, then take off to gather nectar. Clarence's bees fly to nearby chicory, white asters, and purple asters he planted especially for them. He points out what a short time—just a fraction of a second—a bee stays on each flower. "When she's working on one kind of blossom, she won't go to a different kind until she's finished."

The bees keep up their busy pace all summer, in the fields and in the hive, nature's honey factory. The nectar they carry back to the hive isn't honey until the bees process it—and that method is a secret only the bees know. They make enough honey to keep alive and warm over the winter—and that still leaves half a ton each autumn for Clarence to extract and sell.

While they are hard at work gathering nectar, bees live only about six weeks. In that period, a hive loses a thousand bees each day. "Their wings wear out, and they die in the field. Bees are social insects," Clarence says. "They wouldn't dare drop their load of nectar and fly home empty-handed. So they don't come home." (The reason a whole hive doesn't die off in short order is that a queen bee can lay 2,500 eggs per day.)

When the temperature drops, the hum of the hive quiets considerably. The bees cluster together into a tight ball to pool their body heat. All they need for survival is some of their own honey. The temperature inside a hive in the summer is about 92° or 94°F., and in the winter the bees can maintain a temperature close to that in the center of the cluster. "The bees on the outside are a little stiff. . ." Clarence says. But together they make it through until spring warms the air.

Extracting time, usually in late September, is a big event of the beekeeper's year. Clarence starts by blowing all the bees out of the hives with a gasoline-powered air blower. Because the bees cap each honey cell to keep moisture out, he must carefully slice that outer layer off the comb. Then he puts the honey-laden frames, 20 at a time, into his extractor, which spins the honey out by centrifugal force. He screens the honey to remove any lumps of wax, but won't filter it because that would take out the pollen, which he considers good protein. He warms the honey to 110°F., then bottles it.

A small "HONEY FOR SALE" sign outside his home invites regular customers and passersby to enjoy the natural sweetener made by the bees they can see across the lane.

Clarence calls the honeycomb one of nature's construction marvels. Working in harmony, the bees build thousands of perfectly interlocked six-sided cells on each frame. They put two layers of cells back-to-back on one foundation. Instinctive bee engineering tells them to offset the centers of the cells—a construction method that makes the comb very strong. Just as important, the cells also tilt slightly upward so the honey doesn't fall out.

Bees seem proud of their workmanship. They keep a tidy hive, and quickly repair any damaged cells in the comb. If a beekeeper is careful during extracting, there is no reason a honeycomb cannot remain in good condition for many years. That's important, because bees who don't have to spend their time making wax and building comb can devote more of their energy to making honey. Some of Clarence's bees are using comb that is 10 or 15 years old.

Eventually, though, honeycomb will blacken from the many cocoons

Smoking the hives.

(Continued on page 272)

THE ABC AND XYZ OF BEE CULTURE

A. I. Root. 1974.
The A. I. Root Co., Medina, OH
44256. 712 p. hardcover $11.25.

This encyclopedic approach to bee-keeping doesn't follow the logical pro-gression of other books, but it sure is fascinating to wander through its pages of miscellany. *ABC and XYZ of Bee Culture* sports a purple embossed cover that gives it the appearance of a sawed-off high school yearbook.

Photo of page courtesy of
A. I. Root Co.

The "HONEY BEE" is "NUMBER ONE" of the insects exploited by "MAN"

What a volume of valuable products we receive from "HER"!

First is Pollination (USDA estimates $7 billion worth of farm produce)

Second is Honey, a valuable natural food

Third is Wax, which has many commercial uses

Fourth is Pollen, a valuable protein supplement

Fifth is Propolis, now being used medicinally

Sixth is Bee Venom, for the treatment of arthritis

Seventh is Royal Jelly, a beauty cream at $200.00 per pound

Eighth is Dead Bees, a good fertilizer

Ninth is Drone Brood, makes good fish bait

Clarence R. Yahn

Clarence's solar wax melter.

EXHIBITS OF HONEY 243

Harry Vandenberg, La Grange, Ohio, has made a beard of bees by hanging a queen in a queen cage from his chin and waiting for the bees to cluster around it, which took about 15 minutes

Earl Kellogg of Iowa, with a hat and chin full of bees. The face can be rubbed with re-pellent to keep the bees out of mouth, nose, and eyes. This sort of stunt should not be attempted by a novice.

the hive. After pulling out the frames he shows the bees and queen on the comb, then he calls to every-body to wait to see the next stunt, for he is going to make a swarm. Into a large newspaper spread out flat, which he has previously pro-vided, he shakes the bees from two or three combs. Then he takes it up and turns to the crowd, saying, "The bees are not real cross yet, so I'll be-gin to shake them up to make them so." He now gathers the four cor-ners of the paper together making a fold in the middle. With a quick jerk downward a couple of times he shakes the bees down in the fold of the paper. Next he turns the paper so that the fold will stand vertically over a straw hat. Two more quick jerks downward will send the bees in the fold of the paper pell-mell in-to the hat. There will be from a half pound to a pound of buzzing bees crawling all over each other as in a little swarm. If the work has been done right they will be so dis-concerted that they will be perfect-ly docile and can be picked up by

proceeds to take off his coat and vest and roll up his sleeves, after which he puts on bicycle pants guards or slips his trousers into his socks. The crowd will quickly appreciate this part of the performance because the operator tells them the bees will sting if they get inside of his cloth-ing. With a lighted smoker he opens

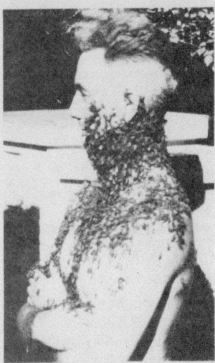

John Jessup of Perry, Iowa, proving that bees won't sting when one knows how to handle them.

GUIDE TO BEES AND HONEY

Ted Hooper. 1976.
Rodale Press, 33 E. Minor Street,
Emmaus, PA 18049. 260 p. hardcover
$10.95.

The author and original publisher are British, but this edition has been edited for a North American audience. So, while the spelling is Anglican, the measurements and geographical references are not.

Many bee culture books on the market have been around for decades—check the copyright page. *Guide* is up-to-date, and the illustrations are clearly of this century.

Most hives today are direct descendants of a prototype developed by a Philadelphia minister, Rev. L. L. Langstroth. He observed that in a natural comb bees allow a fixed space between their combs, amounting to about ¼ or ⅜ inch. If he placed comb-filled frames in a box and allowed at least ¼ inch and not more than ⅜ inch between one frame and the next, bees would use these as walks and leave them open. If the space was less, the bees would fill it with propolis, and if more they would build rows of cells in the interstices.

Murphy's Law of Beekeeping

A complete knowledge of beekeeping is impossible. The more you learn the more you learn that there is more to learn.

From the *American Bee Journal.*

(Continued from page 270)

that remain in the cells after young bees emerge. Beekeepers install a new frame with a thin wax foundation on which the bees build new comb. What happens to the old, discarded comb? On a sunny bank off to the side of Clarence's bee house sits a white wooden box on legs that looks something like a pinball machine. It's a solar wax melter. Running on the sun's heat, it doesn't cost anything to operate as it efficiently separates

valuable beeswax from waste material. The shallow, glass-topped box, about six feet long and two feet wide, slopes downward. Clarence puts old, blackened comb inside, at the top, then lets the sun go to work. Aromatic molten beeswax flows down the tray of the melter in thick, slow blobs, and the dirt stays behind. Clarence sells the beeswax for $1.85 a pound. Of the commercial buyers, the biggest one is the cosmetic industry.

Bees provide a few other ser-

THE HIVE AND THE HONEY BEE

Dadant & Sons, editors. 1975.
Dadant & Sons, Inc., 51 S. Second
Street, Hamilton, IL 62341. 740 p.
hardcover $10.95.

In addition to telling how to raise honey bees, Dadant's book tells a good deal about the animal itself—the history of its long relationship with man, anatomy, life within the hive, nutrition, genetics, behavior, nectar production, and more.

Orientation during flight involves the use of landmarks as well as the position of the sun and polarized light (von Frisch and Lindauer, 1954). Closely located continuous landmarks (forests, shore line, highway) are used more than the position of the sun. If bees have to fly around an obstacle (big rocks, high mountains) to the source of food, they will indicate by their dance the direction toward the feeding place in the straight line which they have never flown. "That they should be able," writes von Frisch (1952), "(without protractors, slide rule, and drawing board) to work out so exactly the direct line between the hive and the feeding place from the detour they have made is surely one of their most amazing achievements."

Bees can perceive the sun even when the sky is completely covered with clouds. This ability is due to the extreme sensitivity of a bee's eye to ultraviolet light, which penetrates through the clouds directly in front of the sun (von Frisch, 1958).

Less than ½ percent of the bees in the hive are likely to sting, even when the colony is greatly stimulated by careless manipulation. The more defensively inclined bees are located primarily near the entrance and can be identified by the typical guard bee position. These bees are sensitive to vibration, odors, and visual stimuli such as movement. Combinations of these stimuli are especially effective in triggering defensive behavior. Once stimulated, the defensive bees fly near the entrance where they are further stimulated by movements and color. Darker colors seem to be more likely to stimulate stinging behavior. After the bee flies toward a moving object, such as one's arm, odor becomes more significant in stimulating the final step in the defensive behavioral pattern, actual grasping of the flesh and deposition of the sting. A very powerful stimulus is exhaled breath, which contains chemicals that elicit stinging. Hair and leather have similar odors that elicit the sting response.

Once the sting is deposited, alarm pheromone is suddenly liberated from the structure, and it lingers at the sting site after the bee has departed, thus tagging the victim and exciting further stinging responses by other bees in the immediate area, until the pheromone dissipates or is removed by washing (water is effective).

The zone of defensive behavior extends only a few meters from the nest in the case of the Italian bee, but much farther in other races, such as adansonii. Hasty departure from the hive area is one effective defense against further stinging, providing one leaves at a speed that exceeds the flight speed of bees!

vices no man-made method can readily duplicate. "Even more than honey, pollination is the most important function of the honeybee," says Clarence. He calls bees the angels of agriculture, because through pollination they render a valuable service to plants, animals, and mankind. They are needed to pollinate about one-third of all the food we eat.

Inspecting a super.

As a member of the Lehigh Valley Beekeepers' Association, he and other veteran beekeepers answer questions and offer advice to members just starting out. "It used to be that everyone at the meetings had gray hair," he says. "We're trying to stimulate younger people to get into the business."

CB

Periodicals on beekeeping. The best-known periodicals are *American Bee Journal,* Dadant & Sons, Inc., Hamilton IL 62341; *Canadian Beekeeping,* Box 128, Oronon, Ontario, Canada L0B 1M0; and *Gleanings in Bee Culture,* The A. I. Root Co., P.O. Box 706, Medina, OH 44256.

Supplies. Dadant & Sons, Inc., Hamilton, IL 62341; A. I. Root Co., P.O. Box 706, Medina, OH 44256.

Maple syrup

You don't need elaborate equipment. You don't even need your own sugar maple trees, if you have neighbors who are willing to exchange the use of their trees for the maple syrup you make. And there are thousands of hard maple trees along roadsides just waiting to be tapped.

A tree with a great number of branches gives the most sap compared to a tree of the same diameter but with few branches. Trees that are not crowded by their neighbors generally produce sap with a higher sugar content. Fence row trees are perfect to tap. They are easy to get to and are the ones that also produce a high sugar content sap. The past two years I measured the amount of sap from fence row trees required to make 1 gallon of syrup. It was 27 gallons. Depending on the tree, it takes anywhere from 25 to 50 gallons of sap to make 1 gallon of syrup.

I use spiles my father-in-law gave me. However, you can make your own from ¼-inch galvanized iron pipe cut four inches long. At one end cut down 1¼ inches in the form of a cross (†). Gently tap the four sections together to form a slight taper. One inch from the other end cut a notch one-third to one-half the way through the spile. This is done so you can hang the sap bucket.

It is best to use two- to three-gallon buckets. If smaller ones are used sap may need to be collected more often. Cut a ¾-inch diameter hole near the top of one side of the sap bucket. This will fit over the ¼-inch spile. Use a 7/16-inch diameter wood bit and drill a hole 2 inches deep into the tree. The hole must be sloped upward slightly so the sap flows down easily. The larger the tree, the more taps you can make. For a 12-inch tree, use one tap; for a 16-inch tree, use two taps, three taps for a 24-inch tree; four taps for a 30-inch tree; five taps for a 36-inch and over tree. If you hang 20 to 30 buckets early in the morning you can collect sap late in the afternoon. It is best to start tapping when the nights are freezing and the days have temperatures above freezing. This makes the sap flow. Start watching for this combination of temperature starting about the last week of February. Once the sap starts to flow, your sea-

son can last two to six weeks.

Now you either start boiling sap or just pasteurize it by bringing it to a boil to prevent bacteria formation. This pasteurized sap can be held in a cool place for a day. Clean 25- to 30-gallon cans, lined with double plastic bags, are good storage containers. The sap must be collected every day or it will sour.

Now, how about an economical sap evaporator? All you need is a sheet of 20- or 22-gauge galvanized iron, 54 by 34 inches. Follow the accompany sketch. In order to bend the sheet iron, clamp it between two 2 x 4s using a large C clamp. Then, using a large wooden mallet, or metal hammer against a thick block of wood, fold the sheet to a right-angle fold. Do this on each side. At the corners, fold the corner back along the side of the pan. (Try this type of fold with a piece of heavy paper to get an idea of how you should fold the sheet iron.) Using a 2 x 4 as a form against which to hammer your mallet, bend the corner into shape. Drill an ⅛-inch hole ½ inch down from the top of the pan near the point of the fold. Make a rivet from an eightpenny nail and rivet each corner to the pan, or use a ¼-inch stove bolt. A strap handle made from ⅛- by 1-inch strap iron stock should be fastened about 12 inches from one end of the pan. In the center of the other end near the top edge, fasten a heavy wire handle or drawer handle. These handles are necessary when pouring the nearly finished syrup out of the pan.

Now comes the construction of the fire box. Lay eight-inch concrete blocks, dry set and perfectly level, spaced so one inch of the pan will rest on the blocks. On one end set up three lengths of six- or seven-inch stovepipe from a chimney. Use one concrete block and enough old bricks set on edge on the block to support the stove pipe. Set the bricks in at different distances so the flame can go up the pipe. Support the three lengths of stove pipe between post supports. These can be either wood or metal. The stove pipe should be held eight to ten inches from the posts by 14-gauge soft iron wire so there is no chance of the wood posts burning. If the ground is not frozen the posts can be driven into the

(Continued on page 274)

(Continued from page 273)

ground. If the ground is frozen, four concrete blocks can be used as anchors and the top of the posts guyed with baler twine or soft wire to the blocks.

Place four old bricks on the ground inside the front of the fire box to support grating material made of heavy steel mesh. This is necessary so the fire gets plenty of air. A piece of sheet iron the width of the space between the concrete blocks will make a suitable fire door.

A three-gallon pail with a $^3/_{16}$-inch hole drilled near the bottom is then placed on a block so sap will stream into the evaporator. This completes the equipment needed to make maple syrup.

The boiling process is where the real fun starts. Put ½ inch of sap into the evaporator. Start a fire with fine kindling and keep adding dry wood frequently enough to keep the fire fast-burning. With practice, you can get the sap boiling in 15 to 20 min-utes. Then make sure you keep your sap supply pail full at all times because with a good fire you can evaporate about five gallons of sap per hour. Keep skimming the foam off. When you have been evaporating sap for four or five hours and the sap in the evaporator is down to ½ inch to ¾ inch and is a little tan with bubbles hanging together, it is time to strain the sap through a flannel cloth. Do not let pan get scorched. Pour the sap into a large kettle, and finish the process in the house on your kitchen stove.

This last phase of syrup-making is critical. Boil the sap rapidly, but do not turn your back on the kettle because as the sap gets to the point of being syrup it will boil over very easily. If you use a thermometer, boil until the temperature is 219°F., which is 7° above the boiling point of water. (Water boils at a lower temperature with increase in elevation, so check the water boiling point where you live.) Strain the finished syrup. One gallon of syrup should weigh 11 pounds.

You can keep syrup for several years by sealing it in clean hot jars.

Louis J. Dushek.
The Conservationist.

For more information.
Back Yard Sugaring (flyer). Roger Sloane, New Hampshire Extension Forester, Durham, NH 03824. *Maple Syrup Digest* (quarterly, $2/year). Editor Lloyd Sipple, RD #2, Bainbridge, NY 13733. *Production of Maple Sirup and Other Maple Products,* Information Bulletin 95, Cooperative Extension, Cornell University, Ithaca, NY 14850.

Equipment suppliers.
Reynolds Sugar Bush, Aniwa, WI 54409; G. H. Grimm Co., Pine Street, Rutland, VT 05701; Leader Evaporator Company, 25 Stowell Street, St. Albans, VT 05478; Robert M. Lamb (Maple Sap Plastic Tubing Gathering System), P.O. Box 368, Route 49, Bernhards Bay, NY 13028; Forestry Suppliers, 205 W. Ranking Street, Jackson, MS 39204.

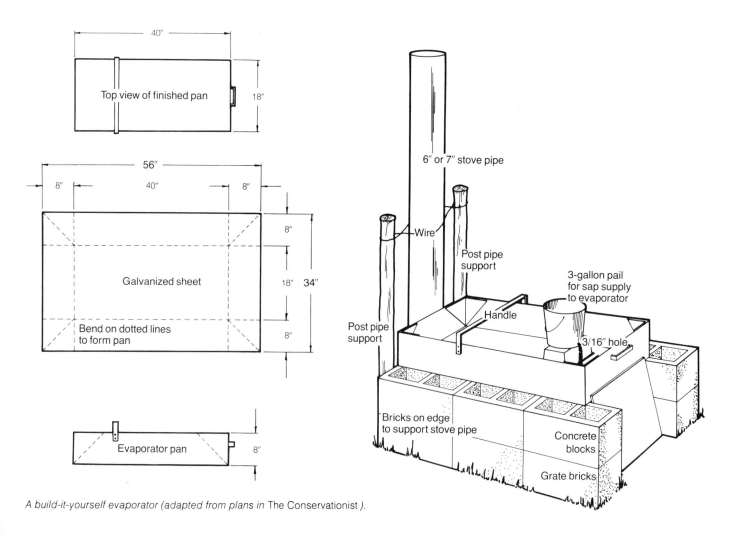

A build-it-yourself evaporator (adapted from plans in The Conservationist *).*

Jackwax and hogwallers: upstate New York

There is frost on the ground this morning, but the winter's snow has all but melted. Only a few traces of white linger in the shade under the evergreens on the hillside. Instead of snow, the path is covered with mud—now firm and frozen under one's boots but later in the day to turn ankle deep and cumbersome in the sun's warmth.

It is mid-March: a season of change in upstate New York, a liaison between winter and spring. It is a time of unpredictable weather, with temperatures which may fluctuate from 20°F. at night to 50°F. the next day. It is this alternation of temperature which is responsible for a springtime phenomenon that is only partly understood: "sap rising." In this transitional time before the leaves begin to bud on the maple trees, unseen forces exert pressures within the trees' capillaries to propel gallons of sugar water through the tree trunk.

Frances and Lewis Edwards are at work on their land. Their home is a small trailer overlooking 26 acres with a running stream and dozens of pine, ash, maple, and shaggy bark hickory trees. The winding dirt road leading to their home is so steep that ice and snow make it impassable for a good part of the year.

Maple sugaring is a satisfying but arduous task. It is work done with patience and love, and the Edwardses do all their work by hand. Their energy and endurance belie their near-70 years.

The sugaring apparatus used by the Edwards is unique because they have built it themselves. After the sap is collected from the trees, it is boiled for many hours in a cast-iron bathtub cemented atop a woodburning fireplace. A spigot built into the tub allows them to pour off the maple syrup when it has reached the desired consistency. The two holding tanks which sit in back of the bathtub were in former days refrigerator backs.

"We make about 25 gallons of maple syrup every year," says Lewis. "And that's a lot of pancakes, believe me."

Maple sugaring is done only in North America, where the sugar and black maple trees grow. Vermont produces the most maple syrup per year—410,000 gallons in 1978. New York is second with 330,000 gallons, and Wisconsin a distant third with 110,000 gallons. New Hampshire, Michigan, Ohio, and Pennsylvania produce sizable amounts of maple syrup, and the activity is carried on as far north as Quebec, as far south as the Virginias, and as far west as Iowa.

The sap that the Edwardses collect each day from their maples during the four week sugaring season is mostly water. Sugar maple and black maple trees have an exceptionally high sugar content in their sap—about 2 to 3 percent—and so are perfectly suited to man's sugaring process.

The mystery of the maple sap rising in the spring is not well understood. It is the daily rise and fall in temperature that starts the sap flowing, but to say the sap rises is only partly true. Sap flows from below, above, and across the taphole.

Hardwood trees are composed of a pithy core surrounded by transport vessels which make up the woody tissue. In these tiny transport vessels the forces of *cohesion* (water molecules clinging tightly together) and *adhesion* (water molecules clinging to the walls of their container) are responsible for making the sap rise from the roots towards the tree top. The temperature fluctuations between freezing and thawing cause pressure to force the sap toward the point of least resistance—the taphole.

As Frances Edwards climbs the steep hillside to tap one of the young maples, her dog Pandora trails behind her. She uses a ½-inch hand drill to bore carefully past the bark and into the woody part of the tree. As the precious sap leaks out, she scrapes out the wood shavings and inserts either a metal or wooden spile which will conduct the sap from the inside of the tree to the pail outside. "You shouldn't tap a tree at the spot a limb comes into the trunk," she explains. "It'll flow better if you tap it either higher or lower. And never tap in the same spot as last year."

Forty gallons of clear sap, continuously boiled, will yield one gallon of sweet, dark maple syrup. For the Edwardses, who must frequently replenish the sap in the boiling pot as the water evaporates, the process for a single gallon can take almost a day.

At the Cayuga Inlet Farm, about 30 miles away, 1½ or 2 gallons of syrup are produced in a single hour. Owner Millard Brink and his assistant collect 1,000 gallons of sap per day and boil it in a modern, two-level evaporator. Instead of collecting buckets, five miles of plastic tubing is used to tap nearly 900 trees. Visitors are welcome and busloads of schoolchildren often visit the Brink sugarhouse.

A float in the evaporator keeps the maple syrup at a constant level, with no guess work involved. Mr. Brink uses a hydrometer to measure the weight of his finished maple syrup. Maple syrup should weigh exactly 11 pounds per gallon, as opposed to water which weighs 8 pounds per gallon. Mr. Brink strains the hot maple syrup carefully through a two-layer cloth bag and pours it while still warm into blue and white containers.

In contrast, Frances and Lewis Edwards use milk to remove the impurities from their maple syrup. Whole milk is mixed with the syrup and it is heated on the stove until the milk, carrying with it dirt and impurities, rises to the top. The milk is skimmed off, leaving pure syrup which is strained again, then bottled.

"My father did it this way for years," explains Frances. "When he didn't have milk, he used eggs."

The fragrant maple steam rises from the boiling pot and mixes with the smoke from the wood fire as Lewis splits kindling to feed the blaze.

"Have you heard of maple rock candy?" he asks, pausing in his work. "After the maple syrup cooks you can put it in the deep freeze. It'll cool down so quick it'll turn to solid rock. Why, years ago at the big fancy dances, they'd always serve maple rock and rye to drink. That rye whiskey's got kind of a harsh taste to it, and that rock candy smooths it."

"Another thing you can make is
(Continued on page 276)

As a guide to following old recipes, 20 grains equal 1 scruple; 3 scruples equal 1 dram; 8 drams equal 1 ounce. In liquid measure, 60 minims equal 1 fluid dram.

GRADING

One-hundred-percent pure maple syrup is graded by quality by the U.S. government. The old standards were US Grade AA Fancy, A, or B. Today's standards are Grade A amber, B dark amber, B or utility, and C or unclassed. The Canadians grade syrup Fancy (ver light amber color with milk flavor), Light (light with mild flavor), Medium (dark and stronger), and Dark (dark and even stronger). Fancy, or light amber, syrups have the best maple flavor. The darker-colored syrups have a natural caramel flavor that overpowers the maple taste.

Dorothy Behler. *American Forests.*

SOUR SAP—BUDDY SAP

At the beginning of the season the sap is water white, clear, and transparent. It has a sweet taste and practically no odor. In strong early runs, the danger of microbial action and souring is not great; but when warm weather comes and the flow is intermittent, the bacteria become active. Microorganisms in the sap cause formation of invert sugar, resulting in darker and lower-grade syrup. Mucous formation in spiles and buckets is a visible sign of sour sap. When found, clean and scald (or wash with chlorine solutions) the buckets and spiles.

True buddy sap comes during bud swelling and indicates the end of the season. The sap has a very unpleasant odor. Buddy sap is caused by physiological changes in the tree as it starts its spring growth and has nothing to do with the microorganisms of maple sap. Producers who specialize in high-quality syrup pull their buckets and spiles before the buds swell.

From *Production of Maple Sirup and Other Maple Products* (see information listing, page 274).

(Continued from page 275)
jackwax," added Frances. "When we have snow—and we're liable to have some more snow this month—you can pour the syrup right over the snow and let it set. Those are jackwax, and they're really good too."

It is nearing midday and time to gather the sap that has been collecting from the trees on the hillside. Lewis walks from tree to tree, emptying the clear contents of the plastic jars and tin cans into his pail. This he will carry to the old washtub which serves as a holding tank. A garden hose connected to the tank siphons the sap to another tank closer to the fireplace. Lewis calls this his "lazyman's method" because it saves him the work of carrying each pailful down the hill to the fireplace.

"You'd think on a nice day like this the sap would run real good'," he says, pausing to pick a butterfly-shaped mill bug out of the foamy sap water. "But there was a south wind blowing last night, and these trees are real particular about which way the wind goes."

He throws a chunk of ice out of the pail and onto the ground spilling some of the precious maple water on his boots.

"The Indians found out about maple syrup first," he explains. "Indians always lived on what comes off the land."

Indeed, the early French and English explorers wrote of the "sweet water" that the Indians drew from the trees. The Algonquin name for maple sugar was *sinzibuckwud* which means "drawn from the wood." The method used by the Indians and early pioneers was somewhat crude: the tree was tapped by slashing it with an axe, and the sap was caught in troughs hewed out of logs.

Later pioneers bored holes in the trees with 1¼-inch augers and boiled the sap to syrup in potash kettles suspended by chains over a blazing open fire.

During the 1700s and 1800s maple syrup was an important food item. Gradually, white cane sugar became less expensive, and so by the late 1800s cane sugar largely had replaced maple sugar.

The maple syrup sells "like hotcakes" during the summer at the farmer's market in town. "We sell both maple syrup and honey," says Lewis. "And if somebody comes along and wants to swap something for a jar of syrup, why, we'll swap and have that to sell too."

Frances surveys my mud-caked boots and shakes her head. "We've got a hogwaller here now," she says, stepping with some difficulty herself through the entrapping mud.

Split-pea soup and homemade rhubarb sauce are heating over the fire. Lunch is a long and relaxing time for joking and storytelling. Overhead some geese fly northward and Frances smiles. "I like to see geese flying in that direction. Gives me goose bumps when they're going the other way." When their day's work is completed, Frances and Lewis will wash their boots off in the creek and carry a pailful of dark amber syrup across the bridge and up the hill to their kitchen.

Kathleen Schmidt.
The Conservationist.

Because of the very large amount of sap needed to make the finished syrup, and the brevity of the season itself, maple syrup is the most expensive of the natural sweeteners. The marvelous natural flavor components of maple syrup—exotic compounds such as ethyvanillin and alpha-furanone—make maple syrup difficult to synthesize artificially.

Phil Levy, Talking Food Co., Salem, MA 01970.

Sorghum syrup

Sorghum syrup is a sweetener boiled down from the sap of sweet sorghum stalks. Gene Logsdon tells how to grow a patch of the plant and press the sap in *Small-Scale Grain Raising* (reviewed on page 85). Eliot Wigginton has chronicled North Carolina sorghum syruping in *Fox-*

fire Three (Doubleday & Company, New York, NY 10017, 1975). South and west of sugar maple country, sorghum was sometimes the only sweetener to be had.

The syrup is available at natural food stores and co-ops. Gene Logsdon says in his book that "you can, if you are very lucky, locate an old sorghum press rusting away in a southern barn." A hand-operated sorghum press is made by CeCoCo, Chuo Boeki Goshi Kaisha, P.O. Box 8, Ibaraki City, Osaka Pref. 567, Japan.

If your climate is too warm for maple sugaring, you might consider pressing sorghum for a thick, flavorful syrup.

Malt

Malt is the sugar isolated from sprouted cereal grains, such as barley, wheat, and rye. See the section on malting in the Grains chapter, page 54. Commercially this sugar is used in brewing and baking.

These grain sprouts are amazingly sweet. Wheat sprouts taste as sweet as raisins—almost too sweet.

The homemade malt described in the Grains chapter can be used to sweeten baked goods and cereals. In fact, the enzymes in malt have the power to convert starch in flour to sugar. Dissolved in hot water, it makes a coffee substitute, although the malt flavor won't come within a mile of that drink unless the malt has been toasted.

THE NEW WAY FOR A NATURAL PURE FOOD

SYRUP TREES

Box Elder · Big Leaf · Silver, Red, Black, and Sugar Maples

By Bruce Thompson

SYRUP TREES
Bruce Thompson. 1978.
Walnut Press, P.O. Box 17210,
Fountain Hills, AZ 85268. 164 p.
paperback $6.95.

Thompson disabuses the notion that syrup is something that necessarily is boiled down at the foot of Vermont sugar maples. Sweet-sapped maple varieties can be found from California north to Alaska, and over most of the eastern half of the United States and southern Ontario and Quebec.

Syrup Trees gives a good account of backyard sugaring, along with tips on taking your product to market.

The most important requirement for the production of maple syrup—one not mentioned by any of the authorities and experts—is the mutual desire of men and women, husbands and wives or whomever, to make the project succeed. Occasionally a male college professor, for example, will run a successful mini-operation during his spring vacation, but almost always the successful operation calls for a joint effort by a man and a woman, and possibly their children, and a dedication to sharing the work and the fun and the challenges.

The syrup you buy in a store or have served to you in a restaurant, reading "maple" on the label, must by law contain at least 2 percent pure maple syrup. That's all. So if you think that store-bought doesn't taste like the real stuff, now you know why. But if you didn't grow up in or close to a maple-syrup-producing region the way I did, consumer tests show that you are likely to prefer commercial syrup.

Dorothy Behler. *American Forests.*

JACKWAX, SOFT SUGAR, AND MAPLE CREAM

Curious things happen to syrup that is raised from 18 to 24° F. above the boiling point of water. If the syrup is poured at once over snow or cracked ice, a chewy sweet called *Jackwax* is the result. With pickles and a fork for twirling up a layer of wax, the makings of a jackwax party are at hand.

If the syrup is stirred for a few minutes after it is removed from the stove, it soon crystallizes into soft sugar and can be poured into molds or containers. Each degree higher the syrup is allowed to boil, the stiffer the soft sugar becomes.

If the syrup is poured hot into a flat dish, allowed to cool rapidly to 60 to 70°F., and then stirred continuously for 10 to 20 minutes, maple cream will result. This product, a fondantlike product, is made of microscopic-sized crystals covered with a thin coat of saturated liquid syrup. This gives a smooth, nongritty texture. The demand for this product is very great. Maple cream is made by using high-quality syrup, low in invert sugars, and boiling it to a temperature of 20 to 23° F. above the boiling point of water. It is then poured into pans in which it will be stirred and rapidly cooled by setting the pans in a running cold water bath to 70°F. or, better, 60°. Then the syrup is stirred, preferably mechanically, in a room at 68°F. or above. It will first become fluid, then begin to stiffen and tend to set, and later lose the shiny appearance on the surface. It is then ready to pour.

After maple cream, soft maple sugar candies are most popular. The sugars are slightly stiffer than cream and about eight pounds of cakes can be made from one gallon of maple syrup.

Cook the syrup to a temperature 25° F. above the boiling point of water. Cool the cooked syrup slowly in pans to a temperature of 155° F.; then stir the syrup with a wooden paddle by hand or mechanically. While it is still semi-liquid, the sugar can be poured into rubber molds of various shapes and sizes. Cakes formed by pouring are more attractive than those made by packing.

F. G. Winch, Jr., and R. R. Morrow, *Production of Maple Sirup and Other Maple Products* (see information listing, page 274).

Suburban sugaring: Macungie, Pennsylvania

The Dill family harvests maple sugar each fall with the simplest of equipment. The buckets are distributed on the trees of obliging neighbors, and collected each afternoon after school. The buckets are any containers that are handy; the evaporator is a roasting pan placed atop a small woodstove. Only the spiles were purchased.

THE MAPLE SUGAR BOOK
Helen and Scott Nearing. 1970.
Schocken Books, 200 Madison Avenue,
New York, NY 10016. 273 p.
hardcover $4.95.

The Nearing's sugaring book is different than others on the subject because, for the authors, this harvest is one answer to the question posed in their forward: "How should one live?" The how-to is supplemented with chapters on maple lore, tools, and marketing. And throughout the book, the authors are aware of their role beyond the sugarbush—as members of society and creatures on this planet.

Originally, we had no intention of going into maple production. On the contrary, we were unaware that maple offered us an opportunity to try out the life pattern we had in mind. Broadly, the purposes we had in mind when we turned our backs on the city and our faces toward the country might be summed up in this manner:

First, we wanted to control our own source of livelihood. . . .

Second, we wanted to get away from the cities, which seemed to us more and more hectic, disorganized, and disorderly. We had lived in cities all over the world, and with minor exceptions, the story was the same everywhere. The city was artificial from top to bottom, imposing upon its victims a life pattern based upon superficialities and upon an endless grind of routine that had as its chief purpose the fleecing of the poor and weak for the profit of the rich and powerful. . . .

Third, we wanted to get our feet on the earth and to get our hands into it–to make and keep that incomparably important contact with nature which balances life at the same time that it cleanses it, rejuvenates it, and keeps it sane. . . .

Fourth, we wanted to live simply, doing as much good as possible to our fellow humans and fellow beings, and at the same time doing them as little harm as possible. . . .

Fifth, we wanted to live solvently. That is, we did not want to beg or borrow or steal, and we did want to produce our living with our hands and in the closest possible contact with the earth. That meant an annual budget that would cover a simple but adequate livelihood and show a surplus rather than a deficit.

Sixth, we wanted, in one sense most important of all, to make a living in about half of our working time—say four or five hours a day—so that we would be freed from the livelihood problem and enabled to devote the other half of our time to study, teaching, writing, music, travel. . . .

Last, we wanted to demonstrate a pattern that might be followed by those who felt with us that self-respect could properly be maintained only at arms' length from the centers of exploitation and only under conditions where the able-bodied individual was doing his share of the necessary social labor at the same time that he was satisfying his own creative urges in his chosen fields of the sciences, the arts, and social intercourse.

Beverages

A great pastime of those who eat healthfully is peeking at the carts of fellow supermarket shoppers—carts piled with snacks for television watching, long loaves of white bread draped here and there, and worst of all, the drinks. Even those who are solicitous of what they chew are often careless about what they drink. It's easy to see why. Good drink is hard to find, and when you do find it—pure fruit juice, bottled water, herbal teas—the prices tend to be exorbitant. It's hard to rationalize paying more for water than beer or soda. The only way out is to count on the market for no beverages whatsoever. There's nothing there that you can't make.

Water

Oddly, the simplest beverage is the toughest to come by: pure water. As occasional newspaper dramas are making clear, tap water likely contains a good deal more than water. You don't have to live downstream from a chemical dump to be concerned. As the water section of the last chapter makes clear, pollutants come from a great number of sources too homely to make the papers, including salt from water softeners, metals and asbestos from pipes, and, ironically, carcinogens formed by the chlorine added to make water safe to drink.

Juice

Store-bought juice costs a lot, unless it has been watered down and adulterated to supplement the fruit. Co-op discounts help some, but the word "natural" on the label is apt to jack the price up several dimes.

You can make your own juices for next to nothing, from a great variety of crops. Obviously, the juicier and larger the crop, the easier the task. That's one reason juices of apple, tomato, and citrus fruits are most popular. See the Juicing chapter.

Teas

Once you've found a commercial herbal tea that tastes especially good, note the ingredients and compound your own. Be warned that an innocent mug of tea may be adminis-

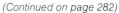

tering a medicinal dose: the properties attributed to herbs is more than dusty folklore.

Unless you plan on sewing up little tea bags of muslin or filter paper, you can keep the bits of herb out of your mouth with a tea egg or a spoon with a perforated clam shell at the business end. These are carried at knick-knack and gourmet shops. One gimmick you don't need is a so-called solar tea-maker. While the principle is venerable and valid, there's nothing fancy about it: just place cold water and herbs in a glass jar, seal, and set it out in the sun for a few hours.

Fermented milk drinks are covered in The Home Dairy, page 296.

You can make excellent carbon-ated drinks with a seltzer bottle. Those sold as antiques are usually of the kind that must be charged by a bottling plant. For about 40 cents you get a fizzy refill of whatever water is piped into the plant—and that may be dank, chlorinated stuff from the city's pipes. It's better to do your own charging with a seltzer bottle that

(Continued on page 282)

Homemade teas: Days Creek, Oregon

During warm weather, our family drinks a lot of iced herb tea. I'd fix a half-gallon pitcher, and before the day was half over, it would be gone, and I would have to interrupt whatever I was doing to prepare more.

One day I got an idea to save some work. Using a cup of herbs to eight cups of water, I made a tea concentrate. The two-quart jug in the refrigerator now holds this concentrate, labeled "ICED TEA—DILUTE." Whenever someone wants iced tea, all that is needed is to put some ice cubes in a tall glass, add about one-quarter cup or less of concentrate and fill with cold water. Now, it lasts for a few days instead of a few hours.

Some of the herbs I like to use for iced tea are comfrey, nettles, lemon balm for its flavor and calming, soothing effect, either borage or sweet woodruff which are stimulating herbs, and some of the mints for a cooling flavor. Although lemon balm has the opposite effect from borage or sweet woodruff, when combined they make one feel calm and relaxed and at the same time alert and active.

During cooler weather, we drink hot herb teas and these we like freshly brewed. The various teas and mixtures are made up ahead of time. The herbs that we've collected and dried are put in the blender to cut them up fine and mix thoroughly.

We like the convenience of tea bags—no mess, no straining, no sediment in the cup. I get the little do-it-yourself tea bags. A teaspoon full of herbs goes in each of the little bags, which come in a strip. They are then sealed with a warm iron and separated along the perforations. For our beverage teas, I do enough to fill a large jar with the tea bags, label the jar, put the lid on tight and store on the pantry shelf. Our tea stays fresh-tasting and fragrant to the last bag.

The special purpose herbs are made up in smaller quantities and kept in small labeled jars in the herb cabinet. By doing them ahead of time in my leisure, they are always ready and convenient when needed. There's always hot water in the glass tea-pot, and it's so easy to just drop a tea bag in the cup and enjoy the flavor and aroma of the brew. I know we drink a lot more tea now than when I made it by the pot.

Marion Wilbur

Some tea drinkers prefer to let the leaves soak in cool water in the sun's rays.

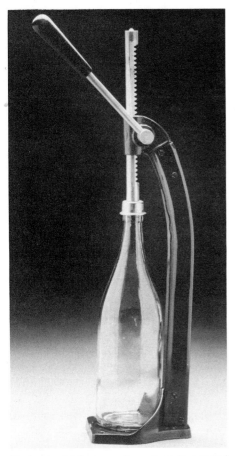

Pasteurized juices can be put up in capped bottles. This is one of several models of manual bottle cappers. Hardware stores carry them, along with the caps.

(Continued from page 281)
uses a CO_2 cartridge. The price per charge is roughly halved, and you can use good bottled, filtered, or well-drawn water. These seltzer bottles—identified by metal safety mesh

A siphon bottle, powered by carbon dioxide capsules, makes additive-free soda water.

—are available from Hammacher-Schlemmer, 147 E. 57th Street, New York, NY 10022.

Try squirting a good shot into a glass of fruit juice for a light, refreshing drink after exercising. The bubbly water makes fruit juices more fun and takes off any cloying edge.

A healthful soda can be made by adding carbonated water to a few tablespoons of fruit syrup (page 256).

Mint Cooler

Here's a change from lemonade.

2   medium oranges
1   small lemon
1   teaspoon dried mint leaves
2   tablespoons honey
6   ice cubes
1   cup water

To prepare oranges and lemons quickly, wash and peel and then cut across the equator. All pits will be visible for picking out. Put into the blender and process at high speed until liquified. Add the mint leaves and honey. Process at low speed. Add the ice cubes and blend at high speed until well chopped. Add the water and blend at low speed until smooth.

Yield: 3 cups

Nuts

Each year unknown tons of food drop out of the sky, ignored by people who walk past or drive over this harvest.

Walnuts, hickory nuts, butternuts, beachnuts, pinyon nuts, and more are left to the deer and squirrels. Even acorns can be eaten, if processed a bit. This crop was a staple of many native Americans, and was relied upon in hard times by Europeans. New England Indians even boiled the pounded nuts for oil.

Acorn meal can be made by milling the shelled nuts. Those of the white oak group (look for rounded lobes on the leaves) are most palatable, but any acorns must be relieved of their bitter tannin for palatability.

Line a colander with muslin or other porous cloth, place the meal upon the cloth, and slowly pour through several quarts of hot or boiling water. The meal is then dried on cookie sheets in the oven. Use as you would cornmeal in bread and muffin recipes, alone or in combination with cornmeal and standard flours. Be prepared for a dark color.

Perhaps its difficulty in preparation is compensated for by its long storage life: the acorn is thought by some to be the most storagable of all foods. North American Indians buried acorns in the cold mud of springs, where they would keep in good shape for 30 years.

Nutcrackers

The variety of nutcrackers on the market reflects the centuries of thought that have gone into tricking the shell into giving up its prize.

We'll deal with the simplest first. The familiar hand-held nutcracker works fairly well for most soft-shelled nuts* and usually comes packaged with a pick for digging out English walnut nutmeats. It is simply constructed of texture, zinc-plated steel arms with a spring-loaded flat pivot. The nut is placed between the arms

*Hard-shelled nuts include black walnut, butternut, and hickory; soft-shelled nuts include almonds, Brazil nuts, English walnuts, filberts, and pecans.

near the fulcrum, against surfaces ribbed for traction. If you aren't careful, you can pinch the palm of your hand between the tips of the two arms.

English walnuts crack easily in most cases, but a small hand could have a problem getting a good grip for the squeeze.

Home nut processing

Many nuts are ready for use as purchased. Others require a bit of work.

Shelling. Everyone seems to have a favorite way of cracking nuts. Some prefer using a nutcracker; and others, a hammer on a hard surface.

Hard shells are easier to crack and nutmeats break less often if nuts are first soaked in warm water several hours or overnight. Spread nutmeats and let them stand a few hours to dry before storing.

To open a coconut, pierce the eyes with an ice pick or large nail, and drain the liquid from the coconut. To remove the shell easily, bake the drained coconut at 350° F. (moderate oven) for 20 to 30 minutes, or put it in the freezer for an hour. Then place the coconut on a firm surface, and tap the shell lightly with a hammer in several places until it cracks. Separate the meat from the shell.

(Continued on page 284)

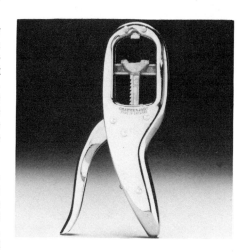

While on an assignment, two Home Food Systems *staffers came across Nelson Boyer, who forages for walnuts and hickory nuts on a bike.*

Black Jack Cracker

$29.95. Cracks ¹/₂ lb. English walnuts in 2 min. 45 sec. Handles all nuts, including hard shell.

The Black Jack is a brute when it comes to cracking hard nuts. Ruggedly constructed of cast-aluminum and chrome-plated steel, it works much like a bench vise. Nuts are cracked between anvils. Close threads on the feeding screw and good leverage from the T-handle give it enough power to crack very hard hickory nuts and black walnuts.

The unit does adapt to the many different sizes and shapes of nuts, but not easily. The anvil's flat cracking surface can adjust to a concave grip. The ram has a flat bolt head for crushing. Remove the bolt, and the ram has a second concave grip. Since making adjustments is time-consuming and because some nuts slip right off the slick, flat surface, it's more convenient to operate the cracker using both concave surfaces. However, Brazil nuts tend to jam in the bolt hole and have to be removed with a pick.

Operating the Black Jack can be awkward if you're used to working with tools like C-clamps and vises which tighten clockwise and loosen counterclockwise. The Jack's handle turns in the opposite direction. Also, we found the adjustment screw for the anvil worked very hard; once in a while the T-handle binded in one spot

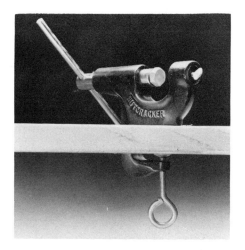

of the stroke; oil leaked from around the ram at the T-handle, getting onto both cracked nuts and hands; and it took a frustrating 11 turns of the handle to move the ram just 1 inch.

The Black Jack has the power for cracking hard nuts, but it is not quick at changing from one nut size to another. It clamps to a table edge up to a 2-inch thickness, but its rough casting could scratch a soft surface.

Garden Way, Charlotte, VT 05445.

Cracker Jack

$14.50. Cracks ¹/₂ lb. English walnuts in 2 min. 30 sec. Handles all nuts except for hard shells such as hickory, butternut, walnut.

The Cracker Jack is a hand-held cracker that operates like a bumper jack. A hand-pumped ratchet controls the movement of a rack which squeezes the nut against a blade at the top of the unit. The blade catches the top, and as the rack applies pressure from the bottom, the blade initiates the cracking action. Put the nut in place with one hand and pump the handle with the other.

The Cracker Jack works nicely for almonds, especially. It slices the shell open and the nut falls out whole. The nutmeat is coaxed out of pecans and English walnuts with some crushing. Crack these nuts only slightly with the first squeeze, then turn them 90 degrees to loosen into quarters. The Brazil nut is often sliced right in half, with the shell still intact. Rather than one hard crack, give two or three to help free the nutmeat from the shell. This takes some practice.

The Cracker Jack is a well-built machine with a chrome finish. Good instructions are supplied for cracking various nuts.

Garden Way, Charlotte, VT 05445.

Potter Walnut Cracker

$34.50. Cracks ¹/₂ lb. English walnuts in 2 min. Handles all nuts, especially hard shells.

It's billed as the "Grandaddy of Them All," and we have to agree. The Potter, with its cast-iron frame and steel base, is the largest, heaviest, most powerful nut-cracker we tested. It's also the ugliest, but cracks all kinds of nuts with finesse, even Brazil nuts.

The Potter cracker uses a simple rack-and-pinion system for its leverage. A rack of grooves is mounted 7 inches above the place where the nut is held. Pulling down on the lever engages it in one of the grooves for leverage, and supplies considerable pressure for cracking. You remain in control. There is no sudden burst of energy and uncontrollable impact. Instead, you pull on the lever until the nut has cracked sufficiently. With some practice, you can tell when everything is about to give and let off the pressure before the nut crumbles.

A safety shield of sheet metal hinges over the cracking chamber to keep shells from flying. It's a good safety feature but slows things down for those who can't quite feel

when a nut is cracked and have to take a peek.

Operating the rack-and-pinion arrangement can be a little awkward at first, but directions are clear and helpful.

Fred Klingensmith, Inc., Sapulpa, OK 74066.

Texas Native Inertia Nutcracker

$14.98. Cracks ¹/₂ lb. English walnuts in 2 min. 15 sec. Handles all soft shells.

This wooden nutcracker is one of the cleverest we tried. We had more fun with it than any other. The Texas Native Inertia gets its power from rubber bands, which propel a ram. It is the only cracker that produced a whole English walnut in one piece. The machine regularly produced whole halves of pecan nutmeats with one shot and whole Brazil nuts after two or three knocks.

Place a nut between the aluminum ram and a steel weight. Pull the aluminum shield over the nut to keep shells from flying. Pull back on the wooden plunger, stretching the rubber bands, then release the plunger. When the weight and shield are removed, the cracked nut rolls down an incline, out of the way.

Trial and error will determine how many rubber bands to use and how far back to draw the plunger for each kind of nut. (Note that this cracker is definitely not for hard-shelled nuts.)

There is no clamp. A wooden block hangs over the table edge to hold the cracker as

the plunger is pulled back. Because there's no protective felt on the bottom, it's a good idea to protect the table surface with newspaper. Keep an eye on the weight if you're prone to dropping things on toes.

Charter Co., Inc., Bickett Boulevard, P.O. Drawer 10938, Raleigh, NC 27605.

(Continued from page 283)

Removing skins. Removing thick skins from some shelled nuts gives them a delicate flavor and improves appearance. This may be done by placing nuts in boiling water (called blanching) or by roasting.

To blanch shelled almonds or Virginia or Runner peanuts, put them into boiling water and let stand three minutes. Drain. Slide skins off with your fingers. Spread nuts on absorbent paper to dry. Roasting also will loosen the skins of peanuts, either shelled or unshelled.

Spread filberts in a single layer in a shallow baking pan. Bake at 300° F. (slow oven) for 10 to 15 minutes or until heated through, stirring occasionally. Cool slightly, and slip skins off with your fingers.

Blanch chestnuts by putting them in boiling water and let stand two minutes. Remove a few at a time, cool slightly, then peel with a paring knife. If nuts are difficult to peel, return them to the hot water for a minute or two. Roasting also will loosen the skins of chestnuts (see below). Remove shells and skins while nuts are warm.

Use a vegetable parer to remove the brown skin from coconut meat.

Buy untoasted sesame seeds in bulk from a natural food store or co-op, and toast them just before adding to salads or baked goods.

Roasting or toasting. Roasting or toasting enhances the flavor and color of nuts. Roasting is done in the oven. Peanuts and chestnuts are commonly roasted in the shell. Toasting may be done in the oven or on top of the range.

To roast or toast shelled nuts, mix one teaspoon cooking oil with each cup of nutmeats, if desired, for richer flavor and even browning. Spread nuts on a shallow pan or baking sheet. Heat at 350° F. (moderate oven) for 5 to 12 minutes, or until lightly browned, stirring occasionally.

Like sesame seeds, sunflower seeds are used toasted in salads and breads. Grow your own, or shop around for a bulk supply – seeds sold as a snack in little plastic bags are overpriced.

Toasting may also be done in a heavy pan on top of the range. Heat nutmeats slowly for 10 to 15 minutes until lightly browned, stirring frequently. Nuts continue to brown slightly after removing from heat, so avoid overcooking. Cool nuts on absorbent paper.

To roast or toast processed coconut, use the method for shelled nuts without added oil.

To roast peanuts in the shell, spread peanuts in a shallow pan and roast at 350°F. (moderate oven), stirring occasionally, for 15 to 20 minutes.

To test doneness, remove a nut from the oven and shell it. The skin should slip off easily and the kernel should be lightly browned and have a roasted flavor.

To roast chestnuts in the shell, first slash through the shells on the flat side of the nuts. Place chestnuts, cut sides up, on a baking sheet. Roast at 400°F. (hot oven) until tender—about 20 minutes. Insert fork through cut in shell to test tenderness.

Cutting. Chop or cut nuts on a board with a knife that has a long, straight cutting edge, or use a nut chopper.

(Continued on page 286)

New Dynamic Nutcracker

$70. (approx.) Cracks ¹/₂ lb. English walnuts in 1 min. 45 sec. Handles soft shells only.

This one is constructed just like the Texas Native Inertia nutcracker, but uses electricity instead of rubber bands for power. It plugs into a regular 110-volt outlet. No electricity is used until the moment contact is made for cracking. It's fast, but it's not designed for hard shells like hickory or black walnuts.

When the Plexiglas shield is pushed over the nut in the cracking chamber, a relay switch triggers an electric solenoid that sends the aluminum piston slamming into the nut. Since the piston travels no more than ⅛ inch, the impact is subtle. The guard keeps shells from spewing all over and there's an incline provided for the shelled nut's escape.

Is it safe? Its quick action will startle you on the first trial, but it's unlikely you'll catch a finger since the shield must be in place to set off the charge. There's also a leather cover over the electric solenoid to protect you from shock.

We enjoyed the New Dynamic for its speed and novelty. However, the force of impact is the same no matter what nut is being cracked. Brazil nuts generally can be extracted whole after two or three cracks, while pecans and English walnuts usually break into pieces.

Dynamic Nutcracker, Box 29153, Dallas, TX 75229.

Lee's Nutcracker

$12, hardwood base; $19.95, mahogany bowl. Cracks ¹/₂ lb. English walnuts in 2 min. Handles soft shells only.

The two Lee's nutcrackers can be quickly switched from one nut variety to another. The less-expensive model has a flat hardwood base, while the other offers a mahogany bowl to catch shells as they fall—a handy feature. Both have felt buttons on the bottom to protect table tops.

The crackers themselves are identical, closing like a C-clamp. The screw pitch has fewer threads per inch than a clamp, so the cracker travels quickly. Some strength is lost, but the Lees are sufficient for the common soft-shelled nuts. They're not guaranteed for extremely hard ones.

Concave surfaces on both the anvil and ram hold any of the common nuts in place. The cup at the end of the ram is free-floating so it doesn't turn the nut as the clamp tightens.

The Lee Manufacturing Co., P.O. Box 20222, Dallas, TX 75220.

Handy Andy Nutcracker

$12.95. Cracks 1/2 lb. English walnuts in 2 min. Handles Brazil nuts with difficulty, won't do hard shells.

The Handy Andy is an attractive, sporty looking nutcracker. The nut is placed between an anvil and a cast-aluminum cylinder. The 9-inch-long handle acts as a lever for the cracking piston, which travels through the cylinder to make contact with the nut.

Although it efficiently cracks one variety of nut at a time, you may find it troublesome to crack a bowl of filberts, pecans, and Brazils. Although the anvil is threaded to accommodate different-sized nuts, the adjustments must be made through trial and error, and this takes time. We had some particularly small filberts that were too small for the cracker.

Handy Andy Industries, Inc., P.O. Box 2456, Jonesboro, AR 72401.

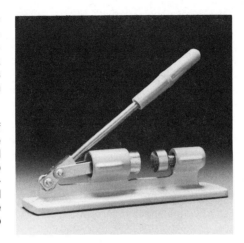

(Continued from page 285)

Sliver or slice nuts, while still warm and moist from blanching, with a thin, sharp knife blade.

Most nuts need protection from oxygen in the air and from high temperatures that may cause the fat in the nuts to become rancid.

Nuts in the shell retain high quality longer than shelled nuts. Whole nuts become rancid less quickly than nuts in pieces. Unroasted nuts keep better than roasted ones.

Nuts in the shell keep well in a nut bowl at room temperature for a short period of time. For prolonged storage, keep them in a cool, dry place. Avoid storing nuts in a damp place.

Shelled nuts will keep fresh for several months stored in tightly closed containers in the refrigerator. Shelled nuts in unopened cans keep well in a cool, dry place but will maintain good quality longer in the refrigerator or freezer.

Shelled or unshelled nuts can be frozen in tightly closed freezer containers at 0°F. or lower. Fresh coconuts in the shell retain good quality up to a month in the refrigerator. Ready-to-eat coconut keeps best refrigerated in a closed container.

Peanut butter will keep its quality

Pumpkin seeds make a good snack when toasted with a shot of tamari. You may want to grow your own, as they are expensive even at co-op prices.

longer in the refrigerator than at room temperature. Chestnuts are perishable at room temperature but will keep several months in the refrigerator in loosely covered containers or in ventilated plastic bags.

Shelled, blanched chestnuts (whole or chopped) may be frozen for longer storage. Pack them in tightly closed freezer containers and freeze immediately at 0° F. or lower. Use in cooking without defrosting.

From "Nuts in Family Meals," USDA Home and Garden Bulletin No. 176.

A Tribute to the Acorn

Any balanced presentation of the economics of the acorn must point out its great nutritive value and its great use as human food. It may be possible that the human race has eaten more of acorns than it has of wheat, for wheat is the food of only one of the four large masses of humans, the European-North American group. The other three groups, the Chinese-Japanese, the Indian (Asiatic), and the tropical peoples, pay small attention to wheat; hundreds of millions of their people have never heard of it. Meanwhile those humans (and possibly pre-humans) who dwelt in or near the oak forests in the middle latitudes—Japan, China, Himalaya Mountains, West Asia, Europe, North America—have probably lived in part on acorns for unknown hundreds of centuries, possibly for thousands of centuries.

It is almost certain that wheat has been of important use only in the era of man's agriculture, while the acorn was almost surely of importance during that very, very long period when man was only a food gatherer.

J. Russell Smith, *Tree Crops.* Reprinted by permission of the Devin-Adair Company, Inc., Old Greenwich, CT 06870. Copyright © 1950 by the Devin-Adair Company.

Hand-pressed sunflower oil

In 2,500 square feet, a family of four can grow each year enough sunflower seed to produce 3 gallons of homemade vegetable oil suitable for salads or cooking and 20 pounds of nutritious, dehulled seed—with enough broken seeds left over to feed a winter's worth of birds.

The problem, heretofore, with sunflower seeds was the difficulty of dehulling them at home, and the lack of a device for expressing oil from the seeds. The job was to find out who makes a sunflower seed dehuller or to devise one if none were manufactured. And to either locate a home-scale oilseed press or devise one. No mean task.

Organic Gardening magazine searched from North Dakota—hub of commercial sunflower activity in the nation—to the files in the U.S. Patent Office, with stops in between. The magazine staff turned up a lot of big machinery, but found no small-scale equipment to dehull sunflowers or press out their oil.

They found that grain mills could do the dehulling: especially good results could be had with the hand-powered C.S. Bell #60 and the electric Marathon Uni Mill. By starting at the widest setting and gradually narrowing the opening, almost every seed was dehulled. The stones crack the hulls open, then rub them to encourage the

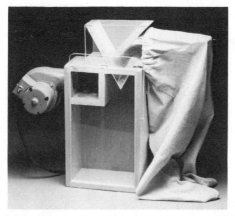
The winnower.

seed away from the fibrous lining. As long as the stones open at least as wide as the widest unhulled seed, any mill will work.

Seeds must be graded to size first, however, and this involves screening them. Since the smallest unhulled seeds are about the size of the largest hulled kernels, the grading prevents these undersized seeds from passing through unhulled. Processed together at a closer setting, the smallest seeds hull out successfully.

Rodale Press next developed a custom winnowing machine propelled by a vacuum cleaner. This pro-

This many sunflower seeds yield this much oil.

totype succeeded in removing all but one hull for every ten kernels, (the mistakes must then be picked out by hand). Fortunately, bug-eaten seeds usually blow off with the chaff.

The homemade oil press shown here is relatively simple to make, but it involves welding. It consists of a welded tubular frame which accepts a three-ton hydraulic jack. You may already have one. They are available at most auto and hardware stores for about $16. A metal canister, with holes drilled in its sides and one end welded shut, holds the mashed sunflower seeds. A piston is inserted in the canister and then inverted and slipped over a pedestal on the frame. With the jack set in place, the pressure is gradually increased over half an hour. Oil drips from the sides of the canister into a tray—the bottom of a plastic jug slipped over the pedestal works fine—which empties the oil into a cup. You can pour the oil through a coffee filter to remove pieces of seed and other fine particles that would burn if the oil were used for cooking. If it's for salads or mayonnaise, filtering isn't necessary.

At this point, plans for the winnower and press are not available from Rodale Press, but the homemade oil project serves to show how ingenuity can be applied to processing one's own food.

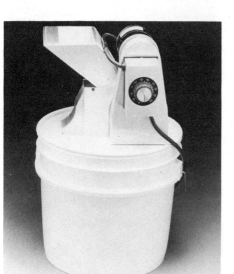
The huller.

You can make your own oil with this press.

The home dairy

It's arguable, but we'll say here that milk is the most processable of foods—that is, more good things can happen to milk than squash or pears or even meat and soybeans. These processes have been around a long time, and are suited to home-scale production. With heat or refrigeration, enzymes or yeasts, you can turn this beverage into forms creamy, rubbery, or rock solid.

What's more, milk is often improved in nutrition and digestibility through these traditional processes. Not all processes, old or lab-inspired, are so kind to our bodies. This chapter first looks at pasteurization and the foods that milk makes: butter and buttermilk, yogurt and other fermented products, and cheese soft and hard. We then introduce the two mammals most likely to be milked in North America, the goat and the cow, and give a glance at a distant third, the sheep.

RY

Store-bought dairy foods

Store-bought dairy products are often not as clean and innocent as the smiling cows on the plastic containers. Consider the basic raw ingredient—milk. Pesticides, antibiotics, and other medications can wind up in dairy foods. Drugs are sometimes fed to cows to control disease, as an alternative to observing strict hygiene in the barn. Other drugs promote growth and milk production. Even if you aren't upset by this news you may be disappointed in using such milk for cheese-making, since antibiotics are apt to interfere with the process.

Store-bought milk is pasteurized—cooked to kill any deleterious life and to prolong shelf life. According to raw milk fans, this process also destroys 20 percent of the vitamin C present, as well as the natural antiseptics in milk. This means that pasteurized milk will not naturally sour and separate into curds and whey. It will only spoil. Sour *raw* milk, however, can be used in baking and cooking.

Homogenization does more than make a quaint memory of the thick layer of cream that capped old-fashioned milk bottles. It is said to destroy naturally occurring lecithin; and according to a renowned Connecticut cardiologist, Dr. Hurt A. Oster, homogenized milk is a major cause of heart and artery disease. This is due to an enzyme called xanthine oxidase. So be happy you can skim off that lovely top milk for ice cream and butter; use the skimmed product for drinking and cottage cheese production.

Ultra-pasteurization, a relatively recent procedure, greatly extends the shelf life of milk by heating it to 210° or 220° F. for one second, but denatures the proteins and harms more nutrients than old-fashioned pasteurization. Ultra-pasteurized cream may be months old by the time you buy it, and will not whip properly.

Sterilization is the most destructive procedure used in milk process-

'Twixt teat and tumbler.

288

ing. Sterilized milk requires no refrigeration, but tastes just terrible. It is marketed extensively in Europe.

Low-fat supermarket milk sometimes hides surprises. Read your labels. Besides added milk solids, you will often find monodiglycerides and other unpronounceables included to provide the necessary mouth feel for customer acceptance. A novel kind of non-milk milk is being marketed in some areas: to lure consumers on low-cholesterol or low-saturated fat diets, the dairy fat is replaced with vegetable fat. Ironically, the vegetable fat used most often is coconut oil, which is even more saturated than butter and more difficult to digest. This "filled" milk also contains corn syrup solids (sugar), sodium caseinate (bad for low-salt diets), other additives, and water.

What happens to all that unsold milk that remains in the dairy cases? Some is reclaimed to reprocess into so-called cultured buttermilk. This is a completely different product from the healthful by-product of your home buttery.

The cheese you make at home is simply milk and a curdling agent. The cheese you buy may contain additives, preservatives, flavorings, and coloring agents. Hydrogen peroxide is added to destroy bacteria and bleach the milk. Other bleaching agents include benzoyl peroxide and potassium alum calcium sulfate, hardly staples in a dairymaid's pantry. Food coloring may be added legally without mention on the label. Dye from wax coverings is apt to penetrate through to the cheese. Anti-caking agents and mold inhibitors are frequently present even in so-called natural cheese.

Cottage cheese is not free from insult. The commercial stuff may contain additives to mask poor quality and staleness, stabilizers, mold inhibitors, plus a whopping overload of salt. There are 918 milligrams of sodium—approximately ½ teaspoon of table salt—in just one cup of lowfat cottage cheese, as compared to 19 milligrams in your own unsalted dry-curd cottage cheese.

Return the manufacture of cottage cheese to its proper place: this fresh cheese originally was the product of cottage industry.

Susan Asanovic

Butter

According to the USDA, ten quarts of milk should yield one pound of butter, but this varies with the fat content of the milk. I have found that two cups of whipping cream (31 percent butterfat) yield approximately ¾ cup (180 grams) of butter; two cups of heavy cream (40 percent butterfat) yield approximately one cup of butter; and four cups of gravity-separated top milk (18 to 20 percent butterfat) yield approximately ⅔ cup of butter. The above figures depend on how well you work the butter after churning, and to some extent upon churning conditions. (It is interesting to note that European countries set the minimum fat content of butter at 85 percent, while ours is only 80 percent. Therefore, the foreign product tastes more buttery, as can yours.)

(Continued on page 290)

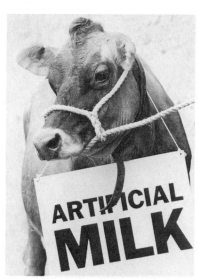

FRANKENCHEESE

What do you call a canary-yellow plastic mass that melts between 400 and 700° F. and turns stringy when tugged?

If the mass crowns a pizza from the supermarket freezer, you'd be forgiven for calling it mozzarella. But it's likely to be a manufactured product.

Artificial cheese weds such things as sodium and calcium salts of casein, hydrogenated soybean oil, artificial flavors, aluminum phosphate, adipic acid, dicalcium phosphate, tricalcium phosphate, magnesium oxide, zinc oxide, minerals, vitamins, and mold inhibitor. What to name the rubbery bricks? The National Cheese Institute suggested Golana, which happens to be "analog" spelled backwards. Anderson Clayton Foods Co., manufacturers of Unique Loaf, have come up with Cheesana. The Fisher Cheese Company likes Pizza Mate; Universal Foods sells Pizza Pal. "Cheese alternate products" is the phonetic disaster used by the USDA. These names are bland and even purposely uninformative or misleading; we propose Frankencheese.

Frankencheese has been approved by the USDA for use in school lunches when accompanied by at least 50 percent natural or processed cheese. The department defended its action by stating that Frankencheese is "made totally from conventional ingredients." The same could be said of Dr. Frankenstein's laboratory analog, of course.

What does the food industry have against milk? Well, it happens to be the most costly ingredient in cheese, and much thought has gone into making products that look like real cheese but are cheaper to crank out. "That's the American way," you might be forgiven for grumbling, and indeed the food industry's best-known invention came to be called American cheese. Any cheese labeled American is a blend of such things as cheddar, colby, stirred curd, wash curd, and an emulsifier to hold it all together in one obedient clump. Process cheese may contain any of 13 chemicals as well as artificial flavor and color. It is pasteurized, and all enzymes and bacteria are destroyed; therefore, process cheese is dead and can't ripen into something marvelous.

Other mutations of natural cheese are pasteurized process cheese food, and pasteurized process cheese spread. Cheese food has less of the cow and more moisture than process cheese. Cheese spread has still less dairy product and more water content, and it's saltier than natural cheese.

RY

(Continued from page 289)

How to make farm butter

Separating the cream. There are several methods of accomplishing this preliminary chore, the easiest of which is to use a mechanical cream separator. These are quite expensive and few homesteads produce enough butter to warrant the purchase of one. I have found that gravity separation is the best method for a cottage industry. The only drawback is that this yields a butterfat content of approximately 22 percent, a bit lean for optimum butter yields. Whipping cream has the recommended fat content of 30 to 33 percent. If you purchase cream at the market, buy whipping cream, not heavy cream. The latter runs about 40 percent fat and will clog in the churn. You will have to mix it with equal parts of water beforehand. Chill freshly drawn milk in wide-mouth containers. Usually 12 hours is enough for the cream to rise. I generally skim a little over three cups of cream from one gallon of rich Guernsey milk. Carefully ladle off the top milk and allow it to warm to churning temperature, in the neighborhood of 58 to 62° F.

Cream separation by the freeze-and-thaw method is an odd method—a bit cumbersome, but useful in

---

Transforming milk: Wilton, Connecticut

A bountiful supply of milk from the homestead cow taught our family that milk isn't just another good or service to be taken for granted. The milkman no longer cometh. We now know that milk is dependent on nature: the cows' reproductive cycles, weather, feed, and the moon. Even the breed of cow you own influences your milk products. For example, I learned to appreciate the high quality of Guernsey milk. In my city-dwelling days, milk was just another standardized item with no subtle taste or nutritional variations. Now I have found that milk's taste, fat content, color, and suitability for cheese-making varies with the cow.

Our entire family becomes involved in milk processing chores. While I seem to make the best yogurt, my husband prides himself on his cheese and has devised ingenious and effective cheese presses. My oldest daughter insists that she finishes the butter better than anyone else in the household.

Because of the versatility of milk products, family meals have become almost exclusively vegetarian. With main dishes, fresh and hard cheeses do wonders for vegetables, grains, beans, and pasta. We use a lot of skim milk, low-fat yogurt, buttermilk, and cottage cheese.

The biblical cooking fat, butter, and its Indian variation, ghee, need no cheering squad—only the *caveat* to use them in moderation. Changing milk into butter extends its storagability to weeks and even months. The children sell extra butter to neighbors for pocket money, or barter it for fresh produce.

Ice cream isn't the best food for you, but it's among the most fun to make. Depending on your supplies and your diet, you may vary the proportion of cream to milk. Ice milk and frozen yogurt have a shorter freezer life than ice cream, but can be nutritionally superior.

Among the cultured milk products, the type of bacterial culture used and the temperature and processing time determines whether the milk will turn out to be buttermilk, sour cream, yogurt, or one of dozens of cheeses. To culture yogurt, kumiss, and the lactic type of kefir, a small amount of the previous batch is used as a continuing culture. Exotic piima (which you either love or hate) begins with a freeze-dried culture from Finland. It will be self-perpetuating for months, especially if you are meticulous about dairy cleanliness. The original Finnish culture comes from the milk of cows who have grazed on the taette flower. As it stands, this milk clabbers into piima with all its intrinsic, health-building qualities. Cultured buttermilk and sour cream are also prepared with special starters, available from dairy supply houses.

The milk used for cheese-making may be raw—right from the cow—or pasteurized. Pasteurized milk has been heated to kill harmful organisms. That means no bacteria remain to help the milk curdle. Before starting to make cheese with pasteurized milk, it must be ripened with the necessary bacteria. That can be done by mixing ½ cup buttermilk with each gallon of milk.

The fat content of milk—and, consequently, cheese—varies widely. Cheesemakers often choose milk with a fat content of 3.2 to 3.5 percent. If you prefer foods lower in fat, skimmed milk makes a successful cheese. Some cheeses, cream cheese and Neufchatel, for example, are traditionally made with rich milk and cream.

A fresh cheese like cottage cheese curdles naturally as clean, raw milk sours. You can also make cottage cheese simply by acidifying it with buttermilk, lemon juice, or rennet. Most cheeses are created with some combination of heat, acid, and rennet.

Rennet, the substance used to cause milk to coagulate, is an enzyme from from the stomach lining of an unweaned calf. Rennet is sold in tablets for making cheese and junket. Vegetarians who wish to avoid rennet use lemon juice, sour whey, fig juice, or an extract from nettles or the herb lady's bedstraw. Each substance yields a cheese with a distinctive flavor.

Hard, aged cheeses begin the same way as soft, fresh cheeses, but then are drained more completely and pressed to remove as much moisture as possible.

As anyone who remembers little Miss Muffet knows, the liquid that separates from curd is whey. Don't throw this mineral-rich by-product of cheese-making down the drain. With high heat and acid you can precipitate the milk protein in whey and get easy, fresh ricotta. Pets and farm animals gladly lap up warm whey. If nothing else, feed your whey to a compost pile.

Susan Asanovic

some situations. You simply freeze the milk in wide-mouthed containers, then thaw it in the fridge, or out, depending on your work schedule. The rich butterfat will lump together and rise to the top. Simply scoop it out, let it warm to churning temperature (60° F.), and churn away. This works very well for small quantities of goat's milk.

Goat's butter is a luxurious delicacy, or delicate luxury, as you prefer. It is also non-existent on the retail market and could fetch a pretty price for you from the gourmet customer seeking exotica. The fat molecules in goat's milk are smaller than in cow's milk—you could say this milk is naturally homogenized. Therefore, gravity separation is not very successful. Either beg or borrow a cream separator, or try the separation-by-freezing method described above. It is possible, if not very practical, to churn unseparated goat's milk into butter. Expect low yields. At least you will have a taste of how heavenly it can be. If you like fresh goat cheese (something like $6 per pound) you will *love* goat's butter.

Ripening the cream. This is an optional step which produces an interesting continental flavor. Ripen the cream at room temperature overnight, then chill it briefly to churning temperature. Or, let it ripen in the fridge several days.

Churning the cream. Temperature is very important to successful churning and for getting a maximum yield from the cream. A simple dairy thermometer is essential. The cream should be about 58° F. in summer and up to 62° F. in winter. If it is too cold, butterfat will be lost to the buttermilk; too high a temperature will result in greasy, soft butter. All the books direct us to fill the churn only half full for best results. This is certainly valid for a standard churn. However, since I still depend on my five-cup blender, I cheat and fill it three-quarters full, with excellent results. Indeed, if you do not want to purchase one of the many models of butter churns now on the market, a strong blender or even a powerful mixer is very efficient for a small buttery. Run the churn, blender, or mixer until globules of butterfat have definitely separated out. Remember that feed composition can affect color and churning time as well as final texture. Expect seasonal variations, which also are governed by the cows' lactation cycle.

With a wooden paddle or rubber spatula, pat the butter into a ball, lift it out of the buttermilk, and proceed to work it. This step is not absolutely necessary if you are going to devour the butter right away or use it the same day in baking. However, for longer keeping you must press out as much buttermilk as possible. You can wash it under cool running water, or run it through the churn with cool water. Press out all the water and prepare the butter for storage.

Storage. Large quantities may be weighed out into half-pound lumps, wrapped tightly, and frozen up to two months. Butter made from sweet cream has a longer shelf life than that churned from ripened cream. Unsalted butter has a short fridge life, so plan to use it within a few days, defrosting what you need as you go. I never salt my butter, preferring its sweet natural flavor. Since salt is only a preservative, careful handling will be as effective. Before universal refrigeration, or even the weekly iceman, European farm women stored butter in salt brines, or tightly covered earthenware crocks, buried in the cool earth. This is just for the historical record, as most of us do not have to rely on these methods today.

You can make whipped butter simply by beating air into soft 60° F. butter until it's light and fluffy and doubled in volume. Served this way, it spreads easily and goes further. Why pay for air at the marketplace?

Clarified butter and ghee. Butter can be converted as clarified butter and ghee. These processes involve separating the perishable milk solids from the pure butterfat or butter oil. Ghee, as prepared in India, will keep sealed in full containers for 9 months at room temperature and for 18 months at 40° F. Clarified butter, the French version, must be refrigerated, but has excellent keeping qualities. Clarified butter is better for cooking, as it can be heated to higher temperatures without burning. Using ghee in cooking imparts the nutty flavor distinctive in Indian dishes. Use at least a pound of butter when preparing these products.

Clarified butter is made by gentle heating just to the melting point,

(Continued on page 292)

Here are three options for churning butter. A blender works for small batches; the British Blow butter churn features a wooden paddle and holds 4 quarts ($49.95, from Garden Way, Charlotte, VT 05445); for another few dollars, Garden Way will sell you the electric Gem Dandy electric churn and 3-gallon jar.

(Continued from page 291)

COMMERCIAL BUTTERMILK

Factory production of butter began about 1860, but most butter was churned on farms until the 1920s. Creameries began to dry the churned buttermilk in the 1940s for more efficient use in baked goods, candy, ice cream mixes, and other dried mises. Milk processors then started producing cultured buttermilk to meet consumer demand.

Today's buttermilk is made by fermentation, a process known to the ancient populations of southern Russia and the eastern Mediterranean countries.

The bacterial cultures and the fermentation controls determine whether buttermilk, sour cream, yogurt, or cheese will be the end product. Salt is normally added in quantities ranging from 0.01 to 0.15 percent to enhance the flavor.

Liquid butter is sometimes added to cold cultured buttermilk. It forms butter granules or flakes that give the product a churned buttermilk appearance. The labels used by some processors of this type of product read "contains butter particles" or "churned."

In most states, however, laws prohibit use of the term "churned" unless the buttermilk is actually the liquid remaining after butter is made. Little if any buttermilk left from butter production is available now.

"Buttermilk," a pamphlet by the National Dairy Council.

DR. KROGER'S RECIPE FOR TRUE BUTTERMILK

Add 1 ounce of cultured buttermilk to 1 quart of heavy cream, stir well, and let it stand at 72° F. overnight (use a light bulb in a Styrofoam cooler for cold weather). The next morning pour the thickened mixture into a blender and stir on low until the mixture separates into liquid and blots of butter that have a moonlike terrain. Pour the mixture through a strainer. You'll end up having 3 cups of buttermilk and 1 cup of butter. The butter can be refrigerated and later kneaded to get the excess buttermilk out.

Dr. Manfred Kroger,
Associate Professor of Food
Science, Pennsylvania State
University.

about 90° F. Let it stand undisturbed until the solids have settled and a rich yellow butter has risen. Skim off any white solids from the top and reserve them to add a delicious flavor to whatever you're cooking that day. To discard the solids would be nutritional heresy. Drain and wash off the milk solids, then refrigerate the clarified butter, tightly covered.

Ghee is prepared by simmering raw butter as gently as possible until the milk solids turn brown and sink to the bottom of the pan. The butter oil will turn a nut brown. Skim, bottle, and store ghee in a cool spot.

This hand-operated cream separator can be adjusted to control the amount of butterfat removed from whole milk. Garden Way (Charlotte, VT 05445) sells it for $449.

Butter will store longer if you press out as much buttermilk as possible, using a wood spatula.

Buttermilk

Genuine churned buttermilk is a delicious by-product of buttering. It cannot be found on the market these days for love or money. Commercial buttermilk is a factory-processed, cultured milk product, often salted. The real stuff is a highly perishable product, understandibly ill-suited for our modern centralized food system. If your buttermilk is made from ripened cream, it should be either used or frozen the same day (although freezing may impart a bitter taste). Sweet cream buttermilk will keep two or three days; after that it still can be used in baking.

It is almost worth making butter just for the buttermilk. This product is very close to whole milk in protein content, is almost fat-free, and can be digested easily. I use it in equal parts with whipping cream in my homemade ice cream mix because it seems to aid the churning.

Buttermilk is, of course, a traditional ingredient for baking pancakes, muffins, waffles, and all quick breads. Combined with baking soda, its natural acidity produces gas and breads rise quickly. Homemade buttermilk varies in acidity, so go easy on the soda until you establish a personal guideline. I also use it, when on hand, in slow-rising yeast breads. It aids in raising heavy whole grain doughs and keeps bread moist.

Cold buttermilk soup is a good way to spark summer appetites. You can substitute buttermilk in any soup recipe calling for yogurt and stock.

More buttermilk can be removed by washing the butter in a blender with cool water.

Since buttermilk is thinner than yogurt, replace the stock or broth with it. In case you lack ideas for soup, here are two favorites.

Vegetable Potage a la Creme

2   cups mixed vegetables in season, chopped (peas, carrots, leeks, celery, spinach, zucchini, onion)
2   cups freshly churned buttermilk
1   tablespoon fresh herbs, chopped
    freshly ground black pepper

Steam any mixture of vegetables until just barely tender. Puree the vegetables, preferably with a hand food mill, or spin a few seconds only in a food processor. Stir in the buttermilk and pepper. Garnish with fresh herbs. Chill before serving.

Yield: 4 servings

Cool Cucumber Buttermilk Bisque

3   cucumbers (2 cups chopped or 1 pound)
1   tablespoon fresh dill, chopped
1   tablespoon fresh parsley, chopped
1   large clove garlic, minced
1   cup buttermilk

Peel and halve cucumbers. Scrape out and discard the seeds from the cucumbers. Chop cucumbers very fine with a sharp knife and combine all ingredients. If a finer consistency is desired, all ingredients can be blended in an electric blender. Cover and chill. Serve in clear glass bowls, with ice cubes, garnished with fresh dill.

Yield: 4 servings
Susan Asanovic

YOGURT IN COOKING

For yogurt that won't separate and curdle when cooked, stabilize it first. Beat the yogurt to liquid consistency, add a whipped egg or a little cornstarch or flour dissolved in water. Bring slowly to a boil, stirring in one direction with a spoon. Reduce the heat and simmer until it returns to a creamy consistency (about ten minutes). Try to add it toward the end of the cooking time.

Fold, don't beat yogurt into recipes. This will keep it from getting runny.

Pat Donahue

Yogurt

Yogurt. It has a funny name and a taste to match and it's not the kind of food you'd expect sweet-toothed American consumers to enjoy. Few did enjoy yogurt when it first arrived in North America in the 1920s. Immigrants from Europe and the Middle East continued to eat it as they had in their homelands, but for decades most Americans and many nutritionists dismissed yogurt as a fad food—the province of health nuts and hippies. Yet yogurt has suddenly become as American as Mom and apple pie. North American consumers are eating their way through hundreds of millions of pounds every year.

Unfortunately, yogurt's newfound popularity has been largely built on the transformation of this basic, wholesome food into a confection, a fabricated food, and a nutritional abomination. Yogurt made in the traditional way is nothing more than milk fermented by the appropriate bacteria. Its custardlike texture and tart taste are a staple in the diets of people all over the world.

But in the United States, fast food joints, supermarkets, and vending machines now offer yogurts laced with an arsenal of additives: sugar, corn syrup, stabilizers, emulsifiers, artificial colors, flavor enhancers, modified food starch, and more. In its various incarnations what passes for yogurt appeals to a wide audience: it's sold as a dieter's delight, a dessert, and a natural food. Doctors prescribe it for convalescents; cosmeticians recommend it as a skin-restoring facial. And behold the sign of yogurt's ultimate acceptance in American life: discarded yogurt cups litter the roads. Swiss Style Blueberry Delight Yogurt cartons nestle in the ditches with beer cans and hamburger wrappers.

Not surprisingly, mass marketing and media hype have obscured the truth about this ancient and remarkable food. Pure, unadulterated yogurt and similar fermented milks have health and dietary values far beyond those of milk alone. Chocolate-covered frozen yogurt on a stick is a joke food, but *real* yogurt—easily made at home—is a boon to sound nutrition, a source of infinite mealtime variety, and a step in the direction of

(Continued on page 294)

The Solait cultured-food kit is a simple insulated jar with freeze-dried starters for making soft cheeses such as a boursin, cream cheese, and neufchatel, as well as yogurt. It can be had for $23 from Crayon Yard Corporation, 75 Daggett Street, New Haven, CT 06519.

(Continued from page 293)
food independence.

Home yogurt-making is easier than baking bread. And though you'll apply some modern knowledge and technology, the overall steps in yogurt making are similar to those followed by our ancestors. Fermentation, after all, is not man's invention: it's a natural process that was discovered, studied, and brought under control.

Fermentation explained

When certain bacteria find their way into milk they feed on elements in the milk, multiply, and produce enzymes that alter the structure, taste, and appearance of the milk. This process is called fermentation, though we may also refer to it as the culturing or souring of milk. Fermentation produces not only a variety of yogurt-type products but also sour cream, milk beverages, and hundreds of different cheeses.

There are many types of fermentative bacteria and each is affected by specific temperature ranges and acidity levels. There are also many kinds of animal milks and numerous methods of handling them. Put all these variables together and you have not yogurt but *yogurts,* fermented milk that varies considerably in taste, texture, and nutrition. The Turkish villager, the Bulgarian farmer, and the New York apartment dweller may all ferment milk to make yogurt, but each product will be different from the others because of differences in both production methods and environment.

Left to itself under the right conditions milk will ferment naturally. So the first notice of milk's curdling properties dates from whenever humans domesticated animals, milked them, and attempted to store the milk in vessels—perhaps 10,000 years ago when wild goats and sheep were first tamed by neolithic humans.

Even before the use of pottery, the stomach sacs of slain animals were used to store food. Fresh warm milk in such a bag would curdle rapidly because of bacteria and enzymes naturally present in the intestinal material. Milk kept in earthenware pots would also be curdled by bacteria present in the air. And though the milk may have been discarded, the bacteria responsible for curdling stayed lodged in the pores of the pottery jars.

In a warm climate, then, milk could not have been kept for more than a few hours before it changed from a fluid into a thickened mass. Eventually, some brave New Stone Age humans tasted the curdled milk and discovered a pleasing taste and consistency. People learned in time which procedures and vessels produced a palatable food and which yielded a gassy, bad-smelling mess.

No one group of people discovered this phenomenon. The fermentation of milk was discovered over and over across the world by many racial and tribal groups. That's why there's such a wide variety of fermented milks. Some are thin enough to drink,

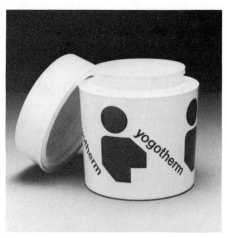

Like the Solait, Yogotherm is little but a well-insulated jar; rather than plug in an appliance, you just place warm milk into the container. Yogotherm sells for $23.95; write International Yogurt Co., 628 N. Doheny Drive, Los Angeles, CA 90069.

In time, yogurt eaters accustomed to the presweetened store-bought stuff will usually find their tastes changing—in favor of unsweetened yogurt, with fresh fruit and other toppings.

some are alcoholic, others are semisolids.

A glass of milk is a staple on North American tables, but milk was more likely eaten in ancient times than consumed as a liquid. (This is still true in most of the world, where drinking milk has never been as popular as in the United States.)

The discovery of the spontaneous fermentation of milk allowed people to both diversify and safeguard their diets. Instead of having only fresh milk to drink, fermentation yielded foods with a variety of consistencies, flavors, and uses. Fine curds could be eaten as yogurt or stirred into a liquid, or cooked with, or dried in the sun. Coarse curds could be strained off and eaten as soft cheeses. And once curds were stored in woven baskets and pottery urns, people discovered that the curds could be pressed and converted to hard, long-lived cheeses.

Early humans didn't know as we do now that fermentation actually produced one of the safest prepared foods they might eat. When milk sours, an acidic medium is created in which many pathogenic organisms dangerous to humans either cannot survive or are inhibited. Tuberculosis, typhoid, paratyphoid, typhus, and coliform bacteria that can infect fresh milk are killed off as milk ferments. So even without refrigeration and under unsanitary and primitive conditions fermented milk served as it still does today as a safe and wholesome product.

Milk will ferment anywhere in the world, of course, but what you get in one place isn't what you'll find in another. Conditions in relatively warm countries will generally convert milk to an acidic food with a firm curd. This is because heat-loving microbes like *Streptococcus thermophilus* and *Lactobacillus bulgaricus* predominate in that environment. In the cooler mountain regions of Europe and Asia, however, bacterialike *Streptococcus lactis* and its variants *Leuconostoc citrovorum* and *Lactobacillus casei*, which thrive at lower temperatures, produce a less-solid, less-acidic food.

In addition to the variables in native bacteria, differences in the types of milk used, and climatological diversity, milk fermentations may also be invaded by wild yeasts that pro-

duce alcohol. Thus what you get depends very much on where you are and what you do with the milk at hand. A bowl of milk left standing in a kitchen in Switzerland may turn into a thick and tasty fermentation. But a bowl of unattended milk in Los Angeles may end up a fetid mess.

Fortunately, modern science now knows enough about fermentation (although many aspects of the process are still mysteries) to isolate some of

From the Cold Teats of Science

Investigation is in progress to obtain knowledge of the chemical compounds which are produced by the culture bacteria and which are responsible for the flavor of the buttermilk. . . . One major compound essential for flavor is diacetyl—and other chemical compounds involved are carbon dioxide, acetic acid, formic acid, pyruvic acid, lactic acid, acetadehyde, ethyl acetate, and ethyl mercaptan. Radioactive tracers are being used to follow the fermentation changes and to establish the relative importance of the chemical compounds to flavor quality and stability.
The consumer has a delightful and versatile product in modern cultured buttermilk.

W. J. Harper and I. A. Gould. "Buttermilk, Then and Now," *Ohio Report*, Ohio Agricultural Research & Development Center.

the principal microbial agents. So it's now possible and practical to make in your own kitchen, no matter where you live, the firm yogurt of the Near East, the low-acid Bulgarian yogurt of Europe, or the kefir of Central Asia.

All the variations in taste and appearance in fermented milk are the result of microbial life—either the activity of long, rod-shaped bacilli or round, seedlike cocci or combina-

tions of the two. And while there are dozens of different types of milk fermentations, all of them fall into three broad divisions: butter cultures; yogurt cultures and related strains; and alcoholic beverages.

Butter cultures demystified

Streptococci are at work here —*S. lactis*, *S. cremoris*, *S. diacetilactis*, and *Leuconostoc citrovorum* favor the production of buttermilk from soured cream, cultured sweet cream buttermilk, cultured skim milk (sold as cultured buttermilk in the United States) cultured cream, and sour cream.

The latter is a favorite food in Central Europe and the Slavic countries, though weight-conscious North Americans tend to shy away from it. Buttermilk—*real* buttermilk—is a popular drink in Belgium, Holland, Finland and Denmark, but is unavailable in U.S. stores.

Acidophilus milk—skim milk inoculated with *L. acidophilus*—is another type of cultured milk popular in Europe but not in North America. Its sour taste is too tart for most Americans raised on sweetened food and drink. A research breakthrough, however, led to the 1975 introduction of so-called "sweet acidophilus milk." Sweet refers not to the addition of sugar but to the fact that this type of acidophilus milk tastes just like plain milk. The idea is that you can get the benefits of lactobacilli without the sour taste of cultured milk.

This is apostasy to cultured milk aficionados, of course, for whom the astringency of fermented milk is a gustatory delight. A sweet- or plain-tasting cultured milk is in the same league as plastic flowers and Astro-Turf. Actually, sweet acidophilus milk isn't a true fermented milk. The lactobacilli are added directly to cold milk. There is no incubation and no fermentation. The milk is held at below 40° F. and acts as a mere carrier for the bacteria.

The bacilli reportedly remain active at this temperature, but whether they are as beneficial to the consumer as fermented milk and whether the bacilli in sweet acidophilus milk can survive digestion and implant
(Continued on page 296)

(Continued from page 295)

themselves in the lower intestine is a debatable question among scientists.

Plain soured milk is popular in many European countries. Milk is left in pans at room temperature to sour for 24 to 48 hours. The thickened product, often called "long milk" because of its viscous, ropy consistency, is eaten with a spoon or stirred into a liquid for drinking.

Long milks are popular in the Scandinavian countries and Finland. The Finns enjoy *viili*, a long milk with a thick layer of cream topped by a covering of *Oospora lactis* mold.* Similar long milks are *ymer* (Denmark), *taettemelk* (Norway), *filmjolk* (Sweden), and *pitkapiima* (Finland).

The principal microbes involved here are *S. thermophilus* and *L. bulgaricus*. These grow best at relatively warm temperatures, 98.6° F. and 113° F. respectively. Yogurt remains a staple food in the areas in which it originated—the Middle East, the Near East, and the Balkans—and is widely used in the Caucasus, Bulgaria, Rumania, Hungary, Greece, and Turkey. Though yogurt is currently popular in North America, even its highest figures for consumption don't approach normal European use of fermented milk products.

The United Nations Food and Agriculture Organization (FAO) estimates that soured milk products make up at least half of all milk consumed in Mediterranean and Near Eastern countries. The yogurts of these regions share some general characteristics. Whole milk from cows, goats, sheep, or buffalo is used, though some peoples use skimmed milk. In most cases the milk is boiled for a long time to reduce the proportion of water, then cooled and inoculated with some yogurt from a previous batch. The incubating yogurt is kept warm in various ways—wrapped in a blanket, set on the back of a wood stove, or warmed by the sun. Some people dilute the thickened milk with water, using it as a beverage. In Afghanistan and Iran

*Eating moldy food sounds terrible, but only because we have—forgive the pun—a cultural bias. We tend in North America to associate molds with pathogens. But beneficial fungi play an important role in food fermentation. Tempeh, miso, and similar Asian staples are the result of mold activity, as are several kinds of cheese.

this drink is called *doogh* or *abdoogh*. The Turks add salt to their diluted yogurt and call the drink *eyran*. Yogurt may also be sun-dried to form sheets of *jub-jub* (Lebanon), *kurut* (Afghanistan), and *kastog* (Iran). And India's traditional yogurt, *dahi*, may be used to make ghee (clarified butter), a food stored for months without refrigeration.

Kefir and kumiss

Kefir and kumiss are two related alcoholic milk beverages. Kefir is also known as kephir, kiaphur, kaphir, kefyr, kefer, khapon, kepi, and kippe. All may be roughly translated to mean "pleasant taste." The peoples of southern Russia, Turkey, and the Balkans have enjoyed kefir for many centuries where it is known as "the champagne of milk."

Kefir contains carbon dioxide, enough to make it effervesce, tickling the tongue and palate. It also contains up to 3 percent alcohol. Traditionally made from camel's or cow's milk, kefir's production was a lucky accident.

Nomadic people stored their milk in leather or goatskin sacks or bottles hung outside in summer and inside in winter. The milk fermented and was used as such, but fresh milk was continually added to the sacks. This caused continuous fermentation and a constant renewal and refreshing of the fermentative organisms.

People eventually noticed the delightful change the milk went through and they also discovered that a unique substance had formed in the sacks—kefir grains. These are small, irregular shaped, yellowish grains resembling miniature cauliflowers or popcorn. The grains, ancient people discovered, could be stored for future use and used to start new batches of kefir.

The spontaneous appearance of these granules and their regenerative properties must have startled their discoverers. Muslim tradition calls kefir "the drink of the Prophet" and maintains that kefir grains were given to Muhammad as a gift from Allah.

Scientists analyzing the grains say they are conglomerated masses of bacteria, yeast, and coagulated milk protein built up in layers. But the traditional practice of regularly adding fresh milk to continuously fermenting milk, plus variations in local conditions and climate, makes it difficult to single out the exact microbial agents responsible for creating this food.

Kefir is the only non-plant food that produces its own seeds. Merely souring milk won't produce the grains, however, so don't run out and buy a goatskin sack. But several laboratories specializing in food cultures will sell you a supply of kefir grains. Unlike yogurt or cheese cultures, you need only buy kefir grains once. The grains will multiply if you maintain the proper condtions. You'll be able to make kefir for years while producing enough excess grains to share with friends.

Kefir culture is also available in packets, but this isn't the same as kefir grains. The culture alone will produce a yogurtlike mixture rather than the traditional beverage.

Kumiss is a variation of kefir, though you can't easily make it at home. Also known as koumiss, kumys, or kumyss, this liquid fermented milk is made only from unheated mare's milk. Kumiss is probably a Mongol word or a name derived from an ancient nomadic people (the Kumanes, Kumyks, or Komans) who originated this drink.

Whoever these wandering herdsmen were, they ranged across Central Asia from China to Rumania during the first century A.D. Their lives revolved around their horses. A later people—the Mongols—also relied on fermented foods made from mare's milk. They made kumiss, of course, and folklore says they made yogurt in leather sacks slung over their saddles. The horses's body heat incubated the yogurt as the rider galloped across the steppes.

Like kefir, kumiss has been described as having the fizz of champagne and the consistency of buttermilk. Its alcoholic content, however, is higher than kefir's. And its flavor, due to the unheated mare's milk, is considerably different. Kumiss is still a traditional drink along the southeastern steppes of the European side of the U.S.S.R. and in western Siberia.

Inoculated with *L. bulgaricus, S. lactis*, possibly *L. acidophilus*, and wild yeasts, fermented mare's milk will yield a brew of upwards of 3 to 4 percent alcohol. (One of the fermented milks Marco Polo described was supposedly 8 percent alcohol.) In Siberia and points east, kumiss is distilled into a brandy with a kick like the horse it came from.

Kumiss varies from place to place. Some varieties are low acid, some are not. The final beverage is the result of a symbiosis of lactic bacteria and alcohol-producing yeasts. The actual microbial activity may involve a mixture of organisms or only a single strain.

The original Mongol method for making kumiss isn't the sort of thing you'd want to do in your kitchen. Soured milk was placed in a leather sack and beaten into a froth with a stick. No quick and easy process, the flailing continued on and off for three to four days. (The Kazaks of Russian Turkestan reportedly still make kumiss by this method.)

Similar products are kurunga, a drink made in Western Siberia from cow's milk; chal, made in Turkmenistan from camel's milk; and arrig (or arrian or arrag), made in Mongolia from yak's milk.

The benefits

All these foods, from buttermilk to kumiss, have a common folklore that claims fermented milk possesses special health benefits far beyond ordinary food. It's not surprising that a food with so long a history and one so intertwined with the growth of civilization should have collected a body of myths and lore. Indeed, the most intriguing and controversial questions surrounding fermented milks have to do with their fabled medicinal and therapeutic values.

But are the claims made for fermented milk only legend? According to tradition, eating soured milk—especially as yogurt or kefir—is variously supposed to nourish, cure, immunize, and help you to live almost as long as Abraham.

Over the centuries, fermented milk has been used as a treatment for diarrhea, constipation, loss of appetite, dysentary, flatulence, anemia, colitis, nephritis, nervousness, and fatigue. Soured milk has also been used as a "blood purifier," and for relief of gastro-intestinal disorders. It has external applications, too. Persian women have traditionally used yogurt as a beauty cream, a use that's caught on in modern circles as a chic facial.

In recent years, regular consumption of fermented milk has been suggested as a means of improving your digestion: the lactic and carbonic acids plus the traces of alcohol found in cultured milk reportedly act as an intestinal tonic. Yogurt has also been used as a douche to cure or relieve vaginal yeast infections. And many surgeons and internists now routinely recommend eating yogurt as a way to help restore intestinal microflora destroyed by surgery or antibiotic therapy.

Making yogurt

There are four steps in making yogurt: milk preparation, inoculation, incubation, and refrigeration.

Milk Preparation. Any type of milk can be fermented, even plant "milk" made from soybeans or nuts, though this is a bit tricky, as we'll see. Let's concentrate first on the more traditional animal milks, either from goats or cows. Each produces a yogurt with a different consistency and taste. These qualities are also dependent on whether the milk is fresh or store-bought.

Pasteurized, homogenized, and containerized whole cow's milk will give a thick custardlike yogurt. But fresh cow's milk right from the milking parlor will be even thicker, with the cream separating to form a buttery cap on the yogurt's surface.

Goat's milk yogurt from fresh milk is rich and creamy. Some batches come out so thick you can almost invert the jar without spilling the contents. Goat's milk has smaller fat globules than cow's milk, so the cream doesn't rise or separate easily. Thus yogurt from fresh goat's milk tends to be more uniform than that made from fresh cow's milk. Many people also find goat's milk easier to digest.

Fresh milk from either cows or goats will vary in butterfat content according to the animal's breed and individual genetic history. A milk richer in butterfat means a thicker yogurt. Nubian goats and Jersey cows give the richest milk, while Toggenburg goats and Holstein cows give milk of a lower fat content.

Store-bought milk in containers generally yields a considerably different yogurt and kefir than that made from fresh milk. Store milk is a blend of milk taken from many different animals living and feeding under a variety of conditions. The best milk—thus the best fermented milk—comes fresh from the farm. It's worth the trip to the small farm or homestead that sells surplus milk. Prices are usually lower than store milk and you can see where your food comes from. Look for organically minded people who put the health and welfare of their animals ahead of profit. Ask about their feeding practices and use of drugs and antibiotics. Inspect the premises. Be sure to ask if milk from animals receiving antibiotics or deworming agents is discarded for at least 72 hours after the animal's *last* injection or treatment.

Potential disease organisms in the milk are no problem for yogurt-makers because the milk must be pasteurized anyway. But drinking raw milk is another matter. Many homesteaders produce safe raw milk because they have few animals and concentrate on keeping them healthy. But for your own safety—unless you're raising your own animals and are experienced in dairy animal management—drink raw milk only if you're sure of its purity.

(Continued on page 298)

DIFFERENT MILKS MAKE DIFFERENT YOGURTS

For *thin,* slightly *bland* yogurt	Use nonfat dry milk reconstituted with water or skim milk
For *rich,* smooth, *custardlike* yogurt with a *slightly tangy* taste	Use whole milk (⅓ cup nonfat milk may be added to thicken the finished product)
For *thick, rich, sweet* yogurt	Use evaporated, half-and-half, or 1 cup of heavy cream added to 1 quart of whole milk

(Continued from page 297)

You can also make yogurt with skim milk powder and get good results. Use only *spray dried* skim milk powder, however. This is available in natural food stores and is the same type of powder used by yogurt and cheese manufacturers. Instant-type powdered milk is all right for drinking, but it produces yogurt with a gummy or lumpy consistency. Spray dried milk can be used either to drink or ferment. It's inexpensive and has a long storage life if kept cool and dry. Dried milk differs nutritionally from whole milk. Lower in calories (because the fat has been removed), powdered milk also lacks the fat soluble vitamins—A, D, E, and K. When dried milk is added to whole or skim milk to make a thicker yogurt, however, the overall protein, calcium, phosphorus, and caloric content is boosted.

Yogurt can be made from nonfat milk powder and water alone, but the activity of lactobacilli is improved when even a little milk fat is present. You'll have a firmer and better tasting yogurt if you add about ½ cup of whole milk to each quart of skim milk.

Store-bought skim milk in containers also needs some thickening, otherwise fermentation will give you a cultured skim milk (the so-called cultured buttermilk sans butter flecks). This is a pleasant drink, but if you want thicker yogurt add whole milk as above or four to eight tablespoons of spray dried milk powder. Add the milk powder prior to heating the milk and blend in thoroughly. Don't add too much milk powder in search of a super-thick culture or your yogurt will develop a gritty consistency. You can also add some sweet cream to skim milk instead of milk powder, but be conservative in its use and experiment with small batches. Too much cream means the microbes will deliver sour cream instead of yogurt.

And here's a novel way to get a firmer yogurt: add a teaspoon of carrot or tomato juice to every eight ounces of milk. Add the juice to the prewarmed milk when you add the yogurt starter. Evaporated or condensed milk may also be used in yogurt-making but neither product is recommended to readers interested in natural foods. Evaporated milk is canned whole milk with 50 percent of the water removed. The processing

destroys much of the milk's vitamin B6 (pyridoxine) content. And the product is laced with additives: disodium phosphate, sodium citrate, carrageenan, and others. Condensed milk is also canned whole milk but it has 60 percent of the water removed, with sugar or corn syrup or both added as sweeteners. Thus eight ounces of whole milk contains 160 calories. The same amount of reconstituted condensed milk has 275 calories.

Soymilk may be fermented, but the process is less predictable in yogurt-making because of the biological differences between soy and animal milk. You can buy canned soymilk—dry or liquid—but read the labels. This product is also intensively "enhanced" by sweeteners and additives. You can make your own soymilk at home, of course. You'll have a natural product and one free of the beany flavor soymilk is infamous for. (See recipe section in Beans chapter, page 102.) The people most usually interested in soymilk either as a drink or as yogurt are vegans who use no milk or persons allergic to animal milk. These readers should note that soymilk is not the nutritional equivalent of animal milk.

Soymilk is low in riboflavin (B2) totally lacking in B12, and has drastically less calcium than dairy milk. On the other hand, soymilk is lower in carbohydrates, has 12 percent fewer calories, 25 percent less total fat, no cholesterol, and contains 15 times more iron than cow's milk.

We've discussed the use of raw milk, and some people prefer this to pasteurized milk because they believe that heating milk destroys much of its nutritional value. Pasteurization does reduce the presence of some vitamins—primarily B1 (thiamine) and ascorbic acid—but only by about 10 percent according to many dairy scientists. Supporters of raw milk also claim that pasteurization denatures milk protein and alters its biological effect on the animal consuming it.

The argument has raged for decades, but the point for yogurt-makers is this: you cannot make yogurt without pasteurizing milk. Fresh milk contains certain bacteria and enzymes that inhibit the type of fermentation associated with yogurt. These agents must be destroyed or altered before yogurt can be made.

Fresh raw milk will sour, of course, and many products can be made from curdled milk. But sour milk isn't yogurt; neither are curds or clabber.

Pasteurization. There are two methods of pasteurization. The so-called flash method involves rapidly heating milk to 160 to 165° F. and holding it there for 30 to 60 seconds. In the holding method you heat the milk to 140 to 145° F. and maintain that temperature for 30 minutes. Use a candy-making or deep-frying thermometer for true readings.

A less technical method is to simply bring the milk to a point just below boiling. When the milk begins to simmer along the edges remove the pot from the heat. Milk scorches easily, though, and the lower temperatures monitored with a thermometer lessen your chances of ruining a gallon of milk. Once burned even slightly the milk is no good for yogurt or anything else but compost. The burned flavor passes into the yogurt.

Pasteurization doesn't sterilize milk, as many people believe. It kills 98 to 99 percent of the microbes present, but those 1 or 2 percent can multiply quickly if the milk isn't properly handled. Chill the milk immediately after pasteurization to prevent the re-entry or renewal of bacteria. Heating milk to 190° F. or boiling it does sterilize it. But even in this case, milk can be invaded by ambient bacteria if not kept chilled and clean. Milk needn't be brought to a boil for yogurt-making.

If you're only making a quart of yogurt at a time, heat the milk in a double boiler or use an insulating pad and a two-quart saucepan. (A one-quart pot holds *exactly* one quart, right to the brim.) Stir frequently at first, then continuously as the milk's temperature rises. For larger amounts of milk—one to two gallons—it's best to start with the burner set at low. Turn it up a notch every five or ten minutes. Raising the temperature slowly minimizes the chance of scorching, but you'll still have to stir continuously once the heat gets in the upper ranges.

You can use a pressure cooker to heat large amounts of milk. It's faster than other methods and uses less energy. The thicker bottoms of well-made pressure cookers are also added insurance against the milk scorching. Raise the heat slowly as above and watch for signs of moisture escaping from the cooker's vent. Remove the lid, check the temperature, and stir until pasteurization is complete.

Inoculation. After pasteurization, the milk is cooled and inoculated with lactobacilli. The simplest way to do this is to add some yogurt from your last batch. But this process cannot go on indefinitely for reasons to be explained later. And what about the first time yogurt-maker with no previous culture to rely on? Most instructions for yogurt-making recommend that you inoculate your first pot of milk with a few tablespoons of store-bought yogurt.

This works but it can be a hit-or-miss procedure—especially for the beginning yogurt-maker. Many store-bought yogurts are pasteurized after fermentation and have nonexistent or severely weakened bacterial populations. Others have additives or stabilizers that also inhibit lactobacilli.

What about being careful to use only those yogurts made with live bacteria and without additives? Several brands of yogurt take care to state on the label: "Contains live yogurt bacteria." You can use these products as starters—most people do—but remember that you're dealing with unseen microbial life. Using a store-bought yogurt means several factors are unknown. First, how old is the yogurt? That is, when was it actually manufactured, not shipped from the factory? Lactobacilli don't remain at peak efficiency indefinitely. Time, pH changes, and invasions of ambient bacteria affect their life cycles and ultimately the yogurt you make.

What types of bacteria does the manufacturer use? Most use *S. thermophilus* and *L. bulgaricus* in a 1:1 ratio, but this information never appears on the label. Besides, the 1:1 relationship can become altered as the yogurt ages and increases in acidity. This condition favors the growth of *L. bulgaricus* at the expense of other lactobacilli. So you could use some *acidophilus* yogurt as a starter and eventually end up with *L. bulgaricus* without knowing it.

Purists, this writer among them, prefer the control and uniform results gained by using pure freeze-dried bacterial starters. These are available in packets and vials and can be purchased in most natural food stores or by mail from the manufacturers of dairy cultures. The starters are not cheap at first glance—about $2.50 a packet—but you'll amortize this after you make your first few weeks of yogurt. Use the dried culture to make one quart of yogurt. Don't eat this. Use it as a "mother culture" to start succeeding batches. Most manufacturers suggest you use the mother culture and its generations as starters for about 30 days. Then you buy another pure culture and begin again. This is good advice. After a month or so the original culture will have changed in pH and will probably have been contaminated by foreign bacteria. You'll know this when later generations of yogurt lack the flavor or consistency of earlier batches.

Unopened packets of freeze-dried culture have a long shelf life when stored in a cool place. Their use makes sense for both novice and experienced yogurt-makers who want good yogurt every time.

Follow the directions on the packet if you're making a first-time starter using dried culture. (All you do is mix the powder with warm milk.) If you're making a batch using either store-bought yogurt or yogurt from your last effort, use two to three tablespoons to each quart of warm milk. Too little starter means the yogurt will take longer to incubate. This increases the acidity and produces more tartness—something you may enjoy or wish to avoid. A relatively rapid incubation generally produces a milder tasting yogurt with more active lactobacilli. Too much starter, however, produces yogurt with a lumpy curd (almost a watery cheese) best used in cooking. Two table-

(Continued on page 300)

HOW HOMEMADE YOGURT COMPARES WITH COMMERCIAL YOGURT
(1 CUP BLUEBERRY-FLAVORED YOGURT)

	Homemade	Dannon lowfat	Light 'n' Lively lowfat	Maya
Calories	158*	260	240	280
Price	15¢**	49¢	43¢	65¢
Ingredients	1 cup whole milk, purchased from store; 1½ teaspoons starter from last batch; 2 tablespoons fresh blueberries; 1 teaspoon honey	lowfat milk; blueberries; sugar; corn sweeteners; nonfat milk solids; pectin; lemon juice	lowfat milk; sugar; skim milk; corn syrup; water; blueberries; food starch; modified yogurt culture; sodium caseinate; gelatin; natural flavor; citric acid; artificial color	whole milk; nonfat milk; cream; crushed blueberry and puree; fructose; honey; carrageenan; vegetable gums (carob bean, tragacanth, and guar); blueberry extract with other natural flavors; red beet juice and grape skin extract as color

*calories calculated from chart in *Joy of Cooking* (Bobbs-Merrill, 1964).
**prices calculated from August 1979 survey of supermarkets in the Philadelphia area. This includes only the price of milk and blueberries since the initial cost of a starter and a teaspoon of honey are negligible.

(Continued from page 299)

spoons to the quart is an approximation. You'll have to experiment to find your optimum. Here again, using a mother culture prepared from freeze-dried lactobacilli yields the most uniform results.

Incubation. Most yogurt-forming bacteria flourish within a range of temperatures—90 to 118° F. Below 90° they become inactive. Above 118° they multiply too rapidly and crowd themselves out of existence. At 120° and above they are killed by the heat. (Kefir cultures prefer lower temperatures, about 70°.)

Incubation is the most critical of all the steps in yogurt-making, almost approaching an art. Commercial yogurt-makers, who can't afford to make a mistake with several hundred gallons of milk, pay close attention to maintaining an optimum incubation temperature. So should you.

Though the range is 90 to 118° F., it's best to keep your culture in the middle, at about 110°. If you can maintain closer tolerances in temperature try for 105 to 107° F., a range some manufacturers and dairy scientists believe produces a yogurt with better body. If your incubation temperature is going to stray, it's better to have it fall slightly than rise. A drop will cause the culture to grow more slowly (commerical makers have successfully held yogurt cultures at 90° F. overnight) but a sharp rise may destroy your microbial helpers.

If you're making your initial or mother culture with freeze-dried bacteria you'll have to maintain the proper incubation temperature for five to ten hours or longer. If you're using a starter from a previous batch you'll need to maintain the temperature for only two to four hours.

There are many ways to produce enough heat for incubation, none of them expensive or difficult. Almost all of them will work better, however, if you use a water bath as an insulator for the incubating yogurt. Let's assume you're going to use wide-mouthed Mason jars. They're the easiest vessels to handle and keep clean. After filling the jars (pre-warmed) with inoculated milk, cap them and place in a large pot or deep broiler pan. Add warm water—110 to 115° F.—up to at least the mid-point of the jars. (To the neck

Electric yogurt-makers may assuage the fears of a novice, but they don't do anything that common pots and jars cannot.

is best, but a large pot may get too heavy to handle safely.)

Now you must maintain the water bath's temperature which will vary less rapidly than that of the individual jars. You can also use the water bath alone as a heat source, checking the temperature frequently and adding warm water as needed, but this method keeps you tied to the kitchen. There are several other heat sources that will give you more leeway.

• Insulated picnic cooler: Place jars in the cooler, fill with warm water, and check the temperature every hour. Add and subtract water as needed.
• Oven: An electric oven set on a very low heat will do the job. So will a gas oven with only a pilot light. But in both cases (especially the latter) use a thermometer and a pot of water to be sure the temperature remains constant.
• Central heating: Set your water bath on or near your oil burner. Check to insure that temperatures don't get too high or low as the burner cycles on and off.
• Solar: Set the water bath in a cold frame, greenhouse, or south-facing window by day. This is tricky, though. A passing cloud may end your efforts. Keep the milk itself sheltered from the direct rays of the sun. Vitamin B_2 (riboflavin) in solution is destroyed by light.
• Blankets: Some yogurt fans have reported excellent results by simply wrapping their jars or other vessels in heavy blankets, sleeping bags, or fleeces. This method is not reliable for very long incubations, as with a first culturing.

• Food dryer: Temperatures used to dry food are usually considerably higher than incubation temperatures but you should be able to adjust for the difference. Should your dryer use light bulbs as a heat source, however, heed the warning about riboflavin under the solar heading above.
• Heat tray: A thermostatically controlled tray used to keep food warm works well. Use a thermometer and jar of water to find the right setting.
• Thermos: Good only for a quart. Pour the inoculated milk into a pre-warmed wide-mouthed thermos, cap it, and let it sit undisturbed.

Refrigeration. When is the yogurt finished? There are three stages of viscosity in yogurt incubation: liquid, thickened liquid, and set.

The third step yields what we generally want as finished yogurt. But knowing just when to stop the incubation process is a knack that comes only with experience. Pull the milk too soon and you'll have a fluid. Wait too long and the yogurt will increase in sourness and the whey—a yellow liquid—will separate from the milk solids.

Here are some suggestions for assessing the incubation process. Are you making a mother culture from freeze-dried bacteria? Relax. Don't even check the yogurt's progress for at least five hours. If you're making a succeeding batch, check it after two hours. Capped Mason jars work well here because you can inspect the development more easily than with bowls or opaque containers.

Watch for the milk to develop a

thickened core with a thin layer of liquid milk surrounding it. In bowls, watch for the milk to become custardy or puddinglike, pulling away from the bowl sides. In both cases tilt the vessels *gently* when making your inspection. Rough handling will rupture the delicate coagulum and cause wheying off.

Don't wait for the yogurt to become totally firm. Watch for that semi-solid central core to develop, then remove the jars and refrigerate them. The yogurt will continue to thicken in the refrigerator. Experience will teach you just how much coagulation will occur after you remove the yogurt from the incubator.

Refrigerate the yogurt, don't freeze it. Store it at 40 to 42° F. Avoid agitating the jar or bowl until the yogurt is chilled and firmly set. This usually takes about 12 hours.

What happened?

Home yogurt-making usually goes on without a hitch. But just as the occasional loaf of bread fails, so too are there gremlins that affect fermentation. Repeated yogurt disasters generally fall into a pattern.

Milk never thickened. If after an appropriate time passage—up to 24 hours—the milk fails to coagulate, look for several possibilities. It could be the milk was not pasteurized, or was heated below optimum levels. Perhaps you forgot to add the yogurt starter. (Don't laugh. It happens.) Or, you used store-bought yogurt as a starter; the lactobacilli were either dead or inhibited by post-culture pasteurization or additives. Another explanation is that the milk got either too hot or too cold during incubation. If the temperature accidentally drops to 90° F. or below during incubation, the lactobacilli will become inactive. You can sometimes save a batch by raising the temperature again and adding more starter.

If the heat rises above 118° F., however, the microbes will be killed. Forget making yogurt at this point and convert the milk to cottage cheese.

It's best to make yogurt during the day when you can keep an eye on the incubation process. Resist the temptation to let the culture work overnight unless your experience tells you the process always takes as long as you sleep.

Fermentation may also be curtailed or inhibited by residues of antibiotics in the milk. As little as 0.005 international units of penicillin per milliliter of milk can adversely affect the aroma of finished yogurt. *S. thermophilus*—one of the two major lactobacilli in yogurt—is exquisitely sensitive to antibiotic residues. Unfortunately, antibiotics are in wide use in the dairy business. (Half of all the antibiotics produced in the United States are used in the rearing of meat, egg, and dairy animals.) Milk producers are supposed to withhold milk contaminated with residues, but the economics of agribusiness make it more costly to discard milk than to

(Continued on page 302)

YOGURT IN SEVEN STEPS

- *Pasteurize* milk.
- *Cool* to 105 to 110° F.
- *Prewarm* jars or other vessels.
- *Add* starter:

   One packet freeze-dried culture to the quart for mother culture.

   or

   Two tablespoons of yogurt from a previous batch.

   or

   Two tablespoons of store-bought yogurt—plain, unflavored, free of additives, with active lactobacilli.

- *Incubate* at 105 to 110° F.

   5 to 10 hours or more for the first quart from freeze-dried culture.

   or

   2 to 4 hours for any amount inoculated with yogurt from a previous batch.

- *Remove from incubator* when thickened.
- *Chill* 12 hours or more at 40 to 42° F.

The equipment for making yogurt is simple—you probably have everything you need on hand.

Milk is pasteurized, then cooled to 105 to 110° F. Culture is added in the form of a freeze-dried packet, or yogurt with live culture.

The cultured milk is now allowed to incubate, here in a kettle of warm water.

When thickened, yogurt is ready to eat. It should be stored in the refrigerator.

(Continued from page 301)

risk the relatively minor penalties associated with trying to slip tainted milk past the inspectors.

Other similar factors can affect milk's fermentation: residues of quaternary ammonium cleaning compounds used in dairying can inhibit lactic bacteria, as can milk taken from animals suffering from mastitis. Short of a laboratory assay, there's no way you can tell whether the milk you buy contains such residues or agents. But if your yogurt continually fails or its progress appears inhibited (very long incubation times are an indication), change brands of milk if you're buying from a store or make some sharp inquiries if you're buying fresh milk at a farm.

Wheying off. (A yellowish liquid separates from the solids.)
● Incubation was too long.
● Agitation during incubation.
    Wheying off is a minor problem in most cases. Don't discard the whey by pouring it off. It's mostly water, but it also contains lactose, protein, and minerals.

Gassy or bubbly yogurt.
● Wild yeasts or coliform bacteria invaded your yogurt mixture. Discard the batch or add it to your compost pile. Be sure to work in clean surroundings. Sterilize your utensils and jars if necessary.

Lumpy or cheesy curd.
● The incubation temperatures were too high or too much starter was added causing overpopulation and overripening. Don't discard. Use in cooking.

Ropiness or stringy curd.
● Invasion of bacteria during handling of milk after pasteurization or during incubation.

Weak or thin curd.
● The incubation time was too short.
● Not enough starter was added.
● Reconstituted or skim milk used with no additional solids or milk fat added as thickeners.
● Some cows in early lactation may give milk that forms a poor curd.
    Don't discard. Use in cooking and baking.

Gritty curd.
● Too much milk powder added as a thickener.

Fermentation may also go astray or be halted by the presence of bacteriophages. These are ultramicroscopic particles containing genetic material. They're thought to be either enzymes or substances related to viruses. When reproduced by a host bacterium—lactobacilli in this case—their multiplication soon supplants the hosts and wreaks havoc with the fermentative process. Bacteriophages are an especial bedevil-

ment in cheese-making.

The best protection against their invasion is cleanliness during all stages of yogurt-making, even to the point of sterilizing jars, spoons, and other utensils if necessary.

If you're reconstituting milk powder and consistently getting poor quality yogurt, look to the source of your water. The activity of lactobacilli may be disrupted by chlorine and other chemicals in city water supplies. Use an activated charcoal filter to remove these substances before adding milk powder to the water.

Bacteria from contaminated well or spring water may also affect the fermentative process. Milk made from milk powder normally doesn't need to be pasteurized because it's been heat-treated to begin with (though your milk mixture can still become contaminated by ambient bacteria). But your water supply may contain sufficient bacteria to throw fermentation off. Have the water checked by the appropriate county agency.

The time of the year may also influence milk fermentation. Fresh milk drawn in the summer is sometimes more difficult to culture than in other seasons, perhaps because warmth and humidity favor increased microbial activity.

Vic Sussman

THE AUTHOR CONFESSES

I struggled through an entire summer once when not a single batch of yogurt cultured properly. I was making (or attempting to make) three gallons of yogurt at a time, something I'd done successfully for years. But that summer every attempt ended in unthickened milk no matter how long the incubation was.

The milk was from healthy, drug-free goats. The culture was freeze dried. The kitchen was clean and all utensils had been sterilized. But no yogurt.

Was it summer milk? Bacteriophage? The evil eye? I never determined the source of the trouble. And thank goodness I've never had a yogurtless summer like that since.

Vic Sussman

Sour cream

Following is a simple recipe for making your own sour cream. If you use raw cream, it's important to pasteurize it. One way to pasteurize cream is to place it in a pan over boiling water, insert a thermometer, and heat until the thermometer registers 185° F., stirring constantly. Remove the cream from the heat, place it in a container of ice water, and stir until cold. Cover and refrigerate until needed.

If you use your own cream, you can save money by making sour cream. If you have to buy commercial whipping cream and buttermilk, it may cost about one-fourth more.

Sour Cream
2 cups pasteurized whipping cream (or light cream for fewer calories)
2 tablespoons buttermilk

Pour the cream and buttermilk into a sterile glass jar. Screw lid on tightly and shake well. Let jar stand in warm place until the cream is thick and sour. This will take about 24 to 48 hours, depending on the temperature of the room.

After the cream has soured, chill it for several hours or overnight for a smooth texture.

Phyllis Olson and Bill LaGrange,
Cooperative Extension Service,
Iowa State University.

Soft cheeses

Cheese-making: the first try

There's not much difference, once you know enough about it to understand the mechanisms, between baking bread, making beer or wine, treating sewage, composting and cheese-making. In bread baking, you are concentrating on the CO_2 (carbon dioxide) involved, which you want; in beer-making you are concentrating on the CO_2 and alcohol involved, both of which you want; in wine-making, you want the alcohol but not the CO_2; in present-day conventional sewage treatment you don't want anything; in composting, you want to destroy the proteins and fats, and keep the cellulose (sugars); in cheese-making, you want the fats and proteins, but not the sugars.

Cheese-making can be very complicated and requires extreme care, especially if you are making it commercially and want to produce a product with predictable flavor and texture. But if you just want to make a pound or two for yourself, then try this way:

If your milk will turn sour in the ordinary way, let it. If it has been efficiently pasteurized, and either "goes bad" horribly, or else doesn't sour at all, then start if off with some ordinary yogurt lactobacillus strain ("starter" bacteria). When it has set a

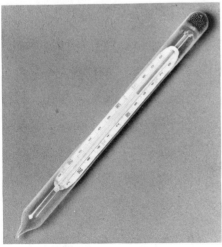

Inexpensive floating dairy thermometers appear in the catalogs of Garden Way (Charlotte, VT 05445) and Cumberland General Store (Route 3, Crossville, TN 38555).

little beyond the point where you would normally wish to eat it as sour milk, break it up, and hang it in cheesecloth overnight. Collect the liquid; use if for soups and gravies, or for cooking cabbage and other vegetables in it.

The curd now in your muslin is mashed up, with herbs if you wish, and eaten straight away. If it is rather dry, you can mash it with some cream. This is known variously as cottage or cream cheese, or farmers' cheese, or fresh cheese.

To make matured cheese, pat this cottage cheese into a roundel (circular shape), and lay it down on a cloth over a wire mesh; the idea is to let air flow over it on both sides as evenly as possible. When it is firm and can be turned without using the cloth as an intermediary, put it into a closed dish in a cool place, and forget it. Since you have covered it up, it cannot lose moisture too fast; and since you have not sealed it with paraffin wax, oxygen can get at it. Some fungus may grow on it; either let it grow, or sprinkle lightly with salt, which will kill it. But the fungus makes a pleasant flavor all of its own, and you may like it. Turn the cheese over every now and then, and eat it when you want to—anytime from a few days to a few months later. The taste will keep on changing all the time, and so will the texture, as it gets steadily harder.

You don't have to make your cheese up into a flat cake; you can make it as a ball or cone. You can coat the surface with herbs, or spices, or seeds—poppy, grape, peppercorns, cumin. You can make it any size you like, from about an inch thick upwards. (Less than an inch tends to dry out too quickly, regardless of how well you keep it covered.) If you keep your dish with its contained cheese in your refrigerator to mature, then you had best put it in the fresh vegetable compartment, which will help stop it from drying out too quickly. So much for the instructions. Now what is actually happening? Your milk is a mass of microscopic-size fat globules (butterfat) held up in water—an emulsion.

Dissolved in that water are also some simple sugars, especially lactose; and protein, especially milk globulins. Originally, of course, these came directly and indirectly from the bloodstream of the cow; they got there from her rumen and intestinal tract, by conversion from the grass she ate. Those conversions have taken place all over her body, with the new products being taken to the udder. Other products are made in the individual glands of the udder—excretory cells—which break open and die, releasing their contents into the milk stream. That is how milk contains some components that are not found in the cow's blood.

So there is milk—water with protein and sugar dissolved in it, bespattered with little fat globules suspended in it. When the milk goes "sour" that is because the lactose sugar is being converted into acetic acid by bacteria—lactobacillus. They obtain energy by this conversion, and so can multiply and continue their life cycles. The rising acidity from this lactic acid coagulates the protein in much the same way as heat coagulates the white of an egg; this is an example of physical coagulation.

You can think of proteins as being rather like cotton lint—long thin fibrils, but of course much smaller; the fat globules are trapped in the gaps between them. As coagulation proceeds, these fibrils get shorter and shorter, squeezing the water out and trapping the fat globules in their mesh. Even when all the lactose has been converted to lactic acid, the contraction of the protein fibrils continues. It is somewhere in the middle of this stage that you will hang the mix up to drain.

During the next stage of maturation, the lactic acid is further broken down by other organisms to carbon dioxide, alcohol, and water. Since the cake is now fairly dense, oxygen gets in only slowly, and anyway, you have more or less sealed the outside, so that the program will be a mix of activity by organisms which require oxygen and those which prefer not to have it. The different proportions will

(Continued on page 304)

(Continued from page 303)

decide what the cheese tastes like, and of course many other variables play their part, too—that is why it is always such a delightful anticipation to find out what your cheese will taste like. Of course, you now understand that you can eat your cheese whenever you want to, even (or especially) if it has gone moldy. The longer you wait, the stronger and the drier it is likely to become.

At the far end of the line, you may find yourself faced with a hard, dry cake, with an exceedingly sharp taste, which may even be gritty. If you examine the grit particles, you will find that they are a perfect rhomboid (slanted rectangle) shape—they are just crystals of pure thioglycollic acid. Lovely. That is what gives "sharp" cheeses their characteristic sharpness. In fact, I have had some success maturing supermarket stan-

dard Baby Edam cheese—those are the red waxed ones. Kept in a cool place and turned over once a week for some weeks, they become excellent eating, quite unlike the soapy stuff that they were when you bought them.

Geoffrey Stanford

Sources for rennet and other cheese-making supplies.

New England Cheesemaking Supplies, P.O. Box 85, Ashfield, MA 01330.

Home Cheesemaking, 2017 Danbury, Madison, WI 53711.

The cheeses

Cottage cheese. Cottage cheese clabbered with rennet makes mild, large-curd or "California style" cheese. A tangy, small-curd version can be made without rennet. Raw skim milk will curd naturally if cov-

ered with a thin cloth and left at room temperature for about 48 hours. The milk will clabber more quickly (12 to 24 hours) if you add ½ cup butter or four tablespoons yogurt for each gallon of milk.

Mozzarella. Mozzarella is one of the most popular cheeses in America—we use literally tons of it on pizza. Your own mozzarella will be a taste revelation when you compare it to the tasteless little rubber balls sold in plastic or the fake cheese melted on commercial pizza. In Italy and in rare Italian-American groceries, you can find genuine mozzarella, stored in water-filled pans.

Blocks of homemade mozzarella keep well for about one week in the refrigerator. The cheese may be frozen with some expected loss of flavor and texture. To use frozen mozzarella for pizza, shred it while it's still partially frozen.

Cream cheese. The way cream cheese turns out depends largely on the butterfat content of the milk or cream mixture. The most famous French cream cheese (Gervais), made with one quart of cream for every gallon of milk, has a 60 percent fat content and is indecently creamy.

Neufchatel. Neufchatel is a rich French version of cream cheese. The less rennet you use and the longer the custard takes to set, the finer the finished product will be. Proceed as for cream cheese, but pack the Neufchatel into little molds with lots of drainage holes. A cheese molded and drained in a fancy ceramic heart is called coeur

Milk on its way to be cheese may have to be encouraged to sour with a bit of yogurt.

Conventional recipes call for rennet to be stirred in.

Place the curds in a colander lined with a piece of cheesecloth.

Hang the balled curds to dry above the sink or a bowl.

The cheese can be eaten after hanging, or you can let it mature in the form of a cake, ball, or cone.

Yogurt cheese is simple to make. It's already cultured and only needs to dry a bit.

BUTTER CHURNER'S CHANT

Need a diversion to make the time go faster? You might like to try the traditional chant that the churner said in time to the up and down movements of the dasher. The arrows indicate the dasher movement.

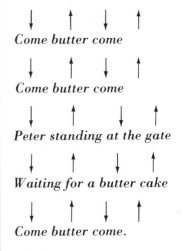

↓ ↑ ↓ ↑

Come butter come

↓ ↑ ↓ ↑

Come butter come

↓ ↑ ↓ ↑

Peter standing at the gate

↓ ↑ ↓ ↑

Waiting for a butter cake

↓ ↑ ↓ ↑

Come butter come.

Eliot Wigginton. Foxfire (Doubleday & Company, 1972).

à la crème. It is traditionally served with strawberries on top.

Yogurt cheese. No cheese could be simpler to make than yogurt cheese. Tie fresh yogurt up in cheesecloth and hang at room temperature for six to eight hours, and you'll have a soft, spreadable cheese. It can be used like cream cheese, but has the characteristic sharpness of yogurt.

Myost. Myost is made with sweet whey left over from making hard cheese. Goat's whey is the most authentic, but cow's whey can also be used. Simmer the whey and cream (about ¼ cup per quart) until the mixture turns thick and brown, like apple butter. Like mozzarella, smooth myost is a kneaded cheese. It is always served unsalted.

Ricotta. In Italian, ricotta means "cooked once again." Whey drained from curd of other cheese is reheated and mixed with fresh milk and vinegar. Home cheese-makers usually don't have the large quantities of whey needed for ricotta, however. A modern method of preparing the cheese uses fresh milk and dried whey powder, available in natural food stores.

Creamy ricotta is much like cottage cheese, but tastes sweeter. It can also be pressed and dried for grating.

Susan Asanovic

The oldest curdling method of them all

The oldest and simplest way to curdle milk for cheese is to do nothing at all to it. Fresh whole milk left to stand in a warm place for a few days will ferment and curdle.

The problem with allowing milk to ferment naturally is that the quality of the cheese so produced may or may not be good. Some bacteria produce off-flavors and may be present in the milk. Most cheese-makers start with fresh, sweet milk and add a starter solution of bacteria of a known type, use rennet, or use the extract of certain herbs. Lemon juice and vinegar, due to their high acid content, are also used to curdle milk. The choice is yours.

Low-Fat Cottage Cheese Dip

2 cups dry curd cottage cheese, unsalted
3 tablespoons blue cheese or Roquefort
2 to 3 tablespoons pimento or fresh red pepper, chopped
2 tablespoons onion or scallions, finely chopped
    yogurt, as necessary

Blend the cheeses, peppers, and onions by mashing them through a sieve or using a food processor. Add yogurt to obtain the desired consistency.

COTTAGE CHEESE: WHAT WENT WRONG?

Sour, acid flavor:
- too much acid developed before and during cooking the curd.
- too much whey was retained in the curds.
- the curd was insufficiently washed and drained.

Yeasty, sweet, or unclean flavors:
- unclean utensils or an impure starter introduced yeasts, molds, or bacteria into your cheese, etc.
- the milk was not completely pasteurized.

Soft, wet curd:
- too much moisture in the cheese.
- development of too much acid while cutting the curd.
- heating the cut curd at too high or too low a temperature
- allowing overly large curd particles to form.

Tough, dry curd:
- insufficient acid development in the curd before it is cut.
- cutting the curd too fine.
- too high a heating temperature.
- too long a holding time after cooking and before dipping off the whey.

From *Countryside.*

BASIC STEPS FOR MAKING SOFT CHEESES

	Cottage cheese	Farmer cheese	Soft white cheese	Mozzarella
Basic ingredients	skim or whole milk	as for cottage cheese	skim or whole milk	skim or whole milk
Ripening (for pasteurized milk)	4 T. yogurt/gal. or ½ c. buttermilk/gal.	"	½ c. buttermilk/gal.	½ c. buttermilk/gal.
Warm	room temperature	"	warm to 92 to 94° F.	warm to 90° F.
Rennet (¼ tablet/gal.)	add rennet for milky, large-curd cheese	"	add rennet	add only ½ rennet solution
Curd forms	cut curd after 12 to 18 hrs. at 75 to 85° F.	"	cut curd after 30 min.	cut curd after 20 to 30 min.
Heat	stir ½ hr. at 95 to 110° F.	"	stir for 30 min. at 92 to 94° F.	stir 15 min. at 90° F.
Drain	drain in cheese-cloth-lined colander, then hang; rinse	" (becomes pot cheese at this step)	pour off whey	pour off and reserve whey
Press		wrap in several layers of cheesecloth, press with 1- or 2-lb.-weight	packed into weighted press 2 to 4 hrs.	pack curds, cut into 3″ x 3″ blocks
Cool				bathe in cool water 15 min.
Refrigerate				tie in cheesecloth, refrigerate in whey 1 to 3 days
Reheat				bring curd to room temperature heat whey to 180° F.
Knead				knead curds in whey until smooth and plastic
Shape				form small balls, cool quickly in cold water

Cream cheese	Neufchatel	Yogurt cheese	Myost	Traditional ricotta	New World ricotta
whole milk and cream	whole milk and cream	yogurt	whey and ¼ c. cream/qt.	whey (2½ gal. yields 1 lb.)	skim or whole milk plus ¼ c. whey powder/qt.
½ c. buttermilk/ gal. (optional)					
60 to 65° F.	warm to 86° F.	room temperature	simmer until thick cream forms	heat until cream rises; add 1 c. milk/gal. at 200° F.	room temperature
add rennet	add only ½ rennet solution			add 6 T. vinegar/gal.	add 2 T. buttermilk OR lemon juice OR vinegar/qt.
custard forms after 12 hrs.; do not cut	custard forms 18 to 24 hrs.; do not cut			remove from heat, dip out curds	leave undisturbed 24 hrs.
			continue to cook to apple butter con- sistency		heat very slowly to 200° F.
drain in cheesecloth- lined colander, then hang overnight	pack into molds, drain 12 to 24 hrs.	hang in cheesecloth 6 to 8 hrs. to drain		drain in cheesecloth- lined colander	dip out curds, drain through cheesecloth
				knead 10 min. at room temperature	
				pack into molds to harden	

An oval cheese form

A cheese form, or mold, is used to shape the curd into cheese. Forms are hard to find, especially in a homestead size. Here is a do-it-yourself plan for a wooden oval form. The form measures 6½ by 8½ inches, and is 3½ inches deep.

The first step is to cut out the bottom piece, and drill ⅛-inch drainage holes. Next cut the thin wood for bending the sides. For bending, the best wood is green oak or ash, but other woods will do. The oval shape is easier to bend than the customary circle. The sides should be a piece 3½ by 27 inches, ¹⁄₁₆ inch thick. To make them pliable, hold over steam or soak in warm water. Once the wood is flexible enough to make the bends easily, wrap it around the bottom and fasten with brads every 2 inches. As soon as the bottom is secured, fasten the top together where the two pieces overlap with a single rivet, as shown.

Cut out a top to fit in the mold, and attach small legs to allow excess whey to drain out the bottom of the mold.

The mold should be left untreated, as finishes may be toxic and, at least, are apt to impart a flavor to the cheese.

Louise Langsner

A simple press

You'll need:
a large juice or coffee can
a can or jar (slightly narrower than can)
2 circles of muslin the diameter of the larger can
a wooden disc that fits snugly into the larger can
a brick or two

Punch eight or ten holes in the bottom of the large can from the inside out. Scald the can. Place a piece of muslin in the bottom and spoon in the curds. Place the second muslin circle on the curds, followed by the wooden disc, then the smaller can or jar. Place a brick on top of the whole set-up. (Two bricks may be tricky to balance; they should be given some support so that the press doesn't topple over around midnight, causing a frightening bang and a lopsided cheese.)

The coffee-can press.

Hard cheese

Equipment

The equipment you will need is modest.
- large enamel or stainless steel pot
- larger pot to serve as the bottom of a double boiler arrangement; a wide basin, roasting pan, or canning kettle will work well (this utensil need not be enamel or stainless steel)
- long spoon for stirring
- dairy thermometer or stainless steel dial thermometer
- long knife or spatula for cutting the curd; it must reach to the bottom of the pot
- large colander or strainer
- several yards of cheesecloth, a clean, well-worn pillowcase, or a square of thin muslin
- cheese press (see suggestions for a simple press)

Procedure

Sterilize all untensils and cloths with boiling water. Cleanliness is of utmost importance if only the friendly cheese-making bacteria are to be present.

If possible, put 1 gallon of evening milk in a cool spot (50 to 60° F.) to ripen overnight. In the morning, add 1 freshly milked gallon. Other- wise, ripen 2 gallons of raw or pasteurized milk by stirring in ⅓ cup cultured buttermilk. Warm milk to 85°F.

Prepare a rennet solution with ¼ of a Hansen tablet (or 2 Salada rennet tablets or 1½ cc. of liquid rennet) dissolved in a cup of cold water. If you use only 1 gallon of milk, mix the rennet solution, use half of it and discard the remainder. The exact amount you will need may vary with kitchen temperatures, the characteristics of the milk, and how much cheese you are making. Some people have found that goat's milk requires more rennet. Experience will be the best guide. Remove the milk from the heat. Gently stir the rennet solution into the milk, then let it sit undisturbed in a warm place. The milk should curd in 30 to 45 minutes.

When the milk appears compleley clabbered, stick your finger into the curd at an angle, then lift straight up. If the curd breaks clean over your finger, it is ready to be cut. With the long knife, cut the curd as you would fudge, into half-inch squares. Then, holding the knife as horizontal as possible, cut diagonally through the depth of the curd, trying to make ½-inch cubes. The whey should run clear. Stir the curd with your hand, and cut any cubes that are not of uniform sizes.

Slowly heat the curd to between 85 and 90° F. The process should take 15 to 30 minutes. Stir the curd

(Continued on page 310)

This homemade press uses barbell weights.

The simplest of presses: a couple of boards and a couple of bricks.

SALT

Salt is used in commercial and home cheese-making not only to add flavor, but also to remove excess moisture from the cheese and to inhibit the growth of organisms that could spoil the cheese. People who choose to limit their salt intake can make fresh homemade cheeses with little or no added salt. Their shelf life is short, however.

Remember that even salt-free cheese contains some sodium naturally present in milk.

A STEP BEYOND

Mold-ripened cheeses are risky for the home cheese-maker. Blue, Roquefort, Camembert, Brie, Port-Salut, and the like require closely controlled ripening conditions and special inoculations as starters. Swiss and Gruyere also require a special bacterial culture and careful ripening.

IS RAW MILK DANGEROUS?

Raw milk should ideally be purchased from certified or approved dairy farms. However, some states do not have certification programs. In Pennsylvania, where producers are state-certified, a bimonthly inspection and laboratory analysis are conducted. Levels for allowable microorganisms are low, and dairymen can be warned well in advance of problems if bacterial levels are rising.

Milk is an excellent medium for the growth of microorganisms because of its good nutritional value. However, sanitary handling procedures control most milkborne diseases, including foot-and-mouth disease, streptococcal infections, typhoid fever, diphtheria, and the Q-fever rickettsia. Cattle must also be tested for tuberculosis and brucellosis. Diseases are rarely a problem on certified farms.

Properly handled raw milk will be cooled rapidly after milking, and even if some microorganisms are present, they will not grow. Keep milk refrigerated. Avoid leaving it out of the refrigerator for any length of time, especially in warm weather.

(Continued from page 309)

CAN THE COTTAGE CHEESE-MAKER SAVE MONEY?

Making cheese with store-bought milk may not have economic benefits at today's milk prices. One pound of cottage cheese takes about three quarts of skim milk. For a pound of hard cheese you need about five quarts. But with your own milk supply, or fresh milk purchased in bulk from a nearby dairy, cheese-making can be a money-saving hobby.

FRESH CHEESE

If you find yourself unable to make cheese at home but appreciate a fresh, low-salt taste nevertheless, scout your area for small Italian-style cheese-makers. They may sell unsalted mozzarella straight out of a bucket of water. The flavor likely will strike you as bland at first, but soon commercial cheeses will seem grossly over-salted in comparison.

Use unsalted cheese in dishes to counteract salted commercial pasta and tomato sauce or puree. The moisture of such cheeses makes them especially suited to otherwise dry sandwiches.

Unsalted cheese won't keep fresh for long, so plan on using it within a few days after purchase. Try storing it underwater, or wrap very securely to keep out air.

Cutting the curd into cubes.

frequently to keep the cubes from matting together.

Remove from heat and ladle out about half the whey. (Don't discard the mineral-rich whey. Save it for making ricotta, feed it warm to pets or farm animals, or pour it onto a compost pile.) Gently pour the curds and remaining whey into a colander lined with three or four layers of cheesecloth, or one layer of thin, worn sheeting. Drain for about 15 minutes, stirring occasionally.

Mix about ½ tablespoon salt with the curds, if desired. At this point you may substitute fresh chopped herbs and spices, such as garlic or crushed caraway seeds. Tie the four corners of the cloth together and hang the cheese to drain for about an hour. Keep an eye out for cruising cats and dogs.

Spoon the curds into a cheese press. Weight with a brick for one hour, then add another brick. After eight hours or so you may increase the weight by another couple bricks. Leave the cheese in the press overnight to expel as much whey as possible: this is essential for a cheese that will keep well.

Remove the cheese from the press and dry it in a cool, airy spot, turning the cheese a few times a day. Leave the cheese out for several days, until the rind is no longer moist to the touch. Rubbing salt on the outside of the cheese helps to draw out moisture and develop a hard rind.

Aging the cheese improves its flavor. Brush the cheese with melted paraffin and place it in a clean, cool spot with good air circulation. Turn the cheese daily at first, then every few days after that. Try to wait at least 60 days before sampling your cheese.

The cheeses

Cheddar. Cheddaring is a step in cheese-making that removes additional whey from curd that has already been heated and drained. Other cheeses besides cheddar are cheddared—Monterey Jack and longhorn are two examples. Because cheddared cheeses have a lower moisture content than fresh or semi-soft types, they can usually be aged for a longer time.

For cheddar cheese, follow directions for basic hard cheese up to cutting the curd, but add the starter and rennet at 88° F. The curd should be ready in half an hour. During the next half-hour raise the heat very, very slowly (2° every 5 minutes) to 100° F. Hold it at this temperature for 45 minutes, stirring occasionally. Press a strainer against the curds and drain off the whey. Let the curds mat together, then cut them into chunks about four by four inches. Pile the chunks on top of each other: this is the cheddaring process. Place the cheese in a pan and keep it at 100° F., over the next two hours, turning every 15 minutes. (An oven heated by just the pilot light is fine, or a warm nook behind a wood-burning stove.) Mill the curds in a meat grinder or similar gadget. They should be in very fine pieces. Salt them with a minimum of one tablespoon per two gallon batch. Press and dry the cheese at 60° F. for about four days or until the rind is no longer moist to the touch. Cheddar should be cured at 36° F. with high humidity for 9 to 12 months (try wrapping it in a moist cloth and storing on the bottom shelf of the refrigerator).

Colby. Colby is a variation of cheddar, but softer and moister. It is also more quickly made. After draining the whey with the strainer, cool the curd to 80° F., stirring continuously, salt, and press. Dry the cheese, then cure in the refrigerator as above, but only two or three months.

Susan Asanovic

The Wheeler Cheese Press is made in England and is available for $120 from New England Cheese Making Supply Co., P.O. Box 85, Ashfield, MA 01330.

Thistlehead cheese: Balearic Island, Spain

The fields are blooming with curdling herbs and somewhere a gallon of milk is waiting. The product is a natural cheese made with a milk-curdling herb, purple thistle, in place of the usual rennet. Rennet is an enzyme, extracted from the stomach of a calf, undesirable to the legions of vegetarians, of which I am one.

There are of course other vegetable milk-curdling substances which can be used in cheese-making, such as the nettle, the sorrels (lemon and common), fumitory, the sap of the unripe fig, and I have tried them all. But without a doubt the giant purple thistle, used in the Balearics but also found throughout the United States, is the best of them all. It is speedy in action (usually curdling milk overnight) and it is sure.

The Balearic cheeses are usually made from a mixture of goat and sheep milk. These are concentrated milks and excellent for cheese. When a cow is kept on a farm, some cow milk is also used, but it is usually enriched with goat or sheep milk.

For the curdling, the milk should be tepid warm. The ideal is warm from the udder, straight into a deep earthen curdling crock—which is how most of the island cheeses are made. If cold milk has to be used, then the crock should be warmed until the milk is moderately warm to the touch. An earthen cooking pot is best for making cheese curds, but an enamel pot can be used, with a piece of new blanket-cloth tied around it to hold the heat. New straw can also be packed around the pot.

Choose a warm place for making curds, as sudden chilling is as bad for cheese as it is for bread dough put to rise. Cheeses are easily spoiled by lack of care, and that is why there are beliefs in dairies concerning good and bad spirits, and trouble is taken to keep on the fair side of the good spirits.

The herb used to curdle the Balearic cheeses is the giant purple thistle. American readers are advised that all species of the thistle section of the compositae family possess milk-curdling properties, especially the giant and thorny kinds. If thistles are not available, then the heads of the globe artichoke (Cynara scolymus Compositae) can be used. The cynara species of the Compositae family are quite common in the United States, especially Cynara cardunculus, commonly called cardoon.

Flowerets are gathered when the thistle-head and the flowerets of which it consists have turned brown. When the thistledown begins to appear, it is getting too late for the gathering, and the flowerets will soon be carried over the countryside by the wind.

The flowerets are air-dried in shallow baskets and then stored in jars to last for the whole cheese-making time, which is usually eight months of the year.

The herb is prepared by pounding until very well crushed in a mortar and pestle. The farms where I learned cheese-making use a wooden mortar and pestle for the cheese-herb pounding—beautiful objects in their carving and the aged quality of the wood. The herb is well-pounded, then a little warm water (I prefer whey) is added to merely cover it. After soaking for five minutes, it's pounded again for five minutes. Soak again, repounding three times at least, until a dark brown liquid forms. The mass is then strained and the herbal liquid is added to warm milk. For every gallon of milk used, five small heaped teaspoons of the herb are required.

If too much herb is used, it dominates and may cause indigestion, being very potent. Therefore, carefully control the quantity.

When I want to eat curds and whey, I do not trouble to pound the herb. I merely dip it in the milk to make it adhesive, press the flowerets together, and bind them with cotton string into a bundle. The string is left hanging out of the crock, so I can pull it out when the milk curdles (usually overnight). Do not add the herb loose into the milk, it's like having hairs in one's mouth when eating the curds and whey.

The cured cheeses are no longer white, but have turned the attractive creamy-gold of honeysuckle flowers.

Juliette de Bairacli Levy

CHEESE VARIETIES AND DESCRIPTIONS

Dairy Laboratory, Eastern Regional Research Center, Agricultural Research Service, Philadelphia, Pennsylvania. 1978. Superintendent of Documents, U.S. Government Printing Office, Washington, DC 20402. 151 p. paperback $1.

This USDA publication has been in print, in various forms, since 1908. It's an encyclopedia of world cheeses, giving cryptic descriptions of hundreds of varieties, many with a hurried suggestion of how the cheese is made. Although there are no illustrations, the sheer number of varieties will stimulate the would-be cheese-maker, and suggests the poverty of our supermarket dairy sections.

RY

Caerphilly.

Caerphilly is a semi-soft, cow's milk cheese made in Wales, and is especially popular among Welsh miners. It is circular and flat, about 9 inches in diameter and 2½ to 3½ inches thick, and weighs about 8 pounds. The cheese is white and smooth, lacks elasticity, and is granular rather than waxy when broken.

Fresh whole milk is inoculated with starter and ripened slightly. The milk is heated to 84 or 90° F., and rennet is added. After a curdling period of 40 to 60 minutes, the curd is cut into small pieces, stirred carefully for 10 minutes, and (if the rennet was added at 84° F.) warmed gradually to 90° in another 20 minutes.

Then the whey is drained off, and the curd is ladeled into small cloth bags which are hung up for draining or pressed lightly. After about an hour the curd is broken up or it may be cut into small cubes, and it is repressed. Again the curd is broken up and salt is mixed in at the rate of 1 ounce of salt to 3 pounds of curd. The curd is then placed in cloth-lined forms and again pressed. The cheeses are redressed and repressed after several hours, and then daily, with gradually increasing pressure, for about 3 days. They are cured on shelves in a damp curing room at 65 to 70° F., for 2 to 3 weeks. During that time they are kept clean and are inverted frequently. A thin, white layer of mold forms on the surface, which is considered desirable in proper ripening. The cheese is perishable and must be eaten soon after curing. The yield is relatively high.

Nieheimer.

Nieheimer, a sour-milk cheese, is named for the city of Hieheim in the Province of Westphalia, Prussia, where it is made. Like Hop cheese, which it resembles, it is packed with hops for curing. Although the two cheeses are not identical, Nieheimer is known as Hop cheese in some localities.

The sour milk is heated to a temperature between 100 to 120° F. The curd is collected in a cloth, and the whey is drained from it for a 24-hour period. Then the curd is worked until it is mellow, after which it is shaped into cakes. The cakes of curd are ripened in a cellar for 5 to 8 days, during which time they are turned frequently. When they have ripened sufficiently, they are broken up, and salt, caraway seed, and sometimes beer or milk, are added. The mixture is molded into small cheeses, shaped like flattened spheres, that weigh about 4 ounces. The cheeses are covered lightly with straw, and when they are sufficiently dry they are packed in casks with hops to ripen.

Sage.

Sage cheese is an American-type, spiced (sage-flavored) cheese made by either the Cheddar or Granular or Stirred-curd process; it is pressed in any of the shapes and sizes in which those cheese are pressed. The curd has a green, mottled appearance throughout. At one time, green sage leaves were added to the curd before it was hooped. Now, sage extract is added for flavor; and the green, mottled appearance is produced as follows: Succulent green corn is cut fine, and the juice is pressed out. This juice is added to a small part of the milk, which is made into curd in the usual way; the rest of the milk is also made into curd, and the two lots of curd are mixed just before hooping.

Sapsago.

Sapsago cheese has been made in the Canton of Glarus, Switzerland for at least 500 years and perhaps more; it is made also in Germany. It is known by various other names, including Schabziger, Glarnerkase, Grunerkase, Krauterkase, and Grunerkrauterkase. It is a small, very hard cheese that frequently is dried. A powder prepared from clover leaves is added to the curd, which gives it a sharp, pungent flavor, a pleasing aroma, and a light green or sage green color. The cured cheeses are cone-shaped, 3 inches thick at the base, 2 inches at the top, and 4 inches tall, and weigh 1 to 2¼ pounds. The fully cured, dry cheese can be used for grating.

Sapsago is made from slightly sour, skim milk. The milk is put into a round kettle and stirred while it is heated to boiling temperature. Cold buttermilk is added slowly as heating and stirring are continued. The coagulum that appears on the surface is removed, set aside, and added to the curd when it is put into the forms. Then enough sour whey is added to precipitate the casein, as in making Ricotta, and stirring is stopped. If too little whey is added, the curd will be too soft and moist; if too much or too sour whey is added, the curd will be too firm and dry. The curd is collected in a cloth or strainer and spread out to cool as the whey is drained off. Then the coagulum that was set aside is mixed with the curd, salt may be added, and it is placed in perforated wooden forms, covered with a press lid, and pressed under heavy pressure at a temperature of 60° F. The curd is ripened (cured) under light pressure at this temperature for at least 5 weeks. At this stage, it is ready for use in making the cheese. In many cases it is sold and transported in large sacks or casks to a distant factory where the cheese is made.

The ripe, dry curd is ground, and about 5 pounds of salt and 2½ pounds of dried, powdered leaves of the aromatic clover, Melilotus coerulea, are added to each 100 pounds of curd. The mixture is stirred into a homogeneous paste, then packed in the small cloth-lined, cone-shaped forms.

About 10 or 11 pounds of fresh curd is obtained per 100 pounds of skim milk and about 65 pounds of Sapsago is obtained per 100 pounds of fresh curd.

Alternate sources for dairy products

Do you want to know if you're getting chemicals through the cow's milk you drink? At the Bethesda Co-op in suburban Maryland, customer questions about milk after the Three Mile Island nuclear plant spread radioactivity produced a sign over the dairy cooler: "The cows that produced this milk are being fed stored grain and well (not river) water. Milk has been tested for radioactivity." A further note told buyers that the cows were Guernseys, a good butterfat-producing breed.

Co-ops and natural food stores (see Grains chapter for description and how to find them) often carry rich creams and unhomogenized milk, unsalted and vegetable rennet cheeses, natural yogurts, additive-free ice cream, and brown eggs.

Of course, buying directly from a farmer is the surest way of knowing about the animals and ingredients. A call to your county agricultural agent (usually listed in the phone book under your county government) will tell you about area farmers who sell directly to the public. A farmers market (extension agents know about these too) will often include eggsellers.

Milk can be a bit tougher to get by the direct route unless you live in one of the 12 states where certified raw milk is legally sold (again the county agent can lead you to sources). In the other 38 states, a farmer who wants to sell directly would have to pasteurize the milk and that's pretty rare. The normal route is for the milk processor to truck out the milk from a farm to a center for pasteurizing, homogenizing, and bottling; the milk then travels to supermarkets and other retailers, usually under the processor's name. There are a lot of health checks on the process, but the identity of the milk source is lost.

If you need quantities of milk for making cheeses or yogurts, one source (besides a farmer if you're lucky) is the milk processor's outlet. (Find them in your Yellow Pages under "milk" or "dairy.") Green Spring Dairy in Maryland, for example, has a cash-and-carry outlet selling milk and cream by the case at prices below the retail level (a case means 4 gallons of milk or 12 to 16 quarts of cream). If you buy directly from a supermarket, it's worth knowing that the processor who marks the expiration date on the milk gives it ten days of shelflife, so getting fresh milk means choosing something that still has over a week's "selling life" to go.

If you need to find goat's milk, your extension agent can lead you to local goat-owners and goat clubs. Health food stores often carry goat's milk for customers allergic to cow's milk. Goat's milk comes under the same regulations as cow's milk, and can't be sold raw legally in many states.

The following chart gives you an idea of some comparative prices for selected dairy foods. Of course, prices fluctuate weekly, and what's true of our sample won't likely hold for yours. Still, the chart serves to show the significant range in prices within an area.

Sara Ebenreck

SHOPPING AROUND FOR DAIRY FOODS

	Co-op	Discount grocery	Supermarket	Farmers' market	National food store	Processor outlet
Eggs, large (per doz.)	$1.15 brown	$.99 brown .79 white	$.85 white	$ .85 brown	$1.05	$.95 white
Natural yogurt (per lb.)	.62	.67	.63 store brand .73 name brand		1.29 2 cups	
Sharp cheddar cheese (per lb.)	2.07	2.19	2.55	2.45	4.42 raw milk, vegetable rennet	
Butter (per lb.)	3.65		1.85	1.69	2.95 lightly salted	.95
Milk, whole (per gal.)	1.48 +1.00 deposit		1.89	1.89	2.00 raw	1.58

Ice cream, frozen yogurt, and sherbet

Back in Grandma's day, making ice cream was a big event, heralded by the arrival of the iceman at the back door with a specially ordered block of ice. The kids eagerly turned the hand-crank on the freezer, and fought for the delight of licking the dasher. Waiting that two hours for the ice cream to harden seemed like a lifetime, but all was forgotten the moment that creamy concoction was in your mouth.

Grandma made the basic ice cream batter with cream, milk, eggs, and sugar. The only additives in those days were a vanilla bean, or maybe some fresh fruit, nuts, or chocolate.

Ice cream never was the ideal health food, rich as it is in milkfat. But it does have some redeeming food value, and besides, who can avoid such a delightful deviation?

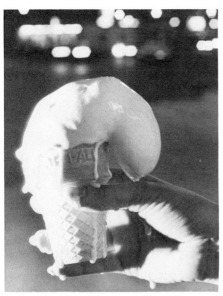

Commercial ice cream is fun on the way down, but the stomach is left with the unenviable job of handling the strange ingredients.

What you can avoid, though, is the growing number of additives in today's commercial version of ice cream. The industry has all but taken the cream out of ice cream, along with eggs, and has replaced them with emulsifiers, stabilizers, artificial colors, flavorings, and fillers. A typical commercial ice cream contains milkfat and nonfat milk (not necessarily cream), corn syrup, sugar, whey, mono and diglycerides, natural and artificial flavor, carob bean gum, cellulose gum, guar gum, polysorbate 80, carrageenan, and artificial color. Mouth-watering, isn't it?

The sweeteners make up 15 to 18 percent of the total volume of ice cream. Whey is a watery milk residue left over from cheese-making, used in ice cream as a cheap substitute for the more expensive cream. It's nutritious but changes the taste and texture of ice cream. By FDA regulations, no protein in ice cream can come from any source other than the milk products.

Stabilizers and emulsifiers simulate the creamy texture of ice cream and hold it together whenever the product thaws and refreezes en route from plant to freezer case. Artificial flavors substitute for the real thing. Artificial colors are purely cosmetic—the most unnecessary evil of all.

Why such heavy reliance on additives? A simple rule suggests the answer: The more additives present, the cheaper the ice cream is.

Adding insult to the possible injury from additives is the fact that as much as 50 percent of the volume of ice cream is air! While a half-gallon of ice cream made exclusively with cream weighs four pounds, the six commercial brands I checked—including the most expensive—weighed no more than 2½ pounds.

The best ice cream money can buy—for texture, taste, and price combined—is homemade. The accompanying table shows a comparison of commercial to homemade

"I scream, you scream . . ."

ice cream on the basis of price, ingredients, and food value. The homemade using purchased milk was made with half-and-half (cream and milk), honey, vanilla bean, and salt.

As for price, you may or may not save money by making your own. If supermarket cheapo brands are fine with you, making ice cream from scratch would mean throwing money down the dasher. If you like real ice cream, without additives, you'll save (especially if you have a cow or goat, of course).

As for price, the homemade varieties fare quite well. The one using fresh cow's milk is cheapest of all, as there was no cost figured in for the milk and cream.

The homemade using purchased milk costs more than the supermarket brand, but less than advertised brand's "ice cream parlor taste" variety; this is interesting, since both commercial brands contain the same ingredients. The commercial "all natural" ice cream was closer in quality to the homemade, and there the savings are considerable.

As for food value, the table shows the homemade version with purchased milk is better overall than the tested supermarket and advertised brands, and at least equal to the commercial "natural" brand in almost every category. The homemade using fresh cow's milk and cream does have a high fat content because of the cream; this may be a negative point for some people.

The richest homemade ice cream contains eggs, which boost nutritional value as well as enhancing flavor and texture. (The commercial brands with eggs in them are labeled "French vanilla" by the FDA.)

As for texture and taste comparisons, it's like comparing fresh-squeezed orange juice to Kool-Aid in most cases. The varieties you can make at home are limited only by your imagination; they can be as slimming or as sinfully sweet as you want; and you know exactly what you are getting.

Cranking it out

The basic ice cream freezer consists of a large bucket, a smaller aluminum can, and a cranking mechanism which fits over the can and turns a "dasher" or paddle inside the can that churns the ice cream.

To make ice cream, you place the can inside the bucket, fill up the bucket with ice and salt, and turn the crank to freeze the cream mixture inside the can. For my money, the old-fashioned hand-cranked freezer offers the best, cheapest, and most enjoyable way to make ice cream. I feel it produces the best-textured ice cream with a smoothness well worth the 20 minutes or so of easy cranking. I have a hand-cranked model with a cedar bucket, which I prefer to a cheaper plastic bucket because it stands up longer to wear and tear. Also, the wooden model is easier to steady if the can gets caught on an occasional piece of ice while turning.

There's an electric version of this ice cream freezer which will do all the cranking for you. But it takes slightly longer to freeze the ice cream, and you still have to keep it packed with ice and salt while it churns. The most effortless machine is one you simply put into the freezer section of your refrigerator. Plug it in, and come back two hours later for finished ice cream. There's no need for ice and salt, but you must clear a flat space in the freezer for it. Also, not all refrigerator freezers can reach the optimum temperature of 0° F.

Recipes

Here are a few tried-and-true recipes I use. They're basic and lend themselves to variations.

French Vanilla

4 cups half-and-half (or light cream)
½ to ¾ cup honey
3 eggs
1 vanilla bean

Heat half-and-half in a double boiler just until boiling (don't let it boil). Stir in honey. Beat eggs in separate bowl, then add slowly to the half-and-half, beating vigorously so the eggs don't cook. Add the vanilla bean and cook on low heat until thickened enough to coat a spoon (about 15 to 20 minutes). Remove bean.

Yield: 2 quarts

No-Cook Vanilla

2 cups milk
1 cup honey
4 cups half-and-half (or light cream)
1 vanilla bean, powdered

Mix all ingredients together and chill. Freeze.

Yield: 2 quarts

Chocolate Ice Cream

1 cup honey
¼ cup water
¼ cup carob syrup
4 cups light cream
½ to 1 vanilla bean, powdered

Mix honey and water in double boiler over boiling water. Add carob syrup. Add cream slowly and heat till just below boiling. Chill thoroughly, add vanilla, and freeze as directed. Makes 1 quart. For carob syrup, mix ¼ cup carob powder and ¾ cup water in a small saucepan. Bring to a boil over very low heat, stirring constantly, until syrup is smooth.

WHAT YOU GET FOR YOUR MONEY (AND LABOR)

	Supermarket brand	Advertised brand	"All natural" brand	Homemade (purchased milk)		Homemade (own milk)	
Price per ½ gal.	$1.39	$1.99	$2.45	$1.80 (no eggs)		$.80 (no eggs)	
Ingredients	Milkfat and nonfat milk, corn syrup, whey, natural and artificial flavorings, mono and diglycerides, carob bean gum, cellulose gum, guar gum, polysorbate 80, carrageenan, artificial color.		Milk, cream, sugar, natural flavor.	Cream, milk, honey, vanilla.		Cream, milk, honey, vanilla.	
				no eggs	eggs	no eggs	eggs
Calories	160	140	140	153	169	269	252
Carbohydrates	17	16	17	13	13	35	35
Protein	2	2	3	3	4	3	4
Fat	9	7	7	11	11.7	22	23

TOWARDS PERFECT ICE CREAM

- Be sure batter is well chilled before freezing. For best results, make batter a day ahead.
- Use too much honey and it won't freeze: use too little, and it freezes hard as a rock. (When converting sugar to honey, cut the amount in half.)
- Don't fill the can more than two-thirds full—it expands.
- Always use crushed ice and keep the ratio of ice to salt six to one. Too much salt makes the ice cream grainy.
- When hand-cranking, turn handle slowly until you feel it thickening, then speed up. Add fruit, nuts, or whatever when the mixture is nearly frozen, being careful not to get salt into the can.
- After the ice cream is frozen, put it into your regular freezer, or drain the water out of the bucket and pack it with ice to freeze it hard.

Homemade ice cream is for people who really appreciate ice cream. Factories can make sweet, cold, semi-hard blocks cheaper than you can. But the White Mountain freezer used here can stir up better ice cream for those who are willing to crank and pay for cream. This model is available in three capacities, with prices starting around $50. Prepare the ice cream mixture according to the recipe. If you're adding fruit, use crushed or pureed fruit, or cut it into small pieces. Ice crystals tend to form around large pieces of fruit. A few tablespoons of lemon juice help prevent the fruit from darkening. Chill the mixture in the refrigerator for several hours.

Dairy products are ideal mediums for bacterial growth, so be sure to carefully wash and scald the ice cream maker can with boiling water. Pour the cooled mix into the can. Don't fill the can more than two-thirds full.

Position the dasher in the can, put on the lid, and place the can in the freezer bucket. Lock the cranking mechanism in place.

Pack the freezer bucket with alternating layers of crushed ice and rock salt. The salt enables the ice to melt more quickly, so it can absorb more heat from the mixture. For best results, the ice should be crushed as finely as possible. (You can put ice cubes in a burlap bag and smash them with a hammer.) Fill the bucket about one-third full with the first layer of ice. Cover that with a thin layer of rock salt, then continue to add thin layers of ice and salt until the bucket is full. Pour in a cup or two of water.

Start cranking slowly, and increase speed as the mixture starts to freeze. Water will flow out the drain hole in the freezer bucket. Make sure ice doesn't clog the drain. Add more ice and salt as necessary to keep the can covered. The ice cream is done when cranking becomes difficult. The ice cream can be eaten right away, but it will still be quite soft at this point. It's best to freeze the ice cream for 2 or 3 hours, either in your freezer, or in the freezer bucket repacked with ice. Remove the dasher from the can, cover the ice cream with foil, plug the hole in the lid, and replace it on the can before freezing.

Other frozen desserts

While ice cream may be one of America's most popular desserts, there are other frozen dairy desserts not as calorie-rich and just as good.

Frozen yogurt. Here is the health food industry's answer to ice cream. But many manufacturers offset the benefits of yogurt's lower fat content by filling it with sugar, which adds calories and cancels out some of yogurt's beneficial effects. There are also no Federal standards for the ingredients in yogurt.

Aside from doubling the nutritional value of yogurt, making your own frozen yogurt is cheaper and better tasting. Dannon's Danny-in-a-Cup frozen yogurt is 59 cents for an eight-ounce container of vanilla. Buying your own and freezing brings the cost down to about 27 cents. When making it from scratch, eight ounces costs but a dime.

Yogurt can be frozen in an ice cream freezer, or you can simply put it in the freezer section of your fridge. Just take plain yogurt, mix in a little more honey than you think it needs (freezing lessens sweetness), add fruit and freeze till slushy.

Remove it, beat in a blender or with a spoon to remove ice crystals, and return to freezer. Repeat this in a half-hour. I find yogurt freezes quite hard, and suggest you take it out of the freezer at least ten minutes before serving.

COMPARING SHERBETS

Homemade ($2/half-gallon)	Commercial brand ($2.20/half-gallon)
Milk, orange juice, honey, lemon juice, grated orange rind.	Whey, sugar, corn syrup, orange puree (concentrated orange juice, pulp, and oil), cream, citric acid, mono and di-glycerides, carob bean gum, cellulose gum, guar gum, polysorbate 80, carrageenan, artificial color.

Sherbet. This is the kind of dessert which, if you've never tasted homemade, you've probably stopped eating altogether because the commercial stuff is a tasteless, sugary slush.

But you'll be pleasantly surprised at how smooth and intensely flavorful sherbet is when it's made the right way; I now consider it a very close second to ice cream in richness and texture. Just don't expect it to turn out a bright neon orange or loud lime. With all that milk, and no artificial color, homemade sherbet may surprise you with its paleness.

The most pleasant surprise is the price. Making your own sherbet is cheaper than buying it, even though you're using superior ingredients.

To make orange sherbet, combine 1½ teaspoons rind, 1 cup honey, ¼ cup lemon juice, and 1½ cups orange juice. Stir them slowly into 4 cups very cold milk. Freeze as you would ice cream, or use the same method as for frozen yogurt, beating the half-frozen mixture at half-hour intervals to remove ice crystals. This recipe makes 6 cups.

For lemon sherbet, substitute ¼ cup lemon juice or to taste. For lime, use ½ cup lime juice, 1½ teaspoons lime rind, and the same amount lemon rind.

Ices. Although not dairy desserts, ices can satisfy your craving for a frozen goody, and are a refreshing low-calorie alternative. They call for nothing more than water, honey, and fruit.

For orange ice, cut a strip of orange peel into tiny pieces, add ½ cup honey, and 1 cup water. Stir well, add 1½ cups orange juice and 2 tablespoons lemon juice. Blend and freeze as for sherbets or ice cream. This recipe yields 3 cups.

For lemon, use ⅔ cup lemon juice and peel; for pineapple, use 1 cup water and 2½ cups fruit; for raspberries, ½ cup water, and 4 cups fruit; for strawberries, ½ cup water, and 3 cups fruit.

Pat Donahue

Butter from the churn

Once you've experienced the clean, fresh taste of home-churned butter, you won't want to use store-bought any more. What's the rap against commercial butter? First, most butter in this country contains between 2 and 3 percent salt and 80 percent fat. Often the extra saltiness masks off-flavors from stale or recycled cream (and even recycled rancid butter). Another trick used to recycle rancid butter is to wash it with diacetyl to disguise the flavor. Sour, stale cream can be transformed into butter with the help of neutralizing agents. Artificial dyes may be added to simulate springtime yellow; even though natural dyes from vegetable sources are available, the synthetic colors are preferred because they are cheaper and more uniform. None of these unsavory procedures need to be mentioned on the label, so *caveat emptor.*

New York City gourmet shops are selling imported French *beurre* at $4 per pound; the "bargain basement" at Macy's department store only charges $3.45. Your own sweet butter, which is much fresher, will cost almost nothing if you use your own cream, or about $2.60 using purchased whipping cream. This is little more than the current price for commercial butter.

Goats

If you were to count goats (instead of sheep), as does the American Dairy Goat Association, you would find that the number of goats registered in its files in 1975 was four times the number registered just five years before. This country is in the midst of a goat boom. Worldwide, goat's dairy products feed more people than cow's milk products. It's estimated there are 65 million goats in India and 55 million in China. And the reason?

The goat, with needs decidedly more modest than those of a dairy cow, produces milk which is not only delicious, but easier to digest than cow's milk. With its smaller fat globules, goat's milk passes through the human stomach in about 20 minutes, processed in a fraction of the time which moo juice requires. This is why doctors often recommend goat's milk for babies and for people with allergies and digestive problems.

Yet goat's milk is similar enough in flavor to cow's milk so that you can learn to enjoy it. The milk of Saanen and Toggenburg goats resembles that of Holstein cow's in its composition of water, fat, lactose, protein, and minerals. The milk of the Nubian contains about 5 percent fat, similar to that of Jersey cow's.

Of course, human consumption is not the only consideration. Extra goat's milk can be fed to other livestock, such as chickens and pigs.

The milk is just for starters. Very likely you've already tried goat's milk cheese and know how good it can taste. Yogurt, butter, and other goat dairy products also rank as delicacies—not to mention practical additions to your larder.

A good example of economy of scale, a goat produces an average of about one or two gallons per day, much closer to the needs of the average family than the milk production of a cow. Based on the amount of food cows and goats manage to put away, their yields are equally efficient, although cows lactate for 12 months after calving, while goats give milk for 10. The big advantage with goats is that you can raise at least seven or eight of them on what it takes to feed one cow. Several goats may often be kept in a location where a single cow cannot.

Goat meat is tasty, tender, and lean. Even if it sounds exotic, it has become popular in our own Southwest. A 12-week-old kid will give 20 to 25 pounds of meat; a six- to nine-month-old "buckling" will yield about 60 to 80 pounds. While large-scale production of chevon (goat meat) uses different breeds and methods from those associated with dairying, male kids from the small homestead dairy herd can provide meat for your table.

The manure and used bedding from a goat can be thought of as a valuable by-product for soil building. If you think of the goat as an organic fertilizer factory, this "factory" can excrete 12 times its own body weight in manure per year. Its composition will be approximately 1.44 percent nitrogen, 0.5 percent phosphoric acid, and 1.21 percent potash.

Yet with goats, the sum is more than all of its parts. For very concrete reasons there are the intangible rewards—the pleasure of self-sufficiency, the opportunity to learn from your own hard work (make no mistake, goats are a commitment), and the affection of your animals. Because you make intimate contact with them twice a day, there will be a deeper bond between you and your goats than with any other livestock you could raise.

The human factor

Keeping dairy goats means work, and if you have never been responsible for maintaining any sort of livestock, goats are probably not the place to start. Rabbits or chickens are much simpler for the beginner. Also for goats, you'd better like

being a real homebody since that's what you'll need to be, with a reliable stand-in for those times when you do go away.

Here's why. Your does will require milking twice daily at 12-hour intervals, the same time every day, in order to keep them comfortable and maintain milk production. For just a few goats the milking process is by hand, into a bucket. But whether by hand or machine, scrupulous concern for sanitation is required, beginning with brushing the doe to remove loose hair and dust and washing her udder. Then the milk must be immediately cooled for storage.

Plan on at least a half hour a day for feeding and milking three goats, once you become proficient.

Periodically your herd will have other needs. Of course, long before anything else can happen, you've had to make provision for breeding. Then there is delivery of the kid(s), usually dehorning at an early age, and eartagging or tattooing after birth. Kids must be weaned at the proper time to allow humans to take over and to prevent the weakening of the doe's udder which occurs when kids nurse too long. Occasionally hooves or hair need trimming, temperatures need taking, and medicines must be administered. Fences, pens, and shelter need looking after.

Anyone maintaining a breeding herd must have official milk production records. In fact, it's good to keep records for the most modest goat venture to find out how much your milk is actually costing you.

Goats are sociable animals by nature, so your herd should start with two females, as a goat alone is a lonely bleater and may refuse to feed. This, however, is just the beginning: lactation (milk production in the doe) is possible only with a small population explosion. To give milk, the doe must bear offspring. The milk flow then continues for about ten months, as a rule. Toward the end of this period, when she goes into heat, the female is bred again, and if the breeding is successful, five months later the new kids are delivered and the milk flow begins again.

Expect to pay at least $200 for a lactating doe, and $100 to $200 for a female kid.

A doe who is over a year-and-a-half old typically gives birth to twins, so you have a small herd in no time. You don't have to keep them all, of course. Male kids are usually butchered or sold off when they reach about 20 pounds. If you have an unflinching attitude, you can learn to do this yourself by assisting someone who knows how at several slaughtering and dressing sessions. Or you can find someone to do it for you at minimal cost.

Breeding

Goats breed seasonally, from September to March. Breed two does at opposite ends of the season, and you will have milk year-round.

Does should not be bred until they weigh 85 to 90 pounds and are about ten months old. In this matter dairy goat owners have two options. One is to keep a buck on the premises and put up with his obstreperous personality and with his smell, which may be described as somewhat less offensive than a skunk's but worse than a locker room after the game. Usually, keeping a buck is not an economical choice if you have fewer than 15 females. A buck must have separate quarters, partly because his musky smell would otherwise spoil the taste of the milk and partly because he has a one-track mind. As William Blake wrote, "The lust of the goat is the bounty of God." (Male goats to be kept as pets or raised for meat are usually castrated shortly after birth. They are called wethers.) Males and females should

(Continued on page 320)

(Continued from page 319)

be kept separated after two to four months of age.

For dairy herds without a buck in residence, a date is arranged with somebody else's billy. This is not a simple matter of any coupling will do. Milk production can vary greatly from one breed to another and from individual goat to individual goat. So the wise owner will research bloodlines and milk production records in the prospective buck's family tree. You may choose to travel some distance for advantageous breeding to build up a good herd.

Then the trick is to know when the doe is in heat. She may be quite vocal about it, losing her appetite and her usual bright, attentive personality, persistently wagging her tail above a swollen pink vulva. Or she may not. The successful goat keeper comes to know what to expect from each doe, though it may take some watching and possibly some trips to the stud only to discover she's not in heat after all. Return engagements because of failure to conceive are not unusual in this business, and usually the owner of the male will not charge for these extra services.

No tin cans, please

Goats are cud-chewers, like cows, and can be fed much the same—alfalfa, clover or other hays, silage, pasture, plus a concentrated grain mixture with protein supplement. Depending on her size, a milking doe will munch her way through 1,200 to 2,000 pounds of dry matter a year. The general rule is that for each pound of milk produced, ¼ to ½ pound of grain should be fed.

Feed is usually the factor most subject to cost management and manipulation. It will cost about $125 a year to feed a lactating doe and feed her well.

Proper feeding is also crucial to the quality and amount of milk a goat will produce, as well as the health of your goats. A commercially prepared grain ratio can be fed, or once you have informed yourself thoroughly, you can mix your own ration from a large variety of possible goat feeds. But do choose one alternative or the other, and don't assume that a goat, because of her popularly held reputation, can be fed just anything and thrive.

Once you know what you're doing, you can supplement your herd's diet with surplus vegetables from the garden, overripe or damaged produce, trimmings, and outer leaves and husks—things which otherwise would go to waste and can thus be considered a source of free food.

It is not necessary to pasture a goat if your space is limited. You may prefer to use your small, extra acreage, if you have any, for growing hay or other feed. Or all feed can be

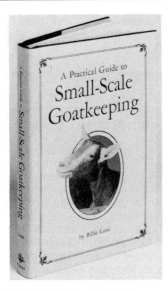

A PRACTICAL GUIDE TO SMALL-SCALE GOATKEEPING
Billie Luisi. 1979.
Rodale Press, 33 E. Minor
Street, Emmaus, PA 18049. 208
p. hardcover $8.95.

Ms. Luisi provides a straightforward and lucid guide for the aspiring goat owner. She begins with the information you must have and the decisions you must make when you buy your first dairy goat, and describes the way a milk animal works to provide us with milk. In a series of informative chapters, she covers milking, breeding, birthing and kid raising, and feeding. She explains the shelter and pasture required, how to manage goats for their own safety and the goatkeeper's sanity, and the few health problems to which they are occasionally prone. One information-packed chapter discusses the many ways goat milk and cheese can enrich the family diet, and another the value of goat manure for assuring a fertile soil in the organic garden.

A final chapter deals with the problems and requirements for operating a commercial-scale goat dairy.

Pam and Richard Darling

My first method of putting the goats in the garden had nothing at all to do with manurial analysis. One bright November morning, I was struck with the innovative idea of turning the goats into the fall garden remnants. They could graze amid the rubble of broccoli stems and corn stalks, clean up the stubby, resistant leaves on the site, deposit their pellets—and we would all mutually benefit. The goats were delighted with the new twist in their regimen. Pasture had been drying up all around them and they had begun a seasonal turn to evergreen needles and bark. Once in the garden, they chewed and chomped away at the succulent garden crop residues with relish and delight. They did an excellent job of clearing. Little did I know that I had created an indelible imprint on their minds.

All the next spring I had to struggle to lead the goats past the garden gate. Mackenzie would stop, plant all four feet firmly and not yield an inch once she was near the gate. She would stand there craning her neck and gazing into the wire mesh squares, perhaps seeing in her mind's eye the rows full once more of cabbage stumps and cauliflower leaves. At night I would dream of the goats escaping the pasture enclosure and flying over the garden fence to gobble up the tender shoots of lettuces, turnips, and peas that were, in reality, coming along fast in bursts of fine spring energy. As my morning conflicts with Mackenzie lengthened, aggravated by the necessity of leading the goats past the garden, dreams continued over the passing nights. I vowed never again to graze the goats among the stumpy leavings of the fall garden.

bought. In some instances local farmers with surplus may offer you the chance to cut feed costs. Here again you need to be well informed yourself. For feedstuffs which must come from the mill, you may save by asking that friendly local farmer to include your requirements as part of his regular order.

Since they are browsers (though it's the glue on the labels they like, and not tin cans) goats will eat some things most other domesticated animals won't touch. This includes some plants which may be toxic or taint the goat's milk. The owner must take care that such vegetation does not grow where goats are to be pastured. For each goat a mineral lick is also a must.

central passageway.

If your choice is loose housing, then a controlled feeding arrangement is necessary so a bossy doe doesn't gorge herself on somebody else's grain ration, plus her own, as that can make for a very sick goat. Goatkeepers who live in year-round warm climates may find that with common penning in the barn they can get away with as little as 12 square feet of housing space for each animal, since much of their time will be spent outdoors.

In either case, a separate milking room facilitates sanitation although many owners of small herds do with-

out. Electricity and running water are desirable.

Storage space for hay in the main barn is a convenience, but hay can be kept in another outbuilding. The main shelter should have a small storage closet for pitch fork, hair clippers, hoof trimming tools, and disbudding iron. Add a hanging scale, and you're well on your way to being set up.

Milking equipment—stainless steel pail, strainer, strip cup, glass jars or metal milk cans for storage—should not be stored in the barn, but in some cleaner place. For a few

(Continued on page 322)

Housing and equipment

For even a small dairy herd the initial investment includes far more than simply buying the goats.

One major concern is a roof over their heads during the winter months and at other times when it is wet or windy. If your animals have access to some outside exercise space, you can make do with 20 square feet of housing per goat, as a rule. A simple, unheated shelter that is draft-free, tight, and dry is adequate, even in northern climates, although milk production will drop in response to the cold. The other alternative, maintaining a heated barn, is generally too expensive for the small-scale goat farmer.

Before building an elaborate structure, it's best to have a few years of goatkeeping under your belt so you know exactly what your requirements will be. If possible, begin your goat venture by converting an existing structure for housing. A garage, shed, old chicken coop, or existing barn will serve.

There is some division of opinion as to whether goats should be confined to individual pens or allowed to run loose in the barn. Penning usually requires more of a building investment. Each pen (minimum size, 20 to 30 square feet) needs to have provision for holding food and water. Usually this takes the form of a rack for hay or other roughage, and buckets fixed in place to hold water and the grain ration. Pens should open onto a

HOME PASTEURIZATION

You'll need a double boiler, wooden spoon, dairy thermometer, glass jar with a lid or another suitable container, and a large bowl or bucket of ice water for cooling the milk.

Put enough water in the lower part of double boiler to touch the bottom of the upper part. Heat the water to boiling. Scald the upper part of the double boiler and spoon. Pour milk in upper part of double boiler.

Stir milk as the water boils in the lower part of the double boiler. As milk heats, test it with dairy thermometer. Heat milk to 160° F., and keep it at this temperature for 15 seconds.

You must cool the milk quickly. Remove upper part of the double boiler and set it in the ice water. Stir the milk as it cools to 50° F. Add more ice if necessary.

Pour milk into clean bottles or jars that have been scalded and cooled. Cover with caps or lids and refrigerate at 50° F. or lower.

Both home electric pasteurizers of 1¼- to 2-gallon capacity and dairy thermometers may be purchased from some mail order farm supply firms. These are batch-type pasteurizers that may use lower temperatures, usually 150 to 160° F. and longer holding times to achieve the desired results. It is possible to achieve adequate pasteurization by holding milk at 145° F. for 30 minutes. Follow the directions provided with the pasteurizer

This small-scale pasteurizer can process 2 gallons of milk in 24 minutes. It's available through Garden Way and Cumberland General Store.

you purchase. Pasteurizers are sold by NASCO, Fort Atkinson, WI 53538; they're listed in Sears' "Farm and Ranch Catalog" and Montgomery Ward's "Farm and Garden Book."

The pasteurization heat treatments suggested here have essentially no effect on the nutritive value of milk. Boiling milk, as was frequently practiced many years ago, did affect its nutritive value; also, excessive heating or oversterilization, as sometimes practiced when milk was canned, affected its nutritional properties. The only nutrients affected by pasteurization are vitamin C and thiamine although milk, raw or pasteurized, is not an adequate source of vitamin C and thiamine.

(Continued from page 321)
goats your own kitchen can serve as a milkhouse to wash and store utensils and refrigerate the milk.

If outside exercise space is to be provided, good fencing is important. Electric fencing is effective, cheap to buy, easy to put up, and takes minimal maintenance.

Finally, how much total land space you need for goats depends on how you want to run the operation. You can make do with a small exercise yard, period. But if you want to grow all the hay, cereals and other food for the goats yourself, something like three-quarters to one acre per goat are required. Many goat-keepers compromise somewhere in the middle and buy the hay, grain concentrates, and straw they need.

Problems or pleasure?

With good management, goats can be a delight. Whether or not you are looking for pets, it doesn't hurt that goats are friendly, curious, and eager for attention. Their backyard gambols will charm even the most pragmatic. Suddenly a kid shoots two feet into the air as if on springs.

But a goat is not a responsibility to be taken on lightly, and you should not assume that because you like the idea of goaty companionship and a daily milk supply that your neighbors will be of the same mind. Some good public relations work in advance is helpful. Promise that the goats will be well controlled so that they don't break out of their exercise yard and tear up the neighbors' Florabundas.

Assure them that the goat barn will be kept neat in appearance, that the manure pile will be kept discreetly out from under their noses, and that the smell goats are reputed to have is largely the property of a stud buck, which you won't have. Hint that you will occasionally share the wealth of your milking with them.

Susan Weaver

For more information:

Dairy Goat Journal, Box 1908, Scottsdale, AZ 85252.

Raising Milk Goats the Modern Way by Jerry Belanger (Garden Way, 1975).

The family cow

If you decide to tell your milkman to look for new customers because you have just decided to get your own cow, you'd better do it with a smile on your face. There is a lot more to keeping a family cow than meets the eye.

Having a milking cow around will change your life every bit as much as having your first baby. In fact, dairymen sometimes describe themselves as nursemaids to the cow.

The family cow really ought to be a family project where several members of the household can do all the things necessary in keeping her. If only one person knows how to milk her, if only one person knows how to feed her, to process her milk, to wash up the "milk things," to clean out the barn and bed the cow, then that person is confined to a daily routine akin to being the only one able to change diapers on the new baby. Farmers have said to me, "You know, John, that barn full of cows is my jail." One cow, dependent on one person, is no different.

The family cow is most practical where there is a family of milk drinkers who are bankrupting the bread-winner with milk bills and where those members are willing to pitch in and share the load. Families with several milk-drinking children are fairly well confined, anyhow, as I recall. You sort of have to make old Bess a member of the family and work her into the family routine. That is why we

Curiosity killed the cat, but it apparently left the dairy cow untroubled.

speak of her as the family cow.

Thank somebody for little favors: keeping a family cow is not a project that is immersed in red ink. If family members are willing to accommodate themselves to her daily needs, she'll do her part in the milkpail; she'll give you a calf every year and a few tons of manure for the garden, and when the sad day comes, she will become meat on the table.

However, even if everybody in the clan is willing to adopt old Bess and reap the benefits, most families don't live in neighborhoods where a cow is welcome or even practical. Zoning laws restrict her. She ought to have a pasture to roam in, although she could be kept confined just as dogs can be hitched. You ought to be able to make use of her valuable manure in a big garden, or be well acquainted with somebody who can. The cow makes a good member of the family only if you live out in the country with a little space that the kids and the cow can use jointly. The rest of you who want to take a cow into the family, I fear, will have to think about becoming country folk first.

Shelter

The cow barn, or the cow shed, must be made ready before your cow arrives in the cattle truck or is led over from a neighboring farm. It would be a lot easier to satisfy the requirements of old Bess if you didn't have to satisfy the whims of the people who are presumably going to take care of her. She would be happy with a three-sided shed, open to the south, a well-bedded place to lie down, a feed rack for the hay and a grain box for the grain she is to eat, and a source of water that doesn't turn to ice. You just keep adding bedding—sawdust, shavings, straw, mulch, hay—to the point where she builds up a pack. As the manure and urine go to work on this foot-deep bedding, heat is created, and she has a warm place to lie down.

When it drops below zero, however, and you have to go out there and milk her, you aren't likely to appreciate her tolerance for cold weather as your fingers freeze. So most folks try to give her housing that both the cow and the rest of the family can tolerate. This can be made

from a building you already have, or one you'll have to erect.

The cow herself does very well in a stanchion. She needs a space 4 to 4½ feet wide; ahead of her is a manger 3 to 4 feet deep. She stands and lies on a platform 60 to 64 inches long from the stanchion back to the gutter. The gutter should be 9 to 10 inches deep, and no more, or she may try to broad jump it and get hurt. It can be 16 to 18 inches wide. This has the capacity to hold a day's manure. If there is no gutter, she'll be lying in manure. You have to bed her well and keep the area she lies on as dry as possible.

Since most cows stalls are made of concrete, the bedding serves as insulation from the cold, and as a mattress between her bony projections and the hard surface. If used in sufficient quantity, the bedding also soaks up the wet part of the manure. Bed her well under her front feet. In the process of doing her thing, she works the bedding back toward the gutter with her feet, and most cows will stay pretty dry and clean. Some cows by nature are

(Continued on page 324)

There are many possible layouts for housing your cow. You can rearrange things to fit the shape of your area. Space for you and the cow to move around is necessary, as is handy storage of feed. If the barn is too big, the cow's body heat will be insufficient to keep the water from freezing.

(Drawings by Margot Apple.)

FAMILY COW LOGIC

The economic value of one family milk cow is most impressive. I emphasize *one* cow, because the value of the second cow will be a great deal less. Given the nature of the larger economy, the value of the produce you eat is much greater than that of the produce you sell.

For example, our own family cow is a small Jersey, who gives very rich milk in return for comparatively little feed. Her real worth to us is a little difficult to determine because, if you have a cow, you are apt to use (and give away) more dairy products than you would if you did not have one. But if (as I estimate) my family uses three quarts of milk a day, and two pounds of butter and a quart of cream a week then, our cow was supplying us products worth something like $10 per week. If we milk her 44 weeks a year, her gross annual "earning" is $440. But in the past year, in addition to keeping us abundantly supplied with milk, cream, butter, etc., our cow has produced a slaughter calf that dressed out at 270 pounds, and our surplus of skimmed milk made a significant part of the diet of a meat hog that finished out at about 400 pounds.

In addition to pasture, our cow ate perhaps 75 bales (2,500 to 3,000 pounds) of hay and perhaps the equivalent of 20 bushes of ear corn. I do not think the money value of her feed could possibly have exceeded $175. (I want to emphasize that the figures in the two preceding paragraphs are estimations.)

On the basis of the dairy budget alone, our cow has put us ahead this year by something like $265. In addition to that, her contribution to the making of our beef and pork must be said to have been substantial (certainly well above whatever her calf would have brought as a veal). And, of course, there is the value of her manure. But that is not all, for after last year's calf was weaned, our cow raised two Holstein "bucket calves" which cost $120 and brought $377.26, leaving $257.26, which I believe easily paid both for the cow's feed, and for the grain and hay fed to the three calves and the hog.

Wendell Berry

(Continued from page 323)
"humpers" when they feel the call and will cause the platform to become wet and dirty back near the gutter. When they lie down they get dirty. You have to spend more time cleaning them off and washing them before you milk them. You haven't lived until you have got a wet tail across the face, so let's hope you don't get a humper. You can buy a little device, called a "cow trainer" which hangs over her shoulder and is charged with electricity. When she humps, she hits it. You must always use it on natural humpers, although you don't have to keep it charged all the time—just often enough to let her know it may be alive. You have to buy a special transformer for the cow trainer before you plug it in. The cow trainers also train people fairly rapidly not to touch them.

In addition to the stall space, you have to have room for a 3- or 4-foot alley for the cow to get from her stall to the great outdoors and enough room around her for you to work, to feed her, to store some of her feed, and to hitch her calf in when it comes. If you can find a space 12 feet by 12 feet, you can usually build a little one-cow stable into it. It should be reasonably windtight and insulated if possible.

You see, the whole object is to make the overall stable space large enough for the cow and for you to work in, small enough so that she can heat it some with her body heat, and tight enough so that the heat doesn't all escape before the temperature of the stable rises to where it is comfortable for you to work in. Try to keep the temperature of the stable between 45 and 50° F. during the winter months. When it is higher, open the doors and the windows.

If you allow winter temperatures to run much higher than this, you'll have a damp, drippy stable. Cows breathe out two to three gallons of water a day as vapor. Hot air holds more water than cold air, and as moisture-laden air comes in contact with cold walls and ceilings, it condenses and makes a wet, gummy mess as the dust from the hay hits it. A cow that has to breathe moist air can have respiratory problems. You

can put a small fan in stables, which will blow warm out when the cow heats it up to 50° F. and shut off when it drops to 45° F. The cow wouldn't mind 10° but the water would freeze and you wouldn't like it that cold. The ceiling shouldn't be more than 7½ feet high or you'll have too much cubic space for old Bess to heat to your satisfaction.

The barn should also include enough room to store the winter's hay supply. The cow will eat three to four tons of hay per year, which occupies 1,500 to 2,000 cubic feet of space if it's baled hay and perhaps double that for loose hay. You can store the hay above her or in a separate area, but when building you should build big enough for both the stable and the storage.

If you can keep the stable from freezing, you can run water to the barn and put in a water bowl that she can drink from at any time. Otherwise you'll either have to carry water to her or let her go to an ice-free water tub or brook. Twice daily, the cow needs all the water she'll drink.

A couple of windows and an electric light or two in the stable are desirable. Be sure to put the doors where she can get in and out, and where you can get the manure out easily and bedding and hay in easily. All in all, a cow doesn't need a mansion to live comfortably.

Some folks keep a family cow in a box stall 12 feet square and let her run free both inside and outside. This way she'll waste more hay and require more bedding, and it is harder to clean the stall daily. Some cows will stand while you milk them in those pens; others have to be hitched. You also have to provide a separate area for a calf when you have removed it from its mother. If a cow is kept in a stanchion, she should be allowed to go free and exercise an hour or two every day.

Finding a family cow

So you have her room ready. Now you need to go out and find old Bess. She is probably living on a dairy farm in the area, or perhaps somebody with a family cow wants to go back to buying milk, or has raised a heifer calf and doesn't need two family cows.

There are four common breeds of dairy cows. There are the black-and-white Holstein-Friesians, large cows weighing between 1,200 and 1,500 pounds, as a rule. They give milk that tests about 3.6 percent butterfat. (A few Holsteins are red and white, for you nitpickers.)

The red and white Ayrshire is also a fairly large cow, weighing in at 1,100 or 1,400 pounds and producing about 4 percent butterfat milk.

The yellowy orange-and-white Guernsey runs about 1,000 to 1,200 pounds and produces a slightly pigmented, golden colored milk testing between 4.5 to 5 percent.

From Jersey, whose color varies from almost black through browns and tans to almost yellow, with assorted white markings, is the smallest, weighing 900 to 1,100 pounds. A Jersey gives the richest milk, usually between 5 to 5.5 percent butterfat.

Generally speaking, nonfat milk solids go up as butterfat test goes up, both within and between the various breeds. Also the larger breeds give more milk than the smaller breeds. There is wide variation in production between cows within a breed. And large cows eat more feed to produce more milk. You'll find extremely efficient producers in any breed, and you'll find klunkers, too—cows with big appetites and little production.

One of the lighter breeds might work out better for a family cow. First, they are smaller and easier for all the members of the family to work with. Second, they eat less and give less milk, and unless you have a very large family, even a small cow will give enough during most of the year to keep the milkman away. Also, the higher-fat, higher-solids milk of the lighter breeds means that you have to handle less milk per pound of butter or cottage cheese or whatever you decide to make with surplus milk, and there will be surplus milk.

Let's say you have decided to buy a cow. You are dealing in one of the great no-man's-lands of agriculture. Farmers who work with cows all their lives seldom feel comfortable when they have to buy one, because you never quite know what you are buying. Farmers are always looking for higher and higher producers. Sellers never want to sell their top producers, at least not unless they get top dollar. Top dollar? How many dollars?

Although the price for cattle goes in cycles, you should be able to buy a decent Jersey for $500, give or take a hundred or two, when she is ready to have a calf. You should look for and get some things that a dairy farmer wouldn't consider to be very important.

A family cow should be gentle, one that you can walk up to and pet. You should be able to put your hands on her udder and legs without getting kicked. You are going to have to work very close to her. Most cows can be worked with and calmed down if you stay calm and don't try to rush them, but there is that occasional cow that doesn't make a family cow because of her temper or nasty disposition.

You are most likely going to milk your family cow by hand. In fact, you may be learning to milk by hand while she is learning to be milked by hand. Not all cows have the proper mammary development to make this hand-milking chore tolerable. The floor of her udder should not hang below her hocks—the curved part of her hind legs. If it is too low, you can't get a pail under her, and you won't enjoy milking into a cake pan.

The single most important thing is the shape of her teats. When you hand-milk you need something to grip. A short-teated cow that you have to milk with two fingers and a thumb provides twice a day misery until she becomes family beef.

You don't want what's called a hard milker, one that you have to squeeze with all your might and all your soul in order to get a stream of milk to come. Some cows have perfectly well-shaped teats for hand-milking, but you don't want to have to put your back into it at milking time.

Neither do you want a string-teated cow, one that has a nice long teat that resembles a dangling angleworm. You squeeze and out comes a little squirt of milk. Milk may come easily, but it takes a lot of squirts from a string-teated cow to fill a pail, and a lot of time, too.

There also are bottle-teated cows, which have teats so large up near the udder that it's like trying to get milk out of the business end of a baseball bat. Such cows are a legitimate excuse for the more active members of a family to play soccer instead of helping with the milking chores.

What you want are teats that are firm and full of milk when you squeeze and somewhat limp after the milk has been squeezed out. They should be about as long as a man's large finger and perhaps a little greater in diameter. The teats should be within walking distance of each other. Some cows have teats so widely spaced that it is hard to reach all four of them from one side of the cow.

You'll find more cows with satisfactory teat size, shape, and placement than cows with flaws in one department or another. Don't buy one of the bad jobs.

I would not buy an old cow, but a two-year-old ready to have her first calf or a three-year-old that has been through one lactation. You can sometimes buy a cow at a farm auction. If you don't know how, take somebody who has a family cow with you, or get a farmer to help. You may find it difficult to get anyone to help, because anybody who knows something about buying cattle also knows that the procedure has some of the elements of pooling ignorance. If you can convince your advisor that you understand this, and won't hold a mistake against him, and that you wouldn't know whether you got a good one or an ordinary one, that though his judgment may be flawed, yours is downright nonexistent, then you may come up with a tolerable critter to enlarge your family.

For a family cow, don't put too much money into blue blood papers. You aren't in the breeding-stock business. There are a lot of good cows out there with foggy pedigrees. If she is registered and has papers, get them, as it might be possible to sell one of her heifer calves for a little more money to somebody who does

(Continued on page 326)

(Continued from page 325)

have a reason for having registered cattle.

At the same time, if you get a cross-bred scrub or a dairy-beef cross, don't expect much of her. Buy a cow that is a reasonably good representative of her breed. With artificial breeding and the overall upgrading of cattle that has resulted, good cows are much more common than poor ones these days. This wasn't true in my hand-milking days three decades ago.

You now have the cow and a place to house her. It is desirable to have two or three acres of pasture you can fence for feed from mid-May to early October. The pasture season lengthens as you go south. If you don't have pasture, then you can figure on feeding her almost twice as much hay in a year's time. If you have good pasture, one acre, with a little lime and fertilizer properly applied, might furnish sufficient roughage. If the pasture is rough, brushy, stony, or swampy, you might have to supplement the grazing from three acres with some hay. In the hot dry periods of midsummer it is often necessary to supplement pasture with hay anyhow. If she'll eat the hay, she probably needs it, but don't give her so much that she wastes it.

A small cow will consume about three tons of good hay in the seven-month barn season. A large

cow may take four tons. You should buy early-cut—I repeat, early-cut—hay and give her all she'll clean up every day she isn't on pasture. It is a good idea to sock in a year's supply when you find good hay. Come spring, good hay has been devoured by somebody else's cows, all too often, and you'll be paying good money for poor late-cut hay. This is why storage for a winter's supply of hay is desirable.

Hay and grain

Milk cows need grain while they are milking. The normal ratio is 1 pound of 16 percent protein dairy ration for every 2½ to 3 pounds of milk she gives. On a Jersey use the 2½ figure. On a Holstein use the 3-pound figure, and you won't be too far off. Store the grain for old Bess in a metal barrel with a cover on it to keep mice and rats out.

If you have poor hay, grain a little more heavily as soon as the cow calves. Work her up on grain as fast as she'll eat it and taper off as she tapers off in production.

You might ask why you don't grain old Bess to get just the amount of milk the family needs. The cow's lactating habits aren't designed for our needs, but for hers. After calving she gives a lot of milk and peaks in about a month and stays right up there in production for two or three months. This, you see, is just the time her calf would need it if it were sucking her. Then she tapers off in production and at the end of ten months is ready to dry off. She stops milk or you stop milking her. Now if you hold back on her grain early in her lactation, then she won't reach her optimum peak. She'll go downhill from a lower peak, and you'll end up with the whistling milkman coming for several months until she freshens again.

If you should have to feed poor-quality, late-cut hay, you'd better switch from a 16-percent protein ration to a 20-percent. It is possible, and costly, to feed too much grain, which your cow will eat in lieu of less expensive hay. At the same time, with hay alone she can't do her best in the milk pail.

Also needed are salt-mineral blocks, which you can put where she can lick them when she wants to. One block will last a long time if you keep it out of the weather.

As she gets down to 10 to 15 pounds of milk a day we stop milking her, and grain her lightly—3 to 4 pounds a day—but not at all if she is fat. About two to three weeks before she freshens again, work her up to 10 to 12 pounds of grain a day. She'll be in better shape to milk well by this lead-feeding. Do not lead-feed a

We usually grain the cow at each milking time and then give her enough hay to last until the next milking. Don't give her more than she'll eat; she doesn't like stale feed.

young animal before she has her first calf, as you may get into udder problems with first-calf heifers.

Your family cow, like all productive dairy cows, must be kept pregnant 9 months out of 12 in order to keep the milkman away. You can milk her 10 months out of 12, and give her a 2-month dry period before she calves. After waiting 60 days from date of calving, try to breed her the first time she comes in heat. She should come in heat about every 21 days. One of the major signs of heat is cows climbing on each other. Inasmuch as you have only one cow, you have to be a little more observant. She may try to ride you, or be overly friendly. She may bellow more than usual. She will commonly drop a little milk production, and stamp around and be uneasy. She will commonly switch her tail more than usual and may show redness and swelling around the vulva, and there may be a mucous discharge from it. If you suspect heat before you are ready to get her bred, mark the date down and be on the lookout for her next period.

Fortunately you don't have to lead old Bess over to see the neighbor's bull anymore, once a common sight along our country roads. You call the local artificial-breeding technician. His batting average is as good as the bull's, and he uses bulls that have a possibility of continued genetic improvement. Cost per breeding runs from $7 to $10 as a rule.

Cows aren't always free of health troubles. You should have them tested by a veterinarian for tuberculosis and brucellosis before you drink their milk. The test for brucellosis involves drawing a blood sample. The test for tuberculosis requires two visits from the vet, since he has to inoculate her and then come back to read the reaction. It is important to have both of these tests, as you will be drinking raw milk, and the diseases can be transmitted to people from the milk.

If you can't get her bred, the breeding technician may tell you to have the veterinarian check her. If she stops eating, often the first sign of trouble, something is wrong. Better have the vet check her out. If she goes down and can't get up just before, but more commonly right after,

having a calf, get the vet there quickly as she may have milk fever. If she has trouble giving birth after three to four hours, better call the vet.

The most common problem with cows is mastitis, or inflammation of the udder. You spot it by seeing flakes in the milk, watery milk, or swollen udders. You can sometimes get rid of it with antibiotics, which you can buy at the feed store. If it persists, call the vet. When using medicines, read and follow directions. If you milk by hand and milk the cow out clean, keep her well bedded, and avoid udder injury, you won't have much mastitis trouble. Much of the mastitis problem is due to milking machine misuse or malfunction—a good reason to milk your family cow by hand.

There are a lot of other little things you will have to do for old Bess occasionally—spraying her so that the flies won't bother her in fly season, trimming her hooves if they get too long, brushing her once in awhile to keep her looking neat, and delousing her if she starts to lick and rub at lice that may have gotten into her hair.

The milk

Well, so far all you have done is spend money, and it's about time you got something from old Bess in return. She has had her calf. We let the calf (or make the calf) suck her immediately after birth. The first milk is called colostrum, and the calf needs it immediately because it contains defense mechanisms against digestive and respiratory problems.

Once on its feet, a calf will zero in on a free meal. You may have to help it to stand up. Get the calf to suck immediately after it is born: new mother's milk is a defense against disease.

Cow comfort is a dry, well-bedded stall, with fresh dry air and clean water. Light is also desirable.

The cow's milk for the first five milkings is used only to feed the calf. Don't save the rest for the house. If the milk looks normal in the sixth milking, you are in business.

Generally, you should take the calf off the cow after a few hours and start milking the cow by hand. To teach the calf to drink from a pail, put your fingers into the milk and let the calf suck your fingers. In the process it will get some milk. After several such feedings, the calf will drink from the pail. Calves should be fed warm milk twice a day. With Jerseys, mix one part of warm water and three parts of milk and feed this to them, as the rich Jersey milk may cause scours, or watery manure.

If you are going to veal the calf, give it only milk, increasing the amounts as it grows. Calves are ready for veal in about eight to ten weeks. If you are going to raise the calf for a dairy cow, keep it on milk for four to six weeks, while giving it all the calf-starter grain it will eat—up to three to four pounds a day—and all the good legume-grass mixed hay it will eat as soon as it will eat it.

Those of the family who are going to milk the cow by hand are going to have to keep their fingernails short. Cows don't appreciate long nails, and they are apt to tell you so by eliminating you from their presence. When you milk by hand you use the end of your fingers, and long nails driven into your cow's teats justify some action on her part, in my opinion, and in hers.

(Continued on page 328)

(Continued from page 327)

I will attempt to tell you how to milk by hand, although you must master the technique yourself. Wash the cow's udder and dry it. Paper towels are commonly used for both. Normally you approach the cow from the right side. Sit on a milking stool with a milk pail between your knees or on the floor and held by your knees. Start doing your thing with both hands. Most hand-milkers start on the two front teats, and then do the rear teats. Very commonly when you start milking, the cow gives a little milk, and then for a minute or so she won't give any. Then she lets it down, as we say. Milk let-down is due to the release of a hormone into the blood stream, and as you start to milk you stimulate the release of the hormone.

You squeeze first one teat and then the other. Once you have the hang of it, you'll find that you are using the large and ring fingers to squeeze the milk out and your index finger to stop the milk from flowing back up the udder. The little finger is sort of in the way, and on short-teated cows, it is useless. Always take all the milk the cow will give. "Milk her dry."

With milking machines readily available, why milk by hand? Milking machines cost money to buy and keep up. It takes as much time to milk with a machine as by hand. It takes a lot of time to clean and sanitize the milking machine daily, and it doesn't pay to rig one up just to milk one cow.

Cows have to be milked twice daily, but the milking times don't have to be 12 hours apart. You can vary the morning-night milkings anywhere between 12 and 12 hours to 10 and 14 hours, but don't jump from 12-12 one day to 10-14 the next day. Cows are creatures of habit and don't like change. They'll gear themselves to your routine, but they like to have you follow your routine, too. You can normally let them eat their grain as you milk, keeping them contented and quiet. Cows don't dislike being milked, insofar as I can tell. In fact, they seem to like it, as it releases the pressure on their udder.

Sometimes when you are breaking a young heifer to hand milk, she doesn't know what's going on. Don't rush her, be calm, calm, calm. Talk to her, handle her, grain her. Let her know you aren't going to hurt her and never strike, never. If she won't let you milk her, have somebody hold her tail up by grabbing it at the base where it meets the backbone and hold it up there tightly. She can't kick you if her tail is up. As she gets used to your milking her, you can slowly release the tail and after a few milk-ings, she gets more interested in her grain than in your milking.

When a cow that is normally quiet while being milked starts to act

Let the calf suck your milk-covered fingers. Then put your hand in the milk, raising the pail closer to the calf's head and let it suck your fingers and some milk at the same time. Pretty soon the calf will drink without your fingers in the pail.

Clean off the cow's teats and udder to prepare her for milking. Washing her just before milking causes her to "let down" her milk.

up, she often has a reason for it. She may have chapped teats, or a cut teat, or maybe you need to shorten your fingernails. You may have to use a compound containing lanolin to soften her teats a little if they get rough, especially in cold weather. Find out what's wrong, if you can. If she starts to lift her foot and put it in the pail, and cows do, crowd her. Keep your left forearm against her leg above the hock joint. You'll soon learn to push your head against her belly just ahead of her stifle joint. A lot of times, speaking sharply to a cow will have the same effect it does with a dog. "So, Bess" is acceptable English when conditions warrant.

You should have a couple of good-sized stainless-steel milk pails, a stainless-steel strainer, and a box of strainer pads that fit it. You milk into one pail and strain into the other. You rinse the utensils in warm or cold (not hot) water both morning and night after milking. You wash and scrub and sanitize all of the utensils once a day. If you don't rinse them, the washing will be harder. You can get the special cleaners and sanitizers for milk utensils at your grain store. Read the directions and follow them. The once-over that you or your automatic dishwater may give your household dishes isn't good enough for milk utensils. If you keep milk in contact with only stainless steel and glass, your washing job will be easier.

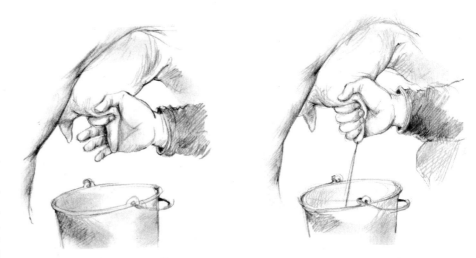

You'll have to learn to milk by yourself, but when you do you'll find that you keep the milk from going up with your index finger and you squeeze it out with the other fingers in descending order.

When milking, do it from the right side. Keep your knee against the cow's left hind leg to discourage her from stepping in the pail. If she gets frisky, crowd her with your head in the back part of her tummy.

Veal, butter, and cheese

Assuming you've got the cow milked, finally, how much milk have you got? At peak production your little Jersey may give you 45 to 50 pounds per day, and the heavier breeds 70 pounds of milk a day. Since a quart of milk weighs about 2 pounds, this may be more than you bargained for. What do you do with 25 quarts of milk a day? For a few months, chances are you'll have more than most families can drink.

Some folks sell raw milk to neighbors while they have extra. Some states have a limit on how much you can sell without undergoing sanitary inspections and meeting a lot of the same requirements that your milkman has to live up to.

Some folks raise a veal calf or two with the extra milk. You'd better get some good information from your extension service on raising a veal calf, because you can easily kill it by overfeeding it. It isn't uncommon for a good cow to give you enough milk for the family and for two veal calves at once. Even as her production tapers off, she might support a veal calf along with the family. Veal calves are readily salable at a good price at auction markets. (Remember: once you start to use the milk, feed the calves from a pail. It isn't commonly accepted sanitary practice to let the calf suck a cow you are drinking from.)

Some people get into the butter-making business, and with the current price of butter, butter-making shouldn't be sneezed at. Often neighbors will buy butter from you; if you let them try it first, they will probably prefer it to creamery butter. When you make butter, you also get buttermilk that is true buttermilk (not the cultured commercial drink). If you have too much buttermilk, it makes good pig feed, along with the skim milk left over from separating. You can separate the cream from the milk with a small separator, or you can use set-pans and skim it off.

A few people get into the cheese-making business. Cottage cheese is easy to make, and I can't imagine your making cottage cheese that isn't better than what you buy.

Now, no matter how you dispose of the milk that the family doesn't drink, you'll have to spend some time getting rid of it, and it really comes at you, every day including Sundays and holidays. The business of keeping a family cow takes time, perhaps an average of an hour a day for much of the year just to take care of the surplus milk and wash up the milk dishes.

Not calculating the time that members of the family spend, it costs around $700 out of pocket per year, *(Continued on page 331)*

Dairy sheep: Stow, Massachusetts

Goat milk lovers brag about their productive animals. As a sheep raiser, I feel sure sheep are even more productive. Besides fertilizer, hides, wool, and meat, they can also give us milk. I knew that Friesian, Tunis, Arabian, and French sheep are famous milking breeds, so last April when a yearling mother ewe had her seven-week-old lamb sold out from under her, I decided to use her milk instead of drying her up. "Tigger" is a yearling Hampshire, not a milking breed, so I did not expect too much from her. But she was healthy, young, and full-bagged, with no lamb sucking on her.

Milking her was not half as hard as overcoming my qualms about catching dread diseases from drinking her milk. Phone calls to local public health officials and to goat dairies did not help. No one had specific facts about keeping sheep clean for milking in a stall with nine others. I was not about to invest in the stainless steel milking parlor recommended by the University Extension.

I read all the books on dairy goats. Little help. None of the bag-cleaning chemicals in the books were available in Stow, so I finally said to myself, "Think of all the families who drank from a family cow before modern antiseptic chemicals were common. Think of all those who still do in underdeveloped countries. Picture Heidi milking her goats in a high Alpine pasture." The words of a friend who sold legal raw cow's milk down the road came to mind. She had quoted statistics which showed that most contamination enters the milk when exposed to air between udder and storage container.

So if I could milk into a sterile jar, strain it within four minutes and cool it in another sterile jar immediately, the risk would be minimal. As far as worrying about the mastitis strip cup, well I simply hoped for the best. The vet said mastitis is not always visible in the milk for days anyhow. Even after all the psychic pep talk, I still cooked the first few days' milk in creamed soups and vegetables, to play it safe. When we did not fall on our backs and curl up our toes, we drank it in coffee. Better than cream!

Then I sipped a little raw warm milk. I remembered that mushroom aficionados take only a tiny bit of questionable mushrooms, to lessen the effects of possible poisoning. Following this line of reasoning, I figured if Tigger's milk would make me sick, a little would cause only small twinges. There were no effects at all . . . ex-

---

Someday, A Home Milk-And-Wool System

Neither East Friesland nor any of the other breeds of so-called milk sheep used abroad for milk product production are likely to become commonplace until the present bureaucratic rulers of the sheep industry's destiny retire or get fired, or, as one innovative student has observed, human female "carrier-tourists" are employed to transport test-tube fertilized ova to recipient ewes on U.S. farms, possibly under state or local veterinary supervision, where states are ill-disposed to follow the hard-linists of the federal bureaucracy.

The Shepherd.

---

cept radiant, blooming well-being. Gradually I began drinking glassfuls. And the milk came. It poured out morning and night, a gallon a day at first. After a month it dwindled down to half a gallon. Remember this was not a milk breed—just an average healthy Hampshire, who had already milked a lamb for two months.

Since I am the only one in this family who likes milk, I had to find ways to use so much milk. I made yogurt, so creamy-thick that no extra powdered milk was needed as thickener. I made gallons of ice cream to

keep in the freezer: coffee, carob, mocha, banana. I made farmer's cheese and cheddar cheese. The two-pound cheddar round was mild and firm, light yellow, with a sprinkling of small holes. It was so easy to make I was amazed. For a recipe I followed the chapter on Cheese-making in *Country Women—A Handbook for the New Farmer* (Anchor, 1976), for aged soft cheese. I oiled the outside instead of dipping it in paraffin. Curiosity forced me to eat it after six weeks, although one is supposed to wait three months for a parmesan-type cheese.

How did the sheep milk taste? Sweet and smooth, like velvet. After a month of using sheep milk, yogurt, soup, ice cream, frappes, and cheese, I gained five pounds. But I felt really healthy.

By mid-June I grew tired of milking twice a day, although it only took ten minutes. By the third day, Tigger had trained herself to stand near the hayrack to which I tied her while I milked. She chomped contentedly away on delicious morsels of last year's hay. Obligingly she shifted sides at my slightest touch, so that I could milk the opposite udder half. At first the other ewes were jealous of her special relations with me and wanted to be milked also. They would rub their heads under my elbow at the most precarious moments, talking to me in conversational tones. I cut down to milking only mornings, and within a week her milk supply halved. I do not know how long she would have continued to produce well had I kept up two daily milkings.

In conclusion, sheep milk tastes marvelous. It is good for very active people doing hard physical labor, and for underweight, sickly types, to build up their health. We do not need to import special sheep for milking or to fight the USDA quarantine laws. Just go out to your barn and milk what you have—whether Dorset, Montadale, Columbia, or a friendly crossbreed. If you follow sensible cleanliness rules, I guarantee you will get enough milk for three months for the average family of four. Bottoms up!

C. G. Schwarzkopf. *The Shepherd.*
Sheffield, Massachusetts.

(Continued from page 329)

which includes buying or raising the 3 tons of hay, and the 2½ tons of grain, the annual cost of the cow herself less salvage for beef, veterinary bills and supplies, housing costs, extra electricity, and the little extras you didn't expect, like a birthday card for old Bess. I could make up a cost budget for you, and we'd not be too far from this figure, providing you aren't a spendthrift.

On the income side, she should give you 10,000 pounds or 5,000 quarts of milk a year. At what farmers sell milk for, this comes to about $1,100. At $11 per hundredweight which a farmer would get for Jersey milk, this comes to about 22 cents per quart. If you were to buy the milk delivered to your home, it would be just about double that, and the milk would be standardized down to 3.5 percent butterfat, besides. Its value to you is what you save on the milk bill, what you earn by selling in flush seasons, the value of the products you make from the milk, or of the veal you raise (to eat or sell), and the value of the manure if you have a use for it. Your income from old Bess will vary with your family size and what use you make of the extra milk, but it certainly would be difficult to get less than the $1,100 that a dairyman gets for a like amount.

Whether or not the keeping of your family cow makes economic sense to you depends on whether the members of the family look at it as employment at so many dollars per hour. If you charge off the labor at minimum wage, then some families will be in black and some in red. If you look at value in and money out, then the family with the cow will have more money at the end of the year than they would have had if they had spent their time watching TV or looking for something to fill idle moments.

The confining aspects of keeping a family cow are something that I fear each family will have to reckon with. I suspect that the unwillingness to be tied down is what causes people not to take old Bess into the family rather than the economics of the thing.

Beyond the practical advantages we have mentioned, I think we should not miss one of the most valuable reasons for keeping a family cow. It is the experience of learning to care for a living thing—in the case of the cow, a fairly complicated living thing. There is the experience of watching a cow give birth to the calf; developing the discipline to care for an animal every day, come measles or chicken pox; knowing where a significant part of the diet comes from, and actually taking part in its production; processing one food into another and finding out that your own production can be better than what you buy. There is the joy of knowing an animal as a member of the family, as an individual, because one cow is as different from another as members of the family are. And inherent in all this is the need to assume responsibility, which a family cow will require of each member of the family.

Keeping a family cow, for a few years at least, might well be a good opportunity for families in a position to do so to add to their lives a dimension that they will cherish in later years. Certainly old Bess will help teach the virtues of constancy and perseverance.

John C. Page. Reprinted by permission from *Blair & Ketchum's Country Journal*. Copyright © 1979 Country Journal Publishing Co., Inc.

Backyard animals

Raising the bacon at home

The greatest value of home livestock production can be summed up in one word: involvement. When you go into the supermarket and load a shopping cart, your involvement is minimal. You choose the items and pay for them, you cook them and eat them. The most important part of your life becomes the job that provides the money to buy the food.

If you plant a garden, on the other hand, you become involved. The soil must be prepared, seeds sown, plants tended, and vegetables harvested fresh from the earth. And if you decide to raise animals, your involvement increases dramatically.

Livestock demands a commitment in terms of financial investment, time and labor, knowledge and skill, and emotion. Let's consider each of these.

Garden seeds don't cost much. If the seeds don't produce a full crop, or even if they don't grow at all, the loss isn't very great. Even a partial crop may return the investment.

The livestock producer has a far greater financial commitment. Even if you happen to acquire the leftovers of someone else's disillusionment or mistakes and find yourself with some chickens or rabbits for little or nothing, your investment will be considerable because the original cost is almost always less than the upkeep. This is one reason it pays to start out with good stock: there are tremen-dous differences among animals, and a gift may be no gift at all, in the long run.

Animals take up space. They need housing, food, and vessels to eat and drink from. The food must have the proper amount of protein, vitamins, minerals. These things cost money.

Livestock raising also requires a commitment of time and labor. There is the daily routine of feeding, watering, cleaning, and other care. You can take a vacation and the tomatoes and beans won't even miss you. But if you take off a single day, who's going to water the rabbits, milk the goats, and see that the chickens have feed? This commitment becomes particularly apparent on mornings when you're running late, or when there's a blizzard howling around the barn and the water pipes are frozen and you have a miserable case of the flu.

In addition, it takes a long time to produce an animal product. You can walk into the grocery store for to-night's dinner without even knowing what you're going to buy, but if you want to produce rabbit stew from scratch, you have to plan at least 3 months in advance. It takes 4½ months for a chick to start laying, and as long as 1½ years to grow a ham-burger or steak. Livestock raising goes against the demand for fast foods and convenience foods.

Livestock raising also requires special knowledge and skills, far be-yond knowing which package of meat to pick up from the supermarket display case. Which is the best breed for your purpose? How do you select healthy animals, with the required genetic potential? What is the best type and arrangement of housing and equipment? The animal caretaker must be a geneticist and nutritionist, midwife and nursemaid, veterinarian and carpenter. Those who take their commitment seriously—who are successful—are soon set apart from the supermarket shoppers by their familiarity with ketosis and coccidiosis, TDN and digestible protein, gestation and castration. They are more deeply involved in life itself.

These considerations can be either negative or positive, and for most people who raise livestock they are both, depending on whether the sow just had a fine litter of beautiful piglets, or whether she just broke out of her pen and ravaged the garden. But I don't think any one of them by itself is enough to explain why so many thousands of people are engaged in raising livestock on a small scale today.

A time to die

The most important commitment of all is the emotional one. Animals are living, breathing creatures, with personalities and intelligence. It's impossible to have them be dependent on you without a bond forming. It may not always be a bond of friendship—as when that hog roots up the garden—but it's a bond nevertheless. There's great satisfaction in assisting a lamb into the world, helping a newborn pig find an unclaimed nipple, or having a goat nuzzle your neck as you milk her. But the trials and pleasures of working with living creatures is only one part of the commitment. Eventually, every farm animal must meet its ultimate destiny. And the act of transforming a living creature into a swinging carcass hones our regard for all of life and opens new vistas on the universe. It can't easily be cataloged or put into words. You have to experience it for yourself.

Perhaps you raised Petunia Pig on a bottle, and she used to follow you around like a puppy. You recall that as she grew, she grunted in contentment when you scratched her behind the ears—and how you

thought that 200 pounds of contented pig is a lot of contentment. You think you couldn't possibly hold a gun to her head and pull the trigger, slit the jugular and listen to the gurgle of gushing blood, clean and scrape the carcass. How can you see a thick chop with applesauce or a succulent ham in something that is so full of life and personality?

Slaughtering a pig at home may require some help from the neighbors, but you'll end up with the best pork you ever tasted.

You could, of course, keep Petunia and buy pork in a supermarket. Or you could ship her off to a custom butcher, and later pick up neatly wrapped flash-frozen packages labeled "bacon" and "spare ribs." Many animal raisers do. But they're missing something; they're stopping short of total commitment.

One of the most basic concepts of a home food system is that food is a sacrament, which can and should nourish us beyond our physical needs. Store-bought food represents nothing more than a few minutes' work. It doesn't tie us to life. It makes us machinelike, and something less than human. That's true of a carrot or a loaf of bread, but even more of a piece of meat.

Among the lessons learned by a home livestock raiser and butcher are the reverence for life and an understanding of the awesome scope of nature. We cannot live without symbiotic creatures inhabiting our bodies, and one day we will pass on to another phase of the wondrous

biological cycle ordained by nature, to be consumed by microbes.

There are many good reasons for raising livestock. Animal products provide important nutrition, and they are among our most expensive foods. As with vegetables and other food products, by growing our own livestock we can be certain that it's wholesome food, uncontaminated by the products whose sole purpose is to make food cheaper without regard for the quality. Home-produced meat *tastes* better. We have had the companionship of the animal.

But most of all, we have tasted life. We have seen the cosmos.

Jerome Belanger

Livestock and the law

If you're considering raising backyard animals, your first step—before investing any money in the venture—should be to pay a visit to your city or town hall to find out if what you want to do is legal. Many cities and rural or semi-rural areas have laws on their books concerning the keeping of livestock. In some cases the laws are clear-cut and enforced; or, they can be vague and open to interpretation.

Don't assume that because you live in a rural area, livestock is allowed. It may not be. As the suburbs have sprawled into formerly agricultural areas, many zoning laws have been changed to suit the new inhabitants. These people may have wanted to live in the country, but not next door to a pigsty. And don't take it for granted that because the fellow next door keeps chickens, you can, too. The laws may have changed since he began, and there may be different rules for someone just starting out.

A check with several cities and towns in eastern Pennsylvania proved, if anything, that there is no rule-of-thumb for dealing with the legality of backyard animals. Most of the larger cities have laws outlawing livestock of any kind within their limits. The laws are clearly written and spell out precisely which animals you can't keep. About the only animals not mentioned are cats and dogs, and in several places you can't
(Continued on page 334)

(Continued from page 333)
even keep more than two of them.

Occasionally, though, there is a loophole. When the health officer of one city was asked about the laws governing livestock, she replied, "We don't allow any." But when she finally read the law over the phone, it turned out that chickens were not specifically outlawed, while many other animals were. The health officer acknowledged that there was probably nothing she could do if someone kept chickens in her town. The point is, don't accept anyone's interpretation of the law; ask the officials to read it to you, word for word, or better yet, go down and read it yourself.

In several of the communities surveyed, animals are allowed but the number that can be kept is limited. That means if you're at the legal limit and your rabbits have babies, you won't legally be able to keep the offspring.

Pigs seem to have the worst public relations at city hall. Probably because of their reputation—largely undeserved—for being dirty and smelly, pigs are banished by most towns, even those that permit other animals.

But don't give up just because your town has a law on the books that says you can't raise animals on your property. Find out how strictly the laws are enforced. Many animal raisers say that while what they're doing is technically illegal, they're left alone as long as no one complains.

"We try to use a little shading when it comes to enforcing our livestock ordinance," one officer told me. "If no one makes a noise, we're liable to overlook a few animals in the backyard. But if someone complains, we've got no choice but to step in."

The art of being a good neighbor is especially important if you keep animals. It's a good idea to check with the folks who live around you before you buy animals, particularly if you're treading in a gray legal area. They may well not object at all if you promise to keep things neat and clean. But be warned if you're planning to have a rooster: good neighbors have a way of becoming enemies after being awakened at dawn once too often. You may be able to smooth things over with a peace offering of your leftover eggs, or, if they're avid gardeners, a load of manure come spring.

You can also run into problems if you try to sell any of your livestock harvest. Some towns allow animals provided they aren't kept "for profit." Obviously that's a pretty vague term, but it may mean that while you can raise and eat all the eggs you want, you can't legally sell any. You may be able to circumvent a law such as this by reverting to the ancient system of bartering.

When you go to check out the law, read it carefully. It's also a good idea to talk to the health officer, code enforcement officer, or whoever actually enforces the law, to find out just how it is applied. In some towns livestock laws are covered by zoning ordinances, in other areas by health codes, and in some places by both.

Finally, don't call it quits just because there's a law that says you can't raise animals. Laws can be changed, or at least waived. If your town has a zoning board of appeals, it may be possible to get a variance that will allow you a special permit. In most cases, however, you'll need to get your neighbors to sign a statement saying they don't object. If enough people in your area want the law changed, you can try that, too. Consider circulating a petition and presenting it to the town leaders.

Michael Lafavore

Chickens

You'd expect to find chickens with pigs and cows and goats down on the farm. But they make just as much sense in the backyards of suburbia or on back porches in town. If you have trouble picturing that, just think of chickens as very productive—and lovable—pets.

They take little space, time, or effort to raise. A cat or dog costs more to feed, and won't give you eggs and meat. And all chickens aren't alike. They come when you call, they delight you with their antics, and they dazzle you almost daily with the miracle of a perfect egg.

To get a dozen eggs a week, all you need are three or four hens, a makeshift coop, a feeder and pan of water, and some chicken feed. You'll get fresh eggs any way you want them—brown, fertile, extra large, organic, even Easter egg colors. Home-fed meat chickens will be the tenderest and tastiest you ever ate, with no hormones or chemicals hiding in them. You'll probably save money too.

Even the manure isn't wasted if you have a garden. Chicken droppings are one of the best animal fertilizers you can get.

The only real obstacle to raising chickens anywhere you want to is the law. Some urban areas may prohibit raising livestock, so check it out before you start. Your neighbors may not be thrilled to hear you're into hens, but good care of your coop should ease their minds.

Read up on chickens, ask friends and local growers about them. But remember that there are few hard-and-fast rules when it comes to chickens. See what works for you, and be sure your new pets will let you know when they're unhappy.

Nurturing the flock

There are lots of options for housing chickens, and none are elaborate. You can make a coop out of a corner of the barn, an old shed, even a packing crate. It's cheap and simple to build one yourself, with plans from a book or your own head.

Chickens have simple needs—four square feet of space each, ventilation to remove moisture and ammonia from the coop, and protection from extreme heat and cold. The only furniture they need are roosts (made from broom handles or dowels) to perch on while sleeping, and a laying nest for every four or five hens.

Chickens don't need a yard, but they thrive on the extra protein they can pick up outdoors, and you avoid problems that arise when they're cooped up. A five-foot fence will keep out most predators—the neighbors' dogs being the most common—with screen over the top to protect them from hawks. Chickens roaming loose may fit some romantic image of farmlife, but they'll get into the garden and wreak havoc, hide their eggs from you, and make easier targets for predators.

You'll need electricity in the coop if you want eggs over a cold winter. Hens need 14 hours of daylight to stimulate their glands, and a light bulb can make up those hours. In colder climates, the bulb also provides heat and keeps their water from freezing.

Feed is the biggest expense in raising chickens. They eat about 100 pounds a year apiece, at about ten cents per pound. Commerical feed is premixed to provide good nutrition, usually along with medication to prevent poultry diseases, but you can buy it with no additives.

You can also grow your own feed if you have the time and land. Just remember you will have to grind up all the grain yourself or pay a feed mill to do it. To cut down on feed costs, toss your flock table scraps, weeds, grass clippings, even seaweed. Plant some leafy vegetables for them, or build a simple flytrap to catch some extra protein. Even city folks can get free leftover produce from a supermarket. Water is just as critical as feed, and six hens can drink as much as a gallon a day.

Every chicken raiser has his pet theory about when they have to be fed. My farmer neighbor insists on feeding his flock twice a day at exactly the same time, while other friends make it once a week. Just make sure your chickens always

A fenced cage like this provides enough space for a small flock of chickens to roost and roam while protecting them from dogs, hawks, and other predators.

have enough food and water, especially in hot weather.

It takes ten minutes a day to check on feed and water. But chickens also need good sanitation. Loose litter such as pine needles, chopped straw, or sawdust, spread thickly on the coop floor, absorbs the ammonia and moisture from chicken wastes. Turning this over often keeps the birds dry and out of the droppings.

One simple way to use the manure is to spread peat on the coop floor, and as it gets damp, shovel it right onto the compost pile. Never put it fresh onto plants. At least twice a year, change the litter and scrub the coop with a mild disinfectant.

Hatching eggs, usually done in the spring, is an effortless task if you let a "broody" hen do most of the work. (All hens can mother, though some breeds are more broody than others.) Otherwise, you need an incubator and it's hard work. Whoever coined the phrase "don't count your chickens before they hatch" knew the frustrations of incubating eggs.

Disease is rarely a problem for the small-scale poultry raiser. Cannibalism and egg-eating among the flock are more common, but all three hazards can be avoided if you keep your chickens well fed, clean, and uncrowded. If disease does crop up,

consult your local agricultural agent or a knowledgeable neighbor. Odors aren't a problem with good sanitation.

Choosing your birds

Before you buy, decide what you want chickens for—meat, eggs, or both. The best laying hens make poor meat, and meat birds are often lousy layers. It's best to select breeds geared for one or the other, and both types will live happily together.

Don't try to save money by buying poor stock. Chickens who bring disease into the coop are a costly bargain. Buy from someone whose stock you can look over firsthand, or from a licensed hatchery, or from a reputable mail-order house.

You don't need a rooster to get eggs. If you want fertile eggs for eating or for increasing your flock, then you need a rooster for every 10 to 15 hens.

We all know roosters rouse the world at dawn, but don't think they stop there. If something startles Chanticleer at midnight, he'll let you and your neighbors know about it. He doesn't make his hens noticeably happier, but you'll be amused at his strutting style, his protectiveness with

(Continued on page 336)

(Continued from page 335)
his ladies, and the fascinating courting ritual.

Raising meat birds is simple and the results are tastier than the Colonel can offer. But don't expect to save lots of money, unless you have a cheap source of high-protein feed. Meat birds eat more feed than laying hens do in the same amount of time.

You can raise chickens to the stage of broilers, roasters, or capons, depending on your meat preference. They do well in confinement, since too much exercise tends to toughen their meat. They're butchered at anywhere from ten weeks to seven months. It's a simple, humane process requiring a sharp knife or ax, a strong stomach, and a philosophical approach.

Choose your layers for such reasons as egg size or color, how much they eat, how they like the cold. Don't forget good looks and personality! One friend won't have the flighty Leghorns in her flock, even if they are great layers. Bantam hens usually make model mothers, and Rocks are calm and quiet.

Brown eggs often bring a higher price, if you plan to sell the eggs. Hybrids and crosses are not suitable for breeding. Try several kinds, and enjoy their unique personalities as you decide which breeds you like the best.

Choosing what age of layers to buy is mostly a matter of cost and convenience. Pullets, 20 to 26 weeks old, cost the most but they're ready to start laying and are ideal for beginners to chicken husbandry.

Day-old chicks are cheaper, less likely to carry disease, and a delight to raise. But they need lots of care during their first few weeks, and you have to add in the cost of five months' feed to bring them to the laying stage. You can buy eggs and hatch them yourself, but this is tricky and often disappointing.

Year-old hens, culled from commercial flocks because they've slowed up egg production, are the cheapest. But with the loving care only the small-scale raiser can give, they'll produce for at least another year—and the eggs get larger.

To give you an idea of costs, in southern Mississippi ready-to-lay hens start at $2.50, chicks at 70 cents and year-olds at $1 for the most common breeds. But that chick costs closer to $2.50 when you add in feed costs. The fancier breeds, like certain Bantams and the "Easter egg" Araucanas, might cost $20 to $25 a pair. Whatever you buy it's best to replace your stock every 18 to 24 months to keep the eggs coming steadily.

The average hen lays 200 to 240 eggs a year, or slightly more than one every other day. So six hens will produce about two dozen eggs weekly. You can sell extras, or store them in the freezer or in waterglass solution for up to ten months. Buy twice as many day-old chicks as you want to end up with, allowing for some mortality. You can butcher any extras for meat.

Start-up costs depend on how big a flock you plan to have, and how much equipment you want to invest in. For less than two dozen hens, you really don't need any equipment other than the shelter, and that's cheap to build. The Sears farm catalog sells feeders for $12 and waterers for $5, a one-time expense.

For more adventurous operations, you can buy a Sears 50-egg incubator for $40, and 100-chick brooder for $15. But you can get cheaper equipment if you buy locally.

Feeding six laying hens for a year will cost about $60. They will lay at least 100 to 120 dozen eggs a year, making the cost of each dozen eggs 50 cents to 60 cents.

Heat lamps can turn a shed into a workable brooder, as long as it's not too cold outside.

Pat Donahue

Ducks and geese

With Mother Goose and Donald Duck firmly entrenched in the culture of America's children, it seems odd that Americans have nearly abandoned their European and rural heritage of raising waterfowl. Roast goose is a traditional English Christmas feast, and duckling is a key ingredient in French and Chinese gourmet cuisine. Yet both are hard to find in many American supermarkets. The United States produces but ten million ducks and one million geese annually, compared to the more than three billion broiler chickens.

That means raising ducks or geese as a sideline business probably won't make you rich, even though both birds command a better price than chickens when there is a market for them. But if you're looking for a source of meat for your own table, waterfowl have so many advantages you'll wonder why the West wasn't settled by ranchers with vast flocks of geese.

They're quite practical, even for the average suburbanite. You don't need a pond to raise ducks and geese, although they'll be happier if you can at least provide them a small swimming pool or tub for splashing and preening. Geese are grazing animals and can live quite well during the summer on only the weeds and crab grass in your lawn. If you feed ducks a little grain, they'll repay you by eating slugs and caterpillars in your garden. And certain breeds of ducks will lay as many eggs as chickens.

Ducks and geese are among the most efficient animals at converting feed to meat, and they're as hardy as larger animals like cattle and sheep. They need little shelter from the elements and suffer from relatively few diseases if properly managed. Many owners of ducks and geese rate them as a nearly foolproof source of tasty meat, eggs, and small amounts of that high-priced clothing insulator—down.

But practicality isn't everything. Many people decide to keep sheep or hogs or rabbits because they like them. Ducks and geese are awkward creatures on land, but the bright plumage of a mallard drake or the swanlike form of a White Chinese goose will enhance the beauty of a pond or shoreline—especially if you don't crowd too many in a small area.

It's overcrowding that has given ducks and geese an undeserved reputation as dirty animals.

Geese and ducks can also form a strong attachment to their owners. Like watchdogs, they'll raise a ruckus if a stranger enters your property. Geese are even used as a kind of alarm system to help guard some businesses, but their honking and hissing is usually worse than their bite. Ganders are good fathers who often mate for life with one goose, sharing the responsibility of protecting young goslings with the mother, so watching a goose family can be a fascinating experience.

Which birds to buy?

The exact costs and benefits you'll derive from owning ducks or geese will vary greatly with the breed you chose. The Khaki Campbell duck, for example, will keep up with any chicken and produce 250 to 300 eggs a year. But a duck that is busy laying eggs has little time to sit on its nest, so you'll need to help out by investing in an incubator to hatch her eggs. (The minimum cost for an incubator is about $25 for a Styrofoam, thermostatically controlled model which holds about 35 duck eggs.)

The Khaki Campbell is a small duck, and adults weigh only about 4½ pounds. The commercial duck industry relies primarily on the familiar white bird with orange feet and bill—the Pekin. It grows quickly to a weight of about 9 pounds, and lays 110 to 130 eggs per year. The Pekin is also a poor nestsetter.

Many homesteaders favor the Muscovy, the only domestic duck not descended from wild mallards. Mus-

Most breeds of waterfowl can survive cold temperatures with only a simple shelter.

covies prefer to roost on a fence or in trees, and are able to get along well without a pond. They like to gather their own grass, seeds, worms, and insects, so they're nearly self-supporting. Muscovy mothers can produce two broods of 15 ducklings each season. Males reach 12 pounds, and the delicious meat has less fat than meat of other breeds. And to top it off, they make almost no noise.

Among geese, the White Chinese are a fairly self-sufficient breed. They thrive on grasses and weeds, although a little grain will fatten them up for your table more quickly. They are small—adults weigh about 12 pounds—but for geese, they are prolific egg layers, producing up to 100 eggs in a year.

(Continued on page 338)

CROWING IN THE MORNING

Where have all the roosters gone?

Since 1950, I've spent much of my life in the country, but in all that time I've seldom heard the clarion call of that posturing male fowl. Crows split the forest air with their cawing; catbirds meow plaintively in the honeysuckle shrubs; calves, dropped from their bovine mothers, bawl piteously in the blue dusk; dogs howl across winter fields under a shimmering moon. But the animating sound that wakes the world from dreams, the tumultuous, soaring cock-a-doodle of the rooster has vanished from much of rural America.

Why? Is the chicken passe? An extinct species like the dodo bird? Every menu, cookbook and supermarket attests to the contrary.

The answer—as with so much out in our agitated society—can be traced to Rampant Technology. Back in the time of social nirvana, The Good Old Days, when the "country" was just over the city line, millions of family farms dotted the landscape. And almost every one of them had a flock of chickens scratching around the barnyard.

Roosters (boys) and hens (girls) lived a life that was elemental and classically simple. As a culmination to the barnyard romance, the hen dropped her fertilized egg into a broody nest and sat on it until a blank-eyed chick emerged into this turbulent world.

As for the rooster, that chauvinist bird blazoned to the world what he was all about. He postured. He strutted. He stalked the barnyard in happy recall of amorous events past and anticipation of more to come.

Alas, the family farm is being replaced by Agribusiness, large factory-farms with single cash crops. Now, the cold ministrations of bottom-line efficiency dominate chicken raising. The fun and games of the family flock are out; the glacial manipulation of Technological Man are in.

Nowadays, thousands of hens bred for egg production are housed in huge buildings from which they never emerge into daylight (chicken factory No. 1). Walk past one of these plants and you won't hear a single clucking hen. Roosters are put in to mate with the hens, and the fertilized eggs are shipped to an incubator plant.

The rooster in this process is a mere assembly-line gigolo incarcerated within the four walls of the chicken works. His glorious cry is muffled, unheard by his fellow creatures in the free world. For him, gone the sun and blue sky, the barnyard Eden of chickendom past.

Now, even most country people get their chickens at the market instead of raising them. In this day of fast-frozen abundance, who needs all that fuss and feathers? All those coops and chicken mess? And who needs roosters? Like juicy tomatoes, creamy milk, and unpolluted air, the rooster has vanished from our daily lives.

Almost. One morning, from somewhere in the spread of the gentlemen's farm bordering my five acres, there floated to my disbelieving ears, a distant, galvanic cry.

Mirabile dictu! Quickly I pelted down the road for a look around my neighbor's place. There among a gaggle of geese and hens was a large handsome rooster. Cocking a brazen eye at the world, he swaggered across the yard, his crimson comb flopping with every toss of his arrogant head. And he crowed his fool head off.

It had taken my neighbor's inspired sensibility to do it: the barnyard flock, with liberated rooster, was back in our neck of the woods.

Now, as the sun moves up over my country haven, I arise each morning to that roistering, primeval cry. That glorious sound of animal exaltation. That vitalizing shout to the world—Cock-a-Doodle-Doo! Awake! It's morning! We're here! Us! Me!

Leo Trachtenberg. © 1980 by The New York Times Company. Reprinted by permission.

(Continued from page 337)

They are also good parents and rear about 20 goslings each spring.

Unfortunately White Chinese geese have the disadvantage of being extremely noisy and not very hardy in winter. If your winters are snowy and cold you'll need to provide them some shelter—although it needn't be more elaborate than a south-facing lean-to with a straw-covered floor. The two most common European breeds, Embden and Toulouse, are more adapted to winter weather. They can give you a lot of meat for your money, since adults often weigh 25 pounds. As a matter of fact, the White Embden goose is considered the most efficient poultry fowl for converting grain to meat. A gosling weighing four ounces at birth reaches 11 pounds in ten weeks—a 46-fold increase.

Naturally a 25-pound bird needs more space than a rabbit or a hen, but geese are still within the limits of a suburban yard. One goose or a pair of ducks can be cleanly kept in a minimum area of about 13 square feet. Besides providing them some water for bathing, shade, and a simple shelter from winter winds, the only other investment you need to make to keep waterfowl (besides the optional incubator) is fencing to keep the birds in and dogs and predators out.

Starting a flock

One of the biggest expenses in starting a flock of ducks or geese will be the birds themselves. Day-old goslings can cost as much as $4 apiece, and ducklings run up to $1.50. And while some breeds of adult waterfowl and their offspring need only a little daily grain and water, raising your own young waterfowl is more demanding.

Baby ducks and geese must be kept warm at first, starting out at about 90° F. when hatched, and they must be fed three times a day. The brooder for keeping goslings and ducklings snug needn't be elaborate—a cardboard box and lamp will do if you can keep the box clean. Unlike waterfowl raised by their parents, however, hatchery birds can't swim until they feather out at about six weeks of age. That means you'll have to keep them away from water for that long.

Commercial chick starter feed is unsuitable for young waterfowl. The medication in it can harm them, and the protein content is too high. It will encourage rapid growth in the young birds' wings and legs and can deform them. A good homemade duckling and gosling ration consists of a breakfast of cooked oatmeal covered with a little water, lunch of scrambled eggs and water, and dinner of homemade whole wheat bread covered with water. The hatchlings should also have some tender greens and no whole grain.

If you're not especially fond of raising baby animals, you might find it simpler just to buy a young pair of adult geese or ducks. In midsummer,

after the spring breeding season and before the Thanksgiving and Christmas demand for meat, adult waterfowl can be almost as cheap as their young. In Eastern Pennsylvania a mated pair of geese sells for as low as $10; a female duck can be purchased for $2 to $3; drakes cost $4 to $6.

Most breeds of duck are ready to eat at about 10 weeks of age, and geese can be butchered at about 15 weeks. Unless you plan to keep them as breeders, ducks should be slaughtered before 17 weeks and geese before they are 10 months old or their meat tends to be tough. Waterfowl do bring a good price where there is a market for them. They sell for 30 cents to 50 cents per pound liveweight, and about $1.50

per pound dressed.

Ducks and geese can be slaughtered like chickens by beheading them on a chopping block. Geese are often suspended upside down, stunned with a heavy blow, and bled by slitting the jugular vein. The real chore in preparing goose and duck for the table involves picking the feathers, especially the small down feathers. If you don't plan to save them for a vest or comforter, you may want to simply dip the bird in wax. When the wax cools you can peel it and the down from the bird.

The final reward for all this work is a savory roast goose or, perhaps, Pekin Duck. You'll find no shortage of waterfowl recipes in gourmet cookbooks.

Dan Looker

Pigeons

It's hard to believe the pigeon which just left its calling card on your windshield is actually a goldmine of food and profit. In fact, unless you grew up during the Depression, you probably think of pigeons as nothing more than nasty, disease-ridden varmints. During the

1930s, however, many families staved off starvation by breeding young pigeons, better known as squabs, for food. Adult pigeons were useful as rapidfire breeders who subsisted on limited rations, usually cold potatoes and sour milk.

Turning cold potatoes into gourmet meat

In later years, as the livestock industry lumbered back to its feet, pigeon-rearing became a rarity. Folks were more interested in eating chicken, a domesticated bird less tasty, but more fleshy, than squabs. Still, many gourmets sought out the delicacy in New York City markets, paying as much as $4 per pound for squab.

Today, pigeon breeding is one of the most inexpensive and convenient projects imaginable. As a USDA brochure pointed out, "Squab growing can be carried on extensively by those who cannot keep livestock of any other kind. There's nothing easier to raise. . .or better to eat." Indeed, as inflation climbs, there may be an upsurge in pigeon breeding for food

once again. After all, there are few types of livestock which can subsist on cold potatoes and milk. In the long run, pigeons may be one of our most valuable natural resources.

Buying birds

The initial investment for breeding pigeons can range from $150 to $300, depending on the type of feed and the number of pigeons you purchase. You won't have much equipment expense, since pigeons can be housed in a garage, shed, or barn, and fed from an easy-to-assemble feeder. Medicinal costs are marginal.

The foremost expense is the birds. It's good to start with four or five mated pairs, which average anywhere from $15 to $30 per pair. Be careful in making your selection; healthy pigeons are capable of pro-

ducing about 10 squabs per year for at least five years. With good planning, you can cover the initial cost of the breeder birds in your first year of operation. A simple formula is to use half the squabs you raise for family food. The other half, somewhere between 20 and 25 young pigeons, can be sold at the going market price, which hovers around $4 a pound. Since squabs only weigh 1½ pounds at the most, you won't make a fortune the first year, but you will pay for the adult breeders.

It is important not to buy diseased livestock. Watch for pigeons with swellings on their necks or at the corners of their mouths—that's a sign of canker disease. As one farmer cautioned, "It takes the same amount of feed to keep champions as it does to keep duffers, so you might as well

(Continued on page 340)

(Continued from page 339)

get the best." Some good breeds are White Carneaux, White and Silver Kings, and Squabbing Homers. Make sure your pigeons are classified as utility (for breeding and food) rather than for show (strictly for exhibition purposes). For more information on where to purchase mated pairs, contact your local feed store or write to the *American Pigeon Journal,* Warrenton, MO 63383.

Finding a place for your pigeons to sleep shouldn't be difficult. A garage, shed, or chicken house will be fine, as long as you allocate about 4 square feet of floor space per pair. The shed should have a cement or solid wood floor for the easy removal of guano for the garden. Adjacent to the pigeons' breeding quarters there should be a flying pen, measuring at least 10 feet wide, 7 feet high, and 15 feet long. The use of one-inch mesh wire for fencing will keep other birds out. The ambitious breeder can even build a narrow plank walkway around the top of the pen for the birds to strut and sun.

Inside the pigeon shed, you'll need to set up nest boxes for breeding. Each mated pair will require two nest boxes, one for the eggs to be hatched, the other for newborn squabs. Adult pigeons have a continuing cycle of reproductivity, mating and producing two eggs which hatch in 18 days, so there's always a new batch of eggs to be set while the young squabs are still being nurtured. These nest boxes can be constructed from egg or orange crates stacked together, and covered at the

bottom with a piece of plywood. Pigeons will use any materials at hand to build their nests, but prefer pine needles, straw, or tobacco stems. Some farmers invest in nest-bowls, which cost about 11 cents and can be inserted in the base of the boxes. After five months or so, the entire nest can be removed and planted as fertilizer in the garden.

Be sure to check the pigeon eggs about seven days after they've been laid to make sure the eggs are fertile. A simple test is to hold each egg up to the sunlight: fertile eggs will show a mass of red veins radiating from a red nucleus, while infertile eggs show only a hint of orange yolk. Destroy the infertile ones.

At 28 days of age, you should separate the squabs for either butchering or breeding purposes. The young pigeons which will be used as breeders should be allowed to nest in a corner of the shed, growing over the summertime, and they'll mature enough to breed by the following spring.

Feeding the flock

While adult pigeons can survive on potatoes, it's better to stick with commercial or homemade mixtures of grain to raise healthy squabs. Commercial feed made from corn, peas, milo, and other grains sells for about $9 per 50-pound bag, and will

Indian Fan Tail squabs not only taste good, they're handsome as well.

last a mated pair at least three weeks. Some economy-minded breeders prefer to buy chicken feed (or pigeon pellets) made from soymeal and cornmeal, which costs less (about $5.50 per 50-pound bag). If you feed your pigeons from grains produced in your own backyard, it's best to feed equal portions of whole corn and wheat, plus a supplement of some type of peas and other small grains. Peas are especially important since they contain a good amount of protein.

During the first weeks, you won't have to worry about feeding the squabs at all. The parents do that job themselves, passing on a creamy substance called pigeon milk which

These White King squabs look rather scrawny, but they'll be ready for slaughter at only four weeks of age.

Pigeons will breed and roost in small boxes, but they need some room for flying.

is produced in the intestinal tract. If you decide to breed a pair of young squabs, you'll need to provide feed after the seventh or eighth week.

Pigeons should be fed twice a day. You can purchase a self-feeder for $5 at the feed mill, construct one from plywood, or use a tray. If feeding from a tray, limit the quantity of food to what the birds can clean up in 20 minutes.

For healthy pigeons, you'll need to keep their sleeping quarters clean, and scrape the guano from pen floors at least once a month. Their water should be fresh, and the nests should be cleaned every five to six months. It's a good idea not to feed your pigeons scraps from the table, since they'll fill up on bread and other starchy foods which don't contain the nutrients essential to growth.

Young pigeons should be butchered at 28 days of age, or around the time the feathers on the side of the body under the wings have opened and matured. It's best not to let the adult pigeons feed the squabs just before butchering. Cut off the head with a sharp hatchet or heavy butcher's knife, bleed the carcass, then pluck the skin dry. Don't scald the skin to remove the feathers. The meat can be dressed and then packed for market.

Although afficiandos swear that squab meat is richer and more tender than chicken, it's doubtful that pigeon will ever be as popular as its larger cousin. Chalk it up to poor press.

Tracy DeCrosta

Quail

They don't take much space, much feed, or much cash, and they'll provide you with eggs, meat, and sweet music. That's why you should consider raising quail, even if you live in town. You can be a poultry keeper without ever having to look into the mean and stupid face of a chicken.

The variety of quail available for domestication includes native-American breeds or the imported Coturnix. Whether the slightly wilder natives or the Coturnix (also called the Japanese quail) are more productive to raise depends on which quail farmer you ask. The Rodale Organic Gardening and Farming Research Center raised Coturnix quail for study, so here's (almost) everything we learned.

In praise of Coturnix

Coturnix quail come from an old and aristocratic family. They have been raised in Europe, Africa, and Japan for centuries, and the Coturnix is the bird pictured in Egyptian hieroglyphics. There are several breeds, denoted by their markings; the Pharoah is the most common. The Coturnix's appeal is based on three features: their rapid maturation, their amazingly low feed to egg ratio, and their small size.

As productive and quickly growing poultry, Coturnix can hardly be bettered. They start laying eggs when they are only 50 to 58 days old, and continue to lay an egg a day for at least one year. The males are mature and ready to be eaten at 6 to 8 weeks. By comparison, the average chicken will be 22 weeks old before she lays and 7 to 10 weeks old before she goes to market. Other quail species are also slower than the Coturnix: the Bobwhite, for instance, doesn't lay until she is 10 months old, by which time the fertile Coturnix has produced four or five generations.

More surprising than the rapid maturation, however, is the efficiency with which a Coturnix hen converts protein. Her speckled brown-and-white eggs are almost identical in flavor and nutrient content to chicken eggs. Yet a quail hen requires only 1.92 pounds of food to produce a pound of eggs, while a chicken requires 2.99 pounds of food to lay a pound of eggs. That's a substantial, money-saving difference.

The emperor's favorite dish

Quail raised for meat will require slightly more feed than their chicken counterparts. But the meat has acquired a gourmet connotation and many chefs say its flavor is more interesting than chicken.

A mature Coturnix hen is slightly smaller than a grapefruit, and will thrive with only one square foot of space (about one-fourth the area a chicken requires). But they will survive and lay with as little as one-quarter square foot apiece, so they can easily be raised in a backyard or even on a back porch.

Quail have another advantage as well. Neither domestic nor imported quail manure is as malodorous as chicken droppings. Simple collection and cleaning techniques will control

(Continued on page 342)

Coturnix quail can live and flourish with only one-quarter square foot of space each.

(Continued from page 341)

any odor so that quail can be good neighbors, even in close proximity to their keepers. As does all poultry manure, their's composts valuably.

In Japan, Coturnix have been kept since the 11th century for their ornamental value and sweet song. Their name is onomatopoetic for their soft, throaty call. The American bob-white male makes the lovely "bob-white" whistle. Any type of quail is a soft-spoken neighbor.

Setting up

If you're attracted to raising quail, you should visit a game bird farm or hobbyist to see the equipment, space, and maintenance the birds need. Bear in mind, however, quail may be raised in a variety of circumstances. Here are the salient aspects of their incubation, brooding, care, and slaughter, based on a small holding of 20 hens.

Fertile eggs cost approximately $2.40 per dozen, chicks sell for 50 cents each, while adult birds are $4.50 each. Figuring on a 60 percent hatch rate with good incubation, and that half of the chicks will be roosters, buy 70 eggs to produce 20 mature hens. Patronize a reputable breeder and if possible, examine the parent stock to assure healthy, productive birds.*

---

*Anthony Mazziotta, co-manager of the Trexler Game Park near Allentown, Pennsylvania, recommends buying adult birds to avoid incubation and brooding on your first poultry attempt. Buying sexually mature birds means you can buy the number of hens and cocks you need—for instance, 20 adults instead of 70 eggs.

Store the eggs at temperatures between 40 and 60° F. before incubation, with the small ends down at a 45-degree angle. Turn them twice a day. Delaying incubation more than a week will decrease fertility. If you are experimenting with poultry for the first time, you may want to send the eggs to a hatchery for incubation. That will save your investing in an incubator before you commit yourself to poultry.

There are two types of incubators: still-air and fan-ventilated. Quail keepers usually prefer fan-ventilated equipment for better temperature and humidity control. These cost more but give a better hatch rate. Small still-air incubators start at about $25; the preferred fan-ventilated models are a bit more, and can include a mechanism to turn the eggs every hour if you want to pay for it. You may need to stagger your egg purchases to fit the incubator.

With either type of incubator, the inside temperature from the first to the fourteenth days should be 99 to 100° F., and 98.5° F. from the fourteenth to the eighteenth days. Humidity for the first two weeks should be about 60 percent (wet bulb, 87), and 70 percent (wet bulb, 90) for the last few days. Follow the incubator's instructions carefully to assure these levels, and test it for a couple of days before you set in the eggs.

The eggs must lie flat in the incubator, not with the points up. Don't touch them for the first 60 hours. After that, turn them (or have the incubator turn them) approximately every 4 hours until the fifteenth day. A schedule of 6 A.M., 10 A.M., 2 P.M., and 10 P.M. is good to the birds and humane to you. A pencil mark on the side of an egg reminds you which side is up. As the embryos grow, they produce carbon dioxide and require oxygen, so open the air vents after eight days.

The eggs should hatch in about 16 days. Some quail keepers put a layer of cheesecloth under the eggs just before hatching so the chicks can stand up more easily. Leave them in the incubator 12 to 24 hours until they are fluffed out and dry before putting them in the brooder. Wash out all down and dirt from the incubator and clean the interior and the grid with a slight chlorine solution.

The brooder should be close to 95° F. the first week and decrease in

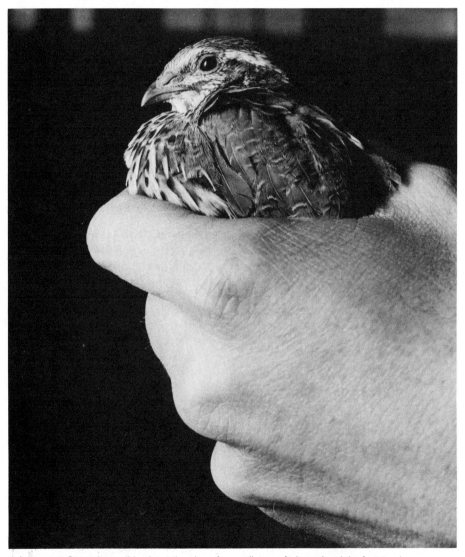

A full-grown Coturnix quail is about the size of a small grapefruit, and weighs four to six ounces.

Under normal conditions, quail eggs hatch in about 16 days.

This incubator has a thermostat and a timed, automatic egg turner.

A newly hatched Coturnix quail is about the size of a bumblebee. The young birds need to remain in the incubator for a day or so until they are fluffed out and dry.

temperature by about 5° a week for the next three weeks. Take the temperature at floor level. The best guide to the chicks' comfort is their behavior. If they're huddled up by the heat source, they're chilly; if they drink frequently and skirt the edges of the brooder, they're too hot. They may also tell you they are cold or hot by chirping or piling.

Besides protecting the chicks from lethal drafts for the first week, you must keep them from drowning themselves. They can do it in as little as ¼ inch of their drinking water. Put marbles in their shallow dishes so they drink from the troughs of water between the marbles. Remember that you are dealing with poultry—and like many creatures they must be protected from their own stupidity. Clean the water jar and marbles with plain water at least daily.

Sprinkle the chicks' food on the brooder floor covering for the first week or so to encourage them to eat, and keep food present all the time.

Use shallow feed dishes filled about ½ inch deep with starter mash. Quail are messy, wasteful eaters. After the first four days or so, put a wire mesh over the mash. They'll pick through the wire and waste less.

Cover the brooder floor with litter to absorb moisture and dilute droppings. Vegetable fibers—sugar cane, dried and crushed corn cobs, peanut hulls, or pine shavings are especially good. (Caution: the dust from *hardwood* shavings may cause nasal and optical inflammation.) For the first week put about two inches of litter on the brooder floor and cover it up with burlap to prevent the chicks from eating the litter.

Watch them closely during brooding. They'll be doing well if they are distributed around the floor with some of them sleeping on the litter. At the end of three weeks, chicks will be half grown and ready for their permanent cages. In another three weeks, you can distinguish the sexes and slaughter some of the males.

Growing up

Coturnix quail are dependent on even temperature and long periods of light for best growth, sexual development, and productivity. It is wise to move the infra-red light from their brooder to their adult cages for heat and light during the fourth to sixth weeks until they reach maturity. Whether the birds are inside or out, constancy of temperature is important to good egg production. Any temperature above 32° F. will do.

Their light requirements decrease from 24 hours a day for the first couple of weeks in the brooder to 8 hours a day until they reach sexual maturity. The hens will then require 14 to 17 hours of light a day for maximum egg production. The light doesn't need to be very bright. If you can read by it, the quail will mature by it.

While lack of light retards their sexual development, it doesn't affect

(Continued on page 344)

(Continued from page 343)

their growth. Therefore, quail cocks that you might want to fatten for slaughter should continue receiving only eight hours of light a day. They will remain sexually immature, and won't waste their energy fighting and mating.

No one knows a 100-percent accurate way to determine a quail's sex before maturity. However, at about two weeks of age most birds will display a sexually distinctive color difference on their breasts. Males are expected to have brick red breasts and hens to have mottled gray. But some birds will show so little color that you can't distinguish them.

At six weeks and sexual maturity, your quails' sex may be determined if you look between their legs into their vents. According to Albert Marsh's *Quail Manual* (see bibliography at end of chapter), the "cock's vent is reddish and points forward. The area behind it is muscular, about the size of a marble. The hen's vent will be bluer, larger, and the opening will be centered. Sometimes you will feel the next egg under the surface. If you notice that a bird has laid an egg, you may assume that bird is a hen."

Small, all-wire cages can be hung on a back porch, or in a garage or shed.

This large cage has a brooder and plenty of room for exercise, but netting must be spread near the top to insure the birds don't break their necks when flying upward.

If you want fertile eggs, plan to keep one male for two or three females. If the birds are in colony pens, one male for every three to five hens is sufficient and will reduce cock fighting.

Food and shelter

When the birds are sexed at five or six weeks, the hens should start eating a layer mash. You may add thrifty supplements of dried greens or

a 15-watt bulb over their cage during the summer to attract mosquitos and moths. The males will plump up on a cheaper, higher-carbohydrate diet which is suitable whether they'll be slaughtered or fertilizing eggs. They will reach their mature weight of ½ to 5 ounces by ten weeks.

The feed should be purchased fresh and stored in covered containers in a cool, dry place. Feed stored longer than eight weeks is subject to vitamin deterioration and rancidity, especially in the summer.

Nothing about quail raising varies as much as their housing. They're adaptable to several arrangements— hence their attractiveness in semi- urban or even urban circumstances. Different cages have different advantages. The more compact cages require a little extra care to keep the birds healthy.

To determine which is best, the Rodale Research and Development Division tested four types of housing units: a tentlike cage over shrubbery, a long, narrow cage-brooder combination, small battery cages, and a large indoor cage built onto an existing structure. The tests showed no significant difference in the birds' weight gain or feed consumption from one cage-type to another. Egg production dropped in the outdoor cage, however, due to fluctuating temperature and lack of lights.

Commercial wire cages require the least amount of space. They are portable and can be set up anywhere in half an hour. Some come with wall brackets for easy hanging. Small units housing four to eight birds start at about $12. These cages usually have rolled paper which pulls underneath the birds for easy cleaning. A small set-up like this can fit in a spare room or on a back porch in warm weather, as well as in a garage or shed if some sort of temperature control is possible.

Instead of buying a cage, you can build one yourself of light lumber and plywood, hardware cloth, and nylon netting. A floor pen or corral can be put in a spare room or garage. Suggested dimensions for 20 hens are 2 feet wide by 10 feet long—a runway design that gives the birds plenty of room for exercise. A brooding compartment as part of the cage would add 9 square feet. A floor pen is inconvenient for cleaning and egg gathering, however. The easiest design for these tasks is one in which the birds walk on a wire cage bottom. The wire makes it easy to collect their droppings on a tray or paper below. Some breeders claim quail lay better on wire, up off the ground.

Rodale researchers designed a portable homemade cage that can be kept outdoors in warm weather.

It's on legs to facilitate egg collection and cleaning, it has a wire floor, and a brooding or shelter compartment. Measuring 3 feet wide, 7 feet long, and 3 feet high, it accommodates 20 birds and another 20 in the box. It took 24 hours to build. (See instructions on page 346.)

Most birds fly upward when they're startled or frightened. If your birds are in a cage higher than seven inches, they can gain enough momentum in their upward flight to break their necks when they bump the top of the cage. Prevent this by draping the cage with nylon netting, mosquito netting, or burlap—anything with enough play to keep them from braining themselves.

If you're raising Coturnix as game birds instead of layers, you can construct a tentlike cage over shrubbery. In an area 6 feet wide by 10 feet long by 5 feet high, 20 mature birds will have enough room for flight conditioning. Brooding must be indoors, however, and the birds should be inside if the weather goes below 32°F. The ground in their cage should be covered with vegetable fiber to prevent their collecting crippling mud-balls on their feet.

These procedures may seem numerous and tedious, but don't be misled. With good equipment and planning, caring for several dozen hens shouldn't require more than ten minutes or so, morning and evening.

Quail are generally hardy. Start with healthy stock, give them proper feed and enough space, keep their cages and equipment clean, control their lighting and temperature exposure, and problems are unlikely. However, trouble might visit in the form of cats, raccoons, or other predators with paws small enough to reach through the cage wire. Some breeders prevent this by putting an eight-inch high piece of plywood along the lower sides of a raised cage and positioning the cage so it can't be jumped on from any nearby perch.

Then there's the chance that the birds will become cannibals, turning on one another. It is easily preventable by putting enough protein in their diet and avoiding over-crowding. Or better yet, allow enough space for the birds to race around a bit—in the runway cage, for instance. Quail are territorial animals and may attack newcomers. Therefore, don't mix any new birds with an established community unless you put all the birds in a new cage.

Quail like to take dust baths to cool off, especially in hot weather. This discourages lice, so put a wide, shallow pan of dirt in their cage.

Inbreeding will rapidly reduce the quality of your quail stock. If you hatch your own fertile eggs, keep careful genealogy records to prevent abnormalities and reduced reproductive performance. For the small family holding of 20 hens, it is probably easier just to buy new stock. Also, use care in returning to the same breeder. One raiser suspects that all the quail in his area are already cousins.

If the cage should be accidentally left open and your stock wanders out, one quail breeder recommends this method for getting them back. Catch one bird and put it in something like a Have-a-Heart trap or an open restraint. Leave the gate ajar and the rest of the flock will likely return on its own.

In some communities, keeping fowl in a residential area may be prohibited by zoning or health regulations. For instance, an Emmaus, Pennsylvania, ordinance would seem to make raising quail emphatically illegal in town, but closer examination showed the real objections were to the construction of visible pens or additional buildings for animals. As Emmaus residents John and Mary Nemeth have discovered, quail live in so little space and are so odor-free (when well tended) that no neighbor has complained. The Nemeths have raised quail for ten years in their suburban backyard, and their neighbors have grown fond of the birds' music and their useful manure. The limited space requirements of quail may make it possible to keep them within city limits when chickens would be absolutely prohibited.

It is difficult to verify that quail and game birds have more interesting personalities than chickens. When you talk to quail owners, however, you may notice they often prefer one or another breed for such imprecise qualities as "intelligence," "camaraderie," or "conversation." Making these observations about quail owners is easier than explaining them.

Reaping the rewards

Quail are ridiculously easy to kill. The only obstacle is a possible Bambi complex prompted by their diminuitive size. If you can't slaughter your own for this reason, you are not alone. Many quail keepers send them off to be killed and dressed. Once that sympathy is conquered, however, all you must do is snip off the birds' heads with pruning shears or scissors. Working with another person, one of you can hold three birds in each hand, tipped down over a container, while the other snips off the heads. When the birds quit flapping and bleeding, dunk them in scalding water and they're ready to pluck. Some people claim that plucking is difficult on such small birds and prefer to skin them, but plucking is fairly easy if they are scalded in water that is close to 148°F.

After plucking and scraping off the pin feathers, wash the carcasses and drop them in chilled water. The quickest method for "drawing" a bird this small is to cut open its back with scissors and draw out the entrails. Drain the carcass, and the meat is ready for your recipe. Dressed cocks should have about four ounces of meat, and hens, about an ounce more.

Quail eggs are the size of giant marbles—five of them equal an average chicken egg. You can cook or bake with equivalents to chicken eggs, or show off their tiny size with pickling. They make a handsome bite-size hors d'oeuvre. They are also attractive served hard-boiled (five minutes) because of their speckled shells, of which no two are alike. Since the quail egg is tougher to open than a chicken's, crack the egg and then use a knife to penetrate the membrane.

Quail meat is just coming to popular attention in America, which shows how much we have to learn. It is a delicacy frequently enjoyed in many other parts of the world, including at the Japanese Prime Minister's ceremonial dinners. It is very similar to dark chicken meat, but a little bit gamier—a description that is useless until you've tasted the meat. Julia Child notes that while chicken and Cornish hen may require marinade in some recipes, quail does not. The

(Continued on page 346)

(Continued from page 345)

birds are so small that two may be considered a dinner serving.

To prepare your birds for the table, roast, fry, stuff, or stew them. Wild game cookbooks provide countless recipes. Here are some more:

Pickled Quail Eggs

12 boiled quail eggs
1 small onion, diced
¼ cup vinegar
½ cup water
  dash of pepper

Place onion, vinegar, water, and pepper in a pan and heat until boiling. Place shelled eggs in a bowl and completely cover with the pickling solution. When the eggs and mixture have cooled, cover the bowl and refrigerate overnight. Store in refrigerator in a glass jar.

Quail Stew

12 dressed quail, water to cover
1¼ pound potatoes, sliced
½ pound carrots, chopped
1 cup celery, chopped
¼ pound onions, sliced
½ clove garlic
  a dash of pepper

Pre-heat oven to 350°F. Cook quail until it is tender enough to be deboned easily. Remove quail from broth and put the broth aside. Separate the meat from the bones and combine it with vegetables in a casserole. Combining pepper, garlic, and quail broth, pour this mixture over casserole. Roast at 350° until tender (about 1 hour).

Yield: serves 8

Sources. Here are a few of the many sources for Coturnix quail eggs, chicks, and equipment. You may find breeders in your own area through pet stores, the county agent in the state agricultural extension service, or possibly the state game bird commission.

Marsh Farms, 14228 Brookhurst Street, Garden Grove, CA 92643.

Ohio Quail Co., Box 117, Xenia, OH 45385.

Ronson Farms, 2019 Wyandotte Road, Columbus, OH 43212.

North Carolina Game Birds, 612 Roundtree Road, Charlotte, NC 28210.

L & L Pheasantry, R.D. 2, Hegins, PA 17938.

McClellans, 4809 Leslie Avenue, Washington, DC 20031.

Iota Quail Farm, Poplarville, MS 39470.

Stromberg's Pets, P.O. Box 717, Fort Dodge, IA 50501.

Plans for a portable quail cage

Here are instructions for a cage three feet wide, seven feet long, and three feet high, which will provide a brooder and living space for 40 quail. The tools are few: hammer, screwdriver, drill, wrench, saw (hand or electric), and staple gun.

Procedure. Saw the two 14-foot 2 x 4s in half to get the four 7-foot boards that make up the lengthwise sides. Two of the 12-foot 2 x 4s should also be cut in half, giving you four 6-foot 2 x 4s for the legs. Connect the sides to the legs using a ship lap joint, nailing the tops and bottoms from the side after you ship lap all the pieces.

Although the width of the cage is 3 feet, the inside of the cage is only 33 inches. Therefore, cut the 12-foot piece of white pine into 33-inch segments and use them for the tops and bottoms across the width of the cage. Nail these cross-pieces to the frame. Then miter the scrap pieces at a 45-degree angle and nail one at the side of each leg to brace it.

Next, build in the brooding box at one end of the cage by nailing four* of the 3- by 3-foot pieces of

*The plywood making the end of the cage is not to be nailed. See following paragraphs

MATERIALS

Quantity	Description
2	14-ft. 2 in. x 4 in. (or 2 in. x 3 in. if cheaper)
2	12-ft. 2 in. x 4 in. (or 2 in. x 3 in. if cheaper)
1	12-ft. piece of 1 in. x 3 in. white pine lath
6	3-ft. x 3-ft. piece of indoor or outdoor plywood, ⅜ in. thick.
4	2½ in. loose pin hinges
1	9-ft. long piece of white pine lath, 3 in. wide
2	16 in. long, 1½-in.-wide strips of white pine
1	7 ft. x 3 ft. piece of ¼-in. hardware cloth
45 sq. ft.	½ in. wide nylon netting
1 box	staples
1 lb.	16-penny nails
1 lb.	6-penny nails
1 piece	Scrap wood, about 1 in. x 3 in.
8 each	no. 10 and no. 6 wood screws
4	3-ft. pieces of 1½ in. wide x ⅜ in. thick plywood (optional)
½ gal.	paint (optional)

plywood to the frame with 6-penny nails. First, nail in the plywood facing the living space of the cage. Cut a small door in it 9 inches high and 6 inches wide. Nail the two strips of 16-inch-long pine on either side of this

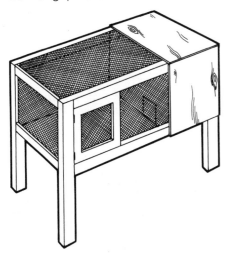

door so it can be made to slide up and down. When lifted, the door will allow the chicks to move in and out of the brooding box.

Next, make the 3- by 3-foot plywood pieces in at the end of the cage into a door by attaching it to two hinges screwed to the upper leg of the cage. To make the door more airtight, you can screw in four 3-foot-long strips of plywood around its perimeter on the outside (optional). (On the inside of the brooding box, nail ¾- by ¾-inch strips of scrap plywood around the top and bottom perimeter in order to keep this back door of the brooding box from being pushed all the way in.)

Nail the third and fourth 3- by 3-foot plywood pieces in as the sides of the brooder box and use the fifth sheet as the top of the box. Like the rest of the cage floor, the bottom of

the brooder compartment will be made of hardware cloth stapled to the frame. To insure heat retention (and allow for removal of wastes) nail some plywood scraps to the long sides of the cage ½ inch or so below the actual floor of the cage so you can slide the sixth sheet of plywood in and out.

A third door, measuring 21½ inches by 31½ inches, should be built for the length of the cage. To make this, cut the 9 feet of white pine into two 21½-inch pieces and two 31½-inch pieces. Miter these sections and using eight screws, one at each side of each joint, fix them together making lap joints. Then hinge the door to the frame of the cage at the end of the side, at the maximum distance from the brooding box. At this point, you may want to paint the cage.

The backyard turkey and beyond: Essex, Massachusetts

Len Putonen flies jets for Eastern Airlines three days a week, and the rest of the time he's home at Boundary Farm raising and processing turkeys. Len sold 2,000 dressed or cooked birds in 1979, so his operation is something more than a backyard thing. Len knows his birds, having raised chicks for a local poultry firm. As a boy, he worked as a turkey plucker. He doesn't really do anything different from the backyard turkey-raiser—he just does it more often.

The turkeys at Boundary Farm are the popular broad-breasted white, of the Rose-A-Linda variety, air-freighted from Jansen's Turkey Farm in Zeeland, Michigan. White-feathered birds traditionally fetch more money than bronze or other feathered birds. Dark feathers, when plucked, leave behind small, dark pluck spots; white feathers leave behind small, light pluck spots.

Raising 300 turkey poults at once, as Len Putonen does, is the same in some respects as raising half a dozen. First, you've got to make sure your poults are smart enough to recognize their feed.

(Continued on page 348)

TURKEY NOTES

If one of the pleasures of turkey raising is seeing the birds grow and develop, the way you pay for that pleasure is in feed. The ratio of feed to weight gain in turkeys is about three to one. Therefore, to raise a turkey to 18 pounds you are going to need at least 54 pounds of feed. At the 1980 feed price of 14 cents per pound, in hundredweight lots, food for one bird will cost about $7.50 (in addition to the initial $2 to $3 cost of the young turkey). This feed to weight-gain ratio will increase as the temperature drops; the birds have to change feed energy into heat and keep themselves warm.

Turkeys like to forage for insects, but also have an eye for ripe vegetables.

One of the old complaints against turkeys, as people moved closer together and backyards began to shrink, was that they would leave the insects and head for the corn. For this reason, you'd do well to learn something about fence building. Turkeys can climb fences, and they can fly, unless one wing has been clipped. So you'll have to keep an eye on them.

The coop is a fact of life in turkey raising. You've got to have some sort of coop, whether it be a converted tool shed, a canvas-covered shelter on poles, or the traditional wood frame building with the low doorway and the high windows. Eight square feet of space is generally satisfactory for one turkey. A six-bird flock could, therefore, be raised in a coop that's roughly six feet wide, eight feet long, and four or five feet tall. The main predators you'll have to worry about are dogs and kids.

Phil Levy

(Continued from page 347)

Poults have been known to stand on their feed, not knowing what it was, and starve. On the other hand, Len says that he fed his first batch of 300 poults by hand, dipping each beak in water and dipping it in the feed. But when the second shipment of 300 arrived soon after, he figured at this rate he'd be dipping beaks all day. So he just dumped the second batch in the coop, with water and feed, and let them train themselves. The second batch had a better survival rate than the first.

On the first floor of the main poultry building at Boundary Farm, the tiny, nearly bald poults are placed within two small "rings," areas about ten feet in diameter, surrounded by a low, chicken-wire fence. The floor is covered with wood shavings and there's a watering trough in the middle and feed bowls around the outside. Lights are kept on at night, so the poults can find water and feed if they need it. The young birds sleep close together, as they will even after feathers develop if it gets too chilly.

Upstairs at Boundary Farm is a group of 400 tom turkeys, living on a wood-shaving covered floor 30 feet by 110 feet. These older turkeys stand or sit quietly in the bright winter sunlight. Len raises turkeys year-round, and the windows stay open year-round as well. The ammonia wafting off decomposing manure would be enough to knock a person out if the coop were unventilated. (Chicken manure is much drier and therefore less smelly.) The ventilation at Boundary Farm is what you'd call natural draft, and it keeps the coop fresh, if a bit cool. Because turkey manure is so wet and because the wood shavings have to be changed so often, the manure is worth relatively little. Too much time is required to sift the droppings from the wood shavings.

Turkeys are high-strung, and they will stampede if frightened. But the calm in the upper floor of the coop is broken only mildly, when one of the tom turkeys suddenly struts a few paces, spreads his feathers, and lets his snood spill down over his beak in the classic mating posture. The snood is that mass of reddish cartilage protruding from the tom's forehead. The snood protrudes, the feathers ruffle, but even if there were

some hens up here it wouldn't do the tom much good. The modern turkey hen has been bred for her exceedingly broad breast—so broad that the tom turkey cannot fully envelop her in his wings as he tries to mount. As a result, nine out of ten turkeys now are conceived artificially.

Large-scale slaughter differs slightly from the small-flock method. The process which will involve several hundred birds over a one or two week period, starts with a line of funnels. The turkeys are placed, upside down, into these funnels with their heads protruding. The funnels may be attached to an overhead trolley system, which moves the birds up to the severing point and beyond to plucking and evisceration. Machinery can also be used in the plucking stage. The Ashley Sure-Pick machine, which looks like a large washing machine, can be used to soften the feathers, and then remove them automatically.

Once you've got the facilities and the birds, and you've raised them and slaughtered them and dressed them, then what?

The modern "traditional" farmer, such as Len, no longer has to fill his wagon with turkey carcasses, hitch up a team and drive to town before sunrise to haggle on the curb with the buyers from the big meat markets. No more arguing on street corners about price while the turkeys begin to warm and ooze in the morning sunlight. Now, thanks to the automobile, the customer is willing to come to the farm to do his shopping. This is how most small turkey farmers sell their poultry—right from the farm. And the local people don't seem to mind the drive to Len's farm.

The more turkeys you sell directly, the better it is for the old cash flow. Len sells everything in his own farm store. His wife Gwen does most of the cooking. He's got a double-door South Bend oven, propane fired, just ten feet from the counter. On a weekend, as the customers crowd in, there are turkeys boiling on the stove, to be used in turkey pies; turkeys roasting in the oven, to be sold whole; and dressed birds hanging in the freezer. There are also fruit pies and apple cider (from the man who lives across the road, an aeronautics engineer).

Phil Levy

Pigs

The neatly hand-lettered sign read "No Trespassing." Naturally my friend and I immediately scrambled over the fence. That summer we had perfected our pitching skills. We settled on a large black rock as a target, which was fine until my friend scored a bulls-eye. The rock grunted, lumbered to its feet, and gave chase. It was the meanest, largest (and only) hog I had ever seen at close range. We narrowly escaped by leaping over a barbed wire fence. The only casualty was my shirt tail, pieces of which may still flutter on the fence.

The experience of raising homestead hogs is far less disastrous than encountering an angry sow in a forbidden field. However, there is *one* similarity: in either venture, unless you're careful, you stand to lose your shirt. While I lost mine to a hog in the most literal sense, naive homeowners may lose their shirts financially while raising one or two hogs to stock the family freezer. You can't expect a windfall of profit from rearing just one pig. All totaled, the cost of feed, labor, and equipment will probably equal, pound for pound, the price of pork found in the grocery store.

But that doesn't take into account the experience you gain, and the satisfaction that comes from eating pure, tasty, home-grown meat. Anyone who has ever eaten the tasteless and textureless product passed off as pork in the supermarket can understand the bonus of rearing and butchering a pig.

The time and labor involved in backyard animal raising can also be an asset, especially if you have children. A man in Phoenixville, Pennsylvania, says that caring for few pigs and chickens has helped his children develop a sense of responsibility. No matter how tired his son may be from basketball practice, he's learned to feed the animals before doing anything else, including feeding himself. "The pigs aren't going to understand my son's excuses for tardiness," the farmer said, "and neither am I. It's good to have the animal's well-being rest entirely on a child."

If you don't grow the food, feeding one or two pigs can be downright expensive. "Raising pigs (on a small scale) is just not a paying proposition unless you have a cheap source of feed," concluded one farmer. Large-scale hog farmers side-step enormous feed costs by either growing their own grain or purchasing grain in ton-lots. For the grower with limited land, however, the first option is virtually impossible. And, the second option is impractical, since one or two hogs won't eat thousands of pounds of feed.

There are a few ways that homesteaders can offset these financial setbacks. If you raise your pig or pigs in the summertime, free vegetables can be culled from the garden. And you won't need to spend several hundred dollars on a self-feeder, since you can devise less-expensive, time-saving devices such as troughs and homemade fountains for watering the pigs.

Home Food Systems *called on the Stevic family in Kosovo, Yugoslavia, the poorest of the country's regions. On not much more than an acre of land, the Stevics grow most of their own vegetables, as well as raise pigs and chickens.*

The initial investment

Your cash outlay will depend on the materials you have at hand. If you're starting from scratch, without any supplemental food sources such as a nearby garden or field, you'll probably spend a total of $125 to $175 per pig. That figures breaks down like this: feeder pig at eight

(Continued on page 350)

In the mountain village of Stavrohouri, Crete, a woman on the way home gets a lift from a neighbor. Animals browse on pods beneath carob trees, a regional cash crop.

PIGS AND MULBERRIES

In one of his plays, J. I. Rodale wrote that there were certain words that always make people laugh. "Pig" was one of them. I'm not sure why.

Now I admit I haven't known pigs for very long—just the last three years—and the pigs I know are not close neighbors. They live in the basement of a big barn about a quarter-mile down the road. But on summer evening walks, I visit the pigs. Last year, as my little girl was learning to walk, she joined me. Along the way there we passed a mulberry tree. I had read that pigs loved mulberries and never quite believed it. But out of a sense of adventure and curiosity I picked a large handful and took them along.

The book was right. The pigs loved mulberries. After that our ritual commonly included an offering of mulberries. As I approached their pen, they'd scamper toward us, noses raised in anticipation of the sweet purple berries. They'd paw and pace at the gate, hoping to be the one to squirm into the front row and receive a treat.

This summer, my little girl is a master walker and insists on her evening stroll. She runs to me after dinner yelling "Pig! Pig!" and soon we're heading toward the barn.

Tom Stoneback

No man should be allowed to be President who does not understand hogs, or hasn't been around a manure pile.

Harry S. Truman

(Continued from page 349)

weeks approximately $35 to $40; commercial feed $65; veterinarian and medical bills up to $25; shelter and livestock items from $20 to $50.

Your most important investment will be, of course, the pig. You'll need a young gilt (a female pig that has never given birth) or a barrow (a castrated male). One or two pigs should be plenty for stocking a year's supply of pork. The best plan is to purchase a pig in the spring, rear it during the summer, and slaughter in the fall. That way you can save money by using garden extras, and be spared the expense of building elaborate insulated sleeping quarters for the pigs. A simple A-frame shed will keep out summer dampness.

Young pigs can be purchased from many sources, including community auctions, graded and association sales, feeder pig shows, and local farmers. They may be graded according to standards established by the USDA. The grades are based on two factors—logical slaughter potential and thriftiness. All thrifty pigs, that is, those that will gain weight rapidly and efficiently, are grouped according to their slaughter potential into either U.S. 1, 2, 3, or 4 grade. Unthrifty pigs are labeled U.S. Utility or U.S. Cull.

Your best bet is to try purchasing a pig from a local farmer. Those pigs are probably less stressed and have received less exposure to disease than their counterparts sold at auctions. Local farmers can also give you the details of the young pig's upkeep in the weeks preceding a sale. All young pigs should have been weaned and received iron shots or iron pellets before being sold.

Follow your instincts when selecting a pig. Think back to the time when you bought your first used car, and the dealer convinced you that poor wheel alignment, a rusted body, and bad brakes were stylish. Don't be fooled this time. Examine the animal carefully, taking note of any peculiar characteristics. At eight weeks, pigs should weigh between 35 and 40 pounds. They should possess deep, wide chests, a large skeletal frame with heavy bones, good length, and adequate muscling.

Don't bother to spend a lot of money on a purebreed pig. Base your purchase on the merits of the individual animal, rather than type of breed. Since your gilt won't farrow, there's no need to insure a blood line for years to come. You just want a pig that can guarantee a quantity of good meat.

Before you take your new purchase home, bear two things in mind: in some states, it is required by law that feeder pigs be inspected by a veterinarian before being transported; and it is essential to purchase a castrated male or have the pig castrated immediately. Boars produce a foul-smelling meat.

Housing the hogs

You need two types of quarters for your pigs: a shed for sleeping and a pen or fenced lot for feeding, watering, and manuring.

A simple A-frame structure can provide shelter for one to four pigs and can be made with salvaged lumber, outside plywood sheeting, or tongue-and-groove siding. Inside the sleeping quarters each pig will need about four square feet of space. Straw or wood shavings are ideal for the floor. If you're not interested in constructing a pig shed, commercially built quarters are available, but they can cost as much as $500.

The pigpen, which is really an exercise yard where the pigs eat and drink, can be made several different ways. Homeowners with a great deal of land can let a hog graze in a fenced-in area. You'll need at least a quarter of an acre, however, since a pig will likely turn a small area into a mudhole. Another option is to construct an apron, a concrete floor that stretches out in front of the pig shed and is surrounded by a woven-wire fence. Each pig needs about eight square feet of space on such a floor. The advantages of a concrete pigpen are that the animals won't tear up your lawn, and the manure can be scooped up easily for the compost heap. Such a floor should be sloped so urine can run off.

You can also build a pigpen with a slatted floor of 2 x 4s, spaced $7/8$ inch apart and set on concrete blocks. You can then lift the floor and rake the manure away. And no matter what type of pigpen you choose, you'll need a sturdy fence.

Tending the herd

The price of food accounts for 70 to 75 percent of the total production cost of pork. Large-scale hog farmers have an advantage over backyard growers, since grain is cheapest when sold in ton lots. A hog eats about 700 pounds of feed from a weaning age of about eight weeks until it reaches a finished (or slaughtering) weight of 200 pounds. Feeding much beyond 200 pounds is an inefficient use of food and results in pork that most of us would judge too fatty.

Pigs, like children, need a well-balanced diet. A protein deficiency, for instance, can stunt their growth and make them susceptible to disease. So while table scraps can supplement commercial feed, it's best not to dump every leftover into the pig's trough. Pigs have a hard time digesting high-fiber substances, such as lettuce, cabbage, and the skins of citrus fruits, but starchy foods such as beans, peas, or leftover corn are excellent for a growing pig's diet. You've just got to keep an eye on the type of table scraps your hogs eat.

No matter what feed you use, you need an efficient method to get it to the animals. Commercial self-feeders retail for $118.50, but you can buy a trough for about $12. Self-watering devices can cost as much as $200, but you can get similar results with a 55-gallon drum and a small spigot.

Rearing pigs can be a pleasant experience. Contrary to public opinion, pigs are not filthy animals. In fact, they're very fastidious, and when provided with a place to manure outside their sleeping quarters, they will do so.

At birth, the newborn pig is sweet-smelling and cuddly, and its only interest in life is to nurse mother and find a warm, dry place to sleep. After a week or so, baby pigs become playful and alert. By weaning time, at six weeks, they're so intelligent they can undo the nuts and bolts that hold their trough together. One farmer said his piglets were so industrious they managed to swarm around his feet and undo his bootlaces during feeding.

As pigs grow older, they lose their playfulness, and eating becomes an activity of paramount importance. By slaughter time, the animal seems to have little in the way of a personality. That's just as well, since some owners tend to treat their pigs like household pets.

During the pig's formative months, you'll not only have to tend to its food and water requirements, but also to its health needs. Most pigs have to be treated periodically for worms and insect infestation, particularly lice and mites. Spray the pigs with a light-weight motor oil to smother the insects. Good pen sanitation will help prevent many health problems.

Pigs have only a small number of sweat glands, so it's important to provide adequate shade areas and sufficient water supplies during the summer. Pigs have been known to die from overheating, and a few hog raisers have even installed electric fans to circulate cool air in the pig

(Continued on page 352)

Pigs can fit in a small space, but in order to keep clean, they need some additional room.

(Continued from page 351)

shed. Covering the shed floor with ground corncobs will cut down on dust, which pigs are especially sensitive to.

Providing that all goes well, you should butcher hogs when they tip the scale at 200 pounds. If they go too much over that, the meat will be fatty. The average hog will yield over half its live weight in hams, picnic shoulders, loins, bacon, and shoulder butt. The hog will also possess about 9 to 13 pounds of lard.

Pigs become so stressed by the impersonal handling and butchering procedures in meat plants that they produce an acidic substance in their bloodstream which affects the taste of the meat. Home butchers don't have this problem since the pig isn't moved or badly treated before slaughtering.

Tracy DeCrosta

Rabbits

The perfect homeowner's meat-producing animal would probably be a slightly modified rabbit. Rabbits take little space; their manure is easy to handle and valuable as a fertilizer; they produce both tasty meat and a fur pelt; they reproduce quickly so herd improvement is easy; they're easy to butcher; and they're highly efficient converters of feed into meat.

Little feed = much meat

The slight modification would be to make rabbits ugly and somewhat repulsive. The biggest problem comes when it's time to eat Fluffy. Rabbits are so cute that they quickly become pets, and a great many potential dinners end up as aged, overweight family friends. Get around this problem, and you can be a successful rabbit raiser.

Estimates vary about the size of the ideal backyard rabbitry. Some experts suggest three females (does) and one male (buck), others say four females and a male. Let's work out the mathematics of this second arrangement.

It is possible to harvest 32 offspring from each doe per year, making a total of 128 young rabbits. These rabbits should be slaughtered at a weight of about 4 pounds, giving a total weight output from your rabbitry of 512 pounds annually. Of that, about 65 percent—333 pounds—ends up as dressed carcass, the rest being skin and waste. About 80 percent of the carcass is usable meat. That gives an edible harvest of 266 pounds, or about 5 pounds of table-ready meat a week. Not a bad harvest from only 5 rabbits.

If you think those figures are fan-

A two-tiered cage arrangement, with the cages against a building, makes a good, basic rabbitry without ambitious construction.

tastic, consider that professional breeders often achieve higher rates of production. An experienced breeder can expect a good doe to be re-bred two weeks after birthing; some will do this as soon as two days after birthing. In the latter case, as many as 100 offspring are possible per year, meaning a yield of 400 pounds of meat per doe. (Consider: a cow often yields only one 400-pound calf per year.)

No other animal produces so much from so little. To keep up that production you need four things: good stock, good housing, good diet, and good management.

Getting good stock

In the past 20 years, several new, improved breeds of rabbits have been developed. There are more than 75 varieties of rabbits registered, each seemingly bred for a particular trait. Flemish Giants often reach weights in excess of 20 pounds, but they take a long time and a lot of feed to reach that weight. Angora rabbits can produce almost 2 pounds of high quality wool a year, but they have very little meat on their bones. For straight-out meat production, New Zealand Whites and Californians are the best. They are both medium-sized rabbits, with extremely fast growth when young. Their carcass is small boned, with a high ratio of meat to bone. Either breed will reach a weight of 4 to 4½ pounds by the time they are weaned from the mother (about eight to ten weeks). This means you can harvest your meat without having to feed the offspring expensive feed, and you'll get the maximum meat output for what you've put into these rabbits. That's why rabbits are such good meat animals—at their peak of efficiency, they are at their peak of quality.

You can buy your rabbits from a breeder or from a commercial operation. Mature animals from good stock will cost $10 to $15 each. Where you get your stock should depend on what you want to do with your rabbits. If you want just meat production, go to a commercial raiser and buy young rabbits. If you ask for "proven stock," you may just be buying culls. Get does and a buck that are somewhat proven, yet not worn out. If you want to experiment with breeds for showing, wool production, or fur production, go to a breeder who specializes in such varieties.

Before buying it is a good idea to go to a local fair or rabbit show and meet a few breeders, talk over the stock options, and see what is available for what cost. No matter how careful you are you may end up with some stock that's less than ideal. But with a little experience, you'll be able to improve the quality of your herd yourself.

Unless you have a personal, long-term use for rabbit pelts, don't let fur color or quality enter into your decision on which breed to buy. Rabbit pelts are just about worthless commercially unless you have a market already established. In most cases, buyers pay less for pelts than it will cost you to ship them.

There are many housing options available for rabbits. As a rule of thumb, each rabbit will need about four to five square feet of living space. A space about 18 inches high is enough to let them stretch their legs. The hutch configuration can vary, from one long row of uniform cages to a two-tiered system, to a two-sided building with an aisle down the center. In most cases, a uniform line of hutches, preferably against the side of a building, will be the simplest. A chicken run under the rabbit hutches will serve two purposes: the chickens will eat any wasted feed as well as any fly eggs in the manure.

If possible, the rabbits' living space should be made from galvanized wire. Wood will quickly become soaked with urine and manure, and may cause health problems. If you make your hutches out of wood, be prepared for extra maintenance to keep them clean. Commercial cages are ideal, but somewhat expensive.

The last ingredient in a good housing system is ample water and a good feeding method. Many small operations depend on ceramic dish-

Rabbits need about four square feet of space, with enough head room so they can stretch their legs.

es for both water and feed, while others use automatic watering devices. The important thing is that the animals get all the water they want, and a regulated amount of food.

The well-fed rabbit

When it comes to feeding, you get out of your rabbits exactly what you put in them. The better you feed your stock, the better they'll produce for you. Most arguments center around whether to feed commercial pellets or homegrown mixtures.

Even though rabbits are vegetarians, they will not produce very well on a diet of garden greens only. The earlier production figures are based on commercial feeds, with some augmentation from garden greens. For a 10-pound doe to produce over 100 pounds of harvestable meat in a year, nutrition must be heeded. In the wild, with their natural diet, rabbits produce nowhere near that level.

For beginners, commercial feeds may be the best bet. They are designed to provide a balanced diet for nursing rabbits, and some of the better meat-producing breeds are

bred with commercial feed in mind. This will leave you free to concentrate on other management problems. Then, as your experience grows, you can begin to blend your own feeds.

The drawback to this plan is cost. Rabbit pellets sell for $10 to $12 per hundred pounds, and it takes about 15 pounds of pellets to bring a young rabbit to the slaughtering weight of 4 pounds. Adults will eat a pound of pellets every three days, while a pregnant doe will consume ten ounces daily. So the more you supplement with home-grown greens, the more you save. Management also includes learning how and when to breed your rabbits, keeping records of birthing rates and growth of young rabbits, and maintaining good production. A well-managed rabbitry should average four litters per year, per doe. Each doe should average eight young per litter, and all should survive to weaning size. To do that, you need to know when to breed what rabbit, when to put in a nest box, when to slaughter, and when to cull a rabbit from the herd. All these things are

(Continued on page 354)

(Continued from page 353)
learned through experience.

Figure on spending about 1½ hours of labor for every dressed rabbit you produce. That includes all the care, from birth to the time they go into the kitchen as dressed meat. So if you expect to produce 100 finished rabbits a year, it should take about 150 hours of your labor.

All that remains for you to be a successful rabbit farmer is to slaughter the animals and eat the meat. Finding ways to use rabbit meat is sometimes a problem. Start slowly and develop a collection of recipes you like. Domestic rabbit meat is all white, and when slaughtered at the right time, has almost no fat.

Ray Wolf

Mr. D. E. B. Clauser grows rabbits in his backyard less than half a mile from our editorial offices. He grew up on a farm, and raised rabbits as a boy, so now he does it as a hobby.

Clauser raises New Zealand Whites, which he believes are best for meat. To kill a rabbit, he gives it a sharp blow on the head, then cuts the throat.

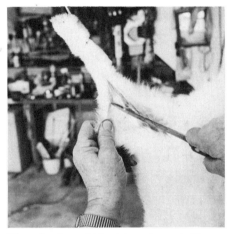

When the flow of blood has stopped, Clauser strings the rabbit up inside his garage, with newspaper underneath to catch any stray drops. The first step is skinning the animal.

Once the skin on the legs has been slit, the entire pelt can be peeled off. There's not much of a market for rabbit skins, but Clauser occasionally has one cured for his grandchildren.

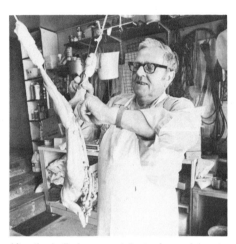

After the belly is opened, just a few quick cuts will clean the carcass. Clauser takes his time, but the whole process–from cage to freezer–requires only ten minutes per rabbit.

Sheep

If you've got some extra land and aren't sure what to do with it, consider sheep. Pound for pound, lamb is much cheaper to raise than beef. You can raise eight sheep on the same amount of land it takes to raise one cow. Those eight sheep will provide more meat through their lambs than the cow does, and do it in 4 to 6 months, while the cow takes 12 to 18 months.

Sheep thrive on pasture that cows and most other critters turn their noses up at, and usually need feed only in winter. They clear land of brush and weeds, leaving it better than when they found it. For this work they ask only a minimum of shelter, care, and equipment.

If you start with two ewes (females), you can graze them all spring and summer, then breed them in their fall breeding season. The next spring, figure you'll get 1½ lambs per ewe, or twins every other time. After grazing the lambs and ewes for five to six months, you'll probably butcher the lambs, keeping the ewe lambs if you intend to expand the flock.

You can buy lambs, adult sheep, or in-lamb ewes who are already pregnant. If you're raising meat, wethers (castrated males) are more docile than rams, and fatten well. Often farmers give away orphan lambs or sell them cheaply because they take more work and loving care to raise without mama's help.

An average 100-pound lamb brings 35 pounds of meat, though some breeds yield more. An average fleece weighs 8 pounds, and it takes 3 pounds of wool to make a pound of yarn, 15 pounds to make a blanket, and 10 pounds to bat a quilt.

If you're an aspiring shepherd, the main limitation on how many sheep you can raise is how much pasture you have, and how much you can afford to fence in. If you have less than an acre for grazing, it may not be worth the effort.

What sheep need

How much land sheep require depends on how good the pasture is. With rich grazing fodder like alfalfa or a clover mix, you can have up to 15 ewes and their lambs on a single acre. Low-grade grasses like timothy can support at least two sheep per acre. Woodland is all right if you plan to make up for the sparse grazing with hay and grain. Swampland is definitely out because it encourages foot rot and other problems.

Sheep are already well dressed for the cold, but they do need protection from drenching rains, and newborns need warmth till they get their wool. If your area doesn't get frost or snow, all the sheep need is shade. Otherwise, a barn or old shed will do, or you can make a simple three-sided structure allowing ten square feet per sheep.

The biggest cost of sheep raising initially is fencing. Sheep may look like clumsy bundles of wool, but they can leap a four-foot fence or shimmy under it faster than you can say "Bo Peep." Even more important, you need a good fence to keep out dogs and coyotes. Dogs simply love to chase sheep, attacking them or harassing them till they die of exhaustion or shock.

Sears catalog prices are generally higher than local prices, but they'll give you some idea of cost: $625 to put woven-wire fence and barbed wire around land 200 feet by 200 feet, just under an acre.

An electric light in the shelter is a good idea. It helps you see at night, and provides newborns with a nonflammable heat source. Sheep

drink several gallons of water a day, so you might appreciate the convenience of running water near your pasture and shelter.

Sheep generally thrive on even mediocre pasture most of the year, eating grasses, brush, even that poison ivy you've been trying to root out for years. When the grass dies off, they'll eat about 20 bales of hay each over the winter.

In addition, each ewe needs about 100 pounds of grain a year to keep her in good shape before and after lambing. It's practical for a homesteader with enough land to raise the corn, oats, wheat, and other grains because you don't have to grind them up for the sheep. Let

them graze in an orchard that otherwise would go to waste. Toss them some root crops now and then, or pumpkins—their favorite treat.

Most of the year, all you need to do for your sheep is say hello now and then, check their water supply, and look for any sign of problems. Whenever a sheep separates from the flock, trouble may be brewing; they're very sociable by nature. Keeping the shelter clean is easy because their droppings are dry.

Once a year, sheep need shearing whether you want the wool or not. It takes a lot of practice to get a fleece that's in one piece and good condition, but you can learn. You can

(Continued on page 356)

(Continued from page 355)
also hire a shearer (if there are any around) or ask a neighbor shepherd for help.

Twice a year, you'll need to trim their hooves and treat your flock for internal parasites. They're especially susceptible to worms picked up in the field because they chew grasses right down to the ground. Ewes mostly manage lambing themselves, but you need to be there for the occasional emergency and to make sure the lambs survive those critical first few weeks.

Some folks find butchering large animals downright distasteful, especially when they've become family pets. You can do it at home if you like. Lamb tastes best when it's aged ten days at 35 to 40°F., which may be hard to do at home, but the trip to the slaughterhouse panics the lamb and toughens its meat.

Counting the costs

One blessing of sheep raising is that you need so little equipment for it. You can make cheap, simple feeders and waterers, and an ordinary heat lamp will do for lambing in cold weather. You can handle the job of shearing with a $10 manual shears (if you can find them) or spend $70 or so for an electric model.

The initial investment in fencing and the sheep can be fairly high: at least $600 for fencing an acre, $75

to $100 for a seven-month-old ewe lamb ready to breed. But after that, the yearly cost of keeping sheep is small. To get an idea, let's take two Suffolk lambs and use southern Mississippi figures to calculate the cost.

At $1.50 a bale, two ewes will eat $60 worth of hay yearly. At $9 for 100 pounds, they'll eat $18 worth of grain. Throw in $12 for a salt block and miscellaneous medications, and the total is $90.

A good shearing is like a good haircut, satisfying to both the shearer and the sheep.

In a year's time they'll produce three lambs or 105 pounds of meat, and 16 pounds of wool. Subtract the value of the wool, $13, from your feed costs, and that makes the cost of raising the lamb less than 75 cents per pound.

The average 35 pounds of meat per lamb usually means money, whether you eat the profits or sell them. Spring lamb, butchered before it's weaned, yields less meat but brings a higher price (especially at Easter) and saves your pastureland. Mutton may not be as tender as lamb, but it's great for stews or ground meat, and many people enjoy its distinctive flavor.

If you don't want the wool yourself, check with local weavers, the local wool co-op, or national weaving magazines for a market. Sheepskin is valuable too, and a single tanning process you can do at home makes it machine washable. You can even use wool scraps left over from trimming the fleece for batting or insulation. Finally, the manure makes great fertilizer, and you can put it right onto the garden without hurting your plants.

The sound of your sheep baaing gently as they wander the fields, their bells tinkling, is a scene of peaceful contentment that's a reward in itself. Sheep are as playful and affectionate as any pets, and among the gentlest creatures you'll ever have the pleasure of knowing.

Pat Donahue

Earthworms

When you think of backyard animals, earthworms are probably not the first that come to mind. If you raise poultry, however, they make a fine alternative feed to soybean meal or corn. Earthworms are a cheaper source of protein and they reproduce themselves regularly, thus replenishing the food supply. Or if you raise rabbits, you can feed their manure to the earthworms. Just place your worm beds under the hutches and worry no more about unpleasant odors. Finally, a rich supply of earthworms will keep any angler happy throughout the fishing season.

But a gardener benefits most from raising earthworms. They're basically eating and composting machines that ingest their weight in earth and organic matter every day. By tunneling into the subsoil layer, earthworms loosen the earth, increase the topsoil layer, and bring up minerals from the subsoil, depositing them as castings. These castings, or manure, are rich in nitrate, phosphorous, potash, and exchangeable magnesium. Earthworms also break down organic matter, aerate the soil, and increase its water-holding capability.

The two species of earthworms most commonly raised on a small scale are the manure (or brandling) worm and the red worm, because they both can live well in boxes or garbage cans. Fisherman prefer night crawlers, but they are best raised on acreage seeded with grass and clover. The red worm is the most common, perhaps because it is quite odorless, as well as being a popular fishing worm.

The initial investment is relatively modest. Mature worms sell for around $12 per thousand and can be purchased from bait shops or through ads in fisherman's or farmer's magazines. In addition, you'll need worm beds and a few tools. A fine-tined fork, such as a hay fork, is essential, as is a shovel, preferably light and broad, a soil thermometer to test manure heat, and lime-testing equipment or litmus paper.

The beds must be made before the earthworms arrive. Worms breed and work best where the temperature ranges from 50 to 70°F., so unless you plan a large enterprise, try to find an extra space in your basement or garage. If you put your beds outdoors, place them in a sheltered area away from falling leaves (which can cause overacidity in the soil).

The bed, or box, should be between 12 and 18 inches deep. Wooden fruit boxes work well. To provide the necessary drainage, drill a few holes into the bottom or make the first layer a two- or three-inch gravel bed.

Hugh Carter of Plains, Georgia, owns the largest worm farm in the world.

For bedding, start with some kind of decaying organic matter, preferably manure mixed with compost or moistened peat moss. If you're using raw manure (other than rabbit droppings), first complete the initial fermentation by watering it heavily and turning it every few days. Add the manure to the bedding when its temperature drops to between 50 and 70°F. Other acceptable types of bedding include torn paper strips or sawdust mixed with manure or compost. When the bedding is ready, just put the worms in and cover them loosely.

Since earthworms prefer a neutral or slightly alkaline environment, sprinkle some agricultural lime or limestone flour over the bed every month or so to keep the acidity down. To be sure you're safe, it's wise to test the pH level of the bed twice a month.

Worm meals

To feed your worms, just spread a layer of food on the top of the bed and cover it with burlap. They'll love coffee grounds or tea leaves. If you use kitchen leftovers, avoid giving them acidic foods, such as onions, citrus waste, or vinegar-soaked garbage. Other garbage, even a small amount of meat, grease, or bones, is safe. Other necessary maintenance procedures include watering the

beds two or three times a week with a sprinkling can and turning the top few inches twice a week.

Earthworms copulate every 7 to 14 days, releasing one or more capsules or egg sacs containing 1 to 20 fertilized eggs. Usually only 1 or 2 will hatch, however, after an incubation period of from two to three weeks. Newborn worms look like tiny white threads, but within a day their color darkens. The worms reach sexual maturity in 80 to 100 days, after which their growth slows dramatically. Mature worms will produce about 50 capsules per year, so your stock should double every four months or so.

The simplest way to harvest the worms is to sift them out with a hay fork. If you feed them heavily the night before, they'll come to the surface. If you are not squeamish, you can sort them by hand. Just be sure to keep immature worms (under four inches) and egg sacs to reseed your beds.

What do you do with your crop of newly harvested adult worms? You can feed them to your livestock, spread them with your compost right into your flower or vegetable gardens, or sell them to anglers or to other gardeners. However, you should think twice about going into earthworm sales in a big way. The marketers and distributors are the

(Continued on page 358)

Two of the most common earthworms are the night crawler (on the left) and the red worm.

(Continued from page 357)

real moneymakers, while the small-scale farmer must create his own markets. Don't count on getting rich quickly.

Earthworms may not have any endearing personality traits, but quietly and inobtrusively these efficient creatures will do their jobs for you day after day. All they ask in return is a little water, a little garbage, and a couple of hours of your time each week.

Krissa Strauss

Husbandry from books and magazines

To read, rank, and review the best livestock books and periodicals would be a huge task, so we list here Rodale Press's several titles and others that came to our attention.

Chickens.
Chickens in Your Backyard, Rick and Gail Luttmann (Rodale Press, 1976).
Raising Poultry the Modern Way, Leonard S. Mercia (Garden Way, 1974).

Ducks and Geese.
Ducks and Geese in Your Backyard, Rick and Gail Luttmann (Rodale Press, 1978).
Successful Duck and Goose Raising, Darrel Sheraw (Stromberg: Pine River, MN 56474, 1975).

Pigeons.
Pigeons, B. Hunt and B. Waldman (Prentice-Hall, 1973).
American Pigeon Journal (periodical), Warrenton, MI 63383.

Quail.
Quail Manual, 8th ed., Albert F. Marsh (Marsh Farms: Garden Grove, CA 1976).
A Quail in the Family, William J. Plummer (Henry Regnery Co., 1974).

Turkeys.
Starting Right with Turkeys, G. T. Klein (Garden Way, 1972).

Pigs.
Raising the Homestead Hog, Jerome D. Belanger (Rodale Press, 1977).

Rabbits.
Raising Rabbits, Ann Kanable (Rodale Press, 1977).
Domestic Rabbits (periodical), American Rabbit Breeders Assn., 1007 Morrissey Drive, Bloomington, IL 61701.

Sheep.
The Shepherd's Guidebook, Margaret Bradbury (Rodale Press, 1977).
The Shepherd (periodical), Sheffield, MA 01257.

Earthworms.
The Earthworm Book, Jerry Minnich (Rodale Press, 1977).

(*Cows and goats* are covered in the Home dairy chapter.)

Buying meat

The United States has been a nation of meat-lovers ever since the Pilgrims feasted on wild turkey. Hot dogs, hamburgers, or steak could be nominated the national food—and the average American eats more than 160 pounds of meat each year, or about seven ounces daily. Many of us spend at least $1 a day on meat. We buy meat partly for pleasure—the juiciness, the taste, the texture—but also for its high protein content. Between 20 and 30 percent of meat's total weight is protein. One ounce of cooked meat, with the fat removed, provides more than one-tenth of our average daily protein requirement. Meat is also rich in other valuable nutrients: iron, thiamine, riboflavin, niacin, and vitamins B_6 and B_{12}.

Most meat in this country also contains less welcome ingredients: the chemicals that contaminate it. Pigs often get sulfa drugs in their food, and chicken feed is laced with antibiotics and arsenic. Cattle still receive ear implants of the hormone DES (diethylstilbestrol), well known for its cancer-causing properties. The grains and silage fed to livestock may still contain residues of banned insecticides such as DDT, and this is passed on to the meat.

There are several options for the wary human carnivore who wishes to avoid these chemicals. One, of course, is to raise your own livestock and give them natural, uncontaminated feed. This solution can also save you a lot of money. One family we know buys a young steer each November, pastures it until the following October, and then fattens it up on grain. After slaughtering the steer, they sell one side and butcher the other for themselves. The profits from the sale finance their purchase of the next year's steer. The cost of their own side of beef works out to a piddling 13 cents a pound, after paying for pasture rent and October's grain ration.

Another approach is to buy organically grown meat. Persuade your local health food store to carry the meats you want, or make arrangements with a local organic farmer. This, too, can save you money, if you buy a whole side of beef from him. Half a steer weighs perhaps 300 pounds, and currently costs a little over $1 per pound. Buying large amounts co-operatively with your neighbors might offer a real incentive to a local farmer to raise livestock organically. Eliminating the middleman (supermarkets) will also help you finance the large freezer necessary for storing your meat.

If these options are not available, try to find a reliable butcher who will not purchase a carcass that shows any signs of disease or damage, and avoid eating the parts of the animal that tend to store contaminants, particularly the liver.

In this country, we've been fed

the idea that prime beef is the very best meat. That's not really true. There are many factors to weigh in choosing what kind, and cut, of meat is best.

For people concerned about world hunger, pork has advantages over beef among the red meats; pigs consume much less grain for every pound of meat they produce. Pork is also leaner and cheaper than beef. Chicken and fish are low in fat, and usually less expensive than red meat. Lamb and veal are also less fatty, but more expensive.

Animal fat is an important consideration in choosing meats. Today's livestock is usually confined without exercise for most or all of its stay on earth. As a result, the fat is much more heavily saturated than when livestock roamed and grazed. In addition, confinement increases the proportion of fat to meat, as does the use of hormones in livestock feed.

Each carcass is graded at the slaughterhouse, and the different grades contain varying amounts and kinds of fat. In order of the usual preference, the grades of beef available to individual consumers are Prime, Choice, Good, and occasionally Standard. Buying Standard or Good grades of meat gives you less fat marbling than Choice or Prime. Marble fat is the kind to worry about, because you can't remove it with a knife—it runs all through the meat. Good and Standard grade meat may need to be cooked with a little liquid to equal the tenderness of Choice or Prime, but otherwise it is cheaper and better. (The above grades also apply to veal. Pork grades are numbered in order of preference, but they mean little because the meat is almost always young, tender, and juicy. Lamb grades are Prime, Choice, Good, and Utility.)

Color is the best indicator of meat quality. Chickens should be plump and have a rich yellow color. Ducks, geese, and turkeys should be white. Lamb should have smooth, firm, pink flesh, with pure white fat and soft, moist, pink bones. Veal should be a pale pink, almost gray-white, and very finely textured. The bones should be soft, moist, and narrow, with a reddish tinge. (Older veal has darker meat and whiter bones.) Beef should be cherry red, with creamy-white fat.

Avoid buying meat that shows blood spots, or major variations in texture. These may indicate mistreatment or other damage to the animal. Very heavy marbling (like crayon marks) may also indicate damage or extreme age. Unfrozen meat is usually fresher than frozen meat. Frozen liquid in the package means poorer quality, because it is a sure sign the meat was thawed and refrozen. Use the date stamped on the package as a guideline for freshness; the store manager is obliged to explain the code to you.

For the best ground meat, select the cut yourself and have it specially ground to insure freshness and the proportion of fat that suits you. And whenever possible buy young meat. The younger the animal, the fewer the contaminants it has absorbed from its feed. Lamb, veal, and calf have received fewer hormones and antibiotics than older beef and mutton. Buying young is especially important with organ meats, particularly liver, which collects and stores some of the toxic compounds.

The kindest cuts

There are two "levels" of cuts: primal (wholesale) and retail. Both should be given on the label of packaged meat under the identity labeling standard that is becoming more common. Each primal cut tends to have its own particular kind of meat (tender, stringy, fat, lean) and its pre-
(Continued on page 360)

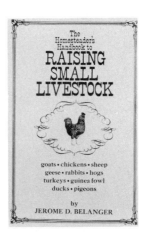

THE HOMESTEADER'S HANDBOOK TO RAISING SMALL LIVESTOCK
Jerome Belanger. 1974.
Rodale Press, 33 E. Minor Street,
Emmaus, PA 18049. 246 p. paperback
$4.95, hardcover $8.50.

This is the book for getting started. It will fuel the city-dweller's dreams, guide the suburbanite's first backyard venture, and inspire the homesteader to diversify his enterprises. Eggs, milk, cheese; fur, feathers, wool, and hides; red meat or white, fresh or frozen, cured or smoked; they're all here. And keeping animals is not seen as just a product-oriented activity, but a way of life attuned to the rhythms of earth and the seasons, the cycles of birth and death, nourishment and growth, labor and rest.

Jerry Belanger, writing from years of experience on a small farm, gives a thorough introduction to the care and feeding, shelter and breeding, of rabbits, chickens, turkeys, geese, ducks, guinea fowl, pigeons, goats, sheep, and hogs.

Richard and Pam Darling

. . . The one word that sums up the reason most people raise small stock today is independence. Freedom from the chemical companies, from price increases, from shortages. Being independent, in this day and age, is something you have to work at, and in keeping with our pioneer traditions, a state worth working for. We . . . have a huge garden, 150 chickens, ducks and geese and guineas, rabbits, sheep and goats, in addition to the hogs and beef cattle. We raise our own wheat and grind our own flour. We make cheese, can hundreds of quarts of fruit and vegetables, and do much of our own butchering.

Work? Of course. But there's a certain grim satisfaction in being "independent." Moreover, just because it's work doesn't mean it isn't fun!

(Continued from page 359)

ferred kinds of cooking. The following chart applies particularly to beef, but also to older lamb, veal, and pork. Older animals, especially beef, have more meat that needs long, slow, moist cooking; younger animals have more meat that can be fried or broiled.

Pick the cut of meat according to cost and the kind of cooking you will do. If meat can be broiled, fried, or dry-roasted, it will be more expensive. Many cheaper cuts must be de-boned, and have the gristle, fat, or tendons cut out, but a little work with a sharp knife can save a lot of money.

In addition to these cuts of meat there are more than 200 types of sausages, lunch meats, and cold cuts made of ground and seasoned pork, beef, veal, and variety meats. Except for fresh sausage, most contain sodium nitrite and nitrate to preserve the meat, kill botulism, and maintain a rosy color. There is substantial evidence that nitrites and nitrates are carcinogenic, and they should be avoided if possible. One unusual meat product is scrapple, a Pennsylvania specialty now available in most regions. Made of pork meat, pork parts, cornmeal, and seasonings, it is cheap, delicious, and nutritious. Slice and fry or bake.

In choosing what cuts to buy, don't neglect the cheaper, tougher, or bonier meats. They are often cooked in ways (like stewing) that fill your home with fabulous aromas and make the very most of flavor. With a boning knife and meat grinder, you can handle anything. Shank, blade, or breast cuts may become your key to inexpensive, high-quality protein.

Once you've bought your meat, keep it in the coldest part of the refrigerator. If it's prepackaged, don't repack; only loosen the ends of the wrapping so it can breathe. If it's not prepackaged, remove the butcher wrappings and re-wrap in foil-type coverings (except for poultry, which is best left in transparent plastic wrap).

When freezing meat, make sure the package is airtight or the meat will dry out.

Cynthia Adcock

BEEF, FARMERS' BULLETIN #2263
LAMB, FARMERS' BULLETIN #2264
PORK, FARMERS' BULLETIN #2265
Superintendent of Documents, U.S. Government Printing Office, Washington, DC 20402. 68, 40, and 65 pages, respectively. $.75 each.

The USDA is a prime source of practical, cheap information about livestock. They've got numerous booklets on how to keep various animals alive: these three publications tell you how to kill them. You'll learn when to slaughter, what equipment to use, and how to preserve and prepare the meat. Each booklet has step-by-step photographs of how to kill the animal and carve its meat. Even in black and white they're pretty graphic, so don't order these booklets unless you're serious.

For a catalog of all USDA publications, write to the Science and Education Administration, USDA, Washington, DC 20250.

COOKING FOR BEST FLAVOR

Primal cut and characteristics	Cooking requirements
Breast cuts (brisket, short plate, bacon, ribs). Alternating layers of fat and lean; many bones and tough tendons; the brisket is leaner and solid, but quite tough; relatively inexpensive.	Large pieces need long cooking with moisture.
Shoulder (chuck) arm cuts (short ribs, shoulder roast, arm roast, stew, ground beef). Much bone, cartilage, and complex muscles and fibers.	Most need long, slow cooking with moisture to prevent dryness.
Shoulder (chuck) blade cuts (chuck eye roast, blade roast, stew, ground beef). Generally more tender than arm cuts; large blade bone to work around.	Chops may often be fried or broiled. A blade roast two inches thick or more can be dry-roasted or braised, likewise the eye roast.
Rib cuts (rib roast, rib steak, rib eye, delmonico steak). The most tender steaks and roasts.	Dry-roast or broil.
Short loin cuts (porterhouse, T-bone, club steaks). No bones except backbone; very tender meat.	Roast, broil, or fry.
Flank meat (flank steak, London broil). Relatively tough.	May be broiled very quickly for London broil, very rare; otherwise must be cooked slowly with moist heat to tenderize it.
Sirloin (hip) cuts (steaks). Leaner than short loin; little waste.	Broil or fry steaks. Sirloin butt may be roasted.
Leg, round and ham cuts (top and bottom round, eye of round, rump roast, cubed steak, stew, ground beef). Still less fatty; tougher than rib, short loin, or sirloin.	Some top round may be dry-roasted; most leg and round cuts benefit from moist, slow cooking. (Pork may be dry-roasted, as may lamb and veal.)

Variety cuts. Fairly inexpensive but need special treatment.
Liver: braise, broil, or fry; tongue: long, slow, moist cooking; sweetbreads (thymus): simmer, then braise, broil, or fry; heart: very tough, cook with moisture; kidneys: in beef, long moist cooking; but fry, broil, braise, or stew veal, lamb, or pork kidneys; brains: soak and then simmer or fry; tripe: precook two hours, then broil, fry, braise; oxtails: tough and bony, need long, slow, moist cooking.

Beating the high cost of beef

With the spectre of $2-a-pound hamburger looming ever greater, and with agricultural prognosticators talking of the worst prolonged North American beef shortage since the late 40s, the neighbors' surplus calves will be drawing calculating looks from more and more of us.

While saving money—or just being able to afford red meat—is sufficient motivation in deciding to raise one's own beef, my own first two calves arrived simply because I wanted the pleasure of their company. I discovered that doing chores in the shadow-filled, fragrant, intimate sanctuary of the cow barn on a winter night worked a powerful, nearly mystical cure on a mind and body ravaged by city work and society.

One pair of the creatures were raised with much pleasure and little trouble and this led, naturally, to another pair, and another—for eight years, a pair of beeves each fall. The system of raising them evolved into a fairly reliable way of putting 400 pounds of free beef into the freezer, but even more important, our home-grown T-bones looked, smelled, cooked, and tasted far better than any commercial feed-lot product.

Don Woodcock. Reprinted with permission of the author and *Harrowsmith* magazine. Copyright © 1979 Camden House Publishing.

COST BREAKDOWN OF RAISING TWO CALVES AND SELLING ONE

Item	Quantity needed	Unit cost	Total cost
Calves	2	$90.00	$180.00
Hay	170 bales	1.00/bale	170.00
Milk replacer	100 lb.	.22/lb.	22.00
Calf starter	600 lb.	.11/lb.	66.00
Calf grower	900 lb.	.10/lb.	90.00
Finisher (feed)	2,100 lb.	.07/lb.	147.00
Salt block	1	2.90/block	2.90
Mineral supplements	55 lb.	7.90/bag	7.90
Veterinary visits			30.00
Miscellaneous equipment			20.00
Slaughter fee	1 steer	8.00	8.00
Butcher and wrap	500 lb.	.11/lb.	55.00
		Total $	798.80

Second steer sells on the hoof for 55 cents per pound; assuming live weight of 1,000 pounds, income is $550.00

Your beef costs $248.95

(Compare this to the current market price of "freezer beef" to determine savings. At $1.29 per pound, for example, a 500-pound carcass would cost $645 after being cut, wrapped, and frozen. You save $396.05. If you sell one side of your butchered steer for $322.50—the freezer beef price—your own beef will be free and leave a modest profit of $73.55.)

Fish gardening
Raising fish in the backyard and basement

Fish gardening may not be the most conventional home food system, but it makes a lot of sense. It seems less intimidating when compared to vegetable gardening: just as a vegetable gardener plants neat rows of seedlings in rich soil, and sees that they get sunshine, nourishment, and regular attention, a fish gardener stocks fingerlings (tiny young fish) in clean water, and gives them food, oxygen, and regular attention. As with a garden crop, rapid-growing species reach eating size within one growing season.

Fish make very efficient use of their diet. Cattle require about 21 pounds of protein to produce 1 pound of protein for the beef-eater; pigs take about 8 pounds of protein for each pound of protein in pork; chickens turn 4 or 5 pounds of protein into 1 pound of protein in eggs or poultry; fish, however, use only 2 pounds of their protein intake for producing each pound of protein the fish gardener nets for his family's dinner. The fish's efficiency is in part explained by the fact that, being in the water, they don't need to keep their body temperature constant and don't have to contend with gravity. Land animals must expend energy just to stay on their feet. Fish retain that efficiency throughout their lives, and keep on growing, no matter how large.

Aquaculture is particularly suited for the suburbs because the quiet and odorless stock aren't likely to offend neighbors.

A home fish harvest makes an important contribution to the family freezer. One pool can furnish at least 50 pounds of good-tasting fish. An average half-pound fish makes a convenient one-serving package of protein. Fish gardeners can be confident that the fish in a closed pool system enjoy refuge from polluted waters other inland and ocean fish must endure.

Embarking on a backyard fish gardening project is not cheap, involving an initial investment of several hundred dollars. The annual operating costs (for fingerlings, food, and electricity) are low, however, and fish are such a productive crop that the cost per pound should be a fraction of the price of fresh fish in the marketplace. If the investment in equipment is amortized over three years, the project becomes increasingly economical.

Nutrition

We'd all be better off if we'd eat more fish. Fish is easy to digest, rich in protein, and low in fat. In fact, it's one of the best sources of high-quality, complete protein, with an array of essential amino acids comparable to that of meat.

Yet fish's calorie count is low. Four ounces of channel catfish has about 130 calories, but four ounces of choice-grade beef has about 370.

Although almost all fish have less saturated fat than meat, the fat content of different fish species varies widely. In general, fish with a darker flesh are more fatty.

Fish is also an excellent source of trace minerals important for health. Zinc helps the body utilize proteins and carbohydrates. Iodine keeps the thyroid gland functioning properly. Iron guards against anemia, and phosphorus helps build bones.

Water quality

Fish need a balanced aquatic habitat. A backyard aquaculturist will sink or swim on the quality of the water he can give his fish.

If fish are to have hearty appetites and gain weight rapidly, the water temperature must be right. Fish are cold-blooded creatures, which means their body temperature matches that of the surrounding water.

Over the ages, certain fish have adapted to specific water temperature ranges. Some, like trout, flourish only in cold water. Others, like tilapia, a fish from the tropics, won't grow well unless the water is very warm. Most species of fish have a narrow optimum temperature range of 10 to 20°F. At high or lower water temperatures they eat less, and consequently gain less weight. As the temperature limit of a particular fish species is reached, the fish become stressed,

stop eating entirely, and may even lose weight or become sick. Rapid changes of water temperature are also stressful to fish. They may die within a few hours if the temperature suddenly drops below their optimum range. Generally, the fish species suitable for backyard aquaculture in most areas of North America do best at water temperatures between the low 70s and the high 80s.

Fish need oxygen to breathe. While air is one-fifth oxygen, water contains relatively little of the life-supporting gas—typically, in a million parts of water only ten parts are oxygen. The level of oxygen that can be maintained in water helps determine how many fish a pond can support.

Temperature affects the oxygen level of water. The cooler the water, the more gas it will hold. Water will hold about 11 parts per million (ppm) of oxygen at 50°F.; at 60° it can hold

only about 9.5 ppm; at 70°, 9 ppm; at 80°, 8 ppm, and at 90°, 7.4 ppm. That can be a problem, because most fish prefer the warmer temperatures. A further complication is that fish are more active in warm water, increasing their need for oxygen. To be safe, fish gardeners try to maintain a dissolved oxygen level of 6 or 7 ppm.

Some oxygen is introduced to water from the air in a process called diffusion. Wind helps diffusion by creating ripples or waves, which increase the surface area of a body of water. The more surface area between water and air, the easier it is for oxygen molecules to move into (or dissolve into) oxygen-low water. Wind is particularly effective for keeping the oxygen level stable in shallow pools, which have a large surface area relative to their volume.

Photosynthesis charges water with oxygen. Algae produce oxygen during the day by absorbing carbon dioxide (CO_2) from the water and breaking the carbon atom from the two oxygen atoms. That may sound like a perfect arrangement, and under ideal conditions, the algae and other plant life, together with moving air, will keep a pond pretty well supplied with oxygen. But photosynthesis works only when there is sunlight. This means that at sunset the conversion of carbon dioxide into oxygen stops, and the algae begin to produce carbon dioxide. Because

(Continued on page 364)

The only disappointing thing about fish gardening may be that kids don't get to take a dip in the backyard pool.

OXYGEN SATURATION AT DIFFERENT WATER TEMPERATURES

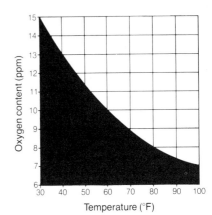

(Continued from page 363)
oxygen use by all pond life still goes on, by the next morning oxygen levels will be low—perhaps dangerously low. Another factor is that on a cloudy day the algae produce considerably less oxygen, so there is less oxygen to sustain the pond population at night.

Fish productivity is affected by the ammonia (NH_3) level in the water. Ammonia is a waste product fish excrete into the water. It also is generated by all decaying organic matter, including algae. Ammonia is toxic to fish at relatively low concentrations: one or two parts per million of ammonia gas in water will kill fish; less than one ppm will reduce fish growth and increase their sensitivity to temperature extremes and low oxygen levels.

PERCENT OF AMMONIA IN THE FORM OF TOXIC NH_3 AS pH CHANGES

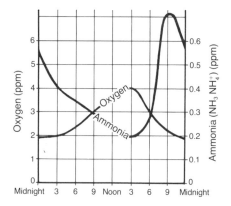

Two natural purification processes filter out ammonia from water. The first is the same process that converts carbon dioxide into oxygen during the day—photosynthesis. Growing plants remove ammonia from the water and use its nitrogen to make proteins for growth. Unfortunately, as more plants grow, less and less light reaches some of them so that the photosynthetic process gradually declines. Some plants begin to die. As they decay, they release their nitrogen as ammonia into the water. A point is reached at which more ammonia is produced from dead plants than assimilated by live plants. Plant use of ammonia also stops at night when photosynthesis stops, and ammonia levels gradually begin to rise until dawn.

Ammonia is also removed by nit-

O₂ AND AMMONIA DIURNAL CYCLE

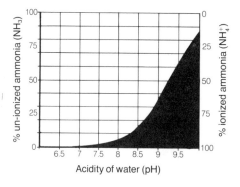

rification, the process by which naturally occurring bacteria convert the ammonia gas (NH_3) into nitrite (NO_2), and then into nitrate (NO_3). At levels found in pool systems, nitrate is not toxic to the fish. It is used primarily as food by algae. In a densely stocked fish pond, nitrification is the main process that controls the ammonia.

Other toxins can also be produced by a fish pond. Methane gas and hydrogen sulfide gas evolve when the growth of aerobic bacteria is stifled. Aerobic bacteria need oxygen to live. When fish wastes and other detritus build up on the bottom of the pond, the aerobic bacteria die off. They are replaced by bacteria that can live *without* oxygen—called anaerobic bacteria—which generate methane and hydrogen sulfide as by-products. Both are toxic to fish. Hydrogen sulfide gas smells like rotten eggs, and you may have smelled it by murky pools. That's the work of anaerobic bacteria. Fish gardeners needn't worry about putting up with rotten egg odors, though, if the oxygen level in the pool remains adequate.

Finally, to raise fish successfully in a backyard pool, you have to consider the acid-alkaline status of the water, as gardeners consider the pH of their soil. The pH scale runs from 1 to 14. A reading of less than 7 is acidic, and more than 7 is alkaline (or basic). Neutral water has a pH of 7.

Fish prefer water that is neither too acidic nor too alkaline—between 6.8 and 8.5 on the pH scale. Fish suffer pH shock if pool water pH changes rapidly. It is essential to know the pH levels when you are transferring fish from one lake to another. Levels of toxic ammonia gas and pH are interrelated, too. The

higher the pH of the water, the more ammonia gas is released. In other words, the more alkaline the pond water, the greater the chance fish will die from ammonia poisoning.

These factors of water quality—temperature, oxygen and ammonia levels, and pH—are all closely interrelated. If one changes, the others change, too. Fish culture, whether large or small scale, is a matter of combining quality fish stock with good water. Fish that are stressed in any way become more susceptible to diseases. Unfavorable conditions such as low oxygen or high ammonia levels, large temperature or pH changes, as well as rough handling, crowding or poor nutrition, can leave a fish prone to infection. And infection might mean a fish kill that will close down your pond for the year.

Anthony DeCrosta

Biological filtration

To control ammonia levels in pool water, most small-scale aquaculture systems rely on biological filtration. A biological filter, unlike a mechanical filter (such as a car's oil filter), doesn't screen out particles. Instead, the work is done by living

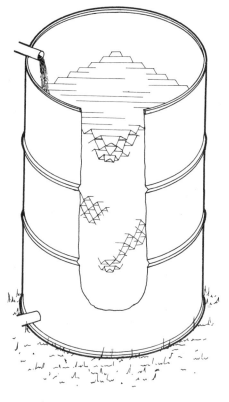

Biodek filter.

bacteria that cling to the surfaces of the filter and purify the water by consuming ammonia. *Nitrosomonas* bacteria convert ammonia (NH_3) to less-toxic nitrite (NO_2). Then, *Nitrobacter* bacteria change nitrite to nitrate (NO_3), which algae can use as food. (This ammonia-to-nitrate process is called nitrification.)

Without these industrious microscopic organisms, fish in a densely stocked pool would soon be poisoned by their own wastes and decaying algae. Most of these bacteria do not float freely in the water, but live attached to surfaces. Therefore, the more surface area on a filter, the greater the bacterial population can be, and the more ammonia can be converted to nitrate.

Biological filters can take many forms. One of the simplest is a gravel bed filter. A bed of limestone rocks furnishes lots of surface area for nitrifying bacteria to colonize. Any hard stone will work as a filter. Limestone is one possibility; it also buffers the water to a pH close to 8.0. Pool water, pumped through the bed, is purified as it passes over the bacteria living on the gravel. (Water must flow over the surfaces to carry ammonia

and oxygen to the bacteria.)

A multi-celled plastic material called Biodek is an effective medium for trickling filters, providing a large surface area and void space for air circulation. Gravel bed and Biodek filters often are installed in a unit adjacent to the pool, through which the water is circulated.

A biological filter can also be floated directly in the pool. Rodale's design maximizes surface area by closely stacking many layers of corrugated fiberglass plates (each about two feet in diameter) on a central axle. Supported by a floating frame and driven by a motor, the filter rotates, half in and half out of the water.

Normally, the primary ammonia purification processes at work in a pool—photosynthesis and nitrification—are dependent on the sun. At nights, the level of ammonia in a fish culture system rises steadily until dawn, as does the level of carbon dioxide. The virtue of an electrically driven biological filter is that nitrification continues throughout the night, so ammonia levels in the pool never reach a point at which the fish are stressed. The major limitation of such a filter is its dependence on energy.

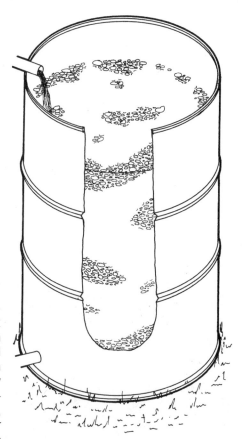

Gravel bed filter.

Fish in the swimming pool

Since 1976, when serious fish gardening experiments first began at the Rodale Organic Gardening and Farming Research Center, our aquaculture researchers have gradually discovered that a component pool system is the cheapest, easiest, and surest way to grow fish in the backyard. The pool system they recommend has three main parts: a vinyl-lined swimming pool, a plastic dome, and a combination biological filter-aerator. Such a system costs less than $400 to build and takes up little yard space—about 150 square feet in all.

A swimming pool might seem an unlikely home for game fish, but we have found that a pool is much easier to maintain—and its water quality much easier to control—than a natural or manmade pond. Ponds often need extra sealing in the form of plastic sheets or bentonite clay granules to stop leaks. Disease or parasites may come from the stream that feeds a pond, and there is the danger of chemical run-off from nearby farm

fields. Wild predatory fish can invade a pond. Stocked fish are also easy prey for raccoons and minks.

You can use a regular above-ground swimming pool from a commercial dealer. Those at the Research Center are 12 feet in diameter and 3 feet deep, and cost less than $80 each. Plastic liners made of polyvinyl chloride (PVC) are not toxic to fish, although new liners should be treated as described below be-

fore any fish are stocked. Ten-mil liners are usually guaranteed only one year, and will likely last no longer. Twenty-mil liners are more durable, though not indestructible.

Pools are easy to put together. Following manufacturer's directions, you should be able to complete the job in one afternoon.

Most fish species suited for backyard pool culture need warm

(Continued on page 366)

(Continued from page 365)

water for rapid growth. In every part of the United States except the Deep South, the optimum range of 70 to 80°F. is reached only for a month during the summer, and one month is too short a time to raise a substantial fish harvest. It's possible to heat a pool electrically, of course, but that is expensive and wasteful. Instead, fish gardeners can borrow a technique from vegetable growers who extend the growing season in solar growing spaces.

Solar-powered pool domes work on the same principle as a vegetable grower's greenhouse. When a pool is

Fish pools at the Organic Gardening and Farming Research Center, Maxatawny, Pennsylvania. At this stage in the experiment the pools have not been covered with solar domes. Aerators float in many of the pools; chicken coops are mounted above some.

enclosed in an airtight, transparent covering, sunlight heats the air and water inside the enclosure and keeps the temperature stable at night and on cloudy days.

The Research Center's domes were built for about $35 apiece. Each was made from a 20 by 20-foot sheet of transparent 6-mil vinyl, readily available from hardware stores. The dome is supported by an eight-spoke framework of ¾-inch plastic PVC pipe. It takes about 100 feet of piping to cover a 12-foot pool. Ideally, some sort of windbreak should be planned for the north side of the pool, both to keep the water warmer and to protect the dome from heavy wind and rain. The thin plastic dome tears easily but it can easily be repaired.

In the Rodale experiments, domes raised the pool water temperature an average of 9° F. By warming the water, and keeping it warmer for an extended period, plastic domes allow fish gardeners to stock fish earlier and harvest them later, lengthening the growing season as much as six weeks. In Pennsylvania, Rodale researchers set up the pools in mid-April—with the water in, the domes on, and the filter operating—to give the water a chance to warm up and to make sure everything is in order.

When you unfold a brand new pool liner you will probably notice a

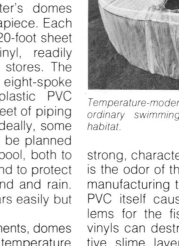

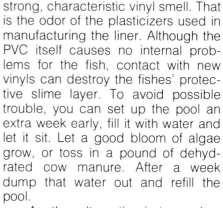

Temperature-moderating plastic transforms an ordinary swimming pool into a cozy fish habitat.

strong, characteristic vinyl smell. That is the odor of the plasticizers used in manufacturing the liner. Although the PVC itself causes no internal problems for the fish, contact with new vinyls can destroy the fishes' protective slime layer. To avoid possible trouble, you can set up the pool an extra week early, fill it with water and let it sit. Let a good bloom of algae grow, or toss in a pound of dehydrated cow manure. After a week dump that water out and refill the pool.

Another alternative is to scrub a new liner with a mild detergent or clay. The entire liner must be scrubbed carefully, and rinsed completely. Then the pool can be assembled and filled, allowing time for the water to warm up.

In a 12-foot pool, 1,600 gallons of water will be about 22 inches deep, which allows clearance for the filter and some splashing room for the fish. If you must use water from a municipal treatment plant, it should sit for a few days so the chlorine can dissipate. To be safe, have the water analyzed for chlorine, fluoride, organic chemicals, and other pollutants. The treatment plants will have some information to help you, and your county agricultural extension agent or appropriate state commission can tell you how to check the rest.

Clean water is essential for thriving fish. As far as they're concerned, though, clean doesn't necessarily mean crystal clear. For a fish gardening pool, green water is best. The green indicates a healthy bloom of algae. To encourage algal growth in your pool water, you can simply add a bucketful of water from a flourishing pond.

Throughout the season you will

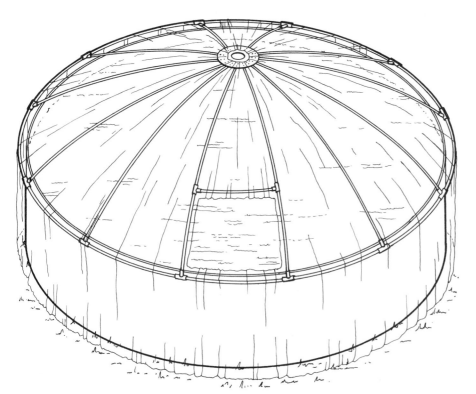

occasionally need to add a little more water to the pool. It's a good idea, especially when using treated water, to keep a 55-gallon holding drum full of water so it can cure.

The filter-aerator

A 1,600-gallon pool of oxygen-hungry fish and algae needs some kind of aeration system to pump plenty of oxygen into the water. Many small air compressors and aerators either mechanically agitate the surface of the water or spray water through the air so it can pick up oxygen molecules. But in a densely cropped fish culture system, biological filtration is just as important as aeration. Rodale researchers developed a rotating biological filter that both aerates water and gets rid of ammonia wastes. You can build this two-in-one unit for $150 or less.

The filter is made of layers of corrugated fiberglass roofing tiles that rotate on a central axle driven by a small ($1/20$-hp) electric motor. The

This Rodale-designed rotating biological filter uses 35 sheets of corrugated fiberglass (cut square, since cutting circles is both wasteful and difficult). A floating framework of 4-inch PVC pipe supports the filter and enclosed motor, on the right. Simple paddles visible on the left help aerate the water as the fiberglass plates rotate.

unit floats half in and half out of the water, bouyed up by a float of tightly sealed 4-inch PVC piping. In just a few cubic feet the fiberglass layers provide more than 300 square feet of surface area for nitrifying bacteria. As the motor turns the plates, paddles churn the water, introducing oxygen to the pool.

For nitrifying bacteria to do their job, they must have oxygen and they must be kept wet. This filter continuously exposes bacteria to fresh water and air. The filter-aerator is run 24 hours a day, at a daily cost of about 10 cents for electricity.

Before stocking, the nitrifying bacteria living on the filter need a few weeks to multiply enough to the point that they can cleanse the water. The filter is seeded with bacteria by simply wrapping a shovelful of ordinary garden dirt in burlap or cheesecloth (so it won't make the water muddy) and tossing the bundle into the pool. The desired bacteria are already living in the soil—providing the dirt has not been treated with chemical fertilizers at some point.

These ammonia-eating bacteria need a little of this compound to start on, so add a little household ammonia—just about five drops a day—and keep testing the ammonia level. It should rise at first, then level off, then drop to less than 0.1 ppm—signaling that the bacteria on the filter are ready to support the fish.

Stocking

Rodale researchers are certain at least a 50-pound harvest can be raised in a pool system this size. Judging from his experiences with different species, polycultures, and stocking densities, biologist Steve

Van Gorder recommends that beginning fish gardeners start with 100 tilapia, or stock a polyculture of 75 tilapia and 25 catfish. If 100 fish all grow to a half-pound each, that will mean a 50-pound harvest of fresh fish at the end of the season.

Selecting an appropriate stocking density for the size and efficiency of the system is critical. Pool systems may be able to support many more fish, but then superior filtration and aeration would be crucial. Naturally you want a substantial harvest, but crowding fish beyond the pool's capacity would risk a big fish kill. Another thing to consider is that as stocking densities increase, growth rates decrease: there are more fish vying for food and oxygen.

A sensible way to manage a fish pool is to stock conservatively, insuring a healthy system that can support rapid growth. After experience makes you confident about maintaining good water quality, it's possible a fish gardening system of this size could support up to 100 pounds of fish.

For fish to reach a nice size (at least ½ pound) in one growing season, you need to start with good-sized fingerlings. Tilapia should be two to three inches long. Catfish do best if they're stocked at five or six inches.

Many hatcheries and fish nurseries sell the species that can be raised in pool systems. One of the best sources of information on suppliers of fish and fish culture equipment is the annual Buyer's Guide published by *Aquaculture Magazine* (formerly *The Commercial Fish Farmer & Aquaculture News*), P.O. Box 2451, Little Rock, AR 72203.

Contact the supplier early to check how much time they will need to process your order. Small orders may not receive priority handling. The fish must not show up before the pool is ready to handle them, but once everything is prepared, you'll want to get started for the longest possible growing season.

When the water temperature hits 65 to 70°F. and remains there for several days, your pool can be stocked. At the Rodale Research Center in Pennsylvania, fish go into the pools in early May.

Unless you happen to live quite

(Continued on page 368)

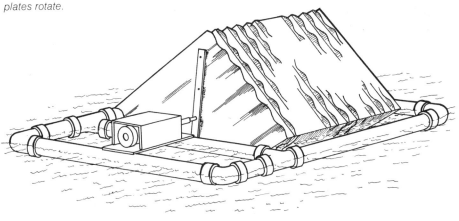

(Continued from page 367)

close to a fish supplier, many of whom are in the southern United States, it's likely that your fingerlings will be shipped by air. Arrange to pick up the fingerlings as soon as they arrive. The water in which they are packed should be kept cool during the trip home from the airport. If the fish are transported without oxygen or an agitator device, splashing the water a little will help keep it aerated.

When you get the fish home, don't put them directly into the warm pool water—the sudden temperature change would be a shock. First, place the plastic bag full of fish into the pool for a half-hour or so, until the water temperatures equalize. Then gradually add small amounts of the pool water to the bag. Before releasing the fingerlings into the pool, count and weigh them for your records. The safest and most convenient way for fish that small is to weigh them all together and subtract the weight of the container and water.

At harvesttime a pool is drained and the fish are gathered in dip nets.

Bluegills ready for stocking.

Feeding

Once the fish are stocked, you can begin feeding them right away. A balanced diet is insured by feeding them commercial fish pellets labeled "complete." The pellets are made in different sizes; and some will sink, while others float. Fingerlings need small (⅛-inch) sinking pellets. As the fish grow they are able to take the larger floating feed. Floating feed is somewhat more expensive than sinking, but is advantageous for pool management. When the fish come to the surface for the food you can see how much and how vigorously they

are eating. Active feeding is a sign of good water quality and healthy fish.

Fish grow at a healthy rate when they're fed a daily meal that is 5 percent of their body weight. Tilapia, however, if they have a lush bloom of algae on which to graze, may need only about 1 percent of their weight in supplemental feed. And because algae are an important food source, it is generally safe to feed tilapia "incomplete" commercial feed. Tilapia also accept small quantities of vegetable scraps.

Feeding takes just a few minutes, but it should always be done at the same time of day and in the same spot. Two feedings are best—late morning (10 or 11 A.M.) and late afternoon. Of course, if you feed them twice a day, they should get only *half* their daily ration each time. If that is

impossible, though, once a day is fine.

Two 50-pound bags of commercial pellets from a farm supply store, costing around $25, should last an entire season.

Some fish may die soon after stocking because of the stress of transportation and handling. This is to be expected, so don't worry. Remove any dead fish from the water, and subtract their weight from the total used for calculating food rations.

Should a fish die later in the season, quickly test water quality to make sure the fish aren't endangered by insufficient oxygen or a buildup of toxins. If you suspect the dead fish was diseased, wrap it in plastic and refrigerate it until you can give it to the state's fish commission for their inspection.

Week after week, bacteria on the biological filter will continue to colonize. There should be a film on the fiberglass plates. After awhile some old growth sloughs off and is replaced by new. That indicates the purification process is working well. A spotted filter is not dirty; do not rinse it off with the garden hose.

For most of the summer, you need only feed daily and check the water regularly. On very hot days the dome must be vented. In the event of an emergency, such as a power failure, stop feeding the fish. Feeding would only magnify the problem by increasing activity and thereby increasing oxygen demand.

Harvest

It's time to harvest the fish when the water temperature drops below 70°F. In Pennsylvania, this occurs in late September or early October. For better-tasting fish, hold them for a week without food before harvest.

The simplest way to catch the fish is to gather them up with dip nets. You're going to have a lot of fish on your hands that must all be cleaned quickly. The job is less rushed if you remove just a few fish at a time for cleaning. Gradually drain the pool to make it easier to scoop up the fish. The nutrient-rich water makes great fertilizer for a garden.

Weigh the fish as you harvest them and keep them on ice (as fisherman do) until they're cleaned. Fish that won't be eaten right away must be wrapped for freezing.

Polyculture

Of the many fish species that can be grown in a backyard pool, carp, catfish, and tilapia hold the best promise of high yields at low cost. They are perfectly suited to the water temperature and quality of a closed, small-scale pool system. These fast-growing fish can reach an edible size after a single growing season.

Of course, the fish should taste good, and the three species all have a mild, pleasant flavor.

Fortunately, the three species make wonderful pool-mates. Stocking more than one species, each with different feeding habits, is called polyculture. Polyculture ideally results in a greater harvest than a like number of a single fish species (called a monoculture), because of the fishes' compatability. Different fish—like different plants in a companionate garden—not only can live together amicably, but also can benefit from each other.

The best known polyculture involves three types of Chinese carp—the grass carp, silver carp, and bighead carp—which feed on different biological levels of a pond. They also utilize more naturally occurring foods than other fish. In many states, though, Chinese carp have been outlawed.

Rodale's experiments in pool polyculture included various combinations and stocking densities of tilapia, bullhead catfish, channel catfish, Israeli carp, and buffalo fish (a type of sucker that looks like a carp).

Tilapia are plant-eating fish, some species of which feed on microscopic algae near the surface of the water, usually the warmest part of the pool. Tilapia also eat water plants and garden wastes. Because they feed so low on the food chain, tilapia are a very productive and inexpensive fish to raise. They efficiently convert free food sources into high-quality protein.

Carp are scavengers, with mouths adapted to feeding off the bottom

(Continued on page 370)

An oxygen meter suitable for home use. The retractable probe measures the dissolved oxygen content of the water.

In polyculture, two or more non-competitive species living together benefit from each other. Fish can live "on top of each other" by occupying different levels of the water, so they require little room compared to land animals. Here, the catfish, buffalo fish, and carp (top to bottom) grow well together because of complementary feeding habits.

(Continued from page 369)
of the pool. They scrounge for anything at that level, including wastes of other fish. When they are raised with the top-feeding tilapia, the tilapia help feed the carp by excreting semi-digested algal wastes.

Bullhead, channel catfish, and buffalo fish are also bottom feeders. They and the carp stir up the pool bottom, increasing the nutrient level of the water. The nutrients stimulate growth of algae which, in turn, are food for the tilapia.

Water testing

Because there are many inter-dependent variables in fish gardening, frequent checks are important. Get into the habit of recording water quality factors—temperature, oxygen level, pH, and levels of ammonia, nitrite, and nitrate. Also note stocking and harvesting dates, the change in the weight of the fish, and feeding rates.

You don't need a lot of expensive equipment to keep tabs on the progress of your fish, but several things are essential. For temperature checks you need a thermometer. You might want to mount one on Styrofoam so it can float in the pool all the time.

An inexpensive bottle of litmus paper, for pH readings, will seem to last forever.

Water test kits are marketed for monitoring oxygen and ammonia levels. A simple, reliable colorimetric test can usually be purchased where tropical fish are sold.

THE FEEDING FORMULA

A daily ration of commercial pellets guarantees fish a balanced diet. This carefully regulated feeding method also helps keep the water quality high.

Fish in a closed pool system should daily consume food amounting to about 3 percent of their body weight. To calculate how much food that is, the fish first must be weighed all together before you stock them. If you start out with 10 pounds of live fish, multiply that amount by 3 percent, or 0.03.

10 lb. × 0.03 = 0.3 lb. food daily

At that rate, in one week you'll feed your fish 2.1 pounds of food.

0.3 lb. × 7 days = 2.1 lb. food

Of course, at the end of the first week your fish won't still weigh 10 pounds, so they'll need a bigger ration. You must correct for the continuous growth of fish, but weekly weigh-ins would be an enormous task, and particularly stressful for the fish. There's a much easier way to estimate their growth, using a mathematical formula.

Hypothetically, every pound of food you feed the fish causes them to gain a certain amount of weight. That relationship can be expressed as a food conversion ratio.

$$\frac{\text{amount of food (lb.)}}{\text{amount of fish weight gain (lb.)}} = \text{food conversion ratio}$$

Suppose you start with a 1-pound fish and feed it until it weighs 3 pounds. If it took 4 pounds of food to achieve that 1-pound increase, then the food conversion ratio will be 2.

$$\frac{4 \text{ lb. food}}{2 \text{ lb. fish weight gain}} = 2$$

The food conversion ratio remains constant from week to week—for fingerlings or for very large fish.

A pool of fish growing at the rate of 1 pound for every 2 pounds of added food is definitely productive, but you can probably achieve even better food conversion ratios. You can reasonably expect your pool culture fish to gain 1 pound for every 1.5 pounds of food you feed them. (Food conversion ratios are like golf scores—the lower the better.) For example, if you stock algae-loving tilapia, they will gain weight from eating algae and need less commercial food.

$$\frac{1.5 \text{ lb. food}}{1 \text{ lb. fish weight gain}} = 1.5 \text{ food conversion ratio}$$

Now, back to the mathematical formula. You need to multiply by a conversion factor, but that factor is easy to calculate. It is simply the inverse of the food conversion ratio.

$$\frac{\text{amount of fish weight gain (lb.)}}{\text{amount of food (lb.)}} = \text{conversion factor}$$

Assuming the standard food conversion ratio of 1.5 for fish in a pool culture system, the conversion factor is 0.67.

$$\frac{1 \text{ lb. fish weight gain}}{1.5 \text{ lb. food}} = 0.67$$

Multiply that conversion factor by the number of pounds of food you fed your fish during the first week, and you will know the number of pounds the fish grew.

2.1 lb. food ×
0.67 conversion factor =
1.4 lb. growth weight for first week

Add the growth weight to the original fish weight and you'll get the new total fish weight.

10 lb. stocking weight +
1.4 lb. growth weight =
11.4 lb. fish weight
after one week.

To find out how much food your fish will need for the second week, and each succeeding week, go through the same formula, starting with the new total fish weight each time.

Here are the figures for the second week.

11.4 lb. total fish weight × 0.03 =
0.34 lb. food daily

0.34 lb. food daily × 7 days =
2.4 lb. food for second week

2.4 lb. food ×
0.67 conversion factor =
1.6 lb. growth weight
for second week

11.4 lb. total fish weight +
1.6 lb. growth weight =
13 lb. total fish weight
after two weeks

In a dynamic biological system, of course, all kinds of things can influence feeding activity and growth. Use this formula as a guide, but monitor how actively your fish take the feed. Young fish can eat as much as 5 percent of their weight. As a pool approaches its capacity it may be necessary to reduce feeding to below 3 percent. The fish should consume their ration quickly. If food remains uneaten after about five minutes, the fish probably are receiving too much, and this is important: extra food collects in the pool and decays, disrupting the balance of the aquatic habitat. Whenever fish are stressed (from low oxygen levels or excess ammonia, for example), withhold food for a day.

Check the temperature daily. The most efficient temperature for rapid growth of warm-water fish is 80 to 82°F. It's especially important to watch the temperature during the hottest days of summer. A solar dome captures heat, so it will be warmer under the dome than it is outside. On particularly hot days the dome must be vented. Warm-water fish don't mind 90°F. water, but remember that oxygen demand increases as the temperature rises.

The critical time to check the pool's oxygen content is early morning, when the level is lowest. A dissolved-oxygen level below 3 ppm is dangerous. Tilapia, catfish, and carp grow safely at 4 or 5 ppm. An oxygen level of 6 to 7 ppm is ideal for warm-water fish, and is *necessary* for trout.

A dietary scale can be used to weigh fish one at a time.

Even though your biological filter is operating smoothly, the ammonia level should be checked weekly for a dangerous build-up of toxic wastes. Be on alert if the level approaches 0.1 to 0.2 ppm. Just 1 ppm can be lethal to fish. A rise in ammonia is followed shortly thereafter by a rise in the nitrite level. Nitrates continue to accumulate throughout the season, but that should not harm the fish. Algae will gobble nitrates readily.

Make sure the water's pH stays close to 7.0. As pH jumps to 8.0 and 8.5, ammonia becomes more toxic to the fish.

Chemical test kits.
Hach Chemical Co., P.O. Box 389, Loveland, CO 80537 or P.O. Box 907, Ames, IA 50010. Lamotte Chemical Products Co., P.O. Box 329, Chestertown, MD 21620. Crescent Research Chemicals, 5301 N. 37th Place, Paradise Valley, AZ 85253.

Bass and bluegills*

Bass and bluegills belong to the sunfish family. The group requires clear water with lots of oxygen. In good quality water they are fairly hardy. Bass and bluegills have mild-tasting, flaky flesh. Bass may sometimes be faintly sweet.

Largemouth bass
(*Micropterus salmoides*)

Bluegill
(*Lepomis macrochirus*)

Oxygen: Bluegills best at 3 ppm and more. Bass need 5 to 8 ppm.

Temperature: Bluegills grow well from 68 to 92°F.; best at 86°F. Bass prefer 73 to 77°F.

Ammonia: Below 0.1 ppm.

pH: 6.5 to 8.5.

Feeds: Insect larvae, algae, animal foods

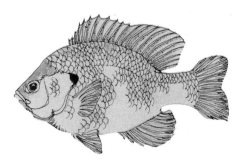

in pond. Bluegills accept pelleted chow; bass must be trained to eat commercial feed.

Stocking: Bass-and-bluegill combination does well with 8 to 10 bluegills per bass. Bluegills should be harvested heavily.

Availability: Hatcheries in almost all states.

Buffalo fish

Buffalo fish were one of the first food fish cultured commercially in the South. When the bigmouth buffalo is crossed with the black buffalo the resultant hybrid grows much faster than either parent species. The flesh has a mild flavor and pleasant appearance, much like carp but with fewer bones.

Bigmouth buffalo
(*Ictiobus cyprinellus*)

Black buffalo
(*Ictiobus niger*)

Smallmouth buffalo
(*Ictiobus bubalus*)

Oxygen: Best growth at 5 to 7 ppm; 3½ ppm is acceptable.

Temperature: Best growth at 68 to 80°F. Growth slows below 55°F.

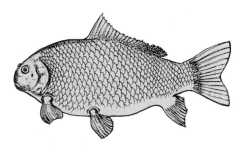

Ammonia: Tolerate 1 ppm, but best growth below .2 ppm.

pH: 6.5 to 8.5.

Feeds: Bottom feeders. Grow well on pond foods: detritus, aquatic plants, plankton.

Stocking: Good in polyculture with catfish.

Availability: Most states in the South.

*Drawings of bluegill, carp, catfish, tilapia, and trout reproduced courtesy of Amity Foundation, Eugene, Oregon.

NETS

Small nylon dip nets with ½-inch mesh are handy for removing fish from a pool. If you raise catfish you'll need a net with coated fibers so the whiskers won't puncture the strands and get caught. For information, write: Champlin Net Company, P.O. Box 788, Jonesville, LA 71343; Delta Net and Twine Company, Box 356, Greenville, MS 38701; J.L. Eager, Inc., P.O. Box 2, N. Salt Lake, UT 84054; Nylon Net Company, 7 Vance Avenue, P.O. Box 592, Memphis, TN 38101.

Dip nets come in a variety of mesh sizes and handle lengths.

*Carp**

Carp have been farmed for more than 40 centuries, starting in Chinese and Roman ponds. They are a practical fish for cultivation because of their rapid growth. Hardy fish, they are disease resistant and tolerant of low oxygen levels and muddy waters. Carp are quite bony, and the oily flesh is both dark and light.

Common carp
(Cyprinus carpio) A specially developed, fast-growing strain is called Israeli Carp or Mirror Carp.

Grass carp or white amur
(Ctenopharyngodon idella)

Bighead carp
(Aristichthys nobilis)

Silver carp
(Hypophthalmichthys molitrix)

Oxygen: Tolerate as little as 0.5 ppm. Grow best with 4 ppm.

Temperature: Growth between 68 and 90°F., but most productive at 80 to 85°F. Won't feed below 50°F.

Ammonia: Tolerate up to 1.2 ppm. Best if kept below 0.15 ppm.

pH: 6.5 to 8.5; prefer a slightly alkaline, around 7.5.

Feeds: Foragers. Eat low on food chain: detritus, fecal matter, aquatic plants, garden wastes, as well as supplemental feed.

Stocking: Grow well in polyculture.

Availability: Most states in the South. Common carp can be captured in natural waters. Certain species are illegal in many states.

PLANS FOR A POOL CULTURE SYSTEM

Complete plans for building and operating a pool culture system will be available in early 1982 from the Rodale Plans Department. *Rodale's Backyard Fish Farm* includes materials lists, blueprints for the rotating biological filter and pool dome, and a schematic diagram of the electrical wiring, with detailed illustrations for their construction.

*Catfish**

Catfish are among the best fishes for home growing. They are hardy and fast-growing. Many people consider the sweet white flesh to be gourmet quality. The few bones are easily removed.

Channel catfish
(Ictalurus punctatus)

Brown bullhead catfish
(Ictalurus nebulosus)

Black bullhead catfish
(Ictalurus melas)

Yellow bullhead catfish
(Ictalurus natalis)

Oxygen: Grow well at 4 ppm or higher.

Temperature: Good growth between 70 and 90°F., but 84 to 86°F. is ideal. No growth below 60°F.

Ammonia: Best below 0.15 ppm; growth severely affected at 1 ppm.

pH: 6.5 to 8.5; prefer a slightly alkaline 7.5 or so.

Feeds: Bottom feeders. Eat low on the food chain. Grow rapidly on supplemental commercial feed.

Stocking: May compete with other bottom feeders, but do well in polyculture with species that graze near the surface (tilapia). Tolerate crowding.

Availability: Most states in the South.

The Eel

I don't mind eels
except as meals.
And the way they feels.

Ogden Nash

*Drawings of bluegill, carp, catfish, tilapia, and trout reproduced courtesy of Amity Foundation, Eugene, Oregon.

Tilapia*

Tilapia enjoy the label "fish of the future" because of their adaptability to small scale aquaculture: they grow rapidly, are inexpensive to raise, and readily breed in captivity. The firm white meat, free of intermuscular bones, has a pleasant taste similar to haddock. Tilapia were common in Bible times, and are sometimes called St. Peter's fish.

Java tilapia
(*Tilapia mossambica*)

Blue tilapia
(*Tilapia aurea*)

Nile tilapia
(*Tilapia nilotica*)

Tilapia zillii

Oxygen: Survive as little as 2 ppm. Best at 3 ppm or more.

Temperature: Grow well between 64 and 90°F., but most productive at 82 to 86°F. Cannot survive in water below 50°F.

Ammonia: Best below 0.15 ppm, although tolerant of high levels.

pH: 6.5 to 8.5.

Feeds: Exploit natural food sources: algae, plankton, plants. Will take supplemental pelleted feeds.

Stocking: Prolific reproduction can crowd the pool, but when stocked densely, reproduction is reduced. Possible, although more difficult and expensive, to obtain sterile hybrids or only males. Males grow faster than females. Have proven successful in polyculture.

Availability: Raised by hatcheries in the Southeast, California, Colorado, Idaho.

GETTING FOOD FROM WATER
Gene Logsdon. 1978.
Rodale Press, 33 E. Minor Street, Emmaus, PA 18049. 371 p. hardcover $11.95.

Logsdon makes the reader feel that the homestead without a stream or body of water is deprived indeed. He describes how to protect these watery environments and use them for homestead aquaculture.

Separate chapters explain cranberry bogs, algae farming, and raising ducks and geese, but most of the book is devoted to the farm pond.

He gives clear and precise directions for building a pond, controlling weeds, maintaining water quality, and raising fish for home consumption.

Richard Darling

Trout*

Those with a good water supply should consider this highly prized food fish. They require clear, cool, flowing water rich in oxygen for efficient growth. Trout are more susceptible to disease than most warm water species.

Brook trout
(*Salvelinus fontinalis*)

Brown trout
(*Salmo trutta*)

Rainbow trout
(*Salmo gairdneri*)

Oxygen: At least 5 ppm for good growth.

Temperature: Prefer 50 to 60°F. Best growth at 58 to 60°F. Severely stressed above 70°F.

Ammonia: Best below 0.025 ppm; 0.3 ppm can be lethal.

pH: 6.5 to 8.5.

Feeds: Animal foods high on the chain and aquatic insects. When stocked densely, need high-protein pelleted trout chow. Will overfeed, which lowers their resistance.

Stocking: Generally not compatible with other species because of extremely high oxygen demand.

Availability: Widely raised by hatcheries across northern U.S.

Sit on the bank of your pond and watch. Watch a water lily. A muskrat feeds on its rhizomes. If it eats all the rhizomes there will be none for next year's muskrats. The muskrat feeds with one eye over its shoulder watching for the mink that would feed on it. A young snapping turtle eats the water lily's leaves. If it makes a mistake and eats too many leaves, the heron will see where it is hiding under the lily and eat the turtle. Sunfish hide among the lilies, too—hiding from the bass—but while they hide, they eat snails that are there eating the algae that is competing with the lily for nutrients in the wastes dropped by the turtle, muskrat, and heron. A frog sits on a lily pad, half-hidden by its camouflage colors and half by other lily pads. The frog doesn't know that the lily pads protect it; it is sitting there waiting to eat the bugs that fly by, attracted by the flowers of the lily.

Basement aquaculture

Backyards aren't the only new places where fish are cropping up these days. A basement doesn't have the benefit of sunshine, but heat from the furnace can keep the water warm. Or cool water can be used to advantage to raise trout. To make up for the lack of photosynthesis, a filtration system must operate full-time. Basement fish pools can operate in both summer and winter, so they can produce two harvests per year.

Rodale biologist Steve Van Gorder says basement fish culture is also a logical complement to an outdoor system. To raise fish to good size in one summer, the fingerlings must be relatively large. Tiny fingerlings are considerably cheaper, though. Steve suggests using the basement system to raise fingerlings up to stocking size during the winter. Experienced fish gardeners may even want to spawn fish at home.

Chickens for fertilizer

Several years ago, Dr. Homer Buck of Illinois, a well-known expert in oriental-style fish culture, showed that fish can be combined with animal husbandry to achieve excellent production. Adapting the methods of the Chinese, Buck fed Chinese carp nothing but algae, which grew in pond water fertilized with pig manure. The manure washed directly into the ponds from pens on the bank and stimulated the growth of algae and plankton.

As part of their aquaculture experiments, Rodale researchers tried this system, using chickens because they're more practical in a backyard than swine. The chickens they used were raised in cages—two to a cage, which provided them with much more room than they normally have in a battery cage. The coops were set directly over the pools to be fertilized. Fish in fertilized pools also got supplemental feed at the rate of 2 percent of their body weight. Fish in unfertilized pools got 5 percent.

The first year, the biggest weight gains occurred in the pool that combined Israeli carp and tilapia. Despite higher supplemental food rations in the trials *not* using fertilization, the

Rodale researchers harvest catfish from a basement pond.

carp and tilapia raised with chickens were significantly larger and had a higher survival rate. Fertilization also cut down food costs by one-third to one-half.

(The other benefit of using chickens to fertilize the pools is the eggs— 1,820 of them were produced by 48 pullets during the 77-day test.)

The following year they again tested chicken fertilization of pool water, with two pullets in each cage above Israeli carp and tilapia polyculture. Since the experiment now could be conducted for the full growing season, almost twice as long as the previous 77-day trial, the researchers figured the advantages and disadvantages of fertilization would become more clearcut.

They were right. Halfway through the test period the chickens had to be removed from the pools. That year, before the research center began using the rotating biological filter of fiberglass plates, the nitrifying bacteria lived in a soil substrate on the bottom of the pool.

These aerobic bacteria converted the poisonous ammonia wastes generated by the fish into nontoxic nitrate. But when the chicken manure built up in excessive amounts, it formed a layer over the soil substrate and created a barrier between the bacteria and the oxygen they needed. That seriously impaired the nitrification process. When the conversion of ammonia into nitrate abated, the ammonia level in the fertilized pools began to rise, threatening the fish.

After the chickens were removed, the nitrifying bacteria were once again supplied with sufficient oxygen and the ammonia level went back to normal. The carp and tilapia began to thrive again.

From the trials, the researchers learned an important lesson: For fertilization of pool water by chickens to work, the addition of manure must be carefully controlled. The biologists will continue to experiment, using only one chicken per cage or introducing the manure in some other way.

The small-scale fish gardener can fertilize the water periodically with a cup of dehydrated cow manure. The advantages are that manure is available at farm and garden centers, it dissolves nicely, and is measurable.

Managing a stream

Fortunate are they who have a spring or stream on their property. Flowing waters can be a reliable source of food for the home. You can harvest this food with relatively little aquacultural management. What's most important is to learn all you can about ecology and gathering wild foods. The work of producing the food becomes a pleasurable pastime and a reward in itself. An hour spent meditating beside a rippling stream while you catch a fish or two can be as healthful to your mind as the nutritious fish protein is to your body.

There's an essential difference between producing food in natural waters and in an artificial backyard pool. In the pool you practice *intensive* production; that is, you try to produce only one or at most several kinds of food, in quantities larger than would be produced naturally in that given amount of water. Producing food from a natural stream or pool is *extensive;* that is, you hope to manage the entire surrounding environment so it will contain a wide variety of foods, in relatively small amounts.

For instance, in a backyard pool, skillful fish gardeners learn how to raise catfish in summer and trout in winter. If the pool is protected by some kind of greenhouse covering, they might also raise year-round vegetables. In a natural stream, the list of possible foods is much more diverse. In our stream, for instance, we have access to a limited supply of catfish, sucker, carp, pike, bullhead, snapping turtle, bullfrog, crayfish, muskrat (yes, this rodent has good, clean meat, since it lives almost entirely on water plants), ducks, geese, and the edible parts of plants like cattails, arrowhead, yellow water lily, and giant bullrush. If our stream were colder and not subject to seasonal flooding, we could have trout, watercress, and perhaps even wild rice.

The first rule of maintaining water quality in natural streams is to protect the land around it. In doing that, you establish a diverse supply of food from plants and animals along the stream or around the pool. For example, suppose you have a clean stream but the water is just barely cold enough for trout (trout like a water temperature of 50 to 60°F.). To insure the survival of the trout you intend to stock, you plant trees along the bank to shade the water on critical hot summer days. The trees and their fruits, nuts, or seeds attract other animals, birds, and insects, which in turn attract other forms of life, all increasing exponentially for your enjoyment or your larder.

Conservationists suggest a protected corridor along a creek, at least 50 feet wide on each side. Even if you have no control over the quality of water coming downstream through your property, you should still maintain these 50-foot strips. Allow no intensive cultivation or grazing of livestock any closer.

In Nebraska, conservationists and ranchers using only this stratagem were able to restore the vitality of an almost-dead trout stream. With the livestock fenced off a mile-long length of creek, the grass, brush, and trees began to grow back, stopping

(Continued on page 378)

Downtown aquaculture: Norristown, Pennsylvania

Dub Smithson's big old house—combination home and photography studio— is right along Main Street of this Philadelphia suburb. All the houses on the block stand shoulder to shoulder. It's nothing like the 20 acres in Texas he and his wife and five sons left years ago.

Still, the Smithson's backyard in Norristown has room enough for a ten-foot wading pool—not for the amusement of visiting grandchildren, but for the crop of tilapia he's raising. In the winter they move inside to a 300-gallon watering trough in the basement.

Dub doesn't act as though it's an unusual accomplishment to have a tub of fish growing in his basement. All it took was a little ingenuity, a lot of common sense, and an assortment of spare parts for the filtration system. Practically speaking, he's raising the fish for food, but the project is just as important to Dub as an unusual learning experience.

Before he started reading about aquaculture, Dub's experience with fish raising went little beyond the family aquarium. "But you haven't gone anywhere if you don't start," he says. "If you have a failure, so what? You learn from it."

The first problem Dub encountered in getting his fish project underway was getting the fish. Fingerlings *are* available, but Dub found that many commercial suppliers were not willing to bother with his small order. It was July 1 when his fish finally arrived.

Not surprisingly, quite a few fish died at first in the outdoor pool. More died while Dub was away from home for several days on a photography assignment. Because he got a late start, the tilapia didn't have a growing season long enough to reach harvestable size. That autumn, Dub rigged up the indoor tank to hold the fish through the cold months and give them another good growing season the next summer. He didn't want to hurry things and try to grow the fish rapidly, since that might have overloaded the system. He was moving the fish from a 1,000-gallon pool to a 300-gallon tank.

With the lower water temperature indoors, the tilapia don't feed as actively or demand as much oxygen. Dub planned to use a 1,500-watt heater, but it proved to be unnecessary. Heat given off by the furnace is enough to keep the water at 70°F.

The only components of the indoor fish system that consume energy regularly are the fluorescent light over the tank, turned on during the day, and the pump for the filter.

Dub's filtration set-up looks like a Rube Goldberg original, but it works. Next to the fish tank is a 55-gallon settling drum. Water and sediment come out the bottom of the tank and flow through a short pipe to the drum. There, particulates are removed with screening and a commercial product, Ammoniasorb.

An inexpensive sump pump propels water from the drum up through an old piece of garden hose. The hose, bolstered by a couple of 2 x 4s, goes back across and hangs several feet above one end of the fish tank. Underneath is a rotating biological filter made from sheets of corrugated fiberglass.

The filter rotates on an axle, and is driven like a water wheel by recycled water pouring out of the hose. Dub chopped the tops off plastic half-gallon containers and mounted them on the filter. As each one fills with water it gets heavy and helps turn the filter. The unit rotates a little faster than two rpm. The falling, splashing water picks up enough oxygen to support the 60 tilapia.

Now that the outdoor season has come around again the fish are already about a half-pound each, and Dub's wife is eager to start eating some right away. (They already harvested a few when the tilapia were smaller, and fell in love with the flavor.) But Dub is holding out a little while longer because he wants to be sure to have a stock of large, healthy fish. Although bringing the tilapia inside was originally a stop-gap measure, now he intends to continue to raise fish all year, both indoors and out. He hopes to spawn the tilapia so he never has to buy more fingerlings.

The tilapia were shipped by air from Florida, packed in plastic bags and cardboard cartons.

Dub at his basement tank.

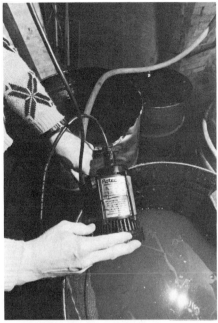

Next to the basement fish tank Dub set up a 55-gallon drum as a settling tank. He uses a sump pump to recycle the water. Water in the other two drums is "curing."

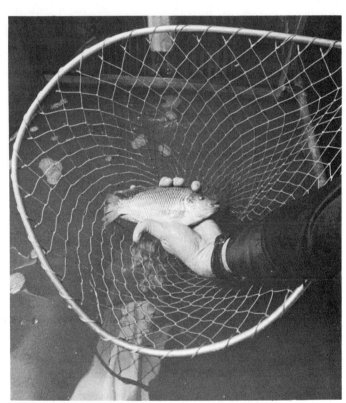

Dub holds a half-pound tilapia. At the far end of the tank water pumped from the settling tank falls into carefully placed plastic containers, turning the fiberglass filter like a water wheel.

Dub fashioned a chicken wire and plastic shelter over the outdoor pool.

(Continued from page 375)

heavy siltration of the creek bed from eroding rains. Since the cattle could not get into the stream along this length, they no longer could stand in the water and make it muddy, so the trout returned and increased in population.

Protected land around your natural water becomes a fringe benefit while it helps improve water quality. Renewing plant cover lures quail, pheasant, rabbits galore, even deer—all wild animal species like the combination of water, cover, and food a wildlife area provides. Bird life will increase dramatically, especially if you provide box nests for those species that will use them, like the beautiful wood duck.

As manager of this wild refuge, your role is that of a referee. You can pick and choose which plants (and how many of each) you want to remain there, among the many that come up as volunteers. Aim for variety—the key to ecological balance and abundance is diversity.

You can sometimes manipulate your landscape to create small bodies of more or less natural water, even where a running spring is not appar-

Mechanical aerators spray pond or pool water into the air so the water can pick up oxygen.

ent. An area of seep springs can be dug out with a dragline to make a pool, or a number of very small trickling springs can be rechanneled together. Small oxbow ponds can be cut off from a meandering river, and replenished each year when the river floods.

Slightly altering a creek or pool itself can increase the fish population. Trout, for instance, like a stream about half riffles and half deeper, quiet pools. Trout enthusiasts have learned to place low (one-foot-high) log dams in their streams, over which the water tumbles, creating little pools below the dams. Just downstream from the pools, sand and pebbles wash out to form shallow riffles. Log and riprap deflectors along the bank can help straighten meanders; weirs can narrow the current, increasing its speed. In both pools and streams, you can sink boulders or watersoaked logs to provide hiding places and perhaps nesting sites for fish. Information from wildlife agencies in your state will show you how to block upstream navigation of undesirable fish in trout streams.

A small dam can be valuable even in a sluggish warm water creek such as ours, where late summer drought can cause the water level to diminish to a mere trickle between the deeper pools. We use rocks piled rather casually across the creek where the banks narrow. We jam the upstream face of the rock dam with clay scooped out of the creek, and raise water level about 8 to 12 inches, which amounts to a considerable increase when figured over a length of 300 feet or so. In winter and early spring, the dam floods out a little, allowing the water to sluice out silt that tends to build up on the stream

floor above the dam. On a warm day in June, we repair the dam and it again keeps the water level up through the critical dry weather of late summer. During spring migration upstream, fish have no trouble getting over the little dam, by swimming over when rains swell the water level above the dam, jumping over with apparent ease, or simply swimming straight up the little waterfalls coming over the rocks.

Because our stock of fish is replenished each spring when the fish migrate up from the river, we don't worry about stocking the stream. The only control we can exercise (or want to exercise) is in removing the large carp that come upstream. They get trapped in the small pools, unless a heavy rain gives them a swollen current in which to swim back to the river. If we wouldn't remove them they'd muddy the water too much for the other fish.

Of course, if you fashion a body of water in which no fish existed before, you must stock your fish. Trout (and watercress in the slow current of protected places) could be your first choice if water is cold and clean enough. In warmer water, largemouth bass, perch, channel catfish, yellow cats, bullheads, pike, and bluegills survive best, from my experience in Ohio. Stocking rates for farm ponds do not apply to streams and springfed ponds unless the fish cannot escape downstream. Talk to wildlife biologists in your area. I would not put more than a dozen of each of the above named species per 500 feet of stream such as mine (average width 6 feet, average depth 18 inches, with pools for winter survival up to 4 feet deep). The fish work out their own balance.

You should not have to provide supplemental food for your fish in a running stream. We sometimes throw handfuls of fish worms into the deeper pools during periods of low water. But because of the dynamic natural populations, the overall in-crease in fish from supplemental feeding is probably slight. In feeding one link in the biological chain you either feed a gradual increase in all the links, which may not mean an appreciable increase in the amount available for you, or you may throw the balance out of whack by encouraging too much of one species over another.

In a stream or small pool, seining will catch more fish faster than hook and line. Seining of some fish (especially trout) is illegal in some states, however, so check with your game warden.

Wherever water occurs constantly throughout the year, ingenuity can figure out a way to divert it into at least a semi-natural pool or stream for fish, plants, and animals. You can enjoy the beauty and peace of your compact wild refuge while you harvest the food you helped the water produce.

Gene Logsdon

Trout pond: Hixton, Wisconsin

Our small farm had a marsh in one field directly below our house. This wet area was a constant irritation. Its boundaries seemed to wander, and every so often the front wheels of our tractor would break through in a new spot, requiring assistance to be pulled out. Marjorie and I decided it would be better to make a pond of the marsh. It could be used for a swimming hole after a hot day of baling and as a source of fresh fish for our table.

Costs were important to us, so we did much shopping for best prices and efficient designs.

To determine a feasible area for the pond, we took soil samples to about a three-foot depth with an ordinary post hole digger. We also made a complete section drawing showing the water source, pond, and drain elevations.

A local contractor dug the pond with a backhoe in just a few hours. It is approximately 30 feet wide by 100 feet long, 9 feet deep at the water inlet and gently sloping to 1½ feet deep at the outlet end. The sides were left as steep as possible, and the dirt removed was piled around the hole about 4 feet high, to compress the surrounding, peaty soil.

A stream fed by a small, year-round spring fills the pond. The water is piped approximately 100 feet through a four-inch plastic sewer pipe buried 3 feet beneath the surface. The outlet from the pond also runs underground, about 20 feet, and empties from the pipe about ten inches above the surface of the natu-

Turning a marshy area into this pond, the Seikmans enjoy "the satisfaction of having made something good out of a nuisance."

ral stream. This elevation is to prevent backup during periods of high water, such as after a heavy rain.

In Wisconsin, the rearing and handling of game fish is strictly controlled by the state Department of Natural Resources (DNR). Recreational fishing is a major tourist attraction, so it is vital that diseases and other sources of catastrophe be avoided. In order to purchase fingerlings, you need a license (for which there is no fee) stipulating the quantity to be purchased, the name of the hatchery, and the date of transportation. We decided to try rainbow trout because of their adaptability to our expected water temperature variation (between 32 and 60°F.) and their local availability.

We bought 100 four-inch fingerlings at 20 cents each, and transported them in garbage pails lined

(Continued on page 380)

A FISH AND VEGETABLE GROWER FOR ALL SEASONS

Robert E. Huke and Robert W. Sherwin, Jr. 1977.
Norwich Publications, Box F, Norwich, VT 05055. 125 p. paperback $4.95.

The ultimate in home aquaculture systems is the solar greenhouse. In this little book, Huke and Sherwin discuss the advantages of raising fish and vegetables together in a completely controlled environment. The book is not a step-by-step guide to fish gardening. It does present a convincing argument for the practice of small-scale aquaculture. The authors also look idealistically beyond the backyard to community greenhouses and food for the year 2000.

We have in mind the rapid expansion of controlled environment enclosures—greenhouses if you like—at the urban fringe, in the suburbs and in the rural non-farm areas. In some locations these would be large structures to provide vegetables and fish for 100 or more families. In the suburbs and more rural areas the structures will be designed to provide part of the needs for a single family. The surface areas in such structures will be one-half water (about 3 feet deep) and one-half soil. The water will provide the habitat for food fish and will also act as a heat sink to help reduce temperature variations and lengthen the growing season. The soil will be fertilized with sludge from the local sewage plant and watered with the waste-enriched effluent from the fish tank filter. The four-day work week will leave ample time for operation of the units. Solar energy will provide the major non-human input and the semi-closed system will allow for the exclusion of major atmospheric pollutants. The microworld of the dome farm and the individual contact with living systems will help to overcome the psychological rigors of a crowded world and will help in a small way to provide an improved diet.

(Continued from page 379)
with plastic bags. The cans floated in our pond for one hour in order to equalize the water temperature before we released the fish. We purchased fine (³/₃₂-inch) granular food from the hatchery as a starter. Since this food was the sinking type, we could not see the fish feeding until they learned to rise near the surface to eat. It took nearly a week before we (or they) found their favorite spot, which was at the deepest end. We feed twice a day, about two hours after sunrise and an hour before sunset. The quantity of food must be carefully regulated. Too little food delays growth. But too much food is wasteful and dangerous: some fish will overeat, and uneaten feed which collects on the bottom will rot. Trout have a much greater appetite in warmer water, and the larger fish obviously eat more, so it is necessary to check the water temperature occasionally and observe the size of the growing fish.

In spite of our very cold winter climate (which occasionally dips below −30°F. with wind chill factors of −70°F. or worse), at least a small circle of water always stays open near the water inlet. It probably helps having the 100-foot inlet pipe located below the frost line. We are thus able to feed once every day throughout the winter (an unpleasant task in bad weather).

Our reward comes in rapid growth. Fish can be thought of as thrifty animals, converting about 50 percent of their feed into body weight. Fingerlings purchased in November, our first year, grew to ten inches by the end of May. Caught at that length, they cost a total of about 50 cents per fish, or $1 per pound.

The rate of growth beyond this length diminishes somewhat, but we prefer to take one fish large enough for two people rather than one for each. We calculate the cost of a one-pound fish at 85 cents, cheaper per pound than the eight-ounce size. After about 12 months the trout are 12 to 16 inches long. With continued feeding, the fish grow to five or six pounds, and 20 to 26 inches long, in three years.

During the summer months we switch to a floating feed, partly because of the pleasure we get watching the trout leap out of the water as

they eat. We have not noticed any difference in weight gain for the larger fish, but the smaller fish seem to prefer the sinking variety.

A license is required in Wisconsin to remove any fish from any body of water, even one's own private pond. This can be a regular fishing license, which requires the holder to use a hook-and-line and follow size and season restrictions; or it can be a special class of commercial (hobby) license which permits seining and open-season removal, but not resale, of fish. The fee is $5 per year plus a one-time inspection fee to insure that the pond design is acceptable to the DNR. Because Marjorie and I see nothing "sporting" about killing the animals we hand-raise for our table, we originally chose to seine those fish ready for harvesting. That practice also permits selective culling, a more efficient process for managing the operation. By the second summer, however, we had to fall back on a hook and line technique because the largest fish broke our net.

We recently spotted some small trout that have spawned and are taking shelter in the weed bed at one end of the pond. We hope they will survive.

The DNR warned us during our planning stages that a muskrat could do considerable damage, and after three years we first spotted one of the critters. It was charming to watch, but we quickly discovered its intent: to dig out all the plants at the water line. Within a few days it had excavated an area three feet long at the shallow end of the pond. The plants roots hold the soil together in our fine sandy silt, and in their absence the water began to percolate through the side of the pond. In just a few days the water was down to creek level.

This was more than a trivial change because the surrounding soils now began to drain into the pond. At deeper levels, these are clays that have a higher pH than our silt loams, and the sudden change in pH stunned the fish somewhat.

The DNR says there is no way to frighten muskrats away. In fact, they will travel long distances over land to find a new pond. One consolation was that our unwelcome visitor arrived during legal trapping season. But because the ground was frozen at the time, pond repairs had to wait until spring.

There are little problems we have not solved. How does one explain to friends that just because the fish are there, they must not all be caught immediately? How does one point out that, like any farm product, they are not simply a free harvest? How do you keep poachers out when rumors of 16-inch trout spread rapidly? To those who exclaim, "Gee, those would look good in a frying pan!" Marjorie retorts that they can have anything they can catch in a frying pan.

It will take many meals of fish to pay for the cost of the pond—about $275 worth—but we think of this as a bargain swimming pool. We also have a lovelier view from the house now, especially when meandering mallards drop in to eat excess algae. Best of all is the satisfaction of having made something good out of what had been a nuisance.

Bill Seikman

In addition to the help they received from the Wisconsin Department of Natural Resources and their county agricultural agent, the Seikmans found two USDA pamphlets very useful in planning their pond. *Trout Ponds for Recreation,* Farmers' Bulletin No. 2249, and *Trout Farming,* Soil Conservation Service Leaflet 552, are available from the Publications Division, Office of Governmental and Public Affairs, USDA, Washington, DC 20250.

Growing fish in cages: East Greenville, Pennsylvania

When Christopher Meinzer bought a rural home outside East Greenville, Pennsylvania, he acquired a good deal of water with his land—enough that fish seemed like an ideal crop to help lower his food budget. Meinzer owns most of a millrace and an island between the millrace and Perkiomen Creek. But like many would-be aquaculturists, he found that his water supply posed problems for fish raising. He could put ponds on his island, but spring flooding might wash away any fish he would stock in them. The dammed up millrace is less prone to flooding, and with its slow-moving current seemed more like a pond. But if Meinzer stocked fish there, he would wind up feeding such uninvited residents as carp and giant goldfish.

In 1978 Meinzer solved his problems with a suggestion from biologists at the Organic Gardening and Farming Research Center. He decided to grow his fish in small floating cages, using the research center's design for a one-cubic-meter capacity cage made of wood, Styrofoam, and plastic netting. In May of that year, Meinzer moored two cages to a bridge over the millrace. In one, he stocked 250 4-inch channel catfish fingerlings; in the other, he placed 250 1½-inch hybrid bluegills. Each day he fed both groups a balanced feed purchased through his local farm supply cooperative.

As with many homesteaders' experiments, not everything went according to plan. In the first two weeks, 40 catfish died from diseases caused by the stress of shipping from the hatchery. Another 40 were lost through attrition during the summer. A few bluegills were lost to cannibalism, and the survivors grew slowly. The surviving catfish grew so well, though, that Meinzer considered his efforts an unqualified success. By November they were 11 inches long.

"The fillets dressed out to about seven inches; one fish was a nice meal for one person," Meinzer says. "I figured that it cost easily under a dollar a pound to raise them, including the cost of buying the fingerlings. It's really an easy, cheap way of supplementing your food budget."

Growing fish in cages is an idea imported from the Orient—an idea that attracted the attention of many American aquatic biologists during the 1970s. To most people, the technique still sounds exotic, but Meinzer thinks the concept is no different than raising chickens in cages or cattle in fenced pastures and pens. "Just because you own a hundred acres of land, you don't throw cows on a hundred acres and ignore them. You manage your herd," he reasons. Similarly, putting fish in cages instead of letting them swim freely in a pond allows more careful husbandry of the fish. "With cages you know exactly where your fish are everyday," he says. "You don't have to keep baiting a hook or trapping them to see if they're healthy and growing."

Management is sadly lacking in most of the small farm ponds that dot America's landscape. One of the most common mistakes pond owners make is introducing too many species

A fish cage of one cubic meter.

of fish into their pond, says Robert Butler, fisheries biologist at the University of Pennsylvania. "A pond is not a lake. It doesn't have deep water and the range of environments needed to support a wide variety of species," he says. Often, one type of fish wins out over the others. Ponds filled with stunted bluegills and nothing else are one result of trying to recreate the bounty of the catch from a Canadian fishing trip in a livestock watering hole.

Cages offer a nice way around that problem. For example, a pond owner who has managed to keep a balanced population and bluegills might want to introduce catfish into his pond. That can throw things off balance, explains Steve Van Gorder, Organic Gardening and Farming Research Center biologist. If you decide to feed the catfish, the bluegills will get some of the feed, too. If they grow too much, the bass, which normally keep the bluegill population in

(Continued on page 382)

(Continued from page 381)

check, will no longer be able to eat them. And your pond may soon be overrun with bluegills.

"If you stock your catfish in cages, you feed only the catfish and you won't hurt the bass-bluegill balance," Van Gorder says. Just about any fish that will accept feed can be grown in cages—catfish, bluegills, perch, carp, and trout in cold water, tilapia in warm and tropical waters, and even bass fingerlings (although mature bass do poorly). During winter months, caged trout have been successfully reared in southern Illinois ponds used for summer catfish production.

Besides allowing aquaculturists to produce a greater variety of fish with more careful management, cages offer other advantages over more traditional methods of growing fish. First, cages protect your fish from predators. Second, if you already have access to a body of water like a large lake or river, cages are a small investment compared to making a pond. The materials for a one-cubic-meter cage cost about $60. A third advantage is that it's easy to harvest fish from cages. Fish can be removed with a dip net, or the cage can be lifted from the water and the fish dumped out. This ease of harvest may be the greatest advantage of cage culture. Commercial fish farmers who use shallow man-made

Rodale aquaculturists check the catfish cages in a pond at the Research Center.

ponds or raceways usually seine their harvest. But most available waters—rivers, large lakes, strip mines, gravel pits, and farm ponds with stumps and snags—cannot be seined. Only cages make growing fish for harvest practical in those waters.

Cages work so well that commercial producers use them on large reservoirs in Arkansas and Oklahoma for rearing catfish. State game and fish agencies use them for growing fish for stocking lakes and streams. Cages aren't trouble-free, however, and anyone who is thinking of trying them for home-fish production should be aware of their disadvantages.

First, since caged fish cannot forage for insects and other natural sources of food, you must buy a nutritionally complete feed, which is more expensive than other available fish feeds. It also must be a floating feed—again more expensive than sinking types, but necessary, because sinking feed would fall through the mesh of the cage and be wasted.

Another problem is that cage fish are more susceptible to bacterial diseases. Further, cages may be vandalized in populous areas. You should know that fish in cages use just as much oxygen as free-swimming fish. That means you can't use cages to crowd more fish into a pond or lake than it would normally support. If your pond is big enough to hold 500 free-swimming fish, it can only hold 500 in cages, or 400 in

cages and 100 swimming free, and so on.

Exactly what results can you expect from cage culture if you can avoid the pitfalls of disease, oxygen depletion, and vandalism? Preliminary research at the Organic Gardening and Farming Research Center is encouraging. Experimenters stocked 84 pounds of channel catfish in a one-cubic-meter cage on June 5. Six times a week the catfish were fed a daily ration equal to 3 percent of their body weight. Once a week the fish were weighed and their feeding rations readjusted. On September 25, 112 days later, 299 pounds of fish were harvested from the cage. The 458 pounds of feed used through the season cost $110. Feeding costs averaged about 37 cents per pound of fish—certainly a bargain when compared to supermarket prices for fish.

In 1970, researchers from the Illinois Natural History Survey successfully grew trout in cages in a southern Illinois pond. On February 17, they stocked two cages with 412 rainbow trout per cage. Each cage had a volume of 1.7 cubic meters. By May 15, when the fish were harvested, the surviving trout had doubled the weight from about ¼ pound apiece to nearly ½ pound. The fish gained poorly in late spring after the water temperature passed 60°F. Because more than a third of the trout died, production costs were about 40 cents per pound of fish. The re-

Harvesting the fish from a cage is as easy as pulling the cage out of the water.

searchers considered that cost too high for commercial production but thought it might be "practical to rear trout in this manner for the home freezer."

Dan Looker

If you decide to try cage culture, you'll need to learn more about such technical details as stocking densities, disease prevention and treatment, and which fish are suitable for confinement (channel catfish, for example, do better than other types of catfish). Here are two excellent pamphlets to get you started. *Catfish Cage Culture: Fingerlings to Food Fish,* Publication No. 13, available free from The Kerr Foundation, Inc., Agricultural Division, P.O. Box 588, Poteau, OK 74953.

Profitable Cage Culture, available for $1 from Inqua Corporation, 18460 S.W. 295th Terrace, Homestead, FL 33030.

Other experiments

Across the country, groups concerned with the well-being of the earth and its peoples are studying ways to increase food production at the home level, and to become less dependent upon large-scale, energy-intensive producers. Several of these groups are turning to aquaculture.

The Amity Foundation in Eugene, Oregon, recently conducted a state-funded project to test the practicality of fish farming in a solar greenhouse. They raised tilapia, carp, oscars, and bluegills in concrete fish tanks enclosed by a steeply gabled passive solar greenhouse. They experimented with two types of biological filters (a trickling filter and an airlift filter) and constructed a Savonius rotor windmill for oxygenation. The design is based on the work of Finnish engineer J. Savonius, who studied the aerodynamic properties of S-shaped vertical axis turbines.

For ten years, the New Alchemy Institute at Woods Hole, Massachusetts, has been developing technologies for providing food, energy, and shelter in self-sustaining communities. New Alchemy emphasizes the use of the sun and the wind. They have raised fish in a round pond capped by a fiberglass geodesic dome, and in a series of three pools through which water flows by gravity. A windmill powers a pump to cycle the water back to the top pool from the bottom.

The Integral Urban House in Berkeley, California, is an ordinary two-story house that the Farallones Institute fashioned into an ecologically stable, resource-conserving living system. Since 1973 the house has been the research and educational center for The Institute, whose engineers, architects, and biologists are working to develop "urban-scale appropriate technology." Bluegills, black bullhead catfish, and Sacramento catfish grow in the 2,000-gallon concrete pond in the backyard. A Savonius-rotor windmill operates a pump that cycles the water through a biological filter. (For a review of The Institute's book, *The Integral Urban House,* see page 458.)

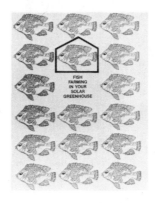

FISH FARMING IN YOUR SOLAR GREENHOUSE
William Head and Jon Splane. 1979. Amity Foundation, P.O. Box 7066, Eugene, OR 97401. 43 p. paperback $5.

Home aquaculture is tricky business. A prospective fish farmer probably will never get a fillet onto his plate unless he understands the complex biological and environmental cycles that are so much a part of any aquaculture system.

This handbook can help. It explains the intricacies of fish farming in totally understandable language. Most useful is the straightforward advice on fish management, including selecting fish, making home-grown diets, and harvesting.

The technical information on solar greenhouses and pool filtration, illustrated in painstaking detail, comes from the experience of a solar greenhouse aquaculture project conducted by the Amity Foundation, a non-profit corporation organized to develop technology for a life that's more self-reliant and less energy dependent.

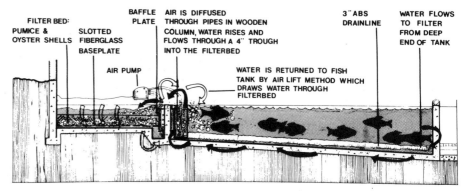

Air-lift filter system, from *Fish Farming in Your Solar Greenhouse.*

COOKING FISH

Few other foods can be prepared in a greater variety of ways than fresh fish. Yet many people shy away from cooking fish altogether, because they are afraid it will come out tough or tasteless. Ironically, the secret to perfect fish couldn't be simpler.

Measure the fish at its thickest point and cook it exactly ten minutes per inch or any fraction thereof. The rule holds true for baking, broiling, frying, sauteing, steaming, poaching—any way the fish is to be prepared.

For whole fish, lay the fish flat and measure vertically at the thickest point. If the fish is filleted or steaked, measure the depth of just that portion.

A fresh fish is naturally tender. Over-cooking will dry it out. When properly cooked, the flesh will be opaque and flaky, and retain its own juices.

Be wary when shopping for fresh fish. Some markets will freeze fish, then thaw it, and offer it as fresh. Do not buy such offerings, for they deteriorate rapidly. Learn to recognize a fresh fish

Clear eyes and bright red gills mean this fish is fresh.

- Eyes are clear and bulging.
- Gills are bright red and free from slime.
- Flesh is firm and elastic; it should spring back when pressed lightly with a finger.
- Skin is shiny, but not slippery or slimy.
- The fish has no "fishy" odor.

A fish with opaque, sunken eyes, flesh discolorations, a thick slime, or noticeable odor is not fresh.

Store fresh fish in the coldest part of the refrigerator, and use it within one or two days.

Tools designed specially for cleaning fish, like the fish scaler and filleting knife shown here, can make you more deft at handling the catch. To scale a fish before cleaning, use a fish scaler or a knife with a serrated edge. Hold the fish firmly by the tail and scrape off the scales from the tail to the head, against the direction in which the scales grow. Filleting knives are thin and flexible so they can maneuver around the bony skeletons of fish.

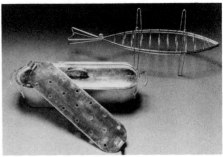

The best way to preserve the succulence and beauty of a fish is to cook it whole. In a fish poacher (foreground) the fish simmers in water or a delicate sauce while the perforated rack protects the fish from the direct heat source. A fish-shaped grill (background) holds a fish within a tinned steel frame, and the grill's short legs allow you to nestle it in the coals of a fire.

Disassembly illustrated

Because catfish have no scales, they are cleaned differently than other fish. The easiest way to skin a catfish is to impale it with a nail through the head and into a heavy board. Beware the sharp spines, or "whiskers." Make a cut in the skin all the way around the fish just behind the head.

Grasp the skin behind the head with a special pliers made for skinning catfish. Pull the skin firmly downward. With practice, you can get the skin to come off in one neat piece, inside-out. (An ordinary pliers will also work—the skin will peel off in narrow strips.) Catfish skinners are available from Champlin Net Company, P.O. Box 788, Jonesville, LA 71343; and Delta Net and Twine Company, Box 356, Greenville, MS 38701.

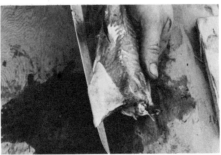

To gut the fish, start at the anal opening and slice along the belly of the fish toward the head. The trick is to cut deeply enough to open the fish, but not so deep that you slit the belly cavity. Cut away the innards, and thoroughly rinse off all blood.

The catfish can be prepared at this point. It has just one large spine, which is easily removed after cooking. If you wish, you can fillet the catfish. Lay the fish on one side with its backbone toward you. Use a thin, sharp filleting knife. Slice through the flesh, following the line of the bones, and peel back the fillet as you go.

Although you may be the first on your block to raise fish in a swimming pool, don't think the idea of farming fish for food is a brand new one. A bas-relief in the Egyptian Tomb of Aktihetep shows someone harvesting tilapia from an artificial pond, cleaning the fish, and laying them in the sun to dry. The year? About 2500 B.C.

The Chinese, Greeks, and Romans were likewise cultivating fish—particularly carp—four thousand years ago. During the Middle Ages feudal lords stocked the moats around their castles. Monasteries raised their own fish to have a fresh, readily available supply throughout the Lenten season.

Cut off the head with a boning knife or fish knife.

To fillet the other side, flip the fish, keeping the backbone toward you, and once again cut along the curve of the spine. Fillets may be cut either head to tail or tail to head.

Eden in a thimble: Norwich, Vermont

Here is Eden in a thimble. Lush vegetation snakes around a steamy green lagoon, fish school beneath a splashing waterfall, and condensation drips langorously from overhead. The occasional insect buzzes through the tropical air. This aquadome is unheated in the conventional sense, but its temperatures remain moderate year-round—despite winter temperatures that may scrape 30°F. below zero.

Viewed from the outside, this geodesic hump seems to fit the season: beehive in July, igloo in January. By night, it glows like the rising moon. Since 1974, Robert and Ellie Huke have tended this homemade Vermont nirvana under a dome, annually raising three crops of vegetables and two harvests of fish.

In the late fall they harvest more than 80 ¾-pound tilapia—a direct descendant of the "fishes and loaves" of parable fame—and stock the pond with 50 brook trout fingerlings. By May Day, they are ready to haul out 40 pounds of gourmet-quality trout and replace them with the warm-water-loving tilapia for another summer.

The aquadome evolved naturally out of Robert Huke's research into global climate trends and their effect on agriculture. A professor of geography at Dartmouth College, Huke is well-aware of the precarious balance of weather and food production.

"Our original idea was to gain a longer growing season," Huke recalls. "We began planning to trap solar energy in a greenhouse structure, but I wanted to avoid the usual wild range of temperatures in a greenhouse—too hot in the day, too cold at night, like a car with its windows raised.

"We had to have a heat sink, something in which we could store heat during the hottest part of the day and which would readily release it at night. Water turned out to be the best heat sink we could think of."

Heat sink is a term that brings to mind expensive solar components, but what Huke turned to was the bottom section of a preformed concrete septic tank, capable of holding 1,500 gallons of water.

Huke might have liked to bring his entire garden indoors, but the original fantasy of a huge dome was soon limited by reality, the energy crunch of 1973. Huke had decided to cover his structure with PVC plastic, and, with motorists lined up at the gas pumps, and supplies of many petroleum-based materials threatened, the best and widest plastic he could find was 10-mil material in 54-inch wide rolls. Accordingly, he decided to build a dome with a maximum span in each triangle of 52 inches, allowing 2 inches for secure stapling.

Having access to a computer and with the aid of colleague Robert Sherwin, Huke determined that—with the available plastic film—they could build a dome 17.1 feet in diameter, with an 8-foot peak. The skeleton of the dome was built of high-stress, clear southern pine, with no knots or weak points.

Close to their house the Hukes had found a level, sunny, well-drained location for the dome. A slope of tall pines buffered the north wind. The setting seemed ideal. Because of its portability, the dome skeleton was built and then moved onto its permanent foundation.

A backhoe operator had been hired to dig a four-foot-deep hole, carved in the circumference of the dome shape. Ten five-foot-long cedar posts were next installed at intervals around the periphery of the excavation, driven one foot into the ground at the bottom of the hole, and used to support the base hubs or the dome.

From ground level down, the Hukes inserted two-inch foam insulation between the anchor posts, to help conserve the heat of the dome's soil. They leveled the pit bottom and covered it with three inches of sand. A truck rigged with a crane delivered the concrete septic tank, setting it in the center of the pit with its long axis running north-south.

Next, the Hukes dug a ditch from the house to the dome, installed an underground electric cable and filled both the trench and the dome pit with soil. A few friends appeared to help carry the dome frame to its cedar post foundation, where it was anchored with bolts and wire and

(Continued on page 386)

(Continued from page 385)
treated with wood preservative.

The nine most northerly panels were covered with 1½-inch foam, its foil facing inside, to insulate against north winds and to reflect back the light coming in from the south. To carry the expected weight of snow on top of the dome, the five panels at the peak were covered with premium-grade fiberglass. The balance of the dome was then skinned with 10-mil polyvinyl plastic, stapled to the outside of the struts. The basic structure of the aquadome was now complete. Material costs in 1974 were about $1,000.

The Hukes filled the 34-inch-deep pond with rain water, which warmed quickly, thereby moderating temperature extremes just as they had predicted.

"It seemed a waste to have all that water inside the dome, and not be able to use it in some way," says Huke. "So we decided to put fish in."

To bring the system to life, the Hukes installed a water circulator. This pump and filter rig is the heart of the aquadome. Without it, waste would quickly accumulate and poison the fish. They linked two 55-gallon oil drums together with a 2½-inch pipe, four inches from the bottom. The tops were removed, the drums thoroughly cleaned, and the exteriors painted black to absorb the heat. A couple of heavy-duty beams were bridged across the width of the dome pond at the south end, and the barrels placed on top. Both barrels were packed with Actifil—multifaceted black plastic matrices, each the size of a spool of thread. A submerged pump capable of circulating 600 gallons an hour was installed to take water from the bottom of the pond, run it through the barrels of Actifil, and return it over a four-foot waterfall.

That first fall the Hukes stocked the dome pond with 50 four-inch brook trout fingerlings. Throughout the winter the fish grew on a daily diet of commercial trout chow in water that never fell below 37°F., although temperatures outdoors dropped as low as 20 below zero.

By mid-April the 12-inch trout average ¾ pound each, and the Huke family set about harvesting them. Then the Hukes stocked the summer pond with tilapia because of

their tremendous growth rate and feed-conversion ratio. "We stocked 87 2-inch fingerlings and harvested 87 9-inchers about 12 ounces each. It's crazy but true!"

When water temperatures cool down to about 58°F., usually in late November, the Hukes drain the pond and prepare the tilapia for the freezer. The septic tank is scrubbed down, rinsed out, and filled with rain water. The filter is thoroughly cleaned, yielding several pounds of rich fish wastes for their outdoor garden. The barrels are scoured and rinsed with rain water before the filter system is replaced. And the pond is ready for stocking with 40 or 50 four-inch brook trout.

"Our last crop of tilapia was beautiful," says Bob Huke. "It took me about a day to clean them up and

wash out the tank, and they sure did smell. But Ellie cooked up a batch with sweet and sour sauce, and our daughter Debbie took the first bite. 'Hey, not bad,' she said. Not bad? Well, they were excellent!"

Since the first year, the Hukes have made a few changes in the aquadome. The fragile PVC dome skin peeled away in a heavy wind during the first winter, and they replaced it with lightweight Kalwall SunLite fiberglass, a rigid greenhouse material with excellent light-transmission characteristics. This raised the cost of the dome by $160, by the time Huke had purchased the fiberglass, bolts, and nails. Today, this would represent an additional investment of about $200—a cost that would be a necessity for a durable structure.

The Huke dome will surely

In spring Huke stocks tilapia fingerlings in his fish tank.

undergo further refinements. For the time being, however, it continues to successfully supply both food and something equally important:

"After a long and busy day at the office with the phone ringing constantly," says Huke, "what could be more relaxing than a few minutes in the dome—no phone, no radio, the damp earth smelling of growing plants and the water cascading from the filter as a miniature waterfall? As one enters the dome, the trout start jumping for their evening ration. Inside the dome one is in a world apart, separated for awhile from the pressures of 20th century society and surrounded by a mini-world of one's own creation."

Article and photography by Tim Matson. Reprinted with the permission of *Harrowsmith Magazine*. Copyright © 1979 by Camden House Publishing Ltd.

Fish and vegetables thrive side by side under Bob Huke's geodesic dome.

Mushrooms

Home-grown protein from an unlikely source

Raising mushrooms shares some of the mystique of wine-making. Lots of people grow grapes, and there is plenty of information on home wine-making and necessary equipment available. But try to find a few wine-makers among the growers. So it is with mushrooms. The key to growing them is good composting and lots of people can do that. But where are the mushroom gardeners?

The mystery associated with mushrooms puts people off. This crop is nothing like garden plants. And unlike other indoor, off-season gardening projects, mushrooms don't provide much of a show: for weeks the boxes of compost sit growing mushrooms, one supposes, but it's hard to tell.

But the people who are involved in it report that mushroom growing is fascinating and has its rewards. First of all, making the necessary compost is a well-known and satisfying job for many gardeners. When the mushrooms finally show, the impressive display of fruiting lasts for weeks. That's not to mention the eating. And, if there's a surplus, fresh mushrooms are as easy to preserve as peas. Finally, when the crop is finished, the material that's left is a rich, thoroughly digested compost.

How to get started

Button mushrooms, the kind in the markets, are the subject of much how-to writing in books and magazines. Many seed companies sell mushroom spawn, along with detailed instructions that should tell you what you need to know.

What does it take?

Mushroom growing doesn't require lengthy research—unless you will only be happy with productivity levels that match commercial growers. After all, cultivated mushrooms are close to their original state, not high-strung and over-bred like many crops. Domestic edible types appear in lawns or on trees. In San Francisco, a wild strain of the common button mushroom has colonized some organic gardens, much like lamb's quarters and other weeds. When the time is right, up pops the fruit. Knowledgeable gardeners make short work of this bonus from their good composting.

Long ago, Asians and Europeans cultivated other varieties than the button which has become synonymous with the mushroom. A few people in the United States have re-developed these old techniques for home production. These exotics include Pleurotus, or oyster mushroom, from Europe, and shiitake from Japan. You may also find spawn for two other oriental types, the velvet stem and the wood far. Cottage industries are now based on growing both shiitake and Pleurotus.

None of these mushrooms has growing requirements as complicated as for the button mushroom. They don't need compost, making it from such hard-to-compost materials as fresh sawdust, straw, and shredded paper. To make a growing medium of these materials requires no manures. After the crop is finished, the stuff has been turned to compost.

In the wild these mushrooms grow on dead or dying trees. Domestically, they can be cultivated on a mix of nine parts sawdust to one part grain bran. They prefer hardwood materials, oak if you can get it. Dry autumn leaves could probably make up part of the growing medium. The mix should be boiled in a large kettle for a time to sterilize it before adding the spawn. Ordinary mushroom beds make good growing containers, but they should be placed in an area with some diffused daylight and the surface should have access to fresh air. The spawn-inoculated growing medium is packed firmly about four to six inches deep.

Temperature requirements aren't complicated. Before inoculation, the growing mix should be cooled to around 75°F. and held there during early growth. The temperature should then gradually drop about 10°, the best level for fruiting. Though not ideal, slightly cooler temperatures will work. The velvet stem mushroom fruits well in the mid-50s. Always keep the beds moist.

The Pleurotus actually will grow better on plain shredded straw. Some shredded paper (up to half) can be mixed with the straw. Straw needs to be heated only to pasteurize it, at least 160°F. for 15 minutes. Of the exotics, Pleurotus spawn is most readily available.

I first encountered this mushroom while visiting a man who made compost for Pennsylvania mushroom farmers. In one mushroom house, uncharacteristically well lit, was a stunning profusion of mushrooms growing out of shredded straw and corncobs.

The chef at a restaurant nearby was buying every one the grower would sell at a premium price. I was honored with a three-pound basket to try, and I felt that I carried away with me a rare gift of the woods. The mushrooms were wonderful: savory, mild, and chewy, much like oysters.

Assuming that a bale of straw which has been steeped in hot water will weigh 90 pounds and that you have provided the right growing environment, you should pick up to 90 pounds of Pleurotus mushrooms over the six-week harvesttime, according to Dr. Ralph Kurtzman, a USDA Research Service scientist who has specialized in developing Pleurotus culture. At a dollar a pound they would be cheap—a handsome return on a bale of straw, some spawn, and the heat for pasteurization.

A few years ago, researchers at the Organic Gardening and Farming Research Center grew a variety of Pleurotus from India on chopped corncobs alone. This worked fine, though the fruits were not as large or abundant as I had seen growing on the straw mix. Whether you grow them in flat trays or on vertical chicken-wire racks does not seem to matter much, as long as a lot of the surface area is exposed to fresh air.

The hardest thing about getting started with these new mushrooms is finding more information. The effort is still in the developmental stages and the literature is still devoted almost exclusively to the button mushroom. One unusual book that may help is *Growing Wild Mushrooms,* by Bob Harris (Wingbow Press, 2940 Seventh Street, Berkeley, CA 94710). This pioneering book contains useful advice on growing species with different habits and habitats, despite Harris's bias toward hallucinogenic rather than culinary mushrooms.

A first-rate introduction to mushrooms can be had by foraging. Once you see how well they grow on their own, growing them will seem less intimidating. Believe it or not, many areas have mushroom clubs. The North American Mycological Association (NAMA, 4245 Redinger Road, Portsmouth, OH 45662) will help you locate a nearby club.

Jack Ruttle

Chapter contributor Jo Mueller, in her mushroom shirt.

Preserving and cooking mushrooms

Mushrooms go through several stages as they mature, and the eating quality of the fleshy cap changes with each stage of development. It is very important to select the correct stage of development to fit the recipe you plan to use.

For instance, small unopened buttons are perfect to use raw for dipping, in salads, or for pickling, while the large fully opened caps simply aren't edible prepared in this way. That's not to say that mature mushrooms lack flavor. In fact, the flavor is richer in dark-gilled, fully opened mushrooms, and for this reason they are excellent to use in sauces and soups. However, mature mushrooms can't be stuffed, as the cap becomes flattened and won't hold a filling. Medium-sized caps are perfectly shaped for stuffing.

Color must also be considered when matching mushrooms to a recipe. As mushrooms mature, the delicate light pink gills on the underside of the cap deepen into dark purple or chocolate brown. The natural dye in the gills is intense and will color any foods with which the mushrooms are mixed. Therefore, don't use mature, dark-gilled mushrooms in creamed recipes or dishes that should be light-colored.

Very few foods have absolutely no waste, and the mushroom is one of them. Both the cap and stem can be used, although the flavor and texture of the cap is preferred to that of the stem. Nonetheless, never discard the stems, for they are excellent in soups and sauces. You can freeze them until there's a large enough batch to puree for a creamy soup.

Some mycophiles claim mushrooms should not be washed—which might explain why many people think mushrooms taste like dirt. Wash mushrooms in cool running water, holding them in the position in which they grew, with the caps up and the stems down. This prevents particles of dirt from lodging in the gills. Don't soak mushrooms in water or they'll become spongy and lose flavor. Snip off the tip of each stem, and if dirt

clings to the cap, gently brush it with a soft cloth. Don't peel the cap, because it will lose both flavor and that velvety mouth feel we associate with mushrooms.

If the mushrooms are not to be used immediately, keep them unwashed in a perforated paper or plastic bag: if mushrooms lack air, they spoil very quickly. Another method is to dip them into water containing a little lemon juice, drain, place on trays, and cover with a moist towel to prevent drying out. As long as air circulates around them and they are kept cold in the refrigerator, newly harvested mushrooms will stay fresh and retain their blemish-free whiteness for about two weeks.

If you grow mushrooms, you'll quickly discover that it's the nature of the beasts to grow in flushes, with the plants producing more mushrooms than can be immediately consumed. Mushrooms can be preserved much like other vegetables—either by freezing, pickling, canning, or drying.

To freeze them, wash the mushrooms, drain, and freeze one layer deep on a cookie sheet, then put them in plastic bags when frozen. Don't defrost them before dipping in batter and deep frying; but if they are to be used in sauces, soups, or casseroles, simply defrost and prepare as directed. Mushrooms that have been frozen raw can be sauteed and added to almost any recipe. Mushrooms sauteed *before* freezing will keep longer in the freezer without losing flavor. And if they are sauteed with onions and peppers, then frozen, you have the beginnings of a beautiful spaghetti sauce, chicken dish, or eggplant parmesan. The lentil and mushroom combination on the next page works as a main dish or vegetable side dish.

Pickled mushrooms will resurrect the dullest meal and, stored in a tart marinade in the refrigerator, will keep for a long time. Chances are they will be eaten long before they can spoil. Use only buttons for pickling and cut the stem just a little

longer than the cap. In the following recipe for marinated mushrooms, tarragon is the flavoring herb, but you can try dill, basil, or marjoram. Marinated mushrooms are great in tossed salads or as hors d'oeuvres.

Mushrooms are a low acid vegetable and should be canned with a pressure canner. Place the mushrooms in cold water to cover for 10 minutes. Drain the water quickly and rinse the mushrooms. Heat them for 15 minutes in a covered saucepan with a small amount of water, just enough to prevent sticking. Pack hot into half-pint or pint containers only and cover with boiled water, leaving ½-inch headspace. Cook in a pressure canner for 30 minutes at 10 pounds. (For extra safety, lemon juice or vinegar can be added to each jar of mushrooms before canning.)

Dried mushrooms retain their flavor for over a year. Slice washed mushrooms, including the stems, and place them evenly on the shelves of an electric dryer. After 36 hours of low temperature heating, they should be completely dry. They can also be dried in an oven with a pilot light, in direct sunlight (bring inside at night so they aren't moistened by dew), over a heat register, or near the fireplace.

Mushrooms strung through the stem and cap and then hung to dry are called fairy necklaces. Use a long needle and heavy cotton thread. Make the necklace quite long as the mushrooms shrink and shrivel as they dry. When dry, tie their ends together and loop them over a hook for a pretty decoration that's good enough to eat. This should take about two to three days in the sun, and less time if hung near the fireplace.

Since mushrooms are nearly 90 percent water, dried mushrooms take up only one-tenth the space of fresh ones. Be certain that the mushrooms are completely dry before packing them away in jars with tight-fitting lids. Just a little residual moisture will cause them to deteriorate. Dried mushrooms can be used in most recipes calling for fresh ones, with exception of salads. Soak dried mushrooms in water before cooking and they will become plump again, although their texture can't compare with fresh ones. In general, dried mushrooms are tougher in texture than fresh or frozen ones.

Jo Mueller and Anita Hirsch

Mushroom recipes

When adding mushrooms to your recipes, use seasonings carefully. The mushroom flavor is delicate and can be overwhelmed by a heavy hand. Nutmeg enhances the flavor of mushrooms, but again, use just a touch. Parsley also complements and adds a bright color contrast to the food. Also, since mushrooms have such a high water content, they must be sauteed over rather high heat or they will simply faint and become surrounded by mushroom liquor. It's easy to ruin mushrooms this way.

Mushroom and Spinach Salad

¼ cup olive oil
¼ cup lemon juice
¼ cup soy sauce
1 pound button mushrooms
1 pound spinach
  several Jerusalem artichokes, thinly sliced
¼ cup toasted sesame seeds

Blend together ¼ cup olive oil, ¼ cup lemon juice, and ¼ cup of soy sauce and pour over 1 pound of button mushrooms, the stems cut even with the caps. Chill. Wash, stem, and break 1 pound of spinach into bite-size pieces. Thinly slice several Jerusalem artichokes and place them in the bed of spinach. Just before serving, drain the mushrooms and add the artichokes. Pour the oil-and-lemon dressing over the salad and sprinkle with ¼ cup toasted sesame seeds. Of, if you prefer, marinate the mushrooms in Italian dressing, eliminate the sesame seeds, and sprinkle the salad with chopped parsley. Or still another variation on the theme, don't marinate the mushrooms but put them absolutely fresh on the bed of spinach with the artichokes. Pour the dressing over the salad and add the sesame seeds. Be sure to use gloriously white mushrooms in this salad. The dark green spinach contrasts beautifully with the mushrooms and emphasizes their whiteness.

Mushrooms and Lentils

4 tablespoons dark miso
4 cups water
1 bay leaf
1 teaspoon allspice
1 pound lentils
¼ cup olive oil
5 cups fresh mushrooms, sliced
1 cup onions, chopped
8 ounces tomato sauce
⅓ cup molasses

In a 3-quart saucepan, bring miso, water, bay leaf, and allspice to a boil. Add dry lentils; cover and simmer 30 minutes. Meanwhile, saute mushrooms and onion in olive oil until just tender. After 30 minutes remove bay leaf from lentil mixture and blend in tomato sauce and molasses. Add mushrooms and onion and stir to mix. Cover and heat for 30 more minutes.

Yield: 10 servings

Tarragon Marinated Mushrooms

1 pound mushrooms
½ cup lemon juice
2 tablespoons apple cider vinegar
1 teaspoon dried tarragon leaves
¼ cup parsley, chopped
  seasonings to taste
½ cup salad oil

Slice mushrooms if large, and use whole if small. Place in a sterilized quart jar. Mix the remaining ingredients and pour over mushrooms. Seal and chill. Fresh mushrooms can be added to the marinade from time to time.

Cream of Mushroom Soup

2 tablespoons onion, chopped
½ pound mushrooms, sliced
2 tablespoons flour
2 cups broth
1 cup light cream
  nutmeg and ground pepper, to taste

If you've made some of the other recipes, you'll probably have some stems left over. They can be put to good use in soup. Saute 2 tablespoons of chopped onion for 5 minutes and add ½ pound sliced mushrooms. Continue cooking for 5 minutes. Blend or grind stems that you happen to have in the house and mix them with 2 tablespoons of flour. Add 2 cups of chicken or beef broth and add this mixture to the mushrooms. Cook till thickened. Cool slightly and add 1 cup light cream. Season with nutmeg and ground pepper. Heat through and serve immediately.

Mushroom Souffle

  large mushrooms
4 tablespoons butter
3 tablespoons flour
¾ cup cream
  ground pepper, to taste
1 teaspoon tarragon
4 eggs

Carefully twist the stems from large but not dark-gilled mushrooms. Cook the caps in 4 tablespoons of butter for 5 minutes and then drain. Chop the stems and saute for 10 minutes. Push the stems aside and add 3 tablespoons of flour to the stem liquor, then add ¾ cup of cream. Stir over a low flame until thick. Season with pepper and 1 teaspoon of tarragon. Slowly add the yolks of 4 eggs. Beat the egg whites until stiff and combine with the yolk mixture. Arrange the mushrooms cap-side up in a greased dish and pour the egg mixture over. Bake at 350°F. for 20 minutes or until lightly browned. Cut between the mushrooms when serving. This souffle billows out of the mushrooms' caps and looks as elegant as it tastes.

Mushroom Tempura

  medium-size mushrooms
½ cup flour
1 teaspoon baking powder
1 egg, beaten
  milk, as needed

Mix together ½ cup of flour, 1 teaspoon baking powder, and 1 beaten egg. Add enough milk to make the batter the consistency of thick cream. Cut stems ½ inch from the cap of medium-size mushrooms with unbroken veins. Dip the mushrooms into the batter, allow the excess batter to drip off, and fry at 350° until golden brown, about 4 minutes. Drain well.

Home-growing the button mushroom

Equipment and supplies

Cultivating mushrooms is considerably different from other gardening techniques and requires a different set of tools. Before beginning a venture with mushrooms you will need the following items.

• *Pitchfork for handling compost.* The type best suited has five or six prongs and a long handle.

• *Trays in which the mushrooms will be grown.* Wooden trays or boxes made from old lumber are quite adequate. The trays should be from 10 to 12 inches deep.

• *Spawn.* Spawn is a pure culture of mycelia that has been grown in specially prepared media and will continue to grow when placed in a suitable environment. Moist spawn is actively growing mycelia that must be used immediately after it is received from the laboratory.

A home gardener cannot be certain that the compost will be at the proper stage for use when the spawn arrives. For this reason, it is more reliable and easier to use dry flake spawn or dry brick spawn. As the names imply, these are dry and dormant, so either may be kept until conditions are conducive to good growth in the mushroom house.

• *Gypsum.* The chemical name for gypsum is calcium sulfate. It can be purchased at building supply companies at very little cost.

• *Watering can.* A pump-type sprayer that gives off a fine spray or mist is needed to maintain the proper moisture content in the trays without adding large water droplets.

• *Straw.* Bales of straw can be bought from farmers or lawn and garden shops. The straw is mixed with manure to form the compost.

• *Manure.* This is the item that makes mushroom growth possible. I use elephant manure simply because it is readily available from the local zoo. Horse manure is more commonly used. This is a "hot" manure and

makes an excellent compost. You must be certain that the manure is not mixed with wood shavings since this would make the compost too acid and would harm the crop.

Other manures can be used for composting. Cow manure is the exception, as it does not make good mushrooming compost.

• *Thermometer.* I always follow quite closely the heating up of the compost pile and this is best done with a stick thermometer or dairy thermometer. Also, when composting is complete, the thermometer should be used to monitor the cropping area; the room temperature will greatly affect the crop and should be checked from time to time.

• *Sieve.* To eliminate clumps of dirt and rocks from the casing soil, a sieve of $^3/_{16}$-inch mesh is very useful. Make the sieve about 15 by 15 inches square and 3 inches deep.

• *Peat.* This is needed for casing the trays.

Manure is the foundation of indoor mushroom culture.

The mushroom house

Let's call it the mushroom house even if it's an unused chicken coop, an old out-building on the back lot, an area under the greenhouse benches, or an old unused coal bin.

It is best if you can regulate the temperature in the house. On one hand, mushrooms grow best at near 58°F., but a room temperature of 52 to 55° will yield a slower-growing but longer-lasting crop. On the other

hand, if you increase the temperature to 62 to 65°, the plants will fruit more abundantly but for a shorter period. A slower-yielding crop is desirable for the home grower, allowing the mushroom production to keep pace with the mushroom grower's appetite.

You can determine the rate of harvest by adjusting the temperature.

Composting

To start a compost pile you will need manure. Just how much manure should you gather? A total of one-half ton of compost is needed to provide 60 square feet of growing surface.

After collecting the manure, you should thoroughly saturate it until the water begins to run out. Keep a spray of water playing over the manure while it is being unloaded. If it has been sitting awhile, it will absorb quite a lot of water, whereas fresh manure requires very little. Be certain to avoid manure that has aged in a pile for weeks or months. Old manure is partially decomposed and important nutrients will have already been destroyed.

If you add straw to the manure, break open the bales and wet it thoroughly. This should be done several days before mixing the straw with the manure to provide ample time for water to soak into the straw. One part straw to one part manure will yield a fine compost.

While building the stack, sprinkle in about 20 pounds of gypsum per half-ton of manure. The gypsum (calcium sulfate) adds a source of calcium to the compost, making it available to the mushrooms, and also guards against a sticky or slimy compost.

After the compost is stacked in a heap, place a long-stemmed thermometer in the pile and check it each day. When the temperature reaches 140°F., *caramelization* takes place. This is a chemical change that induces additional bacterial types to reproduce, forcing the temperature to go as high as 160 to 170°F. When caramelization occurs, you will notice the pile turning a rich brown color and the straw breaking into short pieces.

The pile should be turned every five or six days. At each turning it is important that the heap be literally turned inside out and upside down; the inside goes to the outside, the outside goes inside, the bottom of the pile goes on top, and the top goes on the bottom. If parts of the stack appear dry, water should be added to insure a saturated condition throughout, since bacteria will not function in dry areas.

Also, if the compost has a sticky texture, throw in a little more gypsum when turning the heap.

Have the trays sitting up on something—bricks, blocks of wood, or the like—in such a manner that air can circulate under them. If the trays are sitting in a row, two long 2 x 4s work nicely to keep the boxes off the floor.

Pest and disease control

After the trays are filled with compost (to within 1½ or 2 inches of the top when packed down), most growers insist on a final heating of the entire mushroom house and its contents. They heat the area to 140°F. for four hours so even the organisms in the nooks and crannies are destroyed. This process is called *pasteurization.*

If you plan to crop in an area that can be steamed, you'll have no difficulty, but usually steaming isn't practical on most homesteads. What are your other options? You will have a better chance of maintaining healthy plants if your mushroom house is in

the cellar rather than in an outbuilding. Also, if you can divide your crop, placing a few trays under the stairs, a few more in the wash area, and so forth, then the probability of disease spreading through the trays is greatly reduced. If you must use an outbuilding, cover all windows and openings with a very fine screen to keep out pests.

Because good sanitation plays a big role in pest control, your efforts must be concentrated in this area. By faithfully removing decaying organic matter, stagnant water, and rubbish, you will eliminate breeding places for potential problem-makers.

Adding the spawn

Whether you use dry brick or dry flake spawn, apply about one quart to every 12 to 15 square feet of compost. When using brick spawn, insert a piece about 1½ inches square every 6 inches to a depth of 1 to 2 inches. Cover the spawn with compost and firm down.

In the last few years another method of introducing spawn into the compost has proved to be more acceptable. This technique utilizes flake spawn. The compost is allowed to mature in the stack until all ammonia odor has disappeared. This usually requires one extra turning. At this point, dry flake spawn is sprinkled over the compost and thoroughly mixed into it. The compost-spawn mixture is then placed into trays. Instead of pressing the compost firm, it is left loose for 24 hours.

Since it will be pressed down later, fill the boxes slightly heaping so that when it is compressed the compost will reach to within 1½ to 2 inches of the tops of the trays.

After spawning, the mycelia grow through the compost for two to three weeks. The trays are ready to be cased when a network of the cottonlike mycelia covers the compost surface. Throughout this period the compost will require watering with a fine spray to maintain the correct moisture content. If the compost becomes dry, the mycelia will die, but if it is remoistened the remaining viable strands will continue to grow. The ideal temperature during these two to three weeks is between 65 and 70°F.

Mushrooms will form only if another layer of material, called the *casing,* is placed over the compost. The casing provides the growing mycelia with a layer in which to aggregate and send up mushrooms. The casing material should be placed on the surface of the compost to a depth of 1 to 1½ inches.

A variety of substances can be used for casing the trays. Soil is good, but it should be neither too heavy and claylike nor too sandy. Prepare the soil by first passing it through a fine screen—3/16-inch mesh works well—to remove the clumped soil and pebbles.

Just prior to applying the casing, mix in about one-third peat by volume. This will prevent compaction and also provide nutrients for the mushroom mycelia. Without the peat, the casing becomes crusty after frequent watering and the tiny pinheads have difficulty breaking through the surface crust. Use horticultural rather than bedding peat, because it is less acidic. For good measure, I throw in about a cup of lime for each bucketful of peat used since an alkaline casing is necessary for a successful crop. A pH of 7.4 to 7.6 is ideal.

The moisture content of the casing is rather important to control because fructification cannot take place unless the casing is moist. To apply water directly to the casing after it is added to the compost would injure the developing pinheads, so place several sheets of newspaper directly on the surface of the casing and then twice a day add just enough water to keep the paper moist. The mycelia will grow into the casing layer and become established immediately after it is applied. This will not occur if the casing material is disturbed *in any way* or if it becomes too dry.

After ten days, carefully remove the newspaper covering the trays. At this point the proper watering of the beds requires great care and sensitivity. The goal is to maintain a moist casing, but over-watering can be disastrous. If too much water is applied it will percolate through the casing into the compost and kill the mycelia. If too little water is supplied, the layer of casing next to the compost becomes dry and the mushrooms will not grow abundantly.

Normally beds require watering

(Continued on page 394)

(Continued from page 393)

at least every other day. Sprinkle lightly with a fine mist and then after an hour or so another light sprinkling should be sufficient. Until you get the feel of the amount of water needed, stick a finger into the casing to learn if it is moist all the way through.

Pinheads will appear

If all goes well you can expect to see pinheads appearing over the surface of the beds approximately 12 days after the casing was added. These pinheads will mature into fully formed mushrooms in about 6 to 8 days. The mushrooms will appear in flushes or blooms every 10 to 14 days and will continue to do so for 60 to 120 days depending upon the temperature of the mushroom house.

Between flushes, the moisture content must remain high and this is accomplished by frequent watering

with a fine mist. Mushrooms are over 70 percent water, and this water is taken from the compost and casing. After pinheads appear, and while the mushrooms are undergoing rapid growth, refrain from adding water if possible because water on the caps will cause brown spots to develop, making them appear less appetizing.

When a mushroom is picked, the base of the stem which remains in the casing mixture must also be removed. Failure to do this would provide a substrate for bacterial growth which would reduce the quantity of mushrooms produced.

Jo Mueller. From *Growing Your Own Mushrooms: Cultivating, Cooking & Preserving* (with 200 recipes for home-grown mushrooms). Copyright © 1976 by Garden Way Publishing. Available for $4.95 in local bookstores or from Garden Way Publishing, 509 Ferry Road, Charlotte, VT 05445. (See review in this chapter.)

Pick mushrooms when small if they're to be used raw.

Foraging

Mushroom hunters: Summit Hill, Pennsylvania

John Yurko gathers wild mushrooms where the Pocono vacation lands meet Pennsylvania's pock-marked coal country. His practiced eye knows just where to look. Where a casual observer would see only dead leaves, he can find enough mushrooms for a salad. Even he has to really hunt for the brown variety because they blend in with the leaves. He knows just which ones are safe to eat and which ones are poisonous because his father taught him all about wild mushrooms firsthand.

Using a pocketknife, he slices off the mushrooms near ground level and tosses them into a small basket. He doesn't simply pull the mushrooms out of the ground because he'd end up with the dirty roots in the basket.

"We call these brown ones 'stumpies' because they grow near rotten tree stumps. A few weeks back we were getting 'spongies.' They have a beige top and a tan or yellow sponge. That's just what we call them, you know—spongies and

stumpies. Or povpenki. That's what my mother called them." Povpenki is the Slovak word for stumpies.

John has been hunting mushrooms with his son Bob since the boy was four years old. When you come

from a mushroom foraging family, learning to look for mushrooms is as natural as learning to play catch.

As do vegetables, each mushroom variety has its own growing season. John and Bob find mush-

John Yurko.

Bob Yurko.

rooms from June or July through the middle of October. The brown ones they gather in early October usually aren't plentiful until after a frost. The Yurkos don't know why the frost helps, but the mushrooms seem to come out better then.

Rainy weather is no good for wild mushrooms. "If they sit too long, if it's too damp, they'll rot," Bob said. "And if you get to them before the slugs do you're all right," he added. Slugs love mushrooms, and they're selective: you'll never find one bothering a bad mushroom.

When conditions are right, mushrooms spring up in abundance. "They grow like crazy, when they're gonna grow," Bob said. You can get shopping bags full of a bright yellow, earlier variety. "They'll grow around a tree like this," John says, making a big circle with his arms. "You can fill a basket in five minutes around a tree." That type of yellow mushroom, Bob explains, has a top about the size of a quarter, and a stem almost

as fat. "Boy, they're solid and good eating," John says.

Mushroom pickers aren't always that lucky, though. Sometimes it takes half an hour just to find a handful of mushrooms.

At the bottom of a grassy bank John comes across several clumps of large, firm lavender mushrooms— he nicknames them the Lithuanian variety, because people of Lithuanian heritage often like to gather them. The Yurkos generally don't pick them, though. John thinks they aren't quite as good-tasting as some of the other varieties. "It's just what you learn," he says. "I pick the ones my father taught me to pick. If your father showed you to pick these purple ones, then you would. It's from father to son to grandson."

Then something catches his attention in amongst fallen limbs, leaves, and undergrowth. "See how they hide?" More mushrooms, this time tiny buttonheads.

CB

MUSHROOM GROWING FOR EVERYONE
Roy Genders. 1970.
Transatlantic Arts, Inc., N. Village Green, Levittown, NY 11756. 216 p. hardcover $7.50.

It's been said that the United States and England are two countries separated by a common language. If you keep that in mind as you read the British *Mushroom Growing for Everyone,* it translates into a valuable tool for growing mushrooms. Author Roy Genders is the BBC television mushroom grower.

GROWING YOUR OWN MUSHROOMS
Jo Mueller. 1976.
Garden Way Publishing, Charlotte, VT 05445. 174 p. paperback $4.95.

Growing Your Own Mushrooms is a complete guide to home growing, cooking, and preserving the button mushroom, based on the experience of author Jo Mueller. In simple fashion, Mueller teaches the reader her time-tested methods of mushroom culture and shares over 200 of her favorite recipes.

If you want to grow mushrooms, this $4.95 primer is a good place to start. The preceding section, "Home-growing the button mushroom," was taken from the book.

Foraging guides

The rule for hunting mushrooms is never to eat anything you can't identify beyond the shadow of a doubt. Forget any stories you've heard about how poison mushrooms will turn dimes or silver spoons black,

or how they'll never peel. There are no tricks for telling the poison ones except to know exactly what you're gathering. On the next page are three excellent guidebooks to help you get a foothold in the field.

THE MUSHROOM HUNTER'S FIELD GUIDE

Alexander H. Smith. 1963.
University of Michigan Press, 615 E.
University, Ann Arbor, MI 48106.
Hardcover $9.95.

This is an excellent guide, useful throughout the United States and Canada, but best for the Great Lakes, Western, and Northeastern states. The nearly 200 black-and-white photos are run with descriptions 1 to a page. The back of the book holds a like number of color photographs for more intense study.

Smith seems cautious and reliable. He doesn't recommend eating any *Amanitas,* even with field-guide identification. He issues stern warnings, saying the destroying angel, *Amanita verna,* causes a higher percentage of deaths than rattlesnake bites.

THE SAVORY WILD MUSHROOM

Margaret McKenny and Daniel E.
Stuntz. 1971.
University of Washington Press,
Seattle, WA 98105. 133 p. paperback
$7.95.

This guide is nearly pocket-sized and specializes in the fungi of the Pacific Northwest. It advises going out with an expert for a season before eating any wild mushrooms.

The book includes 156 black-and-white photos, together with complete descriptions, locations, seasons, and helpful remarks on each variety. Another 60 color photographs make identification easier.

With a section on mushroom poisons and an outstanding section on cooking mushrooms, the book is worth reading no matter where you live.

WILD MUSHROOMS: AN ILLUSTRATED HANDBOOK

Linus Zeitlmayr and Otto Gregory.
1976.
Transatlantic Arts, Inc., N. Village
Green, Levittown, NY 11756. 138 p.
hardcover $9.75.

This little book, translated from German, is one of the most interesting mushroom books I've seen. It's a book for those who want to know more about mushrooms than whether a particular variety is edible, although it includes that as well. Included is a list of people who have died from mushroom poisoning—Buddha among them—and nutritional, cooking, and preserving information.

The authors advise caution and suggest eating one kind of wild mushroom at a time to determine your sensitivity. The book concentrates on 70 varieties, with complete descriptions, including spore dust color, and beautifully detailed paintings.

Terri Gabriell

Equipment

Build-it-yourself mushroom shelves

All that's really needed for growing mushrooms is a seed flat or two. But for families with an insatiable appetite for mushrooms, something more elaborate is in order. You can build these shelves in a short time. Construct the uprights and cross-bearers of rot-resistant cedar, redwood, cypress, or pine.

The bed boards can be almost any dimensions that are convenient for you to work with and that will fit the space available. In our prototype, the beds measure 36 inches by 18 inches and are just over 5 inches deep.

Place the bottom bed 6 to 8 inches away from the floor. Leave 18 inches between bed boards, and allow for from 24 inches to 30 inches between the top shelf bed and ceiling. The air will circulate easily. Place away from drafts and sunlight.

You can build this miniature version of a commercial mushroom bed.

Kits

We looked at several grow-it-yourself kits, offering four different varieties: the velvet stem, wood ear, oyster, and the common button.

Some kits are in the form of a compressed log of fertile compost and spawn; the log is simply soaked in water overnight and set under a plastic hood that retains moisture and helps keep the environment semi-sterile. The other type consists of mycelia and compost separated in two plastic bags. Mixing the two with water was a bit messy but certainly not overly difficult. The mixture was dumped into the container provided, and covered.

It was an odd thrill to watch the first baby mushrooms appear. As long as we followed the directions, they kept coming in batches. In the cool environment of the basement, out of direct sunlight, we gave our cultures a fine mist spraying every day or two—just enough to keep the surface moist. In the kitchen, where it wasn't so damp, we watered as often as twice a day.

The kits are fun; as a gift, a learning project for kids, or an introduction to mushroom growing, they're a good idea. But for serious mushroom production you've got to become an expert. Kits are no way to approach serious mushroom husbandry. The harvest we got from one kit came out to about $7 per pound.

DB

You can see here that there's nothing very sophisticated involved in mushroom-growing kits. They're fun, but not a generous source of mushrooms for the kitchen.

Sources.

Spawn for the common button mushroom.

All Fresh Spawn Corp., Toughkenamon, PA 19374

Boy-Ar-Dee Mushroom Co., Milton, PA 17847

Butler Mushroom Co., West Winfield, PA 16950

Farron's Spawn, Kirkwood, PA 17536

Fran Mushroom Co., Central Avenue, Ravena, NY 12143

International Microbiological Products Inc., 3309 W. El Segundo Boulevard, Hawthorne, CA 90250

Krahenbuhl and Son Produce Farm, 4008 Raytown Road, Kansas City, MO 64129

L. F. Lambert Spawn Co., P.O. Box 407, Coatesville, PA 19320

Mushroom Growers Association, Birch Street, Kennett Square, PA 19348

Mushroom Supply Co., Toughkenamon, PA 19374

Oxford Royal Mushroom Products, Inc., Kelton, PA 19346

Sharpless Spawn Co., Kennett Square, PA 19348

Somycel U.S. Inc., Route 1, Avondale, PA 19311

Stoller Research Co., P.O. Box 1071, Santa Cruz, CA 95060

Superior Spawn Co., 7428 Hough Road, Almont, MI 48003

J. B. Swayne Spawn Co., Kennett Square, PA 19348

Utica Spawn Co , 2201 E. Hamlin Road, Utica, MI 48087

Shiitake spawn and complete kit.

Dr. Yoo Farm, P.O. Box 290, College Park, MD 20740

Thompson & Morgan Seed Co., P.O. Box 100, Farmingdale, NJ 07727

The familiar button mushroom.

Various spawn and kits.

Corwin S. Fred, Mushroom Products, Inc., 395 W. Railroad Street, South Lebanon, OH 45065

Kinoko Company, P.O. Box 6425, Oakland, CA 94621

Taylor and NG, P.O. Box 200, Brisbane, CA 94005

Planning
the home
Thoughts on the kitchen, growing spaces, and the garbage can

In assembling a complex machine, one is apt to find there are parts left over—tail ends that look important but somehow aren't really crucial to the mechanics of that machine.

So it is with books. This book, too, is complex in terms of numbers of components, and our final chapter is somewhat of a hood ornament on the essential beast, an after-purchase accessory. Here you will find nice-reading articles and quotes that weren't quite germane to any topic, or were germane to them all and therefore refusing classification by chapter subject.

"Planning the home" shifts restlessly from theme to theme but taken as a whole it expresses the book's message in a way that our title, subtitle, and cover cannot.

RY

The kitchen

The country kitchen is not indigenous to America, but it certainly is as old as the seeds of Independence. The houses constructed by the colonists in New England, known as saltboxes, were generally built with two rooms first. The larger main room was called the common room, a kind of kitchen-living room that functioned as the hub of the house. No matter how many rooms were added on, the common room remained the center of the home. Cooking was done in and over the fireplace, which was very large, as it was the only source of heat. Later, brick ovens were built into the fireplaces for roasting and baking.

These large family rooms were thick with activity. The women did all their sewing, weaving, spinning, dyeing, and candle-making in the warmth of the common room. Children played or studied and the baby's cradle remained near the hearth day and night. At the end of the day's work, the men gathered around the table to discuss local politics and problems. In the evening, both young and old sat on benches in the warm corner of the hearth while the head of the household read from the Bible. It was not uncommon for the master and mistress to sleep in the common room, and quite often the young children slept in a trundle bed which was stored under their parents' bed during the day. It was not until much later, in city houses, that food preparation was moved to a special room separate from the dining and entertaining area. Even then the kitchen remained a gathering spot for all the servants.

Indeed, the country kitchen is American. Wherever new settlements were formed, the initial structures began with and centered around kitchen-living areas. The concept was implicit in the architecture: a beginning, a nucleus, always a common gathering place where the family could gather alone or with friends to eat, exchange news, and discuss future plans.

Prior to central heating, the kitchen, large or small, offered the necessary warmth to draw the family in the morning and evening. Even on the Midwestern prairies, where fireplaces were rare because of the scarcity of wood, the kitchen stove, stoked with soft coal, was kept burning all day in the cold months while doors to other rooms remained closed to prevent drafts. If the kitchen and dining area were one, there was often a separate summer kitchen with a kerosene stove, which did not emit heat as a wood-burning stove did. Or, in some instances, houses contained separate dining rooms where the family could escape the heat of the cookstove during the hot months. The living room or parlor usually contained a coal-burning heating stove, but such rooms were used only for company or on special occasions.

Unfortunately, the last 40 years of American life have witnessed some drastic changes in kitchen space and function. Affluence and changing social patterns made the dining room the proper place for main meals. Kitchen eating became socially less acceptable and, once finished with its meal, the family dispersed to other areas of the house. The living room became the social gathering spot, the place for the radio and subsequently the television. Thus, the idea of the kitchen as a center remained largely in the lower classes.

Then, after World War II, with the enormous birthrate among young marrieds, huge tract home developments built for resale and aimed at middle-income families began to appear all over the United States. A maturing of floor plans and the demand for more living space, compounded possibly by the growing concern and anxiety over the decentralization of larger postwar families, gave birth to the kitchen-family room combination. But, regrettably, in most of these newer homes, the actual kitchen area is very small and sterile, with little more than a cubicle or counter where food is handed through a space that resembles a service window in a hamburger stand.

It is curious, too, that domestic and commercial foods have become so similar. All over America thousands of quick-lunch stands make it possible to get a meal made from precooked, prepared nutritionless products in minutes. In supermarkets, frozen, prepared jiffy breakfast, lunch, and dinner foods have become major stock for the American kitchen. One can drive to the store, select dinner ingredients and, with some quick preparation, finish the whole plastic process in as little as three-quarters of an hour (providing the checkout line is not too long). Quite a change from the time when preparation for the evening meal began after breakfast and there was stirring, kneading, and shaping throughout the day.

Well, given the cultural and industrial changes in our society in the first half of this century, the image of a woman spending the entire day washing, ironing, mending, and fixing the evening meal is not likely to regain popularity. On the other hand, a growing concern about the quality of food consumed by the average American today and a gnawing awareness of the degeneration of the family group necessitate a look toward alternatives to our present family lifestyles.

I suggest we start with a return to the American country kitchen. Let's exchange the food laboratory for the food studio. We all accept the indisputable advantage of gas and electricity in today's home. Stoking the fire in the stove throughout the day and running outside in the rain or snow to the root cellar are among the many chores that no longer absorb one's time and energy. But there is a limit to the validity of labor-saving devices. Many sterile, unattractive kitchen gadgets can be eliminated with no loss of convenience or efficiency, and the avoidance, even rejection, of these items is advantageous on two levels. First, it would make a small, but definite, commitment to the conservation of power, and second, it would help restore a deeply needed aestheticism to food preparation. With so much precooked, processed food being used today, and with so many shortcuts and so many items available for speedy preparation, I feel a decadence emerging in our lives. We have dissociated the means of creating from the end product. In cooking, as in any creative form, attention must be given to the total cycle of creation to produce beautiful, healthful results. This cycle, if reinstated, will have everyone wanting to get into the creative act by questioning, commenting, smelling, tasting, and stirring. And such continuing appreciation will serve as an emotional stimulant for the woman who wants to move into new forms of creativity in her kitchen-studio.

It is difficult to make the commitment to an emotional and functional change in the isolated coldness of the present-day American kitchen, but any kitchen, large or small, can be turned into a studio with a bit of imagination and determination.

After removing some of the superfluous electrical extras (who knows what all the electrical energy passing through the room does to your psyche), you will find you can replace these items with small, inexpensive hand-operated items that take less space. Coffee made in a good old-fashioned drip pot from beans ground fresh each time in a small, inexpensive hand coffee grinder is a great asset to socializing in the kitchen. It is important to mention here, though, that everyone has different cooking techniques and food preferences that are the result of many years of habits. Some items drawing electrical energy may be very important to you, so don't discard them and sacrifice all convenience. It's a matter of priorities. I have an electrical incubator for yogurt and I do appreciate its uniform heating system. The blender is a terrific invention and some of the newer ones have a button that operates only so long as you touch it, which is, I think, a small power and noise saver. I like sourdough waffles, so I keep a waffle iron instead of a toaster and make toast in the oven. Figure out what is really important to you and recycle the rest to people who need and can use them. Consider the aged and the handicapped, or institutions where there is a flow of people and a short-

(Continued on page 400)

(Continued from page 399)

age of labor. There are people and places who could use that electric knife sharpener or can opener.

If the kitchen is large enough, for example, the studio idea can be expanded. It is a perfect place for a piano or sewing machine, a loom or a library table, bookcase, reading chair or a couch, and of course a large table in trestle, round, or mission-plank style, where everyone can gather, eat, and talk.

I do all my writing in the kitchen. I wrote *Whole Earth Cook Book* there, and my sister typed the manuscript at the desk in her kitchen. My neighbor, a creative country-kitchen type, has an alcove in her kitchen where she sews. These arrangements enable you to cook while sewing, reading, writing, painting, or socializing. There are endless personal varieties of country kitchens that one can develop.

The compartmentalization of American life has had an inhibiting effect on the woman in her own home, so that her creativity has been channeled away from her work area. But just as her family's health can be reclaimed through attention to better foods, so should her own kitchen-studio be reclaimed through attention to better explore all levels of her talent. In this way, all of us can gather once again in our common rooms for good, healthful food, rich conversation, and a general inclusive family warmth.

From *In Celebration of Small Things* by Sharon Cadwallader, published by Houghton Mifflin Company. Copyright © 1974 by Sharon Cadwallader. Reprinted by permission.

The baby in the kitchen

A few years ago consumers began to realize that commercial baby food was about as natural as the antiseptic, smiling cherub pictured on the containers it came in. Any parent whose kitchen looks like a slaughterhouse after a lunch of pureed beets knows that kid isn't for real.

When consumers got balky, the manufacturers started advertising the fact that little if any sugar, salt, and preservatives are used in their products. But a typical jar of toddler dinner still doesn't match what you can make at home. One commercial chicken noodle dinner, for example, contains flour and starch. A vegetable and bacon dinner lists modified food starch, defatted soy flour, dehydrated potatoes, and yeast among its ingredients.

When you make baby food at home, you know that because your chicken and noodle dinner contains nothing but chicken, noodles, and water, your child is getting nothing but honest-to-goodness food. Add to that the convenience, speed, and economy of making your own baby food and the choice seems simple.

Most baby foods list the ingredients and the percentage of recommended daily allowances of vitamins, minerals, and protein the contents provide. What they don't tell you is the amounts of each ingredient in the food. Also, remember that the nutritional value of freshly cooked baby food can't be topped.

Modified food starch is added to some baby foods as filler to solidify foods having a high water content. The starch is not completely broken down by saliva, making a burden on the baby's small digestive tract.

Commercial baby food is bland. Desserts, meats, and vegetables all seem to have the same texture. The baby will like your cooking better if he gets a variety of taste and texture to begin with. It will help ease the transition from pureed baby food to table food.

At the beginning, the baby eats only a tablespoon or so of food at a time, and it's a waste to buy a whole jar of food, especially since when opened, it should only be kept for 24 hours. If you gave the baby the leftovers for his next couple of meals, he wouldn't be getting much variety. But if you yourself prepare only a bit of food at a time, you won't waste food and money.

Cook or stew foods until tender, then puree in mill, grinder, blender, or processor. If you use a pressure cooker for cooking, figure ten minutes per hour of standard cooking time. Store in the refrigerator for the next day or two of use or freeze in ice cube trays. Wrap trays in waxed paper and remove when frozen solid (probably the next day). Store in labeled freezer containers or bags. Or save commercial baby food jars, sterilize them, and can food, leaving room at the top for expansion.

Baby Cereal

When you prepare brown rice for your own dinner, set some aside and blenderize it. Use raisin-nut water to thin it to the proper consistency; to make the water, pour a cup of boiling water over eight raw almonds and ¼ cup of washed, unsulfured raisins. Refrigerate overnight. (Test the baby's reaction to raisin-nut water first before you add it to a large batch of rice. The water adds needed iron to the baby's diet.)

Carrots and Cabbage

¼ pound carrots
¼ pound cabbage
1 tablespoon water

Scrub carrots and steam 15 minutes. Add cabbage and steam 15 minutes more. Puree.

Yield: 5 food cubes.

Chicken Dinner

1 cup water
1 piece chicken
¼ cup cooked brown rice
3 whole carrots
¼ pound green beans

Simmer water, chicken, and rice 15 minutes. Add carrots and green beans and simmer 15 minutes more. Blend rice and chicken with enough broth to moisten. Mash the carrots and beans. Freeze the chicken and rice mix and the carrots and beans separately in ice cube trays.

MENU NOTES

- *Meat may be frozen, thawed, cooked, and refrozen.*
- *Millet, brown rice, and barley may be used interchangeably in recipes for toddler dinners.*
- *Egg white, milk, grains, citrus, spinach, starches, and seafoods should be introduced on probation several days in a row; watch for reactions as these are apt to trigger allergies.*
- *Food thickeners: cooked dried peas, beans, lentils, potatoes, rice, egg yolk, chicken, fish, meat, peanut butter, and wheat germ.*
- *Lima beans are generally unpopular with children.*

Avocado Puree

¼ avocado, peeled
1 teaspoon lemon juice

Mash or blend avocado and lemon juice until soft. Don't serve until baby is a year old because it is very rich.

Rebecca Christian

The man in the kitchen

Much as many men now share fully in raising their children, they now are entering the kitchen. One reason is the United States' current fascination with good food, which is paralleled by our interest in home food production and processing and our preoccupation with the health-giving qualities of our food.

Cookware of all kinds—from basic pots and pans to such esoteric utensils as Mongolian firepots or croissant cutters or expresso makers—enjoys phenomenal sales. Cooking
(Continued on page 402)

Root cellar: Payne Hollow, Ohio

Underneath our house is an old-fashioned country cellar, dirt-floored and stone-walled, with dusty timbers close overhead, lighted only by the door through which you enter. According to suburban standards it might seem merely a damp hole-in-the-ground, yet it is a basic part of our establishment. It keeps food cool in summer, protects it from freezing in winter. Moreover, our cellar has a deep significance. We feel more secure with our house firmly rooted in the earth, and the provender stored there is of more value than a pocketful of money.

I like the cellar because it is so honest. It makes no effort to impress anyone and its dust is there for everyone to see who cares to enter.

The cellar is most inviting now in late fall when the garnered harvest makes it a miniature county fair. The wide shelves are loaded with neat rows of full jars, a colorful parade which celebrates the ripening of summer. It was back in April that we gathered the poke in those jars. Next are ranks of spinach and green peas. Farther along are blackberries, tawny gooseberries, golden honey, platoons of redcoat tomatoes, green beans whose uncut pods stand on end—for Anna takes pride in the appearance of her canning. Pale jars of catfish march next to a sanguine block which is goat meat, canned in winter.

Under the shelves are boxes with screens on top to keep mice from nibbling the potatoes which are inside. The cellar is walled on only three sides, the fourth being a sloping earth bank where the squash are laid. Around the edges of the floor are boxes and old buckets containing growing celery and Chinese cabbage whose greenness gives the effect of a conservatory, which is what the cellar really is.

Surely no two people could eat all this accumulation of food in one winter, even with the help of guests, dogs, and goats. Why store up more than we need? I have an idea that squirrels and mice pack away all the seeds and nuts they can lay their paws on. The winter is long, and who knows, the harvest might be scanty next year.

Harlan Hubbard. *Payne Hollow.*
Eakins Press, 155 E. 42nd Street,
New York, NY 10017.

Our buildings in Payne Hollow—house in center, goat stable on left, shed-workshop on right.
Looking north from creek bottom. River off to the left.

BACKDOOR MANNERS

Visitors do not wind up in the kitchen because they use the back door and it just happens to lead to the kitchen; they use the back door precisely because it does lead into the kitchen. For some profound, culture-wide reasons, in 1980 we want to receive our guests in the kitchen. The issue is not a simple one of architecture or transportation, but of attitudes toward food and food preparation.

Roger L. Welsch. © 1979/80 by The New York Times Company. Reprinted by permission.

The story of the Last Supper, the ceremonies of Communion and of Passover, the fact that all major events and milestones in peoples' lives are customarily commemorated by a ritually served meal or banquet is no accident. Such celebrations reflect the recognition that those who gather to eat consume more than food at the table. Symbolically, they take in each other. Through the exchange of words, gestures, smells, sights, and sounds, each participant around the table ingests—in a manner of speaking—the others who share in the meal. Through the sharing of our foods we share in each other.

The process of identification with a group through eating together means that the family's mealtime customs are of great importance to the growing child. This is especially so in contemporary urban life, because the other opportunities for identification through shared activities have been eroded by the absence of the parents from the home for most of the day. Thus, mealtimes assume paramount importance for the creation in children of a feeling of belonging. In many homes, mealtimes provide the only opportunity for the members of the family to relate to each other and counter the fragmentation that afflicts so much of family function in our society.

The importance of sharing mealtimes endures through life. Eating alone remains the ultimate of lonely experiences for a great many people.

Richard Galdston, M.D. Copyright © 1976 Parents Magazine Enterprises. Reprinted from *Parents* by permission.

(Continued from page 401)

classes—running the gamut from natural foods preparation to salad making to fine baking to food processor techniques—are jammed with people and flourish everywhere. And men are heavily involved in this culinary renaissance at every step of the way.

There's no reason why they shouldn't be, although in traditional rural society, men's energies were directed elsewhere, to work the fields. The same social revolution that has urbanized America and drastically altered our food economy has also freed men to discover the rewards of the kitchen.

(In some cases, it's true, men are forced into the kitchen. It's no longer possible for many women to stay at home and assume the role of homemaker, even if they want to: contemporary economic reality forces many households to be two-income households. Besides, many women *want* the stimulation and reward of a career away from home.)

Even those men who are completely untutored in the fine points of food preparation can be extremely helpful in the kitchen. There are many tasks that need doing and don't require any culinary background: peeling and seeding tomatoes, pitting fruit, stirring the contents of a pot. Many simple tasks can be learned easily enough, even by those who don't want to learn more complex cooking skills; the man who adamantly asserts that he can't fry an egg can learn to

chop parsley with a French knife. And there are always cleanup tasks galore that need doing.

For example, my best friend never learned to cook. And doesn't want to. But he's one of the most efficient and ruthlessly thorough cleanup persons I've ever encountered. When we've finally finished one of his wife's dinners, Roy goes into action: clears the table, stacks the dish-

washer, attacks the pots and pans, empties the kitchen scraps on the compost pile, empties the garbage, wipes the crumbs off the kitchen counters and somehow manages to make the kitchen look cleaner than it did before dinner.

This might not sound too glamorous; it isn't. But we're really talking about a working partnership in which both partners pull their weight according to their individual interests annd talents. This is just as important in the kitchen as elsewhere in the relationship. When partners work together, the work gets done more efficiently and, I believe, the relationship is more satisfying too.

There's no reason a man has to be relegated to the role of helper. Many men discover that they begin to feel at home in the kitchen and want to learn more about cooking. My father, for example. My father has always worked long hours at a job away from home and has been a virtual stranger to the inside of kitchen, except, of course, at mealtime. But at 65, he's become quite a cook. He's learned that he not

only enjoys helping my mother prepare a special family dinner or making desserts, but cooking entire meals himself. He's begun collecting recipes and trading cooking tips with his sons, one of whom is a professional cook.

I find that the more I cook, the more I enjoy equally both the *process* and the *result*. What keeps me returning to the kitchen, what makes preparing even the most hurried, routine midweek meal a challenge, is the joy of creation.

I am not a painter, a sculptor, a poet, a photographer, a novelist. I don't throw pots, weave, or do macrame. Although I'm a fair woodcraftsman, I don't have the time or space to do much woodworking. But cooking! I'm a good cook, and I find in the kitchen and in the myriad chores of food preparation a release for my creativity.

There is far more to be found in cooking, of course. The finished product of my creative energies is a source of pleasure for myself, for friends, and family. Cooking is a joy that can be shared; in the ephemeral existence of a pot of soup, a loaf of bread, or the brief expanse of a meal, bonds of friendship and kinship are renewed or forged.

Many food writers recognize this. Anna Thomas opens *The Vegetarian Epicure* by saying, "Good food is a celebration of life. . . ." Marcella Hazan, author of *The Classic Italian Cookbook* and *More Classic Italian Cooking,* describes the satisfaction that Italians find in sharing good meals with family and friends—a feeling that I really believe is at the heart of *all* good cooking:

"The memories associated with the meals of our youth are among the most deeply imbedded in our consciousness. Lin Yutang, a Chinese scholar and epicure, wrote 'What is patriotism, but the love of good things we ate in our childhood.' The joy of unpretentious food, prepared and consumed simply because it is good, leads to moments of pure well-being that are too vital to leave only to guests and to Saturday nights. They should belong to us and to every day.

"Somehow, we must eat. We can do it absent-mindedly, as a matter of routine. We can turn it into an occasion for romance; for taking up

slack in social ties; for displaying our sophistication, or our means; for talking business. Or we can make it into a recurrent source of unhurried, intimate, deep-seated satisfaction. Good Italian cooking can serve most of these ends, but what it serves best is the last."

Anthropologists and folklorists have long struggled to understand the layers of symbolism surrounding food and eating, peeling off layer after layer to find yet another beneath, much as peeling an onion. And many admit they're just beginning to comprehend the complex human reactions to life's most basic ritual.

Michael Stoner

RADICAL HOME ECONOMICS

To Scott Burns, author of *The Household Economy* (Beacon Press, 1977; originally titled *Home, Inc.*), home economics is much like business economics. The home is an important institution with an economy that produces goods and services including "the value of shelter, home-cooked meals, all the weekend-built patios and barbecues in suburban America, painting and wallpapering, home sewing, laundry, child care, home repairs, volunteer services to community and friends, the product of the home garden, and the transportation services of the private automobile." But the value of this work, says Burns, is excluded from the marketplace and the GNP. Its value is invisible outside the home.

There was a time, perhaps a generation or so ago, when the American Home Economics Association (the major voice of home economists) advocated the production of food, clothing, and shelter within the home. But as our food economy has shifted from local and home production to one of large, centralized farms, the AHEA switched to promoting family consumption rather than family production—that is, rather than making. Ironically, some older members of the AHEA have called the theme of *Home Food Systems* "far ahead of its time."

TS

Appliances and gadgets

One cook's pet appliance is another's nuisance. The question of what is necessary and what is expensive frill is asked outside the kitchen of course: photographers, carpenters, and campers, like cooks, go through periodic spells of simplicity and complexity.

This periodicity of fashion made us wonder about the processor and its humble relative, the blender.

RY

Food processors

In 1972, the French Cuisinart made its way into the trendiest kitchens. For the first three or four years of food processor marketing, it was the darling of the culinary cognoscenti, acclaimed as virtually an extra set of hands to chop, grate, slice, mix, whiz, and knead. James Beard, famous chef and cookbook author, declared that it might be possible to live without a Cuisinart, but he'd rather not try.

A backlash ensued. Practical, basic cooks considered the hoopla to be Francophile nonsense and classified the food processor with the electric crepe pan, egg cuber, and automatic peeling wand. Even now, many people still think of food processors as the Cuisinart and a few imitators. Who wants a sleek and sinister contraption like that in the kitchen with its Frenchy name displayed over the front like a smirk? As *New Yorker* magazine remarked, that name "suggests a racy, urbane, sophisticated kind of cooking," and seems alien compared to the "motherly looking" electric mixer.

It's time to jettison all that thinking. If you prepare as few as seven meals a week at home, and especially if you cook from scratch instead of with prepared foods, your cooking enjoyment and flexibility likely will improve with a food processor. Processors are not exclusively for exotic cooking. As chef and cooking teacher Julia Child wrote in a *McCall's* article, the food processor does the "dogwork" in the kitchen. It is the first revolutionary new kitchen appliance to appear since the blender and electric mixer. Although it is as fundamental as those two appliances, it is far more versatile. While it both mixes and blends, it also chops parsley and onions, slices most vegetables and fruits, juliennes mushrooms and what

have you, grates cheese, grinds meat, makes mayonnaise, baby food, pie crust dough, and cake batters. It will even turn coarse coffee to fine grind. As Julia Child recommends, "it is generally about the most useful machine a good cook—or even a seminoncook—can have in the kitchen, since by taking so much of the time and drudgery out of many otherwise arduous operations, it actually makes good cooking possible in a way it never was before."

Because a good food processor will do so many things well, a consumer may expect too much. Even the best food processor now available cannot whip egg whites or cream to anywhere near the volume that a mixer can. The food processor blades spin too rapidly to incorporate very much air into a food, especially inside a closed bowl. A processor can mix up a dough ball, but it will not make as satisfactory or highrising a yeasted loaf as can be mixed by hand. Again, the machine's speed and air flow may be a factor. It will not slice citrus fruits attractively, and some vegetables, such as tomatoes, green peppers, and cabbage, will look better sliced by hand. A food processor won't shuck oysters, shell peas, or change the baby.

However, those basic functions which a good processor performs are done with almost rude efficiency. "It cuts up vegetables so fast that it makes people feel silly," says the *New Yorker*.

If you have a food processor, keep it in mind when you're wondering what to cook and your customary menu will change. You can fix potato or cottage cheese pancakes on a busy morning and breakfast becomes a lot more interesting—and filling. Preparing home sausages and fruit leather seems a possibility instead of a major undertaking. Tasks

like cutting frozen butter into flour lose their intimidating quality. Some recipes—such as those with a pound of grated cheese—become easily accessible instead of discouraging. At best, your menu will widen and your culinary skills improve. At worst, you'll save time in the kitchen. And you may find that specialty foods you used to go out for are now temptingly available at home.

A food processor is an appliance with these basic parts: a base with the motor onto which the bowl fits; a locking lid for the bowl; a pusher to feed food through a chute in the lid; several blades and disks that fit inside the bowl and process the food. The first Cuisinart now has more than a score of rivals, with each generation of machine featuring new refinements. (In addition, several other types of machines perform some of a food processor's functions. We aren't including those "preparation systems" here.)

If you're thinking of buying a food processor, the best advice is, investigate. Either talk to other cooks about their machines or see a thorough demonstration at a retailer. Again, Julia Child has good advice: "Insist, at the store, on trying it out yourself—even bring some raw pork to chop, and hard cheese. (If the demonstrator is snippy, go somewhere else.) Be sure you have a firm guarantee as to its satisfactory operation and as to who pays for return shipping charges if they are necessary." Buy a machine you have confidence in; it means the difference between using it regularly or almost never.

One of the most annoying characteristics a processor can have is loss of balance with dough jobs or when a piece of food is either riding on the disc or stuck on the blade: the machine can dance and vibrate terribly. Watch for good equilibrium.

Check the machine's meat chopping performance (unless you're a vegetarian)! Many machines actually mash or shred meat instead of chopping it. Study the bowl's material. Does it feel solid and have a sturdy handle for tipping out heavy foods? Listen to the machine in operation. Noise levels vary widely from brand to brand.

Once you've picked out your machine, practice with it. Finesse with a food processor is acquired fairly quickly, but still must be learned. Green peppers and onions liquify quickly, but that can be avoided with practice. Soft cheese can turn into a spread before you know it, until you've chopped it once or twice. You will learn how cold a cheese must be to slice well. You will also learn such tricks as thickening gravy with pureed vegetables for enhanced flavor and nutrition. And keep your new purchase plugged in, set to go, and think "processor" every time you have a big chop, slice, or mix job ahead of you. If you do, you'll use it regularly with great success and satisfaction.

With most brands, the food processor manual or cookbook is vital for learning about the machine. A cautious rule of thumb is: Don't assume the processor will do anything that its book doesn't describe. You may have to read between the lines. For instance, the Cuisinart booklet mentions that soft cheese can be sliced, but omits the warning that slicing Parmesan can be fatal for the machine. We learned the hard way. Don't try to chop ice unless the booklet tells you to. Watch out that you don't overheat a machine with bread dough if its booklet doesn't mention mixing yeast breads.

Every manual issues an important warning that should be apparent to anyone with common sense. Food processors move sharp blades rapidly and are dangerous. Even the immobilized blades should be stored out of reach of the kids. Care should be used in washing the blades, particularly the steel cutting blade. Drying it one minute, you may be putting a Band-Aid on your palm the next.

There are little differences in cleanup time for the various processors. Certainly, they require more time than wiping off a knife blade, but usually all that's needed is a simple rinse. Cleanup time becomes negligible for any task larger than a single tomato or a few scallions. Keep your processor handy and ready to go, and you'll use it.

Testing

We tested 12 food processors and 4 food preparation machines. We used one criterion for these tests: based on several representative functions, did the machine do what it claimed it would do?

In the chopping, slicing, and shredding departments, we tested the machines with onions, carrots, green peppers, potatoes, celery, cabbage, mushrooms, and muenster and cheddar cheese. We grated Parmesan cheese, and chopped beef, fish, chicken, and bread for bread crumbs. We mixed a yeast dough and ground-up nuts for peanut butter. We tried (and failed) to slice citrus fruits.

While food processors are somewhat similar in their capabilities and perform most of the basic functions you'd expect from such a machine, we found enough differences on some procedures to make a few machines stand out clearly. Where we don't

(Continued on page 406)

IS IT WORTH THE COST AND CABINET SPACE?

Machine	Functions	Reduce food waste?	Flavor-saving?	Nutrient-saving?	Energy-saving?	Effort-saving?	Time-saving?
Blender	Pulverize, puree, blend	yes	yes	yes	no	yes	yes
Dehydrator	Dehydrate foods	yes	yes	yes	no	no	yes
Food processor	Slice, shred, chop, mix, beat, whip, puree, blend, knead	yes	yes	yes	no	yes	yes
Grinder or mill	Grind, crack, pulverize	no	no	yes	no	yes	yes
Ice cream maker	Whip and freeze	no	yes	yes	no	yes	yes
Juice extractor	Extract liquid	no	yes	yes	no	yes	yes
Juicer	Juice citrus	no	yes	yes	no	yes	yes
Mixer	Mix, stir, beat, knead	no	no	no	no	yes	yes
Oven (convection)	Dry-heat cook	no	yes	yes	yes	yes	yes
Slicer	Uniform slice	no	no	no	no	yes	no
Steamer	Steam-cook	yes	yes	yes	yes	yes	yes
Toaster	Toast bread	no	no	no	yes	yes	yes
Wok (with baskets for steaming)	Pan-fry, steam	no	yes	yes	yes	yes	yes
Electric rice steamer	Steams rice	no	no	no	no	yes	no

American

$59.95

This processor uses lid rotation instead of a switch for on and off. The on-off response was particularly quick, and the American was one of only two brands we tested which had a feed chute long enough that adults couldn't reach down to the disc—a safety feature.

The instruction booklet directed that vegetables should be dropped through the chute for chopping, but this procedure seemed to make the machine labor. When the vegetables were placed into the bowl before the lid was closed and then processed, the machine worked more easily. The American had a slower chopping action than the other food processors, producing coarser products and chunkier peanut butter. The booklet didn't mention making yeast bread, so we substituted pie dough to judge how it cut in shortening. It did fine.

The American had trouble holding still. It tended to dance on the counter with the

vibration of chopping. The bowl's handle felt flimsy, and the open, hollow shaft of the chopping blade easily caught ingredients intended for the bowl. Parmesan cheese repeatedly wedged on the steel blade, causing the motor to overheat and triggering the automatic switch-off.

Cuisinart (model CFP 9A)

$130.

We ran into bad luck with the first Cuisinart we tested. The booklet said that soft cheese could be sliced, but it didn't specifically warn against slicing hard cheese. When we dropped a piece of Parmesan on the slicing blade, it caught, jammed, and cracked the drive shaft. Parmesan is so hard it is lethal to any processing blade except the steel chopping blade.

Proceeding with a much older model, we found that this original, almost legendary brand did perform well. It was the only food processor to produce chopped meat that looked chopped instead of shredded or mashed. It didn't mush up the onions. It formed bread dough into a good ball.

The Cuisinart literature didn't claim that the machine could whip egg white or cream, and, indeed it could not. It was, however, stable, responsive, and easy to assemble and disassemble.

(Continued from page 405)

mention the quality of a machine's performance for specific function, it is because that processor did the job adequately and pretty much as the other machines did it.

Food processors aren't the only versatile appliances that process food. We tested four other machines which perform many of a food processor's tasks, as well as some jobs which a processor can't do. (See page 411.) These machines gain their versatility by adding more parts. But in so doing, they sacrifice one of a food processor's most appealing characteristics: simplicity. ∎

Blenders: further down on the appliance chain

We don't own a food processor. That is, we *do* own one, but we don't call it a food processor. We call it a blender. We bought it eight years ago and have been using it ever since. The same blender. And if a blender doesn't "process" food, what does?

Food processors are handsome gadgets: some of them have all the look of a *Star Wars* prop. And, certainly, the ads make them look attractive, desirable, even make you feel somewhat inadequate as a provider if you have no burning desire to own one. So often, too, they are shown in the hands of professional chefs (or actors in tall white hats) as if to say that without this or that processor, their *cordon bleu* would turn quite *rouge*.

Well, we have never been of the opinion that a machine can make a cook. Just the opposite: a good cook can work with even inferior tools. This is not to say that a food processor can't be both fun and versatile—

more versatile than a blender. But most of the things that a food processor can do, our blender can do (in fact, our blender can do things the manufacturer didn't know it could).

For fine grinding, as with a nut butter or ground sesame or sunflower seeds, we simply put the seeds into the blender and turn the machine on. If things slow down, we give a little help with a rubber spatula. (Don't put a spatula or any hard or fleshy object into the blender while it is working.) The secret of successful grinding with a blender is not overloading the container. For example, we would never attempt to grind more than two cups of seeds at a time.

For coarse grinding or chopping—the consistency you would want for chopped nuts, for example— we load the container, then turn the machine on and off rather quickly, several times, until we have the right texture. Monitoring is essential, of course, but the whole process takes only a few seconds to a minute.

For chopping vegetables (we make a number of soups that cook quickly because the vegetables are chopped before they go into the pot), we add water to the blender container along with the vegetables, then empty both right into the cookpot. Much less of the vegetable comes into contact with air (since it's all under water) and so it retains more of its vitamin content. In this kind of chopping, the cleaning of the blender comes down to a quick rinse, because there is nothing that sticks to the blades or the container.

Slicing is another matter entirely. If we need even slices (say, for bread-and-butter pickles) we use an old-fashioned board slicer (though we have to admit that it is made of plastic), which, as long as we keep the blade sharp, makes very clean slices. For most of our slicing, however, it's simply a matter of Stan and his magic fingers (still intact after all these years of cutting). And a row of well-used, well-sharpened knives looks handsome on a kitchen wall.

We make ices in our blender, and the manufacturer never knew we could. It's noisy but nice. Some ripe cantaloupes cut into cubes, a tablespoon of non-instant powdered milk, and a couple of cups of ice cubes blend into a great dessert for a sum-

(Continued on page 408)

Farberware (model 386)

$140.

This machine was one of several with suction cups for feet, and they held it down when it vibrated and labored with bread dough. But even with the suction cups for stability, the machine produced heavy dough. It had difficulty forming the dough ball and the ingredients repeatedly needed to be scraped down into the mix.

The Farberware features several different speeds and a chart in the manual recommends particular speeds for various procedures. It sliced onions especially well, but took a long time breaking bread into crumbs. Something about the Farberware's blade construction caught food on the underside, making the blades slightly more difficult than most to clean off.

General Electric (including blender; model FP-2)

$99.98

The base unit on this machine accepts either a food processor assembly or a blender, both of which are available. The machine is easy to assemble either way. The food processing blades stopped slowly with the on-off switch, but quickly with the pulse switch. The GE has a short cord, which may or may not be an advantage on your particular countertop, and ran noisily.

It performed most jobs satisfactorily, but vibrated and jumped about when grating Parmesan or forming a dough ball for either yeast bread or pie dough. Its bread dough didn't rise well. Slicing and shredding cabbage was a difficult process for this machine, and the results were not uniform.

GE's *Cooking with a Food Processor* book is sold separately for $7.95. The machine comes with a booklet of instructions and recipes.

Hamilton Beach (model 737)

$99.98

This processor failed during the first test. It began straining with bread dough, emitted an acrid odor, stopped, and could not be revived. (The bread recipe was from its own cookbook.) So we got another machine.

The second Hamilton Beach chopped and shredded well, and had no slicing problems except with cabbage and cheese, two foods which were difficult for all processors to slice perfectly. It vibrated so much when grating hard Parmesan cheese that the suction feet couldn't hold it down; still, the cheese came out nicely grated.

Both of our test machines had lid trouble. It was difficult to snap the top shut, and the problem worsened as the tests went along.

The banana bread ingredients in the accompanying recipe booklet filled up the entire bowl, and got underneath the blade and coated the shaft. It took so long to cut the butter into the dry ingredients that the nuts were obliterated, yet the butter was

unevenly cut in. The second processor did not fail on the yeast bread, although it formed the initial dough ball with considerable difficulty. The dough rose well and the bread was tasty.

This Hamilton Beach comes with both an instruction booklet and a hard-bound recipe book. It has a short cord.

Norelco (model HB 1115)

$70.

The Norelco's short feed tube made it easy to position fruits and vegetables carefully for attractive slicing. However, the food pusher was long enough to extend through the chute and rest on the shredding and slicing discs, a potentially serious problem.

The Norelco made good bread dough, but labored and jumped in the process. In making banana bread, it didn't cut the shortening into the dry ingredients evenly. Carrots tended to jam up on the slicing blade, causing the machine to loose balance and dance. The steel blade jammed and stopped on Parmesan cheese.

(Continued from page 407)

mer's night. It's rich in vitamin A and low in calories and even has some protein and calcium.

Before we bought our blender, we did a bit of research—and we suggest the same kind of research to you. We visited our nearest public library to look at back issues of *Consumer's Report* and *Consumer's Research*. Fifteen minutes of searching told us the recommended models for blenders, and that's what we bought. We haven't regretted it.

The Sunbeam blender we bought was neither the cheapest nor the most expensive, but it has proved to be durable under heavy use—and we would not have chosen it without the help of the consumer-testing organizations. If we were to shop for one today, we would go back to the library and look over their current recommendations.

The blender we bought has eight speeds, but take our word for it that you don't need a lot of speeds to blend well. Three speeds (slow, medium, fast) are plenty, and two speeds will do in a pinch.

The container is made of heavy glass, which is much more durable than plastic. (We have a friend who insists that glass is a "plastic" as it is liquid when molten and will take any shape you mold it to—but what does he know? He's just a physicist.) In eight years of heavy use, it has not even chipped, much less cracked or discolored—both frequent complaints with plastic containers.

To its rim, the container holds about six cups, but the marked scale goes up to only five cups. Actually, you would be better off putting no more than *four cups* in the container because any liquid will just froth over the top if you fill it higher than that. As we said earlier, most things go into the container in much smaller amounts than that—if you want them to process evenly.

We enjoy our blender (we especially enjoy it when we make ices or piña colada in it), but we cooked before we had a blender. Aside from ices, there are very few things that a blender or a food processor or a mixing machine can do that we cannot do as satisfactorily, though more slowly, by hand.

Old-fashioned "knuckle-graters" will grate vegetables. Grinders are still sold: they screw temporarily (or permanently) onto the edge of a counter and, though messy to clean, do a better job of grinding meats than any powered machine.

Whisks will make mayonnaise (we make mayonnaise in our blender —but that is sheer laziness) and meringues and whipped cream. Spend the extra dollar or so and buy a whisk with a dozen or so wires, rather than a skimpy one—the more wires, the faster it whips.

Years ago, television abounded with offers for non-electric kitchen gadgets. One which we have never regretted buying is a spring chopper. (A plastic dome houses some chopping blades and a spring while a handle extends out the top.) A few taps on the handle chops nuts to a perfect texture—a few more does a good job on onions. This is a very handy number if you are afraid of knives.

Stan and Floss Dworkin

Microwave ovens

Knowing that her baking sessions took all afternoon somehow heightened the charm of Grandma's cooking.

"How long should I cook this?" someone would ask about one of her recipes.

"Til it's right," was her inevitable reply.

When a batch of cookies didn't turn out well, she'd shake her head regretfully and observe, "You didn't love them enough."

That's quite a contrast to today's grandmother, who can pop a tray of pizza into the microwave, adjust a precise heat setting, and consult the package to know when it's done. A few minutes later, she can deliver it to the grandchildren sprawled out in front of the tube.

Of course it's not quite fair—or true—to imply that food doesn't taste good unless it requires great effort. And grandma may well have better things to do with her time than cooking all day long. But sentimental arguments are not the only ones against microwave ovens. Many have decided that the dangers and disadvantages of these ovens outweigh their convenience and benefits.

(Continued on page 410)

Nutone Food Center

$287.90

This machine is operated by a power unit that mounts below the kitchen counter, revealing only a dial and cover plate flush with the countertop. We tested only its food processor, although the unit also powers a mixer, can opener, knife sharpener, ice crusher, blender, juicer, and power post for a meat grinder and shredder-slicer. All can be purchased separately.

The Food Center shredded and sliced vegetables satisfactorily, and made well-mixed banana bread, cutting in the shortening nicely with the dry ingredients. Grating Parmesan cheese, however, made the unit vibrate so badly that its top rotated, and the machine labored so much we stopped the test. Making bread crumbs also caused strong vibrations and crumbs seeped out beneath the lid as the top rotated. The Nutone's bread dough was too dense and heavy, and its peanut butter was dry.

Panasonic (model MK-5050)

$159.

This machine was quiet, it held still as it worked, and responded quickly with its on-off switch. It produced a good dough ball quickly and mixed banana bread evenly.

The Panasonic had two features that none of the other machines offered. An addition to the food chute lowers flexible stainless steel fingers which hold food upright for attractive slicing. The fingers worked well. An optional potato peeling attachment also did a fine job.

J. C. Penney (model 8320)

$65.

Sony manufacturers this quiet, smoothly responsive machine with lid rotation instead of a switch for on-off. The blades started quickly but took a second or two to stop. It was one of only two machines with a food chute long enough to make it impossible for a small hand to reach down to the blades—a good feature when familiarity breeds carelessness.

The Penney made particularly attractive french fries (with an optional disc). But it labored on bread dough and failed to form a good dough ball. Some of the shortening didn't cut in uniformly when we mixed banana bread with it.

Sears (deluxe model)

$90.

This machine is produced by Scovill, the company that makes Hamilton Beach products. It features seven motor speeds, controlled with push buttons, which are intended to give greater control for coarse chopping and provide more even results. The Sears did handle most vegetables well, though the cabbage slices turned out uneven.

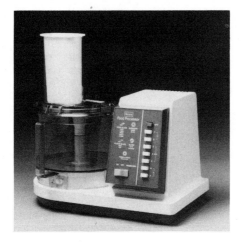

Sears claims the slower speeds enable the machine to whip cream and egg whites because more air can be incorporated. However, cream went from liquid to butter without becoming more than a thin foam. The bread dough failed to rise well, and the processor leaked flour and liquid up under the blade and down the shaft as it mixed the dough. We also found unprocessed peanuts beneath the blade after the Sears had made a somewhat un-

evenly textured peanut butter. The blades were difficult to change, fitting too snugly on the shaft.

Sunbeam

$145.75

This processor has a slightly larger bowl than most other brands. It worked quietly and smoothly, and produced good bread dough which rose well. It made notably smooth peanut butter. Some of the vegetables caught and bunched up on the slice and shred discs and we had to stop the machine and retrieve them. Onions and green peppers were sliced unevenly, but the Sunbeam produced good, well-mixed banana bread batter.

Waring

$179.99

The Waring food processor felt like a heavy-duty machine. It worked quietly and the blades changed easily. It sliced onions and cabbage very well, but didn't do a good job of cutting in shortening to dry ingredients for the banana bread.

The Waring manual doesn't mention making bread dough so we tried pie dough instead. It mixed the dough satisfactorily, but with some vibration.

(Continued from page 409)

When they first came out, microwave ovens sounded like something out of a science fiction movie. Some people feared they would poison food with radioactive particles, but the food prepared in microwave ovens is not contaminated. Curiously, microwave cooking causes slightly lower nutritional losses with some foods than conventional ovens. The problems arise instead from the emissions the ovens make into the home.

The difference between conventional and microwave cooking is in the heat source and the way the heat penetrates the food. Microwaves are

What's cooking? In the microwave age, it may be the cook.

a type of electromagnetic energy, with frequencies just below the infrared region of the electromagnetic spectrum. They have shorter waves and higher frequencies than radio waves, but longer waves and lower frequencies than radioactive substances like radium.

Microwaves cook by vibrating the water molecules in food. The vibrations make the water molecules give up their energy in the form of heat and the heat makes the food cook quickly. The waves can pass right through substances such as paper, cardboard, and plastic without heating them, in the same way the sun can pass through a window and warm a person inside without heating the window. Metal reflects microwaves instead of transmitting them, so metal containers such as aluminum can't be used. When food is subjected to microwave energy in a metal box—the oven—the waves reflect from the metal walls until they strike the food and are absorbed.

In a conventional oven, the entire box and the air within it heats. The heat is then transferred to the food. In the microwave oven, the outer layers of the food are heated and the inside is cooked mainly by the conduction of heat from the outer layers.

Whether ovens and other items that use microwaves can endanger human health is a subject of fierce debate among scientists and consumer research groups alike. Microwave technology is far more advanced than is research on the effects of the waves. There are disturbing indications that emissions of microwaves can be dangerous, even at the "safe" level—10 milliwatts per square centimeter—set by the government. Many people believe this standard is far too lenient. (The Soviet standard for safe emission levels is .01 milliwatts per square centimeter, 1,000 times lower.)

Some scientists contend that the only dangers microwaves present to humans are thermal. In other words, when emissions are too high, microwaves cook human tissue—forming cataracts in the eye, for example—just as they cook food. Areas with poor blood circulation, such as the eyes and the testes, are particularly vulnerable. Especially disturbing is the discovery that such injuries can occur painlessly, with the victim remaining ignorant of them for years.

Still other researchers argue that unsafe levels of emissions have far more wide-ranging effects, particularly on the nervous system. Some serious effects have been seen in animal experiments and in industrial workers who have worked with microwaves. These include: the formation of cataracts; loss of hair, memory, and libido; changes in blood cells; fatigue; headache; aching eyeballs; tickling sensations about the head and ears; sterility; enlargement of the thyroid gland; decreased lactation in nursing mothers; and difficulty in perceiving the color blue.

The validity of these observations, however, is sometimes uncertain. For ethical reasons, humans
(Continued on page 412)

Oster Kitchen Center

$205.95

Oster is a familiar name in blenders. On the Kitchen Center, you can mount a blender, a conventional mixer, a grinder, or a salad maker. The salad maker uses spinning shredder and slicer discs similar to a food processor, but drops the processed food down a chute instead of collecting it in its own bowl. The salad maker sliced and shredded our vegetables satisfactorily, excelling with lovely sliced mushrooms and the best-looking french fries we saw from any machine. Carrots, however, collected above the disc and were sliced unevenly. Shredded food tended to jam up in the exit chute and slow down the operation drastically.

The other Oster attachments—blender, mixer, grinder—were of the usual design for these appliances and we didn't test them. We did, however, grate some Parmesan in the blender to compare that action to a food processor. It did a fair job with the cheese, but threw up pieces with such force that they would have knocked

off the blender lid if it hadn't been held on by hand.

The Kitchen Center's motor did not effectively blend or grind heavy foods. The machine has plastic gears that are prone to breakage.

Kenwood Chef

$275.

This machine looks like a mixer until you add one of its 14 attachments. It will then duplicate some food processor actions as well as hulling peas or opening cans. The Kenwood is a hardware store of parts, and 10 pieces are necessary for the juice separator alone. In our test we passed over the sausage filler and cream maker and tested the high-speed slicer and shredder, and the mincer. If some of those attachments seem redundant, it's because they are.

All of our tests with the Kenwood were hindered by the electrical cord which joins the machine right where a bowl must sit to catch the prepared food. This machine involved so much setting up that it was not an attractive alternative to chopping and slicing mid-sized jobs by hand.

Once assembled for a slicing job, however, the Kenwood sliced and shredded satisfactorily, except that shredded food

tended to jam up in the exit chute. The mincer action was equivalent to the chopping action of a food processor, although the minced food was delivered from a chute rather than processed inside a bowl.

The Kitchen Works

$40.99

This simple machine made by Moulinex looks like a grinder, but it shreds and slices as well. It assembled fairly easily, although the grinding unit has several parts including a sausage filling snout. The Kitchen Works did its work more slowly than a food processor, and its range is obviously less, but it turned out satisfactory slices, shreds, and grinds for all the vegetables, fruits, and meats we fed it.

(Continued from page 411)

can't be dosed with the high number of microwaves used in animal experiments, and scientists disagree on how likely effects observed in animals are to occur in humans.

Meanwhile, microwave ovens are selling in record numbers. As technology improves, the prices drop; they now range from $250 to $700. Industry spokesmen say about 10 percent of American households have microwave ovens now, and they predict that figure will rise to 25 percent in the next few years. In the 1980s, microwaves are expected to outsell gas ranges. They are used increasingly in both restaurants and hospitals.

Clearly, microwave ovens have their advantages. They cook more quickly than conventional ovens, and thus can use less energy. You can cook with paper containers, so they

The Radiant cookstove and grate

When I tested the Radiant cookstove, I got my steak *and* some of the sizzle. I positioned the formed aluminum box about five inches from the fireplace grate to direct radiating heat toward the grill—seven minutes on each side and I was at the table.

To prevent the drippings from staining the hearth, I fashioned a foil guard to capture them and set the oven on another sheet of foil. After the cooking was over, I was uncertain

what to do with the hot stove. If I kept it by the fire the grease would probably burn; if I moved it I might spill the grease. (I eventually ran it to the front porch.)

Another thing that bothered me about using the stove was its dependence on a fireplace. To capture the radiant heat the fire was burning away—an inefficient way to heat your house as so much warm air is lost out the chimney. If you close glass doors and damp the fire to prevent this loss, the heat becomes inadequate for dependable cooking.

The Radiant stove has its greatest value as a cooking alternative when the fire is burning and hospitality abounds, or for recreation. And, of course, you never know when the power will fail.

A quick note about the Radiant fireplace grate. It's neatly designed with a raised front allowing more heat to reach the room than the traditional log holders. Its heavy-gauge steel should withstand prolonged use.

TS

Radiant Grate and Radiant Stove, 11 Reality Drive, Kinnelon, NJ 07405.

The Radiant.

save washing pots and pans. They warm leftovers or convenience foods in seconds. Some models now have browning elements, while others feature a spray that can be used to produce a browning effect.

But there are disadvantages, too. Most microwave ovens cannot crisp or brown foods. Cooking in them requires very precise heat and time settings, so you can't save by cooking at 360°F. when the tuna casserole is supposed to cook at 350° and the gingerbread at 375°, as you can with a conventional oven. A microwave oven usually isn't big enough to hold several dishes at a time. And food doesn't clue you that it's done by turning brown.

Energy costs are increased by defrosting in a microwave. Because moisture contents are different within some dishes, like casseroles, defrosting in a microwave is not as dramatically fast as cooking in one is, and if you try to rush it, you might wind up with a dish that is frozen solid in one place and bubbling in another. Putting food in the refrigerator to defrost all day is more efficient, since the refrigerator is running anyway.

Very few families can get by using only a microwave oven. In a study by *Consumer Reports,* most microwave owners contacted said they used them for re-heating meals, warming snacks and breads, cooking convenience foods, and baking potatoes. They used other appliances for steaks, chops, roasts, hamburgers, and pastries, and thus ended up saving virtually no energy.

As an alternative, a small toaster oven can warm leftovers, breads, and snacks; cook convenience foods; and eliminate the need to heat up the whole oven for a small amount of food. Baked potatoes don't have to delay a meal for an hour if you put them in a low-heat setting in a Crock-pot all day with a little water. Or, they cook more rapidly in a conventional oven if you cut them in half and put a metal stake in them.

With careful use, a microwave oven can save time and energy. But the savings are not as great as the oven manufacturers would have you believe, and they don't take into account the initial cost of the oven or the potential hazards of exposure to errant microwaves.

Rebecca Christian

La Machine

$80.99 ($105. with doughmaker)

This elaborate hybrid is also made by Moulinex. Its three components are a cutting-chopping unit, a Vegetable Chef, and a blender.

The cutting-chopping unit has a capacity of only one cup, but the machine did a good, quick job with the food in that one cup. In fact, it made the smoothest peanut butter of any appliance we tested. A small tool is necessary for removing the cutting blade. (Caution: small tools are easily lost and this one's indispensable).

The Vegetable Chef (a trademarked name) attachment requires elaborate assembly. Once assembled, it shredded and sliced vegetables satisfactorily. But cheese mashed and wedged into the machine instead of being shredded or sliced. Both

the cutting-chopping unit and the Vegetable Chef were very noisy to operate. We didn't test the blender.

Cooking under pressure

Pressure cookers are not time bombs waiting to explode. They're easy-to-use pots with tight lids that can save you energy, nutrients, flavor, and time. At 15 pounds of pressure, the cooker cooks most foods in one-third the normal cooking time at 250° F.

Just follow the directions that come with the cooker. Put the food to be cooked and the recommended amount of water in the pot, close the lid, and place the cooker on high heat with the vent open. Exhaust the air by allowing steam to escape in a solid column through the vent. When the steam flow is steady, put the regulator on.

Once the pressure regulator begins to rock gently, lower the heat to achieve a steady, regular movement. Cook the food for the recommended time, then cool the pot quickly by setting it in cold water. When the pressure is down, the pot will heave a sigh and then you can safely remove the lid. But do so carefully, opening it away from you so you won't get a faceful of steam.

The only problem to watch for is a clogged vent. If food blocks the vent, pressure can build up to a dangerous level, blowing out the safety valve. To prevent this, make

(Continued on page 414)

THE OVEN'S APPETITE

Energy use in the home varies greatly with the appliance and its user. A recent study found that six cooks preparing the same meal on seven days varied by as much as 50 percent in the energy they used, most of the difference being attributed to the way they used their ovens. Although the Department of Commerce hopes for a 20 percent increase in the efficiency of appliances by 1980, 50 percent is a great deal. Mary Rawitscher found that between 4,000 and 5,000 kilocalories is a reasonable cost for an hour's baking in a regular oven. But cooking on top of the stove for an hour uses only 1,800 kilocalories. Using a microwave oven falls in between, about 40 percent less than for a conventional oven and 30 percent more than for stove-top cooking.

It may surprise you (and sadden you a little, as it did me, for I enjoy home-baked bread) that home baking even six loaves of bread at a time takes more energy than a bakery would use for the same weight of bread. The actual figures: About 1,900 kilocalories as opposed to between 1,350 and 1,500. That is an energy trade-off decision your conscience will have to make.

Jean Mayer.

THE HAND

The surest way to save money, of course, is not to spend it. I want to direct your attention to the human hand, a tool that costs nothing at all and is frequently superior to anything else available.

Tasting sauces, for instance, is most commonly done with a spoon. But it is done far more efficiently by holding the index and middle fingers firmly pressed together and rapidly dipping them into the liquid (naturally, you'll want to start with relatively cooler potions and work your way toward greater heat as you get better acquainted with your own pain threshold).

Using the fingers this way, you remove from the pot only a minute amount of the liquid, so you don't have to wait for it to cool, nor do you have to search around for the tasting spoon you keep misplacing, nor do you get the flavor of the spoon in the sauce as you taste it.

The final step in preparing a soufflé for the oven is executed best by the thumb. The way to help a soufflé rise in the shape of a mound and prevent it from toppling over the rim of the dish while it's at it is to create a trough in the mixture around the periphery of the dish. Do it by resting your fist on the rim and drawing your thumb through the mixture until the circle is complete.

In fact, as cooking has become an increasingly technological activity, the closer I can come into contact with the food I'm preparing the better I like it.

The hand is used by every cook to mold chopped beef or fish or fowl into a variety of shapes, and it is used by everyone to initiate the kneading of dough. But not everyone prepares the hands properly for these jobs. When working with chopped meat, wet your hands first with cold water so the meat won't stick to them. With dough, do the opposite, dry them thoroughly with flour before beginning to knead.

A last point: ìf you start out feeling awkward using your hands, don't be disheartened. It is well known that even hands that begin as all thumbs become digitally perfect through experience.

Pierre Franey. © 1979/80 by The New York Times Company. Reprinted by permission.

(Continued from page 413)

sure that the vent is open when you begin cooking, and don't fill the cooker more than two-thirds full. Exercise care with beans, cereals, and dried or pureed fruits—foods that tend to foam when cooked. (These should not be cooked unless your cooker gives specific directions for them.) You can deter beans from foaming by adding one to two tablespoons of vinegar or oil to the cooking water.

If the vent becomes clogged during cooking, the regulator will cease to rock. Remove the cooker from the heat immediately and allow it to cool down. Don't cool the pot by pouring cold water over it, since a sudden pressure drop may cause trouble.

Pressure cooking toughens the proteins in meat and adversely affects flavor. The only advantage to pressure cooking meats is the time savings. There is less shrinkage and better flavor if you cook meat at 10 pounds pressure rather than 15 pounds.

Because of the high temperature, stocks will not be as delicately flavored as those cooked at lower temperatures over longer periods of time. Allow one quart liquid to two pounds of meat in a four-quart cooker. Season very lightly. Cook 30 minutes at 15 pounds pressure to almost finish cooking meat. Allow the meat to cool down before opening the cooker to add vegetables.

A nutrition-saving way to cook vegetables is to use a steamer with the pressure cooker. Boil some water in it, then place the loaded steamer inside. Close the lid and follow regular procedure. Cook one to two minutes longer than the time indicated for normal pressure-cooking of these vegetables.

Mexican Pinto Chili

2   cups dry pinto beans, soaked overnight
2   quarts water
3   medium tomatoes, cut into chunks
1   onion, diced
3   cloves garlic, minced
1   green pepper, diced
2   tablespoons soy sauce
1   tablespoon chili powder
2   teaspoons oregano
1   teaspoon basil
½   teaspoon cayenne pepper
½   teaspoon cumin

Drain the presoaked beans and add to a 6-quart pressure cooker. Add the water. This should be enough to just cover beans and not fill pot more than half full. Add the remaining ingredients and stir to mix through. Put on the lid. When the pressure has built up, time the chili for 25 minutes at 15 pounds pressure. Put to one side to allow pressure to drop; this will take 25 minutes.

Linda Gilbert, Rodale Test Kitchens.

Knives

A good knife is a sharp knife—a simple statement to make, but not easy to put into practice. Centuries of experimentation with materials have gone into developing a kind of metal for knives that will stay as sharp as possible.

By the 1300s, ironworkers in Europe were fashioning strong iron blades that took a sharp edge. They were brittle, however, and would rust when exposed to humid air. So ironworkers learned to produce steel, an alloy stronger and more ductile than iron.

Steel is made from at least 85 percent iron and a combination of several other elements. Carbon makes steel harder, and because the fine edge is easily renewed, many chefs prefer carbon steel blades over any other. The disadvantage of carbon steel, though, is that it can rust from moisture and stain from contact with acid in fruits and vegetables. A car-

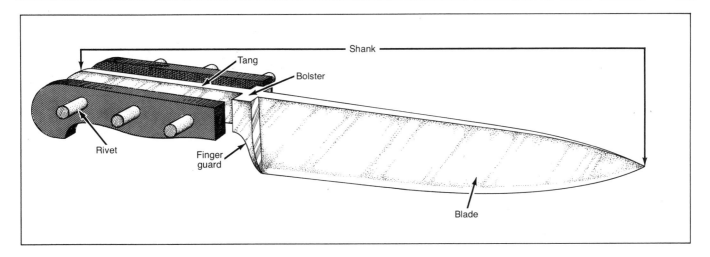

bon steel blade will change color in time. Stainless steel was developed in England in 1912. A high chromium content enables it to resist corrosion and tarnishing. But stainless steel blades are harder to sharpen than carbon steel blades.

A new alloy called high-carbon stainless now combines the best properties of carbon steel and stainless steel. It is by far the best material for cutlery: high-carbon stainless knives are easy to both sharpen and clean.

The quality of steel is affected not only by its mixture of elements but also by the way it is heat-treated. Steel is strengthened by tempering, a process in which it is heated, then immersed in a chemical bath for rapid cooling. Tempered steel is hard but extremely brittle. A second operation, annealing, toughens the steel by reheating and slow air-cooling.

After knife steel is heat-treated, grinding puts the edge on. Grinding also aligns and strengthens the molecular chains on the thin knife blade. Knives hand-ground by craftsmen remain sharp far longer than those done by machine. Machined hollow-ground knives have an abrupt concave profile near the cutting edge, while the rest of the blade is quite thick. Initially, hollow-ground knives are sharp because the lower part of the blade is so thin; but after the edge wears down, the thick part of the blade cannot be properly sharpened.

A long-lasting knife blade in profile appears just barely convex—the curve may not even be noticeable. And for almost all knives the edge should be straight, not serrated. Ser-

rated edges do not actually cut—the dozens of teeth sink into the food and tear it apart. That's fine for bread and tomatoes, since the serrations effectively break through the crust or skin without squashing the soft insides. Roast beef, on the other hand, definitely should not be torn apart. For most cutting chores, a sharp, straight-edged knife is superior. Serrated knives are also difficult to maintain, because you can't sharpen them yourself.

The part of the knife blade that extends into the handle is called the tang. Handles are made of wood, plastic, or wood impregnated with plastic to make it non-porous.

If the handle of a large knife is made from two pieces of wood, the tang should be fully visible around the edge. Inexpensive knives commonly have just a short piece of metal reaching into the handle, held by one rivet. Such weak construction throws the balance of the knife off, and the blade may pull out of the handle. On large knives, look for a full tang held in place with three thick rivets.

Wooden handles should be treated to prevent moisture absorption. If the wood begins to lose its color, rubbing a little mineral oil into the handle will restore its beauty.

On a knife with a plastic handle, the tang is permanently molded into it by heat. The blade will not pull out. Plastics are also sanitary, and their one-piece construction leaves no seams to trap food particles.

Whatever the material, the handle should be large enough for the full grasp of the hand. Some knife handles are formed with a contoured

grip for the fingers. If a grip feels right, it can be a help in keeping a firm hold, because a busy cook invariably has wet or greasy hands. Sometimes the material used for the handle has a texture that prevents slipping.

Many good knives have a thickened section of steel between the blade and the handle that protects fingers from slipping onto the sharp cutting edge. This bolster is forged in one piece with the blade and the tang.

Choosing among these features is a matter of personal taste. Many cooks say they prefer a fairly heavy knife that will do much of the work by itself.

Several brands of quality cutlery most often recommended by experienced cooks are Forschner, Henckels, and Wusthof-Trident. Dexter-Russell knives, also mentioned, are well suited for butchering.

The words Solingen and Sabatier appear on many knives. They are not specific brands themselves. Solingen is a town in Germany, famous for its finely crafted cutlery since the 14th century. Sabatier was once the name of one factory in France, but the term is now used by about a dozen knife manufacturers there.

Using a dull knife is like walking on a sidewalk on ice skates. It's slow going, and it ruins the blades.

Own a sharpening steel and use it often. A sharpening steel, sometimes called a butcher's steel, has a handle much like a knife's and a long, grooved rod, often with magnetic properties. Contrary to its name, a sharpening steel does not actually sharpen a blade. It does hone and

(Continued on page 416)

(Continued from page 415)

renew the edge quickly and easily. The magnetism realigns molecular chains on the edge of the knife, straightening tiny saw teeth which are invisible to the human eye. Ordinary use of the knife puts those teeth slightly out of whack, but a few swipes on the sharpening steel will have them fine-tuned again.

A right-handed person should hold the steel in the left hand, and the knife in the right. The steel should have a guard between the rod and the handle in case the knife slips.

Place the end of the blade nearest the handle flat on the steel, then lift the back of the blade about ⅛ inch. Run the blade along the steel in a smooth arc from heel to tip. Move only the knife, not the steel. Think of it as trying to shave the grooves off the sharpening steel. Do that about five times on one side, then repeat the process for the other side by putting the blade under the steel. Do not overdo it—five times should be enough. Also, be careful to maintain the same angle and pressure on both sides of the knife.

How often you steel your knives will depend on how much you use them. Butchers and chefs do it every time they pick up a knife. In the home, a sharpening steel is helpful whenever you sense you knives becoming a little dull.

Ceramic sharpening "steels" are also available. Their smooth, harder-than-steel surface puts a fine edge on a knife, but the ceramic is brittle and breaks easily.

Eventually, even good knives given proper care need to be sharpened by taking off a small amount of the metal. That thins the blade, which is what makes it sharp. Knives can be sharpened professionally, or you can do it at home with a whetstone. Sharpening stones are often made of silicon carbide. Carborundum is a trademark for various abrasives, including whetstones. The simple rectangular whetstone should have a rough side and a smoother side, and be at least 8 or 9 inches long.

Whetstones should be soaked in oil before using. You can also soak them in a solution of water and enough dish detergent to make the mixture viscous.

Then run the knife across the rough side of the whetstone. Use a few light strokes and, again, be careful to use even pressure and the same angle for both sides of the blade. Don't miss the tip. After sharpening on the rough side, flip to the smoother side of the whetstone and repeat the procedure. Finally, use a sharpening steel to smooth out any burrs made on the knife edge by the coarse stone.

Depending on their wear, you may need to sharpen your knives two or three times a year.

A whole array of knife sharpeners are on the market. One style has two small sharpening rods positioned in a V, so a blade held straight and pulled through is sharpened on both sides at once, at a predetermined angle. Another type consists of two abrasive rollers through which the blade is pulled. Electric can openers often have attached knife sharpeners. (Some sharpening tools also are supposed to be able to sharpen serrated edges, which steels and whetstones cannot do.)

These devices may put an edge back on, but they tend to wear down the blade at the same time. It's better to depend on a steel and a whetstone.

A special cutting technique makes a French knife the ideal tool for efficient chopping and slicing. Keep the tip of the knife on the cutting board and slice back through the food, rocking the blade to its heel. Allow the weight of a heavy French knife to do some of the work for you. Do not, however, repeatedly bang the blade into the cutting board. The edge will suffer. The deep dropped edge of the French knife does make it possible to cut in a rocking motion without knocking your knuckles on the cutting board.

For the jobs you do with a paring knife, its action should be thought of more as an extension of your hand. Sometimes it is helpful to hold the knife with your thumb and forefinger on the blade, close to the handle. You will be able to work with small foods more deftly.

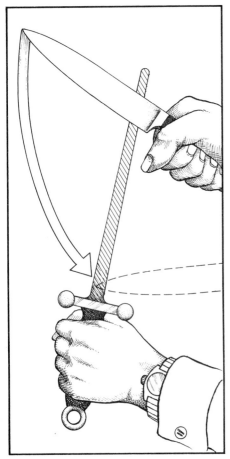
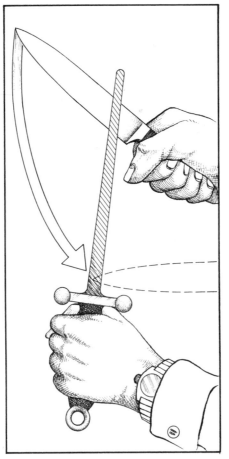

In steeling a knife, use the same pressure and stroke angle for both sides.

CB

Essentials

Professional chefs agree that four or five well-chosen knives are all that are needed for practically every kitchen cutting job. Tom Ney, director of test kitchens and food services for Rodale, recommends the following assortment of knives.

French chef's knife (for home use, 8″ or 10″; $10 to $30).
The indispensable all-purpose French knife with a deep dropped edge is used most often for chopping, slicing, and dicing vegetables and meats.

Paring knife (2″ to 3¾″; $4 to $10).
The smallest and handiest of kitchen knives, the paring knife is likely to be in constant use for coring and peeling small fruits and vegetables, and for fine, decorative jobs.

Slicing knife (10″ to 14″; $10 to $20).
The slicing knife's slender, efficient blade is perfect for making thin slices of cooked meats for serving.

Boning knife (5″ to 8″; $9 to $12).
The thin and narrow blade of the boning knife gives it the necessary flexibility and maneuverability for cutting meat and poultry away from bone, even in hard-to-reach places.

Serrated bread knife (9″ to 10″; $5 to $16).
The saw teeth of a serrated edge break through crust without crushing the loaf. This knife also works well for slicing non-solid vegetables like tomatoes.

Handy extras

Depending on how much use you might have for a particular specialty knife, these additional styles can be helpful in the home kitchen.

Fillet knife (6″ to 10″; $6 to $11).
The extremely thin blade of the fillet knife makes it even more flexible than the boning knife. It can slice thin slabs of fish along the contour of the bone.

Chopping knife (7″ to 10″; $10 to $30).
Absolutely flat cleavers made in the Oriental tradition are designed for rapid chopping using wrist action instead of a rocking motion. These knives prepare vegetables for stir-frying in a wok.

Grapefruit knife (about 4″ to 5″; $2 to $5).
The angled blade of a grapefruit knife and its double-edged serration neatly loosen grapefruit sections from the rind.

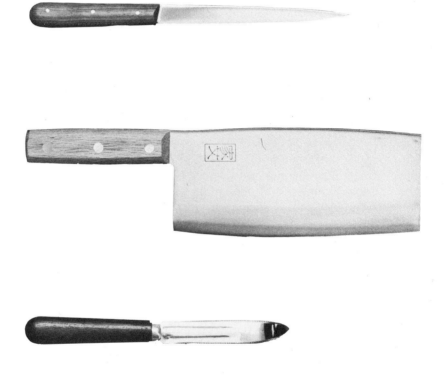

Cookware materials

Perhaps the first pot was nothing more than clay from a river bank, daubed onto freshly killed game by some primitive cook who wanted to seal in the juices and keep the meat from burning. Since then, our culinary goals have changed little, but the equipment is quite another matter.

Modern humans face an array of cookware from unglazed clay to glassware, bronze, tin, aluminum, copper, and many kinds of iron and steel, often coated with porcelain, sprayed-on acrylics, or poly-tetrafluoroethylene (known as Teflon). Choosing among them is difficult, though many of us never properly pick out our pots and pans: they arrive as gifts, hand-me-downs, or stray emergency purchases. Given the chance to choose, however, there are myriad guidelines to follow.

One puzzling factor is whether the pot can poison the food. You may have heard tales of lead poisoning from clay pots, and rumors that aluminum can leach out into our food. Here is a short guide to cookware materials but it does offer reasonable probabilities.

Clay cookware. There is a real danger of lead poisoning when pottery with lead-glazed *interiors* has been fired at too-low kiln temperatures. The lead can then be leached out by acidic foods such as orange juice or spaghetti sauce, to cause damage to the nervous system. Unless your clay cookware has an *unglazed* interior or comes with clear guarantees from a reputable manufacturer, retire it for use as a planter or check it out with the Food and Drug Administration (Consumer Communications Management Staff, Office of Policy Coordination, Food and Drug Administration, 5600 Fishers Lane, Rockville, MD 20852).

Teflon and other non-stick finishes. No-grease, non-stick plastic cooking surfaces are nowadays applied to almost any kind of cookware—frying pans, saucepans, muffin tins, and so on. They are commonly considered safe when used according to manufacturers' instructions. But, as with most chemical compounds developed since World War II, there is no evidence of what happens to people over a lifetime of eating food cooked on such coatings, some of which inevitably gets into the food. Using metal utensils with Teflon scrapes polytetrafluoroethylene (PTFE) from the cookware surface, often into the food. Also, heating PTFE to a very high temperature (above that of smoking oil) releases toxic fumes. If you are the kind of person who sometimes puts a pan of the burner and then forgets it when the telephone rings, Teflon is not for you.

Copper. Copper pans are the very best conductors of heat, and have long been chefs' favorites. Copper salts, however, are extremely toxic and pans should be lined with tin or stainless steel. The moment a tin lining wears thin or is nicked so that copper shows through, the pot *must* be re-lined. Unfortunately, tinsmiths are a rare breed these days and it may be difficult to find the necessary technical knowledge. Big-city gourmet cookware shops sometime still perform this service. Otherwise, it may be necessary to consult the manufacturer for information.

Aluminum. Is it safe to cook in aluminum pots and pans? Aluminum is a natural constituent of many foods and of the human body. While some of the aluminum in cookware does migrate from pot to food (especially with acidic ingredients) the amounts are similar to those in foodstuffs. Very little aluminum is absorbed by the body; most passes right on through. Of what *is* absorbed, most is normally excreted by the kidneys, leaving very little indeed.

For some people, however, the added aluminum may still pose a problem, bringing headaches, indigestion, or noticeable blood pressure fluctuations.

Stainless steel. Stainless steel contains substantial amounts of nickel and chromium. Salt and acidic foods such as mustard, mayonnaise, lemon juice, and vinegar will corrode stainless steel if allowed to remain in contact with the surface for any period of time. The same is true for tin, which is very poorly absorbed by the human body and seems non-toxic, but might possibly irritate some people.

If you or your household suffer from otherwise-unexplained gastrointestinal or skin problems, try avoiding each kind of metal cookware for a month or so in turn, to see if the symptoms improve. Most metals contain some impurities, and if you want to avoid all possibility of metallic irritations, stick to glass or porcelain. For most people, most of the time, however, the best option may be simply to *minimize* the leaching of metals into meals. Use wooden utensils. Don't attack sticky or burned spots with metal scouring pads; instead, soak, then scour gently with brushes or plastic pads. Cook acid dishes in glass, clay, or porcelain (or cast ironware, which may boost your dietary iron).

Aside from the question of health, which work best? That, of course, depends on what's cooking, and where. Copper and aluminum are the best conductors of heat. When put on a flame, they heat quickly and evenly, rather than in hot spots that burn food. Stainless steel, clay, cast iron, and glass are somewhat less reliable this way, but the last two retain heat very well. So, for stove-top cooking, copper and aluminum work best as the pot *base*. This includes stainless steel pots with copper plating on the bottom, or a three-ply sandwich of aluminum between layers of steel, or porcelain-clad metal, and other such combinations. The trick is to combine good conductors with the surface that you prefer.

Oven cookery is less demanding: the oven itself spreads heat to all parts of the baking dish. Cast iron, glass, clay, and porcelain on metal will work well. Indeed, these retain heat beautifully, so that your casserole can stay warm on the table for second and third helpings.

These are the basics of heat distribution, but the subject can get much more complex. Care and cleaning affect the utensil—for instance, darkened tin absorbs heat better than the shiny metal. Moreover, each kind of cookware offers some special advantages and special defects, often related. Glass percolators, for instance, let you see when the coffee has (alas!) begun to boil, but they retain so much heat that you can't quickly stop the boiling. Aluminum is light, easy to heft—and also much easier to dent and warp. Let's look at each kind of cookware in turn for its peculiarities.

Clay (earthenware and stoneware). Clay pots come both glazed and unglazed. Most cannot be used over direct flame, and all are easy to break. When hot they must be set down on a dry surface, or they may shatter. Always guard clay against sudden temperature changes, such as putting hot food in a cold pot. Unglazed ware is porous and should be seasoned before each use by soaking the pot and lid in water. During cooking, this water is driven inside the pot as steam and cooks the food with little or no extra water or grease, preserving the flavors and juices of the food itself. The basic rule with unglazed clay is to put it in a cold oven and to adjust the recipe directions by *adding* 100°F. and one half-hour of cooking time.

Because the clay is porous, it needs special care. Avoid soaps and detergents. Eventually, fat from the food will clog the clay's pores and may turn rancid. This problem is solved by filling it with a strong solution of baking soda in water and baking for one hour at 400°F.

The best glazed clay (stoneware or earthenware, depending on whether higher or lower firing temperatures were used) comes glazed on the inside only. The rough exterior can then absorb the maximum amount of heat. These pots are perfect for long, slow cooking at low temperatures.

Cast iron. Cast iron is very heavy, can crack if dropped, and takes a relatively long time to heat. Pots are superb for long, slow cooking (as Dutch ovens) and for gentle warmups or fyings. Since the heavy metal takes a long time to heat, don't use them for successive quick or deep-fat frying projects: each new food item cools the oil, and before the cast iron can reheat, oil is absorbed, making the food soggy.

Cast iron must be seasoned by heating a layer of oil in the pan before it is ready for use. The oil is then poured off, the pan rinsed, and wiped or heated until dry to prevent rusting. Overheating and over-scouring can destroy the seasoning,

(Continued on page 420)

(Continued from page 419)

which must then be repeated. Ordinarily, a swipe through sudsy water, a rinse, and thorough drying will keep your cast iron clean *and* seasoned perfectly.

Stainless and carbon steel. Stainless steel, an alloy, is moderate in weight, retains its smooth, shiny surface well, and does not dent or warp easily. Stainless steel is not a good conductor of heat, and works best if clad with copper on the bottom, or with aluminum sandwiched between the two surfaces.

Carbon steel is a porous variety that heats quickly, but is usually very thin. It rusts easily, and for that reason is either lined with tin or seasoned with oil. It is superb for use in woks because of its quick heat conduction, but such woks must be wiped out rather than washed.

Porcelain-clad metal. Some of the finest cookware combines a metal base with a baked-on porcelain-glass coating, inside or outside. The metal may be aluminum, steel, or cast iron; the pot takes on the metal's properties, plus the retentive qualities of porcelain. (If the metal is cast iron, make sure the pot or pan isn't too heavy for you to carry safely.)

Cooking at very high temperatures over long periods can melt or crack the porcelain, exposing the metal; and, hard scouring can scratch the glossy surface, making future cleaning more difficult.

Aluminum. Besides its good conducting qualities, lightweight aluminum is easy to handle and easy to dent. Salt or acidic foods may pit and stain the surface. Prevent this by removing the foods immediately after cooking. Stains on aluminum can be removed by boiling a solution of two to three tablespoons of cream of tartar, lemon juice, or vinegar to each quart of water in the utensil, for five to ten minutes. Boiling soda water will remove the grayish oxide that gradually accumulates on aluminum.

Glass. Flavors are *purer* when cooked in glass, as this material does not interact chemically with food. Moreover, glass cookware usually presents a pretty picture: either the colorful collage of the food itself, or the glossy sheen of opaque ceramic. It is, however, monstrously easy to break by dropping, banging, or by sudden changes of temperature (such as setting a hot pot in a countertop puddle). Clear aluminosilicate glass is for surface cooking; borosilicate ovenware is either clear or opaque.

Glass cookware for the stove-top needs a trivet to spread the heat; even so, it may burn food in hot spots. If a glass pot is overheated by boiling dry, discard it for your own safety.

Oven cookware may not normally be used on the surface. Corningware and such can go directly from freezer to oven; *some* of these may be used on the stove-top, but only if the manufacturer so advises.

Harsh scouring is bad for glass. Bleach can remove stains.

Copper. This, the best heat conductor, needs care in use. Avoid scratching the inner lining, and re-line when any copper is visible. Clean with a mixture of flour, salt, lemon juice, and ammonia, or flour and vinegar, or with a manufacturer's recommended cleanser.

Non-stick finishes. It you must use these, avoid high heats and metal

For inexpensive cast-iron ware, try flea markets and restaurant supply stores. (See "Introducing the restaurant supply store," page 422.)

utensils; to remove stains, use a special cleanser or mix ½ cup liquid bleach and two tablespoons baking soda with one cup of water, and simmer. Do not scour with abrasive cleansers or metal.

Cynthia Adcock

Cooking in clay

An ancient technique—cooking in clay—is gaining favor again. People are re-discovering that it enhances and preserves natural flavors. Clay-pot food is cooked alone, with no fat or liquids added to change subtle flavors. As a bonus, all the nutrients remain in the pot and in the food.

Unglazed pots are soaked thoroughly in water before the food is put in.

Glazed clay pots are the original Crock-pots and work best for long-simmered stews, grain dishes, and beans.

There are two basic kinds of clay pots to choose from: glazed and unglazed. An unglazed clay cooker is porous and it is that quality which makes it so efficient at preserving tastes. Before food is put into it, the pot is usually immersed in water for about 15 minutes, and it soaks up some of the water. When the wet pot is put into the oven, steam is driven into its interior. The food inside lightly steams as well as bakes.

Cooking in wet clay takes a little longer than conventional methods. The wet clay pot has to go into a cold oven with the temperature set relatively high. And the cook has to be careful in handling these pots because they break easily and will shatter at a quick change of temperature. If you'd like to try experimenting with clay cooking before investing in a pot (the Romertopf is the most common), try cooking in a new clay flower pot. Get the kind without a drainage hole, soak it, put in the food, and cover the pot with foil. To adapt recipes to clay cooking, the rule of thumb is to add half an hour to the cooking time and 100° F. to the oven temperature. The following recipe is already adapted.

Vegetable Casserole

2   medium-size zucchinis, cut into approximately 1½-inch chunks
2   medium-size carrots, cut into matchstick-sized pieces
2   green peppers, chopped
1   onion, chopped
1   clove garlic, finely minced
2   tomatoes, cut into thick slices
½   pound feta cheese, cubed (or substitute another cheese; a sharp cheddar would work well)
     pinch of rosemary and black pepper

Soak the clay pot in cold water for 15 minutes. Layer the vegetables in the pot, beginning with the zucchini, then the carrots, the peppers, and last the onion and garlic. Sprinkle with rosemary, grind some black pepper on top. Cover with cheese and then the tomatoes. Place the pot in a cold oven, set controls for 450° F., and bake for 30 minutes.

For a change, or depending on what's in season, substitute other vegetables and combinations of vegetables and experiment with different combinations of herbs and cheeses.

The second type of clay pot is earthenware or stoneware, and is glazed. These pots—which are probably the original "crock pot"—are great for dishes that require long, slow cooking. Stews, grain dishes and beans taste great cooked in an earthenware pot. Just cook food for a long time at a low temperature—no higher than 250° F.

There are some inherent problems in cooking with both these types of clay pots. The unglazed pots tend to accumulate fat, which

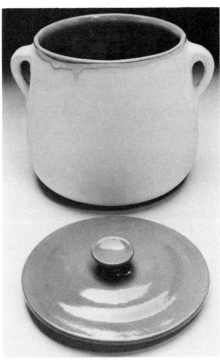

Stoneware pots are good looking—but make sure the glazing is lead-free.

eventually clogs the clay pores. Even worse, the fat can become rancid. To avoid this problem, fill the pot with a solution of water and baking soda and put it in the oven for one hour at 400° F. Another problem is that strong flavors linger in these pots, and make an encore at a later meal. Again, cleaning with baking soda will solve the problem.

The problem with glazed pots is not one of cooking, but buying. Some earthenware pots have been finished with a glaze that has lead in it. So purchase your glazed pot from a reputable source. Avoid cheap souvenir-stand pots.

Plan a cooked-in-clay meal using any of your favorite crockpot recipes, or create your own. Stews, vegetable combinations, desserts, elaborate French cassoulets, and even bread, all lend themselves to clay cookery.

Remembering that this type of cooking was used by primitive man, try using only the foods he might have used—seeds, whole grains, herbs, and fresh vegetables. A primitive meal cooked the primitive way can bring us in closer touch with nature, with early man, and with our own beginnings.

Michael Stoner

Introducing the restaurant supply store

The Ace Hotel and Bar Supply Company in Allentown, Pennsylvania, has almost any kitchen item you could name. Most of their business is with restaurants and bars, but unlike some restaurant supply stores, they welcome walk-in trade.

The pots and pans come in all sizes and materials, from a one-egg cast iron frying pan to a 60-quart stainless steel stockpot. In recent years aluminum cooking utensils have been the best sellers, since they were cheap, lightweight, and distributed heat well. But the price of aluminum has climbed of late, and now more cooks are returning to an old standby—stainless steel. It lasts longer than aluminum, is easier to clean, and there's no problem with any metal migration into your food.

One advantage of a restaurant supply store is the great range of possibilities you have with any utensil. At Ace, you can get a plain hamburger press for $2, or a deluxe, several-patty model for $61.90. A simple plastic butter slicer sells for $1.25, but they've got a heavy-duty version that does two cubes at once for $81. They also stock various potato cutters, tomato slicers, and lettuce shredders.

Along with basic equipment like knives, dishes, hand tools, and electrical appliances, many restaurant supply stores carry picnic and cleaning items. Ace had all kinds of brushes, sponges, gloves, and pails. You can even buy a Magic Mop, a wooden-handled device that looks like a play mop and will sop grease off the top of your soup, stew, or sauce.

So if you're looking for rare kitchen equipment or you simply like a lot of choices, try a restaurant supply store. Even if you don't buy anything, you'll probably enjoy the trip. (And in case you're wondering about that thing that looks like a paper punch with a metal circle in it, it's a lemon press.)

TD

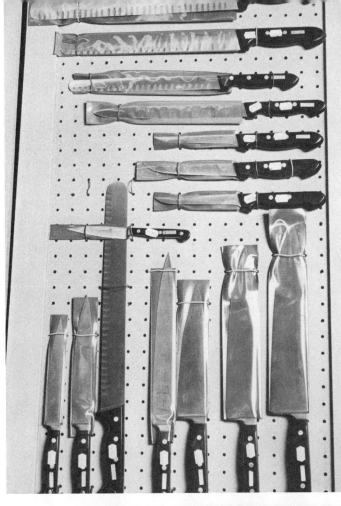

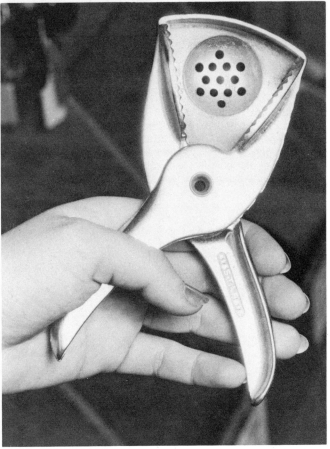

CORN CUTTERS

It's hard to beat the ease, speed, and relative safety of a knife when it comes to taking corn off the cob. We tried a number of different gadgets and in general found them difficult to work with, wasteful, or downright hazardous. There are two types of corn to get from the fresh cob—whole kernels or creamed—and three styles of cutters to give them to you. Using a knife gives you a little of each type.

Corn cutters with a round, expandable, serrated blade are supposed to start at one end and cut all the way around the cob, expanding or contracting to fit its contour. These remove whole kernels fairly well but tend to waste the tiny kernels at the tip. You have to go back and clean up with a knife.

Another type of cutter has a grooved trough for the ear to slide in as you push it into a blade. The blade is curved so that it will remove mostly whole kernels. In general, the blades we tried were not sharp enough to make the going easy. And whichever way you push, you're pushing toward a relatively sharp blade with slippery, wet corn and hands. It would be easy to lose a finger tip to one of these.

To make cream-style corn, you need to score the kernels open with a blade, and then push the innards out with a scraper. The third type of cutter is a hand tool which has one surface for scoring and another for scraping. With this device the insides of the kernels squirt out and the tough outer shells of the kernels are left on the cob. (It seemed like a terrible waste to me, but you can go back and get the remaining corn with a knife.)

No matter which cutter you choose, the job is messy. I'm still partial to using a knife. If it's important to have mostly whole kernels, you can come fairly close by carefully cutting just two rows at a time. If you're in a hurry you can clean the cob in four or five strokes, and then go back for the little nubbins by scraping with the knife. For cream style, run the point of your knife down each row of kernels before you cut them off the cob. This is much slower than the scoring tool, which scores four rows at a time, but works. And you don't leave the outer part of the kernel behind.

DB

Antique cookware

It's hard to imagine how much change has occurred in cooking hardware in the last ten years. Think back to 1971. Can you believe there were *no* food processors back then. A lot has happened since then, what with microwave ovens guided by computer cards and microfilm cookbooks. But if you were to unscramble your junk drawers or to dig deep into your pantry, you might discover at least a half-dozen kitchen tools passed down to you by your mother from her mother from her mother. And that represents a great amount of American kitchen history.

What to do with all these antediluvian hand-me-downs? Your options? Leave them at the back of the drawer; discard them for models of later design and more modern material: transform them into antiques by decorating your kitchen with them; or, use them. What better way to recycle than to give a retired tool a new lease on life. It was designed and built to be used for a good long time.

Let the electric processor share some of the work load. Its blinding speed is impressive, but the elbow-grease-operated mandoline (slaw board) has seen a lot more shredded cabbage in its day. In what follows I shall try to help you identify what may

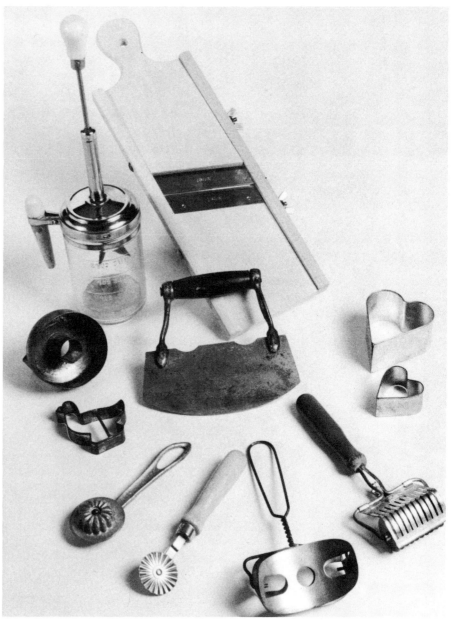

Cutters, choppers, and slicers.

already be lurking in the pantry; show you what I have discovered and, in many cases, am using; and guide you in your new role as collector.

Cutters, choppers, and slicers. The earliest food choppers were knives, and they're still a basic tool today. Food choppers were most often made from one piece of blacksmith-forged iron that joined a native hardwood handle. Lucky you are if you own one with a hand-forged blade—evidence of a village smithy working under a spreading chestnut tree. Later, the iron blade was attached to an additional piece of iron which joined the comfortable wooden handle. Some had straight blades for chopping on a flat board or tabletop. Others had curved blades and fit the roundness of a wooden bowl. Most had a single blade, but really unusual ones (made for impatient people) might have four blades.

My 95-year-old grandmother's walnut-handled chopper (she thinks it was her mother's) is a rare beauty. Family members continue to use it. My mother has a most unusual cast-iron bowl chopper, that was probably her mother's. Instead of four parallel blades, its curved blades intersect each other. Both of these are from the 19th century, but retirement for them is out of the question.

In a cookware shop you find an updated version of an old friend, the chopper enclosed in a jar. The wooden knobs and handles are gone, and the sturdy glass measuring jars have been replaced with modern plastics. But with luck, you might turn up a "blitzhacker" at a flea market, second-hand shop, or in grandma's pantry. It's great for chopping nuts, because the nut meats can't escape the rotating blades.

Other cutters to look for are handmade tin cookie, doughnut, and biscuit cutters. They were made to last. My old ones continue to give me good service.

When I first stumbled upon a strange device with a heavy wire handle in my grandmother's larder, I was mystified. Its twisted body gave no clue to its utility, but once straightened the thing took on the appearance of a pastry cutter. It makes perfectly round cookies or

biscuits with almost no scrap. Two pastry wheels are shown here. One cuts out nine noodles at a time (or minces a bunch of parsley). The other is a cast-aluminum pastry wheel made by Ateco, a firm founded in 1905, which to this day manufactures quality food and cake-decorating utensils.

A slicer popular in the Pennsylvania Dutch countryside where I live is the cabbage cutter, also known as slaw board, kraut cutter, or mandoline. They are still being made, but an old one may please you more. Look for a handwrought blade set diagonally in a rectangular piece of age-darkened wood with a decorative heart carved out for the handle. If you locate one, you will have a rare find. Newer designs may feature several blades, blades that adjust to various thicknesses, blades with a waffled edge for a frilled effect, and with a sliding guard box which holds the cabbage in place and protects the skin on your fingers.

Basher, masher, and press.

Bashers, mashers, and presses. Mashers have long been used for mashing potatoes, pounding meats, and pulverizing herbs and spices. At first there were all-wood food bashers carved from one piece of maple or walnut like the one in the middle. Later models were like the one to the right, with its wooden handle and twisted wires untwisting to form the

(Continued on page 426)

VEGETABLE CUTTERS

You may have heard them called cabbage cutters, slaw cutters, or mandolines, but the one thing they all have in common is that they'll cut firm vegetables quickly and evenly. You just run a vegetable across the wooden chute, into a slicing blade, then watch the slices fall into a bowl below as you draw the vegetable back for the next pass.

The simplest and cheapest versions are made of wood and have a blade set at a height that can't be adjusted. The next step up is a cutter with a guiding box. It's called a slaw box and lets you draw a cabbage head, for instance, across the blade without exposing your fingers to it. I didn't have a slicer with a box to test and wished I had. As you get down to the last inch or so, pushing a wet slippery vegetable into an open blade is dangerous.

Mandolines are made with double blades, interchangeable blades for serrated slices and adjustable blades for different thicknesses. They come in all sizes, but get one wide enough to hold your largest cabbage. No matter what tool you select, be sure the blade is made of metal that won't rust and that can be sharpened.

If you can find a mandoline with a good blade, it will speed up processing time for firm vegetables, and give you more uniform slices than you'll ever get with a knife. This is particularly important if your processed vegetables are heading for a food dehydrator.

DB

SAVE ENERGY: EAT A VEGIE MEAL

Use more vegetable protein—such as legumes, whole grains, and vegetables—rather than animal protein. It takes about 1,450 kilocalories to produce 100 grams of corn protein, as opposed to 80,000 for 100 grams of grain-fed beef protein. Use more milk and dairy products, poultry, and the plentiful varieties of fish.

Jean Mayer.
Reprinted with permission
Family Health Magazine
December 1978 ©. All rights reserved.

(Continued from page 425)
zig-zag base. Both have great character.

Sturdy and direct describes the illustrated cast-iron, factory-made lemon press. Lemonade became a popular drink in the late 1800s. To use this juicer, place a lemon half point down in it and clamp hinged top and bottom together. Juice flows through the eight decorative holes in the bottom. To prevent the acid juice from contacting the iron, the manufacturers gave it a rust-proof tin coating.

Butter-making tools.

Butter-making tools. Butter-making equipment goes back to the 18th century, so collecting it is a challenge. And while you might not get involved in the arduous task of churning your own butter, you might enjoy using decorative molds and stamps.

Butter is made from cream that has been separated from whole milk. Once the milk was strained through fine cloth into a pan, the cream was allowed to rise to the surface where a perforated metal utensil like the one in the picture was used to skim it off. This cream was then placed in a wooden broomstick churn and agitated with a wooden dasher until particles of butter floated to the top. After being washed with cool water to remove any buttermilk, these butter bits were placed in a wooden bowl where they were worked with a wooden paddle to remove excess moisture. (A paddle is shown, but unfortunately

my family cannot remember what happened to the bowl.) Carrot juice was often added to give the butter a brighter color. The butter was finally placed in stoneware crocks or wooden boxes for storage.

One of my prized possessions is the rectangular dove-tailed wooden butter box. About two pounds of butter can be packed into it. Four different designs have been carved into the moveable wood base—an acorn with its oak leaves, a classic sheaf of wheat, an ear of corn with its shucks, and a rose with bloom and bud and leaves. When the mold is inverted, the base imprints the designs and cuts the butter into half-pound sections. Long ago the decorations served a purpose: each dairy farmer had his own symbol to identify his product.

Graters, grinders, and griddles. The nutmeg grater: is it from Colonial days or was it just manufactured? Hard to say, since the design remains unchanged. This one is hand held, a half-round model with a hinged top for storing one nutmeat. How thoughtful of the original designer to assure a nutmeg so close at hand.

The illustrated Royal coffee grinder never has had a vacation in its 70-odd years. The Royal began to grind three generations ago in a general store when the only way you could buy coffee was in bean form.

Grater, grinder, and griddle.

Today, it is still grinding beans,, assured of a job now that we discover that progress doesn't mean canned, preground coffee. We had it right the first time.

Another history lesson resides with the soapstone griddle. Years ago in New England, every kitchen had one: it absorbs heat directly from the woodstove, and since nature provided it with a non-stick surface, it allows for fast cooking without additional fat. The final bonus: after the griddle cakes or crumpets or fried eggs or grilled meats and fishes are cooked, the soapstone provides hours of warmth for cool rooms. (I gave my grandmother a six-inch square of soapstone for Christmas. She remembered what to do with it. After a time on top the woodstove, she wrapped it in a tea towel and took it to bed. So much for her electric blanket.) With the return of wood and coal stoves, these heavy stones—Mother Nature's Teflon pans—are in demand once again.

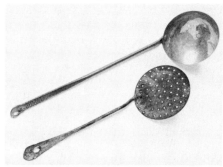

Handmade pieces.

Handmade pieces. The two handmade tools pictured here are probably crafted 150 years apart; both are timeless in their beauty and function. The older of the two is precious for its imperfections. The perforations in the skimmer bowl are unequal in size, and not equidistant from the edge. Some holes are not even punched through completely. The loop of the handle is asymmetrical, and the hanging hole is off-center. The ladle was made just a few years ago by E. J. Frederick of Kutztown, Pennsylvania. Crafted of copper and iron, it has grace and balance. Hand-hammered rivets decorate the bowl, while cross hatching and hook decorate the handle. Years from now this piece will be treasured as an antique.

Mixers and molds. Can you imagine the relief cake bakers must have felt when the hand-operated crank egg-beater was invented, back around the turn of this century. Until that time, cooks had to rely on elbow grease and whisks like these: a miniature birch broom, a blending fork, a metal flat stirring whisk, or a ratchet whisk operated by pushing a handle up and down to activate the beaters. All are still made today, but I feel secure knowing I have my electric mixer to back me up.

The hand-operated beater, designed as a variation of the hand drill, had an arrangement of gears that rotated wire beaters. The earliest models had iron handles, gears, and crank wheels, with tin-plated beaters.

My beater was made to fit over a glass mixing bowl. The tin plate fits over the top of the bowl to prevent the contents from splattering—a necessity when whipping cream or making mayonnaise. My "Merrywhirl" was patented November 28, 1916. Names, dates, places, and patent numbers make items more interesting and valuable.

I would dearly love to add a Universal Bread Maker to my collection. The Universal, made around 1890, uses a pail-shaped container and clamps to a table to make kneading dough a breeze.

Collecting can be a heavy hobby if your theme is iron baking molds. By the early 1800s, factories were manufacturing durable molds for baking breads and cakes. One of the largest producers was the Griswold Manufacturing Co., established in 1865 in Erie, Pa., U.S.A. 6139, weighs a ton, retains heat beautifully, is seasoned for life, and makes six individual scones for cornmeal muffins in half-cylindrical shapes. I also own a current piece by this 115-year-old company—a cast-iron popover pan.

I cannot guarantee that the cast-iron chocolate molds in the photo are American; I suspect they are German. Warmed chocolate is poured into the chilled iron molds—three screech owls and two bunnies—where it sets quickly in the cracks and crevices.

Along came tin and aluminum molds, instantly popular because of their light weight and easier care. The photo shows a tin baking mold with a decorative fluted edge. Its central cone assures even baking in the middle of a dense pound cake or buttermilk bread. The last mold is a plaque of individual compartments in the popular Turk's head design. It is stamped EKCO, Chicago 880. Ekco is certainly a familiar name in kitchen equipment. I wonder if the 880 means 1880?

(Continued on page 428)

Mixers and molds.

COLD AIR AS ENERGY SAVER

Can the frigid air so abundant during our northern winters be used to refrigerate foods?

The idea is being pursued by a Cornell University research team consisting of Professors Robert R. Zall, food scientist, William K. Jordan, food scientist, and David C. Ludington, agricultural engineer. In a pilot study, the researchers have successfully operated a room-sized cooler with cold air from outdoors. The results so far have been highly encouraging.

The walk-in cooler storing milk, vegetables, and fruit was kept at a temperature ranging from 34 to 40° F. When the cold air alone could not handle the cooling within this temperature range, the existing mechanical refrigeration unit took over automatically, but the mechanical unit was unused most of the test period.

New York's Food and Life Sciences Quarterly.

(Continued from page 427)

Collecting your own. I hope I have made you hungry to begin an antique cookware collection. To justify your hunger, consider these facts: you'll find clues to our social, economic, and cultural history; you'll save money by purchasing old equipment—most of which you can use; you'll be making a sound investment, since many tools are now considered antiques.

So visit the flea markets, garage and yard sales, estate sales, and auctions. Don't forget to nose around in your family's junk drawers, pantries, cellars, and barns. There could be an antique tin-and-wire corn popper waiting for you.

Catherine R. Elwell

Hay box cooking

In northern Europe during World War I, fuel was scarce and food even scarcer. One-kettle meals were often packed firmly into a wooden box, surrounded by hay, tucked in with a down pillow, and shut tight for the night to cook on their own steam.

Across the ocean in the tenement squalor of New York, newly landed immigrants found that during the winter an empty ice chest was the perfect fireless oven. All they had to do was boil the ingredients, set a container over hot bricks on a shelf, and let insulation do the rest. The idea caught on and the "fireless cooker" became a household word.

Prewar models were fancy hardwood chests fitted with three compartments, tight-lidded kettles, and iron-radiating plates. By the end of World War I, the U.S. Army had produced asbestos-lined fireless cookers by the millions. But they went out of style, and gradually disappeared from the market.

The common hay box works on the same slow-cooking principle as the old fireleess cooker. In addition, a hay box has an important advantage over the modern Crock-pot—it uses no man-made energy. The cooking is done by retained heat.

Assembling a hay box is easy. First, you need an ordinary wooden box with a lid or a board to cover the top. This box should be as airtight as possible and large enough to hold a

---

FRENCH BEAN SLICERS

I was able to find only two gadgets which profess to turn fresh whole beans into French-style string beans. One worked very well and the other did not. Surprisingly, the less expensive slicer worked best.

When I greeted the two in the mail, I was impressed at the heft of the $19.95 Bean Frencher (Garden Way, Charlotte, VT 05445). Its mouth is wide enough to accept three or four beans at one time. As you crank, six blades turn along with a feeding spool to direct the beans into the blades and drive them through the slicer.

Bean Frencher.

But it just didn't work right. The blades weren't sharp enough to make a clean cut in the beans; they just tore. Also, the beans didn't feed easily through the blades so there was some tugging involved, and that slowed production down to one or two beans at a time. This slicer was faster and safer than using a knife, but it left a ragged-looking bean.

On the other hand, the $3 hand-held bean slicer (The Cook's Collection, Marshfield, WI 54449) worked like a dream. Its seven small blades are razor sharp—in fact, they are razor blades. Four of the blades make the french slices, two skim the bean at its sides so that, if positioned just right, the strings will be removed. The last blade on the top is for nipping the ends off. I had best results with it if I chopped down onto a cutting board.

This slicer handled just one at a time, but because the blades are so sharp the work went pretty quickly. In our house there was nightly bargaining over who got to do the slicing. We did all the beans that way—they're so much sweeter and retain their firmness better in freezing. The bean is slid into a spring-loaded holder which guides it into the frenching blades. To french slice, you just push it on through. Pulling from the other side finishes off the job, with you holding two or three ⅛-inch slices. The strings never make it to the other side of the slicer. They just drop in place, separated from the other slices. Strings don't bother me, and I just let everything fall into one bowl. But for those of you who can't or won't eat them, this little string remover really works.

DB

A hand-held model from The Cook's Collection.

Short-Rib Vegetable Soup

2	pounds short ribs, cut into 3-inch pieces
5	cups water
½	cup onion, diced
1	tablespoon fresh celery leaves minced
¾	cup potato, diced
¾	cup carrot, diced
¾	cup celery, diced
1	cup cabbage, chopped
1	cup fresh whole peeled tomatoes
1	bay leaf
	salt and freshly ground black pepper to taste
¼	cup barley

Assemble and prepare all vegetables. Place short ribs and water in a heavy soup pot. Turn burner to high and boil. Add all the vegetables and bring to a fast boil. Add bay leaf, salt and pepper, and barley. Stir well. Cover immediately with a tight-fitting lid. Cook in a hay box 8 hours. Reheat on top of stove a few minutes and serve.

Yield: 6 to 8 servings

two-quart pot with about six inches to spare on the top, bottom, and sides. Next you will need insulating material. If hay isn't available, sawdust or shredded newspapers make good substitutes.

Pack the box tightly, leaving a nest in the center for the pot. Make sure the pot you choose is sturdy and has a tight-fitting lid. Iron and stoneware are best. Aluminum and stainless steel do not retain heat as well. The receptacle and the ingredients must be boiling hot and covered tightly with a lid so that no precious steam escapes.

In selecting a suitable recipe for your hay box, keep in mind that all ingredients have to be covered with liquid and boiled on top of the stove first. Since the Crock-pot relies on the same slow-cooking principle, you may adapt one-pot stews, soups, porridges, and puddings from slow-cooker cookbooks. If several steps are called for, recipes may be simplified by combining all ingredients at once. The standard boiling time on top of the stove will be a short three to five minutes.

Next, snuggle the piping hot pot into the nest of insulation and cover it with more hay. Stuff a wool blanket or down pillow over the top and into the corners of the box. Some hay box owners recommend wrapping the pot in newspaper.

Meat-based dishes take at least eight hours in the hay box. Traditional baked beans have to be parboiled the usual 40 minutes. However, once they're in the hay box, you're kilowatt-free. Finely diced vegetables cook in four to six hay-box hours. Recipes that ordinarily take two or three hours on the burner will take eight or nine hours. There is no fast rule of thumb. Each individual hay box differs according to its heat retention properties.

Don't peek. The joy of a hay box is to go to work or get a good night's sleep and return to a fully-cooked meal. All you have to do is heat the contents and serve.

Julia Older

The home's green heart

I believe that the garden is the literal heart—the pump, engine, and driving wheel—of the self-sufficient home.

"Self-sufficient" is over-stating the case for most of us. We want to replace dependence on fossil fuel by using solar heat and wood, but we still intend to use electricity for appropriate tasks: running clocks and radios, freezing and cooling our food. Similarly, while a garden provides us with most of our vegetables and fruits, we still visit the grocery store for marinated artichoke hearts, pasta, and tea. The difference is that we have most of our essentials covered by our own efforts, right on our own place. We can get along without marinated artichoke hearts, but not without our home-grown potatoes and squashes. If the lights go out, we eat, we stay warm.

We're not struggling for some degree of self-sufficiency because we're afraid of the dark. Survival insurance in emergencies is merely the pleasant result of having our home systems in place and operating. The real value in managing our own life-support system is what it does to us. How can we feel whole, honest, and real when our very lives depend on such evanescent institutions as the A & P, Exxon, and the local public utility? When our lives depend more on our own muscles and wits, those muscles and wits are worth more to us. Strong concepts of self-worth and self-reliance are two invaluable products of a good garden and good shelter.

The goal is a property organized as a steady-state system, with input minimized, resources cycling rather than flowing through, and enough output to support its inhabitants, human or otherwise. And if this system is organic—operated without the use of chemicals and with the use of natural fertilizers and common-sense techniques—better health for all will result. That's the goal. Let's look at how a garden helps us achieve it. Every pound of food grown and eaten on your own place is a pound less that is brought in from outside the system, trailing clouds of hidden energy costs. Those California lettuces are nourished with fossil-fuel fertilizers and pesticides (and carry the residues of such treatment), moved by truck, cooled with machines that run on fossil-fuel-made electricity, shipped on refrigerated rail cars, moved again by truck, put in supermarket coolers (which account for one-third to one-half a store's energy bill), purchased by someone who has likely driven to the store and back, and again put into a refrigerator before they finally get to the family.

Contrast that with the energy cost of eating home-grown lettuce, planted in soil fertilized with recycled vegetable waste (compost), grown fat on sunlight, and served fresh from the garden. There is no comparison. In such ways do gardens minimize dependence on outside help.

Let's follow our two kinds of lettuce further to see how resources can be cycled within a home system, rather than just flowing through it. The folks who bought the California lettuce obviously don't have a garden, or they wouldn't be paying high prices for stale produce. Since they have no garden, they have no reason to make soil-enriching compost from

(Continued on page 430)

(Continued from page 429)

their wastes. So, the lettuce trimmings and leftover salad go into the garbage can and are whisked away by yet another gas-blowing truck to the local landfill, where more machines bury them so deep they can only putrify and add their bit to groundwater pollution. Or, depending upon on where these people live, this uneaten food might go to the town incinerator, where it adds its bit to air pollution.

Self-worth and self-reliance: products of a good garden and good shelter.

It's only when the leftovers are returned to the top few inches of soil that they do any good, for there they rot away, feeding incredible numbers of soil microorganisms that form the life of the soil, and which are just waiting to be fed some garbage. In turn, these microorganisms use our waste to nourish our growing crops, improve the texture of the soil, destroy disease spores, and support a whole chain of vegetable and animal creatures.

Because the aesthetics of tossing fresh garbage on the ground are questionable, most people who recycle resources put the garbage into a compost pile, where it undergoes a hot rot that preserves its nutrients. Special heat-loving bacteria work not only on your kitchen garbage, but on all your gardan trash: pulled weeds,

leaves, grass clippings, plus manures and other natural ingredients, to create the perfect fertilizer and soil conditioner. The organic garden, with its need for actively decaying organic matter, accepts our leavings and transmutes them back to clean vegetables the next year.

The other by-product of our lettuce is human waste. In the typical home, the nutrients that originated from that California field end up in a septic system where they are partially lost in drainfield leaching, or are pumped out by a septic tank cleaner and dumped into a city sewage system or landfill, again adding to our pollution problems. But let's say our organic gardeners have a composting privy, such as a Clivus Multrum unit or a waste-recovery toilet they've made themselves. When properly composted and exposed to the environment, human wastes are perfectly safe to use to fertilize fruit trees and bushes, or to fertilize soil for vegetables. The nutrients in our home-grown lettuce leave the property, they just keep cycling around, making an appearance in summertime as vegetables and in winter as compost.

If we add a waste-water collection system to this arrangement, and then use the water from tub, sink, and washer to water the garden, we've completed all the cycles regarding the flow of plant nutrients on our property.

But remember: this system is not totally self-sufficient. From time to time, the gardener may want to import some manure or ground rock fertilizers, or toss in leftover marinated artichoke hearts, pasta, and teabags. So nutrients are in constantly increasing supply in the well-managed garden.

Because human activity is powered by food energy, the garden is the driving wheel of the self-sufficient home. This biological engine runs fine on stuff you'd otherwise pay to have hauled away. The establishment of a garden and a compost pile, simple as those acts are, are crucial to reducing our demands on fossil fuels.

This spring I'll begin another year of managing the cycle between my garden and my house. It's a cycle that will keep on functioning through wars, embargoes, and the demise of the oil era.

Jeff Cox

Gardening indoors

Whether you live in an apartment in Queens or a nine-room stucco house in Palo Alto, your indoor environment can be harnessed to produce a successful and bountiful food garden. Gardening skills are relatively unimportant. Just stand near each window in your home and take note of how much light enters from the sun. Try to note how many hours per day the sun enters each room, at what angles and how deep into the room it penetrates. Just like the backyard plot, your indoor garden will require a little advance planning, so write your observations and sketch basic floor plans complete with diagrams of how far the light penetrates during morning and late afternoon hours. Such graphic aids will help you plan for furniture moves, traffic adjustments, and other possible changes in routine that could be forced by garden growth. Plan where you'll situate your plants, taking into consideration such things as drafts, normal room temperature, pet traffic, and access problems. A brightly lighted window area might be more suitable for growing beans or a tomato plant. You'll want to locate spinach or peas where the air is cooler and there is less exposure to the sun.

A good rule of thumb is this: fruit or seed producers (beans, tomatoes, squash, cucumbers) need bright sun and generally more warmth, while leaf and root producers require less light or cooler spots. Variations do occur; peas need good light but cool places.

You won't have to worry about all of the same predators that plague the outdoor plot, but you may encounter a whole new array of them indoors. Forget hail, snow, and sleet. You'll be more concerned with drafts, spider mites, and cats that love windows. Because the same organic balances are in effect, companion planting and soil rotation will be items to note in your record. A spiral-bound notebook will suffice.

Normal summer sunlight ranges between 10 and 18 hours per day. Spring plantings will get along with less. Winter indoor gardens can be helped along with attention to what is planted (try low-light, cool-weather crops) and the addition of artificial

A window ledge can yield an on-going harvest of herbs.

lighting. Drapes and curtains always cause problems. They have a tendency to create a greenhouse effect, raising temperatures considerably. They also can damage foliage when opening and closing.

Selecting containers for your garden can be as simple as visiting a garden shop and selecting from a wide inventory of pottery vessels, wood box planters, or suitable decorative and useful empty things. Imagination can lead to some outrageous discoveries. Old bathtubs offer great habitats for deep roots and provide excellent drainage. Decorative woven baskets lined with plastic bags will match many room decors. Outdoor window boxes can be installed indoors, too. Visit antique shops, scrap yards, and even wine-making supply stores (barrel-halves make great containers for little gardens).

Soil for indoor gardening can be purchased mixed and ready or you can use garden soil. Be careful to use only well decomposed materials. You can sterilize soil by spreading it ½ inch deep on a cookie sheet and baking it in your oven for one half-hour at 200° F.

Depending on your particular planting you may choose to add fine bark, peat moss, sand, vermiculite, gravel, or other materials to either enhance drainage or improve moisture retention. Use the same rules that apply outdoors. Remember that drainage will become a sensitive matter indoors.

If good old Mother Earth is hard to find, use one part each potting soil or packaged soil base (garden or discount store) *and* peat moss or packaged compost; add one-half vermiculite or perlite. Mix in one cup bone meal and blood meal per 50 pounds of the above. Five cups of dried cow manure per 50 pounds of this mixture will complete your recipe. Make plenty and keep it in a covered refuse barrel. Whether you are a fisherperson or not, earthworms added to your soil repository can enrich its quality.

Start your seeds in recycled egg cartons, plastic seedling tubs, or biodegradable peat pots. Following seed package instructions, place your beginners where you intend to grow them as full plants. When they reach two or more inches in height, transplant them into larger containers. Some items such as radishes, beets, and carrots (root producers) don't require transplanting.

Houseplant tools are helpful. Moisture meters will easily inform you about watering conditions below the soil surface. Misting some plants (they

Potted peppers will bloom and fruit indoors.

are accustomed to dewy nights outdoors) will help them adjust to relatively drier in-home conditions. Avoid watering and misting when the sun is shining on the plants. Try to arrange your watering schedule so that you will be approximating out-of-doors watering routines. Above all, be careful that you don't overwater. In hot sun, leaf or bark mulch will help retain consistent moisture levels from top to bottom in your soil.

While you may not subscribe to the use of artificial lighting as a natural tool in gardening, there are available forms that can augment your light in difficult locations and winter seasons. Standard fluorescent fixtures will accommodate lamps designed to provide correct color "temperatures" for plant growth. Houseplant gardeners use them regularly. Most forms of artificial light promote photosynthesis with varying degrees of success. If your garden suffers from low north light or if your kohlrabi fizzles in January, try rigging a plant light about 12 to 18 inches above the topmost foliage. Light timers can be labor savers, too. Try to duplicate adequate daylight hours, adding a half-hour or more per week into the plant's growing season.

Carpenters should note that window greenhouses make handsome additions to small existing windows. You can purchase them ready-to-install at your building and lumber center, or you can fabricate your own. Check with your apartment owner first!

The list of indoor seed varieties grows annually as new hybrids are developed. Some seeds are labeled and advertised as suitable for indoors, and some are better than others because they produce tailor-made plants. Little finger carrots and Short 'n Sweets don't require deep containers. Some tomato plants (Tiny Tim and Presto Hybrid) do better than Beefsteak types. Most peppers thrive indoors. Low-light lettuce seeds are the kind to look for. Slow-bolting spinach that doesn't go to seed quickly (Avon Hybrid, Bloomingsdale) should be grown in low light and cool places. Malabar (actually *Basella alba* and not a true spinach) will grow all year-round. Plant onions as companions to other plants. They'll help keep certain bugs from

(Continued on page 432)

(Continued from page 431)

their companions. Yellow globe and green onions grow very well indoors. Peas and pole beans will require some adjusting. Lead them up trellises and string along window edges. Dwarf Gray Sugar peas make great hanging plants.

Cucumbers and squash need lots of space and may require a helping hand with vine growth direction. Radishes are fast growing and need lots of watering. Turnips and parsnips take longer and need good light. Beets make beautiful houseplants. Try mixing red and golden beets for attractive foliage. Pumpkins? Grow your own "live-in" jack-o-lantern! The entire cabbage family demands long growing seasons, but the indoor results are astonishing. Red cabbage does very well. Strawberries will become your favorite indoor food plant, and they will last many seasons. You can even use them as hanging plants. Use both seasonal and everbearing types. Herbs are a chef's delight to the indoor enthusiast. Follow the package instructions for planting. Preparation after harvest is no problem, either. Cut them when the foilage is heavy, hanging them upside down in a place devoid of light such as a closet or pantry. This retains their color and fragrance. Crumble the leaves in a week and bottle the results for your spice rack.

Jon Adams

The salad-sufficient winter greenhouse

We pick vegetables every day through the winter in our solar greenhouse at the Organic Gardening and Farm Research Center in Maxatawny, Pennsylvania. It is a nice-sized greenhouse for a family: 16 feet long, with about 100 square feet of growing space. Outside, the ground is snow-covered, and nights usually drop to between 0 and 20° F. Inside, the two large raised beds are brimming with tender bunches of greens of all shapes and shades. We grow over two dozen varieties of salad plants, with a few marigolds and nasturtiums scattered here and there for color.

Over the three winter months, harvests average a pound of salad greens every day. From mid-December through January, production slows to about a half a pound a day, which is balanced by the heavier production in late fall and late winter. But that's a lot of fresh food of the most valuable and expensive kind just when you need it most. It's dependable. And we grow it without using any heat or light except what comes from the sun.

We didn't always get such good winter production from the greenhouse. We had to learn how to use it. The greenhouse requires our best gardening effort. It repays the care handsomely, but there is a smaller margin for error in a greenhouse. Missed plantings can leave you with little food in the slowest, sun-poor weeks of January, when stressed plants grow slowly or succumb to

This flier for a fast-food restaurant gives mom reasons for abandoning home food systems: "No cooking, no mess, save utilities, cost less!"

Air movement improves plant health. We channel the hottest air from the peak down over the water-filled heat storage cans. The polyethylene directs the air flow to the fan while allowing radiation to charge the cans.

insect or disease attacks. Beyond good basic gardening practices, here are a few special things that our greenhouse experiences have taught us.

• Grow in large, deep, ground-level beds rather than in big pots up on benches.
• Concentrate on vegetable varieties well adapted to cool-weather growth.
• Provide good air circulation.
• Provide good soil mixes both for the beds and for seed starting, and feed regularly with organic nutrient solutions, such as fish emulsion, liquified seaweed or manure tea.
• Time successions carefully.

Almost any greenhouse you'll see has plants growing conveniently at waist level in pots on benches. Forget that system for growing vegetables in a heat-efficient solar greenhouse. For one thing, putting plants up high shades the heat-storing mass in the back from the direct radiation that it needs. That system also lets plant roots get cold.

Even very large pots can't hold heat for long. Their surface area is too big, and air is free to flow all around the sides and bottom. The year we grew vegetables this way, we measured the temperature in the soil and found that it changed along with the surrounding air temperature. That could mean it ranged from the high 30s at dawn up to the 70s at noon and back down again. Roots tend to cluster in a mat around the inside edges of the the soil ball in a pot, so they get the worst of this swing. The temperature in our 18-inch-deep beds changes very little, typically from about 43° F in the morning up to 47° by afternoon in the coldest weather. Roots are the last thing in the greenhouse to feel stressful cold.

Large growing beds also allow the roots to forage freely for nutrients and water. There is less need to irrigate, less leaching of nutrients, and consequently less need for supplemental feeding. The beds also permit dense plantings and optimum use of greenhouse floor space.

The winter climate in the greenhouse is quite special—well suited to the growth of many leafy crops that wilt in the midsummer heat. But trying to grow heat-loving plants is, by all reports, a losing battle unless you are

High-yielding Chinese greens, grown in succession plantings, mean daily harvests through the winter.

willing to add expensive heat and perhaps a lot of light. Summer vegetables need to do considerable growing before they will flower or make big, fleshy roots. When they do flower, they need to make a lot of sugar for their fruit production. Winter is the worst time of the year to do that. Fortunately, a huge variety of cool-growing salad plants can keep solar greenhouse gardening interesting all winter. And these vitamin-rich greens are just the foods we need most in midwinter.

Of all the plants we've grown in the solar greenhouse, the Chinese brassicas do best. For salads they are as good as lettuce—mild flavored and tender, with crisp, juicy stalks. Food production in our greenhouse doubled the year we started to concentrate on them. Researchers in Louisiana have been growing many Chinese cabbages outdoors in winter, and have found that supermarket shoppers buy these plants eagerly for salads (under the name "Louisiana lettuce") and come back for more. Oriental vegetables seem to be much better adapted to cooler conditions than our lettuces. We like seppaku taina, bok choi, gai choi, dai ga choi, and kyo mizuna.

Of the heat-loving crops, tomatoes do best. They grow well as a

(Continued on page 434)

(Continued from page 433)

late-fall and extra-early summer crop. In midwinter they stop setting fruit, and the ripening process slows to a snail's pace. For year's end tomatoes, start seeds or root suckers in early July, then transplant into the greenhouse after a month's growth. Fruit will start ripening in October. Cherry tomatoes ripen more quickly than large-fruited varieties. Restrict plants to just one or two stems and train them up a stake or string to keep in bounds and the fruit quality high. Top the plants in November to force the last of the fruit to ripen.

Cucumbers, a standard greenhouse vegetable, are tough to raise in a cool greenhouse. They are extremely susceptible to mildew and whiteflies. Lettuces do almost as well as the Chinese brassicas. Some varieties, however, far outperform others. Buttercrunch, Grand Rapids, and Ostinata have all done well in solar greenhouses. Oak Leaf is good, too, but seems to be attractive to aphids. Loose-head types are preferred because they allow early and continuous picking.

Endive and Sicilian chicory (like dandelion) both thrive and make excellent additions to salads. Surprisingly, spinach generally does poorly as a winter crop. It isn't vigorous and the leaves stay small. As an early-spring crop, it does fine in the greenhouse. Monoppa is the best spinach we've found for winter.

Broccoli and cauliflower both produce tender and sweet buds grown as winter greenhouse crops. Plan for picking to start no later than Thanksgiving. However, the plants take a lot of space and do not provide as much food per square foot of bed as do salad crops. Similarly, good carrots will grow in solar greenhouses, but they are not highly productive in the use of space. There is a lot of wasted growth, and summer-grown carrots can be kept through the winter in top condition.

Kale is a more productive cole crop for the greenhouse, and its leaves grow tender and juicy under winter conditions—they're excellent raw in salads. Swiss chard is very reliable in the middle of winter. The best way to grow kale and chard is to bring in fairly large plants from the summer garden. Dig the plants very carefully in September. Be sure to clean the leaves well to remove any insects, especially aphids, which can run rampant during winter. If you have been careful with the roots, the plants will recover from transplanting quickly and provide steady picking through January.

For spicy garnishes, grow a few chervil and parsley plants, as well as Shungiku, the Japanese edible chrysanthemum. Parsley is slow, so start in the fall with mature plants grown in pots (to keep the taproot intact) especially for the greenhouse. Nasturtium leaves have a peppery cresslike flavor, and the flowers brighten up the growing beds. Onions will not form much besides green leaves because of the short day-length. But the tops of some sweet varieties as well as chives are a welcome addition to winter salads.

A fan to circulate air during the day is something nearly all successful solar greenhouse gardeners recommend. We first put a fan in our greenhouse to see if it could reduce the mildew problem on leaves. The moving air helped offset the high humidity and cleared up the mildew on the plants within two weeks. Air movement also helps leaves absorb carbon dioxide (CO_2), a gas plants need for photosynthesis. Furthermore, we found that the fan helped moderate the high temperatures that can occur on bright days. Running a fan can hold the highest temperature to the low 70s, which is ideal.

All your work can be for naught, though, if you don't pay close attention to the soil you provide. It should be porous so that it drains well. Waterlogged soil becomes oxygen-poor and causes root rot. At the same time, the mix should retain water so that irrigation is held to a minimum. That not only saves work and reduces leaching of nutrients, but also helps keep humidity down. A good soil blend is half compost, one-quarter topsoil, and one-quarter sand (or vermiculite).

The large amount of compost in the mix is important not only for its nutrients and beneficial microorganisms, but also as a source of carbon dioxide. Plants in an airtight greenhouse can use up all available CO_2 on a sunny day. But a soil rich in organic matter and microbes will continuously regenerate the CO_2 supply.

Regular feeding with a nutrient solution is essential to good winter production. Many of the microbes that release nutrients from organic matter become inactive at soil temperatures below 50° F. Fish emulsion, liquid seaweed, and manure tea are the best sources of readily available nutrients.

In the middle of winter, beds need watering about once a week, depending primarily on the amount of sunshine. Add nutrients to the water every third or fourth watering; this should be done in the morning, with warm water if at all possible. The drenching should be thorough and deep to flush away any salt buildup.

All fast-maturing leafy plants are best grown from transplants and in succession. To grow lettuces and Chinese brassicas from seed to transplanting size takes about three to four weeks. The first harvests of the fall planting begin about 40 days from transplanting. We transplant young plants into the greenhouse beds in early October and begin picking in mid-November.

In our climate, that's a pretty early crop. Sherry Boutard, who grows plants in the solar greenhouse at the Berkshire Garden Center, plans her first picking to start after Thanksgiving. Before then, she doesn't need the greenhouse, since her outside garden is heavily laden with kale, lettuce, endive, and brussels sprouts. "I want to be able to turn to the greenhouse just as things become frozen outside. I plant my greenhouse with fairly large, good-quality plants, and I don't pick from it for at least a month afterwards," she says.

The first fall planting has so far dominated our greenhouse production right through February. We generally pick only the outer leaves of all the varieties we grow. We thought that relying on well-established, large plants was better than using smaller ones transplanted at a time of low light to replace whole plants that have been removed. However, this year we will also include a second planting of leafy crops in late fall, give it an extra few weeks to come into bearing stage, and see if these younger plants are more vigorous producers in February and early March than those started in early fall.

While the greenhouse provides good growing conditions for leafy crops, it is too cool generally for efficient seed germination in late fall and winter. It is best to set up a special area with bottom heat for germination. You might try small heating pads protected from water with polyethylene.

Start seeds in a small flat. To guard against damping-off, cover the starting soil mix with a half-inch of sterile material such as vermiculite or sand pasteurized in the oven. Plant seed in the sterile layer, and the young root will quickly penetrate to the nutrient-rich layer of soil.

Thin the seedlings as needed, then transplant to small pots filled with a compost-rich soil soon after the first true leaves appear. The young transplants can go into the greenhouse almost as soon as space is ready, but don't hold them so long they become root-bound. When they've got five or six leaves showing, they are ready. Especially with lettuce, planting the seedling any deeper than it grows naturally in its pot invites stem-rot.

Transplanting requires thought and constant attention. But it is the key to high production once you've tuned your solar greenhouse to create the cool climate that a greens crop likes and set up large beds of fertile soil. It is much easier to provide ideal conditions for great numbers of small plants when they are in flats or small pots. You can move them inside, rather than watching them languish in a too-cool climate. When the time comes to transplant, they are at their strongest. A carefully timed transplanting program means that valuable space is not monopolized by tiny, unproductive plants. Likewise, an ample supply of seedlings means that no space need lie idle for long.

The attached solar greenhouse

Heat for the home and production of winter vegetables—the double promise of the solar greenhouse— are in direct competition for the collected energy.

As solar collectors, greenhouses that are well designed and properly oriented are highly efficient. They can deliver almost all of the energy they

A greenhouse is the simplest and most beautiful solar option. Len Meserve's addition delivers heat to the home through existing windows and doors.

Recycled windows and glass are readily available and keep costs low.

collect to the house. But if solar energy is also used to maintain a productive growing climate, as little as 25 percent of the collected energy may be left over for home heating. Why? Because a greenhouse without plants can send all its heat to the house during sunny hours, then be closed off at night. With plants, temperatures need to be kept higher, especially at night, using more of the stored energy from the greenhouse itself.

Let's look first at heat production. Aside from how the greenhouse is managed, its heating potential for your home depends on the amount of winter sunshine and its size in relation to your home. On bright, sunny days, solar greenhouses do a fine job of heating adjacent rooms, even in the coldest weather. On cloudy days, the greenhouse will usually collect enough energy to maintain its own temperature, but may need heat from storage or the home to keep from freezing at night. In predominately cloudy climates, attached greenhouses can be disappointing to people looking for a lot of heat in midwinter. Fortunately, there are few places so cloudy that greenhouses aren't worthwhile.

(Continued on page 436)

SEED SOURCES

Seed companies with good selections of cold-hardy Oriental vegetables include:

Herbst Brothers Seedsmen, Inc. 100 N. Main Street, Brewster, NY 10509.

Tsang and Ma International P. O. Box 294, Belmont, CA 94002.

Kitasawa, 356 W. Taylor, San Jose, CA 95112.

J. L. Hudson, Seedsman, P. O. Box 1058, Redwood City, CA 94064.

Grace's Gardens, Autumn Lane, Hackettstown, NJ 00784.

R. H. Shumway, Rockford, IL 61101.

COLD-WEATHER GARDENING WITH A SOLAR GROW-FRAME

A solar grow-frame works exactly like a solar greenhouse does, using the same principles of heat collection and storage. The solar energy available to plants in April is equal to that in August, so why shouldn't spring be as productive a time in the garden?

April can be very productive. What limits growth in the spring is very cool soil that is warming slowly. And cold nights. The plants are still tiny and unable to capture much sunlight. A solar grow-frame fosters natural heating of the soil by day and prevents freezes at night.

During the spring months a grow-frame will produce very heavy harvests—equal to those you get outside during the summer. A grow-frame also produces abundantly in the fall and through the winter when light levels are low. The food production won't be as high then, because productivity depends directly on the amount of light the plants receive. But even in December there are 6 hours of growing time (compared to roughly 10 hours in late March and September, and 14 hours in June). And a grow-frame creates soil and air temperatures warm enough to take advantage of it.

Jack Ruttle

SOLAR GROWING FRAME
Ray Wolf. 1980.
Rodale Press, 33 E. Minor Street,
Emmaus, PA 18049. 80 p. paperback
$14.95.

For an expanded discussion of modest-sized solar growing spaces, consult this book. Supplementing the patient instructions for construction is a pocket of large, folded plans. The not-so-handy-person will be reassured by the generous use of photos and illustrations.

Whenever we felt a carpenters' trick may help you, we've passed it along. If a cut is potentially dangerous, we've noted it. Likewise, if the chance for error is high for a particular cut or operation, we've noted it and tried to explain it a second time.

Foundation

Of the entire growing frame construction process, this may be the area that intimidates the most people. There is something about the word masonry that sends shivers down the spine of do-it-yourselfers. With that in mind, we have used a foundation technique that does not require traditional masonry skills. We use surface bonding cement for the foundation instead of standard mortar. The concrete blocks are put in place dry. When they are in perfect position (even if it takes you all day to re-adjust them), a coating of the stucco-like cement is troweled on the outside of both walls, bonding the blocks together.

PESTILENCE COMBAT NOTES

In the greenhouse, mealybugs, aphids, and spider mites are the most common invaders. Prepare for battle well. Alcohol and cotton swabs are handy vermin removers for the underside of leaves.

(Continued from page 435)

A solar greenhouse's performance will vary as the winter progresses. In November and March, when solar energy is available on a clear day and the weather is more moderate, the greenhouse will contribute a much greater share of the home's heat than it does in the dead of winter—especially if fans are used to distribute the heat through the home. You'll have a better idea of what living with a solar greenhouse is like if you determine how much sun your climate delivers in each of the winter months. That is easy to do. The best place to start is the *Climatic Atlas of the United States,* found in most libraries. It gives monthly averages for the percentage of possible sunshine. For a more refined idea of the local climate, check with local weather bureaus (usually at airports). Many of them even print summaries of local weather data, including the percent of possible sunshine or cloud

The south side of the house admitted almost no solar energy before the large pit greenhouse was added.

cover. In practice, most people report that their solar greenhouses can't heat the entire home, but only immediately adjacent rooms on sunny days. A rule of thumb is that the greenhouse will heat an adjacent room 1½ times its own size. In Westmoreland, New Hampshire, a cold and often cloudy place during the winter, Len Meserve built a 22 by 11-foot greenhouse against the south wall of a kitchen. The greenhouse heated the 22 by 18-foot kitchen to 80°F. on a clear day. Bob Ericson of Putney, Vermont, has a 16 by 7-foot greenhouse attached to the south side of his kitchen. He says that the greenhouse, plus heat from cooking, provided all the energy needed to

keep the room comfortable on sunny days.

Where the greenhouse collecting surface is large, the energy has a clear path to interior rooms and the weather cooperates, solar greenhouses have heated areas two to three times their own size. David Kruschke designed his home around the solar greenhouse idea. Three hundred square feet of greenhouse glazing collects enough energy to provide half the heat for his 800-square foot home near Wild Rose, Wisconsin. Barbara Eggert of Bucksport, Maine, says that the greenhouse she and her husband added to the south side of their workshop provides all their heat on a sunny day. Their solar greenhouse is only 16 feet long and 7 feet deep and the workshop is 24 by 28 feet. The heating potential is great if the greenhouse is well made and managed skillfully. However, the way the greenhouse is used is the key to how much energy you really get. Night shutters, for example, require attention twice daily. If shutters aren't used, much of the heat that is collected during the day is lost through the glazing at night. Solar greenhouse owners frequently report leaving doors or windows open at night to let heat flow out from the home into the greenhouse to keep plants from freezing. In most cases, though, night shutters and normal heat radiation through the wall will prevent freezing and save all that energy.

Freestanding solar greenhouse experience shows that nearly all of the energy these structures collect is needed to support intensive vegetable production. Rarely is there any extra energy in a freestanding greenhouse in midwinter. Likewise, a lot of the energy that a food-producing, attached greenhouse collects will not be available to the home. The average single-story attached greenhouse, if it's very intensively cropped, will be able to contribute 25 to 35 percent of the energy it collects for home heating. Warm air flows into the home during the day, perhaps assisted by a small fan, and whatever is needed to maintain plant growth is returned at night. For worthwhile growth, freeze protection is not enough. A minimum temperature in the low 40s is better. But if the greenhouse is not well insulated, or is drafty, or is maintained at too high a temperature, the structure can actually use more heat at night than it collects during the day. Many solar greenhouses appear to do little better than break even. The energy flow in the greenhouse must be carefully thought out and watched. Most attached solar greenhouses do not in fact produce much food in the dead of winter. Some owners appear to be very casual about their winter gardening. Others have trouble because they are inexperienced at gardening. Still others choose to run their greenhouses without plants in the coldest months. It is not economical to channel the greenhouse's energy to a handful of plants or crops that are not producing well. That same greenhouse with no plants could be providing a lot more hot air for home heating. At night it can be completely closed off. Used that way for just the coldest months, the greenhouse becomes a very efficient hot-air collector. When the need for home heat declines in spring and fall, gardening can start up again.

Food production from solar greenhouses, though, can be very high. Abby Rockefeller and Carl Lindstrom's attached greenhouse in Webster, New Hampshire, produced 44 pounds of leafy green vegetables and 7 pounds of radishes and turnips in November, 1978—better than a pound of vegetables each day. That's a lot of food, more than many families can use.

A good compromise strategy is to divide the greenhouse into a vegetable-producing zone and a heat-producing zone, at least during the coldest weather. Part of the greenhouse can be set up for maximum vegetable production. Large and deep ground level beds full of compost store heat and allow the roots to forage. Reflective paint or foil on the north wall is less than optimum for heat storage and generating hot air, but excellent for the plants. High-quality thermal curtains at night help hold the heat. The rest of the greenhouse can be handled efficiently and much more simply as a hot air collector for the house and a pleasant, sunny place to sit or work during the day. At night it can be closed and given no extra heat.

Jack Ruttle

The Mechanical Mule

According to the Worldwatch Institute in Washington, D.C., last year some 32 million American households (nearly half of all U.S. families) raised more than $14 billion worth of fruits and vegetables in backyard gardens that totalled nearly seven million acres, some of them carried in paper bags in city elevators to be spread over apartment balconies.

I regret to say, I had little or no participation in the movement. Oh, there is a vegetable garden at our place, one that's typical, I suppose, of the other 31,999,999 plots around the country and in the cities; I got to enjoy much of the produce after it had been harvested, cleaned, and properly cooked for the dinner table. But I made a point of avoiding the garden itself. I left the cultivating to Roger and the Master Gardener, as I have for all the years we have been together. A once-or-twice-a-summer inspection tour—complete with proper admiration for every growing row—was the extent of my on-site presence.

That distance, I may regret to say (for who knows how many blisters there are yet to come) has been obliterated with the start of the 1979 agricultural season. I not only am now involved with the garden on a personal basis, but I helped to plow it. Once that initial wondrous, ground-breaking step is taken, the rest is easy. There must be few people who can do the preparation work a garden requires and then muster the willpower to leave the growing to others. After hours on a plow, the plower wants to see everything that happens as a result of his labor. Every sprout becomes his ward, every

(Continued on page 438)

The Mechanical Mule.

> Gardens, especially food gardens, are tools of freedom. When you grow some of your own food, you begin the process of cutting yourself free from the web of boredom, taxation, high cost of living, and low quality of food that makes so many people frustrated with life in a technological society.
>
> Robert Rodale

WHAT IS YOUR TIME WORTH?

How much time does a garden take? I doubt that there is any sure way to tell. There are too many variables. Doing work at the right time almost always saves time. A push plow is faster than a hoe; a rotary cultivator is faster than a push plow; a horse-drawn single-row cultivator is faster (and a better weeder) than a rotary cultivator. If you want your garden to look picture-perfect you will have to spend more time at work than someone who doesn't mind a few weeds. If you think of gardening as a kind of agony, then perhaps the investment of work may be too great, and you should not do it. If you think of gardening as a pleasure, then you may not think of your work as an "investment" at all; the whole enterprise (less seed) will be profit.

What is your time worth? Though it is often asked, I do not think this question is answerable. It is the same as asking what your life is worth. And I can give it only the same non-answer: it is worth whatever it means. The idea that you cannot afford to raise a garden is based on the assumption that it means money, that if you are not receiving the top dollar for every minute of your life, you are suffering a "loss"—a doctrine that not only would put an end to gardens, but soon drive us all to theft or suicide.

A better question is this: What would you—and your children—be doing with the time "saved" by *not* producing your subsistence? Different families will answer that question in different ways. But if the answer is that you would be doing something expensive away from home, or taking an outside job to buy food, or if you don't know what, you would do well to look carefully into the economics of subsistence.

Wendell Berry

THE SEED-STARTER'S HANDBOOK
Nancy Bubel. 1978.
Rodale Press, 33 E. Minor Street, Emmaus, PA 18049. 384 p. hardcover $12.95.

This book describes, in a step-by-step manner, methods of home plant breeding. It explains enough botany to allow you to deal with plant breeding, hybrids and the other vagaries of seed production.

The basis for selection

Your aims in selecting vegetable plants from which to save seed will depend on the strain with which you're starting, the peculiarities of your local weather conditions, and your personal tastes. Here are some of the qualities you will want to consider in choosing your parent plants:

- *Early bearing.*
- *Flavor.*
- *Size.*
- *Insect resistance.*
- *Disease resistance.*
- *Ability to germinate and thrive in cold weather.*
- *Lateness to go to seed (in leafy crops).*
- *Color.*
- *Storage life.*
- *Yield.*
- *Plant vigor.*
- *Fruit texture—tenderness, juiciness, seediness.*
- *Suitability for purpose; for example, paste tomatoes should be meaty, kraut cabbage should be solid.*
- *Special qualities—absence of thorns, spines, strings, etc.*
- *Resistance to drought, wind, smog, dampness, or other stressful atmospheric conditions.*

Selection won't introduce dramatic new crop improvements, but over the years the process can gradually intensify good qualities.

(Continued from page 437)

green shoot needs his protection and concern.

There is a reason for my plowing, a rather extraordinary one. It would take more than my conscience to get me out there working in the dirt. This year, I had a gadget. I am bananas about gadgets, and this one is so fascinating, no one—much less a gadgetophile—could have resisted.

It arrived in a huge crate sometime in midwinter, sent by friend Robert Rodale of organic farming fame. He, too, is gadget happy, but has managed to convert his interest to producing income rather than outgo: I buy gadgets (remember the lawn-mowing hovercraft?) but Bob builds and sells them. Judging from his permanent aura of contentment, his, I think, is the better system.

What he sent me is known as the Mechanical Mule and, as every successful gadget must be, it is irresistable looking. It has gears, pedals, wheels, and lovely red paint; it is big enough to attract attention from a distance, and it is so original-looking that everyone's first question is: "What's that?" Every gadgeteer glows when asked that question about any item in his collection.

What the Mule does is plow gardens (and of the 31,999,999 average backyard gardens) with pedal power. The Mule is two machines connected by a length of strong and resilient steel cable, some 75 feet of it. At one end is an arrangement that looks a bit like the exercise machines you'd find in any weight reduction emporium—you know, the bicycles that never move, but are driven in place for thousands of puffing miles by folks who pedal like hell not to cover ground, but to sweat away pounds.

Like them, I sat in the Mule's pedalling seat and pedalled like hell, only this time I accomplished more than sweat. The Mule's pedals power a geared winch, and from the winch runs the cable that connects to the Mule's business end: a heavy, gleaming steel plow with two handles, much like the ones that were held by the continent's original groundbreakers two centuries ago. I had Roger play the part of the settler; he held the handles and guided the plow along a straight furrow (well, he tried) while I supplied the energy by cranking away on the pedals. Judging from the crimson that surged through Roger's cheeks as he followed the plow from the far end of the garden toward my flying feet, he had the more muscular of the jobs. My pedalling was eminently bearable; indeed, it was fun.

After some adjustments—like every good gadget, the Mule has wondrous adjustments that kept my pockets bulging with wrenches—we got so we could bury the plow blade in our garden soil and turn over a fine, deep furrow. I was thrilled. There we were, the two of us, doing a job that once required a horse, or a four-footed mule, and of late has seen millions of noisy rotary cultivators come off the assembly lines.

"It's slow," said Roger. "We could have borrowed Alan's rotary cultivator and had this done by now." It was our first effort, and we'd done half the garden. I though that to be remarkable, a miraculous achievement for me, who has learned to expect the worst with new gadgets. I told Roger as much, along with a few words about how wonderful I found the sound of the plow cutting through the soil, and how pleased I was to be doing a job that did not require a gasoline can or an electrical outlet.

"Yes," said Roger, making (for him) a great concession, "and it doesn't chop the worms in little bits the way a rotary cultivator does." I'm certain that not even Bob Rodale has listed that advantage in his Mule brochures. The disadvantage, of course, is that having plowed the garden, I'm hooked for the rest of the season, which is not quite what I expected when I first climbed into the seat of the Mechanical Mule.

John Cole.
Note: At this time, Rodale Press regrets to tell gadgetophiles that it has no immediate plans to produce the Mechanical Mule.

Seeds: our most important inheritance

Part of self-sufficiency in your own food system is ending reliance on hybrids supplied by seed companies, and getting back to nature's way. Surely we'll all continue to plant hybrids—but we should also include open pollinated varieties, from which we'll select out the earliest-bearing and best-tasting plants for seed saving, improving as well as perpetuating our vegetable crops. Home gardens with old varieties and "heirloom" varieties will become the repositories of the germplasm missing from the popular offerings in the seed catalogs. That way, when tomorrow's tomato version of the corn blight works its way through gardens of hybrids, the genetic diversity in our plots will protect us.

Jeff Cox

> Between 12 and 17 percent of total U.S. energy is needed to grow, process, transport, store, sell, and prepare our food. The average American uses three times as much energy simply to bring food to the table as the average citizen of a developing country uses for all purposes.
>
> Mary Rawitscher and Jean Mayer. *Technology Review.*

> Benjamin Franklin exhibited the Newtown Pippin in London in 1759. The apple originated on Long Island, but is known today also as the Virginia, or Albemarle, Pippin. It has a yellowish color. Several apples bear Vandevere in their names, the plain Vandevere being marked with red streaks over a yellow background, becoming deep red. Esopus Vandevere, an old apple that may have originated in Colonial days, is still around but not much grown for market. These are all "snappy" eaters, good for cooking and keeping. (From a letter from Larry L. McGraw, who is keeping old varieties of fruit alive at his Experimental Garden, Portland, Oregon.) In *Apples of New York* S. A. Beach says of the Sweet Bough that "as a kitchen fruit in its honied sweetness and tender flesh, it has no equal of its season," which extends from late July through August. It is greenish-yellow to yellowish-white, and not widely grown today, since it is too soft for shipping.
>
> *John Jay Janney's Virginia.* copyright © 1978 by Edmund Derby Haigler, is published by EPM Publications, Inc., 1003 Turkey Run Road, McLean, VA 22101.

Endangered varieties

With a little searching, people in many sections of the country can still find some of the old [fruit] varieties raised by our grandparents precisely for home processing and preserving . . . . Sadly, selections of vegetables are often much more sparse.

Like their vocational ancestor, Luther Burbank, modern-day plant breeders tend to be pathologically shy.

. . . Few if any new varieties are being bred for taste or nutrition. The home food processor is likely therefore to encounter 'raw materials' inferior to those used a few years ago. People should realize that this is one reason their kitchen productions (and garden vegetables) don't taste as good as when they were children.

A letter from Cary Fowler, co-director of the Frank Porter Graham Experimental Farm and Training Center, Route 3, Box 95, Wadesboro, NC 28170. A booklet of old-time seed sources is available from the Center for $1, postpaid.

Family work

For those of us who have wished to raise our food and our children at home, it is easy enough to state the ideal. Growing our own food, unlike buying it, is a complex activity, and it affects deeply the shape and quality of our lives. We like the thought that the outdoor work that improves our health should produce food of excellent quality that, in turn, also improves and safeguards our health. We like no less the thought that the home production of food can improve the quality of family life. Not only do we intend to give our children better food than we can buy for them at the store, or than they will buy for themselves from vending machines or burger joints; we also know that growing and preparing food at home can provide family work—work for everybody. And by thus elaborating household chores and obligations, we hope to strengthen the bonds of interest, loyalty, affection, and cooperation that keep families together.

Forty years ago, for most of our people, whether they lived in the country or in town, this was less an ideal than a necessity, enforced both by tradition and by need. As is often so, it was only after family life and family work became (allegedly) unnecessary that we began to think of them as "ideals."

I think these ideals are more difficult than they were. We are trying to uphold them now mainly by will, without much help from necessity, and with no help at all from custom or public value. For most people now seem to think that family life and family work are unnecessary, and this thought has been institutionalized in our economy and in our public values. Never before has private life been so preyed upon by public life. How can we preserve family life—if by that we mean, as I think we must, *home* life—when our attention is so forcibly drawn away from home?

We know the causes well enough.

Automobiles and several decades of supposedly cheap fuel have put longer and longer distances between home and work, household and daily needs.

TV and other "media" have

learned to suggest with increasing subtlety and callousness—especially, and most wickedly, to children—that it is better to consume than to produce, to buy than to grow or to make, to "go out" than to stay home.

Another cause, and one that seems particularly regrettable, is public education. The idea that the public should be educated is altogether salutary, and since we insist on making this education compulsory we ought, in reason, to reconcile ourselves to the likelihood that it will mainly be poor. I am not nearly so much concerned about its quality as I am about its *length*. My impression is that the chief, if unadmitted, purpose of the school system is to keep children away from home as much as possible. Parents want their children kept out of their hair; education is merely a by-product, not overly prized. Why should anyone be surprised if, under these circumstances, children should become "disruptive" or even "ineducable"?

We can see clearly enough at least a couple of solutions to this lack of home life.

We can get rid of the television set. As soon as we see that the TV cord is a vacuum line, pumping life and meaning out of the household, we can unplug it. What a grand and

neglected privilege it is to be shed of the glibness, the gleeful idiocy, the idiotic gravity, unctuous or lubricious greed of those public faces and voices!

And we can try to make our homes centers of attention and interest. Getting rid of the TV, we understand, is not just a practical act, but also a symbolical one: we thus turn our backs on the invitation to consume; we shut out the racket of consumption. The ensuing silence is an invitation to our homes, to our own places and lives, to come into being. And we begin to recognize a truth disguised or denied by TV and all that it speaks and stands for: no life and no place is destitute; all have possibilities of productivity and pleasure, rest and work, solitude and conviviality that belong particularly to themselves. All that is necessary is the time and the inner quietness to look for them, the sense to recognize them, and the grace to welcome them. If we consume nothing but what we buy, we are living in "the economy," in "television land," not at home. It is productivity that rights the balance, and brings us home. Any way at all of joining and using the air and light and weather of your own place—even if it is only a window box, even if it is only an opened window—is a making and a having that you cannot get from TV or government or school.

That local productivity, however small, is a gift. If we are parents we cannot help but see it as a gift to our children—and the *best* of gifts. How will it be received?

Well, not ideally. Sometimes it will be received gratefully enough. But sometimes indifferently, and sometimes resentfully.

According to my observation, one of the likeliest results of a wholesome diet of home-raised, home-cooked food is a heightened relish for cokes and hot dogs. And if you "deprive" your children of TV at home, they are going to watch it with something like rapture away from home. And obligations, jobs, and chores at home will almost certainly cause your child to wish, sometimes at least, to be somewhere else, watching TV.

Because, of course, parents are not the only ones raising their children. They are being raised also by their schools and by their friends and by the parents of their friends. Some of this outside raising is good, some is not. It is, anyhow, unavoidable.

What this means, I think, is about what it has always meant. Children, no matter how nurtured at home, must be risked to the world. And parenthood is not an exact science, but a vexed privilege and a blessed trial, absolutely necessary but not altogether possible.

If your children spurn your healthful meals in favor of those concocted by some reincarnation of Col. Sanders, Long John Silver, or the Royal Family of Burger; if they flee from books to a friend's house to watch TV; if your old-fashioned notions and ways embarrass them in front of their friends—does that mean you are a failure?

It may. But the term of human judgment is longer than parenthood, that the upbringing we give our children is not just for their childhood but for all their lives. And it is surely the *duty* of the older generation to be embarrassingly old-fashioned, for the claims of the "newness" of any younger generation are mostly frivolous. The young are born to the human condition more than to their time, and they face mainly the same trials and obligations that their elders have faced.

The real failure is to give in. If we make our house a household instead of a motel, provide healthy nourishment for mind and body, enforce moral distinctions and restraints, teach essential skills and disciplines and require their use, there is no certainty that we are providing our children a "better life" that they will embrace wholeheartedly during childhood. But we are providing them a choice that they may make intelligently as adults.   Wendell Berry

$10 an hour

With food costs rising, even more Americans will discover that their best investment is a backyard garden. Long before food prices began their recent furious climb, home gardens produced food valued at over $10 billion and the average value of home garden work appeared to be over $10 an hour, substantially higher than the average worker's wages. Few investments will be more rewarding in the future.

Scott Burns. *Organic Gardening.*

Home water quality

Perhaps you're perfectly happy with your water supply. Its cleanliness is something you take for granted. After all, doesn't America have the cleanest water in the world? Unfortunately, no. The reasons are more than a compilation of isolated local woes, nasty as they may be. We'll discuss them first, and then suggest several options.

Tap water's many ingredients

Chlorine. Ironically, one of our common problems is rooted in the reason we trusted municipal water so long. Chlorine is the chemical used by municipal water treatment plants to disinfect water before it's pumped to your tap. Stated simply, it's added to your water to kill germs. And it has done a terrific job of killing germs, ever since it became the primary method of water treatment in 1908. Because of its addition to our water supplies, many of the 19th century's deadliest diseases have been passed into legend. Chlorination of water, for instance, is why typhoid dropped from one of the leading causes of death in 1900 to 26th place by 1930.

But in the 1970s, scientific re-

(Continued on page 442)

(Continued from page 441)

search from many quarters confirmed chlorine to be a low-level cancer threat. What happens is this. Untreated water entering a municipal treatment plant generally contains humic acids, which are the by-products of dead leaves, trash, humus, animal waste, and so on. They enter the earth's water systems as run-off from the land or as treated sewage dumped into our rivers and lakes. When the municipal treatment plant puts chlorine in the water, this element combines with the humic acids to form small amounts of chloroform and other chemicals of the trihalomethane family. Chloroform and the other trihalomethanes are linked to cancer.

How do we know? By a multitude of evidence. To give but one example, an epidemiological study of women in seven upstate New York counties showed that those who drank water from chlorinated munici-pal sources (as opposed to those who drank from home wells) had a 44 percent higher death rate from cancers of the gastrointestinal and urinary tract.

A National Cancer Institute study confirmed that injections of chloroform caused high rates of kidney and thyroid tumors in rats, and liver cell cancer in mice.

"If a substance produces cancerous tumors in test animals, as chloroform has done, then regardless of the level of the dosage, we believe

A DROP OF RAIN WATER IS CORRUPTED BY SOCIETY

This diagram is a highly simplified and idealized version of pollutants that end up in drinking water.

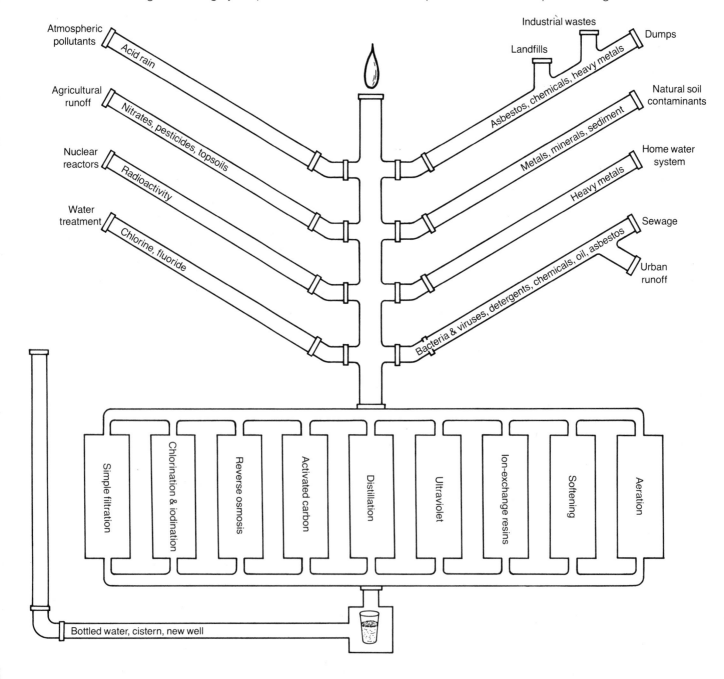

there is the possibility of that same compound producing cancerous tumors in human beings," explained Dr. Ervin Bellack, a chemist for the EPA's Department of Drinking Water Criteria and Standards.

To ascertain whether the chlorination process was indeed resulting in the formation of chloroform, the EPA conducted a massive survey of treatment plants in 80 cities throughout the nation. In each city the EPA found a dramatic increase in the formation of chloroform between the time water entered and left the plants.

In Florida, the chloroform problem has been especially acute, and Miami's drinking water woes have received national publicity. "There's no way to remove organic compounds completely from our water supply," explained Sloan Garrett, director of the Miami-Dade (County) Sewer and Water Authorities, which serves more than a million customers. "The Everglades is our main source of water, and the rainfall penetrates through two or three feet of decomposed matter called muck . . . that's our only source of water in Florida."

The muck creates high levels of humic acids in the water going into Miami's treatment plant, and high levels of chloroform going out: 311 parts per billion, to be exact. The precise effects on health of such an amount are unknown. But a rough guess can be deduced from a study made by Robert Harris, a member of the President's Council on Environmental Quality. The Harris study showed that drinking water containing 250 parts per billion of chloroform carries an estimated cancer risk of between 20 and 300 deaths per million people.

Industrial chemicals. You're probably familiar with cases in your region. You've read about Love Canal, a western New York State dumping ground of highly toxic chemicals. Mothers in nearby houses suffered horribly high rates of miscarriages and children who were born with birth defects. Since then you may have read about EPA efforts to track down hundreds of similar dumping grounds all over the nation. Our industrial society has failed to pick up after itself, and we are beginning to pay the price.

There's no way to gauge accurately the rate at which such chemicals pollute our waters. With approximately 60,000 chemical compounds currently used in industry, and 1,000 more developed yearly, there's no way to keep track of every pollutant entering water treatment plants. Toxic pollutants including heavy metals, pesticides, phenols, cyanides, and polychlorinated biphenyls (PCBs) are dumped every day into our streams and waterways. The National Academy of Sciences has calculated there are at least 309 volatile organic compounds and 55 pesticides found in drinking water. But that's just a rough estimate, because scientists lack the sophisticated equipment to detect every chemical. An EPA report published in November, 1979, pointed out:

"Because the chemicals thus far identified in drinking water account for only a small fraction of the total organic content, the possibility, and indeed the probability, exists that additional substances of equal or greater toxicological significance may be present but remain undetected by present monitoring capabilities . . . ." What we know may be the tip of the iceberg.

"The major emphasis on change in water treatment should be on the industrial polluters," says Richard Miller, Cincinnati Superintendent of Water Works, "and better law enforcement against industries who pollute with toxic substances. We should review all the chemical discharge permits that have been granted over the past few years, and there should be a complete elimination of toxic chemical pollutants at the source. I don't know of any water treatment plant that can handle all those chemical pollutants once they reach the plant. We need to eliminate them at the source."

Cincinnati draws its water from the Ohio River, which serves several municipal water systems. The Ohio has traces of such pollutants as heavy metals and other toxic substances, pesticides, oil and grease. Unfortunately, the Ohio is fairly typical of the northeastern United States, where more than 85 percent of all water basins have been blotted by industrial waste.

In summary, the problem with industrial chemicals is fourfold: the

Some of the material in this section is adapted from Rodale Press's Water Fit to Drink *($5.95 paper, $8.95 hardcover).*

number of chemicals reaching our waterways each year is increasing; we don't know how many of these chemicals are as yet undetected; many of them become more concentrated (hence dangerous to us) as they work their way up the food chain; and they are not being removed from water by conventional municipal treatment technology.

Solids. Solids suspended in water make up the second large category of water contaminants. Little bits of clay, oil, and mineral fibers may be harmless by themselves but can serve as transportation for other pollutants which might be released later within the body. Others, like asbestos, are far from harmless. Inhalation and ingestion of asbestos can cause asbestosis and some forms of cancer.

Trace minerals and heavy metals. Evaluating these elements and their effect on the human body can be tricky. For one thing, some of these trace metals in small amounts can be quite healthful (like zinc) while others in small amounts (like lead) can be deadly. What's more, many of these metals which are good for you in small doses are killers in large amounts (like copper). To further complicate matters, adverse health effects may be caused by a com-

(Continued on page 444)

(Continued from page 443)

bined intake from water, food, and air. If the quantities of an element taken in food and air are unknown, there's no saying how much can be safely drunk in water. Inorganic solutes, usually trace minerals, are picked up naturally by surface or ground water as it flows through soil and rock. Heavy metals, especially mercury and cadmium, are usually products of industry. Other metals leach from home plumbing systems and fixtures.

Soft water with a low pH (acidic water) is most apt to corrode plumbing, so that we drink cadmium and zinc leached from old galvanized pipes, lead from old lead pipes, and solder and copper leached either from pipes or fittings.

Organics. Organics sound harmless enough, suggesting some association with organic gardening and health foods, but in a chemist's terms organics are the antithesis of health food and the nemesis of pure water. Some organics are the things that nightmares are made of—benzene, polychlorinated biphenyls (PCBs), tetrahydroforan, toluene, and chloroform (among others).

Organic compounds in water come from six basic areas: natural sources; runoff; spills and accidents; industrial discharge; sewage treatment plants; and water treatment.

The natural sources include leaves that fall on waterways, waste products of aquatic plants and animals, organic matter from soil, and domestic wastes. As the organic matter decomposes, it can combine with toxic materials such as pesticides and cause problems. As discussed earlier, if chlorine is used to disinfect water, it interacts with the decomposed organic matter to form chloro-organics, some of which are known carcinogens. Other sources of organics are direct and indirect dumping of wastes into our supply of water, from pesticide runoff, urban runoff (such as sewer overflows containing oily automobile residue and street litter), septic systems, animal feedlots, agricultural and home fertilizers and pesticides, manured fields, industrial wastewaters, sanitary landfills, and garbage dumps.

Nitrates. In some areas of the United States, wells that supply fresh water have suffered a sharp increase in nitrate levels.

Nitrates generally show up in well water, especially from shallow-dug or poorly cased wells. While these concentrations are sometimes natural, reflecting the nature of the rock formations in the given area, usually they are caused by human or animal waste that has contaminated either surface water or aquifers. Usually, the deeper the well and the greater its distance from a septic tank, the less is the nitrate pollution. State and local health departments often test for nitrates without charge.

If your water contains nitrates, you are facing two health hazards. Nitrates can affect the blood so that it no longer carries oxygen through the body. This disease most frequently affects infants. The nitrate will also change in the body, becoming nitrosamines, some of which can be carcinogenic. Preliminary studies have suggested a link between high concentrations of nitrates in drinking water and cancer of the stomach.

A homeowner can remove nitrates by distillation, but more common solutions are drilling a new well, or using bottled water.

Radioactivity. Minute traces of radioactive material are found in almost all drinking water. The concentration and composition of these radioactive particles vary from place to place. Some of which is unavoidable results from the natural radiochemical composition of the soil and rocks through which the water has passed. Other radioactivity results from nuclear development. Even the best scientists can't tell us specifically what effect radioactivity will have on our health, and what amount of radiation is tolerable in water. Ion exchangers will remove 99.9 percent of all radioactive materials.

Asbestos. While asbestos naturally occurs in certain water supplies, it is turning up in more and more samples. Perhaps the most frequent single cause of this occurrence is the use of asbestos cement pipe to supply drinking water. Some 200,000 miles of this pipe in use across our country is beginning to disintegrate. As it erodes, asbestos fibers are released into drinking water supplies. In areas where the water is acid, corrosion occurs rapidly.

By adding sodium hydroxide or lime, the pH of the water is raised, slowing the disintegration of the pipes. It is also possible to filter the asbestos out of the water.

Given the state of America's drinking water and prospects for improvement, the homeowner's only real option is to shift for himself. Whether the individual concern is with rusty water, radioactive water, awful taste, awful smell, or the risk of cancer, more and more people are taking water treatment in their own hands.

Testing

Local and state health departments will analyze your water for free (or at a nominal fee) if a problem is suspected. But in most areas they won't make routine checks. They will serve as an information and referral service answering questions and supply names of commercial laboratories equipped for all types of testing.

The first step is to begin locally, at the water authority; then proceed to the county health department or regional department of environmental resources. Finally, approach the state authorities for more complicated testing or if local help is unavailable.

The best way to start evaluating your water is to have it tested for bacteria. The cost is nominal or free. If you want to try testing yourself, the Hach Chemical Company (P.O. Box 389, Loveland, CO 80537) offers a kit requiring little training to get accurate results. However, the initial cost—which includes an incubator—is about $50. In addition to testing for bacteria, many local health authorities and private laboratories will routinely test for turbidity, hardness, pH, and in agricultural states or problem areas, for nitrates.

You now have completed the normal route for testing. But you still don't know if your water contains other pollutants that may be harmful to your health—pollutants like heavy metals, pesticides, and even radioactivity. Tests for such contaminants are not done as part of routine procedure. To detect minute quantities of organic chemicals, for example, technology is so refined and equipment so expensive that the test for each chemical may cost a laboratory more than $100. Some exotic chemicals require two weeks of testing and cost more than $1,000. To have a lab test your water for the presence of all possible organic contaminants, the cost would be staggering.

Yet we know that tap water can contain organic chemicals, lead, copper, cadmium, mercury, zinc, and arsenic. In fact, it is not at all rare to find these elements in household water, although usually in small amounts. How can an individual know if these substances are in his own water supply? You should first make an educated guess about what toxic elements could be in your water. Ask the following questions:

• Where does the water come from?
• Is the original surface water source downstream from an industrial site or landfill?
• What treatment methods does the municipal water works use?
• Does the treated water meet EPA requirements?
• Could contaminated surface waters from landfills, industry, or farms directly enter the home well?
• Could you unwittingly be contaminating your own water with sewage entering an improperly positioned or poorly sealed well?
• Is the water soft and acidic, and therefore apt to leach metals from the plumbing and into the water supply?

Eyeball the quality of your water checking for foam, turbidity, color, taste, and odor. With this information you can guess some of the organic contaminants with fair accuracy. You won't necessarily know the levels at which these chemicals are present, so further tests may be in order.

The heavy metals and other toxic elements can be tested for with considerable accuracy at reasonable cost. You needn't test for all minerals. But you should test for the hazardous ones which could be leaching into your water supply in unhealthy amounts.

Government health offices will often perform tests, but usually for only a select number of minerals.

You can also have your water tested for 17 minerals and heavy metals by the nonprofit Soil and Health Foundation. The cost is $25. (For a sample bottle and instructions, write to the Soil and Health Foundation, 33 E. Minor Street, Emmaus, PA 18049.)

In-house answers

At first, the job of purifying water on your own may seem enormous, requiring the skills of a gifted chemist or engineer. But it's a job almost everyone can do. And there are many units you can build or buy.

Among the devices on the market, you can choose from a variety of filters, distillers, deionization units (such as water softeners), and reverse osmosis membranes. In addition, you can consider disinfecting water with iodine, ultraviolet light, or chlorine.

Filters. Sand filters are simply a column of sand or other porous matter (crushed anthracite, diatomaceous earth, gravel) that strains out particles from water. This kind of filter can remove clay, silt, colloids, and bacteria and viruses if the pores in the filter are small enough. Slow sand filters are commonly used for pond water treatment because the cost is low and the effectiveness high, if they are properly cared for. This cleansing

(Continued on page 446)

The more-expensive carbon filters, like the Everpure THM, are effective at removing trihalomethanes.

First thing in the morning, let your water run for two to three minutes at full force to flush out the system. Water that stands overnight may pick up lead, arsenic, cadmium, and other contaminants.

DO A LITTLE WATER TESTING ON YOUR OWN

Pour a glass of water and take a close look, a good sniff, and a sip. Sometimes just taste, odor, and appearance can tell you that your water needs more treatment than it's getting.

1) Does your water *foam* as it splashes into a glass? Foam can be an indication that detergent residue is getting into your drinking water.
2) Is your water murky looking? *Turbidity* signals clay, silt, metals, synthetic or natural chemical compounds, plankton, or microorganisms. It could also indicate the presence of sewage, industrial waste, or asbestos.
3) Does your water have a *color*? Truly pure water does not. Color can be caused by decaying organic matter—things like leaves and plants.
4) Does your water have a peculiar *taste* or *odor*? The cause of off-taste and smells are many—ranging from industrial solvents to decomposing organic matter. Generally, the most common sources of bad taste and odor are naturally occurring minerals and gases, or chemicals that add a funny flavor even if present in only a few parts per billion (these include formaldehyde, picolines, phenolics, refinery hydrocarbons, petrochemical waste, phenylether, and chlorinated phenolics).

Chlorine can add an unpleasant taste and smell to water. Algae may make your kitchen or bath smell like a stagnant pond. If your water smells like rotten eggs, there is hydrogen sulfide in it. In small amounts, this gas will make your water smell awful.

Taste and odor are difficult to judge because of individual preferences. We become accustomed to the water we drink. But water that tastes bad suddenly is surely in need of testing.

MAKE YOUR OWN CARBON FILTER

Designed by the Environmental Protection Agency, this filter produces high-quality drinking water for three weeks, before the carbon needs changing.

Operate it continuously, 24 hours a day, at a rate of one gallon of filtered water a day.

The carbon filter should be changed every three weeks, or after approximately 20 gallons of filtered water. When changing the carbon, disinfect the rest of the filter by soaking it in a 5 percent solution of liquid laundry bleach.

This homemade unit can be installed beneath the kitchen sink.

For materials, you will need ¼-inch copper tubing, ¼-inch tubing tee, 36 by ¾-inch inside diameter copper of galvanized steel pipe, an ice maker saddle valve, reducing union for ¾-inch pipe to ¼-inch tubing, funnel, paper coffee filter, one-gallon glass bottle, plastic pan, and cotton balls.

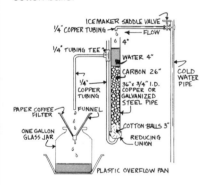

1) Construct the filter as illustrated. A shorter column can be used if there are space limitations.
2) Disinfect the empty column with a 5 percent solution of liquid laundry bleach by filling the column and letting it stand for a couple of minutes.
3) Rinse the column thoroughly.
4) Add the cotton plug.
5) Fill the column with water and add previously wetted washed carbon to a depth of 26 inches.
6) Maintain the water level above the carbon by placing the ¼-inch tubing tee above the surface of the carbon as shown in the accompanying illustration.
7) Operate the carbon column continuously, using the saddle valve to adjust the flow rate.

Remember to change the carbon every three weeks or every 20 gallons.

(Continued from page 445)

will make the water more suitable for further treatment in a distiller, granulated active carbon (GAC) filter, reverse osmosis unit, or deionizing unit. Such filters are simple and can be constructed at home.

To neutralize acid water, you can build a filter that consists of limestone or marble chips. As water passes through the filter bed, the acid is neutralized by the limestone, forming bicarbonates.

Granular activated carbon, often used in aquarium filters, is extremely effective in removing chloroform, chlorine, and bad taste and odor. It can remove or reduce many organic chemicals, including pesticides, industrial chemicals, and most halogenated organic compounds like PCBs and PBBs (polychlorinated biphenyls and polybrominated biphenyls).

Carbon also will reduce the concentration of heavy metals like lead and cadmium, but it also reduces trace minerals thought to be good for health. It will not remove fluoride, nitrates, or other salts, nor will it remove asbestos fibers. In fact, if water is loaded with minute debris, it should be passed through a sediment prefilter before putting it through carbon, or the carbon will become clogged and quickly lose its effectiveness.

Activated carbon filters (mistakenly called charcoal filters) are the most common of the dozens of new water treatment devices on the market. They range from $6 to $600. The most popular—and the least effective—are the small filters that attach to the end of the faucet and filter water as it leaves the tap. Larger units can perform effectively. It is also possible to construct a full-scale GAC water filter on your own.

Most of the unwanted organic pollutants are absorbed in the top layers, and as the water passes downward it progressively contacts cleaner and cleaner carbon. This in-depth filtering "polishes" the water as it passes out of the column and provides a wide margin of safety.

One drawback to the use of GAC filters is that bacteria and other organisms can grow on the carbon surface. Some manufacturers add silver compounds to their filters to kill bacteria. However, silver is not a dependable bacteriostat. It can leach

from the filter, and it may be harmful to your health. Look for one without silver.

Because of the bacterial threat, and the possibility that filtered chemicals may slough off a saturated filter, the carbon should be replaced every three weeks, or after treating 20 gallons of water. In the interim, check for the telltale signs of filter failure:

- A change in taste
- The water pressure is noticably reduced, resulting in slower than normal output
- Sediment appears in the water

Distillation. An age-old method for purifying water still in use today is distillation. The process simply heats water to steam, and then condenses it back to water. In theory, all the debris, bacteria, minerals, and other contaminants are left behind as the water turns to steam. However, in simple stills, chloroform and other organic chemicals, which have a lower boiling point than water, vaporize right along with the water, recondense, and wind up in the finished product. Units employing fractional distillation prevent this from happening. As impurities such as chloroform turn to steam in the boiling chamber they are released through a vent.

Distilled water should be refrigerated in the cleanest container possible. Old hands at distillation say the real trick is keeping good-quality water, not making it. The best storage containers are of glass and stainless

The New World Distiller.

METHODS AND PROBLEMS

Home treatment methods	Asbestos	Bacteria and viruses	Heavy metals	Nitrates	Organic chemicals	Radioactivity
Chlorination/ultraviolet		●				
Fractional distillation	●	●	●	●	●	●
Granular activated carbon (GAC)					●	
Ion exchange			●	●		●
Reverse osmosis	●	●	●	●	●	●
Softening			●			●
Simple filtration	●	●				

steel. Do not use plastic jugs—they can contaminate your water during storage.

Distillation may not be the process for you. Some distillers are very hard to keep clean. Minerals, chemicals, and other contaminants left behind in the boiling chamber build up to form a scale that can interfere with the unit's efficiency and even with the quality of the distilled water. For that reason, the units need frequent cleaning, and often getting access to the boiling chamber is difficult. Moreover, if your water is hard your distiller will clog quickly. Some distillers can be cleaned only with a strong acid such as nitric acid.

Stills can use a lot of water. Generally, it takes five gallons of tap water to produce one gallon of distilled water if water is used as a coolant as well as the source for the distilled water. But the heat and humidity can make a room quite comfortable in the wintertime.

It's important to estimate the energy demand of a still. As the price of electricity rises, the cost of running a still may become too expensive for some households.

The greatest consideration may be the taste of the distilled water. There is none. This lack of flavor may take awhile to get used to, but most people adjust. The reason the water has no taste is that all the minerals have been removed during the process of distillation.

A final word about distillers: they are expensive, ranging from about $150 to more than $500.

Sources: New World Distiller Corporation, Box 476, Gravette, AR 72736; Pure Water, Inc., 3725 Touzalin Avenue, Lincoln, NE 68506.

Deionization. Home water softeners work by a method called ion exchange. What they do, quite simply, is swap particles of sodium for the water's calcium and magnesium.

The amount of sodium which eventually ends up in your water supply depends on the amount of calcium and magnesium it had to begin with. The more grains of hardness in the water, the more salt in the softened water.

Soft water truly is an attractive commodity. It eliminates soap curd and detergent deposit from clothing. You need less soap for laundry and cleaning. White mineral scales no longer deposit on glasses, dishes, and utensils. Soft water rinses away grime faster, and leaves no residue. It allows for luxurious bubble baths and leaves hair clean and shiny. It provides a longer life for household appliances that use water, and it protects expensive plumbing and heating equipment.

Although soft water seems to create a plumbing panacea, it promises few benefits for health. We can't discount the glaring evidence that, for whatever reason, people who drink soft water suffer more heart attacks and strokes than people who live in hard water areas.

Beyond that, softened water also creates an added burden on the ground that eventually receives it.

For these reasons, water softeners should not be installed without careful consideration. But water softeners are not the only kind of deionizing unit that can be used to treat water. There are others that do not use sodium to replace unwanted elements in water. These deionization units have two ion-exchange materials that remove all positively charged ions (called cations) and negatively charged ions (called anions) in exchange for hydrogen ions and hydroxyl ions, respectively. These combine to form water (H-OH or H_2O), a new water that contains no minerals.

While this process may sound like hocus-pocus to those not familiar with it, it actually is a very old procedure, and laboratories that require the very purest water use units that contain special exchange resins that are very effective and efficient.

These resins have not yet been marketed for home units. However, those with pioneering spirit might consider tapping scientific supply houses such as Fisher Scientific, 203 Fisher Building, Pittsburgh, PA 15219; Bellco Glass Inc., P.O. Box B, Vineland, PA 08360; or Continental Water Conditioning Corp., P.O. Box 20018-TR, El Paso, TX 79998.

Deionization will remove fluoride, hardness, soluble iron, nitrates, silica and silicates, sodium, sulfate, total dissolved solids, copper, arsenic, heavy metals, selenium, and just about anything else that goes through it as long as the incoming water meets certain guidelines. Prior to deionization the water should contain little, if any, iron, manganese, turbidity and sediment, organic chemicals, chloride, and hydrogen sulfide. The only limitation with this method is that the purified water is not totally free of microorganisms and still contains organic chemicals.

Reverse osmosis. With reverse osmosis (RO), water flows over the surface of a porous membrane that looks something like cellophane. Under pressure from the tap, some

(Continued on page 448)

(Continued from page 447)

water is also forced through the membrane and purified. The remainder, now concentrated with contaminants, is drained away. It is a kind of filtration under pressure that removes minute pollutants from the water.

RO removes not only the matter in water but also the *dissolved* matter in water. It can make sea water fit to drink.

RO removes turbidity, organic matter, chloride, fluoride, calcium and magnesium, manganese, nitrates, silica, silicates, sodium, sulfates, copper, bacteria, viruses, and pyrogens (fever-causing substances). It will also remove most detergents, pesticides, tannins, chlorinated compounds, and other complicated chemicals. It is an excellent way to remove asbestos from water. But RO will not remove simple compounds like chloroform and phenol.

The membranes have a useful life of one to three years, varying with the pressure, pH, temperature, and quality of the water being treated. If the water is oily, for example, the membrane will be fouled very quickly and need frequent maintenance.

Because RO works slowly, the unit is usually run frequently in order to provide enough drinking and cooking water for the whole family.

A reverse osmosis unit may cost from $300 to more than $500, but needs no electricity or any other source of energy as it operates with water pressure.

Probably the most popular RO unit for household use is made by the Culligan Company. Called the Aqua-Clear, the unit fits in under a kitchen sink and runs the processed water to its own faucet. The entire unit contains a sediment filter, a GAC filter, and offers an option that heats or chills the water.

Aeration. Aeration is a simple, natural treatment of water. By spraying, bubbling, or thinly pouring water, it comes in contact with air. This procedure reduces the concentration of gases—like the rotten-egg-smelling hydrogen sulfide. It also causes volatile organic compounds like chloroform and methane to dissipate into the air. For such a simple procedure, it is quite effective.

Aeration can also be used to oxidize iron and manganese which will then precipitate out of the water as insoluble oxides, to remove odors produced by algae, and to brighten up the flat taste of cistern water.

Probably the simplest and easiest aerator is an appliance you already own—the kitchen blender. Fill it only half full with water, take the glass bubble off the top, and spin.

Disinfection. Several treatments are available for simple disinfection. The most common is chlorination. But water also can be disinfected with iodine, bromine, silver, and even ultraviolet light.

If you decide to chlorinate your water, a system should be developed whereby chlorine will kill germs and bacteria, but then be removed from the water before you drink it. This system provides chlorine's benefits without its related problems. You can buy or make a small mechanical unit called a chlorinator, to add chlorine to your water. The amount of chlorine you need depends on how bad your water is.

Ultraviolet radiation. Ultraviolet radiation is a good disinfection treatment because it does not introduce anything into the water—no chemicals, odors, fumes, or tastes. And it works quickly. Only a short contact time is required for a UV unit, a quartz-mercury vapor lamp, to emit germicidal rays. If you should accidentally

Many local spring water companies have home delivery routes. You can save by using five-gallon carboys, but they necessitate a stand of some sort, such as this tin-and-stoneware antique.

over-radiate the water, no harm is done.

But on the dark side, certain spores and viruses are fairly resistant to UV treatment, and frequent maintenance is required by some units. Ultraviolet rays are not fully effective in cloudy or dirty water or in treating water with dissolved iron or certain organics. Moreover, the unit can become coated and discolored, so install a sediment filter in the line just ahead of the UV unit to trap tiny floating particles that will decrease its effectiveness.

The units are expensive, ranging from $300 to $600. But when a UV light burns out, it can be replaced like any fluorescent tube.

Sources: Trojan Environmental Products, Inc., P.O. Box 2341, London, Ontario, Canada N6A 4G3; Aquafine Corporation, 1869 Victory Place, Burbank, CA 91504; Sanitron, Atlantic Ultraviolet, 250 Fehr Way, Bay Shore, NY 11706.

Iodine. Iodine is effective in destroying bacteria, viruses, cysts, and other contaminants in water, but is not recommended for use by pregnant women, and is not considered safe for long-term use.

Finding a new source. For any number of reasons, you may choose to develop a new water resource rather than tackle the job of cleaning water from the old system. If you are currently drinking water from a surface supply, this switch may be the best way to get good water. Surface water is notoriously polluted, and the simple change to well water—or any underground water—often can provide a dramatic improvement in water quality. If you have a badly located well—too close to septic tanks, an overflowing river, a feed lot, and so on—you may also benefit by finding a new source in a cleaner spot.

Lucky households may find a spring on their property. Others, especially those with a poor or unsteady water supply, may want to install a cistern to catch rain water. No matter what alternatives you choose to explore—dug, bored, driven, or jetted wells, cisterns, springs, or bottled water—you should first calculate your family's water needs.

The average person uses 70 gallons of water each day inside the

Sparkling spring water has become a fashionable alternative to alcoholic drinks, but is too expensive to serve as a long-term replacement for tap water.

home. This figure does not include water used for lawns, gardens, swimming pools, ponds, or car washing but represents the amount used in drinking, cooking, personal hygiene, and household cleaning tasks. Drinking and cooking uses 2 to 3 gallons. For an average family, that's easily 300 gallons every day, 10 to 15 of which are used for drinking and cooking.

For information about constructing a well, cistern, or spring, write the National Water Well Association, Information Officer, 500 W. Wilson Bridge Road, Worthington, OH 43085.

Cisterns. Today, most rural communities have access to a public water supply. Although some people may not understand the need to spend time and energy on an individually maintained water supply, others swear by it. Dave Stevens, for instance, is a free-lance writer in Indiana who decided to construct his own solar home and cistern last year.

"I know where my tap water is coming from, and I've decided I'm a hell of a lot better off gathering water from my dirty roof," he said. "When I finish this house," he added, "I will have no fuel bill, no house payment, and I will be completely free of the water company."

For suburban and rural dwellers who decide to build a cistern, there are many prerequisites to be met. All cisterns have smooth, watertight walls that prevent ground water from get-

ting in and cistern water from leaking out. The inlet and overflow spouts are covered with fine screening. Homes with cisterns often have roof washers or cut-off valves, devices which keep the first rainwater out of the cistern. In addition, good cisterns have an entrance, usually a manhole, to allow cleaning access, and a tight-fitting manhole cover. The ground around a cistern usually is sloped so that surface water cannot enter. Finally, it should hold enough water to meet a family's needs during a drought.

It's impractical, for instance, to build a cistern in an area where there is only slight rainfall. No matter what the size of your roof as a collection area, you'll have a difficult time meeting your water needs in an arid climate. Consult your local weather bureau to find out if there are periods of drought recorded in your locale. When you calculate the size of your cistern tank, bear in mind how much water storage may be necessary to get your through dry spells. A cistern design can be obtained by sending $2.50 to the Cooperative Extension Service, Ohio State University, Columbus, OH 43210.

Tracy DeCrosta and Carol Keough

Add a scant pinch of vitamin C powder or a piece of a vitamin C tablet to a glass of chlorinated water immediately before drinking. Taste and odor will disappear.

The end of the line

Fourteen percent of the food the average family buys is dumped in the trash. This waste can go into building your garden soil instead. Waste paper collected in a paper bag can be compacted by a good foot stamp into a starter for your wood stove. And the ashes produced from the stove will return to the food chain as you recycle them also back into the soil. In the world of nature nothing is a leftover. Even a dead animal's bones are part of a chain of life.

Making the least home waste, then making the most of it

Monitor your waste output for a week. Categorize what's in the trash and think about getting the best use out of it. Watch what kinds of paper, how much plastic, aluminum, glass, and good residues go in. If you count it all by total bagfuls, you may feel some pleasure at your achievement as you cut down on what goes off to the landfill.

Bringing home fewer single-use items is the first way to shrink the size of your trash can contents. Use washable cloth napkins at the table and dishcloths to wipe up spills instead of buying paper napkins and towels. Use china cups and plates rather than paper or plastic throwaways. Packaging counts for 35 percent of all solid waste, so try to avoid packaged foods. Buying in bulk saves on packaging since larger sizes take proportionately less package. (A three-ounce tube of toothpaste takes 50 percent more packaging than a seven-ounce size, for example.)

By shopping at co-ops, you can bring your own reused plastic bags and grocery sacks. Buy drinks in returnable bottles where possible. Reuse lunch sacks until they wear out. Try going back from facial tissue to cloth handkerchiefs.

Most food scraps can be recycled directly back into the food chain. If you have animals, they'll convert some of the scraps back into energy. Chickens eat meat scraps, vegetable trimmings, dry bread, and crushed egg shells. Rabbits munch on fresh salad scraps, and pigs will eat almost anything including chicken bones. Coffee grounds and tea leaves can go directly into your garden soil to feed the earthworms.

With or without animals, kitchen scraps can refuel earth's life by being composted. Mixed together with grass clippings, soil, and other organic materials, your leftovers will decay to form a rich dark humus for the garden, trees, and shrubs. Composting is a planned imitation of nature's own way to handle once-living wastes. In its process, tiny microorganisms get their own life energy by digesting plant and animal debris. As they eat your garbage, eliminate it, and die, they break the matter up into fine particles no longer even recognizable as waste. Their natural magic creates your garden's fertilizer.

At its simplest, composting is a matter of burying scraps outside, mixed with layers of soil, and letting the waste decompose slowly. This will probably take several months to a year, depending on soil temperatures and the composition of the waste. Then you can move the humus to wherever you need it for natural soil fertilizer. Cutting or shredding your scraps will help them decay faster. Turn the soil over to introduce air, and sprinkle as needed to keep the soil moist. Bones, meat scraps, and fats are slower to decompose and also attract dogs, rats, and other animals to the compost pile.

Kitchen pulverizers will grind food, even bones, into a mush for faster decomposition. Commercial or homemade bins and tumblers keep composting matter safe from animals, keep your yard looking ordered, and reduce the labor involved.

All sorts of kitchen throw-aways have multiple uses, so you can save money and landfill space as you recycle them for a second, third, and fourth lifetime. Here are some ideas to spur your creativity:

- Plastic yogurt cups will hold seedlings (punch a hole in the bottom).
- Use mesh bags from store-bought potatoes and onions to hang your own garden produce in a cool, dry place for the winter.
- Bleach bottles with the tops cut off can be used for spring plant protection in the garden or for row-markers (use markers to label). They also make good funnels.
- Banana peels will clean and polish leather (wipe dry with a cloth, and place the peel in the compost).
- Use glass jars to store grains and seeds. Pack lunchbox size jars of fruit or applesauce taken from large jars.
- Plastic bags, washed or turned inside out, can be reused for food storage or for lunch bags.
- Grocery bags make good wastebasket liners.
- Vegetable peels and pieces can go in a slow-brewing soup broth.

Even after you've cut back, composted, and internally recycled as much as you can, a typical home has bags of bottles, cans, and other assorted trash, and community recycling may be an answer.

Newsprint, aluminum, glass, and tin cans are the staples of most community recycling centers. Paper is recycled into insulation, packaging, and recycled paper products from stationery to toilet paper. (You can help the effort by buying those products marked as made from recycled paper.) Aluminum is remelted and rolled into sheets to make new cans, trays, and foil (at a great investment of energy, however). Nonreturnable bottles are crushed and remelted for use as containers, tiles, and insulation. At most recycling centers you should clean and sort the glass into clear and colored for easier handling. Tin cans should be cleaned, top and bottomless, and smashed flat underfoot.

Plastics deserve a word of their own. More than 26 billion pounds of plastic were produced in the United States in 1977, and your kitchen is probably getting a lot of it in packaging. Plastic is made from petroleum, creates poisonous smoke when burned, and is generally not biodegradable—that is, natural microorganisms won't reduce it back to its elements. Some plastics can be recycled but it's technically difficult because the resins in different plastics are not compatible with each other. Industries are slowly coming around to recycling, but the economic incentive is small because manufacture from scratch is still cheaper than recycling.

Organize your kitchen for recycling with separate containers for compostable foods, glass, cans, and burnable or recyclable paper. A covered kettle on the counter will keep scraps from collecting flies; stainless steel won't pit as easily as aluminum, but costs more.

Sara Ebenreck

(Continued on page 452)

WOOD ASH IN THE GARDEN

Of the common materials we add to our gardens, such as manure, hay, and plant wastes, wood ashes boast the highest percentage of potash—up to 10 percent. At one time, when firewood burning was a common practice, ashes were the major source of potash for farms and gardens. Besides potash, several other plant nutrients are found in ash, including some phosphorus.

Garden books advise us to apply five to ten pounds of ash per 100 square feet of ground. I've used this formula with good results. You'll want to keep in mind that ashes from hardwood trees are better; they contain more potash than evergreens.

A good time to apply the ashes is early spring, sometime before planting. It should be mixed well in the soil so that plant roots and sprouting seeds don't touch a concentration of it which could prove harmful. Over winter you'll want to store the ashes in a dry place. Exposing it to rain makes it lumpy and hard to apply evenly, and the rain leaches out the potash.

Plants that are well suited for the ash application include cabbage, broccoli, cauliflower, spinach, tomatoes, corn, and lettuce. These can also be given a side dressing of ash mixed in the soil later in the season, too.

Wood ashes are also very alkaline so when you fertilize with them, you are sweetening the soil. Therefore, if your garden is on the acidy side, you won't have to go out and buy lime. Of course, if your garden is naturally alkaline, you don't want to apply ashes. Generally, a neutral to slightly acid garden is what you want. It's best to test your garden's acidity with litmus paper, available at garden stores. Keep in mind, too, that some plants—notably potatoes and blueberries—prefer a quite acidic soil and don't want ashes whether or not the rest of the garden can use them.

Gary Nelson.
Countryside Magazine.

SINK TOOLS

Most synthetic brush bristles have greater tensile strength (resistance to being broken by stress) than plant or animal fibers such as pig bristles, horse hair, yucca fiber, and broomcorn. Synthetics also make a better recovery from bending: bend the straw on a natural-fiber brush, and if often stays bent, while a nylon or other plastic bristle will usually straighten out when released. Nylon and polystyrene tend to outlast natural bristles, and don't wear down. Synthetic brushes can be much stiffer than natural bristles, especially when wet, because they don't absorb as much water.

And yet I have a brush, a favorite brush, that has lasted for ten years. It's an old wood-and-straw scrubber made from Palmyra straw. It's worn down now, almost all the way on one end. Industrial testing would disapprove: nylon wouldn't wear down that way. What the testers don't seem to understand is that the brush works better when it's worn down. The bristles seem stiffer because there isn't as much length to bend, so my labor power actually goes farther. Very little of it is wasted in bending the bristles. Nylon bristles won't wear down, and so some of your elbow grease is always being used up in bending those everlasting bristles sideways. Eventually they will stay bent, and that doesn't help either.

There is one very effective cleaning tool, however, that *only* comes in plastic: the little mesh balls for dish-washing. They have one advantage over all brushes: you can apply maximum pressure on the mesh *precisely* where you need to scrub hard. Brushes tend to spread the pressure out over the whole surface of the bristles, reducing the pressure in any one spot.

Cynthia Adcock

(Continued from page 451)

Indoor composting

Earthworms can also be used indoors in the winter to produce a small amount of compost from kitchen garbage, dust from a vacuum cleaner bag, even newspapers. Generally, one pound of earthworms will eat one pound of garbage and produce one pound of compost each day, although this varies.

It is best to begin on a modest scale. Construct a wooden box two feet wide, two feet long, and one foot deep. Or, get a vegetable lug box from your local supermarket and, if it has large spaces between the boards, tack in plastic screening to hold the earthworm bedding. If you construct your own box, provide for drainage and aeration by drilling a half-dozen ⅛-inch holes in the bottom, and some more around the sides. A box two feet square, and one foot deep, will accommodate a thousand adult worms (or "breeders," as they are called in the trade) or you can order a pound of pit-run worms which will do as well.

You can prepare a bedding as follows: wet a third of a bucketful of peat moss thoroughly, and mix with an equal amount of good garden loam and manure; add some dried grass clippings, hay, or crumbled leaves, if you wish. (Don't use oak or other highly acid leaves.) Soak this mixture overnight.

The next day, squeeze out the excess water and fluff up the material (which we will now call bedding). Line the bottom of the earthworm box with a single layer of pebbles or rocks. Then, place four inches of the bedding material rather loosely on top of the pebbles, and wait for a day to see if any heating takes place. If initial bacterial action forces the bedding temperature much above 100° F. (38° C.), all the worms will be killed. Any heating that does occur will subside within 48 hours.

When you are satisfied that the bedding will present no serious heating problems, push aside the bedding material, place the worms and the bedding in their shipping container in the center, and cover them loosely. Place a burlap bag and several layers of cheesecloth or wet newspapers over the top of the bedding and moisten it with a houseplant sprayer or sprinkling can.

Keep the bedding moist but never soggy. If the container begins to drip from the bottom, place some sort of container, such as a plastic dishpan, under the box to catch the drippings. Use the drippings to water your houseplants.

Start out by feeding the earthworms cautiously. If you give them more than they will eat in a 24-hour period, the garbage will sour, creating odors and attracting flies, or will heat up, killing the worms. Begin with soft foods, such as cooked vegetables, leftover cereal (including the milk), vegetable soup, lettuce, bread scraps, soft leaves of vegetables, even ice cream. A little cornmeal will be appreciated, and coffee grounds can be added at any time. Do not use onions, garlic, or other strongly flavored foods.

Place the food on top of the bedding and tap it gently into the bedding. After a week or two, your earthworms should have adjusted to their new home and should be on a regular feeding schedule. You can help them along, and build better compost, if you add a thin layer of partially decayed manure from time to time (being sure that it is past the heating stage, but not completely composted).

For two weeks, the bedding in the box should be turned and aerated. Reduce the amount of food after such turnings, since the worms will not come to the surface as readily for a day or two after having been disturbed.

After a month, you can add another two inches of bedding material to handle the increased worm population, and after three months it will be time to start another box.

Jerry Minnich et al. *The Rodale Guide to Composting* (Rodale Press, 1979).

Homemade paper from trash

Kayes and Sonja van Bodegraven seem to have a solution to the "paper war." They use waste writing paper, typing paper, newspaper, envelopes, wrappings, advertising brochures, and greeting cards as raw materials for the manufacture of quality, handmade paper.

The couple came from Holland,

where they ran a news agency and bookstore, Kayes is the son of a son of a printer, "born among paper, printers, and presses." He himself was a professional printer and publisher with his own firm until recently, but decided he's had enough of having people working for him, feeling, "If you want something done, it's better to do it yourself." And so, as seems his way, he's doing just that in a shed in his backyard.

About 18 months or so ago, Kayes was sorting through some books left over from old student days when he came across some details he'd collected on handmade paper. He decided to have a go at it again, but this time instead of using pine, he thought he'd take a short-cut and try substituting waste paper. Now, the van Bodegravens are probably the only professional handmade papermakers in the world to manufacture from recycled waste.

His homemade mill consists mainly of a large barrel, a wooden vat, a press, and plenty of water. Shredded paper and water are put into the barrel which contains an agitator, somewhat like a giant vitamizer, which pulps the paper into a soft mixture. The pulp then flows into a wooden vat and here special effects can be created by adding such things as jute fibers, wood shavings, lichen, leaves, coffee grounds, tea leaves, or whatever, to vary the surface texture of the finished article. Color can be obtained with commercial dyes (vegetable dyes can be used but they tend to fade), however, Kayes doesn't encourage the use of any chemical additives as he considers that there are enough in the waste paper already without adding more.

From the vat, the mixture is strained onto a frame of flywire cut to the size of the paper required. A watermark can be drawn on the sheet before it is tipped out of the mold, like a cake from a tin, and a piece of felt or other absorbent fabric placed on top. The texture of the paper can be further varied by including some other separating sheets: open-weave curtain material or onion bags, for instance. When a stack is formed (the size depending on its weight and your strength), it is shifted to a homemade press which is a combination of wooden blocks,

pulleys, bars, and a hydraulic jack. Here five tons of pressure is exerted, squeezing out the water and adhering the fibers, while forming the characteristic emblem of homemade paper—a torn effect on the edge which is a result of the pulp extruding beyond the mold under the weight. The paper is then pegged on a line to dry, and any buckling rolled out later.

Each batch takes on a quality and essence of its own depending on the materials you care to add; paper can be made to blend with its purpose—in a recent contract, cement-impregnated paper was made for an architect client, which illustrates that the possibilities are nearly unlimited.

Kayes considers that anybody can master papermaking in a short time, just using the kitchen sink.

From Grass Roots, *an Australian homesteading magazine.*

The garbage truck doesn't stop here anymore

On Monday mornings I watch my neighbor across the street struggle with his big garbage cans. Sometimes there is also a large corrugated box filled with smaller boxes. In spring and fall he heaps brush and clippings, carefully tied up with twine and dragged to the roadside for pickup. Does he wonder, as he hauls those cans out and back every Monday, how it is that I sit at my window sipping tea, reading the paper in my housecoat? If he asks, I shall share my system with him.

I used to go through those Monday morning panics, shouting at my children for help as the garbage truck approached. There were wailings from the put-upon: "It's not my turn." "I've got a clean dress on." "It's his fault that the top wasn't put on tight." But no longer.

The weekly charge for a two-can pickup in this suburban community is $1.50, or $78 per year plus tax. While this is not an outrageous sum, it seemed possible that here was a service that I might perform for myself. When I first began saving food scraps for my compost pile, I analyzed the waste materials that left my home and considered how to

(Continued on page 455)

Wastes from a typical person in the United States, crushed together into a landfill, take up 30 square feet each year.

HOW TO FIND A RECYCLING CENTER

Commercial recyclers may be listed in the Yellow Pages under recyclers. Your city or township solid waste department will know about any licensed collection companies who pick up recyclables. The state of Washington has a toll-free hot line (800 RECYCLE) to give information about centers. Other state, county, or city environmental or natural resources agencies may have listings of recycling operations. Locally, check with an environmentally concerned organization such as the Sierra Club.

HAROLD PLUMLEY'S ELEVEN REASONS FOR NOT USING A DISHWASHER

1) Wastes energy used to propel the pump and to dry.
2) Wastes water.
3) Wastes energy used to heat the extra water.
4) Wastes time—requires two to ten times longer to do the job.
5) Wastes space in the kitchen.
6) Involves capital investment.
7) Involves expenses of interest, depreciation, repair, and maintenance.
8) Ruins fine glassware and china by treating every square centimeter as though it were maximally dirty (such as a speck of dried-on egg yolk).
9) If pathogenic bacteria were present, maximizes probability of cross-contamination.
10) Creates acoustic pollution (operating noise).
11) Reduces desirable family participation in a simple manual job (especially by husband and children).

A physicist's dishwashing system: Schenectady, New York

Harold Plumley, retired manager of technical operations at General Electric's Knotts Atomic Power Lab in Schenectady, has applied his analytical mind to washing dishes by hand.

A dishwasher is a foolish appliance, he says, unless the household is very large, perhaps ten people or more. He and his wife Gertrude threw their dishwasher away 14 years ago.

His method is straightforward and his utensils few. All you need is a stream of hot water, a single sink, a bristle brush, and a cup or bowl of soapy water. "You can actually do this without soap," he says, "but Gertrude won't let me."

All is lost if you make the sin of procrastination, because dried food is inimical to speed at the sink.

Harold's first rule is: Don't stack the dirty dishes. Do so and you get food on a side of the plate that could otherwise go without so much as a single pass of the brush. How does he avoid stacking dishes when a lot of people come over for dinner? He washes the dinner dishes while his wife serves dessert.

He simply dips the brush occasionally in the cup of soapy water, and scrubs just enough to supplement the flow of hot water. Use a rack for air drying.

Further coaching from Harold: "People are ridiculously afraid of germs from dishes and silverware"— a curious concern he says, in that kissing is more apt to transmit disease and yet lips are not treated with scalding water and abrasive soap.

RY

THE DUMP AS TEACHER

For five years, University of Arizona anthropology students out in Tucson have been looking at the beer cans and zucchini peels in the daily garbage of over 3,500 Tucson households. Here's what they found.

• Tucson households throw out 9,500 tons of edible food every year, about 15 percent of all food purchased. In some middle-class areas, edible waste is as high as 25 percent.
• More than 80 percent of wasted food is comprised of half-finished cans of beans or fruit, old apples, chunks of meat, and bread rolls.
• As the cost of food rises, more convenience foods and take-out meals show up in the garbage. Speculation is that double-career families have less time to shop, cook, and plan—and waste more food.

Harold Plumley at the sink.

(Continued from page 453)

dispose of them in the simplest and most economic manner. In nature everything is recycled: it's only a matter of time. It's the unnatural sloshing together of all manner of wastes—both manufactured and organic—that causes the problem of unsightly and unproductive landfill dumps and polluted waterways.

Fortunately I live in a town that sponsors a recycling center for papers and glass. There is also a town landfill where I can take my unrecyclable materials as they accumulate. This I do at *my* convenience, perhaps only four to five times a year. I do not have a truck nor even a station wagon but I can manage a large load in my compact car's trunk, tied with a piece of clothesline (recycled).

The system (it is really too simple to call it that) relies on personal attention to waste disposal. Instead of dropping everything into a common receptacle, one must sort and separate. The pure gold of household recycling is the vegetable scraps. These include every bit of organic waste except meat, fat, and bones. If I had a power grinder the bones would make an excellent mineral supplement to my compost, but we eat less and less meat and I grind the few chicken bones and scraps in the disposal. Larger bones from the Christmas turkey or ham hocks from pea soup are stored in the freezer until my quarterly trip to the dump.

Glass is a surprising (and embarrassing) constituent of my waste. Even using returnable beer bottles, I am shocked by the amount of good, and often beautiful, glass I throw away. True, the pickle jars are not as handsome as wine bottles but I am reminded of primitive societies where a glass jar is a prized and cared-for possession, and I wince as I release my cartons of empty bottles (rinsed and sorted by color) into the bins at the recycling center.

Paper is organic matter, and there are those who swear by the *New York Times* for mulching. But with two daily papers and some 25 magazines coming in per month—plus the swelling quantity of junk mail—I can appreciate that paper is the major filler of garbage trucks. I start by trying to reduce the amount coming in. I encourage my husband to take his trade journals back to the office for others to read and dispose of. We have reduced our newspaper consumption and painfully continue to cut back on magazines, no matter how tempting the offer.

The junk mail keeps coming faster and thicker. I realize my first mistake was ever responding to any entreaty that came to my mailbox. That marked us as "reachable" and the lists proliferated. Now I answer none of it. As I sort through the daily influx, I promptly tuck the tiresome, unwanted stuff into a basket lined with a brown grocery bag. As each bag is snugly filled (in three or four days) I store it in the garage ready for mass removal to the school paper drive. The loose paper, opened envelopes, food boxes and wrappings I dump in another bag-lined waste basket, stamping it down firmly with my foot as it accumulates. When it is about half-full, I fold the top down and voila! I have a quick-starter paper log for my Franklin stove. This is not unslightly and is easy to stack near the fireplace. Being firmly packed down, it does not flare up as loose paper will.

The biggest pains in the system are house guests who want to help with kitchen cleanup and dump leftover salad into my scrap paper bag. They soon learn that the extra time it takes to sort before disposing is minimal, and carrying the boxes of bottles and newspapers to the recycling center takes little longer than lugging them each week to the street. (I do admit to qualms, come winter, at emptying the pail of decaying cabbage, egg shells, and banana peels on the freshly fallen snow drift covering last fall's leaf pile.)

Ah, but the satisfactions. Let me list them:
• *Economy*. Since my trips to the recycling center or dump are com-

(Continued on page 456)

REFRAIN FROM PURCHASING PLASTIC WRAPPED ITEMS

During 1971, the consumption of plastic packaging in the United States was 2.5 million tons (out of 10 million tons of plastic produced), [or] 25 pounds per person.

When buried as landfill, plastic does not decompose through bacterial decay. If burned, plastic leaves a non-biodegradable residue and produces toxic air pollutants. . . . These non-degradable properties prevent the recycling of the petrochemicals that compose the plastic. To make one ton of plastic, 72 gallons of refined crude oil, 338 gallons of natural gas liquids, and 37.1 million BTUs of energy in production are required. This represents a total energy cost, fossil fuel for production, and petrochemicals in the plastic equal to 10,500 billion BTUs per year (1971). If projected increases in the production of plastics are met by 1980, this would mean an energy cost equal to 31,500 billion BTUs.

The Food and Drug Administration has proposed banning polyvinyl chloride plastic food containers (50 million pounds per year in the United States) because of studies by Ralph Nader's Health Research Group indicating that vinyl chloride, the chemical used in the production of polyvinyl chloride plastics, can cause a rare type of liver cancer, angiosarcoma. The adverse health consequences result from vinyl chloride, a known carcinogen, leaching into the food from the plastic containers. The critical nature of this threat becomes apparent when one considers the vast number of supermarket and household items packaged in polyvinyl chloride (PVC) plastic coatings. From meat to milk, and bleach to mouthwash, PVC plastic containers are widely used for packaging. Because plastics create environmental pollution problems, deplete our dwinding natural resources, and pose serious health concerns, it is recommended that consumers refrain from buying plastic-wrapped items, especially plastic food packaging.

Way number 47, from *99 Ways to a Simple Lifestyle*, Albert Fritsch, editor (Anchor/Doubleday, 1976).

When it comes to economizing, middle-class Americans are way behind their counterparts in the rest of the world. Even the affluent Swiss and Japanese are appalled by American wastefulness in everything. Not long ago an Italian friend who travels in high fashion circles reported that though the current rage in his home city of Milan is colored paper plates, those used only for cakes or sandwiches are wiped clean with a damp cloth and reused. The idea of throwing away anything that still seems to be in good condition is horrifying to Europeans and Asians alike, in whom a sense of economy is apparently bred in the bone.

Mimi Sheraton. *New York Times.*

(Continued from page 455)
bined with other errands, I don't deduct the cost of gas from my net saving of $78 per year.
• *Better compost.*
• *Paper drive donation to good causes.*
• *Reduced consumption.* I avoid bringing anything into the house which I will have difficulty disposing of: unnecessary wrappings on parcels, fatty meats, plastic dishes, and glasses. I no longer buy the Sunday *Times* unless I know I will have time to read it.
• *Personal satisfaction.* I feel responsible for myself, for the by-products of my living, and see it as a small step against society's excesses.

There are times when the recycling urge precipitates an embarrassing situation. After a dinner party, I gasped involuntarily as the hostess was about to put the petals left from the artichokes vinaigrette into her catch-all waste can. After explaining my feelings for garbage, she put it all into a bag for me to take home, with a few smiles from the other guests.

My husband, who has not reached my enlightened state of environmental consciousness, was uncomfortable at "taking home a doggie bag of garbage," and perhaps that *is* going outside one's recycling system, pirating someone else's waste.

Natalie Mosher

Note: This contribution arrived in a used envelope, with old addresses covered by a return-address sticker at upper left and a larger self-stick label in the center. These labels are available at office-supply firms.

The good-food target

All foods are not created equal. That's a basic truth which has been pushed aside by the idea that eating a balanced diet is the way to achieve good health. The balanced diet concept assumes that there is some good in every food, and that if you eat a little of everything the good of one thing will cancel out the bad of another. In practice, that idea fails to help people work out a way of eating that will do the most for them. A balanced diet is by definition a compromise. The good balances the bad, creating something not good and not bad. My idea is to leave out the bad entirely, causing your diet to be imbalanced in favor of the good foods that can do so much to improve the way you feel, and can help you resist disease.

Balancing a diet (such as by selecting foods from four or more groups) is also too complicated an idea for most people to use effectively. Few people are willing or able to eat according to the dictates of a formula that must be constantly thought about and adjusted. The best way is to eat because you are enthusiastic about your food. You should like everything about your food—its flavor and ability to fit within your budget—as well as its power to improve your health.

I should point out one more weakness of the balanced diet formula. That is, the problem of bad foods becoming more numerous— and of bad qualities being hidden in foods which used to be good. More foods are being processed. They are stripped of their valuable natural nutrients, and then the space those vitamins and minerals occupied is given over to chemical additives and empty-nutrient foods like sugar and salt. Only by applying the food balancing program with extreme care can you pick your way through the booby traps that are hidden in many conventional foods.

A far better diet plan is to focus your attention on the relatively small number of foods that you know are excellent in all respects. In developing that personal list of foods, consider all factors that contribute to quality. That would include not only nutrients like vitamins and minerals, but also whether the food is free of contamination, rich in those badly-needed complex carbohydrates, high in fiber, low in fat, reasonable in cost, and easy to prepare if your time is limited. And, of course, you should enjoy its flavor.

Get in the habit of thinking of all foods as occupying a place on a sliding scale of quality. Then, when shopping, planning meals, choosing recipes, or selecting from a restaurant menu, always lean toward those better foods.

I'll go even further and suggest that you make a habit of eating certain extra-good goods over and over again—day after day. That's getting pretty far from the wide variety of choice inherent in the balanced diet concept, but actually (in a way) it's what most people are now doing. The only thing is that too often, they tend

(Continued on page 458)

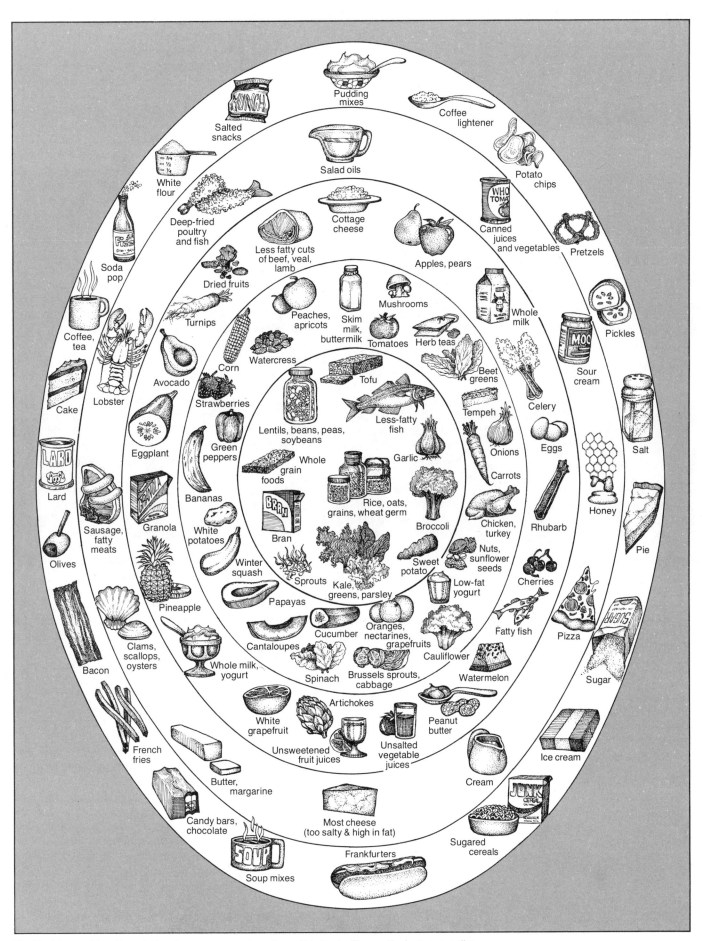

The Rodale good food target: the closer you come to eating within the bull's-eye, the better your diet.

THE INTEGRAL URBAN HOUSE

The Farallones Institute. 1979. Sierra Club Books, 1050 Mills Tower, San Francisco, CA 94104. 494 p. softcover $12.95.

Here are the experiences of a group of San Francisco Bay Area architects, engineers, and biologists who set about to live in the city with fewer resources, and less waste, while looking and smelling acceptable to the neighbors. Nothing, it seems, has escaped their serious consideration: water conservation, noise pollution, aquaculture, solar heating, managing wildlife (an upbeat way of saying "pest control"), gardening indoors and out, and bed bugs. The authors tell you how to do everything but unfold your porch furniture.

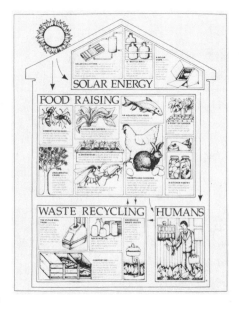

(Continued from page 456)

to eat relatively bad foods like white flour, sugar, and fatty meats at almost every meal. Habits like that seldom disappear. They must be replaced by other habits. So the way to move yourself away from bad food ideas is to fix your mind on a certain number of good foods and eat them regularly.

Eating the same food over and over is far from being a new idea in human nutrition. Many reports of the eating habits of primitive people who enjoyed good health show that they tended to center their diet around a few basic foods. They also kept an eye peeled for special herbs, greens, nuts, seeds, and so forth that they knew from experience made them feel good, but at the core of their diet was often a limited food selection. And many primitive people today who seldom encounter diseases of civilization that afflict Americans (cancer, heart disease, high blood pressure, diabetes, arthritis) do the same thing. As I've said many times in the past, we are all still primitive inside because our genetic make-up hasn't changed for about a hundred thousand years. So what I call the inner historical program, the dictates of our primitive heritage, are an excellent guide whenever we are perplexed about diet or any other aspect of lifestyle that impacts on our health.

Here are some of the reasons why foods are placed where they are on our target diagram. First, the bull's-eye.

Fish is closest to the center. A major reason for giving it that honor is the fact that fish is a steady diet food of many of the healthiest ancient and modern societies. In doing background research for this target plan, we found fish in closer association with the diet of healthy people than any other food. It's low in saturated fat, rich in protein, vitamins, and minerals, and often of excellent flavor. The only problem with fish is that it tends to be expensive. Also keep in mind that the lighter textured fish like haddock and flounder are lower in fat (and therefore in calories) than other types.

Rice is another food like fish— successfully eaten by millions of healthy people. Rice is also good-tasting and cheap. The brown kind (richer in fiber, vitamins, and minerals) is in the bull's-eye, with white rice

further out. That great fiber food, bran, is in the center of the chart, and so is wheat germ, an excellent B-vitamin source. Corn is near the middle, and so are sunflower seeds. Bean sprouts are right there, too—even closer to the middle than beans themselves because sprouts have added nutritional value. Oats are another grain in the bull's-eye because of their high protein, good fiber, and a flavor that can make daily eating enjoyable.

I've put sweet potatoes in the middle, too, because they're so rich in good food elements like vitamin A and protein, and are another healthful food you can enjoy eating regularly. White potatoes are only slightly less healthful—but still are excellent if prepared simply (not eaten as potato chips). Garlic should be in the bull's-eye.

We've put some richer green leafy vegetables in the center (not iceberg lettuce) because they're so needed by many people as a source of calcium. Fruits are great foods, as we all know, but they're slightly away from the middle because of an often high sugar content. Placing fruits on a sliding scale is difficult because varieties can make a difference. A solid Winesap apple—very rich in vitamin A—is a better choice than the more common Delicious, which ranks much lower in that vitamin.

All meats are out of the bull's-eye, to varying degrees. That doesn't mean we think meat is bad, by any means. But since the American diet is already so over-loaded with fat from over-use of lard, oils, and butter and other dairy products, we're almost forced to place most meat visually where you won't think of it as a key to a more healthful diet. Chicken—served without its skin—is the best. Turkey has the same value, for our diagramming purposes.

What about the foods that are completely out of the scoring area of the target? Most are what I call foods of civilization —things like sugar, salt, and fats which simply weren't available to us in any quantity during the period of our genetic development. We can't handle those foods in more than trace quantities without causing a malfunction in one or more of our basic metabolic systems. They can't ever balance a diet—only cause a harmful imbalance.

Robert Rodale

Thirty days in the good food bull's-eye

Writer Pete Roalman came across the target in Prevention *magazine, and recorded his month-long optimum-food diet.*

The strategy was to hit the target by concentrating on foods inside the bull's-eye while avoiding foods in the outer rings. It was such an uncomplicated approach to good eating—the most uncomplicated I have ever seen—that I decided to try it. In fact, I felt challenged by it. Was it possible, I asked myself, for someone living in modern America to abide by such a diet? How difficult would it be to avoid the most easily available foods and live on those, perhaps less easily available, in the inner circle?

I decided to try the diet for 30 days. It was to be an experiment in good nutrition, sort of a 30-day shifting of gears for me to determine if I could get away from such favorites of mine as cheddar cheese (which I have always assumed was a "good" food), peanut butter (which I've always liked and never questioned too much, probably because I didn't really want to), eggs, cottage cheese, butter, veal, and a few other foods that I have fallen into the habit of eating.

I talked with my wife about it. We wondered if it could be done. We weighed the pros and cons of a 30-day experience with the target foods, and finally decided to go ahead.

The experiment started with a problem the first morning. Over the years my wife had gotten into the habit of starting her day with a cup of fresh-brewed coffee. Following well-established patterns, she came down to the kitchen and prepared her coffee. Before she could drink it, I took it away from her.

"You mean coffee is forbidden?" she asked. "Yep," I said, and I pointed out how coffee, tea, soda pop, and other caffeine-heavy drinks were outside the outer ring. She claimed she hadn't noticed that when she agreed to the 30-day test.

But she poured the coffee down the drain, went ahead with the agreement, and survived a 30-day fast from her day-starting drink.

My wife liked one thing about the diet right from the start. "It got me off the hook of preparing elaborate meals," she says. "I came up with a mix of a lot of different vegetables one day and made a big batch of it. Then we had it for three days in a single week. That and a salad and some fruit were all we needed."

But our new diet was never monotonous. Delicately flavored trout and other fish. Lemon. Garlic. Large fruit salads. And salads full of fresh vegetables.

Lunches of thick slabs of home-made, whole wheat bread, tomatoes, and bean sprouts, with a juicy, fresh peach for dessert.

A palmful of sunflower seeds or nuts, plus a cold glass of skim milk made a good midmorning or afternoon snack.

(Continued on page 460)

Page VIII

F. Alcohol

Alcohol, in any form is not acceptable in recipes for use in Rodale Press publications.

Rationale: Alcohol besides containing empty calories has been related to hyperlipidemia, fatty liver, cirrhosis, hepatitis, and beriberi-heart disease (thiamine deficiency). Refined sugar as well as a number of chemical preservatives are used in the processing of wine. Evaporation of 100% of all the alcohol may not occur during cooking.*

Note: Flavoring extracts containing alcohol should be replaced by formulas such as the following or the essential oil of the flavoring in question.

Homemade Vanilla Extract

Cut up a vanilla bean into small pieces and place them in a bowl. Pour ¼ cup of boiling water over them, cover the bowl, and allow the mixture to steep overnight. Grind the mixture in an electric blender. Strain, and return the juice to the blender. Add ½ teaspoon of liquid lecithin and a tablespoon each of honey and vegetable oil. Blend the mixture and pour it into a bottle. Cap it tightly and store in the refrigerator. Shake it well before using. Measure the same amount as any commercial vanilla extract when you use your favorite recipe.

References

Cohen, S. 1978. The Pharmacology of alcohol. Post grad. Med. 64 No. 8, p. 97-102.

Lieber, C.S. 1978. Alcohol nutrition interaction. Contemporary Nutr. Vol 3. No. 9.

McMichael, A.J. 1978. Increases in laryngeal cancer in Britain and Australia in relation to alcohol and tobacco consumption trends. The Lancet No. 8076, June 10, 1978 p. 1244-1246.

Hell, D. et. al. 1977. Vitamin B$_1$, B$_2$ and B$_6$ status in chronic alcoholics. Nutr. Metab. 21 (Suppl. 1): 134-135.

Orlando, J. et. al. 1976. Effect of ethanol on angina pectoris.

Gastineau C.F., 1976. Nutrition note alcohol and calories. Mayo Clin Proc. 51, no 2. p. 88.

Horwitz, L.D. 1975. Alcohol and Heart Disease. J. A. M. A. vol 232. No. 9. p. 959-960.

*Personal Communication - Dr. Ann Noble - Department of Viticulture and Enology, University of California, Davis, California

---

Page XI

I. Miscellaneous

The following are not recommended for use in recipes in Rodale Press publications.

A. Baking Powder - containing aluminum salts (see note).

Rationale: It has been recommended that aluminum salts be withdrawn from use in patients with renal failure and their use restricted in normal persons. Dr. Carl Pfieffer, MD, of the Brain Bio Center, Princeton, New Jersey, claims that inhaled or ingested aluminum accumulates in the brain and over time could cause memory loss and brain deterioration. High doses of aluminum salts are capable of causing gastrointestinal irritation and rickets (by interfering with phosphate absorption).

Note: Baking soda is the chemical sodium bicarbonate. Baking powder contains baking soda plus an acid ingredient such as cream of tartar to release carbon dioxide (leavening gas). An effective baking powder can be made by mixing 2 parts of cream of tartar with 1 part of baking soda. Rumford brand is a commercially acceptable double acting baking powder.

References

Pfieffer, Carl C. New York Times, February 4, 1979.

Berlyne, G.M. et. al. 1972. Aluminum toxicity in rats. The Lancet, March 11, 1972: 654-567.

Speer, Frederic. Food Allergy, PSG Publishing Company, Inc. Littleton, Mass. 1978.

Rodale Press articulated its ideas on healthful eating in a guideline for in-house use. These pages are a sample of that interesting document.

(Continued from page 459)

Rice became a regular part of our diet. Beans also fit in well. Unsweetened fruit juices are delicious, and bran and wheat germ, plus a fresh banana, became a steady breakfast pattern.

Chicken and turkey, both cold and hot, were enjoyable. We were even introduced to tempeh, and we like it.

A problem did develop, though, because I travel a lot. One of the early lessons I learned while trying to convert to the diet is that the food industry tries to load travelers up with useless foods.

For example, during one trip I missed airline connections and had to rent a car. There were a lot of restaurants along the interstate highway I was driving, but most of them were pushing hamburgers or deep-fried fish, steaks or pizzas. I knew from experience that the restaurants where I was headed had a poor selection of nutritious foods, so I was determined to eat something along the way. But I still didn't see, from the highway, a restaurant that looked like it specialized in anything but fat-saturated foods and sugar-water drinks.

Finally, well into the dinner hour, I saw a grocery store with a sign advertising fresh pecans. A solution! I bought half a pound, plus a pint of low-fat milk, and had a satisfying if not ordinary meal. All from the inner circles.

Another surprise I had while on the 30-day experiment was the dis-

covery that sweet potatoes are especially appealing to me. I had hardly ever eaten them before. Oh, one in a rare while, when they happened to be available. But never regularly.

After reading Bob Rodale's article, I tried to make it a point to have a sweet potato at least once a day. After all, he called them an almost perfect food. On a couple of occasions, I put three of four cooked sweet potatoes into my suitcase. They survived well, and they turned out to be a nutritious snack that I ate before going out for dinner—where I ate less than I normally would and easily limited myself to inner-ring foods. No white bread and butter, crackers or sweets to fill me up. (No restaurant I visited during the 30 days had sweet potatoes on the menu.)

Another thing about sweet potatoes I noticed was a distinct increase in energy after eating them. A couple of times I kidded myself by saying I was in a "sweet potato high," but that, in fact, seemed to be the case. The article had described sweet potatoes as being rich in good food elements. They did have a pronounced, positive effect on me.

Another result of the switch was an increased loss of weight. To put that in perspective, I should point out that about six months previously I had been told by my doctor to lose about 40 pounds. I then began a modest running program (I'm now up to two miles a day) and lost about four pounds a month when all was going well.

However, like most weight losers, I slipped back and gained some of the lost weight, so I was nowhere near being 24 pounds lighter six months into the weight-reduction program. More like 10 pounds. After I started eating only those foods in the two center rings, I lost about a pound a week more than I had in the past.

To add a more convincing argument, my wife, who had been trying to get below 123 pounds by watching her food intake and running about six days a week with me, became an excited fan of the diet one morning after about three weeks, when she called down, "I made it! I'm down to 120 pounds!"

What we learned during our

experiment is that there are a lot of meal possibilities—wholesome, satisfying, and tasty—within the two inner circles. We hadn't dreamed it when we started the diet, but it is now obvious. All we had to do was pass by the outer rings of useless food when we went shopping—not as hard to do as you might suspect—and use our imagination with the rich variety of other foods available after the junk was eliminated.

One of the biggest surprises I had during the switch was finding that water is a satisfying drink. Previously, I had been an ice tea freak and seldom ate a meal without a couple of glasses of it. Never with sugar, but always with artificial sweetener, even after the news that such sweeteners might cause cancer. Don't ask me why, except that I usually balanced the remote scare against the more immediate satisfaction, and the sweetened ice tea won.

After reading in the article that tea is a relatively useless food, I switched entirely to water. It also saved me money in restaurants. I developed the habit of drinking straight water and realized that I had been missing a treat.

I also found that I could survive quite well with a tall glass of ice water during social gatherings. Where I used to enjoy a gin and tonic on a warm summer night before, now I'm refreshed by a tall glass of ice water. As it turns out, I can have three or four more ice waters than I could gin and tonics, so I now can have more to drink than before.

Once, during my 30-day switch, I went to a block party, where I knew they were going to be serving a lot of outer-circle foods. I stuck a bag of nuts and dried fruits in my pocket. Washed down with water, they carried me through a pleasant evening and only one person noticed the different eating and drinking pattern I was following. Eventually, she became intrigued and I suspect she will explore the new diet for herself.

Before my wife and I knew it, the 30 days passed. It hadn't been difficult at all. Most important, we had been introduced to a new way of eating, learning healthful lessons we could carry over into the future.

Pete Roalman

FEWER GROCERY STORES, MORE EATING PLACES

While grocery store sales, nationwide, rose 2 percent in constant dollars between 1972 and 1977, in New England (Connecticut, Maine, Massachusetts, New Hampshire, Rhode Island, and Vermont) in the same period, they actually declined about 4 percent, after adjusting for inflation. Nationally, sales at eating places, measured in constant dollars, rose 20 percent between 1972 and 1977.

Michael Van Dress. *National Food Review,* as quoted in *Journal of the American Dietetic Association.*

HOW TO EAT WELL ON $1 A DAY

There is a way to beat food-cost inflation and this is it. We have all read accounts of people living off the land at little cost, but we will now go a different route. This is an exercise in living on purchased food, for a full week, on a dollar a day. We will furthermore live on good things to eat, including some gourmet treats.

This will be a pleasant, non-scientific revolt against highly processed and highly priced foods. Great genius has perfected food preparation, preservation, and packaging. People of faraway pastoral lands come to our supermarkets to stand in awe. For this genius, however, we pay dearly.

The way to beat the cost of food sophistication is to return to unprocessed but hearty foods, which, fortunately, are still in abundance. Not everyone is as well equipped as this writer to enjoy a foray into primitive food. Being a ripe 68, I go back to a period as a farm boy in the Great Depression.

(It's fun to tell the young how really godawful it was.) We had little cash, but marvelous foods in the garden, in the barns, and even in the forests, which were sprinkled with berries, nuts, mushrooms, and, yes, nettle greens, plus squirrels, rabbits, and birds. Had there been TV dinners we could not have afforded them. Nobody ever perfected radio dinners. So I can invoke a degree of nostalgia in an environment of cornbread, baked beans, flapjacks, lusty soups long seasoned atop the wood stove, oatmeal, and ordinary meat and potatoes. Even molasses brings a twinge of homesickness. We used to go to Pleasant Hill where Chet Kennedy had a sugar cane press pulled by a horse caught up in an eternal circle. We bought half-gallon tins from him.

I have another slight advantage for this experiment. I normally eat two meals a day, which cuts start-up cost. My mother used to warn us not to "piece" between meals. I had assumed that to be an obsolete verb, but a check with the big dictionary indicates it is still a good word, although colloquial. I do not have the piecing habit, and in any event, for the coming week there will be no such cheating.

Also going for us is the "fatso factor." It is my understanding that most people eat more than necessary to sustain living. You might loiter in a bakery and note the heft of most customers. The menu that follows would not work for a working man, although I will be spending an hour or two every day splitting great chunks of beech, maple, and oak. Nor would it sustain a growing boy. My best day as a young working man occurred in 1925, when I pitched bundles into a threshing machine for a long afternoon and re-fueled with a beef supper followed by four pieces of pie and three of cake. I will cheat on coffee, for the only time I touch it is at a downtown coffee break, which is not a nutritional experience but a social occasion. I will not count the cost of heat for cooking, since it will be supplied by a wood range that runs all winter in any event. I will list the costs and give some recipes (including one for audacity salad). Only the cost of my portion will be set forth, not the leftovers or what goes to Jason, the golden retriever, who is obliging in the kitchen area. I weigh 160 pounds with the hat on. And so the week begins.

Sunday.

9 cents	Seven ounces of grapefruit juice
7 cents	One medium egg (1¾ ounces)
10 cents	Small glass of milk

That's a cheap egg, fresh from the farm, and we have enjoyed a 26-cent brunch. For the big meal we put a sweet potato (5 cents) in the oven and prepare a rich soup before sinking into the chair to watch the ball game, digestive in hand. The digestive will wipe out our savings, but I didn't promise to give up whiskey.

Soup ingredients:

41 cents	Stewing beef ($1.89 a pound)
4 cents	Carrot
2 cents	Potato
2 cents	Onion
3 cents	Rutabaga
2 cents	Barley
2 cents	Cornstarch

We are skipping the salt and pepper cost as negligible. And so the cost of Sunday would appear to be 87 cents, but wait for the surprise. The soup, with the crunchy barley, filled me with half left over. Half the soup costs 28 cents, and Sunday's total becomes 59 cents and we proceed with money in the bank.

Monday. Today we explore the miracle of corn, which sustained American Indians and Spanish explorers in high style. Since cornmeal at our friendly store costs, in small portions, $24 a bushel, I "borrowed" a few ears from farmer Tom Russell and ground it by hand. We will value it at the Chicago cash market price for No. 2 yellow, namely $2.30 a bushel. Using 70 pounds in a bushel we find that an ounce of our cornmeal costs 2 mils. We allow 19 cents for grapefruit juice and milk, and prepare flapjacks:

1¼ cents	One ounce of store-bought white flour
4/10 cent	Two ounces of cornmeal
7 cents	Egg
5 cents	Buttermilk
2/10 cent	Soda
3 cents	Vegetable oil
8 cents	Blackstrap molasses
3 cents	Butter

This batter rounds off to about 28 cents. I learned two things with these jacks. I always thought blackstrap molasses was something less refined than usual shelf varieties, containing a lot of iron and other character-building things. And it may be. We got a pint in a health food co-op, with no guidance on the container, which was a used tomato juice bottle. My dictionary gives three possibilities for blackstrap, the first being "molasses with rum." The second is a "sailor's term for any wine of the Mediterranean." The third is more fun: "The final mother liquor remaining after crystallization of the juice from sugar from the juice of the cane or beet." I cannot detect a rum taste in this syrup so we may conclude it is the "mother." The second discovery was that, while the taste is fine, the molasses is a powerful dye. I put a dash directly into the batter. It turned the pancakes into a color resembling a walnut antique, to which some squeamish gourmets might object. Furthermore they were a little too crunchy, since my grinder doesn't grind exceedingly fine. So next time I will shoot another penny on white flour and throw a little sugar into the batter, using all the molasses to spread contrast atop the lighter cakes.

It might appear that we have blown 47 cents for breakfast, but here is another O. Henry finish. The recipe made half a dozen cakes, three of them filled me, and Jason, the dog, must pay for the rest. The adjusted cost for a rib-sticking breakfast is 33 cents.

Come Monday evening I get lazy. Remember the stewing beef? It seems

(Continued on page 462)

(Continued from page 461)

a good time for beef fondue, but we are low on cooking oil. So I put the chunks on whole wheat bread and broil them in the electric stove, about 40 cents worth. Then I light into an audacity salad, which is made up of whatever is lying around. This one has a couple lettuce leaves, cottage cheese, half an orange, cheese bits, and a little gooseberry jam sent us by Helen Clayton in Indiana. Count 20 cents for the salad, which makes a total of 93 cents for Monday. We will apply the 7 cents for the electricity, bread, salt, and pepper and call it an even-dollar day.

Tuesday. Here's poached egg on toast, plus juice and milk and a little crabapple jelly, which totals about 30 cents for breakfast, and we proceed to beans.

There are 2,288 navy beans in a pound. I say this confidently because there are 143 in an ounce. A pound costs 39 cents. Soaked overnight they swell up like a snake-bitten thumb. For a dime or maybe 15 cents you can fill up on beans with onion, bacon bits, or anything handy. (We'll bake some later.) A big bowl of beans, small glass of buttermilk, plus bread and butter, did it, leaving room for only about a couple handfuls of popcorn as the evening wore on. It came to about 90 cents for the day.

Wednesday. With a bacon strip to grease the spider, we dice a nickel's worth of Irish potatoes. We do a cottage fry, putting the lid on briefly. I have a bit of milk, but observe we skipped the grapefruit juice, which prevents scurvy at sea. The greatest cost was the bacon, 10 cents. It won't put on much weight, but it is an occasional necessity because of the aroma. On a Wednesday morning, a kitchen filled with fetching whiffs of bacon lifts the spirit and makes high promise for the day. The glorious occasion came to 32 cents.

Claudette Lawton, the egg lady, came by and, informed of our noble experiment to beat the Malthusian peril, said: "Go get yourself a chicken, it's on sale at 59 cents a pound." Some wives are tolerant, but my Bette gets steamed if I mess around the kitchen too much. She commanded: "Place your order, and I'll cook it." So I went out and bought an affable chicken, with gizzard, for $1.52. While Bette started doing chicken with rice, I snuck in another audacity salad to stave off hunger. On a bed of lettuce we have a mixture of sour cream and blue cheese. Blue cheese is a luxury but a little dab makes itself known. The top of the salad was sprinkled with store-bought stuffing. Salad at 23 cents. Although Bette took cooking classes in Paris, she could do chicken and rice before she left the Iowa farm. No French cook could ruin affable chicken with rice. I figured the portion I ate at 35 cents, and a touch of rice pudding, which was redundant, at 10 cents, shooting the buck for the day.

Thursday. Let us now repair to the wheat pit. We have this hand-ground wheat, also pillaged from Tom Russell. For cost we consult Kansas City, where No. 2 hard is $3.42 a bushel. By slide rule we find an ounce costs 3.56 mils, a third of a cent. For pancakes: an ounce of the wheat and an ounce of store flour, with egg, buttermilk, soda, and sugar worked out favorably. I left the molasses out of the batter, but smeared it, with a little butter, atop the light tan cakes. There were a couple left for the dog. I drank the anti-scurvy juice, but skipped the milk. The hearty breakfast cost around 30 cents.

Ground whole wheat as used in the pancakes includes more bran than some might like. That is why I added the white flour.

Leftovers are building up in the refrigerator. Remember Sunday's soup? The other half is still around, but its time is running out. The chicken and rice soup is still fresh and tasty, so we go for 35 cents in that direction. Remember the beans? We subtract a portion from the pot, put it in a small casserole with ketchup, brown sugar, and molasses to bake in the wood stove oven. Out of this world, for 5 cents. Later a dish of buttered and salted popcorn at 10 cents makes it 90 cents for Thursday.

Friday. Poached egg, toast, buttermilk, with a little crabapple jelly for the toast, comes to about 20 cents.

Now for the blast, as Bette takes to the wok to prepare a Chinese dish. I daren't go near the wok. My half of the feast looks like this:

70 cents	Sirloin for strips
18 cents	Water chestnuts
4 cents	Carrot
3 cents	Rutabaga
5 cents	Broccoli
5 cents	Oil
6 cents	Soy sauce
2 cents	Cornstarch

That comes to $1.13, and with breakfast at 20 cents we are at $1.33 for the day, over the buck limit but drawing on credit built up from previous days, so don't worry. The Chinese may cook it better, but we start with better beef than you will normally encounter in Canton. In addition to the naturals, like sugar peas and Chinese cabbage, almost everything that grows in the garden is at home in a wok, so during garden time one can eat Chinese at a niggardly cost by relying on the hoe. And the meat doesn't have to be sirloin, as chicken and pork (how about rabbit?) will do nicely with vegetables fresh out of the sunshine. My little friend from Taiwan, Chiu Mei, says anything a Yankee can do with a wok, she can do with a skillet *ordinaire*.

Saturday. I haven't eaten oatmeal since I left the farm and may never again, but I find that a nickel's worth with cream and sugar (another nickel) is quite filling, if not *haute cuisine*. With 9 cents for the juice we are at 19 cents, and the belt is still tight.

We are not quite finished with corn. I learned of the nutritional value of corn from old Indian Bill, Chief of the Wabigoons.

The tale is worth telling. Many years ago my wife and I spent a wild ten days in a trapper's shack on Big Sandy Lake near Dinorwic, Ontario. The Chief of the Wabigoons didn't live on the reservation. He went afield from time to time, squatted and built a house on a desirable promontory on the lake just across a small bay from our shack, which sat on an elevation with a fine view of Big Sandy. The Chief eventually sold his place to a foreign sportsman at an outrageous profit. There is a smart Indian.

As we awoke the first morning the Chief's canoe approached. He had just pulled his nets and presented to us a fine three-pound walleye. There is a good Indian. We didn't see much more of Bill, but his 18-year-old son Frank virtually moved in with us. He took us to his secret lake where the fish were swimming hub to hub. We gave him cooked meals and took him to the Saturday-night dance in Dinorwic, attended chiefly by the road gang that was building the Trans-Canada highway through there. That was a rich experience, though not gastronomic.

The thing about Frank is that he had three sled dogs that hauled him around his trapline in the winter. They liked Great Northern pike as a diet, and we joined the campaign to catch enough jackfish every day to keep them in good spirits. When we didn't, Frank had to cook them up a corn mash, on which they seemingly could also prosper. What will sustain dogs will in many cases feed people. And so let's make cornbread.

The label on any cornmeal package will provide a reasonable cornbread recipe. It is delicious served warm with butter and honey, though a little dry. There is a better way:

1 cent	Three ounces home-ground cornmeal
2½ cents	Two ounces white flour
7 cents	Egg
12 cents	Cup of sweet milk
7 cents	Half cup buttermilk
3 cents	A little soda and a little baking powder

Now we mix those things, saving out a half cup of sweet milk, and put them into an iron skillet and thence into an oven. After ten minutes, before the bread starts to pick up color, we pour the other half cup of milk on the top. It seeps down, making a moist, prize-winning cornbread. That is a plateful for less than 33 cents and a nickel's worth is about the right slice for any meal. The White Turkey in Manhattan, with a tearoom atmosphere, used to serve, oddly enough, cornbread. But it was dainty, with sugar, something that would never go in the South, where fried fiddler and unsweetened cornbread is considered an alimentary orgy.

Corn is a nutritional gold mine. It runs to spoon bread, omelets, and any number of great Mexican foods. Going from field corn to sweet corn opens a vast world of fritters, casseroles, and soups.

So we will have a slice of cornbread with the last supper of the week, beef casserole unforgettable:

30 cents	Stewing beef
4 cents	Carrot
1 cent	Potato
3 cents	Onion
5 cents	Milk
4 cents	Bouillon cube
1 cent	Cornflakes

I cut the beef a little finer than the butcher left it, to approach a hash consistency. A round whiskey bottle will pulverize the cornflakes. Sprinkled atop the casserole they form an appetizing crust. Cheese is nice too, but expensive. The casserole costs 48 cents.

Now for a big fat apple dumpling, and we can afford it. I simply order one from the chef. She takes a nice Greening, at 15 cents, slices it, adds sugar and spice, wraps it in dough, and bakes, for a total of 20 cents. A dumpling cries out for cream, a lot, like 20 cents' worth. So this noble confection comes to 40 cents, making it $1.12 for the day.

And so, adding the week we get not $7, but $6.84, leaving 16 cents for breakage. The weight, after the week of privation, 160½, up a half pound. I believe one could eat well for 50 cents a day by skipping things like bacon, sirloin, and cream. I further believe that many around the world *are* eating on 50 cents, if not less.

The week, in fact, ended before we got around to several favorite low-priced foods. Fried grits come to mind, and grits are on sale in New England, ground up by a Chicago outfit. It is more ground corn as far as I can see, although some of it is white. Then there is liver, a treat, and macaroni and cheese. Sometimes Bette makes her own yogurt. We could have made some rich wheat bread with our home-ground grain. We skipped spaghetti and meat, best left to a good Italian restaurant. My neighbor makes apple pancakes. Oh well.

What did I miss most during the week? Anchovies. Anchovy salad. Côte d'agneau. Honey. Standing rib roast. (It's out of sight.) Salmon fumé. Café diablo. And, of course, caviar, black or red. Oh well.

Hugh Moffett. Reprinted by permission from *Blair & Ketchum's Country Journal*. Copyright © 1979 Country Journal Publishing Co., Inc.

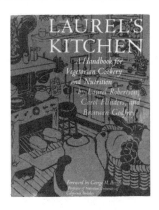

LAUREL'S KITCHEN

Laurel Robertson, Carol Flinders, and Bronwen Godfrey. 1977. Nilgiri Press, Petaluma, CA 94952. 508 p. hardcover $13.95.

A mass-market paperback edition costs a fraction of the hardcover, but *Laurel's Kitchen* is too valuable and too attractive to cram onto stingy, short-lived pages. Laurel's woodcuts and exceptional design make the large hardcover nice to own. Much care is taken to explain the basics of good nutrition for vegetarians, but anyone will benefit from the authors' sane advice.

RY

The real thrills in life are out in the world: life at the office has a certain glamor that your little bungalow is bound to lack. The gist of the typical commercial is, "We know you love your family, but let us free you from its drudgery and give you Time to pursue a more Meaningful Existence."

The tactic is most insidious. For business and industry, the ideal situation is for us to be trying to have a family and a job, for when we do, we spend a lot more money on a lot more things. It's not just because we have more money to spend; a working wife and mother needs a second car (or bus fares), dressier clothing, more nylon stockings, a babysitter, and perhaps a cleaning lady. She's pressed for time in the morning and worn out in the evening, so restaurant meals regularly take the place of bag lunches and home-cooked dinners. Prepared quick-serve foods, far more expensive than basic foods, take another bite out of the budget, along with a dishwasher, a microwave oven, and ready-made clothing for the children—and, in all probability, more money spent on random gifts for them because she feels bad at spending so little time with them. All this on top of the regular operating expenses of the household. "Household" is hardly the word—at this point, when the emphasis falls increasingly on speedy refueling and immediate departure, "pit stop" might be closer to the truth.

Further reading on home, home food systems, and systems

The Mother Earth News, 105 Stoney Mountain Road, Hendersonville, NC 28739. This bi-monthly brought the exclamation mark to the country, where it grows like kudzu. You'll either like or hate *Mother's* aggressively rural prose. Articles in a recent issue included "Don't Miss Out on the Morels!," "Soybeans: Grow'em and Freeze'em," and "Taters in a Barrel!"

Countryside, 312 Portland Road, Waterloo, WI 53594. *Home Food Systems* contributor Jerry Belanger edits this serious homesteader's magazine. Monthly sessions include his philosophizing (Beyond the Sidewalks), The Beehive, The Hen House, The Rabbitry, The Goat Barn, and The Cow Barn.

Country Journal, Box 870, Manchester Center, VT 05255. *Country Journal* is a New England-based magazine that isn't restricted by regional boundaries or by an obligation to mythicize rural America. Still, some issues are almost too pretty to look at, and the fine writing further teases by making you wish you could break out for small-town life.

The CoEvolution Quarterly, Box 428, Sausalito, CA 94965. Editor Stewart Brand's open eyes and restrained blue pencil make this magazine a forum for interesting people.

Harrowsmith Magazine, Camden East, Ontario, Canada K0K 1J0. Canada's fine homesteading magazine describes itself as concerned with "country life and alternatives to bigness." An anthology of articles, *The Harrowsmith Reader* ($8.95), was published by the magazine in 1978.

Organic Gardening, 33 E. Minor Street, Emmaus, PA 18049. Rodale Press's gardening monthly features Robert Rodale's far-ranging editorials, workshop projects, tools for growing and harvesting, and recipes for the harvest.

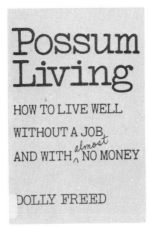

POSSUM LIVING

Dolly Freed. 1978.
Universe Books, 381 Park Avenue S.,
New York, NY 10016. 176 p.
paperback $3.95.

Some years ago, the pseudonymous Ms. Freed and her father quit, respectively, junior high school and a 9-to-5 job. This book is a chatty guide to a comfortable life on little cash. Comfort is important: "Let me re-emphasize that we aren't living this way for ideological reasons, as people sometimes suppose. We aren't a couple of Thoreaus mooning about on Walden Pond here."

People also call them "scavengers" in a deprecating tone. But these same folks will pay exorbitant prices to eat lobster or crab, the most thoroughgoing scavengers of all. (When we lived in Florida, Daddy used to ride to work with a real crab gourmet. Whenever he saw a dead dog along the road he'd stop the car, run over and kick the carcass till the crabs would scurry out and grab them up to take them home for dinner. Daddy was always being docked for lateness because of this guy's predilection for crab meat.)

JOY OF COOKING

Irma S. Rombauer and Marion
Rombauer Becker. 1964.
Bobbs-Merrill Company, Inc., 4300 W.
62nd Street, Indianapolis, IN 46268.
849 p. hardcover $11.95.

The authors delight the browser by going to startling depths in many areas. The excerpt gives an example.

Don't buy the chubby mass-market paperback; invest in the more durable hardcover of this fine cookbook.

—RY

Opposum

If possible, trap possum and feed it on milk and cereals for 10 days before killing. Clean, but do not skin. Treat as for pig by immersing the unskinned animal in water just below the boiling point. Test frequently by plucking at the hair. When it slips out readily, remove the possum from the water and scrape. While scraping repeatedly, pour cool water over the surface of the animal. Remove small red glands in small of back and under each foreleg between the shoulder and rib. Parboil, page 132, 1 hour. Roast as for pork, page 407. Serve with: Turnip greens.

THE END

We laugh with delight at a coffee-drinking dog or a cat that has cultivated a taste for orange juice, but humans are the most adaptable of animals. We inhabit igloo and house boat, mountaintop and bayou, city high-rise and two-pound tent. And our diets—humans are apt to take a fancy to anything they can fit in their mouths, from snails to rock candy, from warm milk to Scotch.

Adaptability is one of our best traits, but it can be a liability as well. Our instincts and cultural traditions undone by advertising, we are suckers for books that tell us what to eat. Now *Home Food Systems* has added its 465 pages of opinion to the discussion.

You can only read so much before you start thinking of your sandwich as fuel and your stomach as a gas tank. If you've reached this sad state, it's time to head for the kitchen and cook up a good-sized plate of your favorite food. Between bites, you might care to ruminate on this: the sensuality of dining has yet to be abridged by law. (Some music lover once claimed that the instrumental is the only uncensorable art form, but this person obviously had forgotten his or her last good feed.) Ultimately, eating is a private act, a dance for one. We may dine with brilliant company, but as we chew, the world recedes: we eat alone, and no one can share our exact sensations. Thoughts of bioflavonoids, amino acids, and such abstractions should gracefully withdraw, leaving tongue and nose to their savage moment.

Index